Pore Structure and Fractal Characteristics in Unconventional Oil and Gas Reservoirs

Pore Structure and Fractal Characteristics in Unconventional Oil and Gas Reservoirs

Guest Editors

Zine El Abiddine Fellah
Jizhen Zhang
Quanzhong Guan

MDPI

Basel • Beijing • Wuhan • Barcelona • Belgrade • Novi Sad • Cluj • Manchester

Guest Editors
Zine El Abiddine Fellah
Laboratory of Mechanics and Acoustics
Aix-Marseille University
Marseille
France

Jizhen Zhang
Key Laboratory of Exploration Technologies for Oil and Gas Resources
Yangtze University
Wuhan
China

Quanzhong Guan
Energy College
Chengdu University of Technology
Chengdu
China

Editorial Office
MDPI AG
Grosspeteranlage 5
4052 Basel, Switzerland

This is a reprint of the Special Issue, published open access by the journal *Fractal and Fractional* (ISSN 2504-3110), freely accessible at: https://www.mdpi.com/journal/fractalfract/special_issues/ZS523FM0B6.

For citation purposes, cite each article independently as indicated on the article page online and as indicated below:

Lastname, A.A.; Lastname, B.B. Article Title. *Journal Name* **Year**, *Volume Number*, Page Range.

ISBN 978-3-7258-5263-5 (Hbk)
ISBN 978-3-7258-5264-2 (PDF)
https://doi.org/10.3390/books978-3-7258-5264-2

Contents

About the Editors

Zine El Abiddine Fellah

Zine El Abiddine Fellah is a Senior CNRS Researcher at the Laboratoire de Mécanique et d'Acoustique (LMA), Centrale Marseille, Aix-Marseille University. His research covers acoustic and poroelastic wave propagation; inverse problems in porous media; fractional derivative modeling; ultrasound-, audio-, and low-frequency-based characterization; and the acoustic modeling of micropolarly structured materials. He maintains several international collaborations and has published a wide range of research on the theory and experimental study of fractal porous materials and sound insulation.

Jizhen Zhang

Dr. Jizhen Zhang is a Guest Editor at the *Key Laboratory of Exploration Technologies for Oil and Gas Resources*, Ministry of Education, and the *Laboratory of Reservoir Microstructure Evolution and Digital Characterization*, Yangtze University, Wuhan, China. His research interests include pore structure, pore heterogeneity, complexity, and the fractal characteristics of reservoirs. He has extensive experience in applying fractal analysis for the characterization of unconventional reservoirs and modeling of fluid transport.

Quanzhong Guan

Dr. Quanzhong Guan is a Guest Editor at the Energy College, Chengdu University of Technology, Chengdu, China. His research focuses on shale gas and oil reservoirs, pore structure and heterogeneity, fractal characteristics, and overall reservoir characterization. He has contributed to both theoretical and applied studies for understanding the transport properties and microstructure of unconventional reservoirs.

Preface

As Guest Editors, we are pleased to present this Reprint, a compilation of pioneering work on fractal characteristics in porous media for unconventional oil and gas reservoirs. The contributions emphasize experimental, numerical, and theoretical approaches to the characterization of pore structures, transport phenomena, and acoustic behavior. Addressing researchers in the fields of geoscience, engineering, and applied physics, this collection offers insights into advanced modeling and measurement techniques for heterogeneous and micropolarly structured materials.

Zine El Abiddine Fellah, Jizhen Zhang, and Quanzhong Guan
Guest Editors

fractal and fractional

Article

Pore Fractal Characteristics between Marine and Marine–Continental Transitional Black Shales: A Case Study of Niutitang Formation and Longtan Formation

Shitan Ning [1], Peng Xia [1,2,3,*], Fang Hao [3], Jinqiang Tian [3], Yong Fu [1,2] and Ke Wang [1,2]

1 College of Resources and Environmental Engineering, Guizhou University, Guiyang 500025, China; 2023311213@student.cup.edu.cn (S.N.); byez1225@126.com (Y.F.); kwang5@gzu.edu.cn (K.W.)
2 Key Laboratory of Karst Georesources and Environment Ministry of Education, Guizhou University, Guiyang 500025, China
3 School of Geosciences, China University of Petroleum, Qingdao 266580, China; haofang@upc.edu.cn (F.H.); tianjq@upc.edu.cn (J.T.)
* Correspondence: pxia@gzu.edu.cn

Abstract: Marine shales from the Niutitang Formation and marine–continental transitional shales from the Longtan Formation are two sets of extremely important hydrocarbon source rocks in South China. In order to quantitatively compare the pore complexity characteristics between marine and marine–continental transitional shales, the shale and kerogen of the Niutitang Formation and the Longtan Formation are taken as our research subjects. Based on organic petrology, geochemistry, and low-temperature gas adsorption analyses, the fractal dimension of their pores is calculated by the Frenkel–Halsey–Hill (FHH) and Sierpinski models, and the influences of total organic carbon (TOC), vitrinite reflectance (R_o), and mineral composition on the pore fractals of the shale and kerogen are discussed. Our results show the following: (1) Marine shale predominantly has wedge-shaped and slit pores, while marine–continental transitional shale has inkpot-shaped and slit pores. (2) Cylindrical pores are common in organic matter of both shale types, with marine shale having a greater gas storage space (CRV) from organic matter pores, while marine–continental transitional shale relies more on inorganic pores, especially interlayer clay mineral pores, for gas storage due to their large specific surface area and high adsorption capacity (CRA). (3) The fractal characteristics of marine and marine–continental transitional shale pores are influenced differently. In marine shale, TOC positively correlates with fractal dimensions, while in marine–continental shale, R_o and clay minerals have a stronger influence. R_o is the primary factor affecting organic matter pore complexity. (4) Our two pore fractal models show that the complexity of the shale in the Longtan Formation surpasses that of the shale in the Niutitang Formation, and type I kerogen has more complex organic matter pores than type III, aiding in evaluating pore connectivity and flow effectiveness in shale reservoirs.

Keywords: marine shale; marine–continental transitional shale; pore structure; fractal dimension; FHH model; Sierpinski model; pore complexity

1. Introduction

Shale gas refers to natural gas existing in mud shale in adsorbed, free, and dissolved states, etc. It is a large-scale, continuous, and low-abundance unconventional oil and gas resource [1–3]. According to different sedimentary environments, shale can be divided into marine shale, marine–continental transitional shale, and continental shale [4,5]. At present, the Sichuan Basin and its surrounding marine shale have been utilized for stable industrial gas flow, and the development and utilization of marine shale have resulted in a strategic breakthrough [6–8]. The production of shales in the Wufeng–Longmaxi Formation in Luzhou, Changning, the Weiyuan and Fuling areas of the Sichuan Basin, and the Niutitang

Formation in Zheng, an area of Northern Guizhou, has achieved commercial breakthroughs and scale-effective development. However, the geological research of marine–continental transitional shale is still in its infancy, and no systematic research on oil and gas has been carried out.

Guizhou is located in South China on the periphery of the Sichuan Basin and has experienced multiple tectonic movements and sedimentary environment changes. From the Sinian to the Middle Triassic period, its sedimentary rocks were marine sediments dominated by carbonates. In the Upper Triassic period, its marine–continental transitional sediments were transformed from a shallow sea platform to an inland lake basin. From the Jurassic to the Tertiary period, its sedimentary rocks were a set of continental sediments dominated by clastic rocks. Therefore, Guizhou has developed a set of thick sedimentary rocks and formed multiple sets of source–reservoir–cap rock assemblages in the vertical direction. The marine Niutitang Formation and the marine–continental transitional Longtan Formation are considered to be shale gas strata with development potential [9–11]. However, in the geological tectonic stage of the collision between the Pacific and the Indian Plate, the Qinghai–Tibet Plateau began to rise sharply, resulting in the formation of the Yunnan–Guizhou Plateau and a large number of folded mountains. The complex pore storage space in shale gas formations is a result of intense structural evolution and substantial heterogeneity. Determining the exact location and associated conditions for shale gas enrichment poses considerable challenges due to these complexities [12–14].

The evaluation methods for pore structure characteristics are diverse, with commonly used ones including image analyses (scanning electron microscopy and image analysis software), experimental techniques (high-pressure mercury intrusion, low-temperature nitrogen adsorption, carbon dioxide adsorption, and small-angle scattering), and geometric fractal methods (the Menger model, Sierpinski model, BET fractal model, FHH fractal model, thermodynamic fractal model, and Langmuir fractal model, etc.). The fractal dimension is an important aspect of geometric fractals and is often utilized to describe the complexity and heterogeneity of solid material pore structures. However, different methods characterize pore diameter distributions differently. Reference [15], based on low-to-medium-grade structural coal, combined with high-pressure mercury intrusion and low-temperature nitrogen adsorption experiments, discussed the applicability of the Menger model, Sierpinski model, FHH fractal model, and thermodynamic fractal model. The authors concluded that the Menger model is not suitable for characterizing the heterogeneity of structural coal pores, the Sierpinski model is suitable for describing the nanoscale pore fractal characteristics of structural coal, and the FHH fractal model is suitable for characterizing the heterogeneity of pores in primary coal and structural coal with pore sizes of 8–100 nm. Reference [16] studied the Longmaxi Formation and Niutitang Formation shales in the southern marine areas of China, using a combination of micro-pore fractal models and FHH fractal models to calculate the fractal dimensions of their pores, and discussed the applicability of the two methods. The authors found that the micro-pore fractal model is suitable for calculating carbon dioxide adsorption data and effectively characterizes the fractal dimensions of shale micropores, while the FHH fractal model is suitable for calculating high-pressure nitrogen adsorption data and effectively characterizes the fractal dimensions of shale mesopores. To summarize previous research on the fractal dimension theory, the use of the Sierpinski model is mostly concentrated in coal and rock formations, with relatively less research on the fractal dimension of shale pore structures based on carbon dioxide adsorption experiments using the Sierpinski model.

Therefore, this study explores the development characteristics of pores in the marine shales of the Niutitang Formation and the transitional marine–continental shales of the Longtan Formation in Guizhou through nitrogen adsorption and carbon dioxide adsorption experiments. We use the FHH model and the Sierpinski model to reveal the complexity of shale and organic matter pores of different sizes and their main influencing factors. Establishing a model to explore the development of pores in the marine shales in the Niutitang Formation and the transitional marine–continental shales and organic matter in the Longtan Formation will help us predict the production of shale gas in Guizhou more accurately and optimize development strategies.

2. Samples and Methods

2.1. Samples and Preparation

In this study, 13 samples were prepared, which were taken from the black shale of the Niutitang Formation in Northern Guizhou and the black shale of the Longtan Formation in Western Guizhou (Figure 1). Among them, 2 samples of Niutitang Formation black shale belonged to platform sedimentary environment and 4 samples from slope sedimentary environment. Seven samples of the Longtan Formation black shale were from the Dahebian syncline structure, representing a transitional marine–continental depositional environment. This type of black shale is typically associated with coal and is also known as coal-bearing shale. Due to their diverse depositional environments and depths, these samples exhibit variations in mineral composition, total organic carbon (TOC) content, and organic matter maturity levels. This allows for a comprehensive analysis of shale gas occurrence states in marine and transitional marine–continental shale formations as well as an exploration of the factors controlling pore structure complexity. Consequently, prior to studying shale and organic matter's pore structures, all black shale samples underwent detailed organic geochemical characterization and mineral composition analyses, including TOC, thermal maturity, whole-rock, and clay mineral analyses.

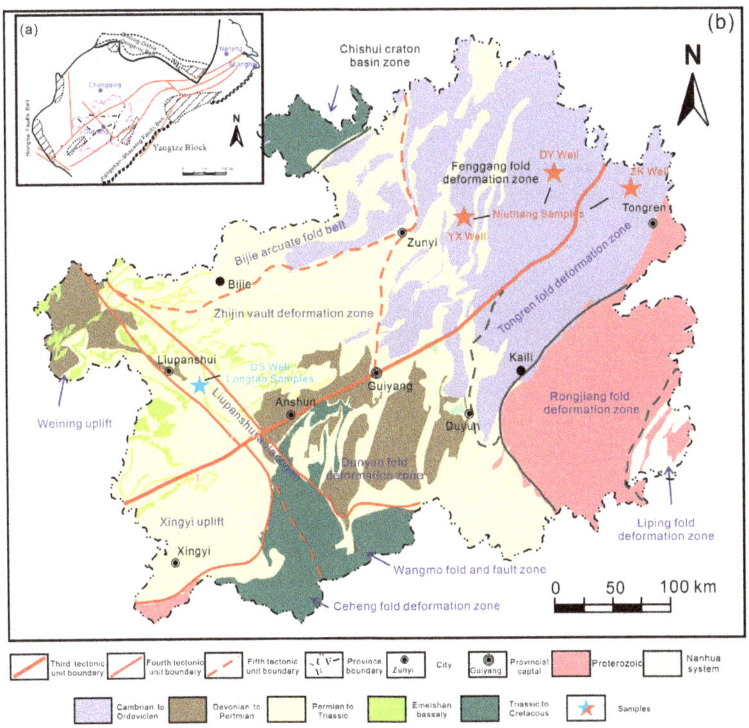

Figure 1. Geological setting. (**a**) Regional location of the Upper Yangtze Platform; and (**b**) a geological map of Guizhou Province, showing its tectonic zones.

Organic matter is usually formed by diagenesis at the same time as shale and evolves into solid kerogen and secondary organic matter in the late diagenetic stage. The main parent material of organic matter is kerogen. In order to study the characteristics of pore development of organic matter and explore the relationship between kerogen pores and shale pores, we isolated kerogen from shale according to National Standard GB/T 19144-2010 [17]. The operation aimed to remove Al_2O_3, CaO, and MgO from the shale by two acid treatments, followed by the removal of the modern sediment formed after the acid treatment by NaOH. At this point, there was still pyrite in the black shale, which was removed by zinc and HCL. Finally, the experimental sample was mixed with heavy liquid (mixture of hydrogen chloride, potassium iodide, and glacial acetic acid with a relative density of 2.01~2.10 g/mL) and centrifuged to separate kerogen. Kerogen can be purified by chloroform treatment of isolated kerogen.

2.2. Field Emission Scanning Electron Microscopy (FE-SEM)

The pore development characteristics of organic matter in black shale were qualitatively observed by field emission scanning electron microscopy (FE-SEM). The experimental instrument was ZEISS SIGMA field emission scanning electron microscope, under which microscopic images magnified 20,000~100,000 times could be observed. Before the experiment, some of the samples were ion polished using argon to remove impurities and form a smooth two-dimensional plane.

2.3. Low-Pressure Gas (N_2 and CO_2) Adsorption

For the analysis of black samples and kerogen samples, we conducted low-pressure N_2 and CO_2 adsorption experiments using the Micromeritics ASAP 2460 automatic specific surface area and pore size analyzer. The N_2 adsorption experiment focused on identifying pores ranging from 1.7 to 300 nm through monolayer adsorption. On the other hand, the CO_2 adsorption experiment aimed to identify pores ranging from 0.35 to 1.5 nm through double-layer adsorption [18,19]. Based on the results obtained from nitrogen adsorption experiments, we utilized the Brunauer–Emmet–Teller (BET) theory and BJH model to calculate parameters such as pore size distribution, specific surface area, and pore volume. To further analyze carbon dioxide adsorption, we employed density functional theory (DFT) to calculate micropore volume and micropore specific surface area [20–24].

2.4. Fractal Theory

2.4.1. Frenkel–Halsey–Hill (FHH) Model

Based on low-temperature N_2 adsorption, fractal models primarily include the BET model, Langmuir model, Frenkel–Halsey–Hill (FHH) model, and thermodynamic methods. Among these, the FHH model has been widely used for analyzing dimensionality in porous media materials. The fractal theory for low-temperature N_2 adsorption in porous media was initially proposed by Pfeifer and Avnir in 1983 [25,26]. This theory suggests that the pore fractal dimension D can be used to characterize the complexity and heterogeneity of porous material pore structures, typically ranging between 2 and 3, where 2 indicates smooth planes and 3 indicates free space. Therefore, when D approaches 2, the pore structure in porous media tends to be simple with low surface roughness, while as D approaches 3, the pore structure becomes more complex. Shale, similar to porous materials, also exhibits typical fractal characteristics in its pore structure [27].

Previous studies have shown that the Pfeifer-corrected FHH model can effectively characterize the pore fractal dimension of shale and kerogen based on nitrogen adsorption experimental results, with a relatively high fractal dimension fitting degree (R2 exceeding 0.90), reflecting high levels of pore structure heterogeneity and complexity [14,28,29]. Thus, this study similarly opts for the Pfeifer-corrected FHH fractal model to calculate pore fractal dimensions, aiming to explore the pore structure characteristics of shale and kerogen, as shown below:

$$\ln V = C + K\ln\left[\ln\left(\frac{P_0}{P}\right)\right] \tag{1}$$

where P is the equilibrium pressure, MPa; V is the adsorption volume at equilibrium pressure P; P_0 is the saturated steam pressure, MPa; C is a constant; and K is the slope of the fitted line. In this study, it was observed that the adsorption curve and desorption curve of the N_2 adsorption/desorption experiment did not coincide. Hysteresis loops were observed at a relative pressure of 0.45, indicating the presence of capillary condensation during the nitrogen adsorption process. Consequently, the fractal dimension of the nitrogen adsorption was calculated using the equation $D = K + 3$.

2.4.2. Sierpinski Model

It is worth noting that the FHH fractal model, considering multi-layer adsorption and capillary condensation effects, is mainly suitable for calculating fractal dimensions within the mesopore range. Errors were consistently found when applying the FHH model to calculate carbon dioxide adsorption experimental data, indicating that the FHH model based on uniform plate-like molecular layer adsorption is not suitable for calculating the fractal dimensions of micropores (less than 1.5 nm) [29]. In micropores, adsorbate molecules typically undergo pore filling, a process distinct from adsorption behavior under mesoporous conditions. Additionally, there is diffusion resistance for liquid N_2 molecules entering micropores at experimental temperatures. Therefore, the applicability of the FHH fractal model is limited, and the model for calculating micropore fractal dimensions should differ from that of mesopores. Considering the existence of micropore filling in CO_2 (273 K) adsorption, a model based on Sierpinski fractal theory was applied by recursively describing the embedding of typical fractal volumes in three-dimensional space relative to their characteristic lengths in the Sierpinski set, as shown below [30–32]:

$$V = L^{(3-D)} \tag{2}$$

where L is length of the yardstick, m; D is volume fractal dimension; and V is the adsorption volume at equilibrium pressure P, mL.

Angulo et al. argued that fluid filling processes, such as mercury intrusion and micropore filling in CO_2 (273 K) adsorption, do not follow a box counting process and adjusted Formula (2) to relate volume to pressure as shown below:

$$V \propto (P - P_t)^{(3-D)} \tag{3}$$

where P_t is the threshold pressure. For CO_2 adsorption at 273 K, the pressure of the first point in the adsorption isotherm was regarded as P_t, MPa. D is the volume fractal dimension. When P is very close to or above P_t, V is the adsorption volume at equilibrium pressure P, mL. Here, if we apply logarithms to both sides of Equation (3), we obtain the following:

$$InV = C + (3 - D)In[In(P - P_t)] \tag{4}$$

from which D can be readily derived through the physisorption data.

2.5. *Calculation Method of Organic Pore Contribution*

Organic matter pores are a distinct type of intergranular pore. In this paper, a calculation method (CRV and CRA) for the contribution rates of specific surface area and pore volume of organic matter is proposed. The CRV formula equation is expressed as follows:

$$\mathrm{CRV} = \frac{\mathrm{V_{org}} * \mathrm{TOC}}{\mathrm{V_{shl}}} * \mathrm{C_1} \tag{5}$$

where V_{org} is the pore volume of 1.0 g of kerogen, V_{shl} is the pore volume of 1.0 g of black shale, TOC is the total organic carbon content of black shale, and C_1 is the pore volume constant of the shale.

Shale pore development exhibits significant heterogeneity, which varies across different formations. Therefore, we suggest that the constant value (C_1) representing the pore characteristics of each formation is essential for analyzing the contribution rate of organic matter pores. The calculation is predominantly contingent on data gathered from prior research, encompassing the average pore volume, average TOC content, and average organic matter pore volume of black shale in the target formation [10,33–39]. These parameters serve as correction constants (C_1) in the study as shown below:

$$C_1 = \frac{\overline{V_{shl}} * \overline{L_V}}{\overline{V_{org}} * \text{TOC}} \tag{6}$$

where $\overline{V_{shl}}$ is the average shale pore volume, $\overline{L_V}$ is the average organic matter pore volume rate, and $\overline{V_{org}}$ is the average organic matter pore volume.

In addition, the contribution rate of the organic matter pore surface area (CRA) of the black shale is expressed as follows:

$$\text{CRA} = \frac{S_{org} * \text{TOC}}{S_{shl}} * C_2 \tag{7}$$

where S_{org} is the pore surface area of 1.0 g of kerogen, S_{shl} is the pore surface area of 1.0 g of black shale, TOC is the total organic carbon content of the black shale, and C_2 is a correction constant.

C_2 has a similar meaning to C_1. The formula equation of C_2 is expressed as follows:

$$C_2 = \frac{\overline{S_{shl}} * \overline{L_S}}{\overline{S_{org}} * \text{TOC}} \tag{8}$$

where $\overline{S_{shl}}$ is the average specific surface area of black shale, $\overline{L_s}$ is the average specific surface area ratio of organic matter, and $\overline{S_{org}}$ is the average specific surface area of organic matter.

3. Results

3.1. Organic Geochemistry and Mineralogical Composition

3.1.1. TOC Content

The TOC can better reflect the hydrocarbon generation capacity of shale and affects the occurrence state and gas content of hydrocarbon gases [40]. The TOC content of the Niutitang Formation shale ranges from 1.0 wt% to 10.0 wt%, with an average of 6.0 wt%. The TOC content of the Longtan Formation shale ranges from 1.2 wt% to 11.3 wt%, with an average of 5.3 wt% (Table 1). According to the classification and evaluation standard of the organic matter abundance of source rocks, it can be judged that the Niutitang and Longtan Formation shales are mainly "good-best" source rocks when only using the TOC index. These indicate that the shale samples of the Niutitang Formation and Longtan Formation are rich in organic matter.

Table 1. TOC content, Ro, kerogen type, mineralogical composition, and clay mineral content of the shales from the Niutitang Formation and Longtan Formation.

Shale Type	Sample	TOC (%)	R_o (%)	Kerogen Type	Mineral Composition (%)							Clay Mineral Content (%)			
					Quartz	Feldspar	Siderite	Dolomite	Calcite	Pyrite	Clay	I/S	It	Kao	Chl
Marine shale (Niutitang Formation)	DY-1	1.0	2.4	I	25.6	12.2	-	-	0.5	11.4	58.1	10.6	88.8	-	0.6
	DY-2	1.1	2.6	I	46.4	8.2	-	1.3	0.7	9.8	40.9	-	73.9	-	0.4
	YX-1	7.7	2.5	I	23.8	18.3	-	14.4	1.9	6.1	33.0	10.0	72.5	1.7	2.9
	YX-2	8.9	3.0	I	54.0	15.0	-	6.0	0.3	6.1	17.2	10.0	77.5	-	-
	ZK-1	6.6	4.1	I	28.2	10.6	0.4	15.1	-	2.0	34.3	13.2	64.4	5	-
	ZK-2	8.8	4.1	I	52.1	5.5	1.4	-	2.0	2.5	29.1	16.5	60.4	5	-
Marine–continental transi-tional shale (Longtan Formation)	DS-1	11.3	0.76	III	54.0	-	-	-	19.1	7.2	19.7	-	-	46.9	53.1
	DS-2	6.34	0.78	III	13.5	-	-	-	28.0	11.2	47.3	-	-	79.8	20.2
	DS-3	7.13	0.81	III	10.5	2.4	-	-	0.0	0.0	87.1	-	-	69.6	30.4
	DS-4	3.89	0.88	III	26.9	18.8	-	-	1.6	0.0	52.7	32.6	5.2	41.0	21.2
	DS-5	1.2	0.95	III	47.1	0.0	-	-	5.3	14.2	33.4	92.0	-	3.2	4.8
	DS-6	1.81	0.94	III	23.9	1.4	-	-	1.3	0.0	73.4	11.5	4.7	55.2	28.6
	DS-7	13.40	0.68	III	36.2	-	-	-	34.9	7.6	21.3	-	-	67.0	33.0

3.1.2. Thermal Maturity

Organic matter undergoes irreversible maturation evolution after deposition, resulting in variations in the occurrence form and structural characteristics of the same type of organic matter at different stages of evolution (the immature stage, mature stage, highly mature stage, and over-mature stage) [41]. The Niutitang Formation samples have an R_o range of 2.40% to 4.10%, with an average of 3.10%, indicating high maturity to over-maturity and passage through the gas window. The Longtan shale has an R_o range of 0.68% to 0.95%, with an average of 0.83%, suggesting that it is in the mature stage and has reached the gas window threshold [42,43].

3.1.3. Mineralogical Composition

Shale minerals were subjected to an XRD analysis, and the results are presented in Table 1. The dominant mineral components of the Niutitang Formation shale consist of quartz and clay minerals, accounting for over 50% of the total mineral composition. The high proportions of clay minerals and quartz in the Niutitang Formation shale suggest that sedimentation processes are primarily influenced by terrigenous clastic, biological, or hydrothermal deposition mechanisms. On the other hand, the Longtan Formation shale predominantly contains high levels of clay minerals, indicating the significant influence of terrigenous detritus during deposition [44]. The mineral compositions within the Longtan Formation shale display significant variability (Figure 2), which is particularly evident when looking at the presence of carbonate minerals and pyrite. This reflects a marine–continental transitional deposition environment in the Longtan Formation characterized by frequent water turbulence. The depositional setting was mainly composed of deltaic deposits, tidal flats, and lagoon sediments, resulting in varying shale mineral compositions [10]. According to the ternary mineral classification (Figure 2), DY-1, DS-4, and DS-6 belong to siliceous-rich clay shale (SRCS). DY-2, ZK-2, and DS-5 belong to clay-rich silicon shale (CRSS). YX-2 and DS-1 belong to silicon shale (SS). YX-1 and ZK-1 belong to a clay/silicon shale mixture (CSSM). DS-2 belongs to calcium-rich clay shale (CRCS). DS-3 belongs to clay shale (CS). DS-7 belongs to a silicon/calcium shale mixture (SCSM).

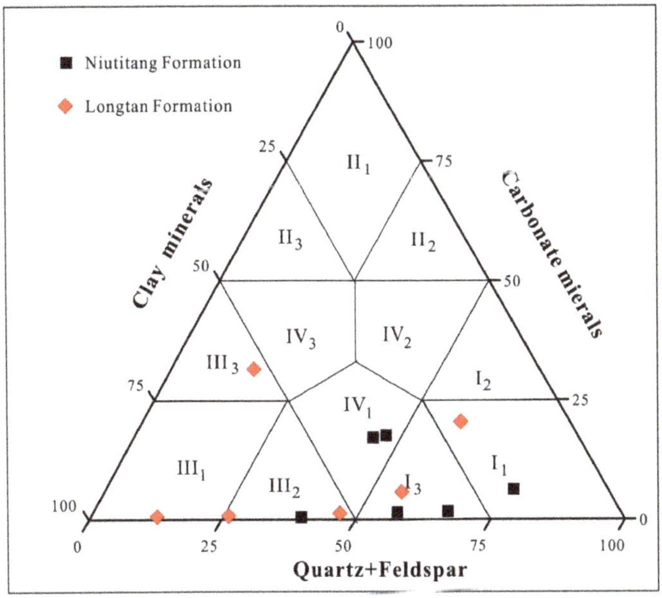

Figure 2. Ternary diagram of mineral composition and classification.

3.2. Pore Morphology and Types

3.2.1. Inorganic Pores

The inorganic pores present in the marine shale and marine–continental transitional shale primarily consist of pyrite intergranular pores, clay mineral interlaminar pores, dissolution pores, etc. (Figures 3a–f and 4a–f). Among these, the most prevalent are the interlaminar pores formed by disordered sheets, slits, or wedges of clay minerals that are susceptible to deformation or closure due to external compression. The strawberry-shaped intergranular pores of pyrite were observed through an FE-SEM analysis. Pyrite dissolution resulted in the formation of dissolution pores which were subsequently filled by clay minerals within the intergranular spaces. While clay minerals contribute a significant number of pore spaces, their high abundance can also impede shale porosity development. Dissolution pores are associated with interactions between organic acids and minerals in coal measure shales; feldspar is commonly dissolved along with quartz, calcite, and pyrite, exhibiting dissolved pore structures. Therefore, in FE-SEM images, it can be observed that inorganic pores exhibit a multitude of pore types, revealing the intricate nature of their pore structure.

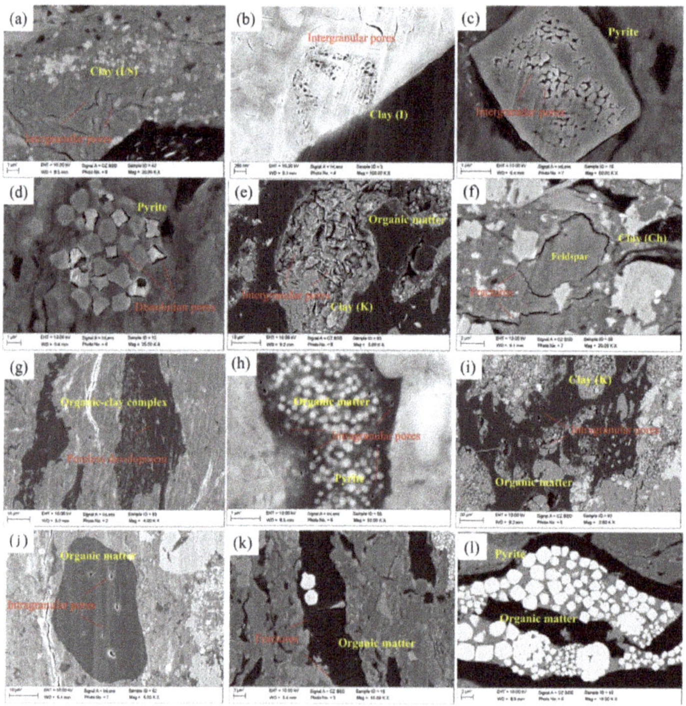

Figure 3. The mineral composition, organic matter morphology, and pore development of Longtan Formation shale under FE-SEM. (**a**) Intergranular pores of illite–montmorillonite mixed layer, Longtan Formation. (**b**) Intergranular pores of illite, Longtan Formation. (**c**) Intergranular pores of pyrite, Longtan Formation. (**d**) Dissolution pores of pyrite, Longtan Formation. (**e**) Intergranular pores of kaolinite, Longtan Formation. (**f**) Fractures of feldspar and intergranular pores of chlorite, Longtan Formation. (**g**) Organic-clay complex with undeveloped pores, Longtan Formation. (**h**) Organic-pyrite complex with intragranular pore development. Longtan Formation. (**i**) Organic-clay complex developed in intragranular pores, Longtan Formation. (**j**) Intragranular pores of organic matter, Longtan Formation. (**k**) Fractures of organic matter, Longtan Formation. (**l**) Banded organic matter wraps pyrite, Longtan Formation.

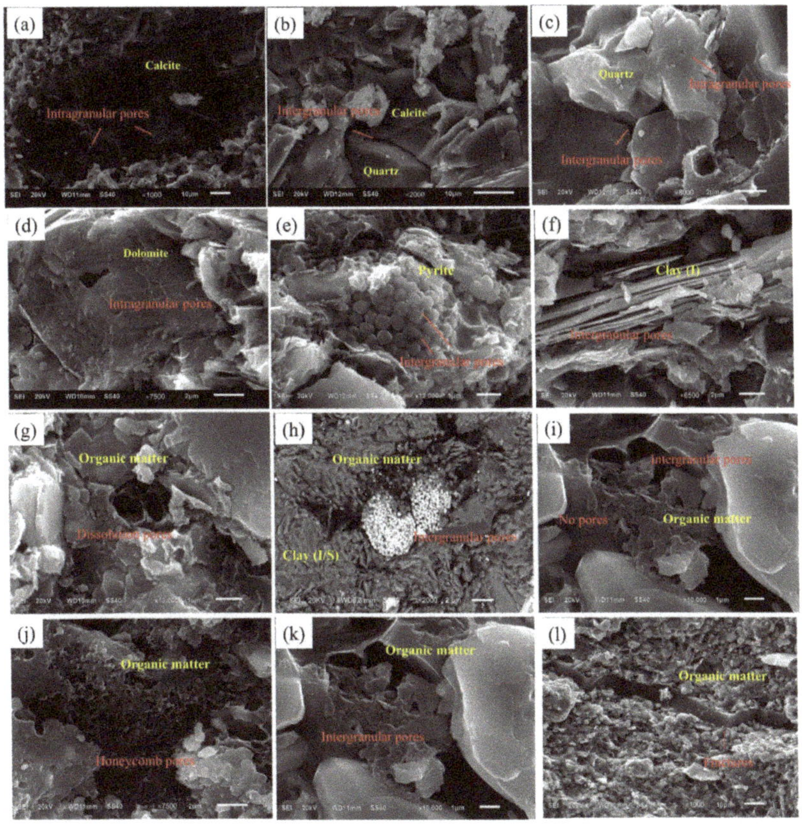

Figure 4. The mineral composition, organic matter morphology, and pore development of Niutitang Formation shale under FE-SEM. (**a**) Intragranular pores of calcite, Niutitang Formation. (**b**) Intergranular pores of quartz and calcite, Niutitang Formation. (**c**) Intergranular and intragranular pores of quartz, Niutitang Formation. (**d**) Dissolution pores of dolomite, Niutitang Formation. (**e**) Intergranular pores of pyrite. (**f**) Intergranular pores of illite, Niutitang Formation. (**g**) Dissolution pores of organic matter, Niutitang Formation. (**h**) Organic-clay-pyrite complex with interparticle pores development. Niutitang Formation. (**i**) Intragranular pores are partially developed in organic matter, Niutitang Formation. (**j**) Honeycomb pores of organic matter, Niutitang Formation. (**k**) Intergranular pores of organic matter, Niutitang Formation. (**l**) Banded organic matter developed fractures, Niutitang Formation.

3.2.2. Organic Matter Pores

Developments in the shape, size, and number of organic matter pores can be effectively observed under FE-SEM. Organic matter pores are the most widely developed pore type in marine organic-rich shale [45,46]. However, the development of organic matter pores in marine–continental transitional shales is suboptimal.

FE-SEM revealed that the pore development of organic matter varied among the different occurrence types (Figures 3g–l and 4g–l). Despite similar TOC contents in the Niutitang Formation shale and the Longtan Formation shale, the number of organic matter pores in the Niutitang Formation shale is significantly greater than that in the Longtan Formation shale [10]. The sapropelic kerogen macerals from the Niutitang Formation primarily consist of 72.00~89.70%, with an average content of 81.45%. Only a few samples contain inertinite, implying that lower aquatic organisms with algae as their primary source contribute mainly to the organic matter composition, while higher plants make up only a small portion. Type I kerogen is predominant in the Niutitang Formation. Our maceral composition analysis reveals that vitrinite constitutes most (76.80~91.00%) of the Longtan Formation shale's macerals; however, there is a relatively low content of inertinite (4.69~15.55%) and exinite (2.06~10.94%), respectively. These results indicate that the Longtan Formation shale predominantly consists of type III kerogen. Different types of organic matter reflect distinct sedimentary environments, determining their maceral compositions [47,48].

Therefore, this discrepancy may be attributed to type III kerogen having fewer pores present in the Longtan Formation's organic-rich shale compared to type I kerogen [49]. Additionally, the lower thermal maturity ($0.68\% < R_o < 0.95\%$) of the Longtan Formation's organic matter relative to that of the Niutitang Formation ($2.40\% < R_o < 4.10\%$) contributes to differences in their respective pore developments. In addition, elliptical and slit pores are the predominant pore types formed by organic matter, characterized by a relatively complicated pore structure. Typically, these pores originate from hydrocarbon production channels and the self-condensation of organic matter itself.

3.3. *Quantitative Characteristics of Pore Development*

3.3.1. The Low-Pressure N_2 Adsorption Analysis

An adsorption analysis of low-pressure gas (N_2/CO_2) can effectively identify pores in organic-rich shale ranging from 0.35 to 300 nm. Additionally, the amount of gas adsorbed reflects the number of pores present and the complexity of the pores. Various classification schemes exist for pore size, and in this study, we primarily adopted the pore classification scheme proposed by The International Union of Pure and Applied Chemistry (IUPAC). According to this scheme, pores are categorized into three groups: micropores with diameters less than 2 nm, mesopores with diameters between 2 nm and 50 nm, and macropores with diameters exceeding 50 nm.

Figure 5 illustrates the N_2 adsorption and desorption curve under low-pressure conditions. Based on the IUPAC's N_2 adsorption and desorption curve classification system, both the Niutitang Formation shale and Longtan Formation shale exhibit typical IV curves. In contrast, their kerogen curves differ significantly from those observed in the black shale samples; they resemble type II or type III curves. Hysteresis rings resulting from the non-coincidence between the adsorption/desorption curves are commonly observed in black shale and kerogen, providing insights into their pore morphologies [50].

The IUPAC divides hysteretic loops into H1, H2, H3, and H4 types (Figure 6). Based on our N_2 adsorption/desorption curve analysis, the pore morphology of the Niutitang Formation shale can be characterized as a combination of lamellar pores (H3 type) and slit holes (H4 type). Similarly, for the Longtan Formation shale samples, a mixture of ink-bottle-shaped pores (H2 type) and slit-shaped pores (H3 type) is observed. These findings highlight the complex nature of pore structures in both of these formations, indicating that a single classification cannot fully capture their morphological diversity. Notably, our examination under FE-SEM revealed further complexity in terms of pore types (Figures 3 and 4). It is worth mentioning that the kerogen samples predominantly exhibited cylindrical pores followed by slit pores; however, only the YX-2 kerogen and DY-2 kerogen displayed hysteresis rings characteristic of H4-type behavior. This suggests that organic matter plays a significant role in shaping shale porosity with a higher adsorption capacity compared to inorganic components at equivalent masses.

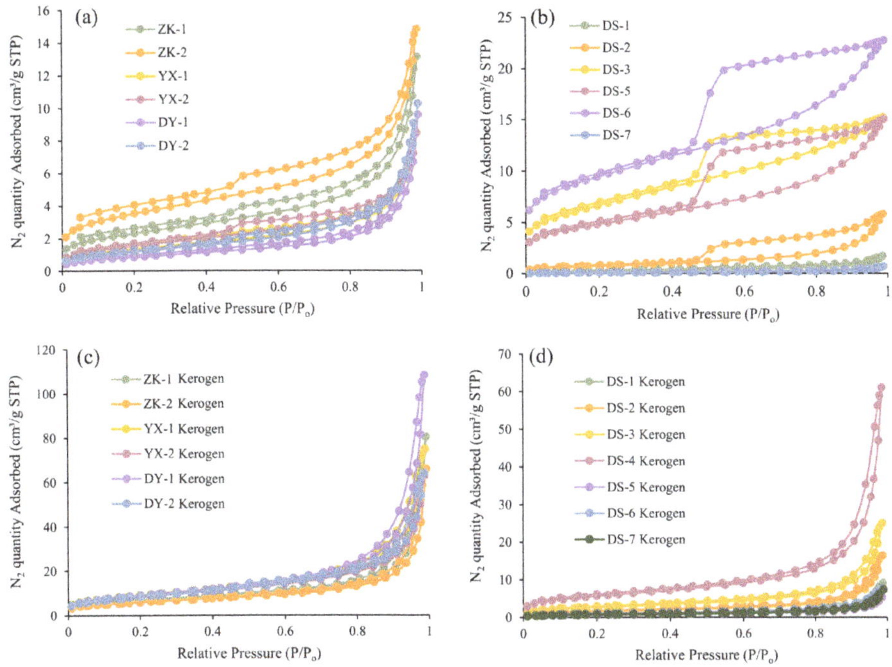

Figure 5. Low-pressure gas N_2 adsorption/desorption of the shales and the organic matter. (**a**) The marine Niutitang Formation shale. (**b**) The marine–continental transitional Longtan Formation shale. (**c**) The marine Niutitang Formation kerogen. (**d**) The marine–continental transitional Longtan Formation kerogen.

Based on the N_2 adsorption and desorption curves, the pore specific surface area and pore volume of the shale and kerogen were calculated using the BJH model [28]. The pore specific surface area of the Niutitang Formation shale ranges from 3.17 m^2/g to 10.38 m^2/g, with an average of 10.38 m^2/g. The shale pore volume ranges from 0.0123 to 0.0222 cm^3/g, with an average of 0.0161 cm^3/g. The pore specific surface area of the Longtan Formation shale ranges from 0.86 to 29.27 m^2/g, with an average value of 14.72 m^2/g. The shale pore volume ranges from 0.0025 to 0.0351 cm^3/g, with an average of 0.0202 cm^3/g. The specific surface area of the Niutitang Formation kerogen pores ranges from 20.26 m^2/g to 33.34 m^2/g, with an average of 20.26 m^2/g. The kerogen pore volume ranged from 0.0954 to ~0.1639 cm^3/g, with an average of 0.1155 cm^3/g. The specific surface area of the Longtan Formation kerogen pores ranges from 2.76 m^2/g to 18.41 m^2/g, with an average of 6.79 m^2/g, and the volume of the kerogen pores ranges from 0.0077 cm^3/g to 0.0916 cm^3/g, with an average of 0.0311 cm^3/g (Figure 7). The pore specific surface area and pore volume of the Longtan Formation shale are higher than those of the Niutitang Formation shale, and the pore specific surface area and pore volume of the Niutitang Formation kerogen are higher than those of the Longtan Formation kerogen. This shows that the inorganic pores of the Longtan Formation shale are more developed than those of the Niutitang Formation shale, and the organic matter pores of the Niutitang Formation are more developed.

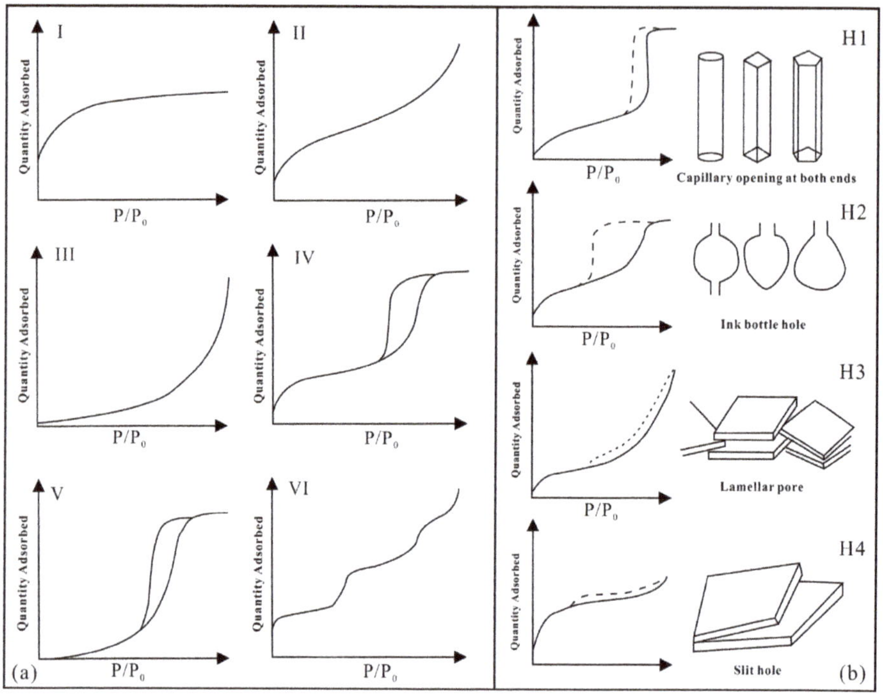

Figure 6. Curve types of gas adsorption and pore morphology (modified for Sing., 1985 [51]). (**a**) Type of nitrogen adsorption curve. (**b**) Hysteresis loop and corresponding pore morphology.

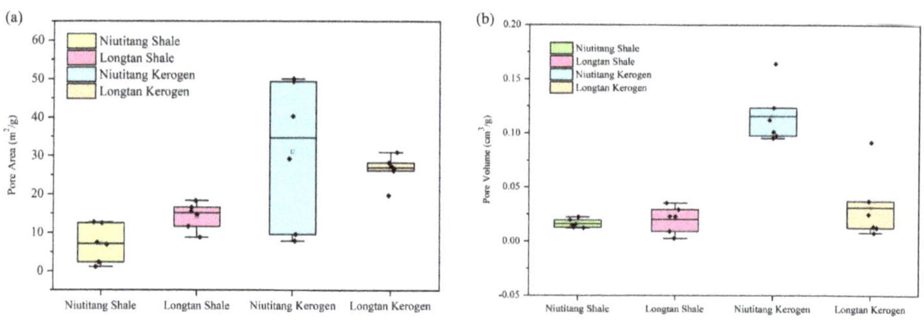

Figure 7. Pore specific surface area distribution/pore volume distribution of shale and organic matter based on N_2 gas adsorption. (**a**) Pore specific surface area of shale and kerogen. (**b**) Pore volume of shale and kerogen.

3.3.2. The Low-Pressure CO_2 Adsorption Analysis

CO_2 adsorption can characterize the pore structure at 0.3~1.5 nm. Based on the BDDT classification scheme [21], the CO_2 adsorption isotherms of the shale and kerogen samples from the Niutitang Formation and Longtan Formation all belong to type I (Figure 8). The CO_2 adsorption capacity of the Longtan Formation shale is higher than that of the Niutitang Formation shale, indicating that the Longtan Formation shale has more micropores. The CO_2 adsorption capacity of the Niutitang Formation kerogen is higher than that of the Longtan kerogen, indicating that the type I kerogen developed more micropores than the type III kerogen.

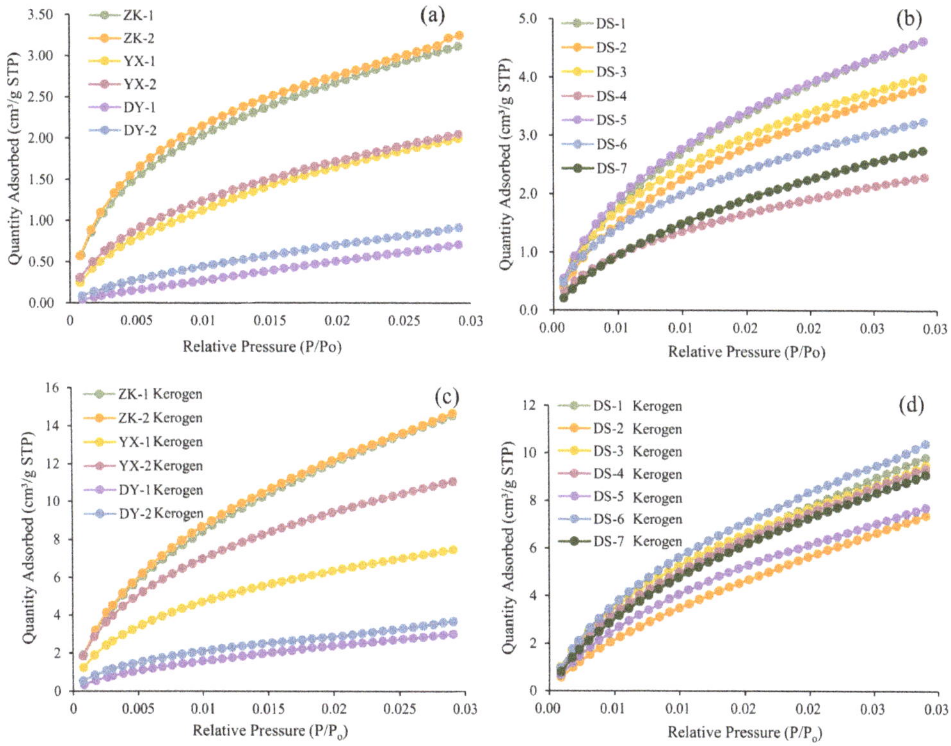

Figure 8. Low-pressure CO_2 adsorption curves of the black shales and the organic matter. (**a**) The marine Niutitang Formation shale. (**b**) The marine–continental transitional Longtan Formation shale. (**c**) The marine Niutitang Formation kerogen. (**d**) The marine–continental transitional Longtan Formation kerogen.

The specific surface area and volume of the micropores of the shale and kerogen were calculated using the DFT model [23,24]. The specific surface area of the micropores in the Niutitang Formation shale ranges from 1.10 to 12.67 m^2/g, with an average of 7.11 m^2/g, and the volume of the micropores ranges from 0.0004 to 0.0047 cm^3/g, with an average of 0.0027 cm^3/g. The specific surface area of the micropores in the Longtan Formation shale ranges from 8.76 to 18.29 m^2/g, with an average of 14.23 m^2/g, and the volume of the micropores ranges from 0.0046 to 0.0077 cm^3/g, with an average of 0.0064 cm^3/g. The specific surface area of the Niutitang Formation kerogen micropores ranges from 7.83 to 50.01 m^2/g, with an average of 31.02 m^2/g, and the micropore volume ranges from 0.0032 cm^3/g to 0.0192 cm^3/g, with an average of 0.0119 cm^3/g. The specific surface area of the Longtan Formation kerogen micropores ranges from 19.65 to 30.94 m^2/g, with an average of 26.43 m^2/g, and the micropore volume ranges from 0.0067 to 0.0116 cm^3/g, with an average of 0.0101 cm^3/g (Figure 9). Similarly, in terms of micropores, the micropore specific surface area and pore volume of the Longtan Formation shale are higher than those of the Niutitang Formation shale, while those of the Niutitang kerogen are higher than those of the Longtan kerogen. This shows that the Longtan Formation shale pores are more developed, and the organic matter pores are mainly micropores.

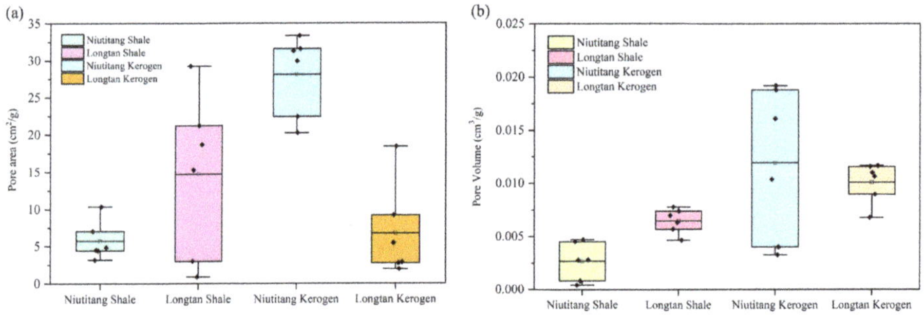

Figure 9. Specific surface area distribution/pore volume distribution of shale and organic matter based on CO_2 gas adsorption. (**a**) Pore specific surface area of shale and kerogen. (**b**) Pore volume of shale and kerogen.

3.4. Fractal Dimension Characteristics

By analyzing N_2 and CO_2 adsorption data, we obtained insights into the pore adsorption properties of these shales. The results indicate distinct differences between the shale pores in the Niutitang Formation and those in the Longtan Formation, with a higher level of porosity observed in the latter while an enhanced development of organic matter pores was evident in the former. Therefore, we employ the FHH model and the Sierpinski model to investigate the structural characteristics of the shale and kerogen pores in marine and transitional marine–continental facies, aiming to elucidate the primary factors controlling pore development.

3.4.1. Fractal Dimensions Calculated from the N_2 Adsorption

In order to enhance our understanding of pore development complexity in the Niutang and Longtan Formations, extensive investigations have been conducted on the pores of shale and kerogen above the mesopore range. These studies have involved the calculation of the fractal dimension using Formula (1) and the examination of N_2 adsorption/desorption results to assess the complexity of shale and kerogen mesopores.

Curve fitting for the N_2 adsorption fractal dimension characteristics is shown in Figure 10a–d, yielding an excellent correlation coefficient (R^2) ranging from 0.9845 to 0.9992, showing a strong fit between them. The fractal dimensions are mainly categorized into D1 and D2 types, reflecting two different adsorption characteristics. Fractal dimension D1 primarily represents the low-pressure region fractality ($P/P_0 < 0.45$), reflecting the van der Waals force phenomenon of monolayer adsorption in micropores with diameters smaller than 3 nm and dominated by adsorption pores containing adsorbed gas. Fractal dimension D2 corresponds to the mid- to high-pressure region fractality ($P/P_0 > 0.45$), indicating the spatial complexity and irregularity of mesopores and macropores with diameters larger than 3 nm, mainly composed of flow-through pores containing free gas [52].

Our calculation results show the following: The fractal dimension D1 of the Longtan Formation shale is observed to be higher than that of the Niutitang Formation shale, while the fractal dimension D2 of the Longtan Formation shale is lower than that of the Niutitang Formation shale (Table 2). This suggests a greater complexity in the micropores within the Longtan Formation compared to that of those within the Niutitang Formation and a higher complexity in mesopore development within the Niutitang Formation relative to that of the Longtan Formation. These findings further highlight differences in the evolution of pore characteristics between marine shales and marine–continental transitional shales. For kerogen, it is noted that both fractal dimensions D1 and D2 are higher for the Niutitang Formation kerogen compared to the Longtan Formation kerogen; however, this difference is not substantial, possibly due to variations caused by the low level of maturity in the

Longtan Formation and the high level of maturity in the Niutitang Formation impacting organic matter pore development.

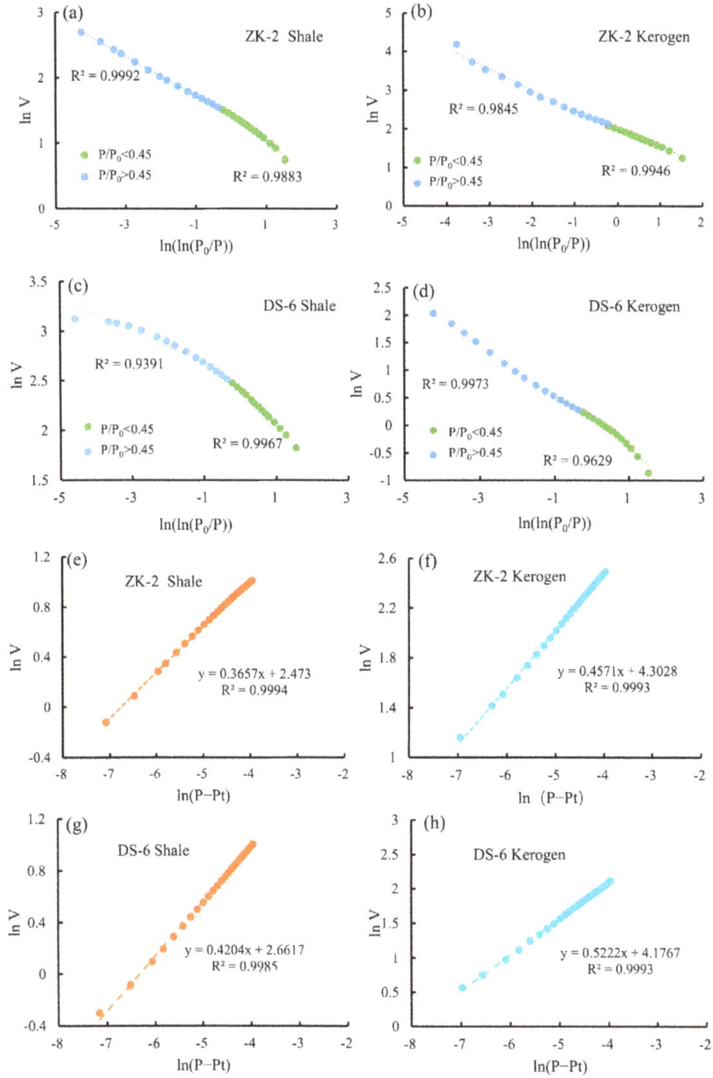

Figure 10. Fractal characteristics of the pore structure from the shales and the organic matter based on N_2 adsorption experiments. (**a**) Fractal dimensions of ZK-2 Shale based on the FHH model. (**b**) Fractal dimensions of ZK-2 Kerogen based on the FHH model. (**c**) Fractal dimensions of DS-6 Shale based on the FHH model. (**d**) Fractal dimensions of DS-6 Kerogen based on the FHH model. (**e**) Fractal dimensions of ZK-2 Shale based on the Sierpinski model. (**f**) Fractal dimensions of ZK-2 Kerogen based on the Sierpinski model. (**g**) Fractal dimensions of DS-6 Shale based on the Sierpinski model. (**h**) Fractal dimensions of DS-6 Kerogen based on the Sierpinski model.

Table 2. Fractal dimensions based on the N_2 adsorption results of the shales and the organic matter.

Samples	$P/P_0 < 0.45$		$P/P_0 > 0.45$		CO_2 Adsorption	
	R^2	D1	R^2	D2	R^2	Dm
DY-1	0.9962	2.4680	0.9986	2.5117	0.9870	2.3232
DY-2	0.9966	2.4523	0.9988	2.5796	0.9974	2.4443
YX-1	0.9969	2.5199	0.9994	2.6160	0.9962	2.537
YX-2	0.9947	2.4894	0.9997	2.6674	0.9989	2.5772
ZK-1	0.9955	2.5495	0.9992	2.6412	0.9993	2.6320
ZK-2	0.9883	2.5854	0.9992	2.7019	0.9994	2.6343
DS-1	0.9900	2.6083	0.9996	2.4357	0.9920	2.4881
DS-2	0.9893	2.5681	0.9955	2.4681	0.9930	2.4968
DS-3	0.9988	2.8743	0.9264	2.5667	0.9990	2.5577
DS-4	0.9985	2.7467	0.9937	2.6514	0.9979	2.5602
DS-5	0.9990	2.7895	0.9548	2.5880	0.9996	2.5284
DS-6	0.9967	2.8421	0.9391	2.6335	0.9985	2.5796
DS-7	0.9978	2.4267	0.9749	2.5147	0.9986	2.4500
DY-1 K	0.9877	2.4490	0.9973	2.4529	0.9957	2.5092
DY-2 K	0.9945	2.4679	0.9996	2.5799	0.9991	2.5891
YX-1 K	0.9939	2.4791	0.9979	2.5558	0.9988	2.5592
YX-2 K	0.9923	2.4782	0.9991	2.5832	0.9988	2.6021
ZK-1 K	0.9928	2.5423	0.9948	2.4883	0.9985	2.5437
ZK-2 K	0.9946	2.5355	0.9854	2.4611	0.9993	2.5429
DS-1 K	0.9672	2.4396	0.9956	2.5210	0.9977	2.4395
DS-2 K	0.9737	2.4362	0.9951	2.4922	0.9973	2.4111
DS-3 K	0.9782	2.4391	0.9976	2.5192	0.9989	2.4494
DS-4 K	0.9848	2.4921	0.9945	2.4702	0.9978	2.4904
DS-5 K	0.9627	2.4558	0.9940	2.5614	0.9989	2.4466
DS-6 K	0.9629	2.4040	0.9973	2.5363	0.9993	2.4778
DS-7 K	0.9593	2.4622	0.9955	2.5278	0.9993	2.4450

3.4.2. Fractal Dimensions Calculated from the CO_2 Adsorption

We calculated the pore fractal dimension under CO_2 adsorption using Formula (4) to investigate the pore complexity of the shale and organic matter micropores. Curve fitting for the CO_2 adsorption fractal dimension characteristics is shown in Figure 10e–h, yielding an excellent correlation coefficient (R^2) ranging from 0.9920 to 0.9996, showing a strong fit between them. The obtained fractal dimension (Dm) values ranged between 2 and 3, suggesting varying levels of complexity in the micropores present in each sample. The similar Dm values observed for the Niutitang Formation and Longtan Formation shales imply a comparable degree of complexity in the micropores across different shale types. However, it is worth noting that the Dm value for kerogen in the Niutitang Formation group was higher compared to that in the Longtan Formation group. This discrepancy indicates that different types of kerogen exhibit diverse micropore complexities, with Type I kerogen displaying a higher level of complexity than Type III kerogen (Table 2).

Shale pores exhibit higher fractal dimensions than those of kerogen pores as they encompass various types within shale pore systems, including the intergranular pores associated with organic matter. Additionally, different mineral composition types may lead to distinct pore types and quantities within shale pore systems, which is particularly evident when examining the marine Niutitang Formation shales and marine–continental transitional Longtan Formation shales, in which their varying sedimentary facies result in differences in mineral composition, thereby influencing shale pore development.

4. Discussion

4.1. Contribution Rate of Organic Matter Pores to Shale Pore System

To investigate the contribution of organic matter pores in the shale pore system, calculations were made based on the previously described method for assessing the contribution of organic matter pores. This allowed for the determination of the contribution rates of organic matter pores in shale under different depositional environments.

According to Formula (6) and previous studies data, the C1 constants of the Niutitang Formation shale and the Longtan Formation shale are 0.93 and 1.13, respectively.

After applying Formula (5) to the gas adsorption data (Figure 11a,b), our results are as follows: the CRV value of the shale in the Niutitang Formation ranges from 6.38% to 60.91%, with an average of 36.08%; and the CRV value of the Longtan Formation shale ranges from 0.93% to 42.53%, with an average of 16.28%.

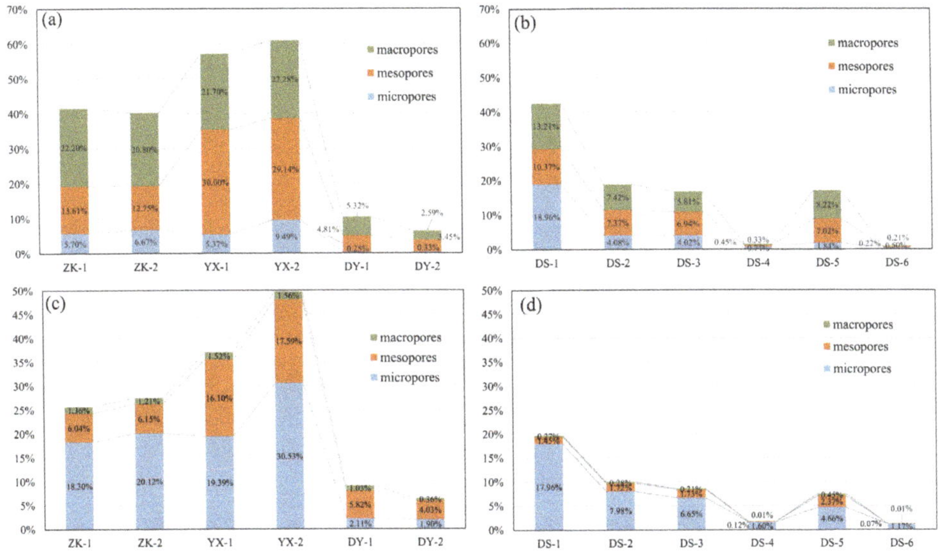

Figure 11. Contribution rate of the organic matter pores to the pore volume (CRV) and contribution rate of organic matter pores to pore area (CRA). (**a**) CRV of Niutitang Formation. (**b**) CRV of Longtan Formation. (**c**) CRA of Niutitang Formation. (**d**) CRA of Longtan Formation.

According to Formula (8), the C_2 correction constant of the Niutitang Formation and Longtan Formation is 0.9301. After applying Formula (7) to the pore specific surface area data measured by gas adsorption, the calculated CRA result is as follows: the CRA value of the Niutitang Formation is between 6.29% and 49.68%, with an average of 36.08%; and the CRA value of the Longtan Formation ranges from 1.26% to 19.68%, with an average of 8.12%.

It is obvious from the calculated contribution rate of organic matter pores that the CRV and CRA values of the Niutitang Formation are significantly higher than those of the Longtan Formation (Figure 11). This indicates that the pores of type I and II kerogen are more developed than those of type III kerogen, so the proportion of type I and II kerogen in the shale pore system increases. At the same time, the contribution of organic pores is also different due to different levels of thermal evolution maturity and the strong heterogeneity of shale.

In order to further discuss the contributions of different organic matter pore sizes, the contribution of micropores, mesopores, and macropores are calculated according to the IUPAC pore partitioning scheme. In the Niutitang Formation, organic matter mesopores are the primary contributors to the shale pore volume, followed by macropores. Organic matter micropores are the main contributors to the specific surface area of the shale pores, followed by mesopores. In the Longtan Formation, organic matter with different pore sizes contributes significantly to the shale's overall pore volume (Figure 11a,b). The micropore specific surface area of the Niutitang Formation accounts for a maximum of 30.53% of the total specific surface area, with an average of 15.39%. For the Longtan Formation, the maximum contribution to the specific surface area by micropores is 17.96%, with an average of 7.64%. This indicates that organic matter micropores are the primary contributors to the specific surface area of the pore systems in both sets of shales (Figure 11c,d).

The contribution of different organic matter pore sizes reflects both hydrocarbon generation and storage capacity in shale gas reservoirs [53,54]. Specifically, organic matter micropores provide a substantial amount of specific surface area and serve as numerous adsorption sites for gases [55–57]. This study reveals that most of the Niutitang Formation samples have an organic matter specific surface area occupying over one-third of total shale porosity, providing a significant amount of specific surface area for gas adsorption primarily through its micropores' and mesopores' networks. Conversely, the Longtan Formation shale, which is characterized by low-maturity marine–continental transitional conditions, shows relatively small specific surface areas within its organic matter pores due to the limited development of this type of formation environment. Pore connectivity plays a crucial role in fluid migration and natural gas development within shale reservoirs; thus, high-porosity connectivity is advantageous for improving natural gas production and recovery under favorable preservation conditions [58,59]. In the Niutitang Formation shale, organic matter contributes to a significant number of mesopores and macropores, accounting for more than 10% of the pore volume and providing ample space for shale gas storage. Although the specific surface area contribution from organic matter pores in the Longtan Formation is relatively small, certain samples from this formation (DS-1, DS-3, and DS-5) exhibit substantial contributions to the organic matter pore volume, indicating a good level of connectivity within these pores. Therefore, organic matter pores play a crucial role in the shale pore system. Building upon gas adsorption experiments and geochemical analyses, we further investigate the fractal characteristics of the shale pores and kerogen pores, as well as the controlling factors influencing pore complexity.

4.2. *Effect of Organic Matter Characteristics on Fractal Dimension and Pore Structure*

Previous studies have shown that mineral composition, TOC, thermal evolution maturity, and kerogen type are the main factors affecting pore characteristics [60,61]. However, the influence of organic matter pore structure needs to be further explored. In this study, by examining the influence of the main factors affecting the pore structure on the fractal dimension, we explore the relationship between the pore complexity of shale with different sedimentary facies and the main influencing factors.

4.2.1. Effect of TOC on Pore Complexity

The fractal dimension (D1, D2, and Dm) of the marine Niutitang Formation shale exhibits a positive correlation with the TOC, whereas the fractal dimension of the marine–continental transitional Longtan Formation shale demonstrates a negative correlation with the TOC (Figure 12a,c). This finding contradicts previous studies [62–64]. This phenomenon suggests that a higher TOC content does not necessarily lead to more complex shale pore development. As the TOC content increases in the shale, the degree of organic matter enrichment rises, but the complexity of organic matter pores varies among the different types. Specifically, at lower TOC levels, the fractal dimensions of the pores in the Longtan Formation shales dominated by type III kerogen are higher than those in the Niutitang Formation shales dominated by type I kerogen. However, as the TOC content increases, the fractal dimensions of the pores in the Niutitang Formation shales dominated by type I kerogen surpass those in the Longtan Formation shales dominated by type III kerogen. This indicates that shales with different types of organic matter exhibit varying degrees of pore complexity as the TOC content increases.

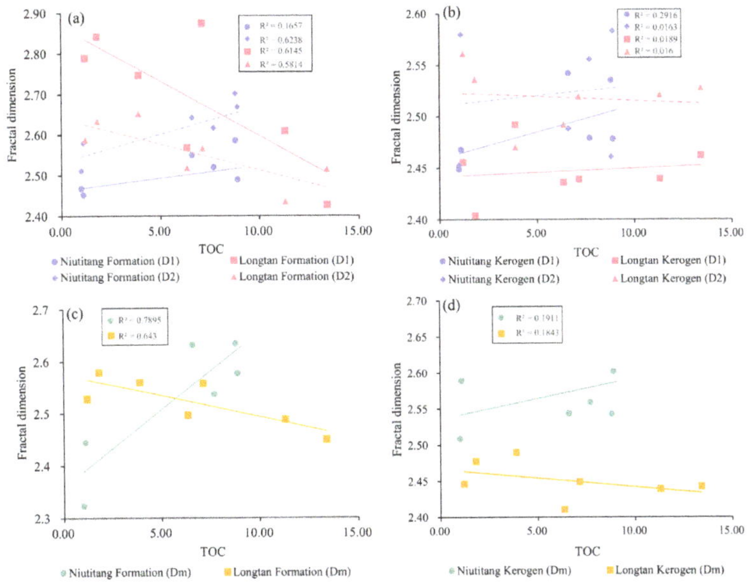

Figure 12. The correlation between fractal dimension and TOC. (**a**) The relationship between fractal dimensions D1, D2, and TOC in shale. (**b**) The relationship between fractal dimensions D1, D2, and TOC in kerogen. (**c**) The relationship between fractal dimension Dm and TOC in shale. (**d**) The relationship between fractal dimension Dm and TOC in kerogen.

Therefore, conducting a separate correlation study of the fractal dimensions of different types of kerogen and TOC contents reveals that an increase in the TOC content does not exhibit a significant relationship with the fractal dimensions of the kerogen pores (Figure 12b,d). This suggests that the increase in TOC content may lead to an increase in the quantity of organic matter pores and exhibit a correlation, but this does not inherently relate to the complexity of pores within the organic matter. From our FE-SEM observations, it is noted that the organic matter pores dominated by type I kerogen in the Niutitang Formation are well developed and exhibit diverse morphological changes (Figure 4g–l). Conversely, in the Longtan Formation, in which type III kerogen dominates, the development of organic matter pores is poor, with only a small number of pores observed under FE-SEM, and their complexity level is low (Figure 3g–l). This also indicates that the complexity of organic matter pores is not significantly correlated with the TOC but rather with the type of organic matter. Different organic matter components possess varying capabilities for pore development.

As for how the TOC content influences the pore complexity of shales from different depositional environments, resulting in opposing correlations between the two types of shales, when considering the relationship between the kerogen pore fractal dimensions and TOC content alongside our scanning electron microscopy results, it is evident that the pore complexity of the kerogen in the Niutitang Formation is consistently higher than that in the Longtan Formation. Additionally, the level of complexity of the organic matter pores observed during our microscopic examination is also higher in the Niutitang Formation compared to that in the Longtan Formation. With an increase in TOC content, the kerogen in the Niutitang Formation positively contributes to the overall pore complexity of the shale. Conversely, in the Longtan Formation, an increase in organic matter content may result in relatively reduced contributions from other mineral components, leading to a lower level of pore complexity.

4.2.2. Effect of R_o on Pore Complexity

During the process of hydrocarbon generation from organic matter burial, the development of organic matter pores undergoes a multi-stage evolution process [45,65]. In terms of the relationship between the shale pore fractal dimension and R_o, the relationship between the fractal dimensions D2, Dm, and R_o in the Niutitang Formation shale demonstrates an initial increasing trend followed by a decrease, with the inflection point occurring at R_o = 3.5%, while D1 initially increases and then decreases with increasing R_o values. Although most of the Longtan Formation shales have a low level of maturity, it still demonstrates a weak positive correlation (Figure 13a,c). This indicates that as thermal evolution progresses, organic matter initiates hydrocarbon generation and develops varying degrees of pores.

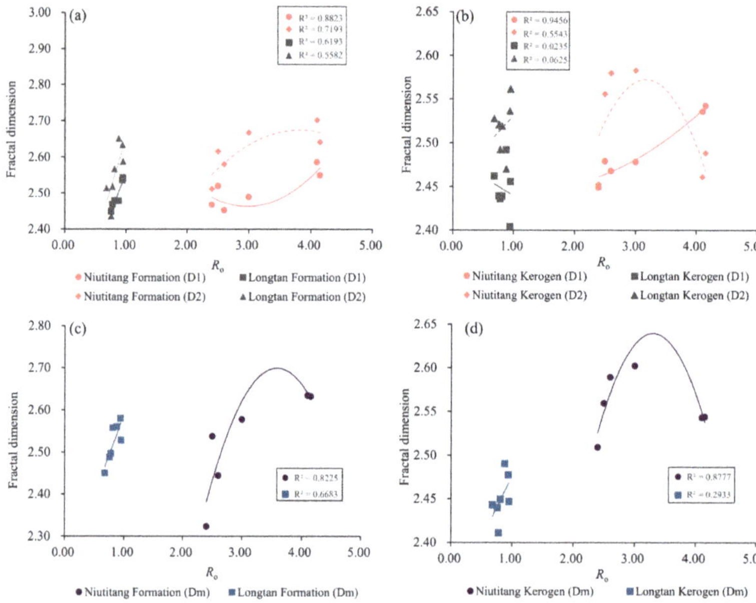

Figure 13. The correlation between fractal dimension and R_o. (**a**) The relationship between fractal dimensions D1, D2, and R_o in shale. (**b**) The relationship between fractal dimensions D1, D2, and R_o in kerogen. (**c**) The relationship between fractal dimension Dm and R_o in shale. (**d**) The relationship between fractal dimension Dm and R_o in kerogen.

A further analysis of the complexity of kerogen pores during thermal evolution reveals a similar effect of R_o on the fractal dimension in the Niutitang Formation kerogen (Figure 13b,d). However, this phenomenon is less pronounced at the stage of the low level of maturity of the Longtan Formation kerogen. Nevertheless, it can be inferred that the pore complexity is influenced by thermal evolution. As the organic matter undergoes thermal maturation, hydrocarbon generation commences, resulting in predominantly dry gas production at high levels of maturation while continuing to develop pores. However, over-maturation leads to graphitization and the collapse of organic matter [34,40]. Therefore, the fractal dimension Dm increases first and then decreases with R_o, indicating that the organic matter begins to generate hydrocarbons and increase micropores at the low level of maturity, leading to an increase in micropore complexity. After Ro reaches 3.5%, the micropores in the organic matter start to collapse and contract, resulting in a decrease in micropore complexity. Similarly, the fractal dimension D2 represents the gradual contraction of the macropores after Ro reaches 3.5%, leading to a decrease in complexity. Meanwhile, due to the macroscopic contraction, macropores gradually transform into mesopores, which is also the reason for the increase in the fractal dimension D1 of mesopores.

4.3. *Effect of Mineral Composition on Fractal Dimension and Pore Structure*

4.3.1. Effect of Clay Minerals on Pore Complexity

The fractal dimension of the Niutitang Formation shale decreases with the increase in clay mineral content, while that of the Longtan Formation shale increases with the increase in clay mineral content (Figure 14a,c). The properties, occurrence, and pore size of different types of clay minerals are obviously different and have different degrees of influence on the pore structure of shale reservoirs [66–68]. According to the XRD clay mineral analysis conducted in this study, it was observed that illite is the predominant clay mineral in the Niutitang Formation shale, whereas kaolinite is primarily present in the Longtan Formation shale, followed by chlorite (Table 2). Previous research indicates that different clay minerals exhibit varying levels of pore complexity and adsorption capacities [69], with their adsorption capacities ranked as follows: montmorillonite > an aemon mixed layer > kaolinite > chlorite > illite [70]. Consequently, having a high illite content within marine shales negatively impacts both their fractal dimension and pore structure. Conversely, in a marine–continental transitional setting, an increase in the contents of kaolinite and an illite–montmorillonite mixed layer positively influences the fractal dimension and pore structure.

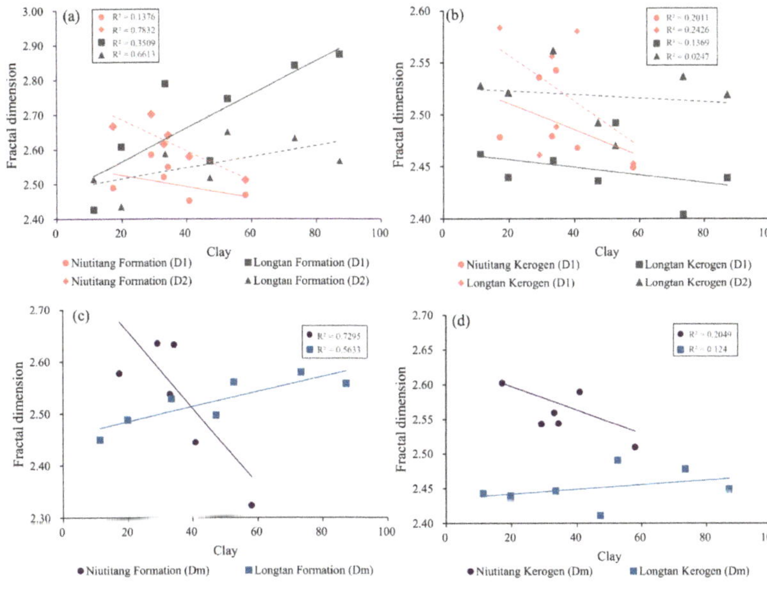

Figure 14. Correlation between fractal dimension and clay minerals. (**a**) The relationship between fractal dimensions D1, D2, and clay minerals in shale. (**b**) The relationship between fractal dimensions D1, D2, and clay minerals in kerogen. (**c**) The relationship between fractal dimension Dm and clay minerals in shale. (**d**) The relationship between fractal dimension Dm and clay minerals in kerogen.

In the relation between the fractal dimension of kerogen and clay minerals, the fractal dimension of the kerogen of the Niutitang Formation shale is weakly negatively correlated with clay minerals, while the fractal dimension of the kerogen of the Longtan Formation shale is not correlated with clay minerals (Figure 14b,d). Under FE-SEM, it can be observed that some clay minerals in the Niutitang Formation shale combine with organic matter in an organic–clay complex, but this is rare in the Longtan Formation shale (Figure 3a–c,f and Figure 4f). The organic–clay complex can develop a certain number of pores with a high degree of pore complexity, but with the increase in clay mineral content, the organic matter content is relatively reduced, and the pores developed in the clay minerals are not enough to offset the decrease in organic matter pores developed with the same large number. However, there is no such problem in the Longtan Formation shale, because type III kerogen has fewer pores and has little relationship to clay minerals.

4.3.2. Effect of Pyrite Mineral on Pore Complexity

Pyrite has two main crystal types: strawberry pyrite and autogenous pyrite. Generally, strawberry pyrite indicates a sedimentary environment, while autogenous pyrite is mainly considered to be a product of burial processes [71,72]. In both the Niutitang Formation shale and Longtan Formation shale, strawberry organic-filled pyrite usually develops a large number of intergranular pores.

The fractal dimensions of the two sets of shales show a tendency to decrease with increasing pyrite content (Figure 15a,c). Among them, the correlation between the fractal dimension of the Niutitang Formation shale and pyrite is the strongest, while the correlation between the Longtan Formation shale and pyrite is weaker. In the relationship between the pyrite content and kerogen pore fractal dimension, the fractal dimension D1 of the kerogen in the Niutitang Formation shows a strong negative correlation with pyrite, while the relationships between the fractal dimensions Dm and D2 and the pyrite content are not significant. On the other hand, the relationship between the fractal dimension Dm of the kerogen in the Longtan Formation and pyrite shows a weaker negative correlation, while the relationships between the fractal dimensions D1 and D2 and the pyrite content are not significant (Figure 15b,d). Pyrite transforms into autopyrite during environmental changes [73–75]. The marine shale of the Niutitang Formation is at a thermal evolutionary stage of high maturity to over maturity, during which the strawberry pyrite transforms into autogenous pyrite, leading to a reduction in porosity and a decrease in pore complexity. Moreover, with the increase in pyrite content, autogenous pyrite becomes the dominant type of pyrite. Apart from providing abundant intergranular pores and a higher level of pore complexity, it is difficult for strawberry pyrite on its own to develop porosity.

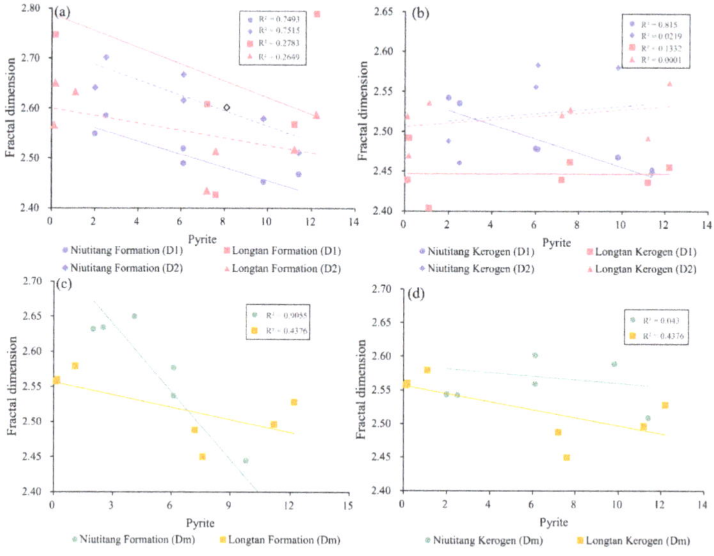

Figure 15. Correlation between fractal dimension and pyrite minerals. (**a**) The relationship between fractal dimensions D1, D2, and pyrite in shale. (**b**) The relationship between fractal dimensions D1, D2, and pyrite in kerogen. (**c**) The relationship between fractal dimension Dm and clay pyrite in shale. (**d**) The relationship between fractal dimension Dm and pyrite in kerogen.

4.3.3. Effect of Brittle Minerals on Pore Complexity

Quartz, feldspar, and carbonate minerals are brittle minerals, but through correlation analyses, they have been found to have no obvious correlation in shale and kerogen. However, brittle minerals also develop intergranular pores, micro-fractures, and secondary dissolution pores, etc., which have an impact on pore complexity, but this impact is limited [76]. This is of great significance for the manufacturing of artificial pore and fracture networks in the mining process, and a certain content of brittle minerals in shale is a favorable condition for the mining of shale gas resources [77].

4.4. Differential Pore Developments of Marine and Marine–Continental Transitional Shales

Due to the differences in their sedimentary environment and diagenetic process, the marine Niutitang Formation shale and marine–continental transitional Longtan Formation shale have great differences in organic matter sources, organic matter content, thermal maturity, mineral composition, and other aspects, resulting in differences in pore structures and fractal dimensions between the marine shale and the marine–continental transitional shale and organic matter. Organic matter, thermal maturity, and clay minerals are the main factors affecting the pore structure and fractal dimension of shale, followed by pyrite [17,78,79]. The type of organic matter along with the content of clay minerals and pyrite that are conducive to preserving organic matter pores predominantly affect the pore structure and fractal dimension of organic matter.

Organic matter plays a pivotal role in marine shale due to its high abundance, primarily comprising solid kerogen and migrated asphalt, along with elevated maturity levels. As a result, marine shale exhibits an abundance of organic matter pores with a substantial contribution rate from such pores while also featuring intricate heterogeneous characteristics (Figure 16a). In contrast, clay minerals assume significance in marine–continental transitional shales in which kaolinite mixed beds provide numerous adsorption sites for pores. Diverging from marine shales due to their lower maturity levels coupled with different types of organic matter compositions mainly consisting of terrestrial plant detritus along with other primary components, secondary pores do not develop within the organic matter framework (Figure 16b). The pore development of secondary organic matter is insufficient, resulting in the poor porosity of organic matter in the marine–continental transitional phase, and its porosity complexity is lower than that of marine shale. However, the porosity provided by abundant clay minerals and pyrite makes the porosity of marine–continental transitional shale more complex than that of marine shale.

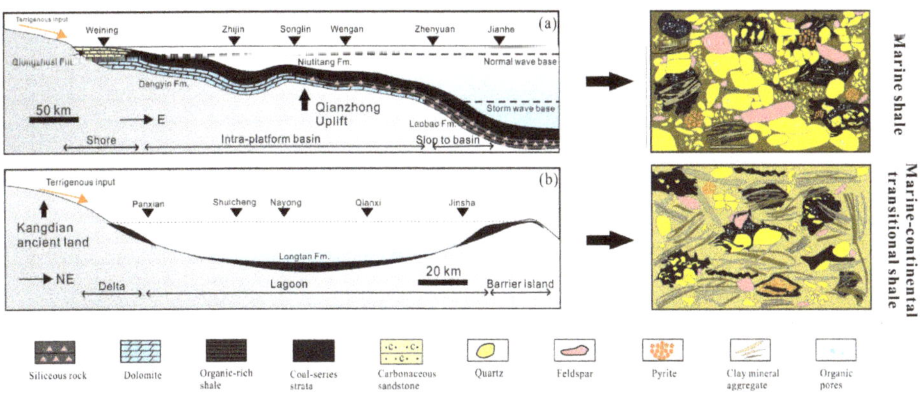

Figure 16. Sedimentary environment and pore development model of organic-rich shale in Guizhou Province. (**a**) Niutitang Formation, (**b**) Longtan Formation.

Considering the comprehensive fractal dimensions Dm, D1, and D2 for the two different sets of shale and kerogen pore complexities, the highest fractal dimension Dm is observed in the Niutitang Formation kerogen, indicating organic micropores have the highest level of complexity in type I kerogen. The highest fractal dimension D1 is found in the Longtan Formation shale, suggesting that mesopores have the highest level of complexity in the Longtan Formation shale. The highest fractal dimension D2 is observed in the Niutitang Formation shale, indicating that macropores have the highest level of complexity in the Niutitang Formation shale. Therefore, based on our N_2 and CO_2 adsorption anaylses, we established two sets of pore fractal models for shale and organic matter (Figure 17). The shale pores of the Longtan Formation in the marine–continental transitional facies exhibit higher levels of complexity compared to those of the marine Niutitang Formation, providing numerous gas adsorption sites (Figure 17a,c). The organic matter pores dominated by type I kerogen in the marine Niutitang Formation show higher levels of complexity than those dominated by type III kerogen in the Longtan Formation marine–continental transitional shale (Figure 17b,d). We have identified the fractal characteristics of these two sets of shale and organic matter, along with their influencing factors, which will aid in the evaluation of subsequent reservoir development. However, further discussions are required to delve deeper into this matter.

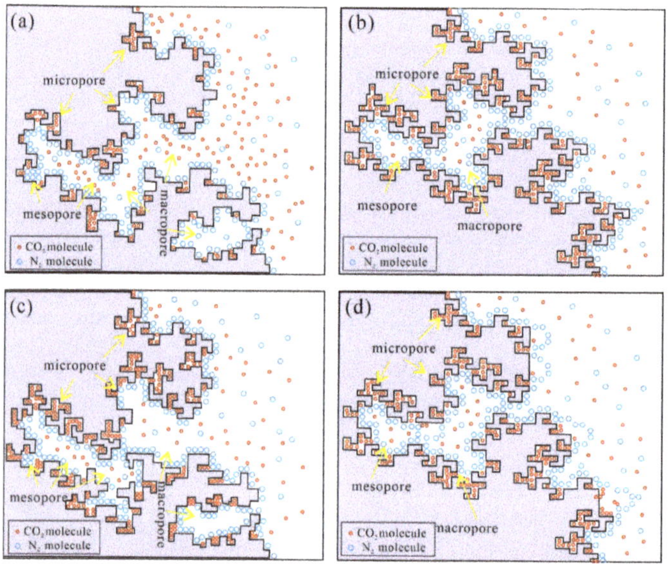

Figure 17. Shale and kerogen pore fractal model of marine Niutitang Formation and marine–continental transitional Longtan Formation. (**a**) Shale pore of Niutitang Formation. (**b**) Organic matter pore of Niutitang Formation. (**c**) Shale pore of Longtan Formation. (**b**) Organic matter pore of Longtan Formation (modified from Seri-Levy, 1993 [80]).

5. Conclusions

In this study, the differences in the pore development of shale and organic matter in different sedimentary environments and the effects of organic geochemical characteristics and mineral components on pore structure and fractal characteristics were studied in the marine Niutitang Formation and the marine–continental transitional Longtan Formation in Guizhou. The results show that the pore complexity of the marine–continental transitional shale is slightly higher than that of the marine shale, and type I kerogen is more complex than type III kerogen. The main conclusions of this study are as follows:

(1) The predominant pore types in the marine shale are wedge-shaped pores (H3 type) and slit pores (H4 type), while inkpot-shaped pores (H2 type) and slit pores (H4 type) dominate in the marine–continental transitional shale. Cylindrical pores (H1 type) are the primary pore structure observed in the organic matter of the marine and marine–continental transitional shale, followed by slit pores (H4 type).

(2) The high CRV value can provide shale gas storage space, and the high CRA value can provide a strong adsorption capacity. In addition to CRV and CRA values, it is shown that organic matter pores contribute more to the marine Niutitang Formation shale pores, which can provide storage space for gas. In contrast, inorganic pores are the main contributors to the marine–continental transitional Longtan Formation shale, with interlayer clay mineral pores being particularly prominent. The clay mineral composition group is primarily composed of kaolinite and emmon mixed layer minerals, which can provide an abundant pore specific surface area and a high adsorption capacity.

(3) The influencing factors of the fractal characteristics of the two sets of shale and organic matter pores are clarified. For the marine shale, the fractal dimensions (Dm, D1 and D2) are positively correlated with the total organic carbon (TOC) and negatively correlated with clay minerals and pyrite. Specifically, D1 and Dm initially increase and then decrease with increasing Ro values, while D2 gradually increases. For the marine–continental transitional shale, the fractal dimensions (Dm, D1, and D2) are positively correlated with R_o and clay minerals and negatively correlated with the TOC and pyrite. Additionally, D2 of the marine–continental transitional shale is negatively correlated with R_o. It is evident that the TOC content, clay minerals, and pyrite are not the primary factors influencing the fractal changes in the organic matter pores. Instead, R_o is the primary factor affecting the fractal characteristics of organic matter pores, indicating that the complexity of the organic matter pores is influenced by their thermal maturity.

(4) Based on N_2 and CO_2 adsorption, two sets of pore fractal models were established for shale and organic matter. It is evident that the complexity of the Longtan Formation shale in the marine–continental transitional facies is higher than that of the Niutitang Formation marine shale. Additionally, the complexity of organic matter pores dominated by type I kerogen is higher than those dominated by type III kerogen. This study will aid in subsequent evaluations of pore connectivity and flow effectiveness in shale reservoirs.

Author Contributions: Conceptualization, P.X. and S.N.; methodology, F.H. and J.T.; formal analysis, S.N.; investigation, Y.F. and K.W.; resources, S.N. and Y.F.; data curation, S.N. and K.W.; writing—original draft preparation, S.N. and P.X.; writing—review and editing, S.N. and P.X.; visualization, Y.F.; supervision, K.W.; project administration, Y.F. and K.W. All authors have read and agreed to the published version of the manuscript.

Funding: This research was supported by the Natural Science Foundation Project of China (No. 42002166, 42162016, 42063009) and Guizhou Provincial Fund Project (Nos. (2022)ZD005).

Data Availability Statement: All of the data and models generated or used in the present study are available from the corresponding author upon request.

Conflicts of Interest: The authors declare that they have no known competing financial interests or personal relationships that could have appeared to influence the work reported in this paper.

References

1. Curtis, J.B. Fractured shale-gas systems. *Aapg Bull.* **2002**, *86*, 1921–1938.
2. Hu, Q.; Ewing, R.; Rowe, H. Low nanopore connectivity limits gas production in Barnett formation. *J. Geophys. Res. Solid. Earth* **2015**, *120*, 8073–8087. [CrossRef]
3. Mahzari, P.; Mitchell, T.M.; Jones, A.P.; Westacott, D.; Striolo, A. Direct gas-in-place measurements prove much higher production potential than expected for shale formations. *Sci. Rep.* **2021**, *11*, 10775. [CrossRef] [PubMed]
4. Liu, Q.; Li, P.; Jin, Z.; Liang, X.; Zhu, D.; Wu, X.; Meng, Q.; Liu, J.; Fu, Q.; Zhao, J. Preservation of organic matter in shale linked to bacterial sulfate reduction (BSR) and volcanic activity under marine and lacustrine depositional environments. *Mar. Pet. Geol* **2021**, *127*, 104950. [CrossRef]
5. Zou, C.; Yang, Z.; Dong, D.; Zhao, Q.; Chen, Z.; Feng, Y.; Li, J.; Wang, X. Formation, Distribution and Prospect of Unconventional Hydrocarbons in Source Rock Strata in China. *Earth Sci.* **2022**, *47*, 1517–1533.

6. Dong, D.; Zou, C.; Dai, J.; Huang, S.; Zheng, J.; Wang, Y.; Li, X.; Guan, Q.; Zhang, C.; Huang, J.; et al. Suggestions on the development strategy of shale gas in China. *Nat. Gas. Geosci.* **2016**, *27*, 397–406. [CrossRef]
7. Ma, Z.; Wang, Y.; Yu, B. Study on Analytical Method for Pore Size Distribution of the Lower Cambrian Niutitang Formation Shale in Southeastern Chongqing. *Rock. Miner. Anal.* **2018**, *37*, 244–255.
8. Zou, C.; Zhao, Q.; Wang, H.; Xiong, W.; Dong, D.; Yu, R. The main characteristics of marine shale gas and the theory & technology of exploration and development in China. *Nat. Gas. Ind.* **2022**, *42*, 1–13.
9. Yi, T.; Zhao, X. Characteristics and distribution patterns of the Lower Cambrian Niutitang Shale reservoirs in Guizhou, China. *Natural Gas. Ind.* **2014**, *8*, 976.
10. Zhu, X.; Cai, J.; Wang, Y.; Liu, H.; Zhang, S. Evolution of organic-mineral interactions and implications for organic carbon occurrence and transformation in shale. *Gas. Bulletin.* **2019**, *132*, 784–792. [CrossRef]
11. Zhao, F.; Sun, Z.; Sang, S.; Zhou, X.; Han, Z.; Zhao, L.; Wu, Z. Progress of coalbed methane exploration and development in Guizhou and the 14th Five-Year Plan development strategy. *Nat. Gas. Ind.* **2022**, *42*, 65–75.
12. Wu, Y.; Fan, T.; Jiang, S.; Yang, X. Lithofacies and sedimentary sequence of the lower Cambrian Niutitang shale in the upper Yangtze platform, South China. *J. Nat. Gas. Sci. Eng.* **2017**, *43*, 124–136. [CrossRef]
13. Jin, C.; Li, C.; Algeo, T.J.; Wu, S.; Cheng, M.; Zhang, Z.; Shi, W. Controls on organic matter accumulation on the early-Cambrian western Yangtze Platform, South China. *Mar. Pet. Geol.* **2020**, *111*, 75–87. [CrossRef]
14. Jia, A.; Hu, D.; He, S.; Guo, X.; Hou, Y.; Wang, T.; Yang, R. Variations of Pore Structure in Organic-Rich Shales with Different Lithofacies from the Jiangdong Block, Fuling Shale Gas Field, SW China: Insights into Gas Storage and Pore Evolution. *Energy Fuels.* **2020**, *34*, 12457–12475. [CrossRef]
15. Song, Y.; Jiang, B.; Li, F.; Yan, G.; Yao, Y. Applicability of fractal models and nanopores' fractal characteristics for low-middle rank tectonic deformed coals. *Ear Sci.* **2018**, *43*, 1611–1622.
16. Dai, F.; Hu, H.; Zhang, A. Suitability study on fractal model of organic shale pore. *Coa Sci. Tech.* **2019**, *47*, 168–175.
17. Xia, P.; Li, H.; Fu, Y.; Qiao, W.; Guo, C.; Yang, Z.; Huang, J.; Mou, Y. Effect of lithofacies on pore structure of the Cambrian organic-rich shale in northern Guizhou, China. *Geol. J.* **2021**, *56*, 1130–1142. [CrossRef]
18. *GB/T 19144–2010*; Isolation Method for Kerogen from Sedimentary Rock. Standardization Administration of the People's Republic of China: Beijing, China, 2010; pp. 1–7.
19. Okolo, G.N.; Everson, R.C.; Neomagus, H.W.J.P.; Roberts, M.J.; Sakurovs, R. Comparing the porosity and surface areas of coal as measured by gas adsorption, mercury intrusion and SAXS techniques. *Fuel* **2015**, *141*, 293–304. [CrossRef]
20. Zheng, F.; Tang, X.; Yuan, K.; Lin, T.; You, M.; Niu, J.; Zi, Y.; Liang, Y. Sedimentation Models and Development Mechanisms of Organic-Rich Shales of the Lower Carboniferous Dawuba Formation: A Case Study in the Yaziluo Rift Trough, South of Guizhou Province, Southern China. *Acs Omega* **2022**, *7*, 29054–29071. [CrossRef]
21. Brunauner, S.; Emmett, S.P.H.; Teller, E.J. Adsorption of gases in multimolecular layers. *J. Am. Chem. Soc.* **1938**, *60*, 309–319. [CrossRef]
22. Wang, T.; Bury, W.; Gómez-Gualdrón, D.A.; Vermeulen, N.A.; Mondloch, J.E.; Deria, P.; Zhang, K.; Moghadam, P.Z.; Sarjeant, A.A.; Snurr, R.Q.; et al. Ultrahigh Surface Area Zirconium MOFs and Insights into the Applicability of the BET Theory. *J. Am. Chem. Soc.* **2015**, *137*, 3585–3591. [CrossRef]
23. Klimakow, M.; Klobes, P.; Rademann, K.; Emmerling, F. Characterization of mechanochemically synthesized MOFs. *Microporous Mesoporous Mat.* **2012**, *154*, 113–118. [CrossRef]
24. Han, H.; Cao, Y.; Chen, S.; Lu, J.; Huang, C.; Zhu, H.; Zhan, P.; Gao, Y. Influence of particle size on gas-adsorption experiments of shales: An example from a Longmaxi Shale sample from the Sichuan Basin, China. *Fuel* **2016**, *186*, 750–757. [CrossRef]
25. Pfeifer, P.; Martin, O.; Milton, W.C. Fractal bet and FHH theories of adsorption: A comparative study. *Proc. R. Soc. London A Math. Phys. Sci.* **1989**, *423*, 169–188.
26. Pfeifer, P.; Wu, Y.; Cole, M.; Krim, J. Multilayer adsorption on a fractally rough surface. *Phys. Rev. Lett.* **1989**, *62*, 1997. [CrossRef] [PubMed]
27. Guan, Q.; Dong, D.; Sun, S.; Hu, L.; Qi, L.; Li, C.; Chen, C.; Chen, X. Fractal characteristics of organic-rich shale pore structure and its geological implications: A case study of the Lower Silurian Longmaxi Formation in the Weiyuan block, Sichuan Basin. *Nat. Gas. Ind.* **2024**, *44*, 108–118.
28. Barrett, E.; Joyner, L.; Halenda, P. The Determination of Pore Volume and Area Distributions in Porous Substances II. *J. Am. Chem. Soc.* **1951**, *73*, 373–380. [CrossRef]
29. Jaroniec, M. Evaluation of the fractal dimension of microporous activated carbons. *Fuel* **1990**, *69*, 1573–1574. [CrossRef]
30. He, X.; Cheng, Y.; Hu, B.; Wang, Z.; Wang, C.; Yi, M.; Wang, L. Effects of coal pore structure on methane-coal sorption hysteresis: An experimental investigation based on fractal analysis and hysteresis evaluation. *Fuel* **2020**, *269*, 117438. [CrossRef]
31. Hou, Y.; Ge, Z.; Zhou, Z.; Guan, Y.; Xiao, C.; Jia, Y. Influence of supercritical CO_2 on fracture network evolution with nanopore structure linkage in high-rank anthracite matrix. *Fuel* **2024**, *357*, 129729. [CrossRef]
32. Wang, H.; Wang, L.; Zheng, S.; Sun, Y.; Shen, S.; Zhang, X. Research on coal matrix pore structure evolution and adsorption behavior characteristics under different thermal stimulation. *Energy* **2024**, *287*, 129677. [CrossRef]
33. Yang, F.; Ning, Z.; Wang, Q.; Zhang, R.; Krooss, B.M. Pore structure characteristics of lower Silurian shales in the southern Sichuan Basin, China: Insights to pore development and gas storage mechanism. *Int. J. Coal Geol.* **2016**, *156*, 12–24. [CrossRef]

34. Wang, Y.; Dong, D.; Yang, Y.; He, L.; Wang, S.; Huang, J.; Pu, B.; Wang, S. Quantitative characterization of reservoir space in the Lower Silurian Longmaxi Shale, southern Sichuan. *Sci. China Earth Sci.* **2014**, *57*, 1348–1356. [CrossRef]
35. Cao, T.; Song, Z.; Wang, S.; Xia, J. A comparative study of the specific surface area and pore structure of different shales and their kerogens. *Sci. China Earth Sci.* **2015**, *58*, 510–522. [CrossRef]
36. Zhang, J.; Li, X.; Wei, Q.; Gao, W.; Liang, W.; Wang, Z.; Wang, F. Quantitative characterization of pore-fracture system of organic-rich marine-continental shale reservoirs: A case study of the Upper Permian Longtan Formation, Southern Sichuan Basin, China. *Fuel* **2017**, *200*, 272–281. [CrossRef]
37. Chen, L.; Jiang, Z.; Liu, Q.; Jiang, S.; Liu, K.; Tan, J.; Gao, F. Mechanism of shale gas occurrence: Insights from comparative study on pore structures of marine and lacustrine shales. *Mar. Pet. Geol.* **2019**, *104*, 200–216. [CrossRef]
38. Zhang, P.; Huang, Y.; Zhang, J.; Li, B.; Liu, H.; Yang, J. Study on shale heterogeneity in western Hunan and Hubei: A case study of the Longmaxi Formation in well Ld1. *Acta Geol. Sin.* **2020**, *94*, 1568–1577.
39. Fu, Y.; Zhou, W.; Wang, H.; Qiao, W.; Ye, Y.; Jiang, R.; Wang, X.; Shu, J.; Li, D.; Xia, P. The relationship between environment and geochemical characteristics of black rock series of Lower Cambrian in northern Guizhou. *Geol. J.* **2021**, *95*, 536–548.
40. Zhang, W.; Hu, W.; Borjigin, T.; Zhu, F. Pore characteristics of different organic matter in black shale: A case study of the Wufeng-Longmaxi Formation in the Southeast Sichuan Basin, China. *Mar. Pet. Geol.* **2020**, *111*, 33–43. [CrossRef]
41. Tissot, B.P.; Welte, D.H. Classification of Crude Oils. In *Petroleum Formation and Occurrence*; Springer: Berlin/Heidelberg, Germany, 1984.
42. Guo, S.; Mao, W. Division of diagenesis and pore evolution of a Permian Shanxi shale in the Ordos Basin, China. *J. Pet. Sci. Eng.* **2019**, *182*, 106351. [CrossRef]
43. Song, Y.; Gao, F.; Tang, X.; Chen, L.; Wang, X. Influencing factors of pore structure differences between marine and terrestrial shale reservoirs. *Acta Pet. Sin.* **2020**, *41*, 1501–1512.
44. Ning, S.; Xia, P.; Hao, F.; Tian, J.; Zhong, Y.; Zou, N.; Fu, Y. Shale facies and its relationship with sedimentary environment and organic matter of Niutitang black shale, Guizhou Province. *Nat. Gas. Geosci.* **2021**, *32*, 1297–1307.
45. Mastalerz, M.; Schimmelmann, A.; Drobniak, A.; Chen, Y. Porosity of Devonian and Mississippian New Albany Shale across a maturation gradient; insights from organic petrology, gas adsorption, and mercury intrusion. *Aapg Bull.* **2013**, *97*, 1621–1643. [CrossRef]
46. Xia, P.; Hao, F.; Tian, J.; Zhou, W.; Fu, Y.; Guo, C.; Yang, Z.; Li, K.; Wang, K. Depositional Environment and Organic Matter Enrichment of Early Cambrian Niutitang Black Shales in the Upper Yangtze Region, China. *Energies* **2022**, *15*, 4551. [CrossRef]
47. Jiang, Z.; Song, Y.; Tang, X.; Li, Z.; Wang, X.; Wang, G.; Xue, Z.; Li, X.; Zhang, K.; Chang, J.; et al. Controlling factors of marine shale gas differential enrichment in southern China. *Pet. Explor. Dev.* **2020**, *47*, 617–628. [CrossRef]
48. Gu, Y.; Li, X.; Wang, Q.; Yang, S. On the Different Characteristics of Organic Pores in Shale and Their Influencing Factors: Taking typical marine, continental, and transitional facies reservoirs in China as examples. *Acta Geol. Sin.* **2021**, *4*, 39.
49. Loucks, R.; Reed, R.; Ruppel, S.; Hammes, U. Spectrum of pore types and networks in mudrocks and a descriptive classification for matrix-related mudrock pores. *Aapg Bull.* **2012**, *96*, 1071–1098. [CrossRef]
50. Thommes, M.; Kaneko, K.; Neimark, A.V.; Olivier, J.P.; Rodriguez-Reinoso, F.; Rouquerol, J.; Sing, S.W.K. Physisorption of gases, with special reference to the evaluation of surface area and pore size distribution (IUPAC Technical report). *Pure Appl. Chem.* **2015**, *87*, 1051. [CrossRef]
51. Sing, K.S.W. Reporting physisorption data for gas/solid systems with special reference to the determination of surface area and porosity (Recommendations 1984). *Pure Appl. Chem.* **1985**, *54*, 2201–2218. [CrossRef]
52. Zou, J.; Chen, W.; Yang, D.; Yuan, J.; Jiao, Y.Y. Fractal characteristics of the anisotropic microstructure and pore distribution of low-rank coal. *Aapg Bull.* **2019**, *103*, 1297–1319. [CrossRef]
53. Jarvie, D.M.; Hill, R.J.; Ruble, T.E.; Pollastro, R.M. Unconventional shale-gas systems: The Mississippian Barnett Shale of north-central Texas as one model for thermogenic shale-gas assessment. *Aapg Bull.* **2007**, *91*, 475–499. [CrossRef]
54. Ma, Y.; Ardakani, O.H.; Zhong, N.; Liu, H.; Huang, H.; Larter, S.; Zhang, C. Possible pore structure deformation effects on the shale gas enrichment: An example from the Lower Cambrian shales of the Eastern Upper Yangtze Platform, South China. *Int. J. Coal Geol.* **2020**, *217*, 103349. [CrossRef]
55. Ross, D.J.K.; Marc Bustin, R. The importance of shale composition and pore structure upon gas storage potential of shale gas reservoirs. *Mar. Pet. Geol.* **2009**, *26*, 916–927. [CrossRef]
56. Chalmers, G.R.L.; Bustin, R.M. Lower Cretaceous gas shales in northeastern British Columbia, Part II: Evaluation of regional potential gas resources. *Bull. Can. Pet. Geol.* **2008**, *56*, 22–61. [CrossRef]
57. Shang, F.; Zhu, Y.; Hu, Q.; Zhu, Y.; Wang, Y.; Du, M.; Liu, R.; Han, Y. Characterization of methane adsorption on shale of a complex tectonic area in Northeast Guizhou, China: Experimental results and geological significance. *J. Nat. Gas. Sci. Eng.* **2020**, *84*, 103676. [CrossRef]
58. Slatt, R.M.; O'Brien, N.R. Pore types in the Barnett and Woodford gas shales: Contribution to understanding gas storage and migration pathways in fine-grained rocks. *Aapg Bull.* **2011**, *95*, 2017–2030. [CrossRef]
59. Hu, M.; Qiu, X.; Hu, Z.; Deng, Q. Current Researches on Shale Gas Reservoirs and Existing Problems. *Spec. Oil Gas. Reserv.* **2015**, *22*, 1–7+151.
60. Löhr, S.C.; Baruch, E.T.; Hall, P.A.; Kennedy, M.J. Is organic pore development in gas shales influenced by the primary porosity and structure of thermally immature organic matter? *Org. Geochem.* **2015**, *87*, 119–132. [CrossRef]

61. Jiao, K.; Xie, G.; Fei, W.; Liu, S.; Liu, X.; Kang, Y.; Deng, B.; Pang, Q.; Liu, W.; Luo, C. The Control Factors and Geological Implications of the Nanopore Morphology of the Lower Paleozoic Black Shales in the Sichuan Basin, China. *Geol. J. Chin. Univ.* **2019**, *25*, 847–859.
62. Mishra, S.; Mendhe, V.A.; Varma, A.K.; Kamble, A.D.; Sharma, S.; Bannerjee, M.; Kalpana, M.S. Influence of organic and inorganic content on fractal dimensions of Barakar and Barren Measures shale gas reservoirs of Raniganj basin, India. *J. Nat. Gas. Sci. Eng.* **2018**, *49*, 393–409. [CrossRef]
63. Han, H.; Guo, C.; Zhong, N.; Pang, P.; Gao, Y. A study on fractal characteristics of lacustrine shales of Qingshankou Formation in the Songliao Basin, northeast China using nitrogen adsorption and mercury injection methods. *J. Pet. Sci. Eng.* **2020**, *193*, 107378. [CrossRef]
64. Chang, J.; Fan, X.; Jiang, Z.; Wang, X.; Chen, L.; Li, J.; Zhu, L.; Wan, C.; Chen, Z. Differential impact of clay minerals and organic matter on pore structure and its fractal characteristics of marine and continental shales in China. *Appl. Clay Sci.* **2022**, *216*, 106334. [CrossRef]
65. Kuila, U.; McCarty, D.K.; Derkowski, A.; Fischer, T.B.; Topór, T.; Prasad, M. Nano-scale texture and porosity of organic matter and clay minerals in organic-rich mudrocks. *Fuel* **2014**, *135*, 359–373. [CrossRef]
66. Loucks, R.; Reed, R.; Ruppel, S.; Jarvie, D. Morphology, Genesis, and Distribution of Nanometer-Scale Pores in Siliceous Mudstones of the Mississippian Barnett Shale. *J. Sediment. Res.* **2009**, *79*, 848–861. [CrossRef]
67. Gao, F.; Song, Y.; Jiang, Z.; Zhang, X.; Chen, L. Influence of Clay Minerals on Shale Storage Space and Adsorptive Capacity. *Spec. Oil Gas. Reserv.* **2017**, *24*, 1–8.
68. Wang, P.; Zhang, C.; Li, X.; Zhang, K.; Yuan, Y.; Zang, X.; Cui, W.; Liu, S.; Jiang, Z. Organic matter pores structure and evolution in shales based on the he ion microscopy (HIM): A case study from the Triassic Yanchang, Lower Silurian Longmaxi and Lower Cambrian Niutitang shales in China. *J. Nat. Gas. Sci. Eng.* **2020**, *84*, 103682. [CrossRef]
69. Tang, S.; Fan, E. Methane adsorption characteristics of clay minerals in organic-rich shales. *Int. J. Coal Geol.* **2014**, *39*, 1700–1706.
70. Ji, L.; Qiu, J.; Xia, Y.; Zhang, T. Micro-pore characteristics and methane adsorption propertiesof common clay minerals by electron microscope scanning. *Acta Pet. Sinica.* **2012**, *33*, 249–256.
71. Wei, H.; Wei, X.; Qiu, Z.; Song, H.; Shi, G. Redox conditions across the G–L boundary in South China: Evidence from pyrite morphology and sulfur isotopic compositions. *Chem. Geol.* **2016**, *440*, 1–14. [CrossRef]
72. Rickard, D. Sedimentary pyrite framboid size-frequency distributions: A meta-analysis. *Palaeogeogr. Palaeoclimatol. Palaeoecol.* **2019**, *522*, 62–75. [CrossRef]
73. Zhao, D.; Guo, Y.; Zhu, Y.; Wang, G.; Liu, J.; Chong, S.; Zhang, J. Micropore Characteristics and Geological Significance of Pyrite in Shale Rocks of Longmaxi Formation. *Acta Sed. Sin.* **2018**, *36*, 864–876.
74. Merinero, R.; Lunar, R.; Martínez-Frías, J.; Somoza, L.; Díaz-del-Río, V. Iron oxyhydroxide and sulphide mineralization in hydrocarbon seep-related carbonate submarine chimneys, Gulf of Cadiz (SW Iberian Peninsula). *Mar. Pet. Geol.* **2008**, *25*, 706–713. [CrossRef]
75. Wang, D.; Zhang, J.; Li, Z.; Tong, Z.; Niu, J.; Ding, W.; Zhang, C. Formation mechanism of framboidal pyrite and its theory inversion of paleo-redox conditions. *Geol. China* **2022**, *49*, 36–50.
76. Zhang, C.; Dong, D.; Wang, Y.; Jiang, S.; Guan, Q. Research Progress on Brittleness of Shale Reservoirs. *Xinjiang Pet. Geo.* **2017**, *38*, 111–118.
77. Ge, M.; Ren, S.; Guo, T.; Wang, S.; Zhou, Z. Identification Method of Marine Shale Gas 'High-Quality Layer' in the Lower Paleozoic Area, Southern China and Its Application. *Rock. Miner. Anal.* **2020**, *39*, 350–361.
78. Milliken, K.L.; Rudnicki, M.; Awwiller, D.N.; Zhang, T. Organic matter-hosted pore system, Marcellus Formation (Devonian), Pennsylvania. *Aapg Bull.* **2013**, *97*, 177–200. [CrossRef]
79. Ilgen, A.G.; Heath, J.E.; Akkutlu, I.Y.; Bryndzia, L.T.; Cole, D.; Kharaka, Y.D.; Kneafsey, T.J.; Milliken, K.L.; Pyrak-Nolte, L.J.; Suarez-Rivera, R. Shales at all scales: Exploring coupled processes in mudrocks. *Earth-Sci. Rev.* **2017**, *166*, 132. [CrossRef]
80. Seri-Levy, A.; Avnir, D. Effects of heterogeneous surface geometry on adsorption. *Langmuir* **1993**, *9*, 3067–3076. [CrossRef]

fractal and fractional

MDPI

Article

Fractal Characterization of the Pore-Throat Structure in Tight Sandstone Based on Low-Temperature Nitrogen Gas Adsorption and High-Pressure Mercury Injection

Taping He [1], Yaoqi Zhou [1,*], Zhaobing Chen [2], Zhenwei Zhang [3], Huanyu Xie [4], Yuehan Shang [1] and Gaixia Cui [1]

[1] School of Geosciences, China University of Petroleum (East China), Qingdao 266580, China; b22010021@s.upc.edu.cn (T.H.); b21010006@s.upc.edu.cn (Y.S.); b22010072@s.upc.edu.cn (G.C.)
[2] College of Earth Sciences & Engineering, Xi'an Shiyou University, Xi'an 710065, China
[3] PetroChina Qinghai Oilfield Company, Dunhuang 736200, China; zzweiqh@petrochina.com.can
[4] School of Energy, China University of Geosciences, Beijing 100083, China; geologyxiehuanyu@163.com
* Correspondence: bdbx465@163.com; Tel.: +86-136-0896-9182

Abstract: The pore-throat structure is a critical factor in the study of unconventional oil and gas reservoirs, drawing particular attention from petroleum geologists, and it is of paramount significance to analyze to enhance oil and gas production. In tight sandstone, which serves as a significant hydrocarbon reservoir, the internal pore-throat structure plays a decisive role in the storage and migration of fluids such as water, gases, and hydrocarbons. This paper employs casting thin section (CTS), field emission scanning electron microscope (FE-SEM), high-pressure mercury injection (HPMI), and low-temperature nitrogen gas adsorption ($LT-N_2-GA$) experimental tests to qualitatively and quantitatively analyze the characteristics of the pore-throat structure in tight sandstone. The results indicate that the pore types in tight sandstone include intergranular residual pores, dissolution pores, intercrystalline pores, and microfractures, while the throat types encompass sheet-shaped, curved-sheet-shaped, and tubular throats. Analysis of the physical and structural parameters from 13 HPMI and 5 $LT-N_2-GA$ samples reveals a bimodal distribution of pore-throat radii. The complexity of the pore-throat structure is identified as a primary controlling factor for reservoir permeability. The fractal dimension (D) exhibits an average value of 2.45, displaying a negative correlation with porosity ($R^2 = 0.22$), permeability ($R^2 = 0.65$), the pore-throat diameter ($R^2 = 0.58$), and maximum mercury saturation ($R^2 = 0.86$) and a positive correlation with threshold pressure ($R^2 = 0.56$), median saturation pressure ($R^2 = 0.49$), BET specific surface area ($R^2 = 0.51$), and BJH total pore volume ($R^2 = 0.14$). As D increases, reservoir pores tend to decrease in size, leading to reduced flow and deteriorated physical properties, indicative of a more complex pore-throat structure.

Keywords: tight sandstone; pore-throat structure; fractal dimension; HPMI; $LT-N_2-GA$

1. Introduction

In the past decade, with breakthroughs in shale oil technology in the United States, unconventional oil and gas resources have received widespread attention [1–6]. Unconventional oil and gas resources typically refer to hydrocarbon reserves that cannot be extracted using conventional methods, including shale oil and gas, tight sandstone oil and gas, coalbed methane, and heavy oil [7–10]. Many researchers define reservoirs with porosity of less than 10% and permeability of less than 0.1 mD as tight sandstone reservoirs [11,12]. A microscopic pore-throat structure refers to characteristics such as the pore-throat size, geometric shape, and connectivity [13,14]. These characteristics play a pivotal role in determining the reservoir's porosity and permeability and at the macroscopic level control the storage and transport of oil and gas [13,15]. Tight sandstone is characterized by fine pores and complex microscopic pore-throat structures, poor pore connectivity, and high heterogeneity, primarily distributed in the nanometer to micrometer range [10,16,17].

Understanding a reservoir's microscopic pore-throat structure is crucial for predicting the "sweet spots" in oil and gas resources and improving petroleum production [18,19].

Nitrogen gas adsorption (N_2−GA) has been widely used to study the micro- and nanopore structures of porous materials [20–22]. In petroleum geology, N_2−GA is typically employed to characterize the pore-throat structures of micro-mesoporous reservoirs [20–24]. Lu et al. [23] conducted a study of reservoir pore-throat structures using N_2−GA and high-pressure mercury intrusion (HPMI). They found that as the number of micro- and mesopores increased and the number of larger pores decreased, the total surface area and total pore volume of shale increased, while the average pore diameter decreased. Wang et al. [20] used N_2−GA to investigate the nanoscale pore morphology and pore size distribution of lacustrine shale in the Yanchang Formation in Ordos Basin. Yuan and Rezaee [24] analyzed the surface fractal dimensions of ineffective water-occupied pores and the volume fractal dimension of effective pores in shale using the results of nuclear magnetic resonance (NMR) and N_2−GA.

Currently, numerous experimental techniques are employed for the study of the pore structure in reservoirs. These primarily include casting thin section (CTS) [25], optical microscopy (OM) [26], field emission scanning electron microscopy (SEM) [26,27], X-ray diffraction (XRD) [27], small-angle neutron scattering (SANS) [28–30], N_2−GA [21], HPMI [31,32], X-ray computed tomography (X-CT) [33,34], and NMR [35,36] testing. However, these experiments have their own strengths and limitations, making it challenging to effectively capture the structural characteristics of various types of pores and throats. Image analysis methods such as OM, SEM, atomic force microscopy (AFM), and X-ray computed tomography (X-CT) provide visual observations of material pores and throats but lack the capability for quantitative analysis of the pore-throat parameters [37]. Techniques like N_2−GA, HPMI, NMR, and SANS can obtain parameters related to materials' pores and throats but have certain technical and methodological defects [37,38]. For example, during HPMI testing, many small pores can expand or even collapse, resulting in measurements that exceed the theoretical values, with a measurement range of 6 nm to 1000 μm [39]. The N_2−GA method, with a testing range of 0.35 nm to 120 nm, cannot adequately measure micron-scale pores [39]. Additionally, these techniques have inherent limitations in their measurement precision. It is essential to select the most suitable experimental method for characterization according to the different types of pores and throats. Accurate characterization of the pore-throat structure, achieved by combining various testing techniques, is crucial in the evaluation and analysis of tight sandstone reservoirs [10].

Fractal theory, introduced by Mandelbrot in 1967 [40], is primarily used to describe chaotic systems of fractal objects [31]. Today, fractal theory finds widespread application in quantitatively characterizing pore parameters in porous materials [31,41–43]. Lai and Wang [44] conducted a fractal analysis of the pore structure in tight sandstone using HPMI, revealing that the fractal correlations are more pronounced in smaller pores than in larger pores and fractures. Yu et al. [45] employed gas adsorption to perform fractal analysis of the micro- and mesoporous structures in coal, unveiling how pore volume and surface heterogeneity respond to structural deformation. Wu et al. [46] evaluated the pore-throat fractal dimensions in tight sandstone using HPMI and NMR, finding that complex pore-throat structures with higher fractal dimensions exhibit lower storage and permeability capabilities.

This paper intends to describe and assess the pore-throat structure of tight sandstone. The analysis is conducted from both qualitative and quantitative perspectives, exploring the pore-throat structure, aperture distribution, and pore connectivity of the samples. Additionally, the utilization of LT−N_2−GA testing is introduced to address deficiencies in the range, precision, and technical methods associated with the quantitative analysis of reservoir pore-throat structures using HPMI. The paper is organized into four sections: (1) Qualitative analysis of the pore-throat types, aperture sizes, pore geometry, and connectivity of tight sandstone; (2) Detailed characterization and quantitative assessment of pore-throat structures using HPMI and LT−N_2−GA; (3) Separate discussions exploring the relationships between pore-throat structure parameters based on HPMI and LT−N_2−GA and reservoir properties; (4) Analysis of the fractal dimensions of tight sandstone pore-throat structures and their relationship with permeability, porosity, and pore-throat structure parameters.

2. Materials, Experiments, and Methods

2.1. Materials

In this study, samples were collected from 18 tight sandstone cores from 12 wells in Chang 6 to Chang 8 members in the Longdong area of the Ordos Basin. The burial depth of the samples ranges from 1500 to 2300 m (Tables 1 and 2). The study area exhibited poor petrophysical properties and a high degree of heterogeneity due to the diagenetic evolution it has undergone.

Table 1. Microscopic pore-throat structure parameters from HPMI analysis.

Sample	Depth	Member	Φ (%)	K (mD)	P_{min} (Mpa)	P_{50} (MPa)	r_{max} (μm)	r_{50} (μm)	S_{Hgmax} (%)	D	Small Pores		Large Pores	
											$R_1{}^2$	D_1	$R_2{}^2$	D_2
H1	2199.51	Ch 6	9.94	0.044	2.41	11.03	0.30	0.067	92.69	2.47	0.969	2.46	0.875	2.9957
H2	2127.4	Ch 6	10.60	0.085	2.21	7.59	0.33	0.097	94.85	2.41	0.977	2.40	0.939	2.9969
H3	2068.3	Ch 6	8.98	0.092	1.52	3.79	0.48	0.194	96.37	2.42	0.964	2.41	0.894	2.9949
H4	2098.4	Ch 6	10.90	0.085	2.76	7.59	0.27	0.097	88.14	2.55	0.984	2.54	0.867	2.9970
H5	2241.6	Ch 6	11.88	0.24	1.10	3.10	0.67	0.237	98.81	2.25	0.970	2.21	0.928	2.9867
X1	1549.5	Ch 7	10.36	0.114	1.38	4.21	0.53	0.175	93.92	2.48	0.988	2.46	0.951	2.9937
X2	1835.4	Ch 6	11.60	0.108	1.38	5.72	0.53	0.128	91.57	2.48	0.989	2.48	0.654	2.9961
X3	1842	Ch 6	11.85	0.13	1.38	4.21	0.53	0.175	91.23	2.51	0.985	2.50	0.893	2.9983
X4	2028.8	Ch 7	12.81	0.262	1.24	2.76	0.59	0.266	96.81	2.37	0.995	2.36	0.911	2.9970
X5	2117.9	Ch 8	9.94	0.112	1.52	3.59	0.48	0.205	97.37	2.42	0.958	2.40	0.975	2.9932
X6	1763.2	Ch 7	11.80	0.23	1.24	3.59	0.59	0.205	97.79	2.35	0.921	2.34	0.993	2.9936
X7	2011.2	Ch 6	8.37	0.025	3.45	16.55	0.21	0.044	84.35	2.62	0.966	2.61	0.920	2.9939
X8	1782.36	Ch 7	11.61	0.055	2.76	7.59	0.27	0.097	88.84	2.52	0.966	2.52	0.849	2.9969

Φ—porosity, K—permeability, P_{min}—the threshold pressure, P_{50}—the pressure at 50% mercury saturation, r_{max}—the maximal pore-throat radius, r_{50}—the median pore-throat radius, S_{Hgmax}—the maximum mercury saturation, r_{ave}—the average pore-throat radius, $R_1{}^2$—correlation coefficient of small pores, D_1—fractal dimensions of small pores, $R_2{}^2$—correlation coefficient of large pores, D_2—fractal dimensions of large pores.

Table 2. Microscopic pore-throat structure parameters from LT−N_2−GA.

Sample	Depth (m)	Member	Φ (%)	K (mD)	BET Specific Surface Area (m^2/g)	BJH Average Pore Diameter (nm)	BJH Total Pore Volume (cm^3/g)	Fractal Dimensions	
								D	R^2
N1	1817.52	Ch 7	12.32	0.54	2.846	17.665	0.0145	2.4633	0.995
N2	2241.60	Ch 6	11.38	0.45	3.846	12.801	0.0108	2.5403	0.9873
N3	2099.63	Ch 6	13.79	0.56	4.846	19.577	0.0106	2.4577	0.9931
N4	1782.36	Ch 7	10.29	0.19	5.846	11.603	0.0128	2.5914	0.9902
N5	2082.00	Ch 6	10.63	0.21	6.846	12.558	0.0176	2.5814	0.9928

Φ—porosity, K—permeability, R^2—correlation coefficient of pores.

Cylindrical rock specimens with a diameter of 2.5 cm were obtained by coring the tight sandstone cores. These samples were subjected to an oil removal process using chloroform solvent and were then cleaned with water in an ultrasonic cleaner, followed by drying. The prepared rock specimens were further processed to create samples tailored to meeting the requirements of various experiments.

2.2. *Experiments*

2.2.1. Casting Thin Section (CTS) Imaging

Samples of tight sandstone measuring 25 × 25 × 0.5 mm were prepared. In a vacuum environment, colored liquid resin was injected into the pores of the samples, and once the liquid resin solidified, the sample was ground to create a rock thin section. The pore-throat structure and mineral composition of the rock were observed under a polarizing microscope.

2.2.2. Field Emission Scanning Electron Microscope (FE-SEM) Imaging

This experiment involved observing the samples using a JSM-7500F field emission scanning electron microscope produced by JEOL Japan Electronics Co., Ltd. (Tokyo, Japan). The samples were prepared as small cubic pieces measuring approximately 1 cm on each side. They were affixed to the specimen holder's stub using conductive adhesive and applied with a gold sputter coating. The samples were observed after being placed in a vacuum within the microscope's chamber. This experiment was conducted at the Xi'an Geological Survey Center of the China Geological Survey.

2.2.3. HPMI

Cylindrical tight sandstone with a diameter and length of 2.5 cm, exhibiting intact surfaces without cracks, was chosen for the HPMI test. A 9220-II mercury intrusion porosimeter (Micromeritics Instrument Corporation, Norcross, GA, USA) was employed to inject mercury into the rock samples. Various parameters, such as mercury saturation, pore radius, and pore surface area, were recorded at different pressures during the mercury intrusion and extrusion processes. Subsequently, curves for each parameter were plotted. The maximum capillary pressure for this experiment was 206.88 MPa, and the minimum measurable pore-throat radius was 3.55 nm. This experiment was conducted at the Analysis and Testing Center of the Exploration and Development Research Institute of Changqing Oilfield Branch Company (Qingyang, China).

2.2.4. $LT-N_2-GA$

This experiment was conducted using an ASAP 2460 fully automatic surface area and porosity analyzer (Micromeritics Instrument Corporation, Norcross, GA, USA). We took a 3 g sample and ground it to less than 250 μm before loading it into the sample tube and recording the weight. The sample tube was then sealed and placed in a degassing station. A heating jacket was wrapped around the sample tube, and the degassing parameters were set. The vacuum pump was turned on to subject the samples to vacuum degassing to remove any gases adsorbed on the material's surface. After degassing, the sample was re-weighed, and the weight was recorded again. The sample tube was then placed in liquid nitrogen, and the analyzer was used to measure the nitrogen adsorption at different pressure points to obtain the adsorption isotherms. This experiment was conducted at Xi'an Shiyou University.

2.3. *Methods*

2.3.1. Fractal Dimension Based on HPMI

According to the basic principles of fractals, the relationship between the number of pores with a radius of r ($N_{(r)}$) and the pore-throat radius (r) can be obtained [47]:

$$N_{(r)} \,\&\, r^{-D} \quad (1)$$

where D is the fractal dimensions.

Based on the Brooks–Corey model [48], the equation for pore-throat volume $V_{(r)}$ can be expressed as:

$$V_{(r)} \,\&\, \pi r^{3-D} \quad (2)$$

The cumulative volume $V_{(<r)}$ of pores with a radius smaller than r is given by:

$$dV_{(<r)} \, \& \int_{r_{min}}^{r} (3-D)\pi r^{2-D} dr = \pi\left(r^{3-D} - r_{min}^{3-D}\right) \quad (3)$$

Therefore, the cumulative volume percentage $S_{(<r)}$ of pores with a radius less than r can be expressed as:

$$S_{(<r)} = \frac{dV_{(<r)}}{dV_{(<r_{max})}} = \frac{r^{3-D} - r_{min}^{3-D}}{r_{max}^{3-D} - r_{min}^{3-D}} \approx \frac{r^{3-D}}{r_{max}^{3-D}} = \left(\frac{r}{r_{max}}\right)^{3-D} \quad (4)$$

where r_{min} is the minimum value of r; r_{max} is the maximum value of r.

The radius of the pore throat (r) and the capillary pressure (P_c) have the following relationship [49]:

$$r = \frac{2\sigma \cos\theta}{P_c} \quad (5)$$

where σ is the surface tension of the material; θ is the wetting angle.

From Equations (4) and (5), we see:

$$\lg(1 - S_r) = (D-3)\lg P_c - (D-3)\lg P_{max} \quad (6)$$

where P_{max} is the maximum value of P_c. $S_{(<r)}$ can be expressed as $1 - S_r$(S_r is the mercury saturation of the reservoir).

The value D is from the slope of the fitted line (6). The range of D is 2 to 3, and the larger D, the more complex the pore-throat structure [44].

2.3.2. Fractal Dimensions Based on N_2−GA

1. Calculation of pore-specific surface area using the BET equation

In the LT−N_2−GA experiments, the specific surface area of the pores is typically calculated using the Brunauer–Emmett–Teller (BET) equation [50]. The calculation equation is as follows [51]:

$$\frac{P}{V(P_0 - P)} = \frac{C-1}{V_m C}\frac{P}{P_0} + \frac{1}{V_m C} \quad (7)$$

where V is the total volume of gas adsorbed at equilibrium pressure; V_m is the amount of gas adsorbed in a monolayer; P_0 is the saturation vapor pressure of the adsorbate; P is the adsorption equilibrium pressure; C is a constant.

Typically, $\frac{P_0}{P}$ data within the range of 0.05 to 0.35 have a good fit with the BET equation, so this study selected seven data points within this range [52]. From Equation (7), it is evident that the BET equation represents a linear relationship, and the slope (m) and intercept (n) of this linear equation can be inferred as follows:

$$m = \frac{C-1}{V_m C} \quad (8)$$

$$n = \frac{1}{V_m C} \quad (9)$$

Based on Equations (8) and (9), it can be derived that:

$$V_m = \frac{1}{m+n} \quad (10)$$

The specific surface area can be inferred from V_m.

2. BJH method for pore size analysis

The use of the Barret–Joyner–Halenda (BJH) method to calculate pore sizes in porous materials is widely accepted among researchers [53]. The pore size (r_0) is calculated using the Kelvin correction equation [21]:

$$\ln\left(\frac{P}{P_0}\right) = -\frac{\alpha\gamma_N V_1}{RT\left[r_0 - t\left(\frac{P}{P_0}\right)\right]} \tag{11}$$

where γ_N is the surface tension of liquid nitrogen; α is the shape factor of the gas–liquid interface; V_1 is the molar volume of liquid nitrogen; T is the absolute temperature (77 K); R is the gas constant; t is the Halsey empirical equation for liquid nitrogen.

$$t = 3.54\left[-\frac{5}{\ln\left(\frac{P}{P_0}\right)}\right]^{1/3} \tag{12}$$

3. Calculation of fractal dimensions using the FHH fractal model

The Frenkel–Halsey–Hill (FHH) fractal model has been proverbially used to characterize irregular porous materials' surfaces [54,55]. By using the data obtained from the $LT-N_2-GA$ measurements and applying them to the FHH fractal model, we can derive the equation for calculating the fractal dimension.

$$\ln V = (D-3)\ln\left[\ln\left(\frac{P_0}{P}\right)\right] + C \tag{13}$$

where C is a constant.

The value D can be calculated from the slope (k) of the straight line in the plot of ln V against $\ln\left[\ln\left(\frac{P_0}{P}\right)\right]$:

$$D = k + 3 \tag{14}$$

3. Results

3.1. Microscopic Characteristics of Pores and Throats

In reservoirs, pores serve as fluid storage spaces, while throats control fluid flowability [56]. Utilizing the CTS and FE-SEM images (Figure 1), four primary pore types are evident in tight sandstone: residual intergranular pores (Figure 1a,c,e), dissolution pores (Figure 1b,d,f), intercrystalline pores (Figure 1g), and microfractures (Figure 1h).

Residual intergranular pores are remnants of the original primary pores in the reservoir, in the shape of triangles or irregular polygons due to compaction and cementation (Figure 1a,c,e). Dissolution pores primarily consist of intragranular dissolution pores (Figure 1f) and intergranular dissolution pores (Figure 1b). They form during the process of minerals transitioning into other minerals due to environmental changes, leading to the creation of dissolution pores. They are commonly found in minerals like feldspar (Figure 1b,f), lithic fragments, and carbonate rocks (Figure 1d). Intergranular pores are typically fine pores formed by the accumulation of clay minerals such as kaolinite, illite, and chlorite. They make a minor contribution to the reservoir's storage and permeability. Microfractures are formed by fracturing brittle minerals such as quartz under pressure, significantly enhancing the reservoir's flowability.

The throat types primarily include sheet-shaped throats, curved-sheet-shaped throats (Figure 1c), and tubular (Figure 1g) throats. Sheet-shaped and curved-sheet-shaped throats are mainly determined by early diagenetic compaction and represent the primary types of throats between the grains in tight sandstone. Tubular throats are mainly formed among clay minerals.

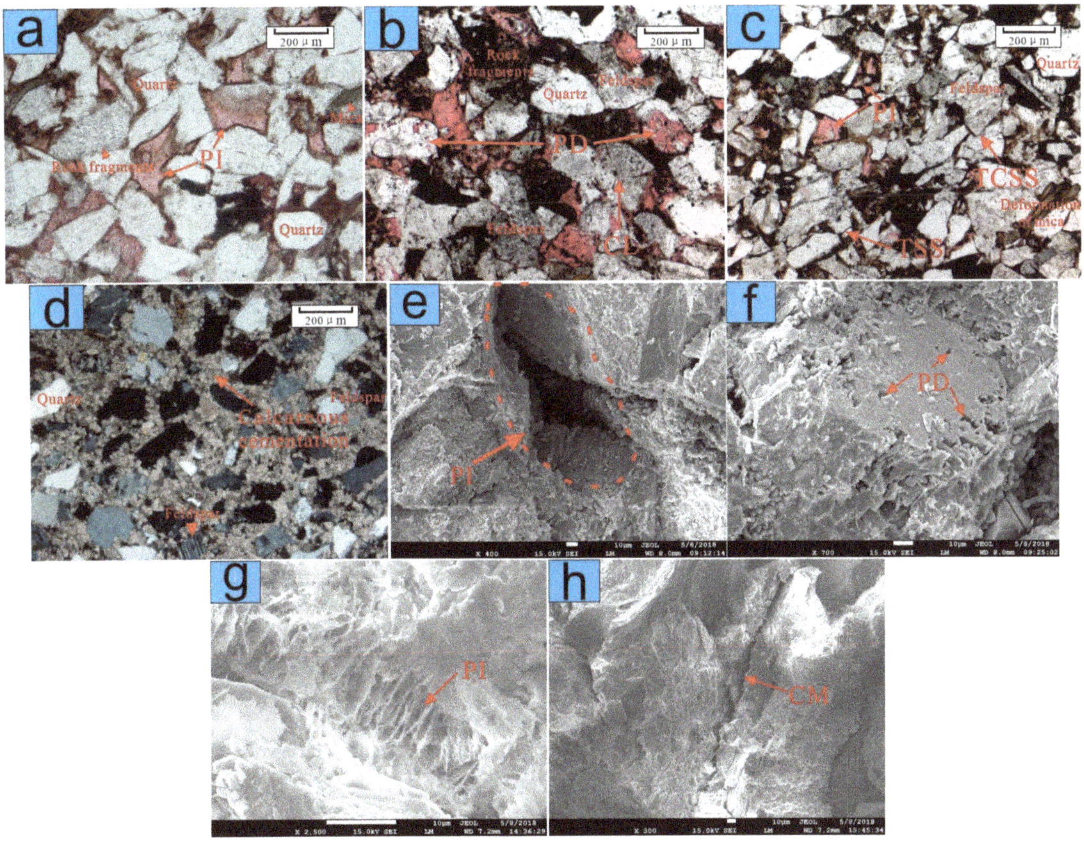

Figure 1. Microscopic pore types under CTS and FE-SEM imaging. (**a**) Intergranular pores (PI), B111, Chang 6_3, 2114.0 m; (**b**) well-developed feldspar dissolution pores (PD), grain-to-grain line contact with irregularities (CL), B143, Chang 6_2, 1923.25 m; (**c**) intergranular pores (PI), sheet-shaped throats (TSS), and curved-sheet-shaped throats (TCSS), B143, Chang 6_3, 1951.40 m; (**d**) calcareous cementation prone to forming carbonate dissolution pores, B239, Chang 6_3, 2105.85 m; (**e**) intergranular pores (PI), B23, Chang 8_1, 2060.2 m; (**f**) intragranular dissolution pores (PD), B23, Chang 8_1, 2060.2 m; (**g**) montmorillonite-filling intergranular pores (PI), intercrystalline throats within montmorillonite aggregates (TI), tubular-shaped throats (TTS), B29, Chang 7_1, 1549.5 m; (**h**) micro-cracks within rock particles (CM), B29, Chang 8_1, 1634.9 m.

3.2. Pore-Throat Distribution Characteristics Based on HPMI

3.2.1. Basic Features

Analysis of the pore-throat structure and pore size distribution using HPMI on 13 samples of tight sandstone reveals that the threshold pressure (P_{min}) ranges from 1.1 to 3.45 MPa (Table 1), with an average of 1.87 MPa. A lower initial slope of the threshold pressure and capillary pressure curve corresponds to the larger porosity and permeability of tight sandstone (Table 1) (Figure 2). Other pore-throat structure parameters include the pressure at 50% mercury saturation (P_{50}) ranging from 2.76 MPa to 16.55 MPa, averaging 6.25 MPa; the maximal pore-throat radius ($r_{max,}$), ranging from 0.21 μm to 0.67 μm, averaging 0.45 μm; the median pore-throat radius (r_{50}), ranging from 0.044 μm to 0.266 μm, averaging 0.153 μm; and the maximum mercury saturation (S_{Hgmax}), ranging from 84.35–98.81%, averaging 93.29%. The pore-throat radius distribution spans from 0.003 to 100 μm, primarily concentrated between 0.025 and 1.5 μm (Figure 3). The HPMI analysis indicates that tight sandstone exhibits complex and highly heterogeneous pore-throat structures.

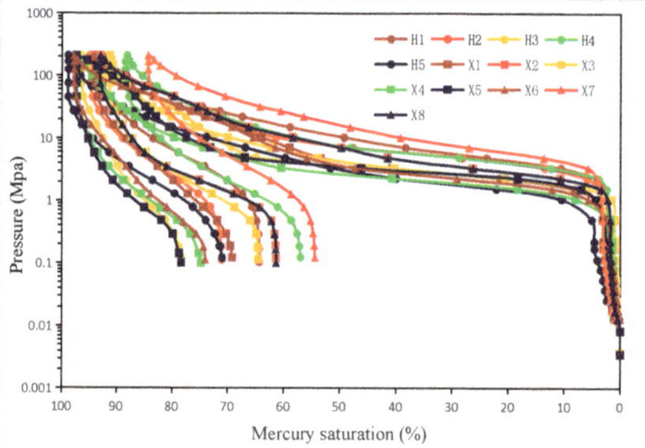

Figure 2. Characteristics of HPMI capillary pressure curve.

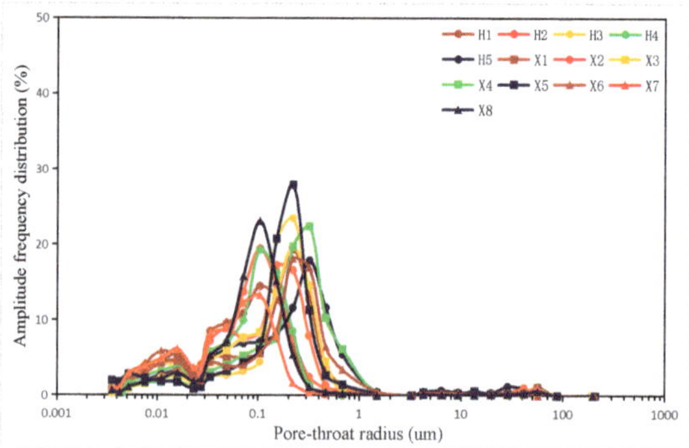

Figure 3. HPMI pore-throat radius distribution features.

3.2.2. Pore-Throat Fractal Dimensions (D)

A fractal analysis was conducted by plotting the $\lg(1-S_r) - \lg P_c$ graph for the 13 samples (Figure 4). The results reveal high correlation coefficients ($R_1{}^2$) ranging from 0.921 to 0.995 (average of 0.972) for small pores and ($R_2{}^2$) from 0.654 to 0.993 (average of 0.896) for large pores. It indicates a strong correlation between $\lg(1-S_r)$ and $\lg P_c$, allowing the application of fractal geometry theory to characterize the tight sandstone samples. Figure 4 shows that all the samples exhibit dual-fractal features, with a transition point at P_c = 1 MPa, corresponding to small pores (radius less than 0.735 μm) and large pores (radius greater than 0.735 μm). The fractal dimensions for small pores (D_1) range from 2.21 to 2.61 (average of 2.44), while for large pores (D_2), the range is 2.9867–2.9983 (average of 2.9949) (Table 1). This suggests that the structure complexity of large pores is higher than that of small pores. The total reservoir fractal dimensions (D) and the fractal dimensions for each pore type are related as expressed by Equation (15) [57]:

$$D = D_1 \times \frac{\varphi_1}{\varphi_1 + \varphi_2} + D_2 \times \frac{\varphi_2}{\varphi_1 + \varphi_2} \tag{15}$$

where φ_1 is the porosity of small pores; φ_2 is the porosity of large pores.

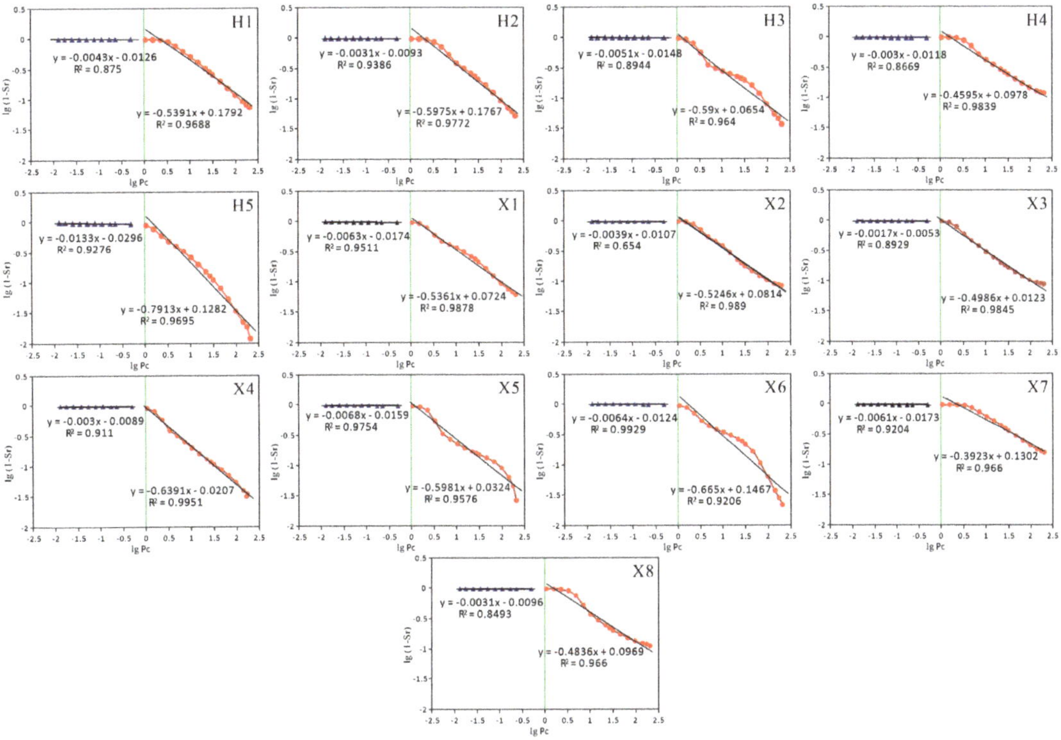

Figure 4. The curve of pore-throat size fractal features intersecting $\lg(1-S_r)$ vs. $\lg P_c$.

As shown in Figure 3, it is obvious that the proportion of large pores is minimal, constituting less than 5% of the total pores. In other words, the fractal dimensions of large pores have a minimal impact on the total fractal dimensions (Table 1). In the 13 samples, the maximum difference between the fractal dimensions of small pores and the total fractal dimensions is 0.04 (Table 1). Therefore, the fractal dimensions of small pores can be approximated as the total fractal dimensions for the analysis of the pore-throat structures in this tight sandstone.

3.3. Pore Distribution Characteristics Based on LT−N_2−GA

3.3.1. Basic Characteristics

LT−N_2−GA experiments were conducted on five samples, and the adsorption–desorption isotherms were plotted (Figure 5). In the classification of adsorption–desorption isotherms by the International Union of Pure and Applied Chemistry (IUPAC) [58], it was found that the curves of the five samples are close to Type IV. In the $0 \leq P/P_0 \leq 0.1$ range, N_2 molecules were adsorbed in the micropores of the samples as a monolayer. At relatively low pressures, the adsorption increased, and the desorption process could not be completed (Figure 5). In the $0.1 \leq P/P_0 \leq 0.47$ range, the mesopores began to adsorb the N_2 molecules as a multi-layer. With increasing pressure, the N_2 adsorption increased slowly, and the adsorption–desorption isotherms overlapped (Figure 5). In the $0.47 \leq P/P_0 \leq 1.0$ range, the mesopores were in the capillary condensation stage. With increasing pressure, the N_2 adsorption increased rapidly, and the adsorption–desorption isotherms formed an H_3-type hysteresis loop (Figure 5). The steeper the capillary condensation segment of the adsorption–desorption isotherms, the more uniform the mesopore distribution. The larger the mesopore diameter, the higher the pressure required for capillary condensation [59]. The capillary condensation segment of the tight sandstone samples is steep. After capillary condensation, there is no adsorption termination plateau, indicating a relatively uniform mesopore distribution with the presence of macropores inside the samples (Figure 5).

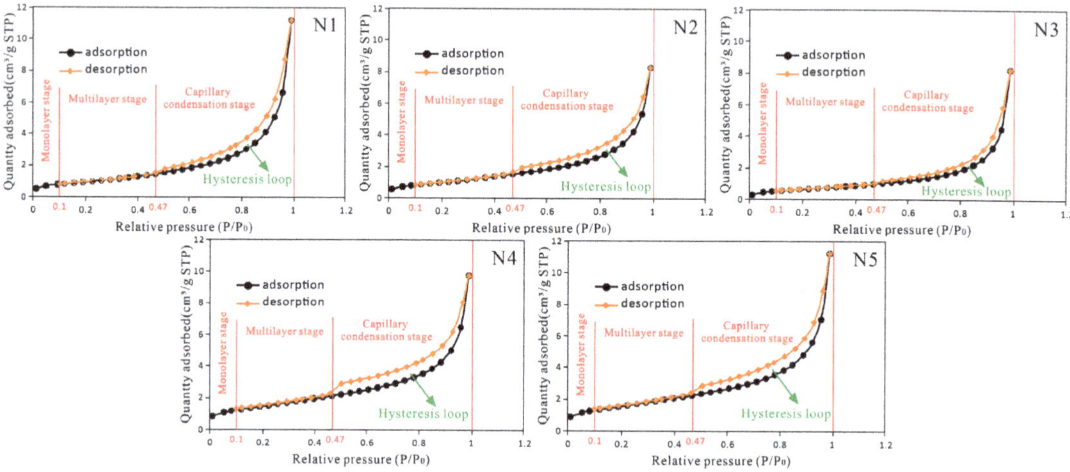

Figure 5. LT−N_2−GA isotherms of different samples.

The use of the BJH analysis method to calculate materials' pore size has been widely accepted by scholars [60–62]. Based on the LT−N_2−GA experiments, this study plotted curves of the pore size distribution using the BJH analysis method (Figure 6). Figure 6 shows that the pore radius distribution of the five samples is highly similar. The N_2 adsorption volume exhibits two consecutive peaks, indicating a relatively high rate with the corresponding pore radius distribution. The two high peaks are in the ranges of 3–22 nm and 25–63 nm, suggesting that tight sandstone is mainly dominated by mesopores, with few macropores larger than 50 nm, and larger pores of an increasing pore size gradually decrease (Figure 6).

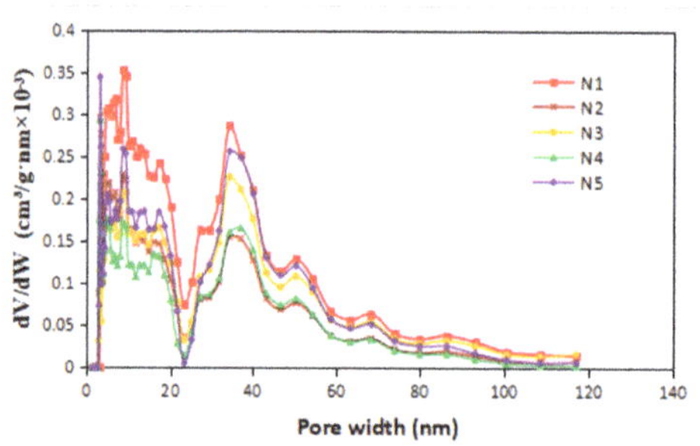

Figure 6. Pore size distribution curves of different tight sandstone samples based on LT−N_2−GA.

3.3.2. Pore-Throat Fractal Dimensions (D)

Utilizing the FHH fractal theory, the $\ln V - \ln\left[\ln\left(\frac{P_0}{P}\right)\right]$ curves for the five samples were fitted (Figure 7). As shown in Figure 7, there is a strong linear correlation between ln V and $\ln\left[\ln\left(\frac{P_0}{P}\right)\right]$, and R^2 exceeds 0.98 for all the samples (Table 2). This demonstrates the pronounced fractal characteristics in the pore-throat structure of tight sandstone. According to the calculation method for D, D primarily ranges from 2.458 to 2.591 (averaging 2.527). This indicates a high level of complexity in the pore-throat structure and strong heterogeneity (Table 2).

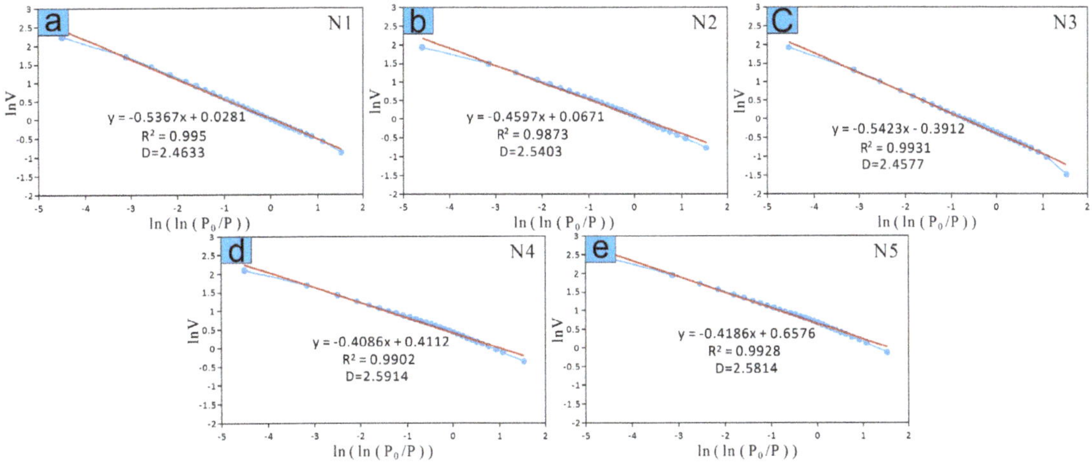

Figure 7. D calculation results from LT$-N_2-$GA experiment.

4. Analysis and Discussion

4.1. Analysis of Tight Sandstone Pores Based on HPMI

4.1.1. Reservoir Pore-Throat Structure Characteristics

The analysis of the reservoir's porosity and permeability through HPMI revealed that its porosity ranges from 8.37 to 12.81%, averaging 10.82%, and its permeability ranges from 0.025 to 0.262 mD, averaging 0.122 mD (Table 1). The observations indicate a certain correlation between porosity and permeability (Figure 8). The correlation between the reservoir pore-throat structure parameters and porosity is relatively weak (all R^2 values are below 0.4), but there is a stronger correlation with permeability (all R^2 values are above 0.5) (Figures 9 and 10). Porosity and permeability show a negative correlation with P_{min} and P_{50} but a positive correlation with r_{max}, r_{50}, and S_{Hgmax} (Figures 9 and 10). This suggests that the reservoir exhibits significant heterogeneity, with throat characteristics being the main controlling factor for reservoir permeability.

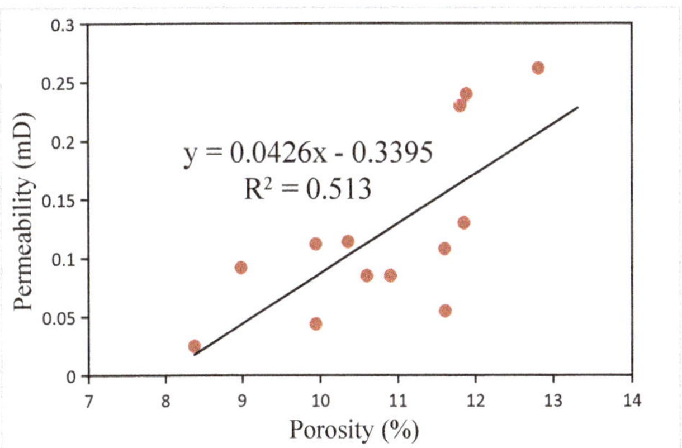

Figure 8. Porosity–permeability relationship chart.

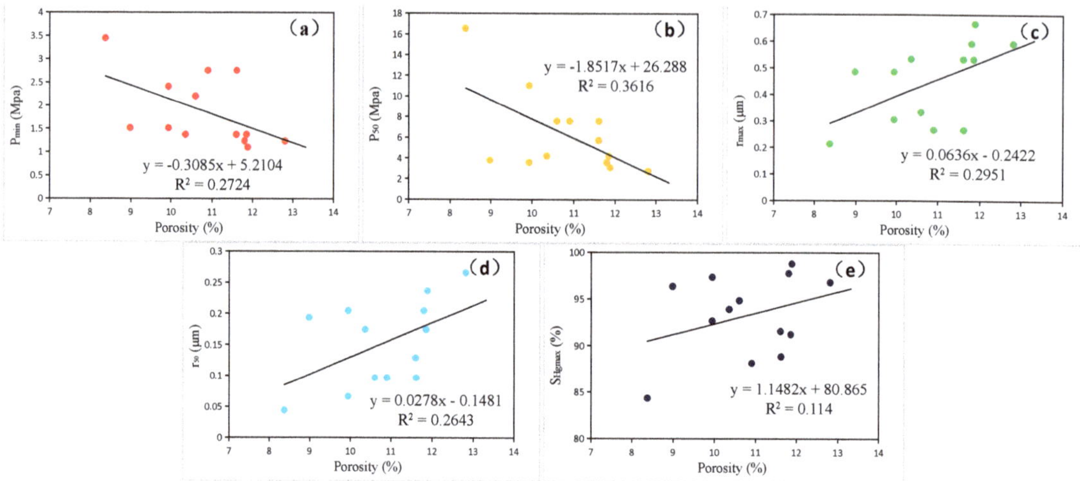

Figure 9. Relationship between porosity and pore-throat structure parameters. (**a**) porosity vs. P_{min}; (**b**) porosity vs. P_{50}; (**c**) porosity vs. r_{max}; (**d**) porosity vs. r_{50}; (**e**) porosity vs. S_{Hgmax}.

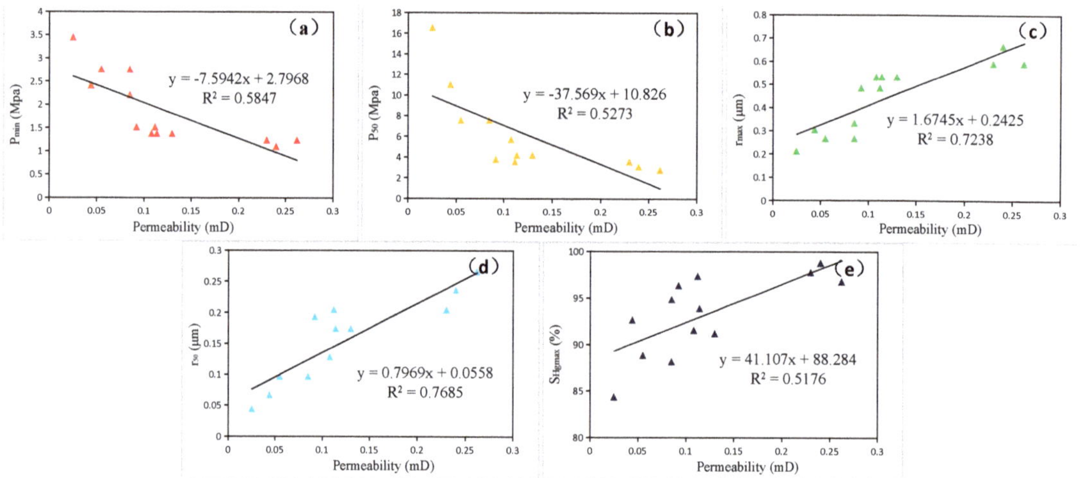

Figure 10. Relationship between permeability and pore-throat structure parameters. (**a**) permeability vs. P_{min}; (**b**) permeability vs. P_{50}; (**c**) permeability vs. r_{max}; (**d**) permeability vs. r_{50}; (**e**) permeability vs. S_{Hgmax}.

4.1.2. Relationship between D and Physical Properties

D shows a clear negative correlation with physical properties, with a weaker correlation with porosity and a stronger correlation with permeability (Figure 11). Ruifei et al. found that porosity is mainly controlled by large to medium pores, while permeability is controlled by small pores and pore-throat connectivity [63]. Therefore, this shows that the reservoirs studied here have fewer large pores, mainly consisting of medium to small pores and good pore-throat connectivity.

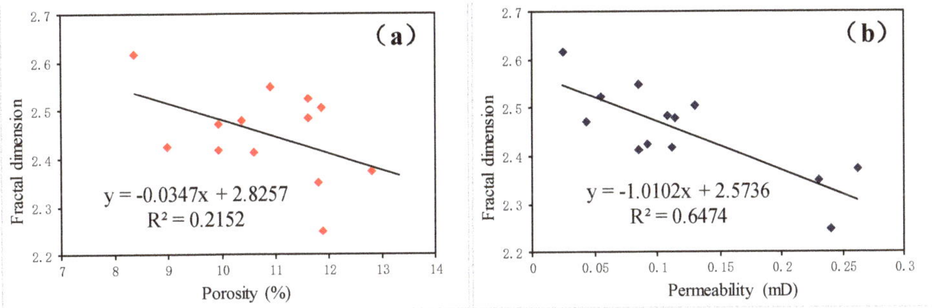

Figure 11. The relationship between D and porosity (**a**) and permeability (**b**).

4.1.3. Relationship between D and Pore-Throat Structure Parameters

Figure 12a,b demonstrate a strong positive correlation between the fractal dimensions and P_{min}, as well as P_{50}. As the fractal dimensions increase, both P_{min} and P_{50} also increase. This indicates that the complex pore-throat structure reduces the permeability of hydrocarbons, requiring higher pressure for fluid flow within the reservoir.

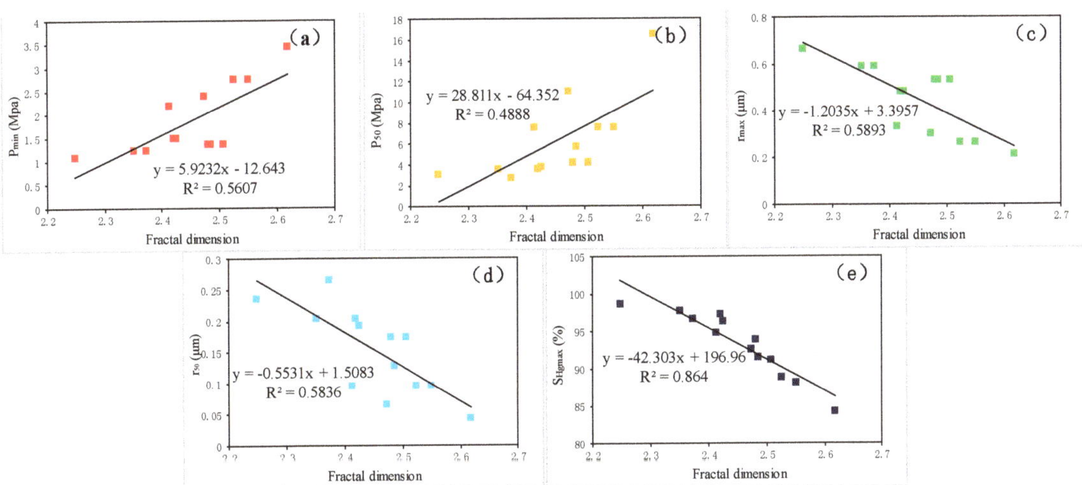

Figure 12. The relationship between the fractal dimensions (D) and pore-throat structure parameters. (**a**) P_{min}; (**b**) P_{50}; (**c**) r_{max}; (**d**) r_{50}; (**e**) S_{Hgmax}.

r_{max} and r_{50} are two physical parameters reflecting the pore-throat size in the reservoir [57]. Both parameters exhibit a strong negative correlation with the fractal dimensions, R^2, close to 0.6 (Figure 12c,d). This indicates that as D increases, the maximum mercury saturation decreases, the pore-throat connectivity worsens, and the heterogeneity of the reservoir properties becomes more pronounced.

The maximum mercury saturation represents the volume percentage of the connected pore space when mercury enters at maximum pressure, indicating the proportion of the connected pore volume to the total pore volume [57]. As shown in Figure 12e, there is a strong negative correlation between D and the maximum mercury saturation, R^2, of 0.864. This suggests that as the fractal dimensions increase, reflecting a more complex pore-throat structure, the maximum mercury saturation decreases, indicating poorer pore-throat connectivity and stronger heterogeneity in the reservoir properties.

4.2. Analysis of Tight Sandstone Pores Based on $LT-N_2-GA$

4.2.1. Relationships between Porosity and Permeability and Pore-Throat Structure Parameters

The pore-throat structure parameters play a pivotal role in determining the fundamental qualities of materials [64,65]. The utilization of pore-throat structure parameters obtained from $LT-N_2-GA$ analysis for investigating oil and gas reservoirs has found wide applications in the field of petroleum geology [66–68]. As depicted in Figure 13a,b, the BET specific surface area exhibits an inverse relationship with the porosity and permeability of tight sandstone. It exhibits a weaker correlation with porosity but a stronger one with permeability. Porosity is determined by pore size and quantity, while the specific surface area is influenced by pore size and quantity and also by factors like pore shape. The difference in the physical properties between porosity and specific surface area accounts for their relatively poor negative correlation. The surface of material pores serves as the boundary for fluid flow, and during the process of fluid infiltration, the fluid interacts with the rock surface, leading to adsorption phenomena that increase the resistance to fluid flow [69–72]. An increase in the specific surface area results in greater resistance, slowing down the fluid flow and reducing permeability.

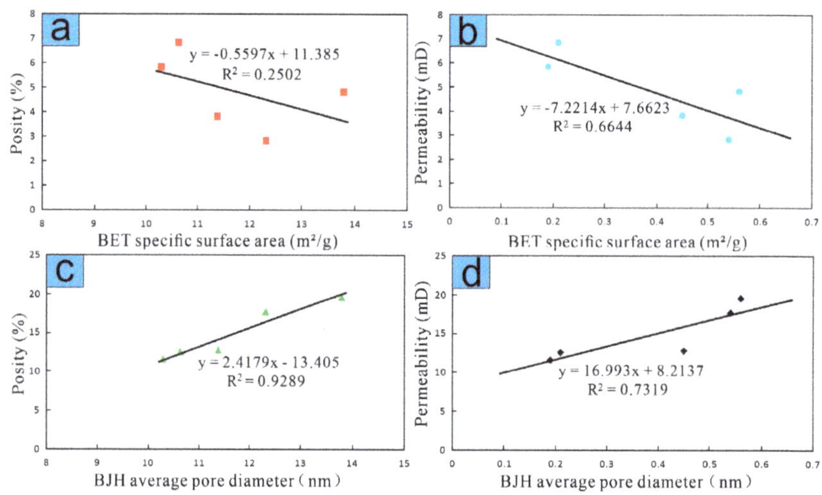

Figure 13. Relationships between porosity and permeability and pore-throat structure parameters. (**a**) Porosity vs. BET specific surface area. (**b**) Permeability vs. BET specific surface area. (**c**) Porosity vs. BJH average pore diameter. (**d**) Permeability vs. BJH average pore diameter.

In Figure 13c, the BJH mean pore diameter exhibits an extremely strong positive correlation with porosity. This suggests that pore size is the primary controlling factor for porosity, indicating that larger pores result in higher porosity. Additionally, the BJH mean pore diameter displays a strong correlation with permeability as well (Figure 13d), implying that larger pore throats are favorable for fluid permeation.

4.2.2. Relationship between D, Porosity, and Permeability

The application of pore fractal dimension analysis to porosity and permeability in porous materials has been universally adopted in the oil and gas reservoir field [37,44,46,57,73]. Fit analysis of the relationships between D, porosity, and permeability revealed a strong correlation among these three parameters (Figure 14a–c). There is a clear positive correlation between porosity and permeability (Figure 14a). This indicates that in tight sandstone reservoirs, the predominant types of throats are sheet-shaped and curved-sheet-shaped.

There are fewer pores such as closed pores or dead-end pores, which can increase porosity but reduce permeability.

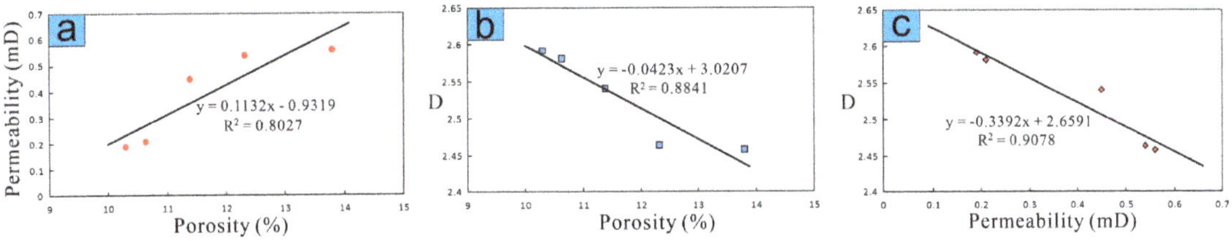

Figure 14. The correlation between D, porosity, and permeability based on LT$-N_2-$GA. (**a**) Porosity vs. permeability. (**b**) D vs. porosity. (**c**) D vs. permeability.

In Figure 14b,c, D exhibits a strong negative correlation with both porosity and permeability. As D increases, the reservoir's porosity and permeability decrease. The value of D of mesopores is significantly negatively correlated with porosity and permeability in tight sandstone reservoirs [57]. As is evident from Figure 6, the pore size distribution in tight sandstone predominantly comprises mesopores. Therefore, the strong negative correlation between D and porosity, as well as permeability, can be utilized to characterize the permeability characteristics of tight sandstone.

4.2.3. Relationship between D and Pore-Throat Structure Parameters

The value of D can effectively characterize the complexity of the pore-throat structure. As D increases, the complexity and heterogeneity of the pore-throat structure also increase [74,75]. As shown in Figure 15a, D has a strong correlation with the BET specific surface area. The more complex the pore-throat structure, the larger the BET specific surface area. In contrast, Figure 15b reveals a weak correlation between the fractal dimensions and the BJH total pore volume, suggesting that the complexity of the pore-throat structure is not related to the pore volume. Figure 15c demonstrates a very strong negative correlation between D and the BJH average pore diameter. This indicates that the pore size is the primary factor in the complexity of the pore-throat structure.

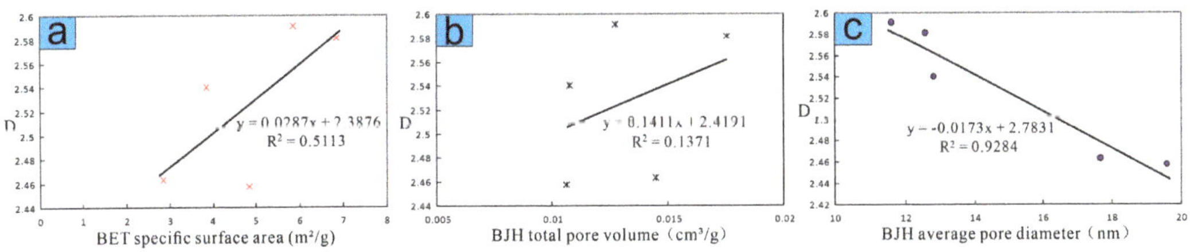

Figure 15. Relationships between D and pore-throat structure parameters based on LT$-N_2-$GA. (**a**) D and BET specific surface area. (**b**) D and BJH total pore volume. (**c**) D and BJH average pore diameter.

5. Conclusions

This study utilized CTS, FE-SEM, HPMI, and LT$-N_2-$GA experiments to analyze the pore-throat structure characteristics of tight sandstone. The following conclusions were drawn:

(1) The pore types in the tight sandstone reservoirs of Chang 6 to Chang 8 members in the Longdong area include residual intergranular pores, dissolution pores, intercrystalline pores, and microfractures. The throat types primarily include sheet-shaped throats, curved-sheet-shaped throats, and tubular throats.

(2) In the studied area, there is a relatively small proportion of pores with a radius larger than 735 nm in reservoirs. For pores with a radius smaller than 735 nm, the fractal dimensions serve as an effective metric to characterize the complexity of pore-throat structures (all the R^2 values are above 0.9). Integrated analysis of the HPMI and $LT-N_2-GA$ tests reveals that for smaller pores, utilizing fractal dimensions is more effective in characterizing pore-throat structures.

(3) D shows a relatively poor negative correlation with porosity but a strong negative correlation with permeability. It exhibits a significant positive correlation with P_{min} and P_{50}, a favorable negative correlation with r_{max} and r_{50}, and a very strong negative correlation with S_{Hgmax}.

(4) The fractal dimensions (D) exhibit a strong negative correlation with porosity, permeability, and BJH average pore diameter; a strong positive correlation with BET specific surface area; and a relatively weak positive correlation with BJH total pore volume. The pore-throat structure of the tight sandstone reservoir in the target area is complex, and the heterogeneity is significant.

(5) There is a strong positive correlation between porosity and permeability. As porosity, permeability, and pore connectivity develop, the reservoir's storage performance and oil and gas recovery efficiency improve.

Author Contributions: Conceptualization, T.H.; methodology, T.H. and Z.C.; software, T.H.; validation, Y.S. and Z.Z.; formal analysis, G.C.; investigation, T.H.; resources, Z.C.; data curation, T.H.; writing—original draft preparation, T.H.; writing—review and editing, H.X.; visualization, Z.Z.; supervision, Y.Z.; funding acquisition, Y.Z. All authors have read and agreed to the published version of the manuscript.

Funding: This research was funded by the National Science and Technology Major Project of China (Grant No. 2016ZX05050006).

Data Availability Statement: The data presented in this study are available from the author upon request.

Conflicts of Interest: Author Z.Z. was employed by the PetroChina Qinghai Oilfield Company. The remaining authors declare that the research was conducted in the absence of any commercial or financial relationships that could be construed as a potential conflict of interest.

References

1. Katz, B.; Lin, F. Lacustrine Basin Unconventional Resource Plays: Key Differences. *Mar. Pet. Geol.* **2014**, *56*, 255–265. [CrossRef]
2. Wang, S.; Qin, C.; Feng, Q.; Javadpour, F.; Rui, Z. A Framework for Predicting the Production Performance of Unconventional Resources Using Deep Learning. *Appl. Energy* **2021**, *295*, 117016. [CrossRef]
3. Slatt, R. Important Geological Properties of Unconventional Resource Shales. *Open Geosci.* **2011**, *3*, 435–448. [CrossRef]
4. Pang, X.; Jia, C.; Wang, W.; Chen, Z.; Li, M.; Jiang, F.; Hu, T.; Wang, K.; Wang, Y. Buoyance-Driven Hydrocarbon Accumulation Depth and Its Implication for Unconventional Resource Prediction. *Geosci. Front.* **2021**, *12*, 101133. [CrossRef]
5. Ashayeri, C.; Jha, B. Evaluation of Transfer Learning in Data-Driven Methods in the Assessment of Unconventional Resources. *J. Pet. Sci. Eng.* **2021**, *207*, 109178. [CrossRef]
6. Davies, R.J.; Almond, S.; Ward, R.S.; Jackson, R.B.; Adams, C.; Worrall, F.; Herringshaw, L.G.; Gluyas, J.G.; Whitehead, M.A. Oil and Gas Wells and Their Integrity: Implications for Shale and Unconventional Resource Exploitation. *Mar. Pet. Geol.* **2014**, *56*, 239–254. [CrossRef]
7. Zheng, M.; Li, J.; Wu, X.; Wang, S.; Guo, Q.; Yu, J.; Zheng, M.; Chen, N.; Yi, Q. China's Conventional and Unconventional Natural Gas Resources: Potential and Exploration Targets. *J. Nat. Gas Geosci.* **2018**, *3*, 295–309. [CrossRef]
8. Rani, S.; Padmanabhan, E.; Prusty, B.K. Review of Gas Adsorption in Shales for Enhanced Methane Recovery and CO_2 Storage. *J. Pet. Sci. Eng.* **2019**, *175*, 634–643. [CrossRef]
9. Alafnan, S.; Awotunde, A.; Glatz, G.; Adjei, S.; Alrumaih, I.; Gowida, A. Langmuir Adsorption Isotherm in Unconventional Resources: Applicability and Limitations. *J. Pet. Sci. Eng.* **2021**, *207*, 109172. [CrossRef]
10. Lai, J.; Wang, G.; Wang, Z.; Chen, J.; Pang, X.; Wang, S.; Zhou, Z.; He, Z.; Qin, Z.; Fan, X. A Review on Pore Structure Characterization in Tight Sandstones. *Earth-Sci. Rev.* **2018**, *177*, 436–457. [CrossRef]
11. Zou, C.; Zhu, R.; Liu, K.; Su, L.; Bai, B.; Zhang, X.; Yuan, X.; Wang, J. Tight Gas Sandstone Reservoirs in China: Characteristics and Recognition Criteria. *J. Pet. Sci. Eng.* **2012**, *88–89*, 82–91. [CrossRef]
12. Gao, H.; Li, H.A. Pore Structure Characterization, Permeability Evaluation and Enhanced Gas Recovery Techniques of Tight Gas Sandstones. *J. Nat. Gas Sci. Eng.* **2016**, *28*, 536–547. [CrossRef]

13. Xie, J.-Y.; Zhang, J.-J.; Xiang, W.; Fang, Y.-P.; Xue, Y.-J.; Cao, J.-X.; Tian, R.-F. Effect of Microscopic Pore Structures on Ultrasonic Velocity in Tight Sandstone with Different Fluid Saturation. *Pet. Sci.* **2022**, *19*, 2683–2694. [CrossRef]
14. Li, G.; Qin, Y.; Zhou, X.; Zhang, Y.; Hu, W. Comparative Analysis of the Pore Structure of Fusain in Lignite and High-Volatile Bituminous Coal. *J. Nat. Gas Sci. Eng.* **2021**, *90*, 103955. [CrossRef]
15. Yongzan, W.; Guanhua, N.; Xinyue, Z.; Yicheng, Z.; Gang, W.; Zhenyang, W.; Qiming, H. Fine Characterization of Pore Structure of Acidified Anthracite Based on Liquid Intrusion Method and Micro-CT. *Energy* **2023**, *263*, 125639. [CrossRef]
16. Xia, D.; Yang, Z.; Gao, T.; Li, H.; Lin, W. Characteristics of Micro- and Nano-Pores in Shale Oil Reservoirs. *J. Pet. Explor. Prod. Technol.* **2021**, *11*, 157–169. [CrossRef]
17. Xiao, D.; Jiang, S.; Thul, D.; Lu, S.; Zhang, L.; Li, B. Impacts of Clay on Pore Structure, Storage and Percolation of Tight Sandstones from the Songliao Basin, China: Implications for Genetic Classification of Tight Sandstone Reservoirs. *Fuel* **2018**, *211*, 390–404. [CrossRef]
18. Lai, J.; Wang, G.; Ran, Y.; Zhou, Z.; Cui, Y. Impact of Diagenesis on the Reservoir Quality of Tight Oil Sandstones: The Case of Upper Triassic Yanchang Formation Chang 7 Oil Layers in Ordos Basin, China. *J. Pet. Sci. Eng.* **2016**, *145*, 54–65. [CrossRef]
19. Lai, J.; Wang, G.; Wang, S.; Cao, J.; Li, M.; Pang, X.; Zhou, Z.; Fan, X.; Dai, Q.; Yang, L.; et al. Review of Diagenetic Facies in Tight Sandstones: Diagenesis, Diagenetic Minerals, and Prediction via Well Logs. *Earth-Sci. Rev.* **2018**, *185*, 234–258. [CrossRef]
20. Wang, Y.; Zhu, Y.; Wang, H.; Feng, G. Nanoscale Pore Morphology and Distribution of Lacustrine Shale Reservoirs: Examples from the Upper Triassic Yanchang Formation, Ordos Basin. *J. Energy Chem.* **2015**, *24*, 512–519. [CrossRef]
21. Cao, Z.; Liu, G.; Zhan, H.; Li, C.; You, Y.; Yang, C.; Jiang, H. Pore Structure Characterization of Chang-7 Tight Sandstone Using MICP Combined with N2GA Techniques and Its Geological Control Factors. *Sci. Rep.* **2016**, *6*, 36919. [CrossRef] [PubMed]
22. Liu, X.; Song, D.; He, X.; Wang, Z.; Zeng, M.; Wang, L. Quantitative Analysis of Coal Nanopore Characteristics Using Atomic Force Microscopy. *Powder Technol.* **2019**, *346*, 332–340. [CrossRef]
23. Lu, Y.; Liu, J.; Tang, J.; Ao, X.; Li, H.; Zhou, J.; Sun, X. Pore Changes of Slickwater-Containing Shale under Supercritical CO_2 Treatment. *Fuel* **2022**, *312*, 122775. [CrossRef]
24. Yuan, Y.; Rezaee, R. Fractal Analysis of the Pore Structure for Clay Bound Water and Potential Gas Storage in Shales Based on NMR and N2 Gas Adsorption. *J. Pet. Sci. Eng.* **2019**, *177*, 756–765. [CrossRef]
25. Kellerhals, R.; Shaw, J.; Arora, V.K. On Grain Size from Thin Sections. *J. Geol.* **1975**, *83*, 79–96. [CrossRef]
26. Lloreda-Jurado, P.J.; Hernández-Saz, J.; Chicardi, E.; Paúl, A.; Sepúlveda, R. Pore Morphology Evolution and Atom Distribution of Doped Fe_2O_3 Foams Developed by Freeze-Casting after Redox Cycling. *J. Mater. Res. Technol.* **2021**, *13*, 1887–1898. [CrossRef]
27. Karin, P.; Chammana, P.; Oungpakornkaew, P.; Rungsritanapaisan, P.; Amornprapa, W.; Charoenphonphanich, C.; Sripapha, K. Impact of Soot Nanoparticle Size and Quantity on Four-Ball Steel Wear Characteristics Using EDS, XRD and Electron Microscopy Image Analysis. *J. Mater. Res. Technol.* **2022**, *16*, 1781–1791. [CrossRef]
28. Hollamby, M.J. Practical Applications of Small-Angle Neutron Scattering. *Phys. Chem. Chem. Phys.* **2013**, *15*, 10566–10579. [CrossRef] [PubMed]
29. Avdeev, M.V.; Tropin, T.V.; Aksenov, V.L.; Rosta, L.; Garamus, V.M.; Rozhkova, N.N. Pore Structures in Shungites as Revealed by Small-Angle Neutron Scattering. *Carbon* **2006**, *44*, 954–961. [CrossRef]
30. Zhang, R.; Liu, S.; Bahadur, J.; Elsworth, D.; Melnichenko, Y.; He, L.; Wang, Y. Estimation and Modeling of Coal Pore Accessibility Using Small Angle Neutron Scattering. *Fuel* **2015**, *161*, 323–332. [CrossRef]
31. Yang, Y.; Wang, B.; Yuan, Q.; Huang, D.; Peng, H. Characterization, Factors, and Fractal Dimension of Pore Structure of Fly Ash-Based Geopolymers. *J. Mater. Res. Technol.* **2023**, *26*, 3395–3407. [CrossRef]
32. Zhang, D.-W.; Sun, X.-M.; Wang, Y.-R.; Li, H. Relationship between Fractal Characteristics of Pore-Structure and Thermal Properties of FA-Based AAM under Different Curing Conditions. *J. Mater. Res. Technol.* **2023**, *26*, 961–972. [CrossRef]
33. Wang, A.; Cao, S.; Yilmaz, E. Quantitative Analysis of Pore Characteristics of Nanocellulose Reinforced Cementitious Tailings Fills Using 3D Reconstruction of CT Images. *J. Mater. Res. Technol.* **2023**, *26*, 1428–1444. [CrossRef]
34. Li, J.; Zhang, Z.; Guo, Z.; Yang, Z.; Qian, W.; Chen, Y.; Li, H.; Zhao, Q.; Xing, Y.; Zhao, Y. Improved Corrosion Resistance of ZrO_2/MgO Coating for Magnesium Alloys by Manipulating the Pore Structure. *J. Mater. Res. Technol.* **2023**, *24*, 2403–2415. [CrossRef]
35. Zhang, C.; Wang, J.; Song, W.; Fu, J. Pore Structure, Mechanical Behavior and Damage Evolution of Cemented Paste Backfill. *J. Mater. Res. Technol.* **2022**, *17*, 2864–2874. [CrossRef]
36. Li, X.; Rao, F.; Song, S.; Ma, Q. Deterioration in the Microstructure of Metakaolin-Based Geopolymers in Marine Environment. *J. Mater. Res. Technol.* **2019**, *8*, 2747–2752. [CrossRef]
37. Zhang, L.; Lu, S.; Xiao, D.; Li, B. Pore Structure Characteristics of Tight Sandstones in the Northern Songliao Basin, China. *Mar. Pet. Geol.* **2017**, *88*, 170–180. [CrossRef]
38. Li, J.; Yin, J.; Zhang, Y.; Lu, S.; Wang, W.; Li, J.; Chen, F.; Meng, Y. A Comparison of Experimental Methods for Describing Shale Pore Features—A Case Study in the Bohai Bay Basin of Eastern China. *Int. J. Coal Geol.* **2015**, *152*, 39–49. [CrossRef]
39. Huang, H.; Chen, L.; Dang, W.; Luo, T.; Sun, W.; Jiang, Z.; Tang, X.; Zhang, S.; Ji, W.; Shao, S.; et al. Discussion on the Rising Segment of the Mercury Extrusion Curve in the High Pressure Mercury Intrusion Experiment on Shales. *Mar. Pet. Geol.* **2019**, *102*, 615–624. [CrossRef]
40. Mandelbrot, B. How Long Is the Coast of Britain? Statistical Self-Similarity and Fractional Dimension. *Science* **1967**, *156*, 636–638. [CrossRef]

41. Ou, K.; Wang, C.; Cao, R.; Lin, H.; Fang, L. Experimental Investigation of the Pore Fractal Characteristics and Damage Degradation Mechanism of Sandstone after Cyclic Freeze-thaw Treatments. *J. Mater. Res. Technol.* **2023**, *23*, 4843–4859. [CrossRef]
42. Zhao, Y.; Wang, C.; Ning, L.; Zhao, H.; Bi, J. Pore and Fracture Development in Coal under Stress Conditions Based on Nuclear Magnetic Resonance and Fractal Theory. *Fuel* **2022**, *309*, 122112. [CrossRef]
43. Kim, J.; Hong, G.; Oh, S.; Choi, S. Application of Various Fractal Models in Characterizing the Morphology of Pore Structures of Hydrating Cement Pastes. *J. Mater. Res. Technol.* **2022**, *20*, 3818–3835. [CrossRef]
44. Lai, J.; Wang, G. Fractal Analysis of Tight Gas Sandstones Using High-Pressure Mercury Intrusion Techniques. *J. Nat. Gas Sci. Eng.* **2015**, *24*, 185–196. [CrossRef]
45. Yu, S.; Bo, J.; Fengli, L.; Jiegang, L. Structure and Fractal Characteristic of Micro- and Meso-Pores in Low, Middle-Rank Tectonic Deformed Coals by CO2 and N2 Adsorption. *Microporous Mesoporous Mater.* **2017**, *253*, 191–202. [CrossRef]
46. Wu, Y.; Liu, C.; Ouyang, S.; Luo, B.; Zhao, D.; Sun, W.; Awan, R.S.; Lu, Z.; Li, G.; Zang, Q. Investigation of Pore-Throat Structure and Fractal Characteristics of Tight Sandstones Using HPMI, CRMI, and NMR Methods: A Case Study of the Lower Shihezi Formation in the Sulige Area, Ordos Basin. *J. Pet. Sci. Eng.* **2022**, *210*, 110053. [CrossRef]
47. Mandelbrot, B.B.; Passoja, D.E.; Paullay, A.J. Fractal Character of Fracture Surfaces of Metals. *Nature* **1984**, *308*, 721–722. [CrossRef]
48. Brooks, R.H.; Corey, A.T. *Hydraulic Properties of Porous Media*; Hydrology Paper; Colorado State University: Fort Collins, CO, USA, 1964; pp. 3–27.
49. Geological Applications of Capillary Pressure: A Review1 | AAPG Bulletin | GeoScienceWorld. Available online: https://pubs.geoscienceworld.org/aapgbull/article/76/6/840/38801/Geological-Applications-of-Capillary-Pressure-A (accessed on 25 May 2024).
50. Dollimore, D.; Spooner, P.; Turner, A. The Bet Method of Analysis of Gas Adsorption Data and Its Relevance to the Calculation of Surface Areas. *Surf. Technol.* **1976**, *4*, 121–160. [CrossRef]
51. Naderi, M. Chapter Fourteen—Surface Area: Brunauer–Emmett–Teller (BET). In *Progress in Filtration and Separation*; Tarleton, S., Ed.; Academic Press: Oxford, UK, 2015; pp. 585–608. ISBN 978-0-12-384746-1.
52. Brunauer, S.; Emmett, P.H.; Teller, E. Adsorption of Gases in Multimolecular Layers. *J. Am. Chem. Soc.* **1938**, *60*, 309–319. [CrossRef]
53. Kruk, M.; Jaroniec, M.; Sayari, A. Adsorption Study of Surface and Structural Properties of MCM-41 Materials of Different Pore Sizes. *J. Phys. Chem. B* **1997**, *101*, 583–589. [CrossRef]
54. Liu, X.; Nie, B. Fractal Characteristics of Coal Samples Utilizing Image Analysis and Gas Adsorption. *Fuel* **2016**, *182*, 314–322. [CrossRef]
55. Chengyang, W.; Shixiong, H.; Wenjing, S.; Wei, C. Fractal Dimension of Coal Particles and Their CH4 Adsorption. *Int. J. Min. Sci. Technol.* **2012**, *22*, 855–858. [CrossRef]
56. Liu, D.; Ren, D.; Du, K.; Qi, Y.; Ye, F. Impacts of Mineral Composition and Pore Structure on Spontaneous Imbibition in Tight Sandstone. *J. Pet. Sci. Eng.* **2021**, *201*, 108397. [CrossRef]
57. Li, P.; Zheng, M.; Bi, H.; Wu, S.; Wang, X. Pore Throat Structure and Fractal Characteristics of Tight Oil Sandstone: A Case Study in the Ordos Basin, China. *J. Pet. Sci. Eng.* **2017**, *149*, 665–674. [CrossRef]
58. Everett, D.H. Manual of Symbols and Terminology for Physicochemical Quantities and Units, Appendix II: Definitions, Terminology and Symbols in Colloid and Surface Chemistry. *Pure Appl. Chem.* **1972**, *31*, 577–638. [CrossRef]
59. Al-Ghouti, M.A.; Da'ana, D.A. Guidelines for the Use and Interpretation of Adsorption Isotherm Models: A Review. *J. Hazard. Mater.* **2020**, *393*, 122383. [CrossRef]
60. Hayati-Ashtiani, M. Characterization of Nano-Porous Bentonite (Montmorillonite) Particles Using FTIR and BET-BJH Analyses. *Part. Part. Syst. Charact.* **2011**, *28*, 71–76. [CrossRef]
61. Ojeda, M.L.; Esparza, J.M.; Campero, A.; Cordero, S.; Kornhauser, I.; Rojas, F. On Comparing BJH and NLDFT Pore-Size Distributions Determined from N2 Sorption on SBA-15 Substrata. *Phys. Chem. Chem. Phys.* **2003**, *5*, 1859–1866. [CrossRef]
62. Bardestani, R.; Patience, G.S.; Kaliaguine, S. Experimental Methods in Chemical Engineering: Specific Surface Area and Pore Size Distribution Measurements—BET, BJH, and DFT. *Can. J. Chem. Eng.* **2019**, *97*, 2781–2791. [CrossRef]
63. Ruifei, W.; Pingping, S.; Ziqi, S.; Hua, Y. Characteristics of Micro-Pore Throat in Ultra-Low Permeability Sandstone Reservoir. *Acta Pet. Sin.* **2009**, *30*, 560. [CrossRef]
64. Wang, Z.; Cheng, Y.; Wang, G.; Ni, G.; Wang, L. Comparative Analysis of Pore Structure Parameters of Coal by Using Low Pressure Argon and Nitrogen Adsorption. *Fuel* **2022**, *309*, 122120. [CrossRef]
65. Yu, F.; Guo, J.; Liu, J.; Cai, H.; Huang, Y. A Review of the Pore Structure of Pervious Concrete: Analyzing Method, Characterization Parameters and the Effect on Performance. *Constr. Build. Mater.* **2023**, *365*, 129971. [CrossRef]
66. Pang, P.; Han, H.; Hu, L.; Guo, C.; Gao, Y.; Xie, Y. The Calculations of Pore Structure Parameters from Gas Adsorption Experiments of Shales: Which Models Are Better? *J. Nat. Gas Sci. Eng.* **2021**, *94*, 104060. [CrossRef]
67. An, S.; Yao, J.; Yang, Y.; Zhang, L.; Zhao, J.; Gao, Y. Influence of Pore Structure Parameters on Flow Characteristics Based on a Digital Rock and the Pore Network Model. *J. Nat. Gas Sci. Eng.* **2016**, *31*, 156–163. [CrossRef]
68. Wei, M.; Xiong, Y.; Zhang, L.; Li, J.; Peng, P. The Effect of Sample Particle Size on the Determination of Pore Structure Parameters in Shales. *Int. J. Coal Geol.* **2016**, *163*, 177–185. [CrossRef]
69. Rabbani, A.; Jamshidi, S. Specific Surface and Porosity Relationship for Sandstones for Prediction of Permeability. *Int. J. Rock Mech. Min. Sci.* **2014**, *71*, 25–32. [CrossRef]

70. Odler, I.; Köster, H. Investigation on the Structure of Fully Hydrated Portland Cement and Tricalcium Silicate Pastes. III. Specific Surface Area and Permeability. *Cem. Concr. Res.* **1991**, *21*, 975–982. [CrossRef]
71. Luhmann, A.J.; Tutolo, B.M.; Bagley, B.C.; Mildner, D.F.R.; Seyfried, W.E., Jr.; Saar, M.O. Permeability, Porosity, and Mineral Surface Area Changes in Basalt Cores Induced by Reactive Transport of CO2-Rich Brine. *Water Resour. Res.* **2017**, *53*, 1908–1927. [CrossRef]
72. Sibiryakov, B.; Leite, L.W.B.; Sibiriakov, E. Porosity, Specific Surface Area and Permeability in Porous Media. *J. Appl. Geophys.* **2021**, *186*, 104261. [CrossRef]
73. Wei, D.; Gao, Z.; Zhang, C.; Fan, T.; Karubandika, G.M.; Meng, M. Pore Characteristics of the Carbonate Shoal from Fractal Perspective. *J. Pet. Sci. Eng.* **2019**, *174*, 1249–1260. [CrossRef]
74. Lyu, C.; Cheng, Q.; Zuo, R.; Wang, X. Mapping Spatial Distribution Characteristics of Lineaments Extracted from Remote Sensing Image Using Fractal and Multifractal Models. *J. Earth Sci.* **2017**, *28*, 507–515. [CrossRef]
75. Wang, F.-Y.; Yang, K.; Zai, Y. Multifractal Characteristics of Shale and Tight Sandstone Pore Structures with Nitrogen Adsorption and Nuclear Magnetic Resonance. *Pet. Sci.* **2020**, *17*, 1209–1220. [CrossRef]

fractal and fractional

MDPI

Article

Pore Structure Characterization and Fractal Characteristics of Tight Limestone Based on Low-Temperature Nitrogen Adsorption and Nuclear Magnetic Resonance

Wei Lin [1,2,*], Xinli Zhao [3,*], Mingtao Li [1,2] and Yan Zhuang [3]

[1] Institute of Digital Geology and Energy, Linyi University, Linyi 276000, China; limingtao@lyu.edu.cn
[2] School of Resources and Environment (College of Carbon Neutrality), Linyi University, Linyi 276000, China
[3] School of Petroleum and Natural Gas Engineering, Changzhou University, Changzhou 213164, China; zhuangy1991@cczu.edu.cn
* Correspondence: ucaslinwei@126.com (W.L.); ucaszxl@163.com (X.Z.)

Abstract: Pore structure characterization and fractal analysis have great significance for understanding and evaluating tight limestone reservoirs. In this work, the pore structure of tight limestone, low-temperature nitrogen adsorption (LTNA), and low-field nuclear magnetic resonance (NMR) are characterized, and the fractal dimension of the pore structure of tight limestone is discussed based on LTNA and NMR data. The results indicate that the pores of tight limestone have H3 and H4 types, the pore size distribution (PSD) of the H3 type is a wave distribution ranging from 2 to 10 nm, and the PSD of the H4 type is a unimodal distribution ranging from 2 to 10 nm. The transverse relaxation time (T_2) spectrum of tight limestone shows a single peak (DF), double peak (SF), and triple peak (TF), and the ranges for the T_2 spectra for micropores, mesopores, and macropores are 0.1 to 10 ms, 10 to 100 ms, and greater than 100 ms, respectively. The LTNA fractal dimension of tight limestone (D_L) ranges between 2.4446 and 2.7688, with an average of 2.5729, and the NMR fractal dimensions of micropores (D_{NMR1}), mesopores (D_{NMR2}), and macropores (D_{NMR3}) are distributed between 0.3744 and 1.1293, 2.4263 and 2.9395, and 2.6582 and 2.9989, respectively. Moreover, there is a negative correlation between D_L and average pore radius, a positive correlation between D_L and specific surface area, and a positive correlation between D_{NMR2} and D_{NMR3} and micropore content, while D_{NMR2} and D_{NMR3} are negatively correlated with the content of mesopores and macropores.

Keywords: tight limestone; pore structure; NMR; LTNA; fractal dimension

1. Introduction

Inspired by the economical and efficient development of North American marine tight oil, China began to explore and develop tight oil and gas resources [1–6]. In China, tight oil resources are mainly distributed in the Ordos Basin, Sichuan Basin, and Songliao Basin [7–11]. For example, the tight oil resources are abundant in the Taiyuan formation in the Ordos Basin and the Jurassic Da'anzhai Member in the Sichuan Basin, which are the typical lacustrine carbonate tight oil resources [12,13]. The tight limestone reservoir space of the Taiyuan formation is mainly composed of dissolution pores, residual organism cavities, intergranular pores, and microfractures, and the development of such reservoirs is mainly controlled by sedimentary microfacies, quasi-syngenetic karstification, and fractures, whereas the tight limestone reservoirs in the central Sichuan Basin are controlled by dissolution and tectonic processes [14], and the porosity and permeability of tight limestone are related to the calcite content [15].

In the development of tight oil, the accurate pore structure characterization of tight reservoirs is not only a comprehensive understanding of oil storage space but also the basis for its efficient development.

At present, there are many techniques for characterizing the pore structure of tight reservoirs, such as scanning electron microscopy (SEM), X-ray computed tomography (X-CT), high-pressure mercury injection (HPMI), constant-pressure mercury injection (CPMI), low-temperature nitrogen adsorption (LTNA), nuclear magnetic resonance (NMR), and so on. Generally, the HPMI test is prone to making microfractures in rocks due to a high mercury entry pressure [16,17]. At the same time, the CPMI test can only characterize a pore size distribution above 120 nm [18,19]. Tian et al. [15] used SEM to explore the pore structure of tight limestone in the Da'anzhai Member in the Jurassic Ziliujing Formation, central Sichuan Basin. They suggested that multi-type nano- to micro-meter micropores or microfractures develop in tight limestone. Additionally, Volery et al. [20] used SEM to study the factors controlling the difference between microporous and tight facies in the tight limestone of the Urgonian Formation of the French Jura Mountains. Liu et al. [21] analyzed the full-scale distribution map of tight limestone by using capillary pressure curve test data and LTNA experimental data and compared it to NMR data. They found that the main peak of tight limestone pores in the central Sichuan Basin was in the range of 10–50 nm; the pore diameter was 27–967 nm, and the average was 235 nm. However, their study did not explain how to establish a full-scale map. Zhao et al. [22] used a CPMI test to study the pore size distribution characteristics of tight limestone and tight sandstone in the central Sichuan Basin and discussed the fractal characteristics of these two types of rocks. They found that the fractal dimension of tight limestone was greater than the fractal dimension of tight sandstone, indicating that the pore structure of tight sandstone was simpler than that of tight limestone.

In this study, LTNA and NMR were employed to characterize the pore structure and pore size distribution of tight limestone. Furthermore, based on the LTNA and NMR data, the fractal theory was used to calculate the fractal dimensions of tight limestone pores, and the fractal characteristics of tight limestone pores are discussed. These comprehensive results help to reveal the pore space characteristics of the tight limestone reservoirs.

2. Experiment and Samples

2.1. Experimental Samples

In this study, LTNA experiments and NMR experiments were carried out on the 35 cores collected from the typical tight oil areas of SW, China. In the LTNA experiment, the core porosity ranged from 0.89% to 3.47%, with an average of 1.72%. In addition, the permeability was between 0.00079 and 0.25×10^{-3} μm^2 with an average of 0.03276×10^{-3} μm^2. In the NMR experiment, the tight limestone core plug porosity ranged from 0.85% to 3.47%, with an average of 1.61%. The core plug permeability was between 0.00098 and 0.435×10^{-3} μm^2, with an average of 0.08009×10^{-3} μm^2. Moreover, all these tight limestone cores came from the same research block. It is noteworthy that these cores are typical tight limestones (permeability < 1×10^{-3} μm^2 and porosity < 10%). According to the previous study, we know that the pores of the tight limestone in the target area can be divided into three types: intragranular pores (Figure 1d), intergranular pores (Figure 1e), and microfractures (Figure 1a–c,f) [23].

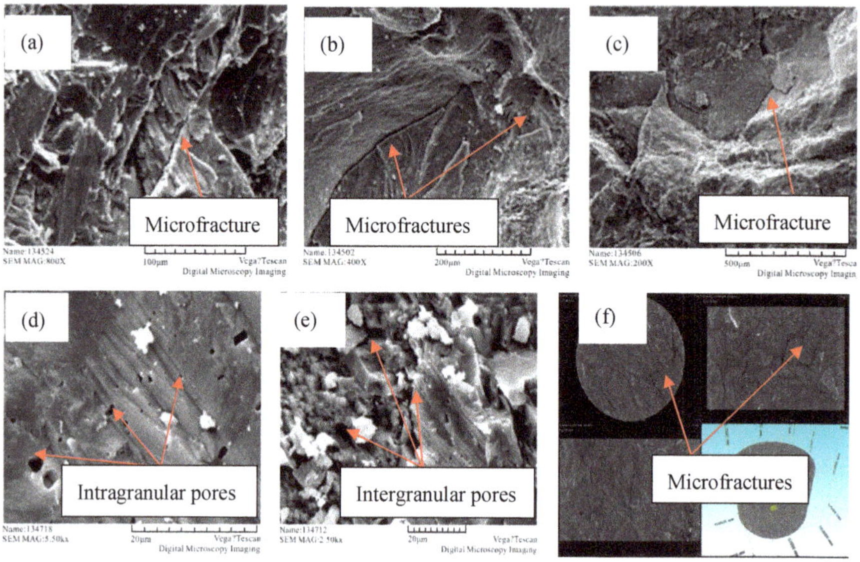

Figure 1. SEM and CT photographs of tight limestone in the study block [23]. (**a**) SEM image containing microfracture. (**b**) SEM image containing microfractures. (**c**) SEM image containing microfracture. (**d**) SEM image containing intragranular pores. (**e**) SEM image containing intergranular pores. (**f**) CT images containing microfractures.

2.2. LTNA Experiment

According to the International Union of Pure and Applied Chemistry (IUPAC) definition, micropores have a pore size of less than 2 nm, mesopores are 2–50 nm in size, and macropores are larger than 50 nm in size. For the analysis of the distribution of micropores and mesopores, LTNA was carried out using a USA autosor-6B automatic isothermal adsorption instrument from Quantachrome under a temperature of 77.35 K and a relative pressure of 0.01–1. In the LTNA test, the critical step is to measure the equilibrium pressure (P) and quantity adsorption (V) of nitrogen from a series of given pressures under a constant temperature (T) for generating the isothermal adsorption curve. It can be applied to obtain the specific surface areas, pore diameter, and pore volume of tight samples by using the BET adsorption isotherm [24] and to determine the pore size distribution curve using the BJH model [25,26].

2.3. NMR Experiment

In the NMR experiment, a small RecCore04 low-field NMR core analyzer (Langfang, China)was used, which was independently produced by the Institute of Porous Flow and Fluid Mechanics, Chinese Academy of Sciences. The specific parameters were set as follows: the main frequency was 3.841 MHz, the echo interval time was 600 μs, the waiting time was 3000 ms, the number of echoes was 1024, the number of scans was 64, the gain was 50, and the experimental ambient temperature was 24 °C.

3. Fractal Dimension Calculation Method

We applied the Frenkel–Halsey–Hill (FHH) model proposed by reference [27] to calculate the LTNA fractal dimension of tight limestone (D_L).

$$\ln\left(\frac{V}{V_0}\right) = A[\ln(\ln\frac{P_0}{P})] + C \tag{1}$$

where V is the volume of gas molecules adsorbed under equilibrium pressure P, in cm^3/g; V_0 is the volume of monomolecular adsorption gas, in cm^3/g; P_0 is the saturated vapor pressure of gas adsorption, in MPa; P is equilibrium pressure, in MPa; and A and C are the coefficients to be determined, which are obtained by fitting using Equation (1) based on the LTNA data. Furthermore, D_L is obtained by Equation (2).

$$D_L = A + 3 \tag{2}$$

In the calculations of the NMR fractal dimensions, we refer to the previous practices [28–35]. First, the relationship between the capillary pressure and the pore radius can be calculated according to the Washburn equation [36]:

$$p_c = \frac{2\sigma cos\theta}{r} \tag{3}$$

where p_c is the capillary pressure, in MPa; σ is the surface tension, in N/m; and θ is the contact angle between the water molecules and the surface of the sandstone.

The relationship between the Sv and the p_c and p_{cmin} can be calculated according to the equation developed by Chen et al. [35].

$$S_V = \left(\frac{p_c}{p_{cmin}}\right)^{D-3} \tag{4}$$

where Sv is the volume fraction of pores occupied by a wetting phase when the capillary pressure is equal to P_c and P_{cmin} is the capillary pressure corresponding to the maximum diameter, in MPa.

The relaxation time T_2 can be expressed as follows:

$$\frac{1}{T_2} = F_s.\frac{\rho}{r} \tag{5}$$

where T_2 is relaxation time, in ms; Fs is the geometry factor, and r is the pore diameter. For spherical pores, Fs = 3. For a columnar pipe, Fs = 2.

The relationship between p_c and T_2 can be calculated as follows.

$$p_c = C\frac{1}{T_2} \tag{6}$$

The conversion coefficient C can be expressed as:

$$C = \frac{2\sigma cos\theta}{F_s\rho} \tag{7}$$

where σ is the surface tension, in N/m and θ is the contact angle between the water molecules and the sandstone surface.

Hence, the following can be obtained:

$$p_{cmin} = C\frac{1}{T_{2max}} \tag{8}$$

where T_{2max} is maximum relaxation time, in ms.

By substituting Equations (6) and (8) into Equation (4), Equation (9) is obtained:

$$V_c = \left(\frac{T_2}{T_{2max}}\right)^{3-D} \tag{9}$$

where V_c is the percent of cumulative pore volume with a transverse relaxation time less than T_2 in terms of the total pore volume.

Taking the logarithm of both sides of Equation (9), an approximate fractal geometry formula for the NMR T_2 spectra can be expressed as follows:

$$\log V_c = (D-3)\log P_c - (D-3)\log P_{c\,min} \tag{10}$$

where D is the fractal dimension (dimensionless) and $P_{c\,\min}$ is the capillary pressure related to the largest pore radius, in MPa.

4. Results and Discussion

4.1. LTNA Characterization

The LTNA experiment can quantitatively characterize the pore size distribution and calculate the specific surface area (A_p) of tight limestone. The detailed LTNA test data of the tight limestones are shown in Table 1. Among them, the specific surface area of the rock sample is between 0.11 and 2.06 m^2/g, with an average of 0.76 m^2/g. Chen et al. [35] measured the specific surface area of shales in southwest China, which was between 7.67 m^2/g and 26.33 m^2/g. It can be seen that the specific surface area of the tight limestone sample is much smaller than the specific surface area of the shale. In other words, the pore throat structure of the shale is more complex than tight limestone. In addition, the pore radius (r_a) of the tight limestone sample is between 8.71 nm and 44.89 nm, with an average of 23.99 nm. Its pore volume (V_c) is between 0.35 and 5.26 mm^3/g, with an average of 2.76 mm^3/g.

Table 1. LTNA test data of different samples.

Sample	Depth	Lithology	φ	K	A_p	V_c	r_a	D_L
S1	2514.37~2516.06	Limestone	1.46	0.00600	1.15	2.87	13.25	2.6698
S2	2517.56~2519.43	Limestone	2.31	0.08300	0.64	3.66	17.63	2.5046
S3	2577.88~2579.35	Limestone	1.42	0.00084	0.11	0.35	23.90	2.6356
S4	2579.35~2580.68	Limestone	1.35	0.01400	0.20	1.79	36.04	2.5007
S5	2825.99~2827.59	Limestone	1.76	0.04000	0.31	2.70	24.29	2.4446
S6	2841.91~2843.56	Limestone	1.64	0.00400	0.25	1.88	27.51	2.4933
S7	2666.12~2667.36	Limestone	1.69	0.00210	1.52	3.68	18.59	2.6648
S8	2672.69~2674.56	Limestone	1.02	0.00160	0.21	1.37	33.79	2.5496
S9	2679.78~2682.04	Limestone	1.29	0.00640	0.39	1.77	26.78	2.6208
S10	2857.60~2858.60	Limestone	1.35	0.07100	2.06	2.24	8.71	2.7688
S11	2874.00~2875.10	Limestone	0.89	0.00084	0.12	1.64	44.89	2.4628
S12	2878.25~2879.43	Limestone	1.50	0.00300	0.32	3.00	29.98	2.4606
S13	2151.53~2153.70	Limestone	3.47	0.25000	0.92	4.81	15.41	2.5464
S14	1723.09~1725.29	Limestone	2.50	0.00790	1.65	4.44	18.01	2.6604
S15	1725.29~1727.32	Limestone	2.08	0.00079	1.51	5.26	21.01	2.6114

Note: φ: porosity, in %; K: permeability, in $\times 10^{-3}$ μm^2; Ap: specific surface, in m^2/g; Vc: pore volume, in mm^3/g; ra: average pore radius, in nm; and D_L: fractal dimension (dimensionless).

A large number of studies have shown that the shape of the adsorption and desorption curves obtained by LTNA experiments can qualitatively evaluate the pore size distribution of tight rock samples [37–39]. At the same time, based on the IUPAC classification criteria, the isothermal adsorption curves of the tight rock samples in this study block can be divided into two categories (Figure 2) [40]. The first type is the H3 hysteresis loop, which mainly includes five rock samples, namely, S3, S4, S8, S11, and S12. In addition, the other 10 tight reservoir rock samples belong to the H4 hysteresis loop. Due to the complexity of tight reservoirs, its pore type is diverse. It can be seen from Figure 3a that the pore radius distribution of the H3 type is mainly between 2 nm and 15 nm. The pore radius of the H4 type is mainly distributed between 1.7 nm and 5 nm (Figure 3b). Simultaneously, H3-type samples have multiple peaks, while H4-type samples have single peaks. This indicates that there is a bimodal type of rock sample (H3) whose micropores are more developed. It is worth noting that the permeability decreases as the average pore radius increases

(Figure 4a). This suggests that when the pore space of tight oil rock exists in the form of micropores, although micropores provide reservoir space, they do not contribute much to rock seepage capability. In addition, there is a positive correlation between porosity and pore volume (Figure 4b).

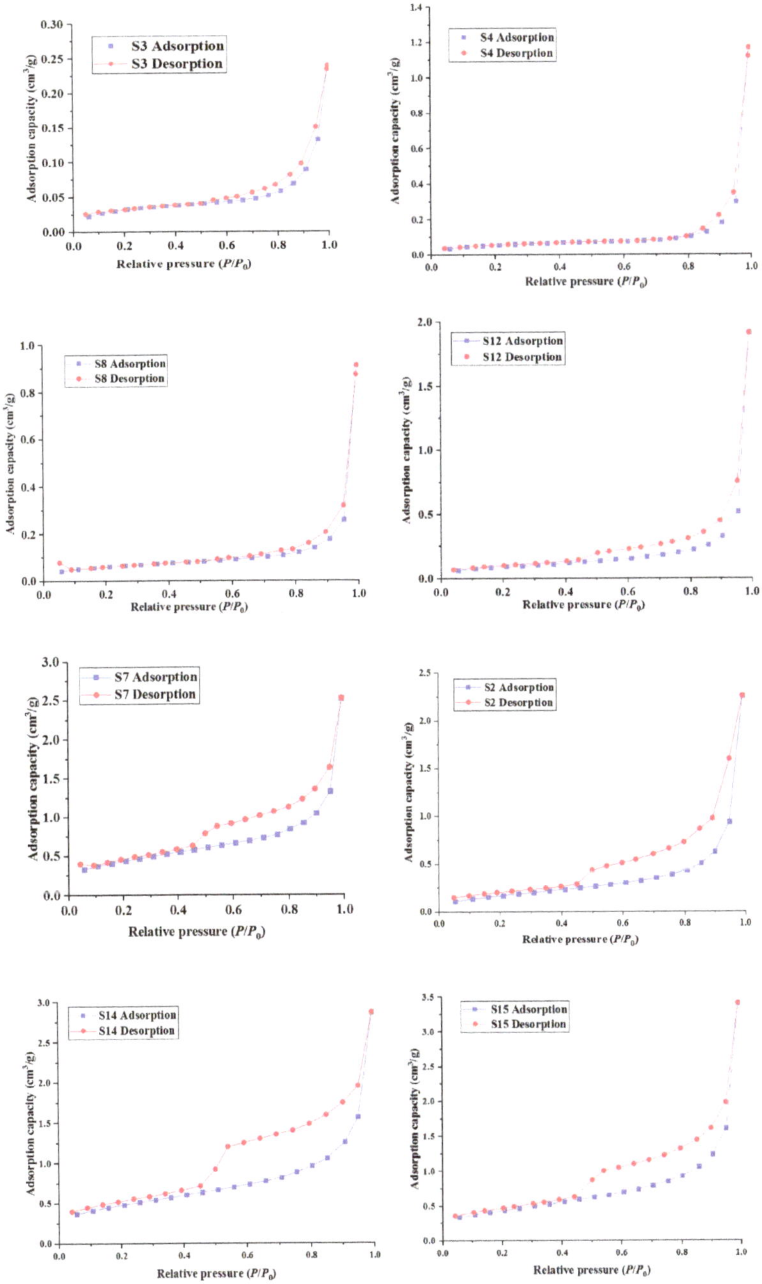

Figure 2. LTNA adsorption–desorption isotherm curve of tight limestone.

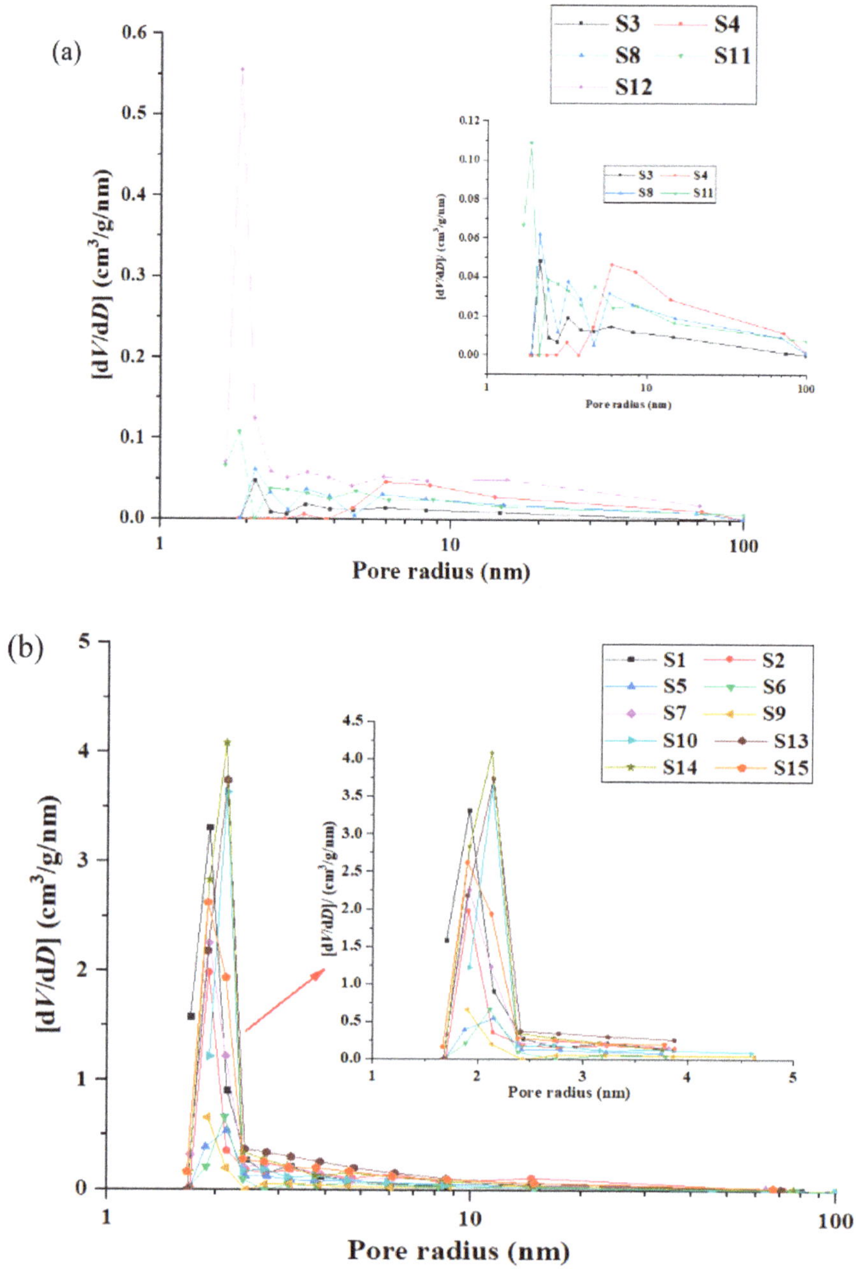

Figure 3. Pore volume difference curve of the LTNA of tight limestone. (**a**) H3 hysteresis loop. (**b**) H4 hysteresis loop.

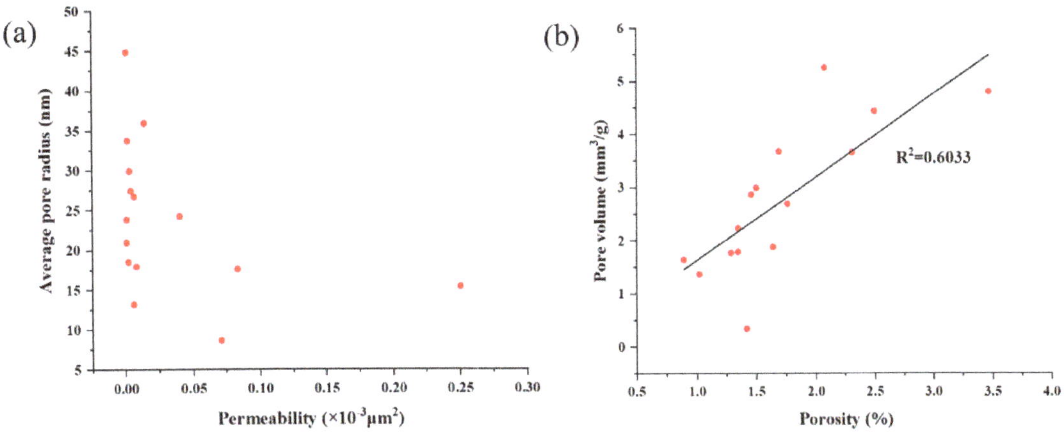

Figure 4. Relationship between permeability, porosity, and average pore radius and pore volume. (**a**) The relationship between permeability and average pore radius. (**b**) The relationship between porosity and pore volume.

4.2. NMR Characterization

NMR technology has unique advantages for studying the pore structure of tight limestones (e.g., non-destructive testing); it obtains the pore size distribution of the core by detecting the relaxation time (T2 spectrum) of the hydrogen protons in the core with 100% saturated water. In this study, NMR experiments were carried out using 20 tight reservoir cores with 100% saturated formation water. At the same time, according to the NMR test principle, the large pores have a long T_2 relaxation time; in turn, the small pores have a short T_2 relaxation time. Therefore, we can characterize the pores with a T_2 relaxation time between 0.1 ms and 10 ms as micropores; the pores with a T_2 relaxation time between 10 ms and 100 ms are characterized as mesopores. Finally, we define pores with a T_2 relaxation time greater than 100 ms as macropores.

According to the experimental test results (Figure 5), the tight limestone cores can be divided into three types: the first type only has one peak in the core (DF); the second type has two peaks in the core (SF); and the third type has three peaks in the core (TF). It can be seen from Figure 5 that in the tight limestone core, these three types of cores contain micropores, mesopores, and macropores, and the pore size exhibits a continuous distribution. In addition, we define three different pores (micropores, mesopores, and macropores) in proportions of Ω_1, Ω_2, and Ω_3, respectively. Among them, the calculation method for Ω_1, Ω_2, and Ω_3 is as follows [41]:

$$\Omega_1 = \frac{\int_{T_{2,min}}^{T_{2,10}} S_i dS}{\int_{T_{2,min}}^{T_{2,max}} S_i dS} \times 100\% \quad (11)$$

$$\Omega_2 = \frac{\int_{T_{2,10}}^{T_{2,100}} S_i dS}{\int_{T_{2,min}}^{T_{2,max}} S_i dS} \times 100\% \quad (12)$$

$$\Omega_3 = \frac{\int_{T_{2,100}}^{T_{2,max}} S_i dS}{\int_{T_{2,min}}^{T_{2,max}} S_i dS} \times 100\% \quad (13)$$

where, Ω_1, Ω_2, and Ω_3 represent the volume ratio of the micropores, mesopores, and macropores, respectively, in %; $T_{2,10}$ and $T_{2,100}$ are relaxation times of 10 ms and 100 ms, respectively; $T_{2,min}$ and $T_{2,max}$ are minimum and maximum relaxation times, respectively, in ms; and S_i is the amplitude value corresponding to each point on the saturated water T2 curve, in %.

Table 2 shows the calculated values of the volume fraction of each tight limestone core. Figure 6 is a statistical plot of the average proportion of different pore types in three types of cores. It can be seen from Figure 6 that the average proportion of micropores of DF, SF, and TF are 48.55%, 42.19%, and 9.26%, respectively; the average mesopores of the three types of cores are 65.52%, 31.21%, and 3.27%, respectively; and the average proportions of macro holes in the three types of cores are 38.43%, 45.68%, and 15.89%, respectively. It can be seen that the content of mesopores and macropores in TF-type cores is more than 60%; so, the pore development of the TF-type core is better than the other two types of cores.

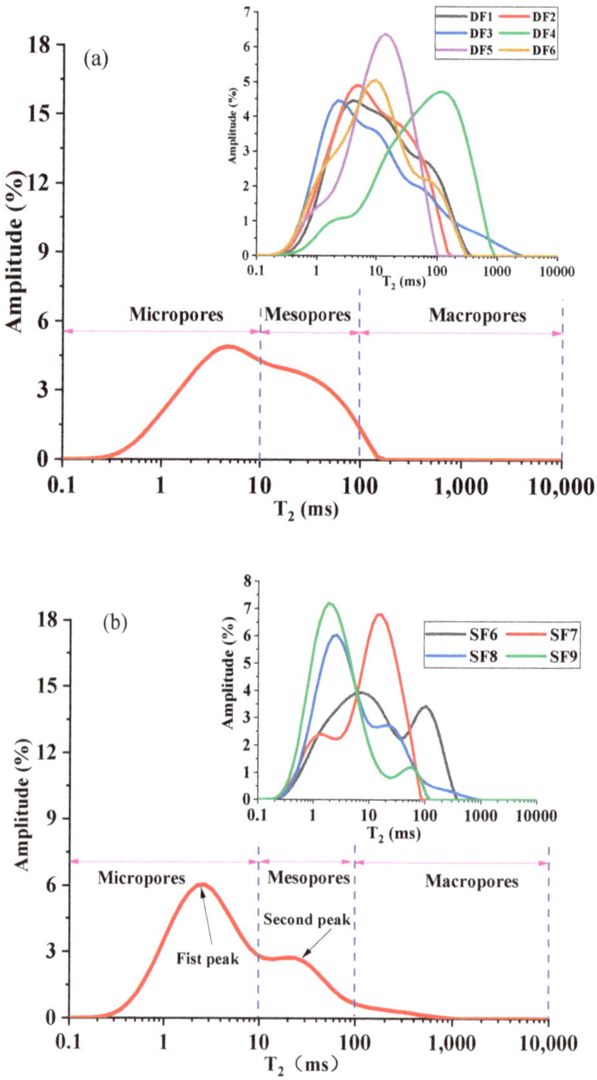

Figure 5. *Cont.*

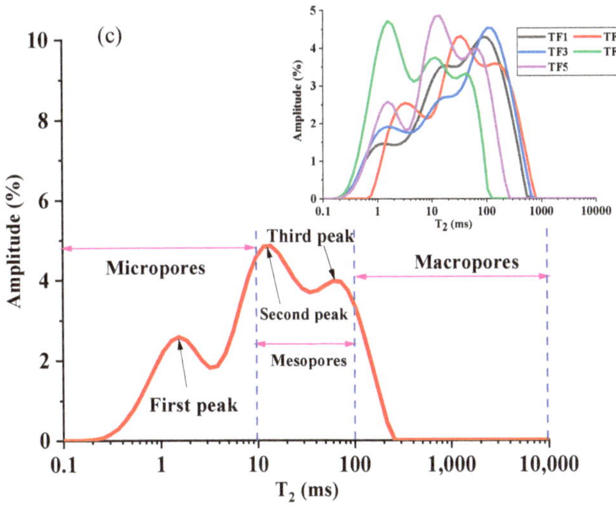

Figure 5. T2 distribution of tight limestone. (**a**) DF type. (**b**) SF type. (**c**) TF type.

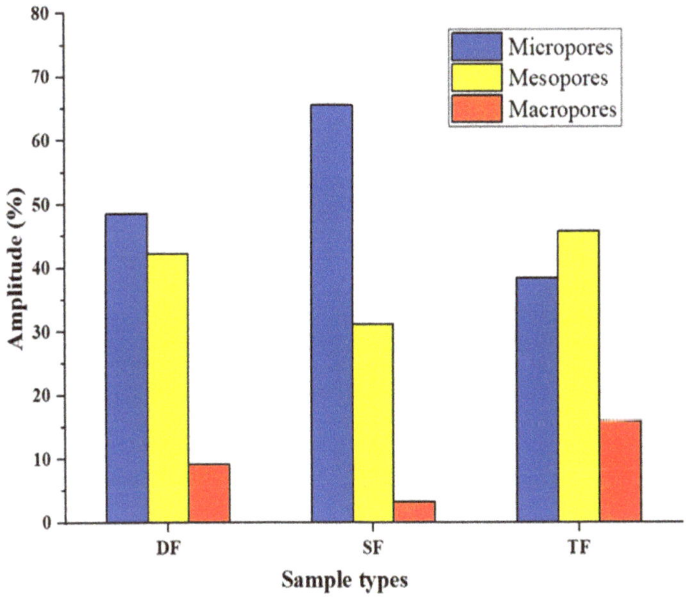

Figure 6. Different pore space amplitudes of three types of tight limestone.

Table 2. NMR test data of different samples.

Sample	Lithology	φ/%	$K/\times10^{-3}$ μm^2	D_{NMR1}	D_{NMR2}	D_{NMR3}	Ω_1/%	Ω_2/%	Ω_3/%
DF1	Limestone	1.25	0.00260	0.5897	2.6529	2.8905	52.71	41.05	6.24
DF2	Limestone	2.14	0.21600	0.4347	2.7603	2.9794	57.87	41.30	0.83
DF3	Limestone	1.08	0.04400	0.6262	2.8420	2.9670	59.18	30.31	10.51
DF4	Limestone	1.01	0.00260	0.3744	2.4263	2.6582	17.66	49.50	32.84
DF5	Limestone	3.35	0.00110	0.5769	2.6783	/	48.89	51.11	0.00
DF6	Limestone	1.30	0.05400	0.5875	2.7735	2.9423	54.96	39.89	5.15
SF1	Limestone	2.05	0.02000	0.5233	2.8709	2.9716	72.61	25.01	2.38
SF2	Limestone	1.85	0.00850	0.5829	2.8517	2.9927	72.42	27.03	0.55
SF3	Limestone	1.14	0.00540	0.6876	2.7863	2.9895	63.46	35.86	0.68
SF4	Limestone	1.10	0.00240	0.4519	2.8449	2.8874	59.97	29.14	10.89
SF5	Limestone	1.01	0.00260	0.4199	2.7429	/	62.19	37.81	0.00
SF6	Limestone	1.62	0.03900	0.5754	2.7721	2.8701	49.79	38.47	11.74
SF7	Limestone	1.15	0.00098	0.7286	2.6752	/	51.36	48.64	0.00
SF8	Limestone	3.47	0.25000	0.4694	2.8573	2.9756	70.90	25.87	3.23
SF9	Limestone	2.85	0.00670	0.5774	2.9395	2.9989	86.94	13.06	0.00
TF1	Limestone	1.27	0.43500	0.8115	2.5930	2.8371	29.86	48.70	21.44
TF2	Limestone	1.07	0.00200	1.1293	2.5509	2.8111	27.38	46.99	25.63
TF3	Limestone	1.20	0.41400	0.7702	2.6460	2.8163	31.37	42.47	26.16
TF4	Limestone	1.44	0.09200	0.6950	2.7859	/	62.61	37.39	0.00
TF5	Limestone	0.85	0.00300	0.6969	2.6839	2.9123	40.94	52.83	6.23

4.3. Fractal Characterization

The LTNA data were analyzed by the fractal theory; we linearly fitted the ln V/V_0 and $\ln[\ln(P_0/P)]$ data. It can be seen from Figure 7 that the correlation curve (R^2) of the two types of cores is greater than 0.9, indicating that the tight limestone reservoir core has obvious fractal features. The fractal dimension of LTNA (D_L) is between 2.4446 and 2.7688, with an average of 2.5729 (Table 1). At the same time, we calculated the multi-scale fractal dimension of the limestone cores by using NMR data (Table 2). The fractal dimension fitting lines of the three types of cores, namely, DF, SF, and TF, are shown in Figure 8. Among them, the mesopore fractal dimension (D_{NMR2}) of the DF-type tight limestone core is between 2.4263 and 2.8420, with an average of 2.6889, and the macropore fractal dimension (D_{NMR3}) is between 2.6582 and 2.9794, with an average of 2.8875. The mesopore fractal dimension of the SF-type tight limestone core is between 2.6752 and 2.9395, with an average of 2.8156; the macropore fractal dimension is between 2.8701 and 2.9989, with an average of 2.9551. The mesopore fractal dimension of the TF-type tight reservoir core is between 2.5509 and 2.7859, with an average of 2.6519; the macropore fractal dimension is between 2.8111 and 2.9123, with an average of 2.8442. It can be seen that the fractal dimension of macropores in the TF-type core is smaller than that of the other two types of cores; so, the pore structure is relatively simple, and the pore surface is relatively smooth.

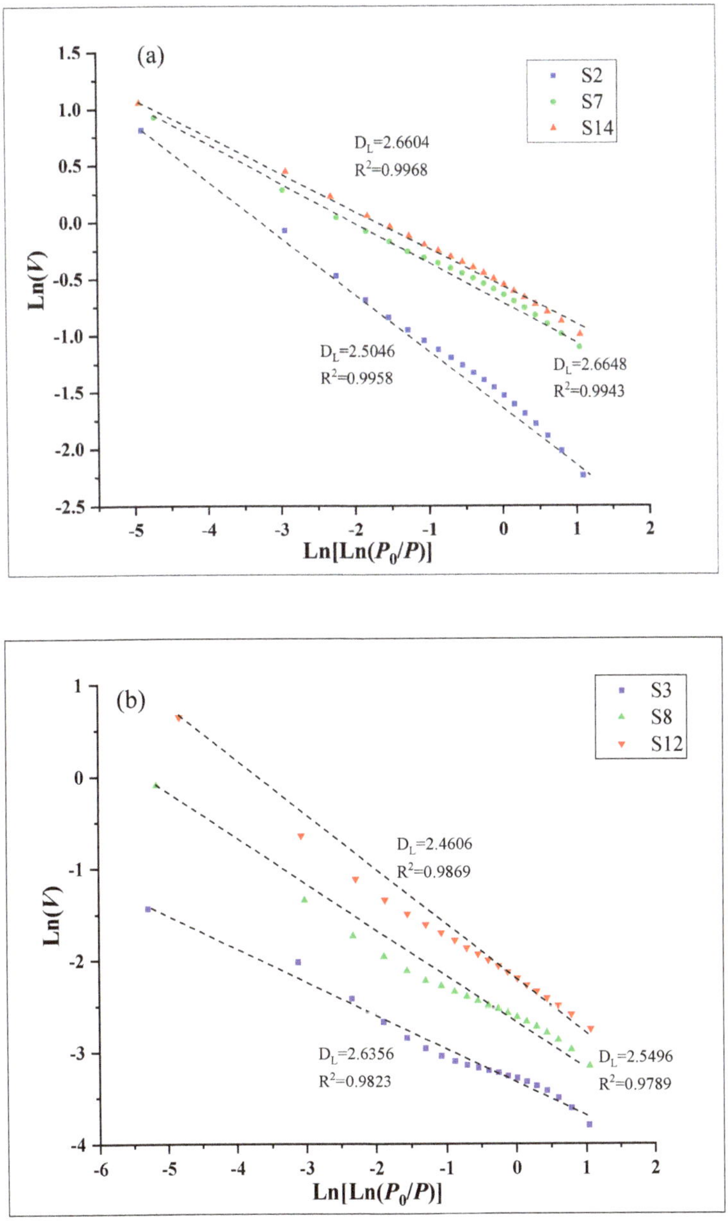

Figure 7. Relationship between Ln(V/V_0) and Ln[Ln(P_0/P)]. (**a**) fractal dimension of H3 type. (**b**) fractal dimension of H4 type.

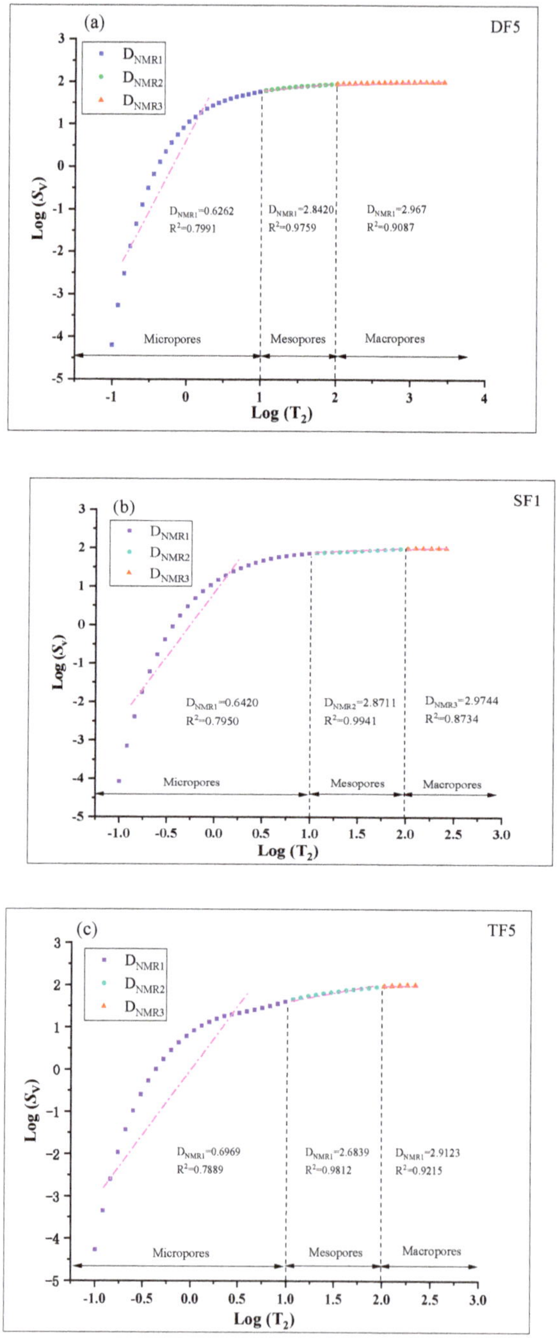

Figure 8. T2 fractal spectrum of tight limestone. (**a**) fractal dimension of DF5 sample. (**b**) fractal dimension of SF1 sample. (**c**) fractal dimension of TF5 sample.

Furthermore, we discuss the correlation relationship between the fractal dimension and permeability, porosity, etc. An obvious phenomenon is that there is no specific relationship between the fractal dimension and permeability and porosity in the tight limestone (Figures 9a,b and 10), which is consistent with the results of other scholars [42,43]. In addition, there is a positive correlation between the fractal dimension calculated from the LTNA data and the specific surface (Figure 9c). This indicates that as the specific surface area of tight limestone becomes larger, the complexity of its pore structure becomes higher. In addition, as the fractal dimension increases, the average pore radius of tight limestone shows a decreasing trend (Figure 9d). At the same time, by analyzing the fractal dimension calculated by the NMR data, it can be seen that D_{NMR2} has a positive correlation with the content of micropores of tight limestone and a negative correlation with the content of mesopores and macropores (Figure 11a). Moreover, D_{NMR3} also has a positive correlation with the content of micropores of tight limestone and a negative correlation with the content of mesopores and macropores (Figure 11b). In other words, with an increase in the micropore content of tight limestone, its fractal dimension tends to increase as the content of micropores is a significant factor affecting the fractal dimension of tight limestone.

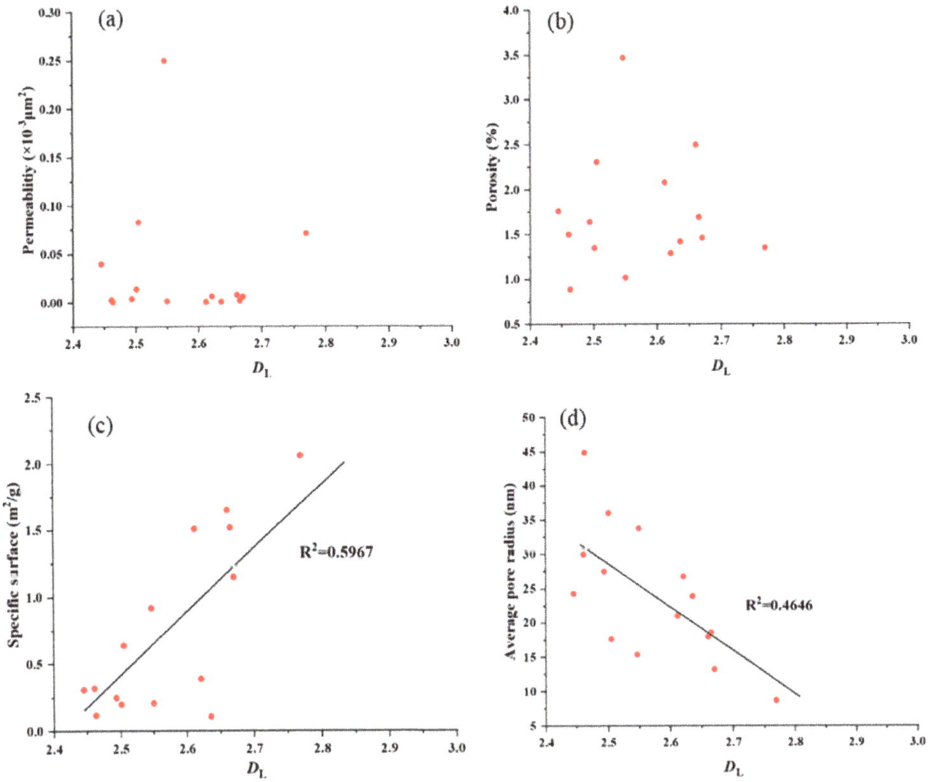

Figure 9. Relationships between D_L and the physical parameters of tight limestone. (**a**) The relationship between D_L and permeability. (**b**) The relationship between D_L and porosity. (**c**) The relationship between D_L and specific surface. (**d**) The relationship between D_L and average pore radius.

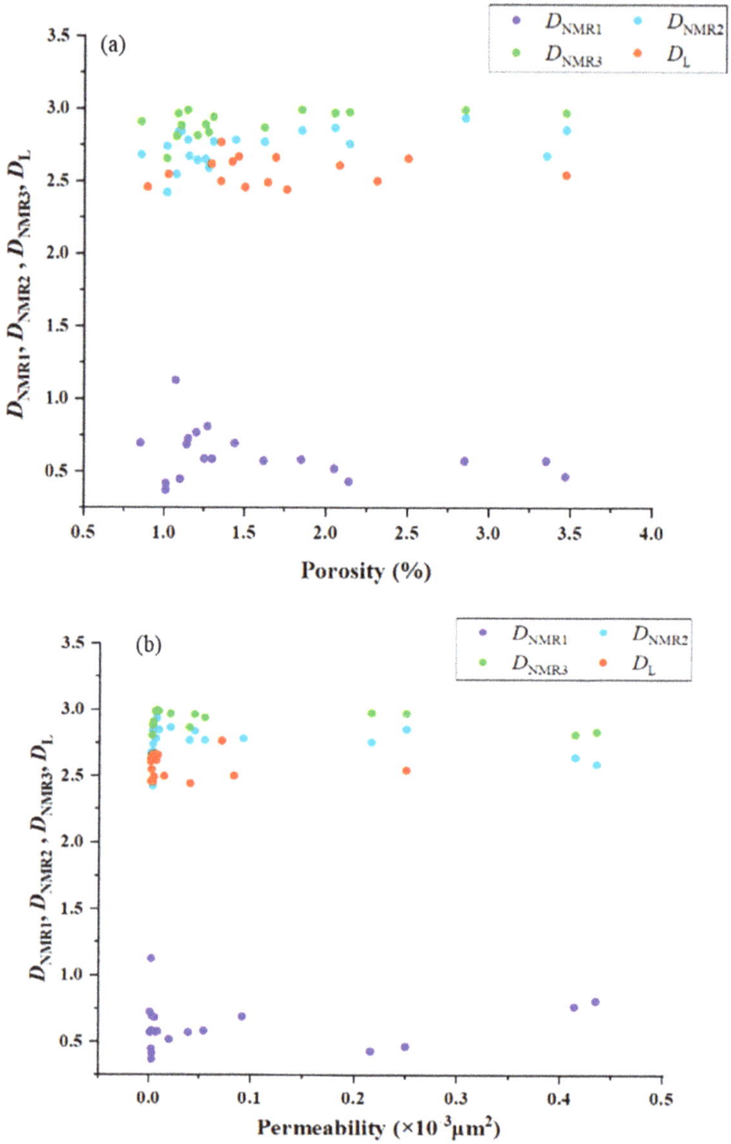

Figure 10. Intersecting graph of fractal dimension and porosity and permeability. (**a**) The relationship between porosity and fractal dimension. (**b**) The relationship between permeability and fractal dimension.

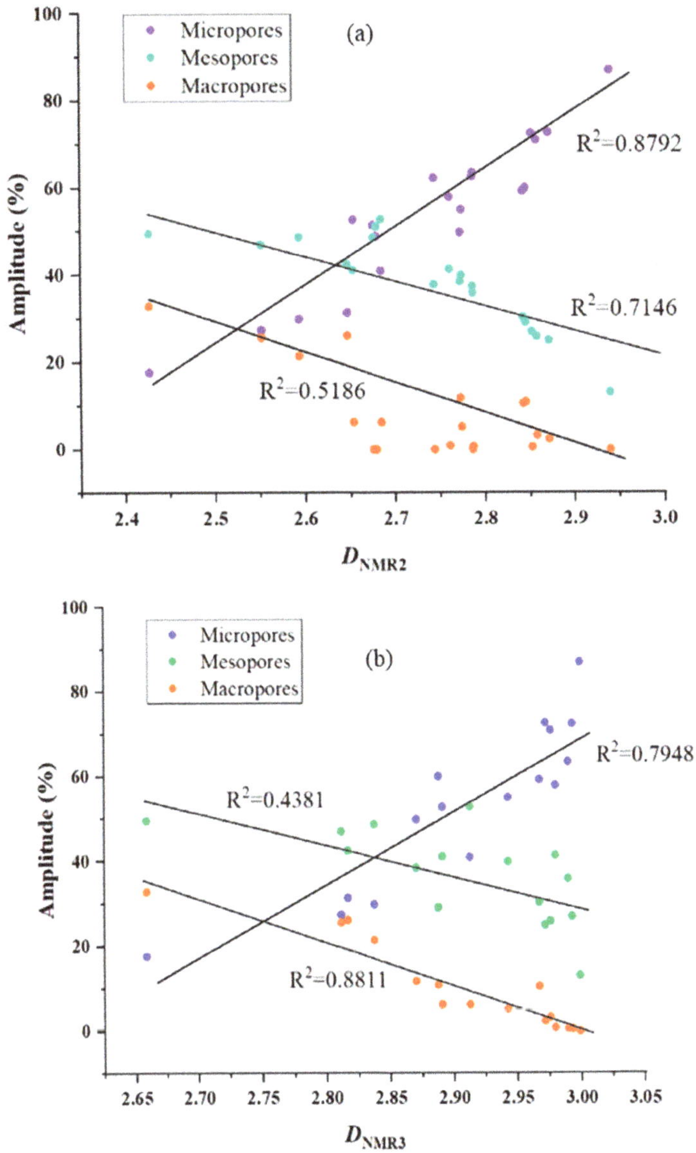

Figure 11. Relationship between the fractal dimension of NMR and different pore amplitudes. (**a**) The relationship between D_{NMR2} and pore space amplitudes. (**b**) The relationship between D_{NMR3} and pore space amplitudes.

5. Conclusions

In this work, LTNA and NMR were applied to characterize the pore structure of tight limestone, and these experimental data were used to analyze the fractal features of tight limestone, from which the following conclusions can be drawn:

(1) The tight limestones have intergranular pores, intragranular pores, and microfractures. Their LTNA test data indicate that the pores of tight limestone have H3 and H4 types,

and the pore radius of tight limestone is between 8.71 nm and 44.89 nm, with an average of 23.99 nm.

(2) The NMR results of tight limestone show that there are three types of T_2 distribution curves, namely, DF, SF, and TF, and the contents of micropores, mesopores, and macropores of the three types of cores are different. Among them, the SF-type tight limestone core has the highest micropore content, with an average of 65.52%. In addition, the average content of macropores in the TF-type tight limestone is 45.68%.

(3) The fractal dimension of LTNA (D_L) is between 2.4446 and 2.7688, with an average of 2.5729. There is a good positive correlation between D_L and the specific surface area of tight limestone, and there is a good positive correlation between D_{NMR2}, D_{NMR3}, and the content of micropores of tight limestone, which shows that with the increase in micropore content, the surface roughness of the pores in tight limestone is more complex.

Author Contributions: Conceptualization, W.L. and X.Z.; Methodology, W.L., X.Z. and M.L.; Software, X.Z. and M.L.; Validation, W.L. and X.Z.; Formal analysis, W.L., X.Z., M.L. and Y.Z.; Investigation, W.L., X.Z., M.L. and Y.Z.; Resources, W.L. and X.Z.; Data curation, W.L., X.Z., M.L. and Y.Z.; Writing—original draft, W.L., X.Z., M.L. and Y.Z.; Writing—review & editing, W.L., X.Z. and M.L.; Visualization, X.Z. and M.L.; Supervision, W.L.; Funding acquisition, W.L. All authors have read and agreed to the published version of the manuscript.

Funding: This work is funded by the National Science and Technology Major Project (2017ZX05013-001) and the CNPC Innovation Fund (2022DQ02-0102).

Data Availability Statement: The data used to support the findings of this study are available from the corresponding author upon request.

Conflicts of Interest: All authors declare that there are no conflicts of interest in this article.

References

1. Hughes, J.D. Energy: A reality check on the shale revolution. *Nature* **2013**, *494*, 307–308. [CrossRef]
2. Wang, S.; Yu, Y.; Guo, Q.; Wang, S.; Wu, X. New advances in the assessment of tight oil resource in China. *Pet. Res.* **2017**, *2*, 1–12. [CrossRef]
3. Hu, S.; Zhu, R.; Wu, S.; Bai, B.; Yang, Z.; Cui, J. Exploration and development of continental tight oil in China. *Pet. Explor. Dev.* **2018**, *45*, 790–802. [CrossRef]
4. Zhou, F.; Su, H.; Liang, X.; Meng, L.; Yuan, L.; Li, X.; Liang, T. Integrated hydraulic fracturing techniques to enhance oil recovery from tight rocks. *Pet. Explor. Dev.* **2019**, *46*, 1065–1072. [CrossRef]
5. Shen, W.; Li, X.; Lu, X.; Guo, W.; Zhou, S.; Wan, Y. Experimental study and isotherm models of water vapor adsorption in shale rocks. *J. Nat. Gas Sci. Eng.* **2018**, *52*, 484–491. [CrossRef]
6. Shen, W.; Song, F.; Hu, X.; Zhu, G.; Zhu, W. Experimental study on flow characteristics of gas transport in micro- and nanoscale pores. *Sci. Rep.* **2019**, *9*, 10196. [CrossRef] [PubMed]
7. Sun, Y.; Deng, M.; Ma, S.; Chen, Y.; Yu, L.; Zhang, Y.; Yan, B.; Zhang, Y. Distribution and controlling factors of tight sandstone oil in Fuyu oil layers of Da'an area, Songliao Basin, NE China. *Pet. Explor. Dev.* **2015**, *42*, 646–655. [CrossRef]
8. Luo, Y.; Wang, Y.; Liu, H.; Wang, G.; Zhao, Y. Overpressure controlling factors for tectonic fractures in near-source tight reservoirs in the southwest Ordos Basin, China. *J. Pet. Sci. Eng.* **2019**, *188*, 106818. [CrossRef]
9. Guo, Q.; Wang, S.; Chen, X. Assessment on tight oil resources in major basins in China. *J. Asian Earth Sci.* **2019**, *178*, 52–63. [CrossRef]
10. Yang, Z.; Zou, C.; Hou, L.; Wu, S.; Lin, S.; Luo, X.; Zhang, L.; Zhao, Z.; Cui, J.; Pan, S. Division of fine-grained rocks and selection of "sweet sections" in the oldest continental shale in China: Taking the coexisting combination of tight and shale oil in the Permian Junggar Basin. *Mar. Pet. Geol.* **2019**, *109*, 339–348. [CrossRef]
11. Pang, Z.; Tao, S.; Zhang, Q.; Zhang, T.; Yang, J.; Fan, J.; Yuan, M. Enrichment factors and sweep spot evaluation of Jurassic tight oil in central Sichuan Basin, SW China. *Pet. Res.* **2019**, *4*, 334–347. [CrossRef]
12. Wang, Y.; Tong, M.; Sun, Y.; Zhang, Y.; Yuan, D. Reservoir and development characteristics of the Da'anzhai tight oil in Sichuan Basin, SW China. *Pet. Res.* **2019**, *4*, 212–226. [CrossRef]
13. Cao, J.; Liu, Z.; He, F.; Liu, X.; Deng, W. Analysis of limestone reservoir forming geological conditions of Taiyuan formation in central and eastern Ordos basin. *Pet. Geol. Eng.* **2022**, *36*, 41–47.
14. Ni, C.; Yang, J.; Chen, W.; Lin, J.; Hou, G.; Wang, Y. Characteristics and Development Patterns of Tight Limestone Reservoirs: A Case Study of the Daanzhai Formation in the Central Sichuan Basin. *Lithol. Reserv.* **2015**, *27*, 38–47.

15. Tian, Z.; Song, X.; Wang, Y.; Ran, Q.; Liu, B.; Xu, Q.; Li, Y. Classification of lacustrine tight limestone considering matrix pores or fractures: A case study of Da'anzhai Member of Jurassic Ziliujing Formation in central Sichuan Basin, SW China. *Pet. Explor. Dev.* **2017**, *44*, 234–246. [CrossRef]
16. Zhao, H.; Ning, Z.; Wang, Q.; Zhang, R.; Zhao, T.; Niu, T.; Zeng, Y. Petrophysical characterization of tight oil reservoirs using pressure-controlled porosimetry combined with rate-controlled porosimetry. *Fuel* **2015**, *154*, 233–242. [CrossRef]
17. Wu, H.; Zhang, C.; Ji, Y.; Liu, R.; Wu, H.; Zhang, Y.; Geng, Z.; Zhang, Y.; Yang, J. An Improved Method of Characterizing the Pore Structure in Tight oil Reservoirs: Integrated NMR and Constant-Rate-Controlled Porosimetry Data. *J. Pet. Sci. Eng.* **2018**, *166*, 778–796. [CrossRef]
18. Zhao, X.; Yang, Z.; Lin, W.; Xiong, S.; Luo, Y.; Wang, Z.; Chen, T.; Xia, D.; Wu, Z. Study on pore structures of tight sandstone reservoirs based on nitrogen adsorption, high-pressure mercury intrusion and rate-controlled mercury intrusion. *J. Energy Resour. Technol.-Trans. ASME* **2019**, *141*, 112903. [CrossRef]
19. Zhao, X.; Yang, Z.; Lin, W.; Xiong, S.; Luo, Y.; Liu, X.; Xia, D. Fractal study on pore structure of tight sandstone based on full-scale map. *Int. J. Oil Gas Coal Technol.* **2019**, *22*, 123–139. [CrossRef]
20. Volery, C.; Davaud, E.; Durlet, C.; Clavel, B.; Charollais, J.; Caline, B. Microporous and tight limestones in the Urgonian Formation (late Hauterivian to early Aptian) of the French Jura Mountains: Focus on the factors controlling the formation of microporous facies. *Sediment. Geol.* **2010**, *230*, 21–34. [CrossRef]
21. Liu, Y.; Zhang, Y.; Wang, Y.; Wang, L. The pore structure of tight limestone—Jurassic Ziliujing Formation, Central Sichuan Basin, China. *Appl. Geophys.* **2018**, *15*, 165–174. [CrossRef]
22. Zhao, X.; Yang, Z.; Lin, W.; Xiong, S.; Wei, Y. Characteristics of microscopic pore-throat structure of tight oil reservoirs in Sichuan Basin measured by rate-controlled mercury injection. *Open Phys.* **2018**, *16*, 675–684. [CrossRef]
23. Wei, Y. *Study on Tight Sandstone Reservoir Characteristics and Development in Sichuan*; University of Chinese Academy of Sciences: Beijing, China, 2018.
24. Brunauer, S.; Emmett, P.H.; Teller, E. Adsorption of Gases in Multimolecular Layers. *J. Am. Chem. Soc.* **1938**, *60*, 309–319. [CrossRef]
25. Barrett, E.P.; Joyner, L.G.; Halenda, P.P. The Determination of Pore Volume and Area Distributions in Porous Substances. I. Computations from Nitrogen Isotherms. *J. Am. Chem. Soc.* **1951**, *73*, 373–380. [CrossRef]
26. Haul, R. Adsorption, surface area and porosity. *Z. Phys. Chem.* **1969**, *63*, 220–221. [CrossRef]
27. Naveen, P.; Asif, M.; Ojha, K. Integrated fractal description of nanopore structure and its effect on CH4 adsorption on Jharia coals, India. *Fuel* **2018**, *232*, 190–204. [CrossRef]
28. Zhang, Z.; Weller, A. Fractal dimension of pore-space geometry of an Eocene sandstone formation. *Geophysics* **2014**, *79*, D377–D387. [CrossRef]
29. Wang, F.; Yang, K.; Cai, J. Fractal characterization of tight oil reservoir pore structure using nuclear magnetic resonance and mercury intrusion porosimetry. *Fractals* **2018**, *26*, 1840017. [CrossRef]
30. Lai, J.; Wang, G.; Fan, Z.; Zhou, Z.; Chen, J.; Wang, S. Fractal analysis of tight shaly sandstones using nuclear magnetic resonance measurements. *AAPG Bull.* **2018**, *102*, 175–193. [CrossRef]
31. Zheng, S.; Yao, Y.; Liu, D.; Cai, Y.; Liu, Y. Characterizations of full-scale pore size distribution, porosity and permeability of coals: A novel methodology by nuclear magnetic resonance and fractal analysis theory. *Int. J. Coal Geol.* **2018**, *196*, 148–158. [CrossRef]
32. Zhang, J.; Deng, H.; Deng, J.; Gao, R. Fractal Analysis of Pore Structure Development of Sandstone: A Nuclear Magnetic Resonance Investigation. *IEEE Access* **2019**, *7*, 47282–47293. [CrossRef]
33. Li, C.; Liu, G.; Cao, Z.; Yuan, W.; Wang, P.; You, Y. Analysis of Petrophysical Characteristics and Water Movability of Tight Sandstone Using Low-Field Nuclear Magnetic Resonance. *Nat. Resour. Res.* **2019**, *29*, 2547–2573. [CrossRef]
34. Chen, S.; Tang, D.; Tao, S.; Ji, X., Xu, H. Fractal analysis of the dynamic variation in pore-fracture systems under the action of stress using a low-field NMR relaxation method: An experimental study of coals from western Guizhou in China. *J. Pet. Sci. Eng.* **2018**, *173*, 617–629. [CrossRef]
35. Chen, L.; Jiang, Z.; Liu, Q.; Jiang, S.; Liu, K.; Tan, J.; Gao, F. Mechanism of shale gas occurrence: Insights from comparative study on pore structures of marine and lacustrine shales. *Mar. Pet. Geol.* **2019**, *104*, 200–216. [CrossRef]
36. Mandelbrot, B.B.; Wheeler, J.A. The fractal geometry of nature. *Am. J. Phys.* **1998**, *51*, 468. [CrossRef]
37. Ojha, S.P.; Misra, S.; Tinni, A.; Sondergeld, C.; Rai, C. Pore connectivity and pore size distribution estimates for Wolfcamp and Eagle Ford shale samples from oil, gas and condensate windows using adsorption-desorption measurements. *J. Pet. Sci. Eng.* **2017**, *158*, 454–468. [CrossRef]
38. Zhao, P.; Wang, L.; Xu, C.; Fu, J.; Shi, Y.; Mao, Z.; Xiao, D. Nuclear magnetic resonance surface relaxivity and its advanced application in calculating pore size distributions. *Mar. Pet. Geol.* **2020**, *111*, 66–74. [CrossRef]
39. Zhao, P.; Wang, X.; Cai, J.; Luo, M.; Zhang, J.; Liu, Y.; Rabiei, M.; Li, C. Multifractal analysis of pore structure of Middle Bakken formation using low temperature N2 adsorption and NMR measurements. *J. Pet. Sci. Eng.* **2019**, *176*, 312–320. [CrossRef]
40. Thommes, M.; Kaneko, K.; Neimark, A.V.; Olivier, J.P.; Rodriguez-Reinoso, F.; Rouquerol, J.; Sing, K.S.W. Physisorption of gases, with special reference to the evaluation of surface area and pore size distribution (IUPAC Technical Report). *Pure Appl. Chem.* **2015**, *87*, 1051–1069. [CrossRef]
41. Chen, T.; Yang, Z.; Ding, Y.; Luo, Y.; Qi, D.; Lin, W.; Zhao, X. Waterflooding Huff-n-puff in Tight Oil Cores Using Online Nuclear Magnetic Resonance. *Energies* **2018**, *11*, 1524. [CrossRef]

42. Sun, W.; Zuo, Y.; Wu, Z.; Liu, H.; Xi, S.; Shui, Y.; Wang, J.; Liu, R.; Lin, J. Fractal analysis of pores and the pore structure of the Lower Cambrian Niutitang shale in northern Guizhou province: Investigations using NMR, SEM and image analyses. *Mar. Pet. Geol.* **2019**, *99*, 416–428. [CrossRef]
43. Wang, J.; Cao, Y.; Liu, K.; Gao, Y.; Qin, Z. Fractal characteristics of the pore structures of fine-grained, mixed sedimentary rocks from the Jimsar Sag, Junggar Basin: Implications for lacustrine tight oil accumulations. *J. Pet. Sci. Eng.* **2019**, *182*, 106363. [CrossRef]

fractal and fractional

MDPI

Article

Fractal Dimension Analysis of Pore Throat Structure in Tight Sandstone Reservoirs of Huagang Formation: Jiaxing Area of East China Sea Basin

Wenguang Wang [1,2,3,*], Chengyan Lin [3] and Xianguo Zhang [3]

1 Institute of Unconventional Oil & Gas, Northeast Petroleum University, Daqing 163318, China
2 Key Laboratory of Continental Shale Hydrocarbon Accumulation and Efficient Development, Ministry of Education, Northeast Petroleum University, Daqing 163318, China
3 School of Geosciences, China University of Petroleum (East China), Qingdao 266580, China; lincy@upc.edu.cn (C.L.); zhangxianguo@upc.edu.cn (X.Z.)
* Correspondence: wangwenguang.0@163.com or wangwg@nepu.edu.cn

Abstract: The reservoir quality of tight sandstone is usually affected by pore throat structures, and understanding pore throat structures and their fractal characteristics is crucial for the exploration and development of tight sandstone gas. In this study, fractal dimensions of pore throat structures and the effect of diagenesis on the fractal dimension of tight sandstone sweet spot in Huagang Formation, Jiaxing area, East China Sea Basin were studied by means of thin sections, scanning electron microscopes, X-ray diffraction analysis, scanning electron microscope quantitative mineral evaluation, and high pressure mercury injection experiments. The results show that the total fractal dimension ranges of type I, type II, and type III sweet spots were 2.62–2.87, 2.22–2.56, and 2.71–2.77, respectively. The negative correlation between total fractal dimensions, porosity, and permeability of type I sweet spots was different from those of type II and type III sweet spots. The negative correlation between total fractal dimensions of type II and type III sweet spots and maximum mercury saturation, average pore throat radius, and skewness were significant, whereas the correlation between total fractal dimensions of type I sweet spots, and maximum mercury saturation, average pore throat radius and skewness were not significant. The positive correlation between the total fractal dimensions of type II and type III sweet spots and the relative sorting coefficient, displacement pressure, and efficiency of mercury withdrawal were significant, whereas the correlation between the total fractal dimension of type I sweet spots and relative sorting coefficients, displacement pressures and efficiency of mercury withdrawal were not significant. The effect of diagenesis on fractal dimensions was investigated. Compaction reduced the pore space of tight sandstone and increased fractal dimensions. Quartz cementation and calcite cementation blocked pores and throats, reduced pore space, and increased fractal dimensions. Chlorite coat can inhibit compaction, protect pore throat structures, and maintain fractal dimensions. Most clay minerals filled primary pores and secondary pores and increased fractal dimensions. Dissolution increased the pore space of tight sandstone and decreased the fractal dimensions of the pore throat structures. The pore throat structures of type I sweet spots were mainly composed of macropores, mesopores, transitional pores, and micropores, and the fractal dimension of type I sweet spots was chiefly controlled by chlorite coat formation, dissolution, and a small amount of compaction. This study provides a reference for pore throat structure and fractal dimension analysis of tight sandstone sweet spots.

Keywords: fractal dimension; tight sandstone; diagenesis; pore throat structure; Xihu Sag

1. Introduction

Tight sandstone usually experiences strong diagenesis, such as compaction, cementation, and dissolution, which results in small pores and complex pore throat structures at the microscopic level [1]. The characteristics of pore throat structures are crucial for a comprehensive understanding of tight sandstone reservoirs. This is because heterogeneous pore throat structures can affect reservoir quality evaluation and final exploration risk assessment [2,3]. Fractal theory is a useful tool for irregular and self-similar phenomena widely present in the natural and social sciences [4]. The pore throat structure of sandstone is proved to be a statistical fractal in a certain range [4]. Therefore, the fractal dimension is a key parameter and an effective means to describe pore throat complexity and microscopic heterogeneity, and has been applied to pore throat structure evaluation, permeability prediction, and classification and evaluation of tight reservoirs [1,5–7].

At present, qualitative and quantitative research on pore throat morphology and size of tight sandstone can be divided into three categories: image observation methods, fluid injection methods, and spectral analysis methods [8–11]. Image observation methods, such as thin section, scanning electron microscope and energy spectrum analysis, and micro/nano CT, can be used to characterize the geometric characteristics of the pore throat visually [10,12]. Fluid injection methods, including high pressure mercury injection, constant rate mercury injection, gas adsorption/desorption measurement, and nuclear magnetic resonance, can quantitatively characterize pore throat size and distribution characteristics [13–15]. Spectral analysis methods chiefly include X-ray small-angle scattering, neutron small-angle scattering, and focused ion beam scanning electron microscopes [11]. However, these analytical methods have their advantages and limitations for measuring pore throat diameters [8,16]. Fractal theory is an effective tool to quantitatively characterize the complexity of pore throat structures in tight sandstone [17,18].

The fractal dimension is an important parameter to characterize the heterogeneity of pore throat structures [19,20]. With the increase in protuberances and concavities, pore surface and structure become more rough and complex [21], and fractal dimensions can quantify the roughness of the material surface. Fractal theory provides an effective bridge between microscopic characteristics and physical properties of tight reservoirs [22]. The fractal dimension of pore throat structures is between 2 and 3, and the increase in fractal dimension indicates a discrete distribution of pore throats and poor connectivity [18]. At present, there is much research on pore throat structures, fractal characteristics, and diagenesis of tight sandstone [16,23,24]. However, there are few studies on the fractal dimension of the pore throat structure of tight sandstone sweet spots and the effect of diagenesis on fractal dimensions.

The fourth (H4) and fifth (H5) members of the Huagang Formation in the Jiaxing area of the East China Sea Basin are tight sandstone reservoirs. The reservoir quality of these tight sandstones is often affected by pore throat structures, and understanding pore throat structure characteristics of sweet spots is crucial for effective development. This work takes the Jiaxing area of the East China Sea Basin as an example to investigate the fractal dimension of the pore throat structure of tight sandstone sweet spots and the effect of diagenesis on fractal dimensions. The structure of this article is as follows. Firstly, reservoir characteristics of tight sandstone were investigated. Secondly, the diagenesis of tight sandstone was analyzed, and the diagenetic evolution sequence was established. Thirdly, pore throat structure characteristics of tight sandstone were investigated. Fourthly, fractal characteristics of tight sandstone sweet spots were elaborated. In addition, a discussion was given, including the relationship between fractal dimensions, reservoir properties, and pore structure parameters, the effect of diagenesis on fractal characteristics, and implications for sweet spot prediction. Finally, the main conclusion was drawn. The main objectives of this work are the following: (1) Outline compaction, clay mineral transformation, quartz cementation, calcite cementation, and dissolution of tight sandstone; (2) Investigate the fractal dimension of pore throat structure of tight sandstone sweet spot, and elaborate the relationship between fractal dimension, reservoir properties, and pore structure parameters; (3) Investigate the effect of diagenesis on fractal dimension. This study provides a reference for pore throat structure and fractal dimension analysis of tight sandstone sweet spots.

2. Materials and Methods

2.1. Geological Setting and Samples

The East China Sea Basin is located in eastern China (Figure 1a) [25]. The Xihu Sag is located in the middle of the Zhedong Depression in the East China Sea Basin (Figure 1a) [26,27]. The length of the Xihu Sag is about 440 km, the width is approximately 70–130 km, and the area is about 59,000 km^2. The Xihu Sag is a basin dominated by Cenozoic strata, and its basement is the Proterozoic metamorphic rock series crystalline basement and Jurassic and Cretaceous inland clastic rock and volcanic rock series Mesozoic basement. East of the Xihu Sag is bounded by the Diaoyudao uplift, the west is bounded by the Haijiao uplift and the Yushan low uplift, the south is connected with the Diaobei Sag, and the north is adjacent to the Fujiang Sag (Figure 1b) [28]. The N gas field is located in the Jiaxing area of the central–northern parts of the central uplift belt in the Xihu Sag, East China Sea Basin (Figure 1b). The Huagang Formation in the N gas field is primarily a braided river delta sedimentary system with a large development range and continuous sandstone distribution (Figure 1b). The upper part of the Huagang Formation can be divided into H1–H5 and the lower part of the Huagang Formation can be divided into H6–H12. The thick sandstone deposits in the upper part of the Huagang Formation are primarily subaqueous distributary channels in the braided river delta front part, which are the main natural gas reservoirs. H1–H2 is chiefly composed of thin interbeds of sandstone and mudstone. H3-H5 develops thick positive rhythm superimposed sandstone, with glutenite and gravel-bearing sandstone common at the bottom. The tight sandstone of the H4 and H5 members is the target interval of this study (Figure 1c). The burial depth of the H4 and H5 sandstones ranges from 3800 m to 4150 m, and the formation temperature ranges from 135 °C to 155 °C.

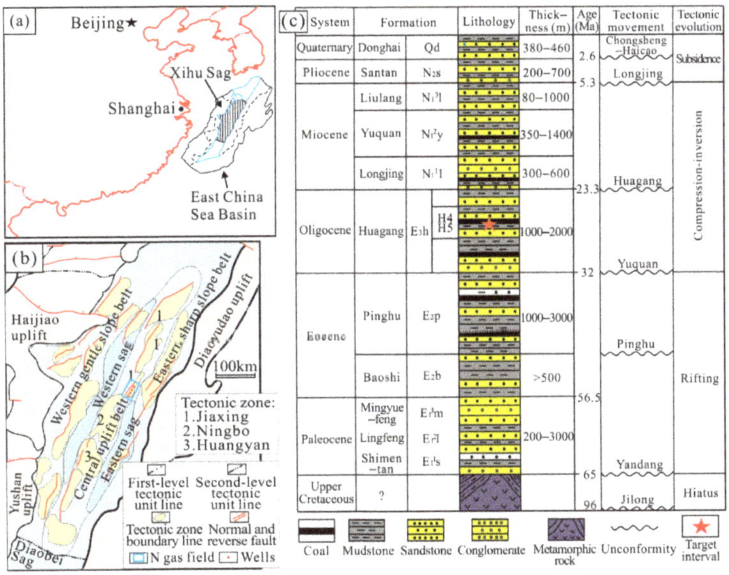

Figure 1. (**a**) Location of East China Sea Basin and Xihu Sag. (**b**) Division of structural units and location of N gas field in Xihu Sag [29]. (**c**) Comprehensive stratigraphic column of Xihu Sag [30]. The red star indicates the studied target intervals.

More than 500 core samples and 2713 conventional core analysis data were collected from China National Offshore Oil Corporation Shanghai Branch. Sandstone samples were taken from core sections of the H4 and H5 members of four wells in the N gas field in Xihu Sag. The total length of the core section in the target interval of four wells is 129.9 m, including 27.1 m in the N1 well, 48.3 m in the N2 well, 45.5 m in the N3 well, and 9.0 m in the N4 well. These sandstone samples were prepared for 296 thin-section analyses, 158 scanning electron microscopes, 64 X-ray diffraction analyses, 6 scanning electron microscope mineral quantitative evaluations, and 16 high pressure mercury injection measurements.

2.2. *Experiment Methods*

X-ray diffraction analysis data were used to analyze reservoir characteristics of tight sandstone. Thin sections, scanning electron microscopes, and scanning electron microscope mineral quantitative evaluations were used to investigate the diagenesis and diagenetic minerals of tight sandstone. Then, pore throat structure of tight sandstone was investigated by high pressure mercury injection experiment, and fractal dimension of pore throat structure of tight sandstone sweet spot was investigated by fractal theory.

2.2.1. Thin Section

A total of 296 blue epoxy-infected thin sections were used for petrographic analysis. Thin sections were stained with alizarin red for the determination of carbonate minerals. Petrographic point counting analysis was performed on thin sections of sandstone samples, and mineral composition, authigenic minerals, primary pores, and secondary pores of each thin section were determined. Point counting was performed on a thin section for detrital composition with at least 300 points, which can provide a standard deviation of approximately 5.5% or less [31]. The pore types contained primary intergranular pore, intragranular dissolution pore, intergranular dissolution pore, and moldic pore. The initial porosity evaluation equation proposed by Scherer (1987) was used to calculate the initial porosity (or initial intergranular volume) [32]. The initial porosity of H4 and H5 tight sandstones calculated by this method ranged from 31.8% to 35.5%. The porosity loss caused by compaction in H4 and H5 sandstone was calculated using the Houseknecht (1987) method [33].

2.2.2. Scanning Electron Microscopy

A total of 158 gold-coated core sandstone bulk samples were analyzed for scanning electron microscopes. The core sample treatment process included oil washing, sample selection, piling, drying, dust removal, and gold plating. A scanning electron microscope was used to describe the pore morphology, pore size, and mineral distribution of tight sandstone [29,34]. High-resolution imaging was performed on the Quanta FEG 450 scanning electron microscopy to determine pore throat space, cement morphology, and mineral composition.

2.2.3. X-ray Diffraction Analysis

A total of 64 core samples were selected for whole-rock X-ray diffraction analysis and clay X-ray diffraction analysis. Whole rock (massive) and clay fractions (<2 μm) were analyzed for whole rock mineral content and clay mineral content using X-ray diffractometer Ultima IV.

2.2.4. Scanning Electron Microscopy Mineral Quantitative Evaluation

Based on petrographic analysis, six representative core samples were selected for scanning electron microscope minerals quantitative evaluation (QEMSCAN). The quantification analysis of mineral composition was performed by scanning the surface of the sample with a high-energy electron beam accelerated along the preset grating scanning mode to obtain a color map of the distribution characteristics of mineral assemblage. By combining backscattered electron (BSE) image grayscale with X-ray intensity, the content of the element can be obtained and converted into mineral phases. By scanning the surface of the sample, almost all parameters related to mineral texture characteristics can be calculated, including mineral grain morphology, mineral distribution characteristics, mineral dissociation degree, element occurrence state, porosity, and matrix density. The quantification analysis results obtained mineralogy images of investigated thin sections.

QEMSCAN analysis was carried out by iRock Technologies company, Beijing city, China, and test instruments were Hitachi FlexSEM1000II desktop electron microscope, Bruker Energy Spectrometer EDSQ80detectors, and AMICS 3.2.0.222 mineral analysis software. The Hitachi FlexSEM1000II desktop electron microscope was operated at an accelerated voltage of 0.1 kV–20 kV, and the highest resolution of the BSE was 5 nm. A resolution of 0.5 μm or 0.83 μm was used for analysis to obtain clear BSE images. Mineral identification was carried out using the Mapping (debris)/Particles (core) of the analytical mineral library (Shale 2021 SY) in the AMICS 3.2.0.222 mineral analysis software. For each region analyzed, the BSE image was converted to a mineral image by merging energy dispersive spectrometer spectra into a single mineral.

2.2.5. High Pressure Mercury Injection Experiment

Based on petrographic analysis, high pressure mercury injection experiment was performed on 16 core samples to obtain pore throat size data. Before conducting high pressure mercury injection experiment, oil and gas should be removed from the sample, and the parameters such as porosity, permeability, and density need to be measured in advance. High pressure mercury injection analysis was conducted at the Exploration and Development Experiment Center of Geological Science Research Institute, Jiangsu Oilfield Branch of China Petroleum and Chemical Corporation. According to the Chinese petroleum industry standard GBT 29171-2012 [35], pore throat size of core samples with a diameter of approximately 2.54 cm and a length of approximately 4.5 cm were tested by using AutoPore IV9520 mercury porosimeter at a temperature of 25 °C and a relative humidity of 35–50%. Mercury was injected to obtain the intrusion curves until the maximum pressure was reached, and then the experiment of decreasing pressure was carried out to obtain the mercury extrusion curves. The maximum intrusion pressure was 116.67 MPa, which corresponded to a pore throat radius of 0.006 μm. During the experiment, the volume accuracy of the entry or exit of mercury was greater than 0.1 μL. The pore distribution characteristics of the cores can be converted from the capillary pressure curve.

The complex pore throat structure is thought to be composed of a series of interconnected irregular pore networks with self-similarity. When non-wetting phase mercury was injected into a porous sample, capillary force prevented mercury from entering the pore throat. Therefore, an injection pressure was required to overcome capillary resistance, and each injection pressure corresponded to a corresponding magnitude of capillary force. Under capillary pressure equilibrium, the injection pressure and injection amount of mercury were recorded, and then the capillary pressure curve characterizing pore throat structures was obtained according to the relationship between injection pressure and mercury saturation. The pore throat radius curve was obtained by the relationship between capillary pressure and pore throat radius (Equation (1)), and the pore structure of tight sandstone was evaluated:

$$P_c = \frac{2\sigma\cos\theta}{r} \tag{1}$$

where P_c is capillary pressure, MPa, θ is the wetting angle, °, σ is surface tension, N/m, and r is pore throat radius, μm. In general, θ = 140° and σ = 0.48 N/m.

2.3. Fractal Methodology

The pore throat structure of a sandstone reservoir has self-similarity, which can be described by fractal theory, and the fractal dimension is related to the surface roughness of the reservoir pore throat [36]. In this paper, based on the advantages of high mercury injection pressure and large pore diameter range, a fractal dimension study of obtained experiment data was carried out to quantitatively characterize the heterogeneity of the pore throat structure of the H1 and H5 tight sandstone in the study area.

In high pressure mercury injection experiments, capillary pressure can reflect the difficulty of mercury entering different pore throat radii. According to the principle of fractal geometry, when the pore throat radius property of tight sandstone conforms to the fractal characteristics, the number of pore throats with a radius greater than r is calculated as N. The correlation between N and r can be expressed as Equation (2):

$$N(>r) = \int_{r}^{r_{max}} F(r)dr = ar^{D} \tag{2}$$

where r_{max} is the maximum pore throat radius, μm, and F(r) is the distribution density function of the pore throat radius, and a is the proportionality constant. Since the minimum value of pore throat radius, r_{min}, is much smaller than r_{max}, Equation (3) can be obtained.

$$S = \left(\frac{r}{r_{max}}\right)^{3-D} \tag{3}$$

where S is the wetting phase saturation corresponding to capillary pressure P_c, %.

Assuming that θ is not affected by pore throat size, the fractal Equation (4) of pore throat radius distribution can be obtained.

$$S = \left(\frac{P_c}{P_{min}}\right)^{D-3} \tag{4}$$

where P_{min} is capillary pressure corresponding to r_{max}, MPa. Take the logarithm of both sides of Equation (4) to obtain Equation (5).

$$\lg(S) = (D-3)\lg(P_c) - (D-3)\lg(P_{min}) \tag{5}$$

The fractal dimension calculation procedure is as follows. Firstly, according to the experiment data of capillary pressure, the correlation between lg(S) and $\lg(P_c)$ can be determined. In the high pressure mercury injection experiment, the wetting phase is air, and the non-wetting phase is mercury, so lg(S) can be expressed as $\lg(1-S_{Hg})$, where S_{Hg} is mercury saturation. Secondly, the slope of the $\lg(1-S_{Hg})$ and $\lg(P_c)$ correlation trendline was obtained by linear fitting, which is the Hurst index (H). Finally, the fractal dimension is calculated based on Equation (6).

$$D = 3 + H \tag{6}$$

According to fractal theory, pore throat structure fractals can be divided into total fractals and segmented fractals [37]. The segmented fractal curve of the $\lg(1-S_{Hg})$ and $\lg(P_c)$ plot curve is not a straight line and has obvious turning points. The curve can be divided into several segments, and the fractal dimension of different sizes of pore throats needs to be calculated. Many classification criteria for pores were proposed [38,39]. In this study, the Xoaoth classification criteria was used to classify pore types, namely micropores (<10 nm), transitional pores (10–100 nm), mesopores (100–1000 nm), and macropores (>1000 nm). These four types of pores had obvious turning points on $\lg(1-S_{Hg})$ and $\lg(P_c)$ curves, and then fractal dimensions of micropores, transitional pores, mesopores, and macropores were calculated by linear fitting according to above-mentioned fractal dimension calculation procedure.

The weighted average method can consider the fractal dimension and pore space content of different pores, and the total fractal dimension obtained is accurate. Finally, based on the fractal dimension and porosity of macropores, mesopores, transitional pores and micropores, and the total fractal dimension (D) of the entire pore throat can be obtained by using the weighted average method (Equation (7)).

$$D = D_1 \times \frac{\varphi_1}{\varphi_1+\varphi_2+\varphi_3+\varphi_4} + D_2 \times \frac{\varphi_2}{\varphi_1+\varphi_2+\varphi_3+\varphi_4} + D_3 \times \frac{\varphi_3}{\varphi_1+\varphi_2+\varphi_3+\varphi_4} + D_4 \times \frac{\varphi_4}{\varphi_1+\varphi_2+\varphi_3+\varphi_4} \quad (7)$$

where, D is total fractal dimension, D_1 is fractal dimension of the macropore, D_2 is fractal dimension of the mesopore, D_3 is fractal dimension of the transitional pore, D_4 is fractal dimension of the micropore, φ_1 is the porosity of the macropore, %, φ_2 is the porosity of the mesopore, %, φ_3 is the porosity of the transitional pore, %, φ_4 is the porosity of the micropore, %.

3. Results

3.1. Reservoir Characteristics

The rocks of the H4 and H5 members are chiefly feldspathic lithic quartzose, with a small amount of feldspathic litharenite and litharenite (Figure 2a). The content of detrital quartz, feldspar, and rock fragments ranges from 53% to 75%, from 8.0% to 20.0%, and from 14.0% to 37.0%, with an average value of 65.04%, 16.29% and 18.68%, respectively (Figure 2b). The types of rock fragments are primarily metamorphic rock (average 8.87%) and volcanic rock (average 7.97%), followed by sedimentary rock (average 2.54%) (Figure 2c). The pore space is mainly composed of primary pores and secondary pores. The content of the argillaceous matrix ranges from 0.0% to 28.0%, with an average value of 5.14%. The cement content ranges from 3.2% to 34.6%, with an average value of 12.58%. The clastic grains were medium or well sorted, and the roundness was subangular and subrounded. The content of clay minerals in the H4 and H5 sandstones ranged from 2.1% to 35.2%, with an average value of 7.45%. Illite is the predominant clay mineral in the tight sandstone, followed by chlorite and illite–smectite mixed layers (Figure 2d). The illite content in the H4 and H5 sandstones varies from 0.231% to 9.37%.

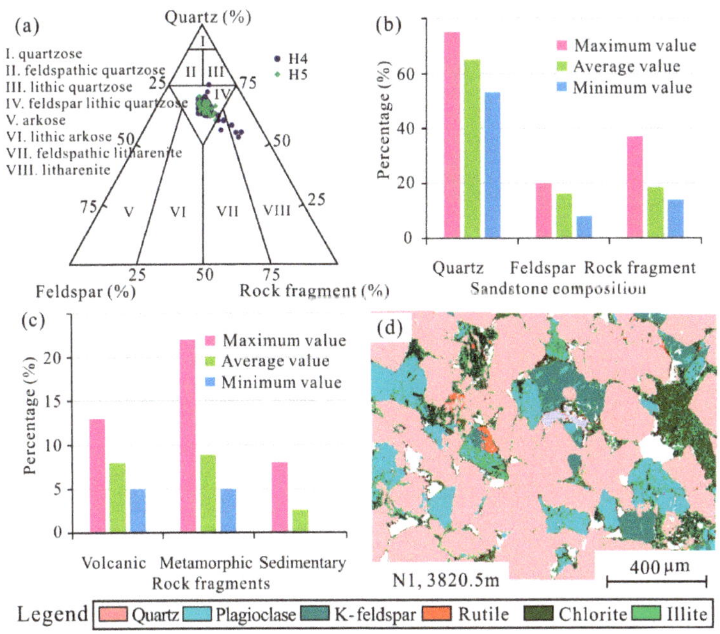

Figure 2. (**a**) Ternary plot of rock types in the H4 and H5 members of N gas field in Xihu Sag [40]. (**b**) Histogram of framework mineral composition. (**c**) Histogram of rock fragment type. (**d**) Scanning electron microscope mineral quantitative evaluation data.

3.2. Diagenesis

3.2.1. Compaction

The average burial depth of H4 and H5 in the N gas field is greater than 3780 m. Point-to-line contact, line contact, concave–convex contact, and a small number of point and sutured contacts can be observed under petrographic observation (Figure 3). The clastic grains of the H4 and H5 sandstone experienced mechanical compaction, which is mainly evidenced by the point and line contacts between mineral grains, bending of mica grain, and breakage of quartz and feldspar grains (Figure 3a–d). Evidence of chemical compaction is concave–convex contact and sutured contact between mineral grains (Figure 3e,f).

According to calculated values, the compaction porosity reduction rate of H4 and H5 sandstone ranges from 27.48% to 96.57% and from 47.93% to 93.27%, respectively, and the average values are 83.93% and 85.12%, respectively.

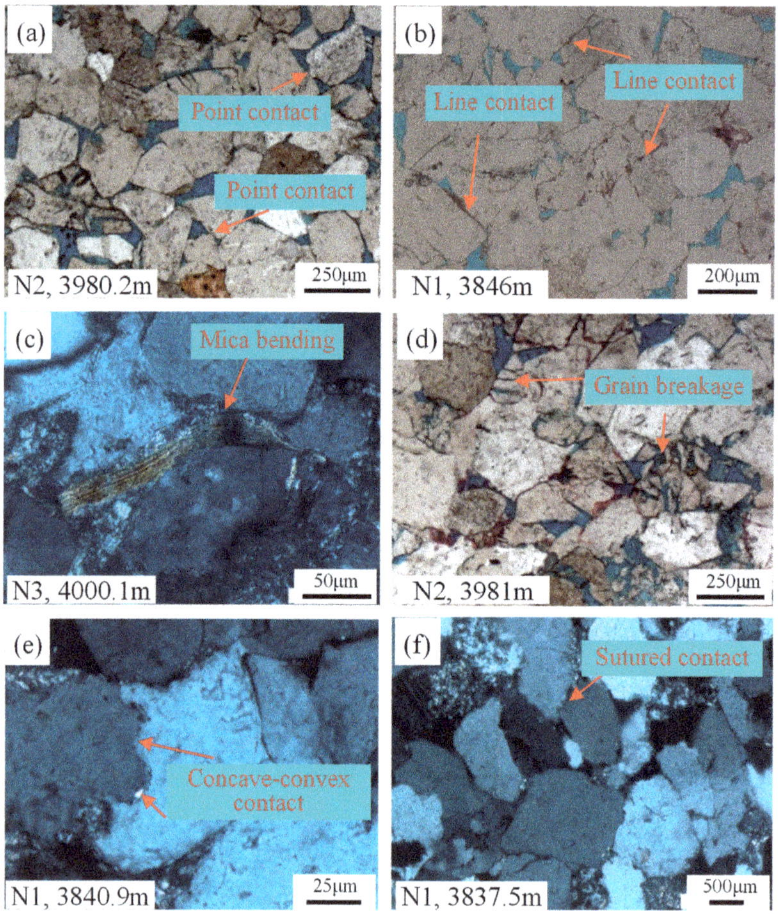

Figure 3. Compaction evidence of the H4 and H5 sandstones. (**a**) Point contact and point-to-line contact. (**b**) Line contact. (**c**) Mica bending. (**d**) Breakage of quartz and feldspar grains. (**e**) Concave–convex contact. (**f**) Sutured contact.

3.2.2. Quartz Cement

Quartz cement is a common authigenic mineral in H4 and H5 tight sandstones (Figure 4a–d). Quartz cement primarily contains two forms of quartz overgrowth and authigenic quartz crystals (Figure 4a,b). The thickness of quartz overgrowth is primarily distributed between 10 μm and 100 μm (Figure 4a), and the size of authigenic quartz crystals is usually less than 35 μm (Figure 4b). Authigenic quartz crystals are chiefly distributed in the intergranular pores. Two-phase quartz overgrowth can be easily identified by dust line (Figure 4c). The H4 and H5 sandstones developed quartz overgrowth and authigenic quartz crystals (Figure 4a–d).

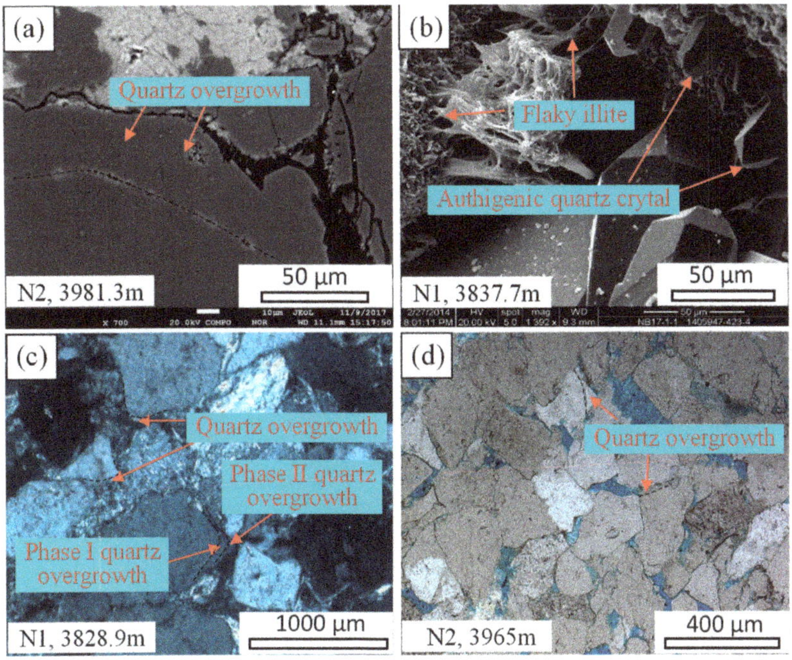

Figure 4. (**a**) Quartz overgrowth. (**b**) Authigenic quartz crystals and flaky illite in intergranular pores. (**c**) Two-phase quartz overgrowth. (**d**) Quartz overgrowth.

3.2.3. Illite

Illite is one of the main clay minerals in the H4 and H5 tight sandstones. Detrital illite and authigenic illite can be observed under a scanning electron microscope (Figure 5). Detrital illite is characterized by thick flake, dark color, tight packing, and distinct squeeze features (Figure 5a), whereas authigenic illite is distinctly different. The morphologies of authigenic illite are inner illite coat, honeycomb, flaky, hair-like, filamentous, and pore-bridging (Figure 5b–f). The inner illite coat can be observed on mineral grains (Figure 5b), and the inner illite coat is not common in H4 and H5 tight sandstone. Flaky and honeycomb illites chiefly appear in residue intergranular pores (Figure 5c,d). Filamentous and hair-like illites are widely developed in intergranular dissolution pores and intragranular dissolution pores (Figure 5e,f). Pore-bridging illite is distributed at the narrow pores and throats (Figure 5f). Hair-like and filamentous illites are common at the end of flaky illite on and around the mineral grain. The length of flaky illite is primarily 2–15 μm, and the length of filamentous illite is mainly 5–20 μm.

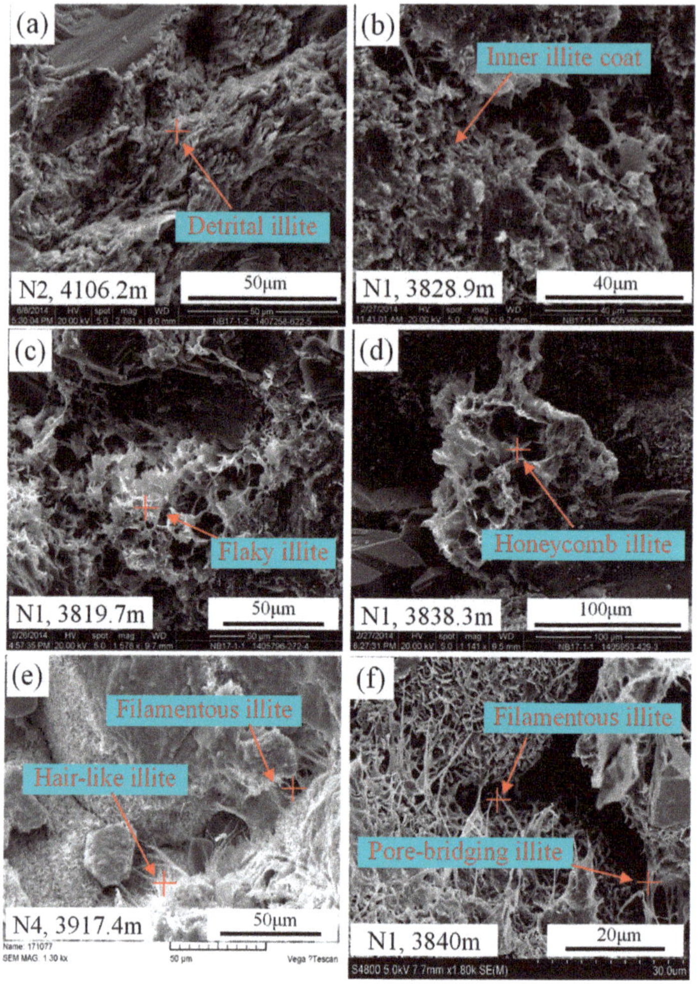

Figure 5. Evidence of illite in H4 and H5 tight sandstone of N gas field in Xihu Sag. (**a**) Detrital illite. (**b**) Inner illite coat. (**c**) Flaky illite. (**d**) Honeycomb illite. (**e**) Intragranular dissolution pores of plagioclase and filamentous illite. (**f**) Filamentous illite and pore-bridging illite.

Based on the morphology of illite and the relationship between illite and mineral grains, the illite distribution was analyzed. Inner illite coats are mainly distributed on the grain surface (Figure 6a,b). The inner illite coat is mainly developed in sheet sand sedimentary facies and the upper part of subaqueous distributary channel sedimentary facies. Flaky illite and honeycomb illite mainly appear in residue intergranular pores (Figure 6c). Flaky and honeycomb illites chiefly develop in the upper part of channel bar sedimentary facies, the upper part of subaqueous distributary channel sedimentary facies, and sheet sand sedimentary facies. Filamentous and hair-like illites mainly distribute in intergranular dissolution pores and intragranular dissolution pores, indicating the formation of secondary pores to a certain extent (Figure 6d–f). Filamentous and hair-like illites mainly develop in the middle and upper parts of channel bar sedimentary facies and the middle and upper parts of subaqueous distributary channel sedimentary facies. Illite distribution has a certain influence on reservoir quality. Authigenic illites are associated with the presence of primary pores and secondary pores (Figure 6d–f).

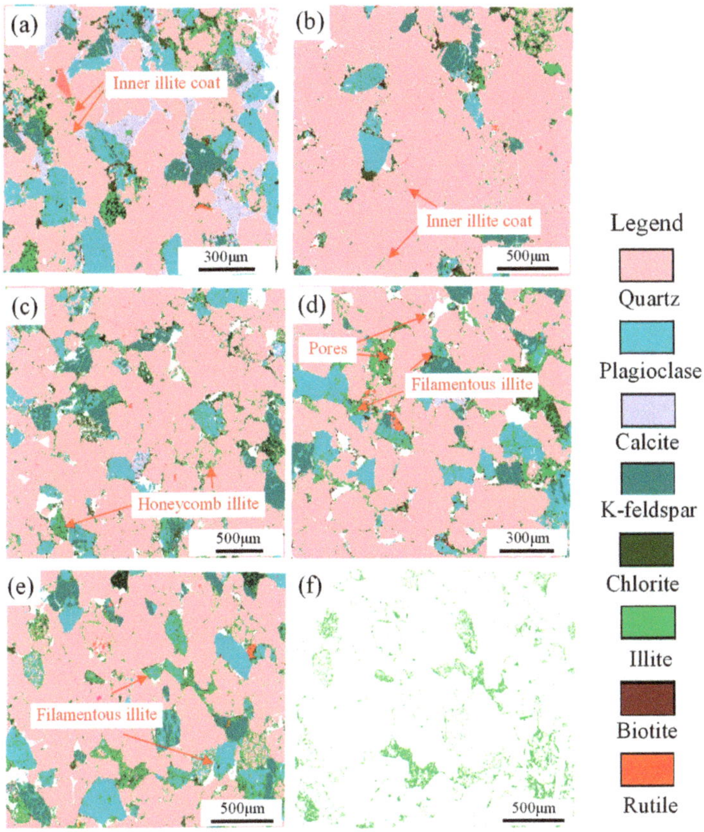

Figure 6. Mineral quantification evaluation images showing the presence of different illites. (**a**) Mineral composition and inner illite coat in sheet sand sandstone. (**b**) Mineral composition and inner illite coat in the upper sandstone of subaqueous distributary channel. (**c**) Mineral composition, flaky and honeycomb illite in the upper sandstone of subaqueous distributary channel. (**d**) Mineral composition, primary pores, secondary pores, filamentous, and pore-bridging illite in the channel bar sandstone. (**e**) Mineral composition, filamentous and pore-bridging illite in the channel bar sandstone. (**f**) Filamentous and pore-bridging illite in the channel bar sandstone.

3.2.4. Chlorite and Illite–Smectite Mixed Layers

Chlorite is an important clay mineral in the H4 and H5 sandstones. The morphologies of chlorite are primarily detrital grain coat, authigenic grain coat, and pore filling. The size of the detrital chlorite coat is chiefly 1–2 μm, whereas the size of the authigenic chlorite coat is primarily 6–8 μm (Figure 7a,b). It is common that an authigenic chlorite coat grows on a detrital chlorite coat (Figure 7b). Pore-filling chlorite fills the pores and blocks the throats (Figure 7c). The H4 and H5 sandstone primarily develops pore-filling chlorite and minor chlorite coat (Figure 7a–c).

Illite–smectite mixed layer is an intermediate transition mineral in the diagenetic evolution from smectite to illite. The morphology characteristics of the illite–smectite mixed layer are between smectite (honeycomb and mesh) and illite (flaky, filamentous, and hair-like). Illite–smectite mixed layers appear in the form of flaky and honeycomb crystals (Figure 7d), and the size of curled edges ranges from 2 μm to 15 μm. The illite–smectite mixed layer is primarily distributed in the intergranular pores.

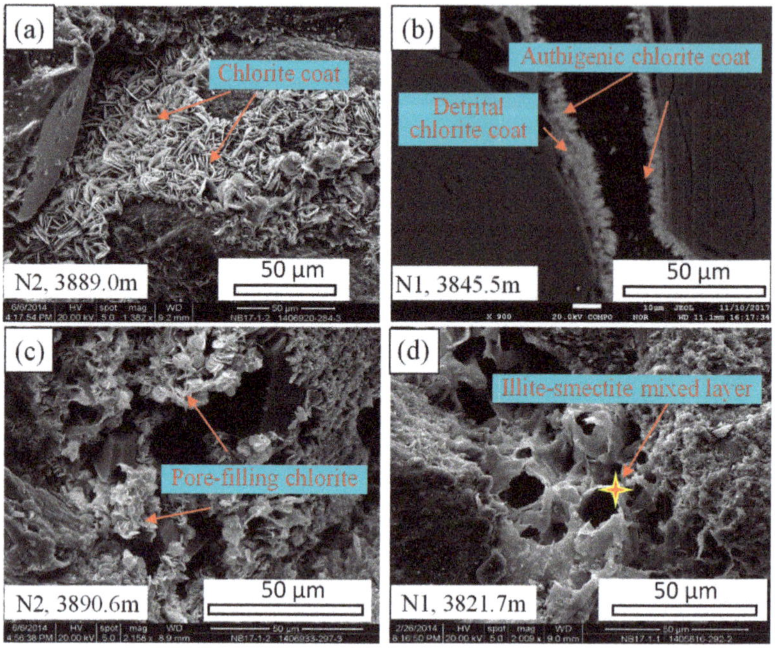

Figure 7. (**a**) Authigenic chlorite coat. (**b**) Detrital chlorite coat and authigenic chlorite coat. (**c**) Pore-filling chlorite. (**d**) Illite–smectite mixed layer.

3.2.5. Carbonate Cement

Under thin sections, calcite is a red pore-filling poikilitic bulk crystal, and ferrocalcite appears purplish red (Figure 8a). Under cathodoluminescence thin sections, red calcite can be observed surrounded by light to dark purple ferrocalcite (Figure 8b). The development of calcite cement is uneven, showing basal distribution and clustered distribution (Figure 8c,d). The cathodoluminescence of calcite is orange to orange-yellow (Figure 8d). The ferrocalcite cement is primarily developed at the edge of calcite cement or dispersedly distributed in residual primary pores and secondary pores (Figure 8d), and the cathodoluminescence of ferrocalcite is orange-red (Figure 8d).

3.2.6. Feldspar Dissolution

Feldspar in the H4 and H5 tight sandstones in Xihu Sag chiefly includes k-feldspar and plagioclase (Figure 8e,f). Partially or completely dissolved plagioclase and k-feldspar are common in the H4 and H5 sandstones, and pore spaces are chiefly intergranular dissolution pores, intragranular dissolution pores, and moldic pores. K-feldspar grains are usually partially dissolved and filled with authigenic illite, usually with intragranular dissolution pores and moldic pores (Figure 8e,f). The dissolution of k-feldspar and potassium-rich rock fragments may produce authigenic illite. Authigenic illite is primarily distributed in the bicrystal structure and cleavage plane of k-feldspar (Figure 8e,f). Under certain circumstances, k-feldspar dissolved to form a network structure (Figure 8f).

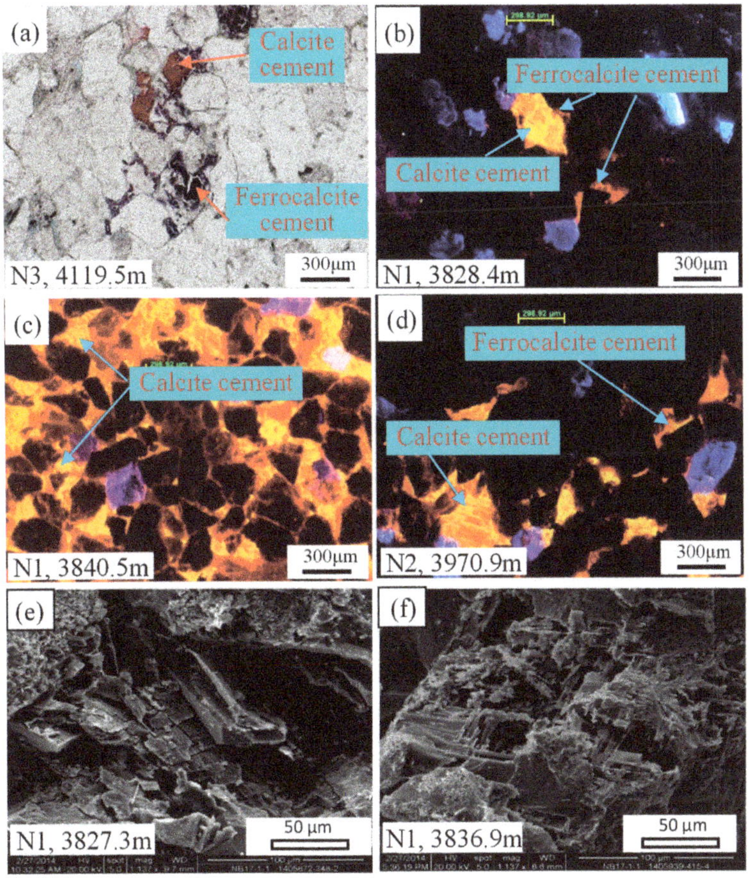

Figure 8. (**a**) Thin section image showing calcite and ferrocalcite cement. (**b**) Cathodoluminescence thin section image showing calcite and ferrocalcite cement. (**c**) Cathodoluminescence thin section image showing calcite cement. (**d**) Cathodoluminescence thin section image showing calcite and ferrocalcite cement. (**e**) Intragranular dissolution pores of k-feldspar. (**f**) Intragranular dissolution pores of k-feldspar.

3.2.7. Diagenetic Evolution Sequence

Based on thin sections, cathodoluminescence thin sections, scanning electron microscopes, mineral quantification analysis, and previous diagenetic evolution sequences, a diagenetic evolution sequence was established (Figures 3–8) [34]. The characteristics observed during early diagenesis chiefly include (1) compaction, (2) formation of chlorite coat and calcite cementation due to connate alkaline pore water, (3) formation of inner illite coat, (4) conversion of smectite to illite, and (5) dissolution of feldspar due to late Oligocene meteoric water leaching, accompanied by the precipitation of kaolinite and quartz cement (Figure 9) (Wang et al., 2021). The middle diagenesis characteristics primarily contain the following: (1) Dissolution of feldspar, rock fragment, and carbonate cement due to organic acid and CO_2 acid fluid and precipitation of kaolinite and quartz cement; (2) Continuous compaction; (3) Illitization of k-feldspar; (4) Illitization of k-feldspar and kaolinite; (5) Formation of chlorite, illite, and late ferrocalcite in alkaline diagenetic environment from late Miocene to present day (Figure 9).

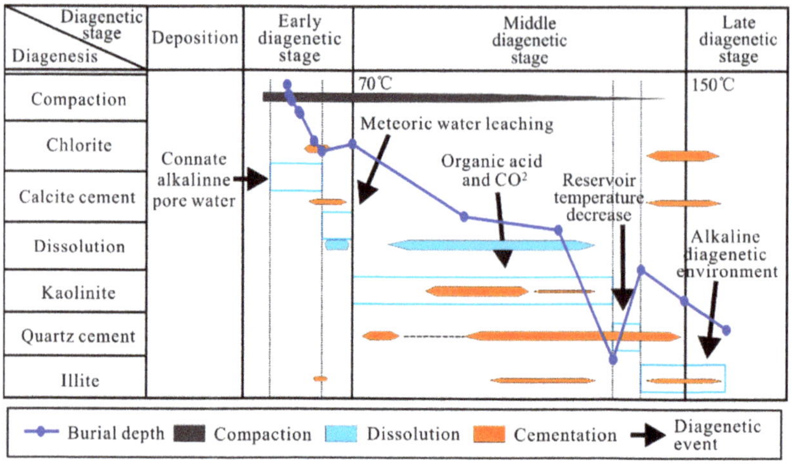

Figure 9. Diagenetic evolution sequence of the H4 and H5 tight sandstone.

3.3. Pore Throat Structure Characteristics

The pore throat structure and pore size distribution of the sample were obtained by high pressure mercury injection experiment. According to high pressure mercury injection experiment data of 16 representative samples, the displacement pressure and average pore throat radius of the samples are 0.02–2.0 MPa and 0.11–4.7 μm, respectively (Table 1). Maximum mercury saturation, relative sorting coefficient, and skewness of the samples range from 65.6% to 99.5%, from 0.16 to 0.53, and from 0.85 to 2.02, respectively.

Table 1. Parameters of the high pressure mercury injection experiment.

Well Name	Depth (m)	Porosity (%)	Permeability (mD)	Sweet Spot	Maximum Mercury Saturation (%)	Relative Sorting Coefficient	Skewness	Average Pore Throat Radius (μm)	Displacement Pressure (MPa)	Efficiency of Mercury Withdrawal (%)
N2	3960.2	7.3	0.261	Type III	75.1	0.34	1.62	0.21	1.5	39.25
N2	4102.5	7.2	0.197	Type III	73.1	0.36	1.57	0.18	1.5	36.27
N3	4113.7	6.1	0.109	Type III	67.6	0.42	1.43	0.11	2	44.59
N3	4123.8	6.6	0.113	Type III	75.1	0.33	1.56	0.16	1.5	38.66
N1	3823.7	9.9	0.316	Type II	87.1	0.23	1.85	0.29	0.8	33.36
N1	3825.7	8.8	0.581	Type II	98.4	0.18	1.17	0.39	0.7	32.44
N1	3829.3	9.8	0.424	Type II	92.2	0.21	1.93	0.43	0.5	26.39
N1	3831.3	10	0.331	Type II	99.2	0.16	1.21	0.54	0.5	26.84
N1	3833.3	8.7	0.257	Type II	94.4	0.17	1.68	0.24	0.8	30.34
N1	3841.7	9.7	0.441	Type II	99.5	0.2	0.95	0.58	0.3	29.52
N1	3843.3	9	0.416	Type II	93.9	0.22	1.53	0.47	0.3	41.11
N1	3821.7	11	1.32	Type I	78.3	0.37	1.94	1.78	0.05	29.93
N1	3827.7	12.3	2.67	Type I	65.6	0.53	1.85	1.06	0.2	25.5
N1	3846	12.6	3.63	Type I	99	0.25	0.85	1.38	0.15	32.08
N2	3963.1	13.4	24.3	Type I	96.1	0.35	0.9	4.7	0.02	31.05
N2	3973.2	12.2	1.76	Type I	83.6	0.28	2.02	0.67	0.3	26.45

The pore throat distribution curves of tight sandstone in the study area were obtained by using a high pressure mercury injection curve. It can be seen from the mercury injection curve (Figure 10a) that the trend of the mercury injection curve for most samples is consistent. The front section of the mercury injection curve is relatively gentle, while the back section gradually rises. The pore throat radius was mainly distributed in the range of 0.01–63.0 μm (Figure 10b), and the curve showed a multi-peak distribution. Small pore throats are mainly concentrated in about 0.03 μm, medium pore throats are chiefly concentrated in about 0.8 μm, and large pore throats are mainly concentrated in about 6.5 μm. Although the variation range of pore throats radius is basically similar, the distribution frequency of different pore throats is very different.

Based on reservoir properties, the classification and evaluation criteria for the sweet spot in Huagang Formation were established, namely, type I sweet spots ($\Phi > 10\%$, $K > 1$ mD), type II sweet spots ($8\% < \Phi < 10\%$, 0.4 mD $< K < 1$ mD), and type III sweet spots ($\Phi < 8\%$, $K < 0.4$ mD). According to the matching conventional core analysis data and high pressure mercury injection experiment data, the experiment samples were divided into type I, type II, and type III (Figure 10c–e).

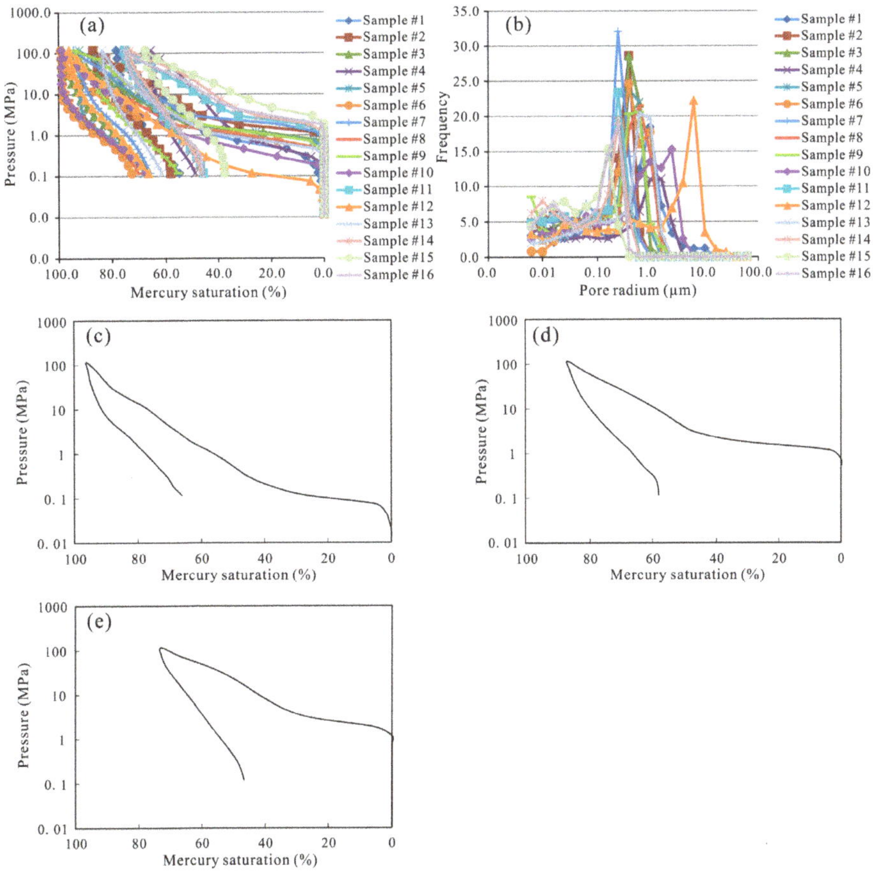

Figure 10. (**a**) Typical high pressure mercury injection experiment curves. (**b**) The characteristics of pore radius of high pressure mercury injection experiment. (**c**) Capillary pressure curve of mercury intrusion and extrusion of type I sweet spots. (**d**) Capillary pressure curve of mercury intrusion and extrusion of type II sweet spots. (**e**) Capillary pressure curve of mercury intrusion and extrusion of type III sweet spots.

3.4. Fractal Characteristics

Figure 11 shows the relationship between $\lg(1-S_{Hg})$ and $\lg(P_c)$ of typical samples. The average fractal dimensions of macropores, mesopores, transitional pores, and micropores of type I sweet spots are 2.93, 2.72, 2.632, and 2.45, respectively, and the corresponding average porosities are the following: 3.24%, 5.6%, 2.6%, and 0.87% (Figure 11a). The average fractal dimensions of mesopores, transitional pores, and micropores of type II sweet spots are 2.46, 2.46, and 2.05, respectively, and the corresponding average porosities are 6.16%, 2.35%, and 0.85% (Figure 11b). The average fractal dimensions of mesopores, transitional pores, and micropores of type III sweet spots are 2.78, 2.75, and 2.64, respectively, and the corresponding average porosities are 2.98%, 2.77%, and 1.05% (Figure 11c). According to the fractal dimension analysis of the macropores, mesopores, transitional pores, and micropores of the experiment samples, the fractal dimension range of different pore types is between 2 and 3. The fractal dimension of the macropore is large, the fractal dimension of the micropore is small, and the fractal dimension of the mesopore and transitional pore is intermediate (Figure 11d).

The total fractal dimension of the entire pore throat was obtained by the weighted average method. The total fractal dimensions of type I, type II, and type III sweet spots were distributed in the ranges of 2.62–2.87, 2.22–2.56, and 2.71–2.77, respectively (Table 2). The average value of the total fractal dimension of type I sweet spots was 2.726, and the high total fractal dimension was mainly caused by macropores, mesopores, transitional pores, and micropores, especially the contribution of macropores. The average value of the total fractal dimension of the type II sweet spot was 2.43, and the low total fractal dimension was chiefly caused by mesopores, transitional pores, and micropores. The average value of the total fractal dimension of the type III sweet spot was 2.738, and the high total fractal dimension was primarily caused by mesopores, transitional pores, and micropores, especially the contribution of micropores.

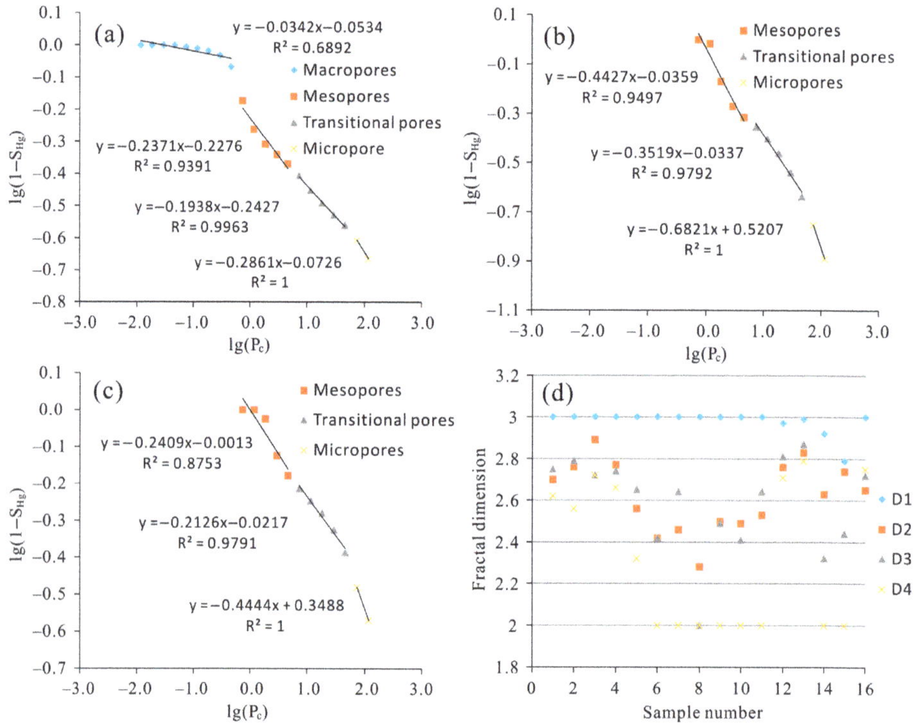

Figure 11. (**a**) Fractal dimension curve of type I sweet spots. (**b**) Fractal dimension curve of type II sweet spots. (**c**) Fractal dimension curve of type III sweet spots. (**d**) Fractal dimension distribution of macropore, mesopore, transitional pore, and micropore of the sample in the study area.

Table 2. Statistical table of fractal dimensions of different pore types of experiment samples in the study area.

Well Name	Depth (m)	Sweet Spot	Macropore		Mesopore		Transitional Pore		Micropore		D
			D_1	Φ_1	D_2	Φ_2	D_3	Φ_3	D_4	Φ_4	
N2	3960.2	Type III			2.7	3.8	2.75	2.56	2.62	0.95	2.71
N2	4102.5	Type III			2.76	3.31	2.79	2.52	2.56	1.37	2.73
N3	4113.7	Type III			2.89	1.77	2.72	3.36	2.72	0.98	2.77

Table 2. *Cont.*

Well Name	Depth (m)	Sweet Spot	Macropore		Mesopore		Transitional Pore		Micropore		D
			D_1	Φ_1	D_2	Φ_2	D_3	Φ_3	D_4	Φ_4	
N3	4123.8	Type III			2.77	3.04	2.74	2.64	2.66	0.92	2.74
N1	3823.7	Type II			2.56	5.84	2.65	2.87	2.32	1.19	2.56
N1	3825.7	Type II			2.42	5.76	2.42	2.25	2	0.78	2.38
N1	3829.3	Type II			2.46	6.76	2.64	2.16	2	0.88	2.46
N1	3831.3	Type II			2.28	7.7	2	2.1	2	0.2	2.22
N1	3833.3	Type II			2.5	5.13	2.49	2.7	2	0.87	2.45
N1	3841.7	Type II	3	0.24	2.49	6.33	2.41	2.27	2	0.85	2.44
N1	3843.3	Type II	3	0.18	2.53	5.58	2.64	2.07	2	1.17	2.5
N1	3821.7	Type I	2.97	2.01	2.76	6.03	2.81	2.15	2.71	0.81	2.8
N1	3827.7	Type I	2.99	2.98	2.83	5.15	2.87	2.76	2.79	1.41	2.87
N1	3846	Type I	2.92	3.87	2.63	5.3	2.32	2.71	2	0.72	2.62
N2	3963.1	Type I	2.79	6.83	2.74	3.08	2.44	2.68	2	0.8	2.66
N2	3973.2	Type I	3	0.49	2.65	8.42	2.72	2.68	2.75	0.61	2.68

4. Discussion

4.1. Relationship between Fractal Dimension, Reservoir Properties and Pore Structure Parameters

4.1.1. Relationship between Fractal Dimension and Reservoir Properties

The relationship between total fractal dimension, porosity, and permeability was investigated. The relationship between total fractal dimension, porosity, and permeability of type I, type II, and type III sweet spots were significantly different. Among them, the negative correlation between total fractal dimension, porosity, and permeability of type I sweet spots was different from that of type II and type III sweet spots (Figure 12a,b). This indicated that pore space types of type I sweet spots were different from those of type II and type III sweet spots. The pore types of type I sweet spots included macropore, mesopore, transitional pore, and micropore, especially the contribution of macropores, whereas type II and type III sweet spots contained mesopores, transitional pores, and micropores. The total fractal dimension of type I sweet spots was large, indicating that the complexity of the macropores had a great influence on porosity and permeability. The total fractal dimension of type II and type III sweet spots had a significant negative correlation with porosity and permeability, and mesopores and transitional pores had a great influence on reservoir properties. The total fractal dimension of type III sweet spots was larger than that of type II sweet spots, indicating that the complexity of micropores had a greater influence on porosity and permeability.

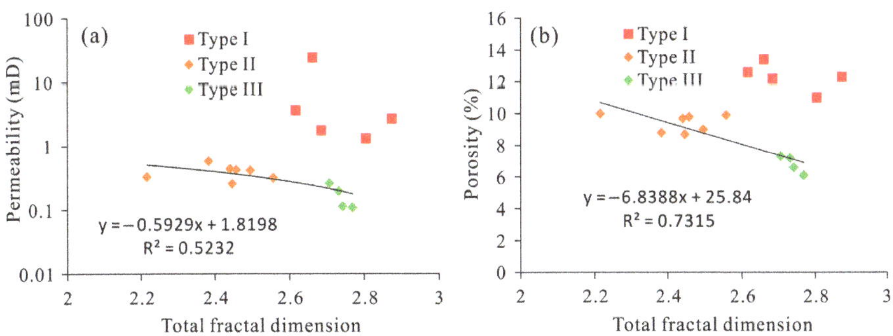

Figure 12. (**a**) Relationship between total fractal dimension and porosity. (**b**) Relationship between total fractal dimension and permeability.

4.1.2. Relationship between Fractal Dimension and Pore Structure Parameters

The relationship between total fractal dimension and pore structure parameters was investigated. Figure 13a shows the negative correlation between the total fractal dimension and maximum mercury saturation. Due to different pore types, the negative correlation between total fractal dimension and maximum mercury saturation of type I sweet spots was different from that of type II and type III sweet spots. The maximum mercury saturation for type I and type II sweet spots was high. This indicated that mesopores and transitional pores had a significant impact on the storage space of type II sweet spots, macropores, mesopores, and transitional pores had a significant impact on the storage space of type I sweet spots, and transitional pores, micropores and mesopores had a significant impact on the storage space of type III sweet spots. Figure 13b shows the negative correlation between the total fractal dimension and the average pore throat radius. The negative correlation between the total fractal dimension and average pore throat radius of type II and type III sweet spots was significant, while there was no significant correlation between the total fractal dimension and average pore throat radius of type I sweet spots. This indicated that the smaller the total fractal dimension of the pore throat structure of type II and III sweet spots, the larger the average pore throat radius. The average pore throat radius of type I sweet spots was large, the average pore throat radius of type III sweet spots was small, and the average pore throat radius of type II sweet spots was between type I and type III sweet spots. The negative correlation between total fractal dimension and skewness of type II and type III sweet spots was significant, whereas there was no correlation between total fractal dimension and skewness of type I sweet spots (Figure 13c). This indicated that the symmetry of micropores, transitional pores, and mesopores distribution had an impact on the fractal characteristics of the pore throat structure of type II and III sweet spots, and the symmetry of macropores distribution had an important impact on the fractal characteristics of the pore throat structure of type I sweet spots.

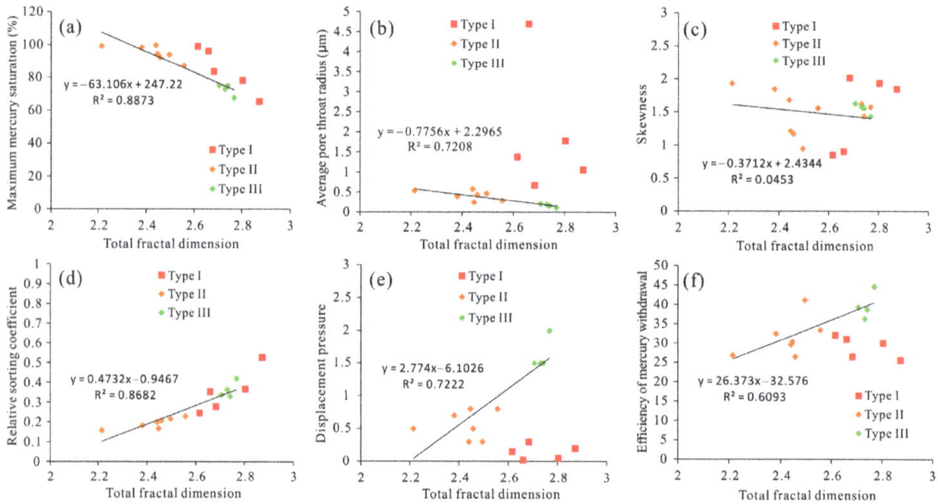

Figure 13. Relationship between total fractal dimensions and pore structure parameters. (**a**) Relationship between total fractal dimension and maximum mercury saturation. (**b**) Relationship between total fractal dimension and average pore throat radius. (**c**) Relationship between total fractal dimension and skewness. (**d**) Relationship between total fractal dimension and relative sorting coefficient. (**e**) Relationship between total fractal dimension and displacement pressure. (**f**) Relationship between total fractal dimension and efficiency of mercury withdrawal.

The positive correlation between the total fractal dimension of type I, type II, and type III sweet spots and the relative sorting coefficient was significant. The relative sorting coefficient of type II sweet spots was smaller than that of type I and type III sweet spots (Figure 13d). The relative sorting coefficient is a parameter that accurately reflects the uniformity of pore throat distribution. The smaller the relative sorting coefficient, the more uniform the pore throat distribution. This indicated that the pore throat distribution uniformity of type II sweet spots was better than that of type I and III sweet spots. Figure 13e shows the relationship between the total fractal dimension and displacement pressure. The correlation between the total fractal dimension of type I sweet spots and displacement pressure was not significant, whereas the positive correlation between the total fractal dimension of type II and type III sweet spots and displacement pressure was significant. This indicated that the larger the total fractal dimension of type II and type III sweet spots, the larger the displacement pressure. The displacement pressure of the type I sweet spot was small, and the displacement pressure of the type III sweet spot was large. The correlation between the total fractal dimension of type I sweet spots and the efficiency of mercury withdrawal was not significant, whereas the positive correlation between the total fractal dimension of type II and type III sweet spots and the efficiency of mercury withdrawal was significant (Figure 13f). The efficiency of mercury withdrawal of type I sweet spots was similar to that of type II sweet spots. Mercury withdrawal efficiency can reflect the pore throat connectivity of tight sandstone. This indicated that the pore throat connectivity of type II sweet spots was the best, the pore throat connectivity of type III sweet spots was the worst, and the pore throat connectivity of type I sweet spots was similar to that of type II sweet spots.

4.2. *Effect of Diagenesis on Fractal Characteristics*

The H4 and H5 tight sandstone in Xihu Sag was mainly composed of quartz, feldspar, and rock fragment minerals and experienced various diagenesis during burial. According to thin section and scanning electron microscope observation, pore space and reservoir properties of tight sandstone were affected by compaction, quartz cementation, clay mineral conversion, carbonate cementation, and dissolution (Figures 3–8). There is a good relationship between fractal dimension and pore space (porosity and average pore throat radius) and its homogeneity (sorting coefficient). Therefore, the effect of diagenesis on fractal dimension was studied through the effect of different diagenesis on pore space and its homogeneity.

4.2.1. Effect of Compaction on Fractal Characteristics

With the increase in burial depth, the compaction of tight sandstone caused the change of contact type from point contact, point–line contact, line contact to concave–convex contact, or mica bending and grain breakage (Figure 3a–d), which reduced pore space of tight sandstone and increased fractal dimension. The concave–convex contact and sutured contact formed by pressure dissolution increased reservoir heterogeneity and fractal dimension (Figure 3e,f).

4.2.2. Effect of Cementation on Fractal Characteristics

Quartz cement is a major type of cement in the H4 and H5 tight sandstones, including quartz overgrowth and authigenic quartz crystals. Quartz overgrowth and authigenic quartz crystals blocked pores and throats (Figure 4), reduced pore space, and increased fractal dimension.

The effect of clay minerals on pore throat structures was quite complex, and the effect of different types and contents of clay minerals on pore structure was different [41–43]. Chlorite coat can inhibit compaction, protect pore throat structures, and maintain fractal dimension (Figure 7a,b). Pore-filling chlorite mainly filled intergranular pores, reduced pore space (Figure 7c,d), and increased reservoir heterogeneity and fractal dimension. Illite and illite–smectite mixed layers were filamentous or flaky and mainly dispersed on the surface of dissolved grains (Figure 5a,b). When the potassium ions produced by complete dissolution of k-feldspar cannot be effectively transported and discharged, the dissolved products of k-feldspar often form illite (Figure 5e). A large number of clay minerals filled primary pores and secondary pores, destroyed sandstone porosity and permeability, and increased the fractal dimension.

Calcite is an important cement in H4 and H5 tight sandstones, including calcite and ferrocalcite (Figure 8a–d). Calcite and ferrocalcite appeared orange to orange-yellow and light to dark purple under a cathodoluminescence thin section (Figure 8b–d). Calcite and ferrocalcite filled residual primary pores and secondary pores, reduced sandstone porosity and permeability, and increased reservoir heterogeneity and fractal dimension.

4.2.3. Effect of Dissolution on Fractal Characteristics

The dissolution of H4 and H5 tight sandstone in Xihu Sag was mainly caused by organic acids and CO_2 acidic fluids [34,44–46]. The organic acids and CO_2 acidic fluids led to the dissolution of feldspar, rock fragments, and carbonate cement to form intergranular dissolution pores and intragranular dissolution pores, increasing the pore space of tight sandstones (Figure 8e,f). Therefore, the dissolution of feldspar and rock fragments improved pore space and reduced the fractal dimension of the pore throat structure (Figure 8e,f).

4.3. Implication for Sweet Spot Prediction

The accuracy of the fractal dimension of tight sandstone sweet spots was analyzed based on analytical test data and fractal dimension calculation procedure. The accuracy analysis of the fractal dimension primarily includes the following three aspects: (1) High pressure mercury injection experiment is to inject non-wetting phase fluid mercury into the pore throat radius distribution of the core sample under a certain pressure. By directly injecting mercury into pore space, three-dimensional pore structure information can be obtained from large core samples, which ensures the accuracy of fractal dimensions of pore throat structures. (2) The fractal dimension calculation procedure considers in detail the turning point determination on $\lg(1-S_{Hg})$ and $\lg(P_c)$ curves and fractal dimension fitting. The turning points of micropores, transitional pores, mesopores, and macropores can be clearly determined on $\lg(1-S_{Hg})$ and $\lg(P_c)$ curves. Considering the fractal dimension and pore space content of different pores comprehensively, the weighted average method was used to calculate the total fractal dimension of core samples, which indicates that the reliability of the fractal dimension of pore structure in this study is relatively high. (3) The average value of the total fractal dimension of type I, type II, and type III sweet spots in tight sandstone are 2.726, 2.43, and 2.738. The fractal dimension is an important index to characterize the heterogeneity of microstructure, and the fractal dimension of the pore throat structure is between 2 and 3. This supports the accuracy of the fractal dimensions of pore throat structures in this study.

The clastic components of H4 and H5 tight sandstone in Xihu Sag are mainly composed of quartz, feldspar, rock fragments, and clay minerals. These tight sandstones experienced complex diagenesis, resulting in differences in pore throat structures and fractal dimensions.

The fractal dimension of H4 and H5 tight sandstone sweet spots in Xihu Sag and the effect of diagenesis on the fractal dimension were investigated. Type I sweet spots developed macropores, mesopores, transitional pores, and micropores, with macropores leading to a more complex pore throat structure, with a total fractal dimension range of 2.62–2.87. The average fractal dimensions of macropores, mesopores, transitional pores, and micropores of type I sweet spots are 2.93, 2.72, 2.63, and 2.45, with corresponding average porosity values of 3.24%, 5.6%, 2.6%, and 0.87%, respectively. The fractal dimensions of type I sweet spots were mainly controlled by chlorite coat formation, dissolution, and a small amount of compaction. Type II sweet spots developed mesopores, transitional pores, and micropores, and the total fractal dimension range was 2.22–2.56. The average fractal dimensions of mesopores, transitional pores, and micropores of type II sweet spots are 2.46, 2.46, and 2.05, respectively, with corresponding average porosity values of 6.16%, 2.35%, and 0.85%. The fractal dimension of the type II sweet spot was chiefly controlled by quartz cementation, clay mineral transformation, and a small amount of dissolution. Type III sweet spots developed mesopores, transitional pores, and micropores, and micropores led to a more complex pore throat structure, with total fractal dimensions ranging from 2.71 to 2.77. The average fractal dimensions of mesopores, transitional pores, and micropores of type III sweet spots are 2.78, 2.75, and 2.64, respectively, with corresponding average porosity values of 2.98%, 2.77%, and 1.05%. The fractal dimension of the type III sweet spot was primarily controlled by compaction, calcite cementation, and quartz cementation. This study indicates that the fractal dimension of the pore throat structure of the sweet spot and the effect of diagenesis on the fractal dimension can improve the understanding of the reservoir quality of tight sandstone.

5. Conclusions

In this study, fractal dimensions of pore throat structures and the effect of diagenesis on the fractal dimension of the fourth (H4) and fifth (H5) members tight sandstone sweet spots of the Huagang Formation in the Jiaxing area of Xihu Sag were investigated by integrating thin section, scanning electron microscope, X-ray diffraction data, scanning electron microscope quantitative mineral evaluation, and high pressure mercury injection experiment. Five main conclusions were obtained.

(1) The H4 and H5 tight sandstone is composed of moderately to well sorted, subangular to subrounded feldspathic lithic quartzose. The average values of quartz, feldspar, and rock fragment grains are 65.04%, 16.29%, and 18.68%, respectively. The H4 and H5 sandstone experienced early diagenesis and middle diagenesis. The early diagenesis stage included compaction, chlorite coat formation, calcite cementation, inner illite coat formation, smectite to illite conversion, feldspar dissolution, kaolinite, and quartz cement precipitation. The middle diagenesis stage contained the dissolution of feldspar, rock fragment and carbonate cement, precipitation of kaolinite and quartz cement, illitization of k-feldspar, and formation of illite, chlorite, and ferrocalcite.

(2) The fractal dimension of tight sandstone sweet spots was investigated. The average fractal dimensions of macropores, mesopores, transitional pores, and micropores of type I sweet spots are 2.93, 2.72, 2.632, and 2.45, respectively, and the corresponding average porosities are 3.24%, 5.6%, 2.6%, and 0.87%, respectively. The average fractal dimensions of mesopores, transitional pores, and micropores are 2.46, 2.46, and 2.05, respectively, and the corresponding average porosities are 6.16%, 2.35%, and 0.85%, respectively. The average fractal dimensions of mesopores, transitional pores, and micropores are 2.78, 2.75, and 2.64, respectively, and the corresponding average porosities are 2.98%, 2.77%, and 1.05%, respectively.

(3) The relationship between fractal dimension, reservoir properties, and pore structure parameters was determined. The negative correlation between total fractal dimension, porosity, and permeability of type I sweet spots was different from that of type II and type III sweet spots. The negative correlation between the total fractal dimensions of type II and type III sweet spots and maximum mercury saturation, average pore throat radius, and skewness were significant, whereas the correlation between total fractal dimensions of type I sweet spots and maximum mercury saturation, average pore throat radius and skewness were not significant. The positive correlation between the total fractal dimension of type II and type III sweet spots and the relative sorting coefficient, displacement pressure, and efficiency of mercury withdrawal were significant, whereas the correlation between the total fractal dimension of type I sweet spots and relative sorting coefficient, displacement pressure and efficiency of mercury withdrawal were not significant.

(4) The effect of diagenesis on fractal dimensions was investigated. Compaction reduced pore space of tight sandstone and increased fractal dimension. Quartz cementation and calcite cementation blocked pores and throats, reduced pore space, and increased fractal dimensions. Chlorite coat can inhibit compaction, protect pore throat structures, and maintain fractal dimensions. Most clay minerals filled primary pores and secondary pores, destroyed sandstone porosity and permeability, and increased fractal dimensions. Dissolution increased the pore space of tight sandstone and decreased the fractal dimension of the pore throat structure.

(5) The pore throat structures of type I sweet spots was mainly composed of macropores, mesopores, transitional pores, and micropores, and fractal dimensions of type I sweet

spots was chiefly controlled by chlorite coat formation, dissolution, and a small amount of compaction. This study shows that fractal dimensions of pore throat structures of the sweet spots and the effect of diagenesis on fractal dimensions can improve the understanding of the reservoir quality of tight sandstone.

Author Contributions: Conceptualization, W.W., C.L. and X.Z.; methodology, W.W.; software, W.W.; validation, W.W., C.L. and X.Z.; formal analysis, W.W.; investigation, W.W.; resources, W.W., C.L. and X.Z.; data curation, W.W.; writing—original draft preparation, W.W.; writing—review and editing, W.W. and C.L.; visualization, W.W.; supervision, C.L. and X.Z.; project administration, W.W., C.L. and X.Z.; funding acquisition, W.W. and C.L. All authors have read and agreed to the published version of the manuscript.

Funding: This work was financially supported by the National Natural Science Foundation of China (42202144), Innovative Research Groups of the National Natural Science Foundation of China (U19B2006), and the Natural Science Foundation of Shandong Province (ZR2022QD048).

Data Availability Statement: The data presented in this study are available on request from the corresponding author.

Acknowledgments: The authors gratefully thank the main editors and anonymous reviewers for their valuable comments and suggestions leading to improvement of this paper.

Conflicts of Interest: The authors declare no conflict of interest.

References

1. Kong, X.; Xiao, D.; Jiang, S.; Lu, S.; Sun, B.; Wang, J. Application of the combination of high-pressure mercury injection and nuclear magnetic resonance to the classification and evaluation of tight sandstone reservoirs: A case study of the Linxing Block in the Ordos Basin. *Nat. Gas Ind. B* **2020**, *7*, 433–442. [CrossRef]
2. Li, J.J.; Jiang, H.Q.; Wang, C.; Zhao, Y.Y.; Gao, Y.J.; Pei, Y.L.; Wang, C.C.; Dong, H. Pore-scale investigation of microscopic remaining oil variation characteristics in water-wet sandstone using CT scanning. *J. Nat. Gas Sci. Eng.* **2017**, *48*, 36–45. [CrossRef]
3. Zhao, N.; Wang, L.; Zhang, W.; Peng, Y.; Sima, L. Investigation of pore structure characteristics in marine deep-water high-temperature and high-pressure sandstone: A comprehensive characterization combining various experimental techniques. *Geoenergy Sci. Eng.* **2023**, *222*, 211433. [CrossRef]
4. Zhou, Y.; Xu, J.; Lan, Y.; Zi, H.; Cui, Y.; Chen, Q.; You, L.; Fan, X.; Wang, G. New insights into pore fractal dimension from mercury injection capillary pressure in tight sandstone. *Geoenergy Sci. Eng.* **2023**, *228*, 212059. [CrossRef]
5. Ge, X.; Fan, Y.; Deng, S.; Han, Y.; Liu, J. An improvement of the fractal theory and its application in pore structure evaluation and permeability estimation. *J. Geophys. Res. Solid Earth* **2016**, *121*, 6333–6345. [CrossRef]
6. Fan, X.; Wang, G.; Li, Y.; Dai, Q.; Song, L.; Duan, C.; Zhang, C.; Zhang, F. Pore structure evaluation of tight reservoirs in the mixed siliciclasticcarbonate sediments using fractal analysis of NMR experiments and logs. *Mar. Pet. Geol.* **2019**, *109*, 484–493. [CrossRef]
7. Dai, Q.; Wang, G.; Zhao, X.; Han, Z.; Lu, K.; Lai, J.; Wang, S.; Li, D.; Li, Y.; Wu, K. Fractal model for permeability estimation in low-permeable porous media with variable pore sizes and unevenly adsorbed water lay. *Mar. Pet. Geol.* **2021**, *130*, 105135. [CrossRef]
8. Liu, W.; Lu, S.; Wang, M.; Zhang, S.; Liu, Y.; Zhou, N.; Guan, Y.; Wang, H. Overall pore characterization of tight sandstone reservoir space and its significance: A case study of the outer expansion area of Long 26 in Longhupao oilfield, Songliao Basin. *J. Northeast Pet. Univ.* **2018**, *42*, 41–51.
9. Guo, R.; Xie, Q.; Qu, X.; Chu, M.; Li, S.; Ma, D.; Ma, X. Fractal characteristics of pore-throat structure and permeability estimation of tight sandstone reservoirs: A case study of Chang 7 of the Upper Triassic Yanchang Formation in Longdong area, Ordos Basin, China. *J. Pet. Sci. Eng.* **2019**, *184*, 106555. [CrossRef]
10. Wu, Y.; Liu, C.; Ouyang, S.; Luo, B.; Zhao, D.; Sun, W.; Awan, R.S.; Lu, Z.; Li, G.; Zang, Q. Investigation of pore-throat structure and fractal characteristics of tight sandstones using HPMI, CRMI, and NMR methods: A case study of the Lower Shihezi Formation in the Sulige area, Ordos Basin. *J. Pet. Sci. Eng.* **2022**, *210*, 110053. [CrossRef]
11. Wang, W.; Zhao, X. Study on characteristics parameters of coal core structure on fractal theory. *J. Henan Univ. Nat. Sci.* **2023**, *53*, 729–737.
12. Mayo, S.; Josh, M.; Nesterets, Y.; Esteban, L.; Pervukhina, M.; Clennell, M.B.; Maksimenko, A.; Hall, C. Quantitative micro-porosity characterization using synchrotron micro-CT and xenon K-edge subtraction in sandstones, carbonates, shales and coal. *Fuel* **2015**, *154*, 167–173. [CrossRef]
13. Clarkson, C.R.; Freeman, M.; He, L.; Agamalian, M.; Melnichenko, Y.B.; Mastalerz, M.; Bustin, R.M.; Radliński, A.P.; Blach, T.P. Characterization of tight gas reservoir pore structure using USANS/SANS and gas adsorption analysis. *Fuel* **2012**, *95*, 371–385. [CrossRef]

14. Huang, H.; Sun, W.; Ji, W.; Zhang, R.; Du, K.; Zhang, S.; Ren, D.; Wang, Y.; Chen, L.; Zhang, X. Effects of pore-throat structure on gas permeability in the tight sandstone reservoirs of the Upper Triassic Yanchang formation in the Western Ordos Basin, China. *J. Pet. Sci. Eng.* **2018**, *162*, 602–616. [CrossRef]
15. Zhang, F.; Jiang, Z.; Sun, W.; Li, Y.; Zhang, X.; Zhu, L.; Wen, M. A multiscale comprehensive study on pore structure of tight sandstone reservoir realized by nuclear magnetic resonance, high pressure mercury injection and constant-rate mercury injection penetration test. *Mar. Pet. Geol.* **2019**, *109*, 208–222. [CrossRef]
16. Qu, Y.; Sun, W.; Tao, R.; Luo, B.; Chen, L.; Ren, D. Pore–throat structure and fractal characteristics of tight sandstones in Yanchang Formation, Ordos Basin. *Mar. Pet. Geol.* **2020**, *120*, 104573. [CrossRef]
17. Mandelbrot, B.B. *The Fractal Geometry of Nature*; W.H. Freeman & Co. Ltd.: San Francisco, CA, USA, 1983.
18. Li, P.; Zheng, M.; Bi, H.; Wu, S.; Wang, X. Pore throat structure and fractal characteristics of tight oil sandstone: A case study in the Ordos Basin, China. *J. Pet. Sci. Eng.* **2017**, *149*, 665–674. [CrossRef]
19. Li, K. Analytical derivation of Brooks-Corey type capillary pressure models using fractal geometry and evaluation of rock heterogeneity. *J. Pet. Sci. Eng.* **2010**, *73*, 20–26. [CrossRef]
20. Giri, A.; Tarafdar, S.; Gouze, P.; Dutta, T. Fractal pore structure of sedimentary rocks: Simulation in 2-D using a relaxed bidisperse ballistic deposition model. *J. Appl. Geophys.* **2012**, *87*, 40–45. [CrossRef]
21. Wang, F.; Cheng, Y.; Lu, S.; Jin, K.; Zhao, W. Influence of coalification on the pore characteristics of middle-high rank coal. *Energy Fuels* **2014**, *28*, 5729–5736. [CrossRef]
22. Hu, Q.; Ewing, R.P.; Dultz, S. Low pore connectivity in natural rock. *J. Contam. Hydrol.* **2012**, *133*, 76–83. [CrossRef] [PubMed]
23. Guo, X.; Huang, Z.; Zhao, L.; Han, W.; Ding, C.; Sun, X.; Yan, R.; Zhang, T.; Yang, X.; Wang, R. Pore structure and multi-fractal analysis of tight sandstone using MIP, NMR and NMRC methods: A case study from the Kuqa depression, China. *J. Pet. Sci. Eng.* **2019**, *178*, 544–558. [CrossRef]
24. Qiao, J.; Zeng, J.; Jiang, S.; Wang, Y. Impacts of sedimentology and diagenesis on pore structure and reservoir quality in tight oil sandstone reservoirs: Implications for macroscopic and microscopic heterogeneities. *Mar. Pet. Geol.* **2020**, *111*, 279–300. [CrossRef]
25. Deng, Y. Analysis on differences of petroleum type and geological conditions between two depression belts in China offshore. *Acta Pet. Sin.* **2009**, *30*, 1–8. [CrossRef]
26. Gao, M.; Lu, Y.; Du, X.; Ma, Y.; Zhang, J.; Deng, K. Multi-factor evaluation for fine grading of tight sandstone reservoirs: A case study from H3 sand group in the upper section of Oligocene Huagang Formation, Xihu Sag, East China Sea Basin. *Pet. Geol. Exp.* **2021**, *43*, 1097–1106.
27. Wang, W.; Lin, C.; Zhang, X.; Dong, C.; Ren, L.; Lin, J. Seismic diagenetic facies prediction of tight sandstone in the offshore sparse well area: An example from the Xihu Depression of the East China Sea Basin. *J. Pet. Sci. Eng.* **2022**, *216*, 110825. [CrossRef]
28. Liu, J.; Zhang, S. Natural gas migration and accumulation patterns in the central-north Xihu Sag, East China Sea Basin. *Nat. Gas Geosci.* **2021**, *32*, 1163–1176. [CrossRef]
29. Zhang, Z.; He, X.; Tang, X.; Zhu, H. Structural trap characteristics and reservoir types in Xihu Sag, East China Sea Basin. *Mar. Geol. Front.* **2022**, *38*, 27–35. [CrossRef]
30. Su, A.; Chen, H.; Chen, X.; He, C.; Liu, H.; Li, Q.; Wang, C. The characterisitcs of low permeability reservoirs, gas origin, generation and charge in the central and western Xihu depression, East China Sea Basin. *J. Nat. Gas Sci. Eng.* **2018**, *53*, 94–109. [CrossRef]
31. Stroker, T.M.; Harris, N.B.; Crawford Elliott, W.; Marion Wampler, J. Diagenesis of a tight gas sand reservoir: Upper Cretaceous Mesaverde Group, Piceance Basin, Colorado. *Mar. Pet. Geol.* **2013**, *40*, 48–68. [CrossRef]
32. Scherer, M. Parameters influencing porosity in sandstones: A model for sandstone porosity prediction. *AAPG Bull.* **1987**, *71*, 1508. [CrossRef]
33. Houseknecht, D.W. Assessing the relative importance of compaction processes and cementation to reduction of porosity in sandstone. *AAPG Bull.* **1987**, *71*, 633–642. [CrossRef]
34. Wang, W.; Lin, C.; Zhang, X.; Dong, C.; Ren, L.; Lin, J. Provenance, clastic composition and their impact on diagenesis: A case study of the Oligocene sandstone in the Xihu sag, East China Sea Basin. *Mar. Pet. Geol.* **2021**, *126*, 104890. [CrossRef]
35. *GBT 29171-2012*; Rock Capillary Pressure Measurement. Chinese Petroleum Industry Standard: Beijing, China, 2013.
36. Liu, S.; Wang, F.; Yu, R.; Gao, J.; Shi, H.; Zhu, Y. Micro pore throat structure and fractal characteristics of tight sandstone reservoir. *J. Jilin Univ. Earth Sci. Ed.* **2024**, *54*, 96–107.
37. Zhu, W.; Zhang, X.; Zhou, D.; Fang, C.; Li, J.; Huang, Z. New cognition on pore structure characteristics of Permian marine shale in the Lower Yangtze Region and its implications for shale gas exploration. *Nat. Gas Ind.* **2021**, *41*, 41–55. [CrossRef]
38. Xoaoth, B. *Coal and Gas Outburst*; Song, S., Wang, Y., Eds. and Traslators; China Industry Press: Beijing, China, 1966.
39. Loucks, R.G.; Reed, R.M.; Ruppel, S.C.; Hammes, U. Spectrum of pore types and networks in mudrocks and a descriptive classification for matrix-related mudrock pores. *AAPG Bull.* **2012**, *96*, 1071–1098. [CrossRef]
40. Folk, R.L.; Andrews, P.B.; Lewis, D.W. Detrital sedimentary rock classification and nomenclature for use in New Zealand. *N. Z. J. Geol. Geophys.* **1970**, *13*, 937–968. [CrossRef]
41. Spencer, C.W. Review of characteristics of low-permeability gas reservoirs in western United States. *AAPG Bull.* **1989**, *73*, 613–629. [CrossRef]
42. Malunic, O., Dypvik, H.; Hammer, E. Diagenetic influence on reservoir quality evolution, examples from Triassic conglomerates/arenites in the Edvard Grieg field, Norwegian North Sea. *Mar. Pet. Geol.* **2018**, *93*, 247–271. [CrossRef]

43. Zhao, D.; Hou, J.; Sarma, H.; Guo, W.; Liu, Y.; Xie, P.; Dou, L.; Chen, R.; Zhang, Z. Pore throat heterogeneity of different lithofacies and diagenetic effects in gravelly braided river deposits: Implications for understanding the formation process of high-quality reservoirs. *Geoenergy Sci. Eng.* **2023**, *221*, 111309. [CrossRef]
44. Wang, W.; Lin, C.; Zhang, X.; Dong, C.; Ren, L.; Lin, J. Effect of burial history on diagenetic and reservoir-forming process of the Oligocene sandstone in Xihu sag, East China Sea Basin. *Mar. Pet. Geol.* **2019**, *112*, 104034. [CrossRef]
45. Wang, W.; Lin, C.; Zhang, X.; Dong, C.; Ren, L.; Lin, J. Discussion of seismic diagenetic facies of deep reservoir in the East China Sea Basin. *J. Pet. Sci. Eng.* **2022**, *208*, 109352. [CrossRef]
46. Wang, W.; Lin, C.; Zhang, X.; Dong, C.; Ren, L.; Lin, J. Diagenesis, diagenetic facies and their relationship with reservoir sweet spot in low-permeability and tight sandstone: Jiaxing area of the Xihu Sag, East China Sea Basin. *Minerals* **2023**, *13*, 404. [CrossRef]

fractal and fractional

MDPI

Article

Logging Evaluation of Irreducible Water Saturation: Fractal Theory and Data-Driven Approach—Case Study of Complex Porous Carbonate Reservoirs in Mishrif Formation

Jianhong Guo [1,2], Zhansong Zhang [1,2,*], Xin Nie [1,2], Qing Zhao [2,3] and Hengyang Lv [1,2]

1 College of Geophysics and Petroleum Resources, Yangtze University, Wuhan 430100, China; 2022730024@yangtzeu.edu.cn (J.G.); xin.nie@yangtzeu.edu.cn (X.N.); 2023710429@yangtzeu.edu.cn (H.L.)
2 Key Laboratory of Exploration Technologies for Oil and Gas Resources, Ministry of Education, Yangtze University, Wuhan 430100, China; 2021710322@yangtzeu.edu.cn
3 Seventh Oil Production Plant, Changqing Oilfield, Qingyang 745000, China
* Correspondence: zhangzhs@yangtzeu.edu.cn

Abstract: Evaluating irreducible water saturation is crucial for estimating reservoir capacity and developing effective extraction strategies. Traditional methods for predicting irreducible water saturation are limited by their reliance on specific logging data, which affects accuracy and applicability. This study introduces a predictive method based on fractal theory and deep learning for assessing irreducible water saturation in complex carbonate reservoirs. Utilizing the Mishrif Formation of the Halfaya oilfield as a case study, a new evaluation model was developed using the nuclear magnetic resonance (NMR) fractal permeability model and validated with surface NMR and mercury injection capillary pressure (MICP) data. The relationship between the logarithm mean of the transverse relaxation time (T_{2lm}) and physical properties was explored through fractal theory and the Thomeer Function. This relationship was integrated with conventional logging curves and an advanced deep learning algorithm to construct a T_{2lm} prediction model, offering a robust data foundation for irreducible water saturation evaluation. The results show that the new method is applicable to wells with and without specialized NMR logging data. For the Mishrif Formation, the predicted irreducible water saturation achieved a coefficient of determination of 0.943 compared to core results, with a mean absolute error of 2.37% and a mean relative error of 8.46%. Despite introducing additional errors with inverted T_{2lm} curves, it remains within acceptable limits. Compared to traditional methods, this approach provides enhanced predictive accuracy and broader applicability.

Keywords: irreducible water saturation; carbonate rock; fractal theory; Thomeer Function; nuclear magnetic resonance; logarithm of the mean transverse relaxation time; deep learning; well logging; Mishrif Formation

1. Introduction

In petrophysics, the distribution characteristics of immovable fluids are predominantly characterized by the irreducible water saturation parameter, which is defined as the volume of immovable water relative to the total pore volume of the rock [1]. Precisely evaluating irreducible water saturation is a crucial parameter in the exploration and development of oil and gas reservoirs [2,3]. The computation of irreducible water saturation directly influences the calculation of related parameters such as rock wettability, movable water saturation, and relative permeability [4,5]. Furthermore, it plays an essential role in identifying reservoir fluid properties, discerning water-flooded zones, and predicting production capacity [1,6].

Currently, the precise irreducible water saturation of core samples is typically determined through core experiments. The primary methods include the mercury injection capillary pressure (MICP) method [7], the semi-permeable membrane technique [8], nuclear magnetic resonance (NMR) testing [9], and phase infiltration testing. However, these core-based experimental methods do not consider economic and scalability factors. The long time needed and financial costs of these experiments, along with the expensive sampling process, limit the number of core samples [10,11]. Consequently, these methods cannot be directly applied to producing wells without coring data. To obtain continuous vertical irreducible water saturation parameters, the prediction model is typically constructed using high-resolution geophysical logging data in conjunction with core data. For conventional logging data, physical parameters such as porosity and permeability are calculated to construct the irreducible water saturation prediction model [12,13]. However, this method is generally inapplicable to unconventional reservoirs with significant heterogeneity. Therefore, some scholars have constructed irreducible water saturation models for each reservoir type by dividing flow units [14,15]. These flow units are categorized by minimizing physical property differences within the units and maximizing differences between units. The core idea is to mitigate the interference of reservoir pore structure variations on fluid evaluation [16].

Subsequently, some scholars have employed the oil column height method to calculate the irreducible water saturation, also referred to as original water saturation [17]. It is posited that the process of oil and gas continuously overcoming the capillary pressure within the rock and displacing water due to driving forces constitutes the transport of oil and gas. The current state of oil and gas distribution is thus seen as the relative equilibrium between driving forces and capillary pressure [18]. Therefore, an intrinsic connection exists between capillary pressure and the volume of displaced water, which corresponds to oil and gas saturation. For Middle Eastern carbonate rocks characterized by complex pore structures and multi-peak distribution, a relatively mature evaluation process has been established: Initially, the physical model is employed to fit the capillary pressure curve. Subsequently, combined with the saturation height function, a comprehensive quantitative evaluation of the original water saturation of the reservoir is performed based on reservoir classification [19]. However, this method involves a significant workload and low efficiency, requiring high accuracy in permeability calculation and oil–water interface determination. Subsequent studies have demonstrated that NMR data from specialized logging series can effectively evaluate irreducible water saturation [20,21]. The most common method is to differentiate between irreducible and movable fluids by determining the cut-off value of the transverse relaxation time ($T_{2cutoff}$), a standard initially recommended by major international oilfield companies [22,23]. Based on extensive experimental samples from the Gulf of Mexico, a fixed value of 92 ms is recommended as the $T_{2cutoff}$ for carbonate reservoirs. As more experiments are conducted, it has been shown that the fixed value is not applicable to complex reservoirs with significant heterogeneity [24]. For instance, in the pore-type carbonate reservoirs of the Mishrif Formation in the Middle East, which feature a complex pore structure, selecting a single fixed $T_{2cutoff}$ leads to substantial errors in practical applications. Subsequently, Zhang et al. [25] determined irreducible water saturation by fitting variable $T_{2cutoff}$ to NMR experimental sample data. However, due to the small number of samples used, the prediction model is difficult to extend to other regions. Wu et al. [26] selected different $T_{2cutoff}$s by dividing the reservoirs based on lithological differences. Nonetheless, it is challenging to reflect the complex pore structure of carbonate reservoirs based solely on lithological differences. The complexity of carbonate reservoirs' pore structures results in varying transverse relaxation time (T_2) spectrum morphologies, making this method poorly applicable in practice.

In the study of heterogeneous carbonate reservoirs, many international scholars have discovered a common phenomenon known as "diffusion coupling" in these reservoirs [27,28]. Consequently, it is challenging to accurately evaluate the irreducible water saturation of the reservoir using the $T_{2cutoff}$. Coates [29] analyzed and proposed the spectral coefficient (SBVI) method for addressing this phenomenon, asserting that each term of the relaxation time incorporates the contribution of irreducible water, and introduced an empirical formula specifically applicable to carbonate rocks. Additionally, some scholars utilized NMR data to construct a pseudo-capillary pressure curve, with Zhu et al. [30] specifically constructing such a curve based on NMR logging. The criterion of calculating the cumulative permeability at 99.9% was adopted, resulting in a continuous and accurate evaluation of irreducible water saturation. In addition to NMR logging data, imaging logging data have also been employed to evaluate irreducible water saturation. Li et al. [31] analyzed the porosity characteristics of irreducible fluids using mercury intrusion capillary pressure curves. Based on the capillary pressure approximation theory, the inverse accumulation curve of the porosity spectrum was derived. The physical significance of its "inflection point" was clarified. Subsequently, the irreducible water saturation at each reservoir depth was predicted using Archie's formula. Table 1 summarizes the advantages and disadvantages of various methods.

Table 1. Evaluation methods for irreducible water saturation.

Method Type	Specific Methods	Advantages	Disadvantages	References
Core experiment	Mercury injection capillary pressure experiment Semi-permeable membrane technique Nuclear magnetic resonance testing Phase infiltration testing	High accuracy	High costs, poor generalization	[7–9]
Conventional logging method	Physical property parameter fitting method	Strong generalization, cost-effective	Low precision	[12,13]
	Division of flow unit construction model method	Raises the accuracy, cost-effective	Non-uniform division standards	[14,15]
	Physical model method	Strong anti-interference ability, cost-effective	Low precision, high operation complexity	[18,19]
Special logging method	Nuclear magnetic resonance logging method	Theoretically strong, diversified forms, high recognition	Difficult to promote, high costs, destabilization	[21–26,29,30]
	Imaging logging method	Theoretically strong, moderate cost performance	Difficult to promote, poor feasibility	[31]

Currently, numerous evaluation methods for irreducible water saturation exist, yet two primary issues persist. The method based on conventional logging data for calculating physical property parameters is relatively popular but suffers from low overall accuracy. Even with the consideration of flow unit division, it is challenging to account for the promotion error of the flow unit and the carry-over error of the calculation. Although artificial intelligence models, such as machine learning [32], are now widely employed in the logging interpretation industry, the limited amount of core data complicates the construction of these models, significantly reducing their reliability. The introduction of a specialized logging series can resolve the accuracy issue, but general applicability remains problematic. Additionally, due to production cost considerations, specialized logging series are rarely collected in production wells. Consequently, evaluation models based on specialized logging data are challenging to apply to oilfield wells. The irreducible water saturation evaluation model is predominantly used in core wells with abundant logging series.

This paper introduces a novel method for evaluating irreducible water saturation to address existing challenges. Initially, an evaluation model for irreducible water saturation is derived from the NMR fractal permeability model, incorporating the logarithm mean of the transverse relaxation time (T_{2lm}) as a key parameter. Subsequently, the study verifies the numerical relationship between the modal parameter of pore throat diameter and the T_{2lm} in the Thomeer Function theory, using data from ground NMR and MICP experiments. The fractal theory is then applied to derive calculation formulas for the three key parameters in the Thomeer Function, establishing a numerical relationship between the T_{2lm} and the Thomeer Function. To operationalize this theory, the paper integrates it with logging data to construct a T_{2lm} prediction model based on conventional logging curves and advanced deep learning algorithms. This approach results in a versatile method for evaluating irreducible water saturation. In wells lacking specialized logging data, the T_{2lm} curve can be constructed using conventional curves and the deep learning model for assessing irreducible water saturation. The efficacy of this method was demonstrated in the Mishrif reservoir within the Halfaya oilfield, Middle East, yielding promising outcomes. Overall, this study introduces a novel perspective and methodology for evaluating irreducible water saturation.

2. Overview of Geological Area

The methods explained in this paper are applied to the Mishrif Formation of the Halfaya oilfield. Regionally, from the Jurassic to Early Cretaceous, the Persian Gulf Basin was generally in a large transgressive cycle [33]. South Iraq was at the west gentle slope of the deepest water area (Figure 1a). Mishrif is a sedimentary cycle deposited in the Cenomanian period of the Early Cretaceous. Some of the wells in the Halfaya oilfield are depicted in Figure 1b. The Mishrif Formation Group can reach a thickness of 400 m [34]. It is overlain by the Khasib Formation and underlain by the Runaila Formation. The Mishrif Formation Group can be subdivided into three major sections: MA, MB, and MC. The MB section can be further divided into MB1 and MB2 (Figure 1c). According to the previous literature, the Mishrif Formation is a porous carbonate reservoir. The primary mineral is calcite, constituting over 90% of the formation. The pore types of the reservoir predominantly consist of connected or disconnected moldic pores, fossil endopores, intergranular pores, intragranular pores, and matrix micropores. However, the influence of sedimentation, diagenesis, and tectonic modification, coupled with the complexity and variety of pore types, results in a complex pore structure, intricate pore-permeability relationships, and strong heterogeneity of the reservoir [35–37]. The reservoir exhibits a complicated pore structure, complex pore–permeability relationships, and strong heterogeneity.

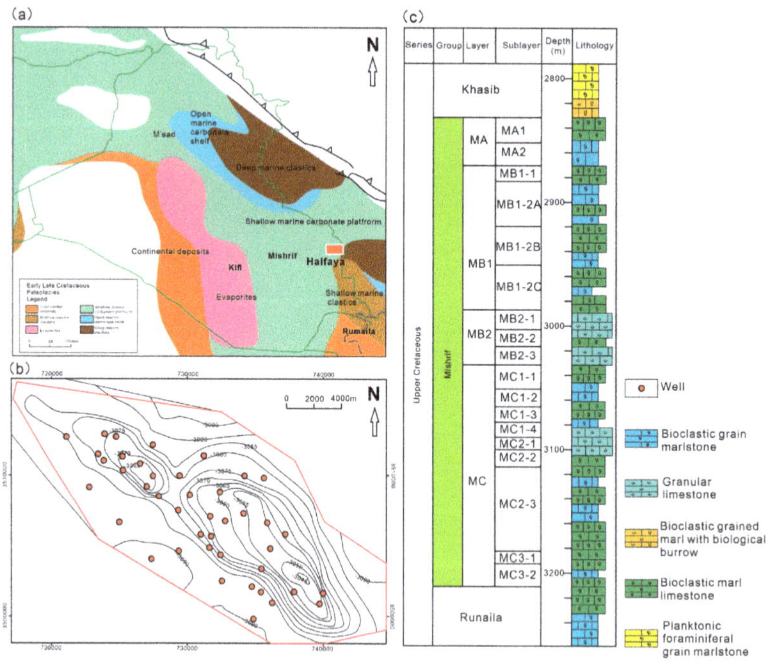

Figure 1. A geological survey map of the study area. (**a**) The geographical location of the Halfaya oilfield; (**b**) the deployment and elevation of important wells in the Halfaya oilfield; (**c**) core section histogram and stratification information of the Mishrif Formation.

3. Material Source and Experiment

The core samples in this study are sourced from the Mishrif Formation in the Halfaya oilfield, with plunger cores extracted from boreholes (Figure 2a). In Figure 2a, ① indicates the core sample, and ② indicates the plunger core, which is taken from the MB2_1 reservoir. For the plunger cores, column samples of appropriate size for the instrument are extracted separately to determine the core's water content and T_2 after centrifugation (Figure 2c). T_{2lm} and other parameters can also be determined post-centrifugation. The T_2 values after centrifugation (Figure 2c) were used to determine the distribution of T_2 in water-saturated cores, calculate the T_{2lm}, and establish the irreducible water saturation and corresponding $T_{2cutoff}$. Concurrently, the same plunger cores were subjected to MICP experiments using a Micromeritics 9320 mercury compression instrument (Figure 2d). The mercury compression curve of the core sample (Figure 2e) was obtained to calculate the distribution of pore throat radius, irreducible water saturation, and other physical parameters. For the samples and experimental processes shown in Figure 2, a total of 21 co-tested samples were obtained.

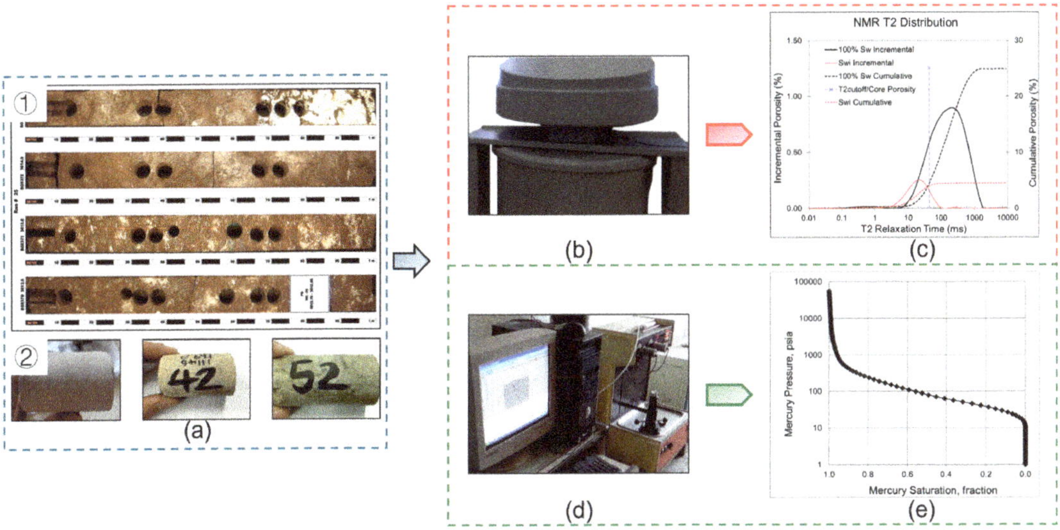

Figure 2. Core material source and core experimental diagram. (**a**) ① schematic diagram of coring sample, ② schematic diagram of plunger sample acquisition; (**b**) nuclear magnetic resonance instrument; (**c**) ground nuclear magnetic resonance experimental results of core samples; (**d**) mercury injection instrument; (**e**) MICP experimental results schematic.

4. Methods and Principles

4.1. Derivation and Verification of Irreducible Water Saturation Formula Based on Fractal Nuclear Magnetic Permeability

4.1.1. Derivation of Irreducible Water Saturation Formula

Previous studies have demonstrated that NMR logging data can effectively calculate permeability. The Timur-Coates model [38,39] and the Schlumberger Doll Research (SDR) model [40,41] are the most widely used, corresponding to Equations (1) and (2), as shown in Figure 3A,B.

$$K_c = a_1 \times \varphi^4 \times \left(\frac{FFI}{BVI}\right)^2 = a_1 \times \varphi^4 \times \left(\frac{1 - S_{wirr}}{S_{wirr}}\right)^2 \tag{1}$$

$$K_s = a_2 \times \varphi^4 \times T_{2lm}^2 \tag{2}$$

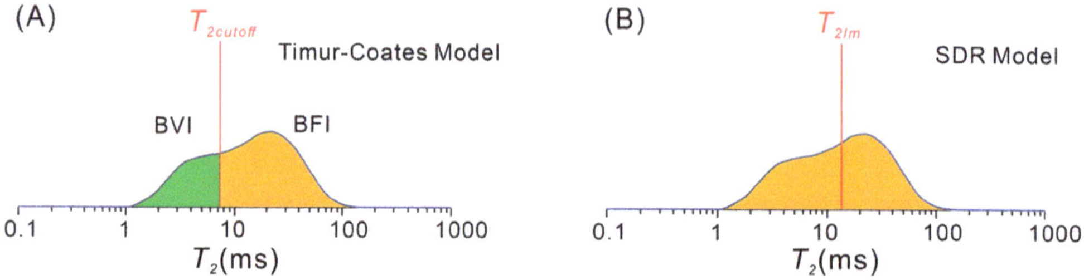

Figure 3. Nuclear magnetic permeability prediction model diagram. (**A**) Timur-Coates model; (**B**) SDR model.

In Equations (1) and (2), φ is the porosity, *FFI* is the free fluid index, and *BVI* is the irreducible volume index. Both can be converted to the form of irreducible water saturation (S_{wirr}, fractional), which can be determined by the $T_{2cutoff}$. T_{2lm} is the logarithmic mean of the transverse relaxation time, and a_1 and a_2 are the fitting coefficients.

Numerous studies have been conducted to improve the classical equations, primarily updating the coefficient solution for reservoirs with different lithologies or introducing correction coefficients. Wu et al. [42] modified the classical NMR permeability formula using fractal theory. Based on Equations (1) and (2), they introduced fractal dimensions and rewrote the formula as Equations (3) and (4):

$$K_{cD} = a_3 \times \varphi^{(b_1 + c_1 D)} \times \left(\frac{1 - S_{wirr}}{S_{wirr}} \right)^{d_1} \tag{3}$$

$$K_{sD} = a_4 \times \varphi^{(b_2 + c_2 D)} \times T_{2lm}^{d_2} \tag{4}$$

Unlike the classical formulas, the permeability formulas incorporating fractal theory replace the original constant exponent of porosity with a linear function of the fractal dimension *D*. The terms a_3, a_4, b_1, b_2, c_1, c_2, d_1, and d_2 are the fitting coefficients.

In this study, Equations (3) and (4) are conjugated to derive an expression for irreducible water saturation, as shown in Equation (5):

$$S_{wirr} = \frac{1}{1 + \left\{ \left(\frac{a_4}{a_3} \right) \times \varphi^{[(b_2 - b_1) + (c_2 - c_1) \times D]} \times T_{2lm}^{d_2} \right\}^{\frac{1}{d_1}}} = \frac{1}{1 + \left(\frac{a_4}{a_3} \right)^{\frac{1}{d_1}} \times \varphi^{\frac{[(b_2 - b_1) + (c_2 - c_1) \times D]}{d_1}} \times T_{2lm}^{\frac{d_2}{d_1}}} \tag{5}$$

In Equation (5), the constant terms in the formula can be replaced. A*, B*, C*, and D* are constants, $A^* = \left(\frac{a_4}{a_3} \right)^{\frac{1}{d_1}}$, $B^* = \left(\frac{b_2 - b_1}{d_1} \right)$, $C^* = \left(\frac{c_2 - c_1}{d_1} \right)$, $D^* = \left(\frac{d_2}{d_1} \right)$, whereas *D* is a variable (fractal dimension). Irreducible water saturation can be expressed as a function of porosity, fractal dimension, and T_{2lm} (Equation (6)).

$$S_{wirr} = \frac{1}{1 + A^* \times \varphi^{B^* + C^* \times D} \times T_{2lm}^{D^*}} \tag{6}$$

4.1.2. Verification of Irreducible Water Saturation Formula

After deriving the expression for irreducible water saturation, it was validated using data from core ground NMR experiments and MICP experiments. Initially, T_2 spectra obtained from the core NMR experiments were utilized to calculate the fractal dimension. Previous studies have consistently demonstrated that the reservoir's pore size distribution adheres to a fractal structure [43,44]. Derivations indicate that the density function of pore size distribution in a reservoir with pore sizes larger than *r* can be expressed as follows:

$$P(r) = \frac{dN}{dr} = -D \times a \times r^{-D-1} \tag{7}$$

In Equation (7), N is the number of pores with a diameter greater than r; r is a specific pore diameter; D is the pore fractal dimension; a is the proportionality constant; and $P(r)$ is the pore size distribution density function.

The cumulative pore volume fraction S_v for pore diameters less than r is simplified and expressed by Equation (8) [45]:

$$S_v = \frac{r^{3-D}}{r_{\max}^{3-D}} \tag{8}$$

In Equation (8), r is the pore diameter, and S_v is the cumulative pore volume fraction for pore diameters less than r.

For the T_2 spectrum from NMR experiments, the superposition mechanism and simplification based on the transverse relaxation rate can be considered:

$$\frac{1}{T_2} = \rho(\frac{S}{V}) = F_s\frac{\rho}{r} \tag{9}$$

In Equation (9), T_2 is the transverse relaxation time, milliseconds; ρ is the transverse surface relaxation strength of the rock, μm/ms; S is the pore surface area, cm^2; V is the pore volume, cm^3; and F_s is the geometrical shape factor, typically 2 or 3 depending on the specific model.

For capillary pressure, this can be expressed by Equation (10) [46]:

$$p_c = \frac{2\sigma\cos\theta}{r} \tag{10}$$

In Equation (10), p_c is the capillary pressure corresponding to pore diameter r; σ is the surface tension and contact angle of the liquid; and θ is the wetting contact angle, °.

According to Equations (9) and (10), the expression between p_c and T_2 can be obtained:

$$p_c = \left|\frac{2\sigma\cos\theta}{F_s\rho}\right| \times \frac{1}{T_2} \tag{11}$$

The correlation between the distribution of NMR T_2 spectra and capillary pressure curves obtained from mercury compression experiments reflects their expressions of the rock's pore structure. Xiao et al. and Zhu et al. [21,30] successfully reconstructed pseudo-capillary curves using these T_2 spectra, as demonstrated in

$$p_{c\min} = \left|\frac{2\sigma\cos\theta}{F_s\rho}\right| \times \frac{1}{T_{2\max}} \tag{12}$$

In Equation (12), $p_{c\min}$ is the capillary pressure corresponding to the maximum pore size in the reservoir, i.e., the mercury feed pressure, and $T_{2\max}$ is the maximum transverse relaxation time.

Equations (7), (8), and (10) are conjugated to obtain S_v, and then Equations (11) and (12) are substituted:

$$S_{vT_2} = (\frac{T_{2\max}}{T_2})^{D-3} \tag{13}$$

At this stage, S_{vT_2} represents the cumulative volume of pores with a transverse relaxation time less than T_2, expressed as a percentage of the total pore volume [47]. Applying a logarithm to both sides of Equation (13) results in

$$\lg(S_{vT_2}) = (3-D)\lg(T_2) + (D-3)\lg T_{2\max} \tag{14}$$

According to Equation (14), the fractal dimension can then be calculated in conjunction with the T_2 spectrum of the core experiment.

In the past, when calculating the fractal dimension of rock, only the large pore part was usually considered, and few scholars discussed it from the physical properties of the rock itself. In this paper, the relationship between the cumulative permeability contribution and pore system is described by using MICP experimental data of a core. The permeability of reservoir rocks correlates closely with their microscopic pore structure [48,49]. Core MICP experiments provide capillary pressure curves that offer insights into this relationship. The cumulative permeability contribution derived from MICP experimental data reflects the physical properties of the rock [50,51]. It quantifies how specific pore throat intervals within a rock sample contribute to its overall capacity for fluid flow. Thicker pore throats contribute to higher permeability, whereas thinner ones contribute to lower permeability. Equation (15) enables determination of the cumulative permeability contribution across pore throat intervals, ranging from larger to smaller sizes.

$$\Delta K_j = \frac{r_j^2 \times Shg_j}{\sum (r_j^2 \times Shg_j)} \tag{15}$$

In Equation (15), r_j denotes the radius of the pore throat corresponding to the jth point, μm; Shg_j denotes the increment of mercury feed corresponding to the jth point, pu; and ΔK_j denotes the permeability contribution corresponding to the jth point, pu.

Figure 4a shows pore throat frequency distribution curves for three rock samples: single peak (green dashed line), double peak (red dashed line), and triple peak (blue dashed line). Figure 4b–d depict how these distributions relate to cumulative permeability contribution. The blue curve in the double logarithmic scale represents the cumulative permeability contribution, indicating how permeability accumulates from large to small pore throats. The red curve in the semi-logarithmic scale illustrates the pore throat distribution, revealing the pore space structure. Analysis of these curves reveals multiple pore systems in the Mishrif Formation. Single-peak samples (Figure 4b) exhibit a 100% permeability contribution. In double-peak samples (Figure 4c), the first and second pore systems contribute 99.7% and 0.3% to permeability, respectively. Triple-peak samples (Figure 4d) show contributions of 97.6%, 2.2%, and 0.2% from their respective pore systems. This finding aligns with Tang et al.s' [52] results, suggesting a negligible impact of secondary pore throats on permeability in double-peak samples, focusing instead on the largest pore throat system. This supports Wu et al.'s [42] premise that calculating fractal dimension should primarily consider the macroporous portion.

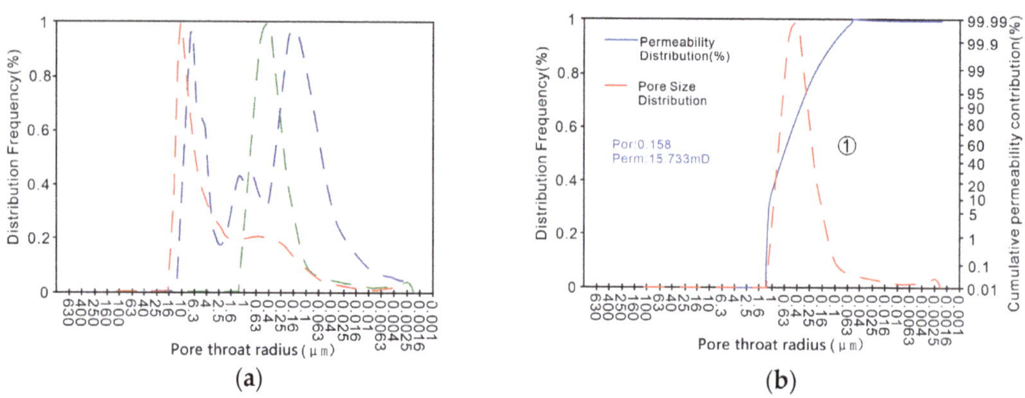

Figure 4. *Cont.*

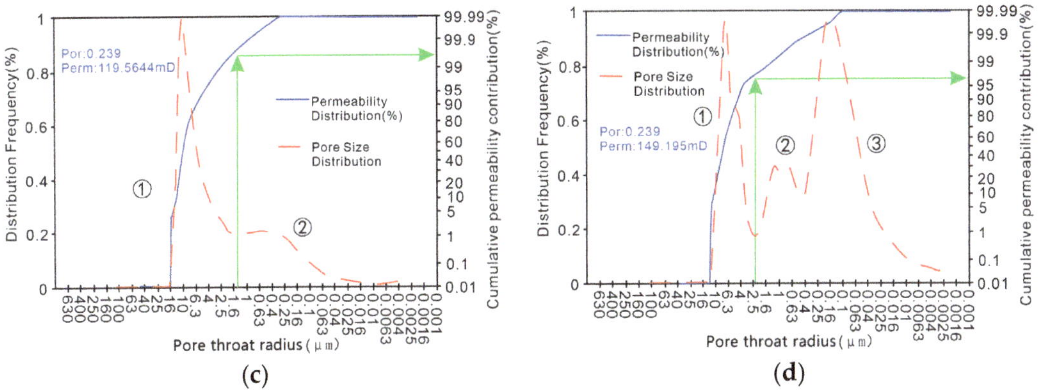

Figure 4. The pore throat frequency distribution curve and the relationship between cumulative permeability and pore throat distribution. (**a**) The pore throat frequency distribution curves of the three rock samples correspond to the pore throat distribution characteristics of single peak (green discontinuous line), double peak (red discontinuous line) and three peak (blue discontinuous line); (**b**) single-peak pore throat distribution and permeability contribution; (**c**) bimodal pore throat distribution and permeability contribution; (**d**) pore throat distribution and permeability contribution of three peaks.

Among the co-tested core samples, the 151VA sample from well X-95 is used as an example. In Figure 5a, the T_2 spectrum of this sample when saturated with water is shown. The black solid line indicates the pore fraction, while the red solid line represents the cumulative pore volume of the core under saturation. Figure 5b outlines the process for determining the fractal dimension of this core based on the NMR T_2 spectra's volume percentage. The fractal dimension is derived from the intercept obtained through Equation (13) and its fit.

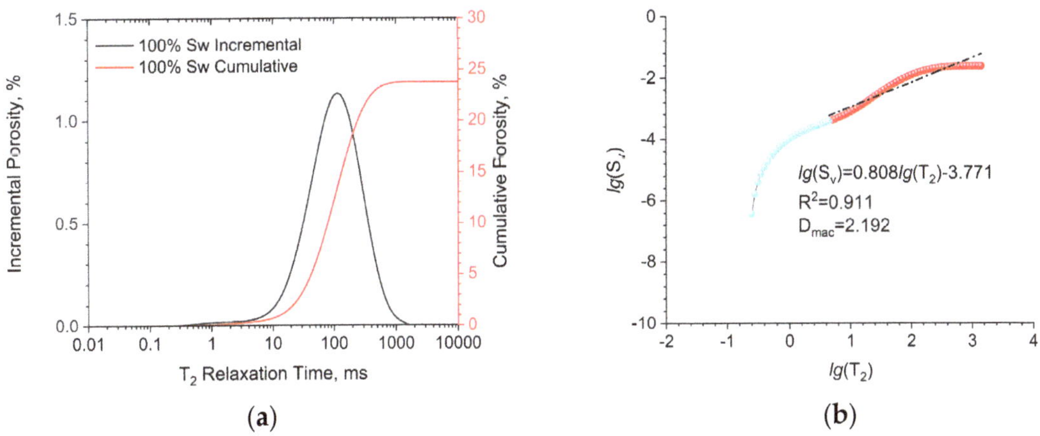

Figure 5. The NMR experimental results of 151VA sample and the calculation process of fractal dimension. (**a**) The T_2 spectrum of the sample is saturated with water, the black solid line is the pore component of the core saturated with water, and the red solid line is the pore accumulation of the core saturated with water; (**b**) fractal dimension calculation diagram of core (The light blue color is the microporous portion and the red color is the macroporous portion).

In previous studies of inhomogeneous carbonate reservoirs, researchers have categorized them into distinct petrophysical types to enhance understanding. In this study, we employed the classical Winland R35 method to classify reservoir types based on the pore throat radius (R35), which corresponds to a 35% cumulative mercury saturation. Martin's [53] criteria were also used due to the scarcity of samples with R35 values exceeding 10 μm. Specifically, R35 values between 2 μm and 10 μm were classified as first Rock-Type (RT1), those between 0.5 μm and 2 μm as second Rock-Type (RT2), and values below 0.5 μm as third Rock-Type (RT3). The study includes NMR T_2 spectra of three fully saturated core samples, X-1 (RT1), X-9 (RT2), and X-14 (RT3). The porosity and permeability of these samples vary: X-1 (25.9%, 54 mD), X-9 (19.7%, 8.4 mD), and X-14 (15.8%, 0.56 mD). Their corresponding R35 values are 2.4 μm, 0.71 μm, and 0.37 μm, respectively. Figure 6a, 6d, and 6g illustrate the saturated water NMR T_2 spectra of samples X-1, X-9, and X-14, respectively. The NMR spectra of X-1 show a notable shift to longer T_2 compared to X-9 and X-14. Pore throat frequency distributions (Figure 6b,e,h) reveal complex porosity systems in X-9, with indications of dual sets of pores. The peak pore throat sizes decrease gradually from RT1 to RT3. The fractal dimensions calculated for the samples are 2.148 (X-1), 2.256 (X-9), and 2.299 (X-14). This trend aligns with existing research, suggesting higher fractal dimensions correlate with poorer reservoir physical properties. In contrast to dense sandstone reservoirs (typically 2.5–3), the porous carbonate rocks in our study range between 2 and 2.39.

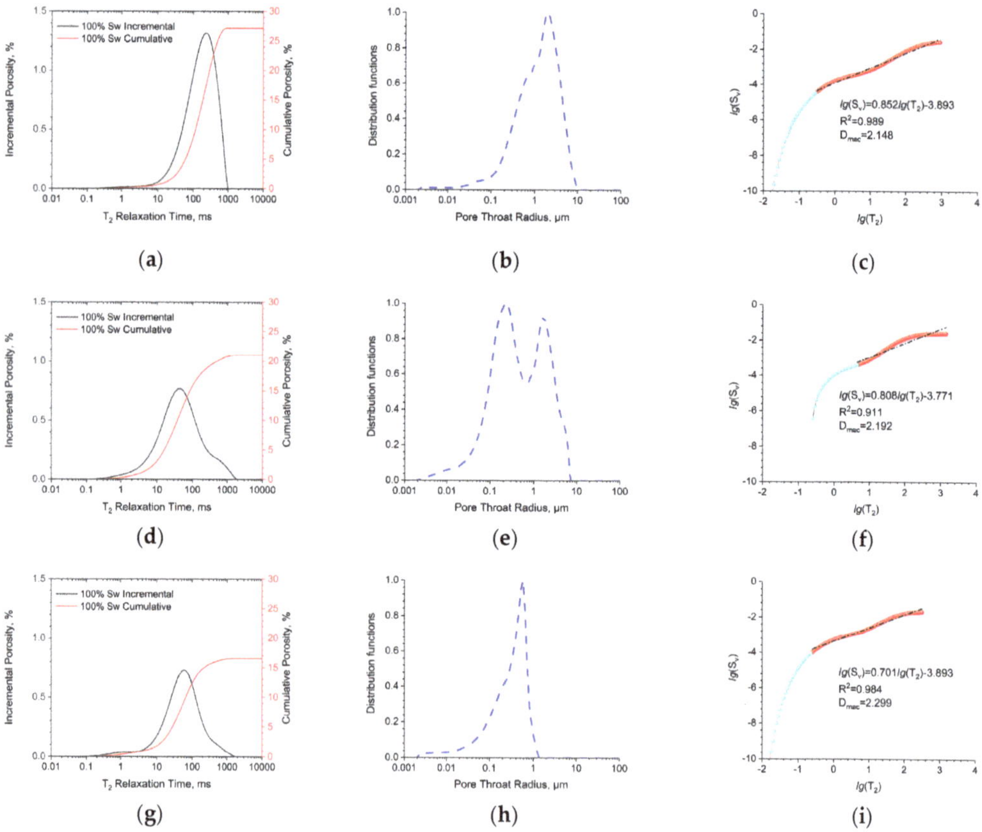

Figure 6. Surface NMR experimental results of core samples of three types of rock types and calculation of pore throat distribution frequency and fractal dimension based on MICP experimental

data. (**a**) Water-saturated T_2 spectrum of X-1 sample; (**b**) pore throat distribution frequency diagram of X-1 sample; (**c**) the calculation process of fractal dimension of X-1 sample; (**d**) water-saturated T_2 spectrum of X-9 sample; (**e**) pore throat distribution frequency diagram of X-9 sample; (**f**) calculation process of fractal dimension of X-9 sample; (**g**) water-saturated T_2 spectrum of X-14 sample; (**h**) pore throat distribution frequency diagram of X-14 sample; (**i**) calculation process of fractal dimension of X-14 sample. In Figure c, f, i, the light blue color is the microporous portion and the red color is the macroporous portion.

For the co-tested samples, the fractal dimension, T_{2lm}, and porosity can be determined for each core sample. There are 21 groups of samples in total. In Equation (6), a total of four groups of coefficients are involved; therefore, the number of samples used for determining the coefficients should be no less than four. In this study, 10 groups of samples were randomly selected and combined with the Least-Squares Optimization method to determine the four groups of coefficients in Equation (6). The other 11 groups of samples were used to test the accuracy of the method. The parameter results and the formula's form are shown in Table 2. Figure 7 plots the predicted irreducible water saturation results against the core-irreducible water saturation. The blue spheres in Figure 7 represent the ten groups of sample points involved in the model construction, and the red spheres represent the other 11 groups of samples used for validation. The calculated irreducible water saturations are in good agreement with the core samples. The goodness of fit for the two groups of samples is 0.947 and 0.920, with relative errors ranging from 0.01% to 18.17%, and the average relative error is 8.44%. These results indicate that the irreducible water saturation formula derived in this paper is valid.

Table 2. The coefficients in the irreducible water saturation model with the final form of the equation.

Coefficient Value			
*A**	*B**	*C**	*D**
0.034	0.607	−0.256	0.903

$$S_{wirr} = \frac{1}{1+0.034\times\varphi^{(0.607-0.256\times D)}\times T_{2lm}^{0.903}} \quad (16)$$

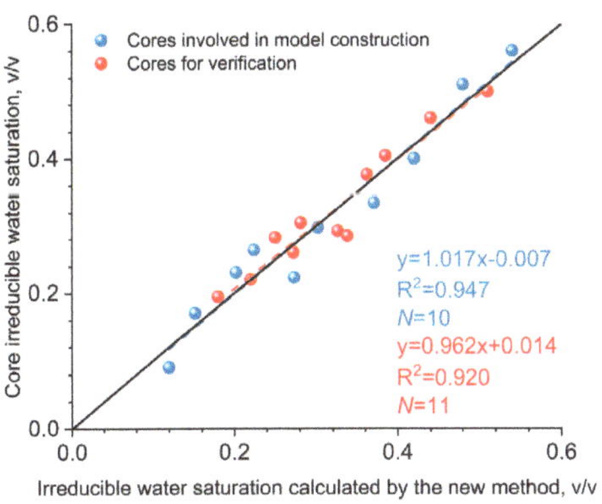

Figure 7. The practical application effect of the irreducible water saturation prediction model in this paper.

4.2. The Conventional Expression Form of T_{2lm} and the Promotion of Deep Learning Algorithm

4.2.1. Numerical Expression of T_{2lm} Based on Thomeer Function and Fractal Theory

The expression for irreducible water saturation was derived and validated in Section 4.1. T_{2lm} in the equation requires NMR logging data, which are scarce and costly to obtain in practical applications. Thus, promoting Equation (16) presents a technical challenge. In this study, a numerical expression for the T_{2lm} is derived using Thomeer Function theory combined with fractal theory.

Thomeer [54] observed a capillary pressure curve in double logarithmic coordinates, which can be approximately characterized by a hyperbolic model (Figure 8a). The model is defined as follows:

$$\frac{(B_v)_{p_c}}{(B_v)_{pc\infty}} = e^{-G/(\lg p_c/p_d)} \tag{17}$$

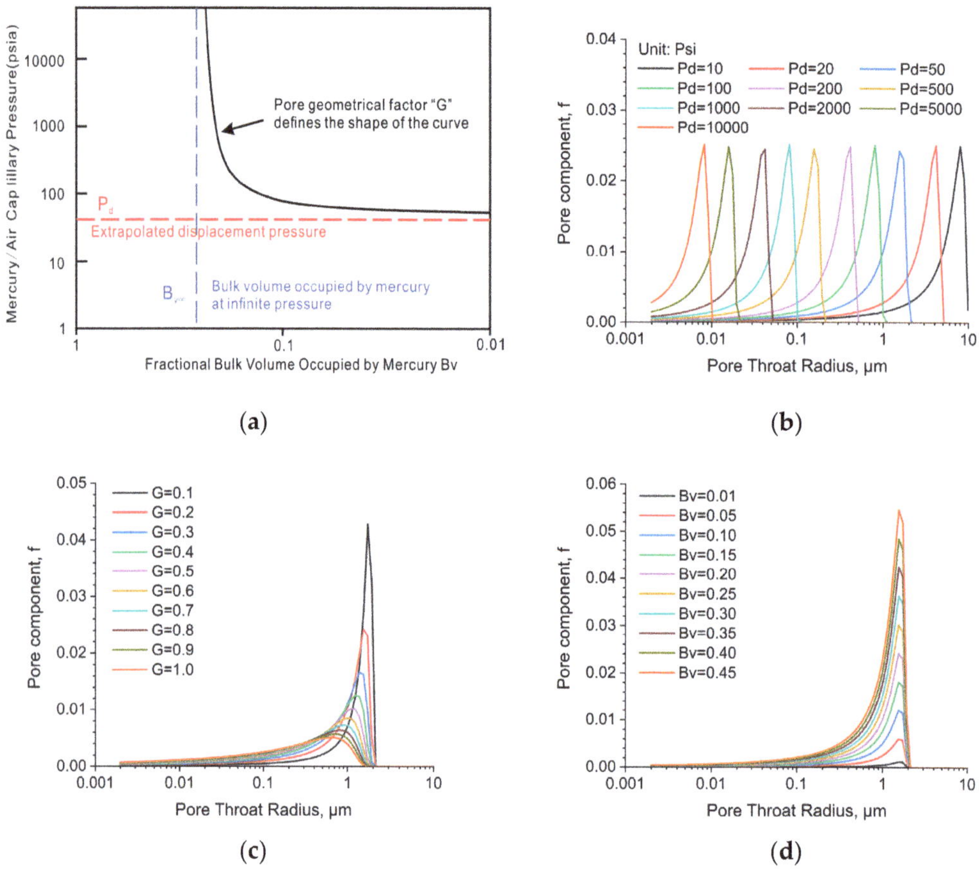

Figure 8. Theoretical form of Thomeer Function and physical meaning characterization of three parameters under control variables. (**a**) In the double logarithmic coordinate system, the mercury injection curve was simulated by hyperbola; (**b**) characterization of displacement pressure (p_d) on pore throat size; (**c**) characterization of pore throat characteristics by geometric factors G; (**d**) invisible characterization of rock samples by maximum mercury injection volume $(B_v)_{pc\infty}$.

In Equation (17), p_c is the mercury–air capillary pressure, psi; p_d is the discharge pressure, i.e., the inlet pressure of mercury, psi; $(B_v)_{pc\infty}$ is the volume fraction of injected mercury at infinite pressure, v/v; $(B_v)_{P_c}$ is the volume fraction of injected mercury at the p_c pressure, v/v; and G is the dimensionless pore geometry factor.

For the expression form of Equation (16), numerical simulations were carried out with controlled variables, plotting the pore throat radius versus pore fraction under different parameters:

(1) Changing only the discharge pressure (p_d); the larger the discharge pressure, the smaller the radius of the pore channel, indicating worse physical properties of the sample (Figure 8b).
(2) Changing only the geometry factor G; the smaller G, the narrower the shape of the pore throat distribution curve, indicating that the pore channel is more concentrated and the sorting of rock samples is better (Figure 8c).
(3) Changing only the maximum mercury feed volume $(B_v)_{pc\infty}$; the larger $(B_v)_{pc\infty}$, the larger the pore volume, indicating better physical properties of rock samples (Figure 8d).

In the context of the Thomeer Function parameters, the minimum mercury injection pressure corresponds to the maximum pore throat radius, marking the boundary of the larger radius in the pore throat frequency distribution. The parameter G signifies the narrowness of this curve. Typically, these two parameters are combined to pinpoint the peak radius (RP) location within the orifice frequency distribution. A multivariate relationship between RP and p_d relative to G can be established to align with the graphical representation in Figure 9, corresponding to Equation (18):

$$RP = \frac{1.244}{G^{0.211} \times p_d^{1.355}} \tag{18}$$

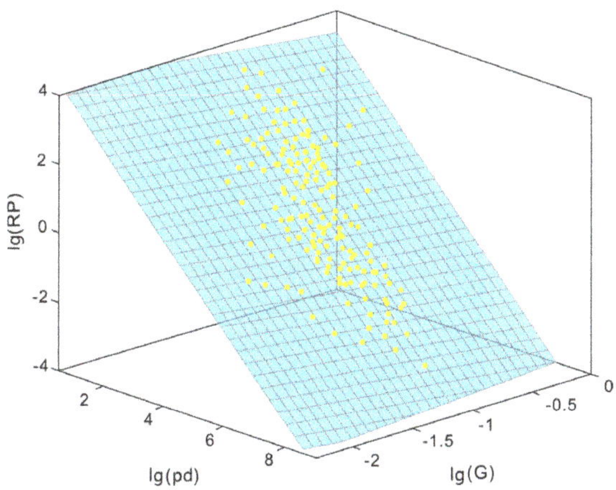

Figure 9. The fitting diagram of the peak of the pore throat radius frequency curve and p_d and G in the Thomeer Function.

Tang et al. [52] proposed a new parameter, Mode, which is highly correlated with permeability. This was based on Equation (18) using a large amount of core data from the Mishrif Formation in the Middle East, in conjunction with the permeability model form. This parameter is defined as

$$Mode = \frac{1.244 \times (0.037Bv - 0.370G)}{G^{0.211} \times p_d^{1.355}} \tag{19}$$

This parameter, Mode, can be regarded as a comprehensive characterization of the pore throat size, and the T_{2lm} has similar physical significance. Chen et al. [55] also showed that this type of element characterizing the pore throat diameter has a significant correlation with the T_{2lm}. For the 21 groups of samples, the Thomeer Function parameters were extracted based on the MICP experiments. The Mode of the samples was calculated according to Equation (18) and correlated with the T_{2lm}. Figure 10 shows the results of the correlation analysis between the two parameters. The goodness of fit between the two parameters is 0.931, indicating a significant correlation. This relationship can be characterized by Equation (20), meaning that T_{2lm} can be characterized using the Mode parameter. Since this parameter consists of the parameters in the Thomeer Function, T_{2lm} can be written as an expression involving these three parameters.

$$T_{2lm} = 424.984 \times Mode^{0.314} \tag{20}$$

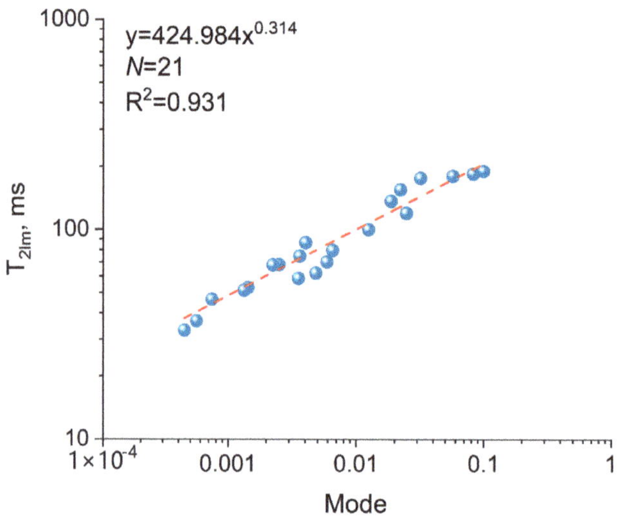

Figure 10. The fitting results between T_{2lm} and the Mode parameter.

Previous calculations of parameters in the Thomeer Function were primarily extracted from core samples and fitted to the response of conventional logging curves, lacking a basis in theoretical derivation [52,56]. This study incorporates fractal theory to propose a new argument. The derivation in Appendix A shows that

$$p_d = \frac{\sigma \cos\theta}{\sqrt{\frac{2K}{f\varphi}\frac{5-D}{3-D}}} \tag{21}$$

$$(B_v)_{p_{c\infty}} = \varphi\left\{1 - \lim_{r\to 0}\left[\frac{\sqrt{(8K/f\varphi)(5-D)/(3-D)}}{r}\right]^{(D-3)}\right\} \tag{22}$$

Equation (21) proves that p_d and $\sqrt{\frac{K}{\varphi}}$ have a correlation. Especially in the same type of reservoir, when the difference in fractal dimension is not significant, the two are linearly correlated.

In Equation (22), since (D-3) is a negative number, $(B_v)_{pc\infty}$ represents a function entirely determined by porosity from a mathematical perspective. Due to the presence of irreducible water, r is not equal to zero, and the porosity φ remains the parameter with the greatest influence on the calculation results.

The geometric factor G, while mathematically simplified, lacks direct physical significance. Numerical simulations indicate that G reflects the concentration of pore throat distribution; a smaller G suggests better sorting. According to Thomeer's theory, G is influenced by both the P_d and $(B_v)_{pc\infty}$ parameters. Given their relationships with φ and $\sqrt{\frac{K}{\varphi}}$, we established a correlation between G and φ, $\sqrt{\frac{K}{\varphi}}$ as expressed in Equation (23):

$$G = -1.215 \times \varphi - 0.154 \times \lg(\sqrt{\frac{K}{\varphi}}) + 0.784 \tag{23}$$

After completing the above theoretical derivation, the demonstration was carried out using the co-tested samples. Figure 11a shows that the P_d has a significant correlation with the $\sqrt{\frac{K}{\varphi}}$, which is consistent with the theoretical derivation of Equation (21). Figure 11b shows a strong linear correlation between $(B_v)_{pc\infty}$ and φ, with an R^2 as high as 0.921, which aligns with the theoretical derivation of Equation (22). Figure 11c,d show a correlation between the geometric factor G with the $\sqrt{\frac{K}{\varphi}}$ and φ, consistent with the theory mentioned above.

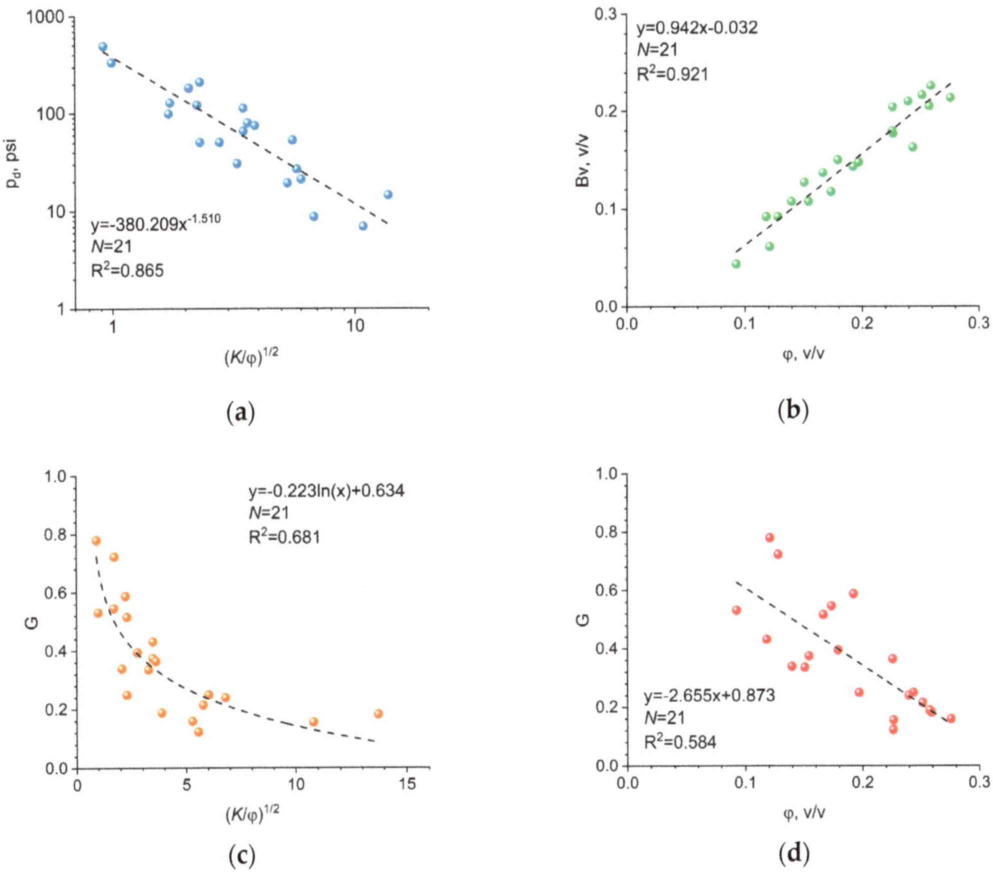

Figure 11. Correlation between the parameters in the Thomeer Function and the physical parameters in the formula derivation form. (**a**) The fitting diagram between p_d and $\sqrt{\frac{K}{\varphi}}$; (**b**) the fitting diagram between $(B_v)_{pc\infty}$ and φ; (**c**) the fitting graph between geometric factor G and $\sqrt{\frac{K}{\varphi}}$; (**d**) fitting diagram between geometric factor G and φ.

Combining Thomeer Function with fractal theory, the functional relationship between each parameter in the Thomeer Function and the physical property parameters and the fractal dimension is deduced. The combination of Equations (19) and (20) shows that T_{2lm} can be expressed by G, $(B_v)_{pc\infty}$, and p_d. The combination of Equations (19)–(23) can be written as

$$T_{2lm} = 28.574 \times \left(\frac{\varphi}{k}\right)^{1.023} \times (-\varphi - 0.062\lg k + 0.062\lg\varphi + 0.645)^{-0.211} \times (\varphi + 0.057\lg k - 0.057\lg\varphi - 0.601)^{0.314} \tag{24}$$

4.2.2. T_{2lm} Inversion Based on Deep Learning Algorithm

In Section 4.2.1, it was theoretically proven that T_{2lm} can be calculated from porosity and permeability, and this was verified using experimental data from actual co-tested rock samples. However, in practical applications, there is a significant error in permeability calculation, and the numerical relationship in Equation (24) also has fitting errors. This leads to an increase in the calculation error of the Thomeer Function parameter, an increase in the error in the T_{2lm} inversion results, and ultimately a decrease in the accuracy of the irreducible water saturation calculation.

Unconventional reservoirs pose challenges in calculating physical property parameters, prompting the use of machine learning or deep learning methods in conjunction with conventional curves to build predictive models. The usefulness of these models has been demonstrated in successful applications. This study focuses on constructing a T_{2lm} inversion model using deep learning algorithms based on conventional curves. Previously, core data were directly integrated with conventional curves to develop parameter inversion models, requiring substantial core datasets that are costly and time-consuming to obtain, thus limiting their availability. In contrast, inversion curves offer a more abundant source of data samples. The NMR logging instrument's small sampling interval ensures each set of points collected represents a dataset. Therefore, sufficient samples are available for constructing deep learning models whenever NMR logging data are gathered from boreholes, overcoming the data sample limitations encountered in traditional model construction.

In this study, data extracted from the Mishrif Formation in the study area were analyzed to compute the Pearson correlation coefficient (Equation (25)) across conventional, derived, and T_{2lm} curves. The conventional curves analyzed include the uranium-free gamma curve (KTH), acoustic time-difference logging curve (AC), natural gamma logging curve (GR), compensated density logging curve (DEN), deep resistivity curve (RD), and compensated neutron logging curve (CN). Derived curves encompass the porosity curve (Por) derived from the regression analysis of the porosity logging series, the permeability curve (K) determined by Winland R35 reservoir classification, the mud content curve (Vsh) calculated from uranium-free gamma curves, $\sqrt{\frac{K}{\varphi}}$, the triple porosity difference ϕ_D, and the triple porosity ratio ϕ_R.

$$Pearson = \frac{M\sum_{i=1}^{M} x_i y_i - \sum_{i=1}^{M} x_i \sum_{i=1}^{M} y_i}{\sqrt{M\sum_{i=1}^{M} x_i^2 - \left(\sum_{i=1}^{M} x_i^2\right)}\sqrt{M\sum_{i=1}^{M} y_i^2 - \left(\sum_{i=1}^{M} y_i^2\right)}} \tag{25}$$

x_i is for each logging parameter, y_i is T_{2lm}, i corresponds to the serial number of each sample, and M is the total number of samples for statistics.

A heat map based on the Pearson index was generated (Figure 12), demonstrating significant correlations between the porosity series logging curves, calculated porosity, and T_{2lm}. Similarly, permeability curves and the $\sqrt{\frac{K}{\varphi}}$ also exhibit strong correlations with T_{2lm}, achieving Pearson indices of 0.81 for both. Furthermore, T_{2lm} shows a robust correlation with depth. The logging signal and depth form a time series within the logging curve.

Consequently, a T_{2lm} curve inversion model was developed using the CNN-GRU–Attention algorithm [57].

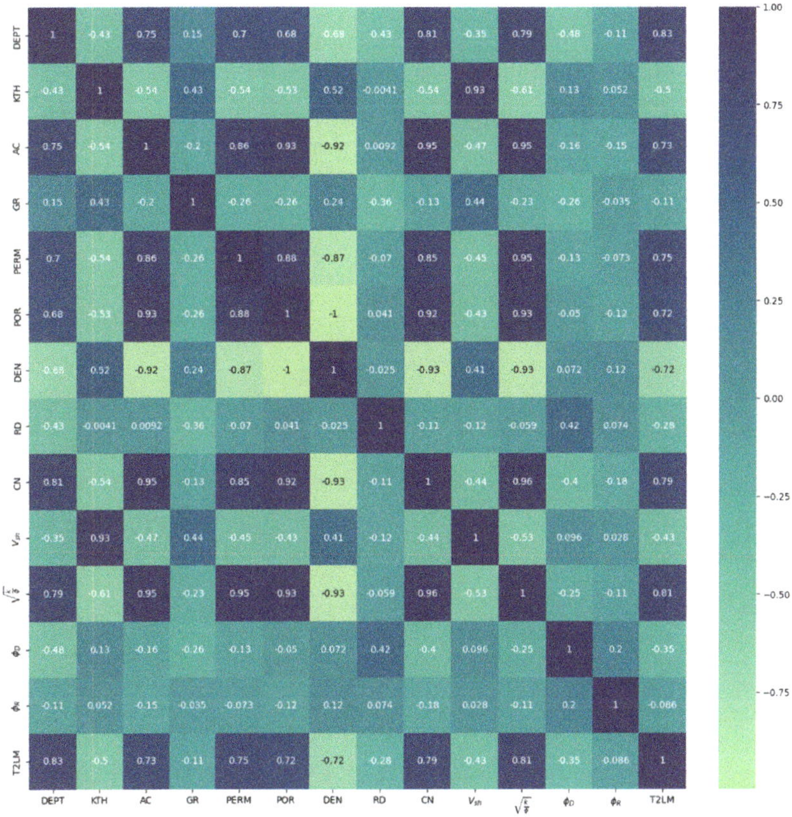

Figure 12. Pearson index heat map between conventional logging curve, calculated curves, and T_{2lm} curve.

This study introduces an innovative CNN-GRU-ATT model, integrating Convolutional Neural Networks (CNNs), Gated Recurrent Units (GRUs), and Attention mechanisms (Figure 13). The CNN component performs feature extraction through its convolutional layers, utilizes pooling layers to reduce the spatial dimensions of the feature maps, and employs fully connected layers for classification or regression tasks. CNNs are characterized by weight sharing and local receptive fields, which enhance their ability to automatically learn data features, thus improving prediction accuracy [58]. Specifically, input data are represented as N-dimensional vectors, and the CNN processes these through sliding window techniques to generate M×N-dimensional feature maps [59], extracting spatial features as illustrated in the first row of Figure 13.

The GRU, an advanced variant of Recurrent Neural Networks (RNNs), optimizes memory retention and mitigates gradient vanishing issues. It uses reset and update gates to regulate the flow of information, enabling the model to capture long-term dependencies more effectively, which is beneficial for sequence modeling tasks such as language processing and time-series forecasting [60,61], as depicted in the second row of Figure 13.

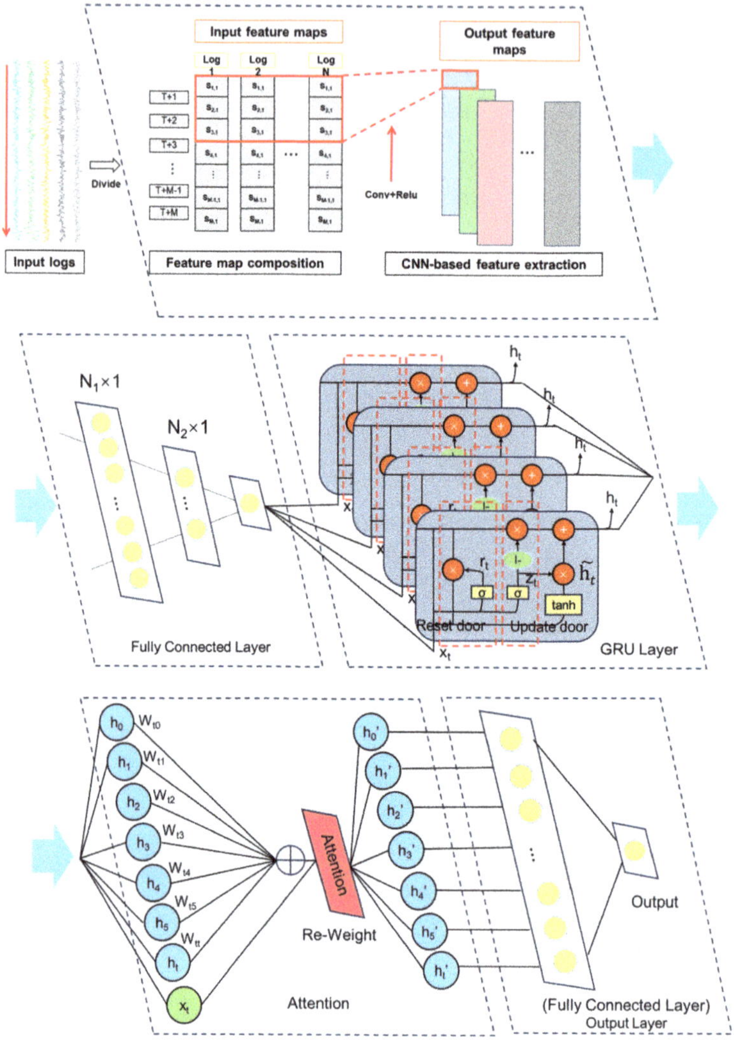

Figure 13. The principle diagram of CNN-GRU-ATT algorithm.

The Attention mechanism enhances model performance by assigning varying degrees of importance to different parts of the input sequence, improving its ability to focus on critical elements of complex data [62]. This dynamic weighting significantly boosts prediction accuracy and generalization, as shown in the third row of Figure 13.

In the figure, x_t is the input at time t; h_t is the hidden layer output at time t; z_t is the update gate; r_t is the reset gate; σ is the sigmoid function; and tanh is the activation function. $h_0, h_1, \ldots, h_t$ are the hidden layer output vectors of the GRU at each time step, and $W_{t0}, W_{t1}, \ldots, W_{tt}$ are the corresponding weight coefficients; $h'_0, h'_1, \ldots, h'_t$ are the weight coefficients of the Soft-max normalized, $\widetilde{h}_t$ are the candidate hidden states, and x_t are the query vectors. By combining CNN's localized feature extraction capabilities, GRU's sequence modeling strengths, and the dynamic weighting provided by Attention, the CNN-GRU-ATT model offers a refined and efficient approach for T_{2lm} prediction. This integrated methodology markedly advances the processing of well log data by effectively capturing both spatial and temporal dynamics. The model architecture consists of an input layer, convolutional layers, max pooling layers, Dropout layers, GRU layers, Attention layers, fully connected layers, and an output layer.

Considering that only a small number of coring boreholes have collected NMR logging data, the T_{2lm} curve cannot be directly obtained for uncored boreholes. Conventional logging data are usually collected from uncored boreholes. Therefore, for such boreholes, combined with conventional logging curves and calculated parameters as input, the constructed CNN-GRU-ATT network is used to invert the T_{2lm} curve, and then the inverted T_{2lm} curve is used to evaluate the irreducible water saturation.

4.3. The Calculation Process of Irreducible Water Saturation

For the evaluation of irreducible water saturation, this paper proposes a new method and process, as shown in Figure 14. The first step is to collect the logging curves and core data of the target reservoir in the study block, calculate and evaluate the porosity curve, and decide the next process based on whether the borehole collects NMR logging data:

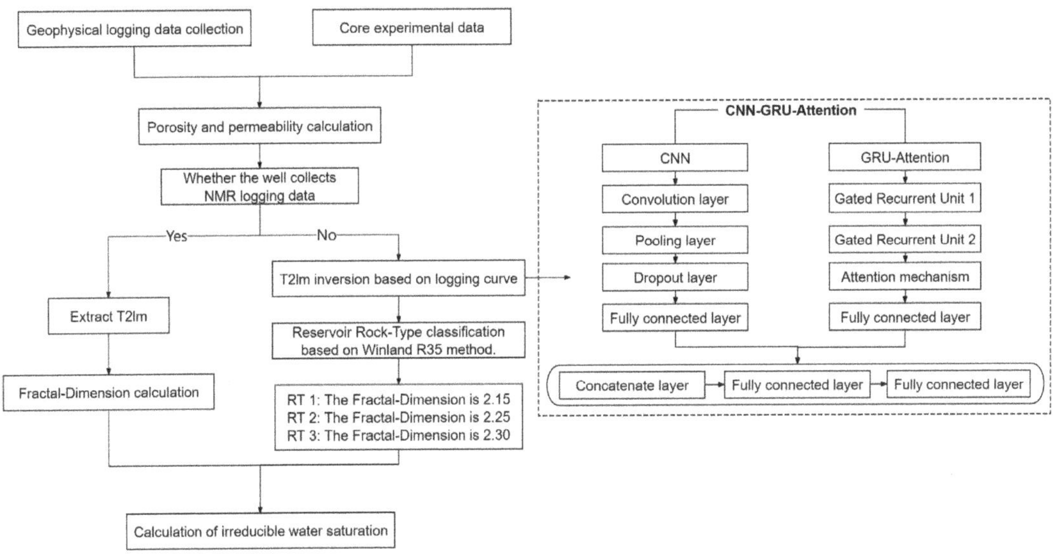

Figure 14. Evaluation process of irreducible water saturation.

1. If the borehole collects NMR logging data, extract T_{2lm} from the T_2 spectrum collected by the NMR logging instrument, calculate the fractal dimensions of each sample point, and ultimately calculate the irreducible water saturation.
2. If the borehole did not collect NMR logging data, it is necessary to perform T_{2lm} inversion based on the logging curve with the deep learning model, determine the fractal dimension for each type of reservoir using the Winland R35 method, and ultimately calculate the irreducible water saturation. This entire process considers both coring and production wells in the study area. Usually, the coring wells have a complete logging series and collect special logging series curves, while the production wells typically contain only conventional curves. The irreducible water evaluation method and process proposed in this paper take full account of this practical situation.

5. Results and Applications

5.1. The Inversion Results of T_{2lm} Curve

The inversion of T_{2lm} curves utilized a CNN-GRU–Attention neural network. A dataset comprising 7060 sets of NMR logging sampling points from the Mishrif Formation in the study area was split into training and validation sets at a 4:1 ratio. Hyperparameter tuning for the CNN-GRU–Attention network involved optimizing the width of the convolutional layer's scanning window, the number and size of CNN kernels, the quantity of GRU neurons, and the batch size, guided by established parameter selection practices. Scanning window widths of 16, 32, 64, and 128, alongside batch sizes of 32, 64, or 128, were evaluated. Training employed the Adam optimizer, with a dynamically adjusted learning rate for enhanced convergence [63]. Hyperparameters were selected through a grid search, prioritizing mean squared error (MSE, Equation (26)) loss to maximize prediction accuracy. The model architecture (see Table 3) featured 32 CNN kernels with a kernel size of 3 and a regularization rate of 0.3. The GRU layer comprised 10 to 20 hidden neurons, varying as per experimental conditions.

$$MSE = \frac{\sum_{i=1}^{N} (y_i - \hat{y}_i)^2}{N} \tag{26}$$

Table 3. The hyperparameter usage table of CNN-GRU-ATT neural network.

Layer	Parameters	Value
CNN Layer	Number of Convolutional Kernels	32
	Convolutional Kernel Size	3 × 3
	Activation Function of Convolutional Layer	ReLu
	Padding Method of Convolutional Layer	Valid
	Pooling Layer Pooling Window Size	2 × 2
	Pooling Layer Activation Function	ReLu
	Dropout Rate	0.3
GRU Layer	Number of Hidden Neurons	10
	Number of Hidden Neurons	20
Attention	Dimensionality	50
Output Layer (Fully Connected Layer)	Dense1	10
	Dense2	1

Based on the parameters in Table 3, the deep learning network inversion T_{2lm} is constructed, and the results of the training dataset and the verification dataset are displayed. Figure 15a shows the application effect of the training data and the verification data. The blue sphere is the result of the training data. With a total of 5491 sets of samples, the goodness of fit between the predicted results and the measured results reached 0.984, and 1569 sets of samples formed a verification dataset. The goodness of fit between the predicted results and the measured results reached 0.986. The samples of the two types of datasets are evenly distributed on both sides of the zero-error line. To further determine the accuracy and generalization of the T_{2lm} inversion network, a blind well in the study area was used for validation. This new well was not involved in the training and construction of the model. There were a total of 785 groups of samples in the new well, with the goodness of fit between the predicted and measured results reaching 0.973 (Figure 15b), showing no significant deviation. This demonstrates the generalizability of the T_{2lm} inversion model.

5.2. Evaluation Results and Application of Irreducible Water Saturation

After validating the general applicability of the T_{2lm} inversion model, we applied and evaluated the proposed method for assessing irreducible water saturation. Figure 16 illustrates an exemplary application well within the study area. The channels are organized as follows: the first for depth, the second for stratigraphy. The third channel logs lithological data, including the GR, KTH, SP, and CAL curves and the drill bit diameter (BS). The fourth channel records porosity (CN, DEN, DT), while the fifth displays resistivity at maximum depth (RD) and microsphere resistivity (RXO). The sixth channel presents NMR logging data, featuring T_2 spectra and the T_{2lm} curve. Porosity calculations and core sample porosity are detailed in the seventh channel. The eighth channel includes permeability calculations and core permeability. Channels nine and ten depict irreducible water saturation evaluation: one is based on the T_{2lm} curve extracted from NMR and porosity data, alongside core measurements; the other uses T_{2lm} inversion results. The eleventh channel is for explanations and conclusions.

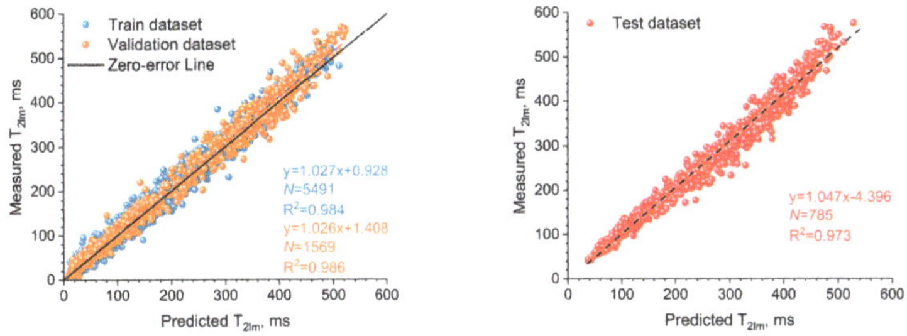

(**a**) (**b**)

Figure 15. T_{2lm} curve inversion effect diagram based on deep learning network. (**a**) The application effect of the training dataset and validation dataset; (**b**) the application effect of the new well (test dataset).

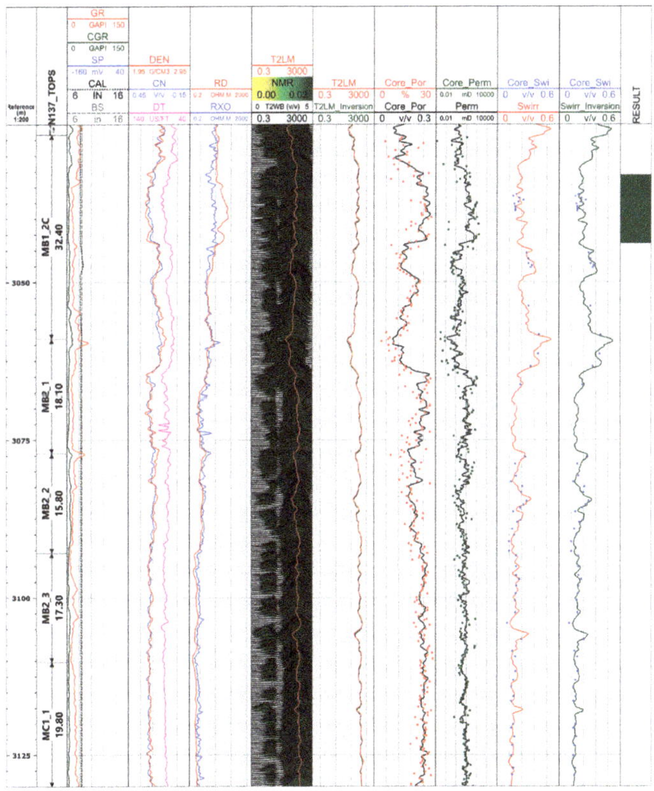

Figure 16. The evaluation effect of irreducible water saturation of the applied well (Well N137) in the block is studied.

The practical application of the proposed method demonstrates its effectiveness in calculating irreducible water saturation, aligning closely with the core results. Furthermore, to underscore the method's generalizability, two scenarios are compared: firstly, when borehole NMR data allow direct extraction of T_{2lm} for saturation evaluation, the average relative error across 64 samples is 9.55%, with a minimum and maximum of 0.10% and 41.06%, respectively. The calculated results exhibit a strong fit with core data (R^2 = 0.897). Secondly, utilizing a deep learning network for T_{2lm} curve inversion (yellow samples in Figure 17a,b) results in an average relative error of 10.92%, ranging between 0.49% and 42.19%. Despite a slightly higher error, this method remains within acceptable bounds. Comparatively, direct T_{2lm} extraction yields higher precision in irreducible water saturation evaluation, with lower relative errors evident in Figure 17b. Importantly, applying the inverted T_{2lm} curve, despite increased error, remains viable for practical applications in new wells within the study area, affirming the method's reliability and broad applicability.

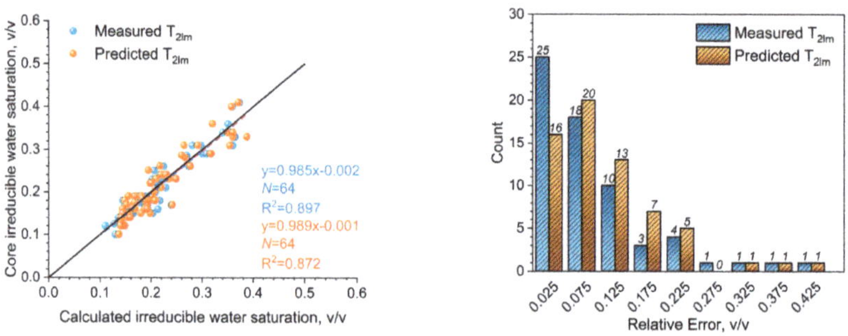

(a) (b)

Figure 17. Prediction effect of irreducible water saturation under two T_{2lm} curve acquisition methods. (**a**) The evaluation results of irreducible water saturation under two T_{2lm} curve acquisition modes and the intersection diagram of core irreducible water saturation; (**b**) the relative error distribution histogram of the irreducible water saturation prediction results under the two modes.

6. Discussion

6.1. Methods Comparison

Previous methods for predicting irreducible water saturation primarily rely on physical properties or the $T_{2cutoff}$ method combined with NMR logging data to construct the evaluation model. In this paper, the new method is compared with traditional methods, using the evaluation results of 21 sets of core samples for multi-method comparison. The mean absolute error (MAE) and mean relative error (MRE) of each method are calculated. Figure 18 shows the application effect of each method. Figure 18a shows the application result of the method proposed in this paper, with a goodness of fit between the predicted irreducible water saturation and the core irreducible water saturation reaching 0.943, an MAE of 2.37%, and an MRE of 8.46%. Figure 18b shows the application effect of the direct evaluation of irreducible water saturation using the porosity parameter [64]. Figure 18c shows the application effect of the evaluation of irreducible water saturation using the $\sqrt{\frac{K}{\varphi}}$ parameter [12]. Figure 18d shows the application effect of evaluating irreducible water saturation using the $T_{2cutoff}$ method [20,21], and Figure 18e shows the application effect of the spectral coefficient method [29]. Comparing the method proposed in this paper with the other four classical methods, the accuracy of this method is the highest. The spectral coefficient method has the second-highest accuracy. Predicting irreducible water saturation directly using porosity parameters has the lowest accuracy, and the results of predicting irreducible water using the $T_{2cutoff}$ method are significantly higher than the core irreducible water saturation. This demonstrates the higher accuracy and generalizability of the methods proposed in this paper.

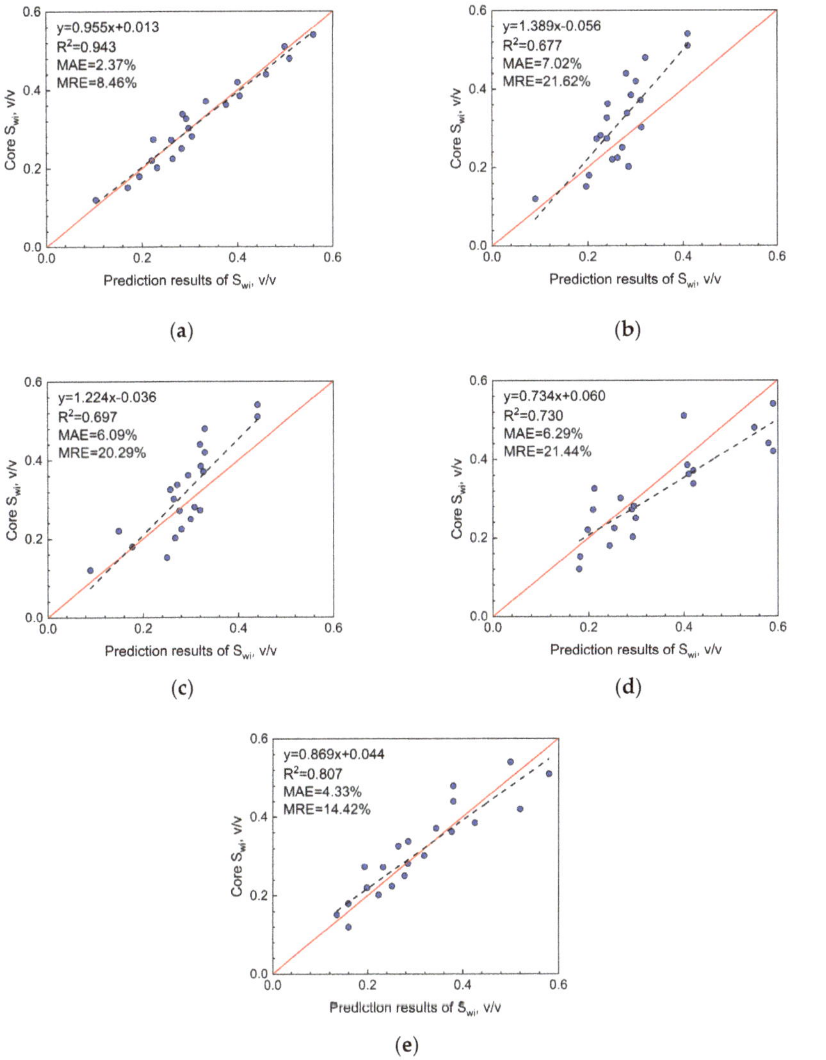

Figure 18. Comparison of the application of this method and the traditional irreducible water saturation evaluation methods. (**a**) crossplot of irreducible water saturation and core irreducible water saturation calculated by method in this paper; (**b**) evaluation results of porosity fitting method; (**c**) evaluation effect of $\sqrt{\frac{K}{\varphi}}$ parameter fitting method; (**d**) evaluation effect of $T_{2cutoff}$ method; (**e**) evaluation effect of spectral coefficient method.

The differences in the accuracy of each method are further discussed in this paper. The sorting coefficient and average pore throat radius can be considered a combination of the degree of mineral weathering, depositional environment, and transport distance. Core samples from the Mishrif Formation were used to obtain sorting coefficients and average pore throat radii from MICP data and were analyzed in conjunction with core irreducible water saturation. As the sorting coefficient changes from large to small, the irreducible water saturation increases (Figure 19a). With the sorting coefficient becoming smaller, the rock particle size tends to be more homogeneous, the specific surface area of the rock increases, and the film irreducible water increases. Additionally, a lower sorting coefficient corresponds to finer reservoir pore throats, and the fluid mobility deteriorates, increasing the capillary irreducible water (Figure 19b). Previous studies have shown that, in general, weaker sedimentary hydrodynamics result in a smaller sorting coefficient, stronger mechanical compaction and cementation, poorer pore throat connectivity and microscopic pore structures, and higher irreducible water saturation [31]. In other words, rock particles and pore radius directly affect irreducible water saturation. This explains why the previous parameters extracted based on the experimental data of mercury compaction to characterize pore throat size can be used to evaluate irreducible water. The T_{2lm} in this paper's method is also proven to be highly correlated with the mode value, as both indicate the pore throat size of the reservoir.

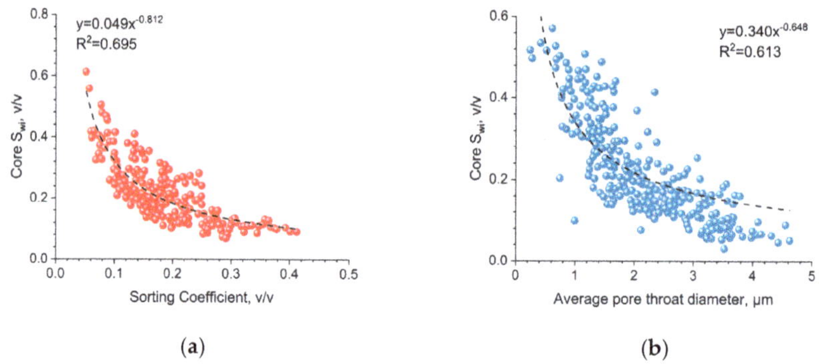

Figure 19. Correlation analysis between sorting coefficient and average pore throat diameter and irreducible water saturation. (**a**) Crossplot of sorting coefficient and core irreducible water saturation; (**b**) crossplot of average pore throat radius and core irreducible water saturation.

The size of pore throats differs significantly from porosity in geological formations. While a larger porosity generally corresponds to a larger pore throat radii in homogeneous reservoirs, the Mishrif Formation studied here exhibits non-homogeneous characteristics. Figure 20a presents two cores from the Mishrif Formation, with pore throat size distributions extracted via Hg-pressure experiments. Despite similar porosities (19.2% vs. 19.3%), core samples M-107A and M-144A display distinct pore throat peaks: 1 μm and 0.466 μm, respectively, resulting in more than a fourfold difference in permeability. This discrepancy accounts for the low goodness of fit (0.184) observed between porosity and irreducible water saturation (Figure 20b). In contrast, the correlation between pore throat parameters and irreducible water saturation is notably stronger, as evidenced by a goodness of fit of 0.613 when correlating the average pore throat radius with irreducible water saturation (Figure 19). In reservoirs exhibiting significant non-homogeneity, the pore throat radius cannot be accurately inferred from porosity alone, which contributes to the highest average relative error (21.62%) observed in direct evaluations of irreducible water using porosity. While introducing permeability parameters improves accuracy to some extent, its effectiveness remains inferior to the method proposed in this study. Moreover, the complexity involved in assessing permeability further restricts the accuracy and applicability of this approach compared to the proposed method.

The $T_{2cutoff}$ method is commonly employed for assessing irreducible water saturation using NMR logging data. Analysis of actual core samples reveals variability in $T_{2cutoff}$ values (Figure 21a), ranging from 31.6 ms to 82 ms. This variability complicates the establishment of a uniform $T_{2cutoff}$ for accurate irreducible water saturation evaluation, despite 92 ms often being used as a standard cut-off value. However, applying this cut-off frequently results in calculated irreducible water saturation values higher than those observed in core measurements. To address this discrepancy, this study integrates analysis from ground nuclear magnetic experiments. Traditionally, fluid relaxation processes in pore throats are assumed to be independent and non-interfering. Yet, recent research has uncovered a "diffusion coupling" phenomenon in carbonate reservoirs. In such formations, significant variation in pore scale leads to interconnected water content across different pore sizes. This diffusion among hydrogen-containing fluids in varying pore sizes complicates the accurate determination of relaxation time distributions [27,40]. NMR data from carbonate reservoirs in the Middle East (Figure 21b) demonstrate that after centrifugation, the long and short relaxation components of T_2 spectra do not correspond to pre-centrifugation readings. This indicates a notable "diffusion coupling" effect, where fluids in different pore throat radii intermix due to diffusion. Ramakrishnan, using the Monte Carlo random walk model, suggests that the surface relaxation rate plays a crucial role in accurately reflecting pore size distributions [65]. Experimental data from the Mishrif Formation in the study area indicate surface relaxation rates ranging from 0 to 5 μm/s, further complicating accurate reservoir irreducible water saturation assessment via $T_{2cutoff}$ predictions. Consequently, the $T_{2cutoff}$ method exhibits significant error in evaluating irreducible water saturation. Moreover, its limited applicability is exacerbated by the fact that most production wells do not gather NMR data, thereby restricting its generalizability.

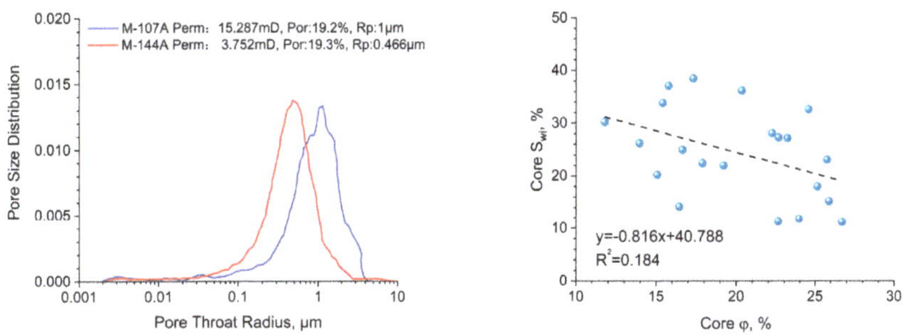

(a) (b)

Figure 20. The difference between porosity and pore throat radius in the study area. (**a**) The porosity of different samples is close, but the difference in pore throat radius leads to the difference in permeability; (**b**) the correlation between core porosity and core irreducible water saturation.

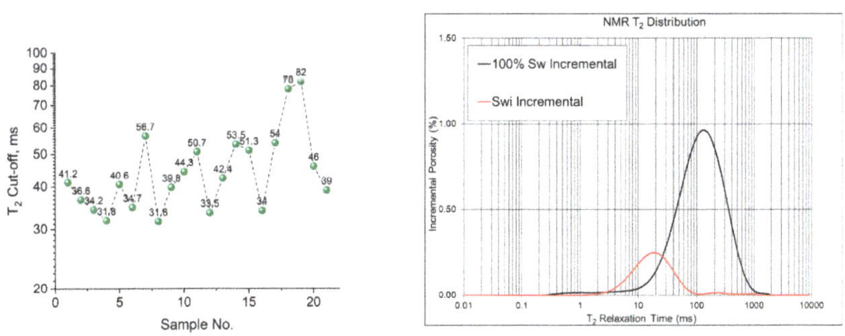

(a) (b)

Figure 21. The application effect of $T_{2cutoff}$ method is illustrated. (**a**) $T_{2cutoff}$ values of different core samples; (**b**) NMR experimental results of typical rock samples with diffusion coupling effect.

Coates [29] proposed the spectral coefficient method as an alternative to the cut-off value approach for determining irreducible water saturation in carbonate rocks. This method assumes a consistent irreducible water content regardless of pore size, although this content decreases as pore size increases. Each T_{2i} signal reflects a specific irreducible water saturation due to its contribution from irreducible water at various T_2 time points. Comparative analysis shows that this method exhibits the lowest relative error among the four evaluated methods. Figure 22, based on experimental core data, demonstrates a strong correlation ($R^2 = 0.73$) between T_{2lm} and irreducible water saturation. This correlation is further supported by this study's examination of the relationship between T_{2lm} and pore throat diameters. The significant association observed between T_{2lm} and irreducible water saturation, coupled with correlations to pore throat parameters, underscores the efficacy of the spectral coefficient method. This finding aligns with the perspective presented by Wang et al. [66], emphasizing the substantial correlation between irreducible water saturation and T_{2lm} in unconventional reservoirs, contrasting with the weaker correlation with porosity. Although Wang et al. did not specify reasons, this study provides empirical support for their observations.

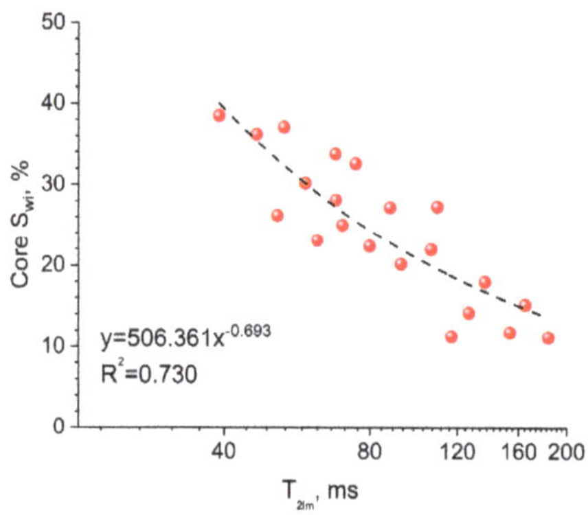

Figure 22. Correlation between T_{2lm} and irreducible water saturation of core.

In summary, the irreducible water saturation formula derived by the method in this paper considers the fractal dimension of the reservoir, porosity, and T_{2lm}, resulting in a higher accuracy. Additionally, this study combines the Thomeer Function with fractal theory and derives the specific expressions for the p_d parameter and $(B_v)_{p_c\infty}$ parameter in the Thomeer Function.

A novel irreducible water saturation model was developed by integrating the T_{2lm} from NMR fractal permeability models. Validation with ground NMR and MICP data confirmed the correlation between pore throat diameter (Mode) and T_{2lm} values. Fractal theory was applied to derive numerical relationships among Thomeer Function parameters, elucidating connections between T_{2lm} and physical properties. Leveraging these insights, a robust predictive model was created that combines traditional logging curves with advanced deep learning techniques, facilitating accurate evaluation of irreducible water saturation even in the absence of NMR logging data. This method enhances the reliability of water saturation assessments and extends analytical capabilities when direct NMR measurements are unavailable.

6.2. Limitations and Errors of the Method

Additionally, this paper discusses the limitations of the methodology, noting that the Mishrif Formation has a very low clay content and is a pure greywacke reservoir. Previous studies have shown that clay mineral content affects the magnitude of irreducible water saturation. Experimental data from clay mineral-bearing cores in the Halfaya field show a positive correlation between clay mineral content and irreducible water saturation (Figure 23). As the clay mineral content increases, the specific surface area of the reservoir rock increases, enhancing its ability to adsorb formation water. The surface of clay minerals is generally negatively charged, which can directly adsorb electrically unbalanced polar water molecules and combine to form hydrated ions, indirectly attracting water molecules and increasing film irreducible water [67]. Additionally, clay minerals fill the pore space, strengthening the binding of pore fluid by capillary force and increasing capillary irreducible water.

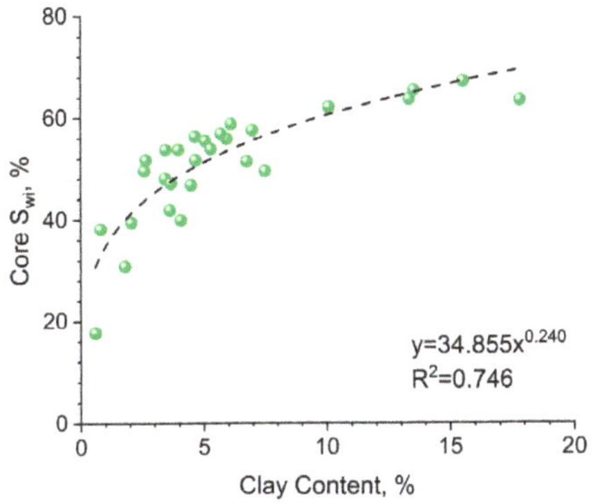

Figure 23. Correlation analysis between clay mineral content and irreducible water saturation.

Since the Mishrif Formation is almost free of clay minerals, the effect of clay mineral content is not reflected in the final form of the equation, which is a limitation of this method. When applying this method to other layer groups or reservoirs with different geological profiles, the form and process of evaluating irreducible water saturation can be followed. The effect of clay mineral content on irreducible water saturation should be considered in conjunction with actual core experimental data for clay-rich reservoirs.

Regarding the application error of this method, although the evaluation process of irreducible water saturation proposed in this paper accounts for the lack of NMR data from production wells, it cannot avoid the generation of carry-over error after inverting the T_{2lm} curves using the deep learning method and then evaluating the irreducible water saturation. This includes the evaluation error of porosity, which cannot be eliminated by constructing a separate error model. This issue is discussed in Section 5.2 of this paper. The error results discussed in Section 5.2 of this paper show that using the inverted T_{2lm} curve to evaluate irreducible water saturation increases the relative error. However, the overall evaluation accuracy can still provide effective data support for the formulation of subsequent production and development programs and the selection of perforated intervals.

7. Conclusions

This study introduces a novel method for predicting and evaluating irreducible water saturation by integrating actual core experimental data and practical applications, leading to the following conclusions:

(1) The nuclear magnetic fractal permeability model effectively predicts irreducible water saturation.
(2) A distinct numerical relationship exists between the modal parameter of pore throat diameter based on the Thomeer Function and T_{2lm}. Utilizing fractal theory, this relationship clarifies the connection between each parameter and the physical property parameter in the Thomeer Function, thereby establishing the numerical association between T_{2lm} and the Thomeer Function.
(3) Employing deep learning, conventional logging curves, and derived parameter curves enables the effective inversion of the T_{2lm} curve for evaluating irreducible water saturation. While the inverted T_{2lm} curve exhibits a slightly higher relative error compared to that extracted from NMR logging data, it remains within an acceptable range, particularly valuable in the absence of NMR logging data. This method proves effective and feasible.

In summary, this paper presents a broadly applicable method for assessing irreducible water saturation. In wells lacking specialized logging data, a T_{2lm} curve can be constructed using conventional curves and a deep learning model, successfully applied in the Mishrif Formation within the Halfaya field, Middle East, yielding favorable outcomes.

Author Contributions: Conceptualization, J.G. and Q.Z.; Methodology, J.G., X.N., Q.Z. and H.L.; Software, J.G. and H.L.; Validation, J.G.; Formal analysis, J.G. and Z.Z.; Resources, X.N.; Data curation, J.G.; Writing – original draft, J.G.; Writing – review & editing, J.G., Z.Z. and X.N.; Visualization, Z.Z. and X.N.; Supervision, Z.Z. and H.L.; Project administration, Z.Z.; Funding acquisition, J.G. and Z.Z. All authors have read and agreed to the published version of the manuscript.

Funding: This work was financially sponsored by the Open Fund of Key Laboratory of Exploration Technologies for Oil and Gas Resources, Ministry of Education, Wuhan, China (No. K2023-02) and the Hainan Provincial Natural Science Foundation of China (No. 421QN281).

Data Availability Statement: All our data have been displayed in the pictures in the paper.

Acknowledgments: The authors would like to express their most sincere gratitude to the field workers in the Halfaya oilfield. We sincerely thank the anonymous reviewers for their valuable comments and suggestions, which led to significant improvements of the manuscript.

Conflicts of Interest: Ms. Qing Zhao was employed by the company Changqing Oilfield. The remaining authors declare that the research was conducted in the absence of any commercial or financial relationships that could be construed as a potential conflict of interest.

Appendix A

In Section 4.1.2, according to Equations (6), (7) and (9), the following can be derived:

$$S_v = \left(\frac{p_c}{p_{\min}}\right)^{D-3} \tag{A1}$$

In Equation (A1), $p_{\min}$ is the capillary pressure corresponding to the maximum pore size $r_{\max}$ in the reservoir, i.e., the inlet capillary pressure; and S_v is the wetting phase saturation in the reservoir at the capillary pressure p_c.

Assuming that the porous medium consists of a capillary bundle (multiple capillary tubes), and using Poiseuille's law [68], the flow rate of a fluid with viscosity μ under a pressure difference ΔP through a single capillary tube i with radius r_i and length L is q_i. The volume of the capillary tube $V_i = \pi r_i^2 L$, when considering the flow rate of the entire capillary bundle $q = \sum\limits_i q_i$, can be written as

$$q = \frac{\Delta P}{8\mu L^2}\sum V_i r_i^2 \tag{A2}$$

Substituting the relational equations for the volume V_i of capillary i, the length L, the fraction S_i of the total pore volume accounted for by capillary i, the porosity of the porous medium, and the cross-sectional area A into Equation (A2), one obtains

$$q = \frac{\phi A \Delta P}{8\mu L}\sum S_i r_i^2 \tag{A3}$$

Assuming that the flow of fluid in a porous medium obeys Darcy's law, we have

$$q = \frac{KA\Delta P}{\mu L'} \tag{A4}$$

In Equation (A4), K is the permeability coefficient and L' is the length of the porous medium.

By associating Equation (A3) with Equation (A4), we obtain

$$K = \frac{f\varphi}{8}\sum S_i r_i^2 \tag{A5}$$

In Equation (A5), $f = \frac{L'}{L}$, is the curvature of the capillary in the porous medium. Since the pore distribution is continuous, Equation (A5) can be written in integral form:

$$K = \frac{f\varphi}{8}\int_0^1 r^2 dS \tag{A6}$$

Substituting Equation (8) into Equation (A6) and integrating and shifting yields

$$r_{\max} = \sqrt{\frac{8K}{f\varphi}\frac{5-D}{3-D}} \tag{A7}$$

Since the discharge pressure P_d corresponds to the pressure when the nonwetting phase enters the largest pore $r_{\max}$ in the pore structure, combining Equation (10) with Equation (A7), we obtain

$$p_d = \frac{\sigma\cos\theta}{\sqrt{\frac{2K}{f\varphi}\frac{5-D}{3-D}}} \tag{A8}$$

Since $(B_v)_{P_{c\infty}}$ is the inlet volume of mercury at infinite pressure, therefore

$$(B_v)_{P_{c\infty}} = \varphi S_{Hg_{\max}} \tag{A9}$$

In Equation (A9), $S_{Hg_{\max}}$ is the maximum mercury feed volume.
By combining Equations (10), (A1), (A8) and (A9), we obtain

$$(B_v)_{P_{c\infty}} = \varphi\left\{1 - \lim_{r\to 0}\left[\frac{\sqrt{(8k/f\varphi)(5-D/3-D)}}{r}\right]^{(D-3)}\right\} \tag{A10}$$

References

1. Su, Y.; Fu, J.; Li, L.; Wang, W.; Zafar, A.; Zhang, M.; Ouyang, W. A new model for predicting irreducible water saturation in tight gas reservoirs. *Petrol. Sci.* **2020**, *17*, 1087–1100. [CrossRef]
2. Gao, H.; Li, H.A. Pore structure characterization, permeability evaluation and enhanced gas recovery techniques of tight gas sandstones. *J. Nat. Gas Sci. Eng.* **2016**, *28*, 536–547. [CrossRef]
3. Liu, D.; Yue, X.; Wang, L.; Hou, J.; Zhang, J. Model study on the irreducible water saturation by centrifuge experiments. *J. Petrol. Sci. Eng.* **2006**, *53*, 77–82. [CrossRef]
4. Zhou, X.; Zhang, C.; Zhang, Z.; Zhang, R.; Zhu, L.; Zhang, C. A saturation evaluation method in tight gas sandstones based on diagenetic facies. *Mar. Pet. Geol.* **2019**, *107*, 310–325. [CrossRef]
5. Zhu, L.; Ma, Y.; Cai, J.; Zhang, C.; Wu, S.; Zhou, X. Key factors of marine shale conductivity in southern China—Part II: The influence of pore system and the development direction of shale gas saturation models. *J. Petrol. Sci. Eng.* **2022**, *209*, 109516. [CrossRef]
6. Neog, D. Effect of temperature on the pH of water flood effluents and irreducible water saturation: A study with reference to the Barail sandstone outcrop of the upper Assam Basin. *J. Pet. Explor. Prod. Te.* **2022**, *12*, 1–17. [CrossRef]
7. Lai, F.; Li, Z.; Zhang, W.; Dong, H.; Kong, F.; Jiang, F. Investigation of pore characteristics and irreducible water saturation of tight reservoir using experimental and theoretical methods. *Energy. Fuels.* **2018**, *32*, 3368–3379. [CrossRef]
8. Dong, X.; Sun, J.; Li, J.; Gao, H.; Liu, X.; Wang, J. Experimental research of gas shale electrical properties by NMR and the combination of imbibition and drainage. *J. Geophys. Eng.* **2015**, *12*, 610–619. [CrossRef]
9. Liu, J.; Xie, R.; Guo, J.; Wei, H.; Fan, W.; Jin, G. Novel Method for Determining Irreducible Water Saturation in a Tight Sandstone Reservoir Based on the Nuclear Magnetic Resonance T_2 Distribution. *Energy Fuels* **2022**, *36*, 11979–11990. [CrossRef]
10. Sun, J.; Wang, K.; Li, W. Development and analysis of logging saturation interpretation models. *Pet. Explor. Dev.* **2008**, *35*, 107–113.
11. Fu, X.; Qin, Y.; Wang, G.G.X.; Rudolph, V. Evaluation of coal structure and permeability with the aid of geophysical logging technology. *Fuel* **2009**, *88*, 2278–2285. [CrossRef]
12. Morris, R.L.; Biggs, W.P. Using log-derived values of water saturation and porosity. In Proceedings of the SPWLA Annual Logging Symposium, Denver, Colorado, 12 June 1967; p. SPWLA-1967-X.

13. Amaefule, J.O.; Altunbay, M.; Tiab, D.; Kersey, D.G.; Keelan, D.K. Enhanced reservoir description: Using core and log data to identify hydraulic (flow) units and predict permeability in uncored intervals/wells. In Proceedings of the SPE Annual Technical Conference and Exhibition, Houston, TX, USA, 3 October 1993; p. SPE-26436-MS.
14. Arifianto, I.; Surjono, S.S.; Erlangga, G.; Abrar, B.; Yogapurana, E. Application of flow zone indicator and Leverett J-function to characterise carbonate reservoir and calculate precise water saturation in the Kujung formation, North East Java Basin. *J. Geophys. Eng.* **2018**, *15*, 1753–1766. [CrossRef]
15. Lala, A.M.S.; El-Sayed, N.A.E.A. The application of petrophysics to resolve fluid flow units and reservoir quality in the Upper Cretaceous Formations: Abu Sennan oil field, Egypt. *J. Afr. Earth Sci.* **2015**, *102*, 61–69. [CrossRef]
16. Mirzaei-Paiaman, A.; Ostadhassan, M.; Rezaee, R.; Saboorian-Jooybari, H.; Chen, Z. A new approach in petrophysical rock typing. *J. Petrol. Sci. Eng.* **2018**, *166*, 445–464. [CrossRef]
17. Gao, C.; Yuan, Y.; Wu, H.; Li, H.; Hu, X. Research on the log evaluation method of irreducible water saturation in Yinggehai Basin. *Nat. Gas. Ind.* **2003**, *5*, 38–40+142.
18. Taktak, F.; Rigane, A.; Boufares, T.; Kharbachi, S.; Bouaziz, S. Modelling approaches for the estimation of irreducible water saturation and heterogeneities of the commercial Ashtart reservoir from the Gulf of Gabès, Tunisia. *J. Petrol. Sci. Eng.* **2011**, *78*, 376–383. [CrossRef]
19. Biswal, D.; Prasad, K.; Nedeer, N.; Lerdsuwankij, T.; Singh, K.H. Petrophysical modeling in the absence of core data in a small, marginal field in the Mumbai Offshore Basin, India. *Interpretation* **2021**, *9*, 1025–1038. [CrossRef]
20. Lyu, C.; Ning, Z.; Cole, D.R.; Wang, Q.; Chen, M. Experimental investigation on T2 cutoffs of tight sandstones: Comparisons between outcrop and reservoir cores. *J. Petrol. Sci. Eng.* **2020**, *191*, 107184. [CrossRef]
21. Xiao, L.; Li, J.; Mao, Z.; Lu, J.; Yu, H.; Guo, H.; Li, G. A method to determine nuclear magnetic resonance (NMR) T2cutoff based on normal distribution simulation in tight sandstone reservoirs. *Fuel* **2018**, *225*, 472–482. [CrossRef]
22. Hossain, Z.; Grattoni, C.A.; Solymar, M.; Fabricius, I.L. Petrophysical properties of greensand as predicted from NMR measurements. *Petrol. Geosci.* **2011**, *17*, 111–125. [CrossRef]
23. Straley, C.; Matteson, A.; Feng, S.; Schwartz, L.M.; Kenyon, W.E.; Banavar, J.R. Magnetic resonance, digital image analysis, and permeability of porous media. *Appl. Phys. Lett.* **1987**, *51*, 1146–1148. [CrossRef]
24. Zhao, Y.; Lin, B.; Liu, T.; Zheng, Y.; Sun, Y.; Zhang, G.; Li, Q. Multifractal analysis of coal pore structure based on NMR experiment: A new method for predicting T2 cutoff value. *Fuel* **2021**, *283*, 119338. [CrossRef]
25. Zhang, W.; Wan, J.; Wu, Y.; Qu, Y. Prediction of nuclear magnetic resonance (NMR) T2 cutoff based on multivariate statistical method. *Nucl. Electron. Detect. Technol.* **2020**, *40*, 193–199.
26. Wu, F.; Sima, L.; Yang, H.; Ling, H.; Tian, C. Research on NMR T_2 cutoff of complex lithology in the west Qaidam Basin. *Well. Log. Technol.* **2014**, *38*, 144–149.
27. Wang, Y.; Tang, H.; Zheng, M.; Tian, G. The limitation of NMR to characterize the pore structure and irreducible water saturation of carbonate rocks: An experimental analysis. *Mar. Orig. Pet. Geol.* **2018**, *23*, 89–96.
28. Kenyon, W.E.; Howard, J.J.; Sezginer, A.; Straley, C.; Matteson, A.; Horkowitz, K.; Ehrlich, R. Pore-size distribution and NMR in microporous cherty sandstones. In Proceedings of the SPWLA Annual Logging Symposium, Denver, CO, USA, 11 June 1989; p. SPWLA-1989-LL.
29. Coates, G.R.; Galford, J.; Mardon, D.; Marschall, D. A new characterization of bulk-volume irreducible using magnetic resonance. *Log. Anal.* **1998**, *39*, 51–63.
30. Zhu, L.; Zhang, C.; Shi, W.; Chen, Y.; Liu, J.; Guo, C.; Cheng, Y. Study on the method of prediction of irreducible water saturation by combining mercury intrusion and NMR logging data. *Sci. Technol. Eng.* **2016**, *16*, 22–29.
31. Li, Y.; Zhang, Z.; Hu, S.; Zhou, X.; Guo, J.; Zhu, L. Evaluation of irreducible water saturation by electrical imaging logging based on capillary pressure approximation theory. *Geoenergy Sci. Eng.* **2023**, *224*, 211592. [CrossRef]
32. Zhu, L.; Zhang, C.; Zhang, C.; Zhang, Z.; Nie, X.; Zhou, X.; Liu, W.; Wang, X. Forming a new small sample deep learning model to predict total organic carbon content by combining unsupervised learning with semisupervised learning. *Appl. Soft Comput.* **2019**, *83*, 105596. [CrossRef]
33. Gao, J.; Tian, C.; Zhang, W.; Song, X.; Liu, B. Characteristics and genesis of carbonate reservoir of the Mishrif Formation in the Rumaila oil field, Iraq. *Acta Pet. Sin.* **2013**, *34*, 843–852.
34. Zhang, M.; Qiao, Z.; Gao, J.; Zhu, Y.; Sun, W. Characteristics and evaluation of carbonate reservoirs in restricted platform in the MB1-2 Sub-Member of Mishrif formation, Halfaya oilfield, Iraq. *J. Northeast Pet. Univ.* **2020**, *44*, 35–45.
35. Alsharhan, A.; Nairn, A. A review of the Cretaceous formations in the Arabian Peninsula and Gulf: Part I. Lower Cretaceous (Thamama Group) stratigraphy and paleogeography. *J. Pet. Geol.* **1986**, *9*, 365–391. [CrossRef]
36. Alsharhan, A.; Nairn, A. A review of the ctretaceous formations in the Arabian Peninsula and gulf: Part II. Mid-cretaceous (Wasia Group) stratigraphy and paleogeography. *J. Pet. Geol.* **1988**, *11*, 89–112. [CrossRef]
37. Sadooni, F.; Alsharhan, A. Stratigraphy, microfacies, and petroleum potential of the Mauddud Formation (Albian-Cenomanian) in the Arabian Gulf basin. *AAPG Bull.* **2003**, *87*, 1653–1680. [CrossRef]
38. Timur, A. An investigation of permeability, porosity, and residual water saturation relationships. In Proceedings of the SPWLA Annual Logging Symposium, New Orleans, LA, USA, 23 June 1968; p. SPWLA-1968-J.
39. Coates, G.R.; Xiao, L.; Prammer, M.G. NMR logging. In *Principles and Interpretation*; Halliburton Energy Service: Huston, TX, USA, 1999.

40. Kenyon, W.E.; Day, P.I.; Straley, C.; Willemsen, J.F. A three-part study of NMR longitudinal relaxation properties of water-saturated sandstones. *SPE Form. Eval.* **1988**, *3*, 622–636. [CrossRef]
41. Morriss, C.E.; Maclnnis, J.; Freedman, R.; Smaardyk, J.; Straley, C.; Kenyon, W.E.; Vinegar, H.J.; Tutunjian, P.N. Field test of an experimental pulsed nuclear magnetism tool. In Proceedings of the SPWLA Annual Logging Symposium, Calgary, AB, Canada, 13 June 1993; p. SPWLA-1993-GGG.
42. Wu, F.; Li, Y.; Burnham, B.; Zhang, Z.; Yao, C.; Yuan, L.; Zhang, F.; Deng, H.; Xi, Y.; He, J. Fractal-based NMR permeability estimation in tight sandstone: A case study of the Jurassic rocks in the Sichuan Basin, China. *J. Petrol. Sci. Eng.* **2022**, *218*, 110940. [CrossRef]
43. Mandelbrot, B.B.; Passoja, D.E.; Paullay, A.J. Fractal Character of Fracture Surfaces of Metals. *Nature* **1984**, *308*, 721–722. [CrossRef]
44. Brooks, R.H.; Corey, A.T. *Hydraulic Properties of Porous Media*; Hydrology Paper; Colorado State University: Fort Collins, CO, USA, 1964; pp. 3–27.
45. He, T.; Zhou, Y.; Chen, Z.; Zhang, Z.; Xie, H.; Shang, Y.; Cui, G. Fractal Characterization of the Pore-Throat Structure in Tight Sandstone Based on Low-Temperature Nitrogen Gas Adsorption and High-Pressure Mercury Injection. *Fractal Fract.* **2024**, *8*, 356. [CrossRef]
46. Geological Applications of Capillary Pressure: A Review1 | AAPG Bulletin | GeoScienceWorld. Available online: https://pubs.geoscienceworld.org/aapgbull/article/76/6/840/38801/Geological-Applications-of-Capillary-Pressure-A (accessed on 1 June 1992).
47. Lai, J.; Wang, G. Fractal Analysis of Tight Gas Sandstones Using High-Pressure Mercury Intrusion Techniques. *J. Nat. Gas Sci. Eng.* **2015**, *24*, 185–196. [CrossRef]
48. Dullien, F.A.L. *Porous Media: Fluid Transport and Pore Structure*; Academic Press, Department of Chemical Engineering, University of Waterloo: Waterloo, ON, Canada, 2012.
49. Han, Y.; Zhang, C.; Zhang, Z.; Zhang, H.; Chen, L. Pore structure classification and logging evaluation method for carbonate reservoirs: A case study from an oilfield in the Middle East. *Energy Source Part A* **2019**, *41*, 1701–1715. [CrossRef]
50. Cheng, Y.; Zhang, C.; Zhu, L. A fractal irreducible water saturation model for capillary tubes and its application in tight gas reservoir. *J. Petrol. Sci. Eng.* **2017**, *159*, 731–739. [CrossRef]
51. Zhang, C.; Cheng, Y.; Zhang, C. An improved method for predicting permeability by combining electrical measurements and mercury injection capillary pressure data. *J. Geophys. Eng.* **2017**, *14*, 132–142. [CrossRef]
52. Tang, X.; Zhou, W.; Yang, W.; Zhang, C.; Zhang, C. An improved method in petrophysical rock typing based on mercury-injection capillary pressure data. *Energy Source Part A* **2020**, *2020*, 1–16. [CrossRef]
53. Martin, A.J.; Solomon, S.T.; Hartmann, D.J. Characterization of petrophysical flow units in carbonate reservoirs. *AAPG Bull.* **1997**, *81*, 734–759.
54. Thomeer, J.H.M. Introduction of a pore geometrical factor defined by the capillary pressure curve. *J. Pet. Technol.* **1960**, *12*, 73–77. [CrossRef]
55. Chen, X.; Zhang, C.; Zhu, L.; Zhang, C.; Zhang, Z.; Guo, J. Evaluation of the saturation of carbonate reservoirs by combining the nuclear magnetic resonance logging and the Thomeer model. *Geophys. Geochem. Explor.* **2023**, *47*, 110–119.
56. Chen, L.; Jiang, Z.; Liu, Q.; Jiang, S.; Liu, K.; Tan, J.; Gao, F. Mechanism of shale gas occurrence: Insights from comparative study on pore structures of marine and lacustrine shales. *Mar. Pet. Geol.* **2019**, *104*, 200–216. [CrossRef]
57. Mei, P.; Li, M.; Zhang, Q.; Li, G.; Song, L. Prediction model of drinking water source quality with potential industrial-agricultural pollution based on CNN-GRU-Attention. *J. Hydrol.* **2022**, *610*, 127934. [CrossRef]
58. Wang, Y.; Sun, Y.; Liu, Z.; Sarma, S.E.; Bronstein, M.M.; Solomon, J.M. Dynamic graph cnn for learning on point clouds. *ACM T Graphic.* **2019**, *38*, 1–12. [CrossRef]
59. Das, V.; Pollack, A.; Wollner, U.; Mukerji, T. Convolutional neural network for seismic impedance inversion. *Geophysics* **2019**, *84*, 869–880. [CrossRef]
60. Jiao, R.; Huang, X.; Ma, X.; Han, L.; Tian, W. A model combining stacked auto encoder and back propagation algorithm for short-term wind power forecasting. *IEEE Access.* **2018**, *6*, 17851–17858. [CrossRef]
61. Chen, T.; Gao, G.; Liu, H.; Li, Y.; Gui, Z.; Yu, Z.; Zhai, X. Rock brittleness index inversion method with constraints of seismic and well logs via a CNN-GRU fusion network based on the spatiotemporal attention mechanism. *Geoenergy Sci. Eng.* **2023**, *225*, 211646. [CrossRef]
62. Li, N.; Xu, B.; Wu, H.; Feng, Z.; Li, Y.; Wang, K.; Liu, P. Application status and prospects of artificial intelligence in well logging and formation evaluation. *Acta Pet. Sin.* **2021**, *42*, 508.
63. Sabour, S.; Frosst, N.; Hinton, G.E. Dynamic routing between capsules. *Adv. Neural Inf. Process. Syst.* **2017**, *30*, 1–11.
64. Buckles, R.S. Correlating and averaging connate water saturation data. *J. Can. Petrol. Technol.* **1965**, *4*, 42–52. [CrossRef]
65. Fordham, E.J.; Kenyon, W.E.; Ramakrishnan, T.S.; Schwartz, L.M.; Wilkinson, D.J. Forward models for nuclear magnetic resonance in carbonate rocks. *Log Anal.* **1999**, *40*, SPWLA-1999-v40n4a1.
66. Wang, H.; Kou, Z.; Bagdonas, D.A.; Phillips, E.H.W.; Alvarado, V.; Johnson, A.C.; Jiao, Z.; McLaughlin, J.F.; Quillinan, S.A. Multiscale petrophysical characterization and flow unit classification of the Minnelusa eolian sandstones. *J. Hydrol.* **2022**, *607*, 127466. [CrossRef]

67. Bai, Z.; Tan, M.; Li, G.; Shi, Y. Analysis of low-resistivity oil pay and fluid typing method of Chang 81 Member, Yanchang Formation in Huanxian area, Ordos Basin, China. *J. Petrol. Sci. Eng.* **2019**, *175*, 1099–1111. [CrossRef]
68. Ren, X.; Zhao, Y.; Deng, Q.; Kang, J.; Li, D.; Wang, D. A relation of hydraulic conductivity—Void ratio for soils based on Kozeny-Carman equation. *Eng. Geol.* **2016**, *213*, 89–97. [CrossRef]

fractal and fractional

Article

Factors Controlling Differences in Morphology and Fractal Characteristics of Organic Pores of Longmaxi Shale in Southern Sichuan Basin, China

Yuanlin Wang [1,2], Denglin Han [1,2,*], Wei Lin [3], Yunqian Jia [4], Jizhen Zhang [2,5], Chenchen Wang [2,6] and Binyu Ma [1,2]

1 School of Geosciences, Yangtze University, Wuhan 430100, China; wangyl.st@yangtzeu.edu.cn (Y.W.); mabinyu2011@163.com (B.M.)
2 Laboratory of Reservoir Microstructure Evolution and Digital Characterization, Wuhan 430100, China; wcc1220@163.com (C.W.)
3 Institute of Geology and Paleontology, Linyi University, Linyi 276000, China; ucaslinwei@126.com
4 School of Earth Sciences and Engineering, Nanjing University, Nanjing 210023, China; yqjia@smail.nju.edu.cn
5 School of Resources and Environment, Yangtze University, Wuhan 430100, China
6 Cooperative Innovation Center of Unconventional Oil and Gas, Yangtze University, Wuhan 430100, China
* Correspondence: handl@yangtzeu.edu.cn

Abstract: Shale gas is a prospective cleaner energy resource and the exploration and development of shale gas has made breakthroughs in many countries. Structure deformation is one of the main controlling factors of shale gas accumulation and enrichment in complex tectonic areas in southern China. In order to estimate the shale gas capacity of structurally deformed shale reservoirs, it is necessary to understand the systematic evolution of organic pores in the process of structural deformation. In particular, as the main storage space of high-over-mature marine shale reservoirs, the organic matter pore system directly affects the occurrence and migration of shale gas; however, there is a lack of systematic research on the fractal characteristics and deformation mechanism of organic pores under the background of different tectonic stresses. Therefore, to clarify the above issues, modular automated processing system (MAPS) scanning, low-pressure gas adsorption, quantitative evaluation of minerals by scanning (QEMSCAN), and focused ion beam scanning electron microscopy (FIB-SEM) were performed and interpreted with fractal and morphology analyses to investigate the deformation mechanisms and structure of organic pores from different tectonic units in Silurian Longmaxi shale. Results showed that in stress concentration areas such as around veins or high-angle fractures, the organic pore length-width ratio and the fractal dimension are higher, indicating that the pore is more obviously modified by stress. Under different tectonic backgrounds, the shale reservoir in Weiyuan suffered severe denudation and stronger tectonic compression during burial, which means that the organic pores are dominated by long strip pores and slit-shaped pores with high fractal dimension, while the pressure coefficient in Luzhou is high and the structural compression is weak, resulting in suborbicular pores and ink bottle pores with low fractal dimension. The porosity and permeability of different forms of organic pores are also obviously different; the connectivity of honeycomb pores with the smallest fractal dimension is the worst, that of suborbicular organic pores is medium, and that of long strip organic pores with the highest fractal dimension is the best. This study provides more mechanism discussion and case analysis for the microscopic heterogeneity of organic pores in shale reservoirs and also provides a new analysis perspective for the mechanism of shale gas productivity differences in different stress–strain environments.

Keywords: Sichuan Basin; Longmaxi Formation; unconventional petroleum resources; shale gas reservoir; pore structure; fractal dimension; tectonic compression

1. Introduction

In recent years, with the success of the "shale gas revolution" in North America and the continuous growth of global energy demand, the exploration and development of shale gas resources have attracted increasing attention all over the world [1–3]. Shale gas is a large-scale, high-potential but difficult-to-exploit unconventional oil and gas system [4]. Organic matter in shale is not only the source of natural gas but also the storage space [5,6], of which, the widely developed organic pores make significant contributions to the shale reservoir space [7]. Well-developed organic pores can form interconnected pore networks that control shale gas content and methane adsorption capacity [8,9]. Therefore, the study of organic pores is of great significance for reservoir quality evaluation and shale gas resource exploration.

The sizes, morphologies, structures, and types of organic pores greatly influence the reservoir capacity and distribution pattern of shale gas [10]. When characterizing the complex pore network of shale reservoirs, FE-SEM imaging technology can qualitatively observe two-dimensional shale pores and quantitatively extract pore structure parameters such as surface porosity and pore diameter [11], while low-pressure N_2 /CO_2 adsorption experiments can quantitatively characterize pore size distribution, pore volume, and specific surface area [12]. These pore structure parameters can characterize the content estimation, transport dynamics, and occurrence state of shale gas [13]. Furthermore, SEM images combined with successive FIB milling of the rock surface can also reconstruct the three-dimensional digital pore network model [14], which can be used to quantify the pore throat distribution and coordination number to evaluate the connectivity and permeability of the shale pore system [15,16].

In view of the complexity and non-homogeneity of pores in shale reservoirs, fractal characteristics have been introduced as a new method for studying shale pore structures that cannot be addressed using Euclidean geometry theory [17–19]. Fractal dimension (D) can be used to quantitatively evaluate pore surface roughness and structural irregularity, and in general, a large fractal dimension value represents a more complex pore structure [20–22]. The fractal characteristics of shale are currently studied using image methods and low-pressure nitrogen adsorption, focusing on linking fractal characteristics to pore structure and rock components [23,24]. The fractal dimension is usually higher for smaller pore sizes and larger porosities, while the effect of mineral or TOC content remains inconclusive, even resulting in completely opposing conclusions in various studies [25–27]. Therefore, it is particularly important to identify the factors influencing pore fractal dimension in the geological background of the study area, which is also one of the focuses of this study.

Organic pore development is restricted by multiple factors such as TOC, organic matter types, maturity, mineral composition and rigid frame, pore fluid, and pressure [28–31], and its formation and evolution mechanism is very complex. The essence of organic pore formation is kerogen decomposition during hydrocarbon generation, but this is non-linear [32]. In the over-maturity stage (the R_o = 2.0–3.5%), the number of micropores increases and micropores will connect with each other to form meso-macropores with larger pore sizes as the thermal maturity increases [33,34]. The thermal evolution is also accompanied by a change in the overlying formation pressure and fluid properties [35,36]. Furthermore, dissolution will provide charging volumes for subsequent bitumen generation [37,38], and mineral recrystallization may also enhance the brittleness of shale, thereby protecting the pores from compression [39,40]. In recent years, there has been abundant progress in identifying the controlling mechanisms of organic pore generation and evolution such as sedimentary environment, mechanical compaction, and burial evolution [37,39–41], while work on identifying the restriction mechanism of stress, especially tectonic deformation, on pore development started relatively late. The influence of tectonic activities on the preservation conditions of shale gas is considered to be the key factor controlling the enrichment and high yield of shale gas [42,43]. Scholars have systematically investigated the pore structures and gas adsorption/desorption properties of shale reservoirs in tectonically complex areas and tried to reveal the influence of tectonic deformation in gas shales on the evolution of pore structures [44–46]. For example, the deformation strength gradually weakens from the center of the anticline to the two wings, so the pore characteristics are different in these areas [47]. From the perspective of pore-fracture configuration, the presence of organic pores in contact with the fractures will lead to formation pressure leakage, resulting in smaller and deformed pores [48]. In addition, the influence mechanisms of ductile folding and brittle sliding on pore structure are also different [49,50]. However, there are still insufficient studies on the morphological characteristics and distribution of organic pores under different tectonic evolution histories and tectonic stress backgrounds.

The southern Sichuan Basin has experienced complex multi-stage tectonic movements such as Caledonian, Hercynian, Indo-Chinese, and Himalayan, and the differences in shale reservoirs in the Basin are closely related to the tectonic deformation mechanism [51]. Therefore, this study focused on a typical Marine shale reservoir of the Silurian Longmaxi Formation in southern China. The Weiyuan block in the southwestern Sichuan depression and the Luzhou block in the southern Sichuan fold belt were selected because of their similar sedimentary environments, composition types, and organic matter abundance. Scanning electron microscopy (SEM) and focused ion beam scanning electron microscopy (FIB-SEM) were used to visually observe the morphology, size, and distribution of each pore, while liquid injection was used to quantitatively obtain pore size distribution, pore volume, and specific surface. Through analysis of the above pore characteristics and parameters combined with knowledge of the tectonic background of the late Mesozoic-Cenozoic differential extrusion, the genetic mechanism linked with storage and the permeability properties of organic pores with different morphologies were revealed, which provides a theoretical basis and case analysis for constructing the microscopic pore structure characteristics of organic-rich shale under complex tectonic backgrounds.

2. Geological Setting

The Sichuan Basin is located in the northwest portion of the Yangtze Plate and has experienced numerous complex tectonic movements including Caledonian, Hercynian, Indosinian, Yanshanian, and Himalayan [52] and undergone two complete geological evolution cycles of weak tensioning-weak compression-weak tensioning-weak compression since the Sinian Dengying period, accompanied by two stages of geoclastic tensioning and two stages of compression [53]. Taking the Huaying Mountain and Longquan Mountain anticlinal belts as boundaries, the Sichuan Basin can be divided into three tectonic zones and six sublevel tectonic units [54,55]. The Weiyuan Block located in the southwestern Sichuan low-steep fold belt and the Luzhou Block in the southern Sichuan low-steep fold belt were selected as the study area (Figure 1), both of which are located in the complex superposition of the south north Daliang mountains-Wumeng mountains tectonic deformation belt on the eastern margin of Qinghai-Tibet Plateau and eastern Sichuan fold belt. Strong composite tectonic deformation since the Indosinian period has formed a complex tectonic framework and differential uplift denudation in the study area [56]. The comb anticline structure arranged in an echelon is developed from north to south in the Luzhou block, the syncline structure is wide and gentle, and the developed fault is small in scale [57]. Meanwhile, the Weiyuan is generally controlled and restricted by the Weiyuan large vault anticline [58].

At the end of the Ordovician and the beginning of the Silurian, two global transgressions and multi-stage large-scale volcanic activities caused extensive sedimentation of the Silurian Longmaxi Formation in the Sichuan Basin and its periphery with siliceous, calc, and clay shales as the main lithofacies [61,62]. The Longmaxi Formation shale in the study area has undergone a multi-stage tectonic and thermal evolution process of shallow burial, slow uplift, secondary deep burial, and rapid uplift since its deposition [63], and the thermal evolution is in the high mature-overmature stage, so it has generally experienced oil generation, oil cracking into gas, bitumen cracking into gas, and kerogen direct gas generation [64].

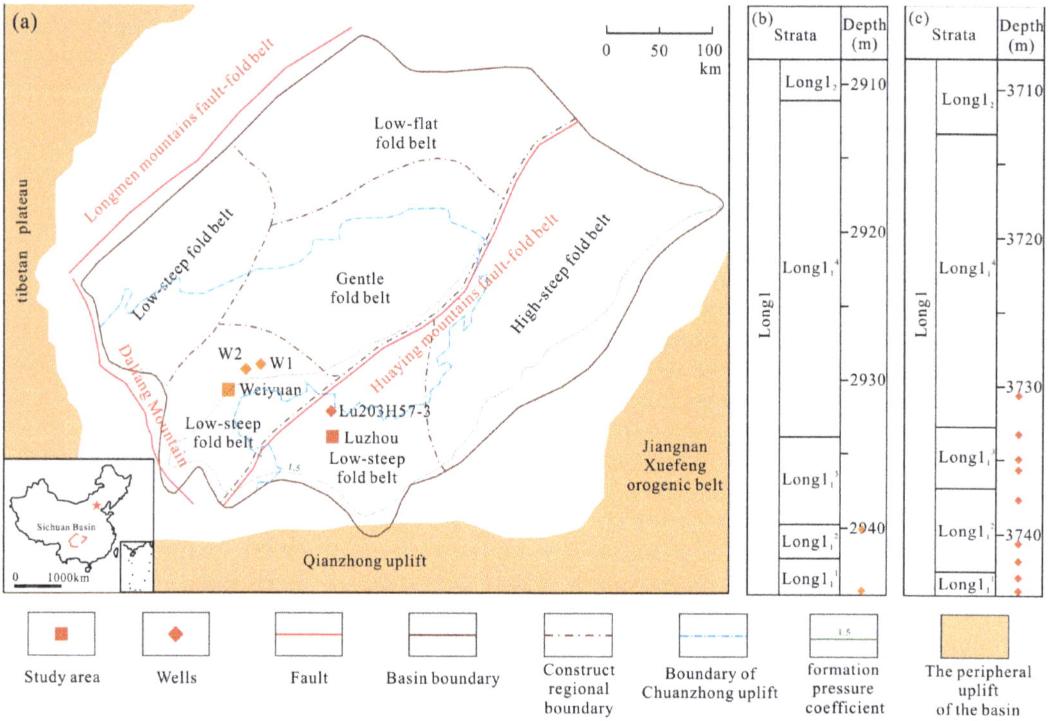

Figure 1. Tectonic location of Sichuan Basin and stratigraphic column of Longmaxi Formation. (**a**) Structural regional map of Sichuan Basin [59,60]. (**b**) Weiyuan block sampling distribution. (**c**) Luzhou block sampling distribution.

3. Materials and Methods

3.1. Samples

In this study, a total of 12 representative samples from Wells W1 and W2 in the Weiyuan Block and Well L203H57-3 in the Luzhou Block were selected for comparison, and the samples were all from 1–4 small layers of Long1_1 sub-member (Figure 1). In principle, the samples of each small layer in the Luzhou are involved, and they all represent the main lithology of the interval. For example, the sample of Long$1_1{}^1$ consists entirely of siliceous shales, which are the most common lithology in this layer, Long$1_1{}^2$ contains siliceous shales and calc-siliceous shales, and Long$1_1{}^3$ is dominated by clay siliceous shales (Table 1). For comparison, we selected well locations (closer to the center of the depositional water body) in the deep-water plain sedimentary microfacies of Weiyuan and selected the siliceous shale in the lower part of the Long1_1 sub-member. Thin sections were prepared by cutting 2~3 mm thick sections perpendicular to the direction of core bedding, and then the images in the visual field were determined to be consistent with the direction of formation deposition according to the direction of mineral laminae and fracture and orientation of clay minerals under scanning electron microscopy. All SEM experiments were accomplished in the Laboratory of Reservoir Microstructural Evolution and Digital Characterization of Yangtze University.

Table 1. Minerals characteristic of $Long1_1$ sub-member.

Sample ID	Layer	Depth/m	Mineralogy Composition/%							Brittleness Index/%
			Quartz	Feldspar	Illite	Chlorite	Calcite	Dolomite	Pyrite	
L203H-1	$Long_1{}^4$	3730.7	48.04	5.65	17.68	4.8	6.87	2.46	2.49	70.6
L203H-2	$Long_1{}^3$	3733.2	64.28	2.65	5.83	0.12	0.93	12.46	2.52	92.2
L203H-3	$Long_1{}^3$	3734.9	64.70	4.84	14.33	0.94	3.33	2.37	4.92	80.2
L203H-4	$Long_1{}^3$	3735.6	49.99	6.85	25.85	1.29	0.92	2.56	4.53	72.1
L203H-5	$Long_1{}^2$	3737.6	58.85	1.52	3.86	0.14	3.28	25.18	2.88	94.6
L203H-6	$Long_1{}^2$	3740.6	59.58	1.33	3.59	0.06	10.71	8.91	9.55	94.1
L203H-7	$Long_1{}^2$	3741.8	76.44	1.31	1.68	0.05	4.46	11.2	1.52	97.6
L203H-8	$Long_1{}^1$	3742.9	69.66	2.44	5.36	0.09	13.55	9.07	3.68	94.9
L203H-9	$Long_1{}^1$	3743.8	73.52	2.16	3.45	0.09	7.99	5.7	2.62	94.4
W1-1	$Long_1{}^2$	2736.8	64.23	4.54	5.38	0.63	4.6	13.33	3.32	92.9
W1-2	$Long_1{}^1$	2747.2	75.0	0.81	2.21	0.18	5.93	3.62	1.47	96.1
W2-1	$Long_1{}^1$	2856.1	83.12	2.08	2.6	1.61	4.07	2.8	2.47	95.2

3.2. MAPS

The morphology and distribution of organic pores were observed using MAPS (Modular Automated Processing System) scanning, which was used to automatically acquire high-resolution large-format mosaic images. The surface area of interest was divided into a grid format and a sequence of BSE/SE images were acquired in real-time [65]. Therefore, compared with ordinary SEM images, MAPS has a wider imaging range, an adjustable field of view, and is more representative. The equipment used for MAPS and FIB-SEM scanning was a HELIOS NanoLab650 dual beam FIB and SEM workstation, with voltage ranging from 1 to 30 KV and current ranging from 0.78 pA to 26 nA. MAPS with 500 nm resolution was first used for overall rough scanning of the sample to observe the fracture and veins in a large field of view. On the basis of this larger scope of vision, a representative region (400 μm × 400 μm) was selected under this field of view for local fine scanning with a higher resolution (4 nm). Then, Avizo software (https://www.thermofisher.cn/cn/zh/home/electron-microscopy/products/software-em-3d-vis/avizo-software.html (accessed on 22 September 2024)) was used to extract organic porosity and the specific parameters of each pore such as length, width, and equivalent radius based on the MAPS scanned images. The number of organic pores extracted from each sample ranged from 10,759 to 57,920, with an average of 28,314.

3.3. FIB-SEM

The two-dimensional microstructure, structural characteristics, and three-dimensional connectivity of the samples were studied using focused ion beam scanning electron microscopy (FIB-SEM) on a HELIOS NanoLab650 operating in high-vacuum scanning mode. The three-dimensional images (10 μm × 10 μm × 10 μm) scanned in this experiment were selected from MAPS fine scanning, and the objects selected in this study were mainly large organic matter with organic pores. The three-dimensional pore-throat model of organic pores was extracted by image analysis and processing, and parameters such as aperture and coordination number of the three-dimensional organic pores were quantified.

3.4. QEMSCAN

QEMSCAN (Quantitative Evaluation of Minerals by Scanning Electron Microscopy) mineral analysis scanning was carried out using a QEMSCAN 650F with a voltage of 1~30 KV, and a current of 0.78 pA~26 nA, following the Chinese National Standard GB/T 20726-2015 [66]. The field of view selected in this experiment corresponds to that of the MAPS scan used to obtain the mineral content and distribution in this area.

3.5. Low-Pressure N_2/CO_2 Adsorption-Desorption

We performed low-pressure N_2/CO_2 physisorption tests using a Micromeritics ASAP 2460 Surface Area and Porosity Analyzer, the experiment was completed in China University of Petroleum (East China). Low-pressure N_2 adsorption was carried out at 77.35 K (−195.85 °C) with a relative pressure of N_2 ranging from 0.002 to 0.993. Low-pressure CO_2 adsorption was measured at 273 K (ice-water mixture) with a relative pressure of CO_2 ranging from 0.006 to 0.029. The isothermal adsorption curve was generated according to the amount of N_2 adsorbed under a certain pressure, then the surface area of shale samples was calculated by using the Brunauer–Emmett–Teller (BET) method, while the Barrett–Joyner–Halenda (BJH) method was used to obtain the pore volume and pore size distribution [67]. Based on the adsorbed quantity of CO_2, a density functional theory (DFT) model was used to calculate the specific surface area and pore volume of the micropores [68].

3.6. Fractal Theory

3.6.1. Frenkel–Halsey–Hill (FHH) Model

Based on the data obtained from the low-pressure N_2 adsorption-desorption experiments, the fractal dimension of the shale was calculated using the FHH model, which has been widely applied and is reported in detail in [27,69]. The FHH model can be simplified as:

$$\ln\left(\frac{V}{V_0}\right) = K \ln\left[\ln\left(\frac{P_0}{P}\right)\right] + C \tag{1}$$

where P is equilibrium pressure, Mpa; V is the volume of N_2 adsorbed under equilibrium pressure, cm^3/g; V_0 is the volume of monomolecular adsorption gas, cm^3/g; P_0 is the saturation pressure, Mpa; C is a characteristic constant; and K is the slope of the fitted line from the plot of $\ln(\ln(P/P_0))$ versus $\ln(V)$. In this study, the fractal dimension based on low-pressure N_2 adsorption (D) is obtained by Equation (2).

$$D = K + 3 \tag{2}$$

3.6.2. Box Counting and Dilation Method

In the calculations of the MAPS fractal dimensions, we used the method of gray extraction and box-counting methods to calculate the fractal dimension of only the organic pores. The method involves regarding the SEM image as a three-dimensional surface, and dividing it into many boxes with side lengths of ε, making the number of divisions in each direction the same, and then calculating the number of boxes $N(\varepsilon)$ covering the organic pore, when the box size is reduced to zero, the fractal dimension can be obtained, the number of boxes can be calculated as follows [70]:

$$\log N(\varepsilon) = -D_L \log \varepsilon + c \tag{3}$$

where ε is the length of the box side; $N(\varepsilon)$ is the number of boxes of size ε containing organic pores; c is the constant that describes the vertical intercept; and D_L stands for length fractal dimension. Linear fractal dimension refers to the fractal dimension obtained by two-dimensional images, which meets $1 < D_L < 2$. On the basis of the assumption of isotropic structure, the fractal dimension of a surface (D_S) in three-dimensional space is obtained by using $D_S = D_L + 1$ [71].

4. Results

4.1. Mineral Composition and Brittleness Index

The mineralogy composition based on QEMSCAN scanning of the shale of the Longmaxi Formation derived from three wells (W1, W2, L203H57-3) is presented in Table 1. In the same small layer, the mineral composition of the two study areas is relatively similar. Taking the $Long1_1{}^1$ and $Long1_1{}^2$ layers as an example, the clay mineral contents in Luzhou and Weiyuan vary from 1.8% to 5.8% (avg. 3.8) and 2.4% to 6.2% (avg. 4.2), respectively.

But there are obvious differences between strata in a single well. For example, the clay mineral contents vary from 6.1% to 27.5% (avg. 17.5) in the $Long1_1{}^1$ and $Long1_1{}^2$ layers of one of the Luzhou wells. To clarify this difference, we introduced the brittleness index to characterize it.

The mineral composition of Marine shale in southern China is relatively complex and quartz, feldspar, calcite, and dolomite are all brittle minerals [72]. Therefore, the calculation method of the brittleness index in this manuscript adopts the mineral composition calculation method, which is expressed by the proportion of brittle minerals as follows:

$$\text{Brittleness Index} = \frac{\text{Quartz} + \text{Feldspar} + \text{Calcite} + \text{Dolomite}}{\text{Quartz} + \text{Feldspar} + \text{Calcite} + \text{Dolomite} + \text{Clay Mineral}} \times 100\% \tag{4}$$

The results show that the brittleness indexes of the $Long1_1{}^1$ and $Long1_1{}^2$ layers are higher, with averages of 94.8% and 96.3%, than those of the $Long1_1{}^3$ and $Long1_1{}^4$ layers, with averages of 81.4% and 70.6%, respectively.

4.2. SEM Observation and Statistics

The extraction of organic pore porosity based on the grayscale of the MAPS images is shown in Table 2. It was found that the porosity of organic pores of shale samples was positively correlated with the brittleness index. The organic porosity of the $Long1_1{}^1$ and $Long1_1{}^2$ layers with higher brittleness index is higher, with an average of 1.66% and 1.56%, while the small layers of $Long1_1{}^3$ and $Long1_1{}^4$ had a low brittleness index and an organic porosity of only 0.89% and 0.48% (Figure 2). It is speculated that the brittle mineral skeleton can improve the mechanical strength of shale such that the organic pores can be preserved under strong external force [73].

In the layer with a high brittleness index (>90%), the organic porosity of shale samples from Luzhou is different, ranging from 1.25% to 2.12%, even though the brittle mineral framework provides protection to the organic pores (Figure 2). At the same time, MAPS also showed that organic pores were distributed in various forms. Therefore, according to the data on pore diameter and length-width ratio of the organic pores, organic pores with a pore diameter greater than 50 nm and length-width ratio less than 2 were defined as suborbicular pores and the organic pores with a diameter greater than 50 nm and length-width ratio greater than 2 were defined as long strip pores. Meanwhile, the organic pores with a diameter of less than 50 nm were defined as honeycomb pores because the small diameter and the deformation under stress were not obvious. The proportion of pores of different morphologies for all samples with organic pore porosities of greater than 1% was calculated and the results are shown in Table 2. Samples with porosities lower than 1% were not counted due to the small amount and size of organic pores that could be observed under MAPS and because these were of little statistical significance.

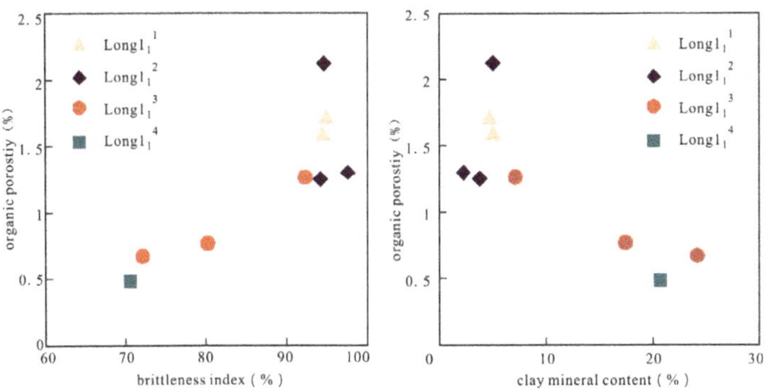

Figure 2. The interrelationship between brittleness index, clay mineral content, and organic porosity of the $Long_1{}^1$ sub-member of Well L203H57-3.

Table 2. Shale reservoir parameters of Long1_1 sub-member from MAPS.

Sample ID	Organic Porosity	The Proportion of Different Morphology Pores/%			Suborbicular Pores/Long Strip Pores	Fine Scanning Region Selection
		Suborbicular Pores	Long Strip Pores	Honeycomb Pores		
L203H-1	0.48	/	/	/	/	/
L203H-2	1.26	41.3	17.4	41.3	2.37	/
L203H-3	0.76	/	/	/	/	/
L203H-4	0.67	/	/	/	/	/
L203H-5	2.12	47.5	22.2	30.3	2.13	/
L203H-6	1.25	35.0	27.1	38.0	1.29	vein
L203H-7	1.30	49.5	13.2	37.2	3.75	/
L203H-8	1.72	35.9	27.0	37.1	1.32	vein
L203H-9	1.59	35.9	26.2	37.9	1.37	high-angle fracture
W1-1	2.08	44.2	25.7	30.1	1.71	/
W1-2	2.25	42.5	29.4	28.1	1.44	/
W2-1	2.16	44.9	29.3	25.8	1.53	/

4.3. Pore Structure Examined by Gas Adsorptions

The obtained N_2 adsorption isotherms of the selected samples conformed to both type II and type IV(a) physical adsorption isotherms defined by IUPAC (Figure 3) [74], suggesting that the samples are dominated by micropores and mesopores with a heterogeneous surface. Adsorption in Longmaxi shale occurs mainly in the 0–0.1 and 0.9–1 regions in terms of relative pressure, and there is no plateau at high relative pressure, which indicates that the micropores and mesopores are well developed. The adsorption-desorption isotherms of all samples have obvious hysteresis loops, indicating that capillary condensation occurs in the mesopores [75]. Based on the IUPAC classification of hysteresis [74], the morphology of the hysteresis loops is H2(a), H3, and H4, that is, multiple [hysteresis loops are superimposed, which reflects the fact that shale samples are disordered porous materials with an uneven pore network structure.

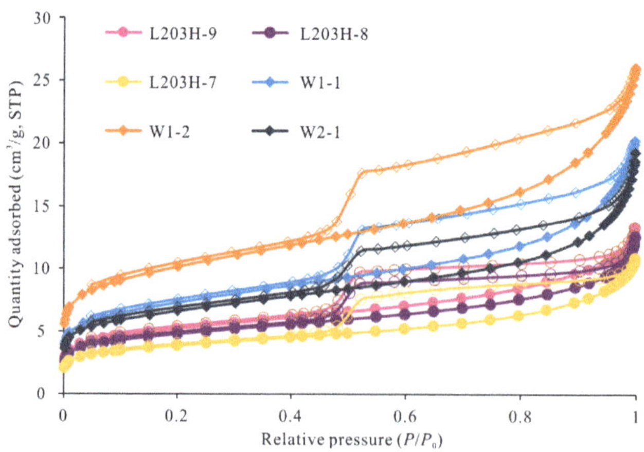

Figure 3. Low-pressure N_2 adsorption-desorption isotherms of shale samples in Luzhou and Weiyuan. The line connecting the filled points is the adsorption curve and the line connecting the hollow point is the desorption curve.

N_2 adsorption isotherms and hysteresis loops can reflect pore shape [28], such as open cylindrical, layered slit, and ink bottle. Comparing the adsorption-desorption isotherms of shale samples between two study areas, the desorption curves of Weiyuan are steeper, and the hysteresis loops are similar to type H2(a), while the hysteresis loops of Luzhou are more inclined to type H2(b), which indicates that the size distribution of pore necks is narrow in Weiyuan shale, while the range of pore neck sizes in Luzhou is wider [74]. Meanwhile, based on the classification of Labani [76], hysteresis loops can be divided into six types (A–F). Our analysis shows that the hysteresis loops in Weiyuan are type B, while those of Luzhou are closer to type E [27]. The results indicate that the pore shapes of Weiyuan consist of slit-shaped pores, while Luzhou may be dominated by ink bottle pores.

Based on the N_2 adsorption and desorption curves, a BJH model was used to calculate the specific surface area and pore volume of shale [77]. The specific surface area of the pores of the Weiyuan shale ranges from 22.64 m^2/g to 34.76 m^2/g (Table 3), with an average of 27.48 m^2/g. The pore volume ranges from 0.0118 to 0.0176 cm^3/g, with an average of 0.0143 cm^3/g. The specific surface area of the pores of the Luzhou shale ranges from 13.36 m^2/g to 18.12 m^2/g, with an average of 16.44 m^2/g. The pore volume ranges from 0.0066 to 0.0090 cm^3/g, with an average of 0.0081 cm^3/g. The specific surface area and volume of the pores of Weiyuan shale are higher than those of Luzhou shale, which is consistent with the higher porosity calculated by MAPS.

Table 3. Pore parameters of $Long1_1$ sub-member from gas adsorption.

Sample ID	CO_2 Adsorption		N_2 Adsorption	
	V_{CO_2} (10^{-3} cm^3/g)	S_{CO_2} (m^2/g)	V_{BJH} (10^{-3} cm^3/g)	S_{BET} (m^2/g)
L203H-5	1.71	3.79	9.07	18.12
L203H-7	1.62	3.66	8.95	17.95
L203H-8	1.56	3.54	7.99	16.36
L203H-9	1.16	2.58	6.67	13.36
W1-1	2.07	3.02	13.62	25.04
W1-2	2.79	3.83	17.67	34.76
W2-1	2.00	4.47	11.83	22.64

Meanwhile, the specific surface area and volume of shale micropores were calculated using a DFT model [78], The specific surface area of the micropores of the Weiyuan shale ranges from 3.02 m^2/g to 4.47 m^2/g, with an average of 3.77 m^2/g. The micropore volume ranges from 0.0020 to 0.0027 cm^3/g, with an average of 0.0023 cm^3/g. The specific surface area of the micropores of the Luzhou shale ranges from 2.58 m^2/g to 3.79 m^2/g, with an average of 3.39 m^2/g. The micropore volume ranges from 0.0011 to 0.0017 cm^3/g, with an average of 0.0015 cm^3/g. The specific surface area and volume of the micropores of Weiyuan shale are, thus, also higher than those of Luzhou shale.

4.4. Fractal Characterization

Based on the fractal FHH model and N_2 adsorption data, the plots of $\ln(V)$ versus $\ln(\ln(P_0/P))$ from the studied shale samples are shown in Figure 4. These show a significant linear relationship with a high correlation coefficient ($R^2 > 0.9$), indicating that the Longmaxi Formation shale has obvious fractal features. Previous studies have commonly divided FHH plots into two regions, reflecting two different adsorption characteristics, corresponding to the fractal characteristics D_1 and D_2 [79,80]. Fractal dimension D_1 corresponds to low-pressure region fractal characteristics ($P/P_0 < 0.45$), reflecting the van der Waals forces of monolayer–multilayer adsorption in micropores. Fractal dimension D_2 mainly represents mid- to high-pressure region fractal characteristics ($P/P_0 > 0.45$), indicating capillary condensation based mainly on the spatial complexity and irregularity of mesopores and macropores [71,81].

The results of fractal analysis based on the N_2 adsorption experiments are shown in Table 4. Although the fractal results of the N_2 adsorption experiments are based on all the pores of the entire shale sample, they still represent the fractal characteristics of the organic pores to a certain extent. The results show that fractal dimension D_1 of the shale samples in Weiyuan is observed to be higher than that of the samples in Luzhou, while the data with respect to fractal dimension D_2 are more complicated. Due to the resolution limitation of a large amount of data collected in MAPS used in the comparison, the minimum pore diameter that can be identified is only 10 nm, and macropores are more affected by stress extrusion pore extrusion deformation [9,82], so D_2 received more attention in this study. In the following discussion, samples from Luzhou were further divided according to whether veins or fractures developed. In general, the samples with developed veins or fractures (L203-8, L203-9) had the highest D_2 values (mean of 2.877), the homogeneous Weiyuan samples had the second highest values (mean of 2.872), and the D_2 values of the homogeneous samples from Luzhou (L203-5, L203-7) were the lowest (mean of 2.859).

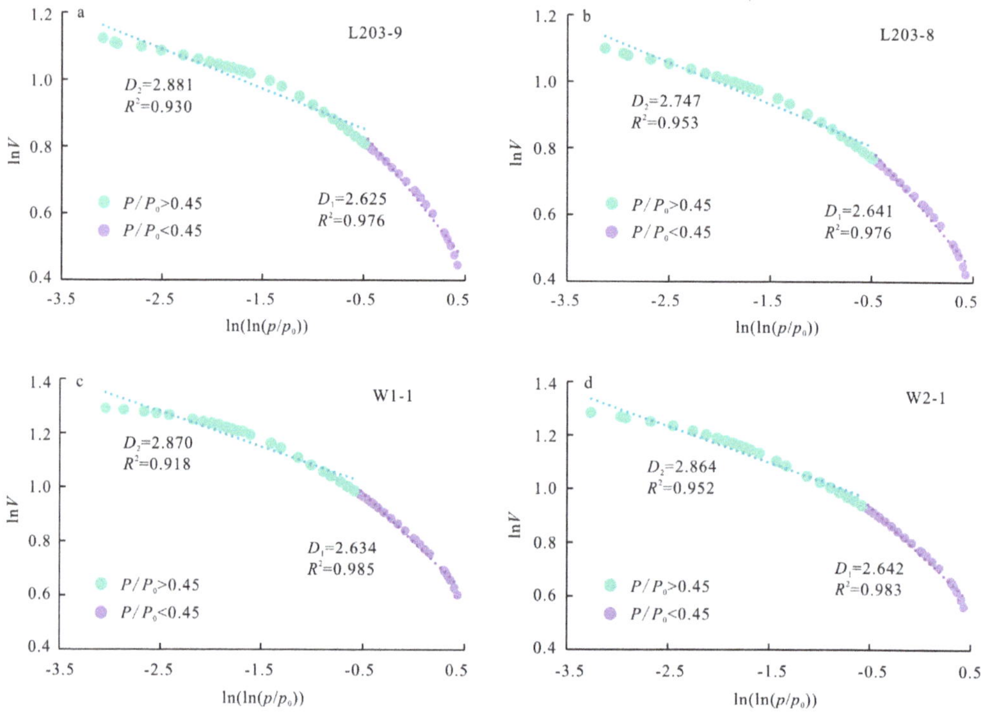

Figure 4. Plots of ln(V) vs. ln(ln(P_0/P)) from N_2 adsorption-desorption experiments using FHH model. (**a**) Fractal dimensions of L203-9 shale. (**b**) Fractal dimensions of L203-8 shale. (**c**) Fractal dimensions of W1-1 shale. (**d**) Fractal dimensions of W2-1 shale.

Table 4. Fractal dimensions calculated by the FHH model from N_2 adsorption experiments.

Sample ID	$P/P_0 < 0.45$		$P/P_0 > 0.45$	
	R^2	D_1	R^2	D_2
L203H-5	0.971	2.629	0.933	2.855
L203H-7	0.975	2.625	0.924	2.863
L203H-8	0.977	2.641	0.953	2.874
L203H-9	0.976	2.625	0.930	2.881
W1-1	0.985	2.634	0.918	2.870
W1-2	0.982	2.646	0.901	2.883
W2-1	0.983	2.642	0.952	2.864

Based on the box-counting and dilation method and SEM images of the organic pores, plots of ln($N(\varepsilon)$) versus ln(ε) for the studied shale samples were generated and are shown in Figure 5. These show an excellent correlation coefficient. The fractal dimension obtained from the measurement of the area of the organic pores extracted based on MAPS is shown in Table 5. The results are similar to the fractal dimension calculated by the FHH model and again show that the fractal dimension of Weiyuan is larger than that of Luzhou in the homogeneous region.

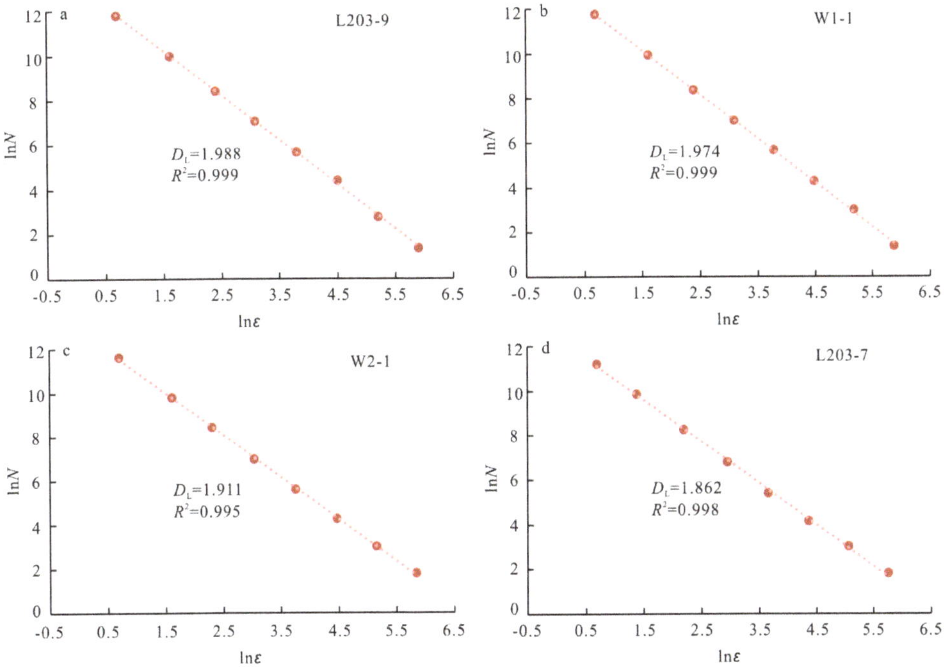

Figure 5. Plots of ln(N) vs. ln(ε) from MAPS images using box-counting and dilation method. (**a**) Fractal dimensions of an image from L203-9. (**b**) Fractal dimensions of an image from W1-1. (**c**) Fractal dimensions of an image from W2-1. (**d**) Fractal dimensions of an image from L203-7.

Table 5. Fractal dimensions calculated from MAPS images using box-counting and dilation method.

Sample ID	No. of Images	D_S	
		Range	Average
L203H-5	10	2.867–2.937	2.907
L203H-7	10	2.858–2.944	2.912
L203H-8	10	2.954–2.989	2.971
L203H-9	10	2.963–2.991	2.973
W1-1	12	2.885–2.981	2.958
W1-2	12	2.901–2.979	2.951
W2-1	8	2.911–2.958	2.939

5. Discussion

5.1. Microscopic Deformation Characteristics

The relatively stable brittle mineral framework provides support and preservation for the pores in the organic matter to a certain extent, but under the action of intense tectonic compression, compaction, mineral recrystallization, and other stresses [49], rigid mineral particles will also locally squeeze adjacent organic matter components. As a result, the outline of organic matter components and internal organic pores are significantly deformed [83], and the originally round organic matter pores are deformed into ellipses by extrusion (Figure 6). SEM images showed the following evidence of deformation of organic pores under stress conditions: (1) The deformation of organic pores was directional, that is, the long axes of elliptical organic pores are arranged in the direction perpendicular to the extrusion of particles (Figure 6a). (2) The deformation of organic pores closer to the edge of organic matter is more obvious, and the roundness of organic pores away from the contact interface between mineral particles and organic matter is better (Figure 6b). (3) The images of 10 nm slices arranged in sequence on a three-dimensional scale (Figure 7a–e) show that long organic pores on both sides of the particles are gradually obvious as the edges of mineral particles become clearer inside the red frame. By reproducing the 3D image of the mineral framework in this scanning area, it can be observed that the mineral particles in the red frame range are obviously raised (Figure 7f), and it is inferred that the mineral particles have an extrusion effect on the nearby organic pores.

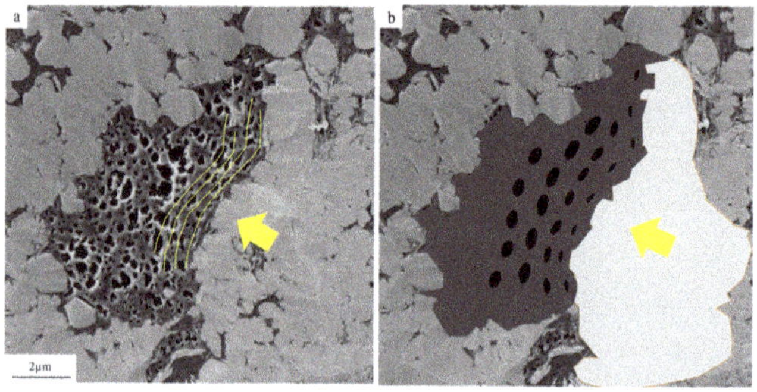

Figure 6. Mineral particles changing shape of organic pores under stress, L203H-7. (**a**) SEM images; the yellow line indicates the line along the long axes of the organic pores, and the yellow arrow indicates the direction of the mineral particles squeezing the organic matter. (**b**) Schematic diagram; light gray indicates mineral particles, dark gray indicates organic matter, black indicates organic pores, and yellow arrows indicate the direction in which mineral particles squeeze the organic matter.

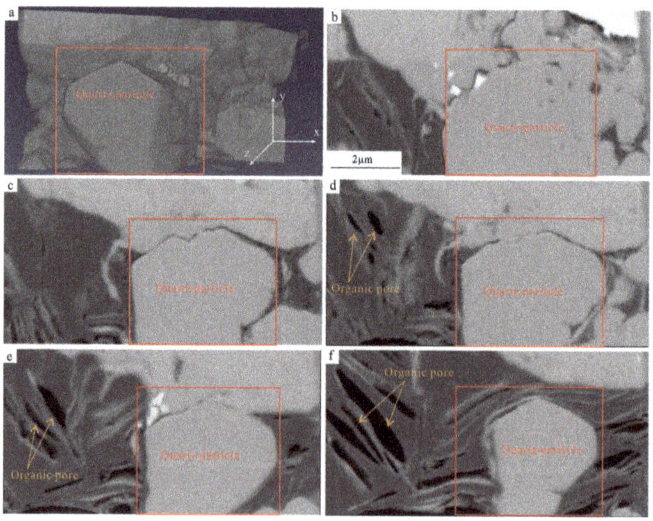

Figure 7. FIB-SEM images, W1-2. (**a**) Three-dimensional mineral diorama reconstructed by Avizo; there are prominent raised minerals in the red frame. (**b–f**) Continuous SEM 2D slice images along the z-axis in the 3D mineral model in (**a**), i.e., the xy-axis slices shown successively, illustrating the change in organic pore morphology characteristics.

The above arrangement characteristics of organic pores are common in the shale reservoir of the Longmaxi Formation in southern Sichuan. It is believed that under the influence of tectonic compression and mineral recrystallization, mineral particles will squeeze the organic matter around them, resulting in the deformation of organic pores and thus, the pore morphology becomes a long strip with a higher length-width ratio.

5.2. Influence of Reservoir Internal Stress

Based on the deformation characteristics of local organic pores, the morphological differences between organic pores in different stress deformation regions are discussed. In this study, the regions of vein development (Figure 8a) and high-angle fracture development (Figure 8b) were selected as examples of stress concentration regions. The selected veins are calcite and quartz veins with a width of 0.5~1.5 mm. Calcite and quartz were deposited and cemented in the tectonic uplift stage [84]. During the formation and later opening of fractures, the two sides of the area are squeezed, and the compressive stress near the vein is stronger [85], so this represents a stress concentration region. Organic matter with undeveloped pores can be observed in the selected high-angle fracture. In the stage of tectonic uplift, the high-angle fracture continued to open and heal and was filled with fluid [86,87]. During the formation and later opening of the high-angle fracture, its two sides are squeezed, and the extrusion stress near the high-angle fracture becomes stronger, so this also represents a stress concentration region. Similarly, two areas with no large micro-fractures and good development of organic matter and brittle minerals were selected to represent non-stress-concentration regions (Figure 8c,d). The fine SEM scans showed that the organic pores in stress concentration regions such as veins and high-angle fractures were deformed significantly, had poor roundness, and were dominated by long organic pores (Figure 9), while the organic pores in non-stress-concentration regions maintained good shape and roundness, and were dominated by round and elliptical organic pores (Figure 10).

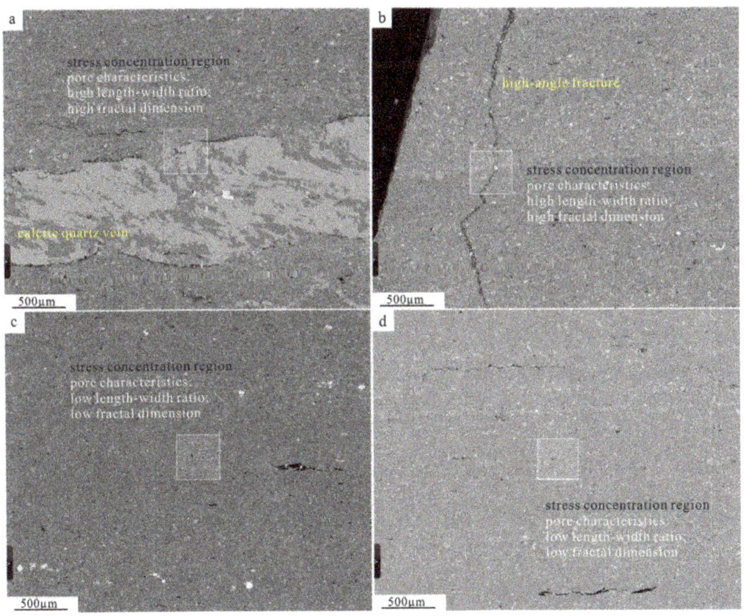

Figure 8. MAPS images of 500 nm resolution. (**a**) Regions of vein development, L203H-8; (**b**) Regions of high-angle fracture development, L203H-9. (**c**) Non-stress-concentration regions, L203H-5. (**d**) Non-stress-concentration regions, L203H-7. The area in the white box is the selected fine scanning region with a size of 400 μm × 400 μm.

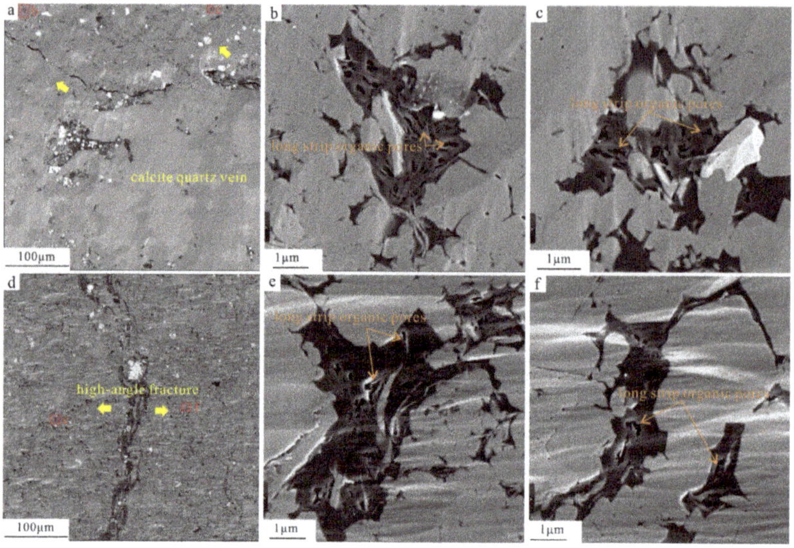

Figure 9. MAPS images of stress concentration region (4 nm resolution). (**a**) Enlargement of the white box area (the region of vein development) in Figure 8a, L203-8. The yellow arrow indicates the direction of extrusion when the vein is consolidated or filled. (**b**,**c**) Partial enlargements of (**a**) showing long strip organic pores with D_S = 2.962 and 2.989. (**d**) Enlargement of the white box area (the region of high-angle fracture development) in Figure 8b, L203-9. The yellow arrow indicates the direction of extrusion when the high-angle fracture opens and heals. (**e**,**f**) Partial enlargements of (**d**) showing long strip organic pores with D_S = 2.986 and 2.991.

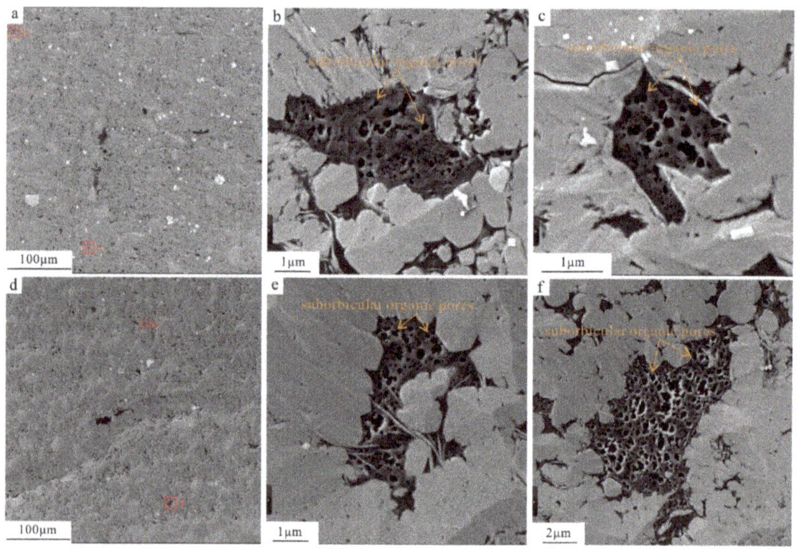

Figure 10. MAPS images of non-stress-concentration region (4 nm resolution). (**a**) Enlargement of the white box area (non-stress-concentration region 1) in Figure 8c, L203H-5. (**b**,**c**) Partial enlargements of (**a**) showing suborbicular organic pores with D_S = 2.911 and 2.912. (**d**) Enlargement of the white box area (non-stress-concentration region 2) in Figure 8d, L203H-7. (**e**,**f**) Partial enlargements of (**d**) showing suborbicular organic pores with D_S = 2.883 and 2.917.

Since organic pores with small diameters generally have better roundness and less obvious deformation characteristics, they are of little reference significance in a discussion of the cause of stress modification [47,83]. Therefore, the long-axes and short-axes data of 138,437 organic pores with diameters greater than 50 nm were extracted from the four fine scanning areas, and the length-width ratios of organic pores in different regions were quantified (Figure 11a). The results showed that the length-width ratios of organic pores (the long-axes/short-axes of organic pores) are higher in the stress concentration region (near the high-angle fractures and the veins), the deformation of the organic pores is more obvious, and their shape is closer to that of a long strip. Meanwhile, the organic pores in the non-stress-concentration regions have a lower length-width ratio and are more rounded in shape.

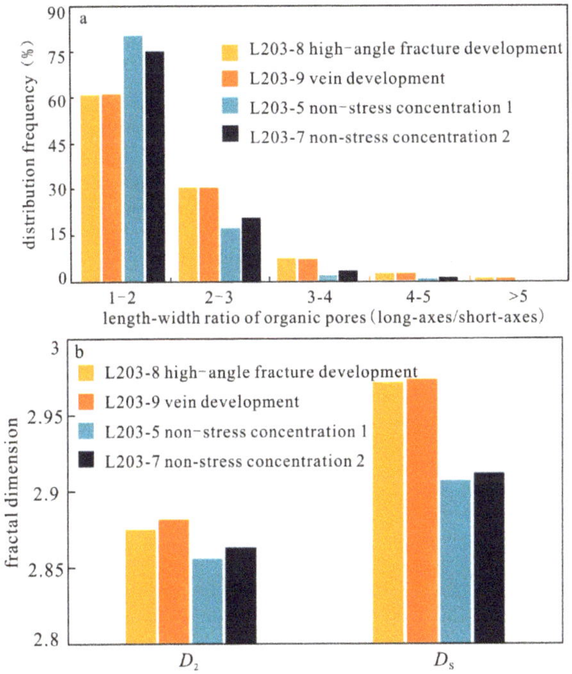

Figure 11. Comparison of deformation characteristics of organic pores in different stress-affected regions in the Long1$_1$ sub-member. (**a**) Length-width ratio of organic pores. (**b**) D_2 calculated by FHH model and D_S measured by box-counting method.

We also compared the differences in fractal characteristics between different stress regions. D_S values derived from the SEM images using the box-counting and dilation method are given in the captions of Figures 9 and 10. The results also show that the fractal dimensions (D_2 and D_S) of the samples in the stress concentration regions are higher (Figure 11b), which indicates that the pore structure of the organic pores around the high-angle fractures or veins is more complex. It is speculated that the compressive stress may lead to pore deformation and enhance pore heterogeneity in a single organic matter region.

Therefore, in the vertical variation of the Long1$_1$ sub-member shale in L203, high-angle fractures and veins are widely developed in samples (8, 9) at the bottom, resulting in obvious deformation of the organic pores and a complex pore structure, which are manifested as high length-width ratios, a high proportion of long strip pores (Table 2), and a high fractal dimension, while the samples in the middle (5–7) are generally less affected by internal reservoir stress, resulting in higher roundness of the organic pores and a lower fractal dimension. Meanwhile, the upper part of the samples (1–4) has higher clay mineral contents (Figure 2), resulting in weak compaction resistance and lower organic porosity.

5.3. *Influence of Tectonic Stress*

The morphologies and fractal dimensions of organic pores in different study areas are different (Tables 2 and 4). In this section, we discuss the influence of tectonic stress on pore deformation in the reservoir. Therefore, we choose to compare samples of undeveloped veins and high-angle fractures in Luzhou (L203-5, 7) with shale samples from Weiyuan to exclude the influence of reservoir internal stress. The results show the following: (1) As pointed out above, strong stress will lead to the extrusion of suborbicular pores into long strip pores. The distribution frequency of long strip pores is higher (Figure 12a) and the ratio of suborbicular pores to long strip pores in Weiyuan is smaller than that of Luzhou (Table 2), indicating a higher proportion of long strip pores in Weiyuan. (2) Based on the type of hysteresis loop found by N_2 adsorption, the porosity of Weiyuan shale consists mainly of slit-shaped pores, while that of Luzhou is closer to ink bottle pores (Figure 3), which is consistent with the results of organic pore length-width ratio calculations calculated by MAPS. (3) The fractal dimension D_2 calculated by the FHH model and the fractal dimension D_S measured by the box-counting method are both higher for Weiyuan (Figure 12b), indicating that the organic pores of Weiyuan are more heterogeneous and the pore structure is more complex. (4) According to the results of CO_2 adsorption and N_2 adsorption (Table 3), it is found that although the pore volume and specific surface area of micro-, meso-, and macropores of the Weiyuan shale samples are larger, the difference between the meso- and macropores obtained by N_2 adsorption experiments is more obvious, while the difference between the micropores calculated by CO_2 adsorption experiment is relatively unnoticeable. This means that there may be fewer micropores in shale samples from Weiyuan, which is consistent with the smaller proportion of honeycombed pores calculated by MAPS (Table 2). Previous studies have shown that brittle deformation in natural shale may cause micropores to connect with each other and become mesoporous, thus reducing the number of micropores [44,88]. All the above phenomena indicate that the pore structure of Weiyuan is more complex and the deformation is more obvious.

In terms of geological background, the sedimentary environment of the two study areas is similar, both are deep-water plain sedimentary facies, and the lithofacies and mineral composition are roughly the same [89]. Therefore, according to the differences in organic pore morphology between the two study areas, the mechanism of the difference is analyzed in terms of the evolution background of Weiyuan in the southwestern Sichuan depression and Luzhou in the southern Sichuan fold belt.

Burial evolution curves for the Weiyuan and Luzhou blocks show (Figure 13) that after the Silurian Longmaxi Formation was deposited in southern Sichuan, the Chuanzhong uplift was uplifted and eroded as a whole at the end of the Silurian, and its movement intensity gradually weakened from the western, central, and southeast Sichuan regions [90]. Therefore, due to the influence of early uplift and denudation, oil generation in Weiyuan was late and the formation underwent pressure relief, resulting in a low formation pressure coefficient [91,92]. Against the background of late stress extrusion, organic pores in the Weiyuan block with a low-pressure coefficient were also more prone to deformation. Conversely, the Silurian Longmaxi Formation shale in Luzhou enters the oil generation window earlier after being quickly buried and stabilized, and the preservation condition of organic pores is more stable in the later period.

During the late Mesozoic to Cenozoic tectonic activities, during the early period of uplift and denudation, the Jiangnan Xuefeng orogenic belt in the Late Jurassic and Early Cretaceous was extruded from the southeastern margin of the basin to the basin, and the Late Cretaceous Qianzhong Uplift pushed northward. In the later period of uplift and denudation, the Daliang Mountain front area in the southwest margin of the Sichuan Basin was subjected to superimposed deformation under the influence of stress in the late Paleogene, and the Qinghai-Tibet Plateau continued to expand eastward from the Cenozoic onward. At this time, Weiyuan was significantly more subjected to stress transformation than Luzhou (Table 6) [53,93]. In the process of tectonic extrusion, the formation pressure coefficient gradually decreases with uplift and denudation, and the formation pressure coefficient is lower when the tectonic compressive stress is stronger in the Weiyuan block. Therefore, the transformation of organic pore morphology by reservoir stress is more obvious, which shows that the organic pores in Weiyuan are more prone to deformation.

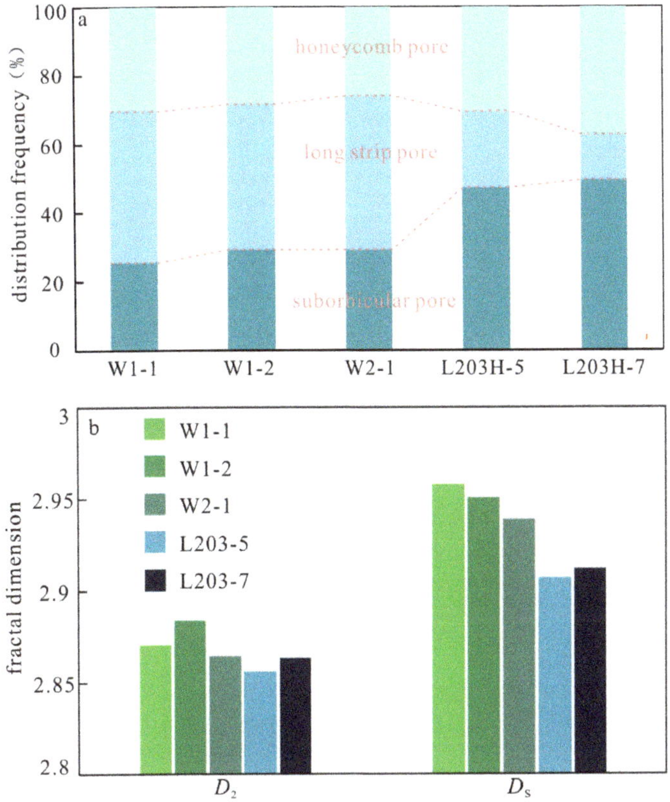

Figure 12. Comparison of deformation characteristics of organic pores in the Long1$_1$ sub-member of Weiyuan and Luzhou shale. (**a**) The proportion of organic pores with different morphological characteristics. (**b**) D_2 calculated by FHH model and D_S measured by box-counting method.

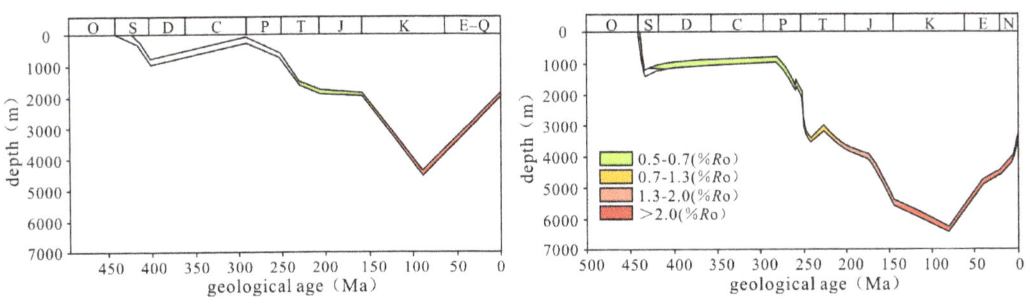

Figure 13. Burial evolution diagram of Weiyuan and Luzhou blocks [56,91].

Table 6. Different tectonic backgrounds in Weiyuan and Luzhou blocks.

Tectonic Activities	Weiyuan	Luzhou
Period of entering the oil window	Triassic, relatively late	The end of the Silurian, relatively early
Formation pressure coefficient	Relatively low	Relatively high, the reservoir sealing condition is better.
Intensity of late Mesozoic deformation	The intensity of structural deformation is relatively weak	The intensity of structural deformation is relatively strong
Intensity of Cenozoic tectonic deformation	The intensity of structural deformation is relatively strong	The intensity of structural deformation is relatively weak

To sum up, the shale of the Weiyuan Silurian Longmaxi Formation has been more strongly reformed by tectonic activities since its deposition, the pore structure of shale changes more obviously, which is reflected in the fact that micropores in Weiyuan connected to each other to become mesopores or macropores, resulting in a reduction in the number of micropores, and, as macropores are susceptible to compressive deformation under stress, this results in a significant increase in the number of long strip pores. The pore structure in Weiyuan shale, modified by compressive stress, is more complex and shows strong anisotropy, which is expressed in the higher fractal dimension.

5.4. Factors Influencing Fractal Dimensions

Based on the differences in pore characteristics in different intervals and study areas, this study also investigated the relationship between shale composition, pore structure, and fractal dimension to explore pore complexity and the main factors influencing this under a stress modification environment.

A comparison of fractal dimension values for different mineral components in shale shows that the proportions of quartz, clay minerals, and pyrite are not correlated or weakly correlated with fractal dimension (Figure 14). This can be explained from two aspects. Previous studies have shown that organic pores are the most important reservoir space in the overmature shale in the study area [31,94]. Therefore, different minerals such as intergranular pores of quartz and interlayer fractures of clay minerals have little contribution to the pore structure, and thus, have little influence on the fractal characteristics of the pores in the samples. Secondly, the content of mineral components also indicates the sedimentary environment—strawberry pyrite reflects the degree of redox state, and, while clays are often indicative of terrigenous inputs [95–97], quartz is relatively complex, and its origin may be the coupling of various geological events [98–100]. As the purpose of this study was to explore pore deformation characteristics under different stress environments, samples with a relatively consistent depositional environment were specially selected for comparison to control variables. Therefore, the weak correlation between fractal and mineral composition was within expectation. It is worth mentioning that in the correlation between fractal and mineral content, the negative correlation with clay mineral contents is relatively more obvious, which may be related to the lower brittleness of clay mentioned above, making the pores more susceptible to extrusion deformation.

The correlation between fractal dimension and pore structure parameters is further discussed. As presented in Figure 15, the fractal dimension shows good positive linear relationships with pore volume and specific surface area, thereby indicating that the pore structure becomes more complex and heterogeneous with increasing pore volume and specific surface area. The results are consistent with the results of previous studies on marine shales in South China and North America [27,101,102].

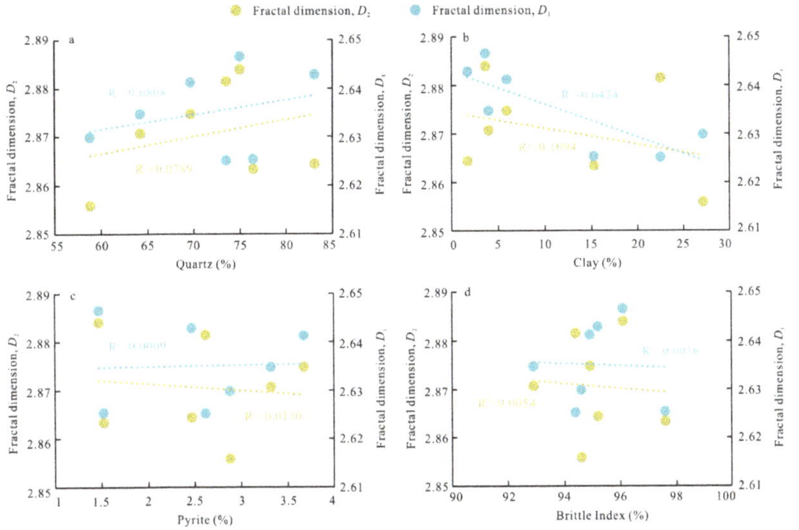

Figure 14. Correlations between fractal dimension and mineral components. (**a**) Quartz. (**b**) Clay minerals. (**c**) Pyrite. (**d**) Brittle index.

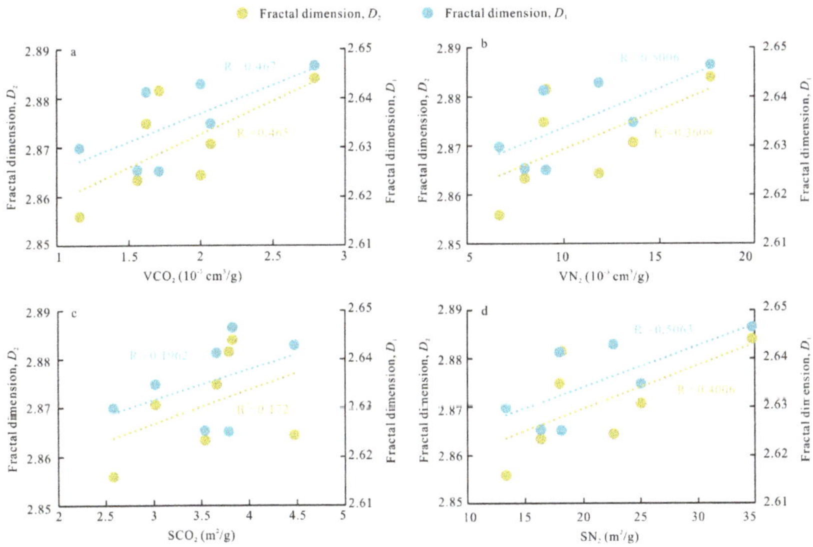

Figure 15. Correlations between fractal dimension and pore structure parameters. (**a**) Pore volume calculated by CO_2 adsorption. (**b**) Specific surface area calculated by CO_2 adsorption. (**c**) Pore volume calculated by N_2 adsorption. (**d**) Specific surface area calculated by N_2 adsorption.

As illustrated in Figure 16, the correlation between the proportion of different pore types and the fractal dimension is completely different: the correlation between suborbicular pores and fractal dimension is negative, long strip pores are positively correlated, and the correlation between honeycomb pores and D_2 is weak. It should be noted that since the abscissa in the graph shows the relative content of pores with different morphology features, which represents the proportion of such pores in the sample, the results of correlation do not reflect the long strip pores have better fractal characteristics than the suborbicular pores. According to the data, the fractal dimension of the suborbicular pores and long strip pores is highest when the proportion is close to 33%, which indicates that the fractal dimension of pores is higher when all kinds of pores in the sample are developed. This result can be interpreted as indicating that the development of different pore morphology types represents stronger heterogeneity [71]. From the point of extrusion stress, a sample with a higher proportion of long strip pores may be the result of stronger stress modification, and brittle or ductile deformation under stress will enhance the complexity of pore structure, resulting in an increase in fractal dimension [103]. The weak correlation between honeycomb pores and fractal dimension may be due to their limited contribution to pore structure. In addition, pores with a small diameter are not significantly affected by stress modification [49,103].

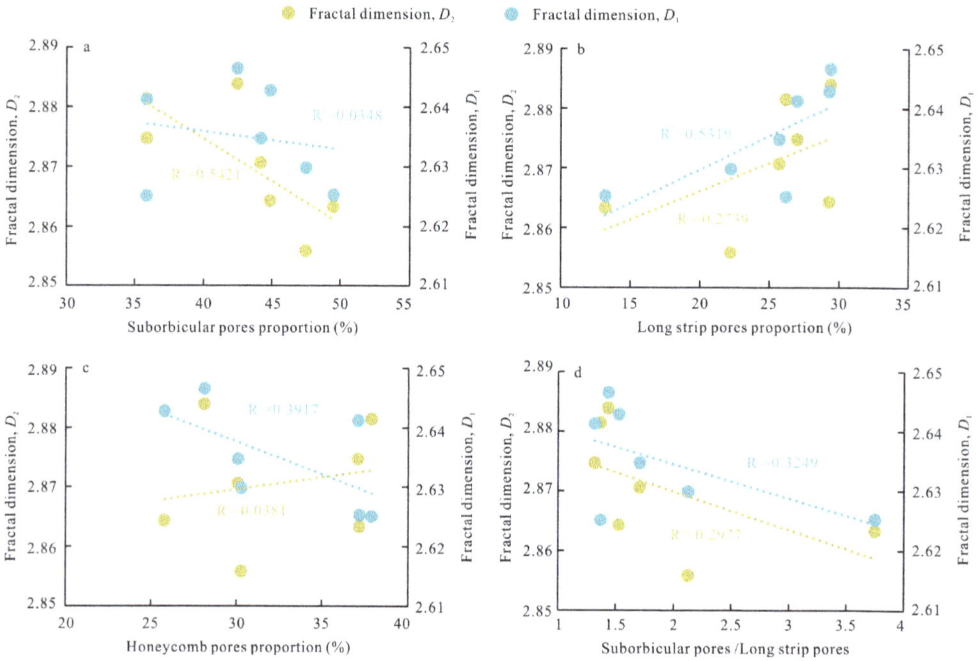

Figure 16. Correlations between fractal dimension and proportion of different pore types. (**a**) Suborbicular pore. (**b**) Long strip pore. (**c**) Honeycomb pore. (**d**) The ratio of suborbicular pores and long strip pores.

5.5. Storage and Permeability Performance of Different Pores

The differences in the storage and permeability performance of various pores were further clarified by FIB-SEM. Honeycomb pores are formed when organic matter reaches a certain maturity and there is sufficient gas expansion force [104]. The aperture of the honeycomb pores is small and the distribution is concentrated, generally in the 2 to 30 nm range, with obvious peaks (Figure 17a,b). The fractal dimension of honeycomb pores calculated by the box-counting method based on SEM images is the lowest, with an average value of 2.922. The data show that low values of the coordination number of pore throat account for a high proportion, while high values are rare, so the connectivity of the honeycomb pores is poor. Moreover, by comparing the pore data in Figure 17a,b, it is found that the connectivity of honeycomb pores gradually becomes better with an increase in aperture size.

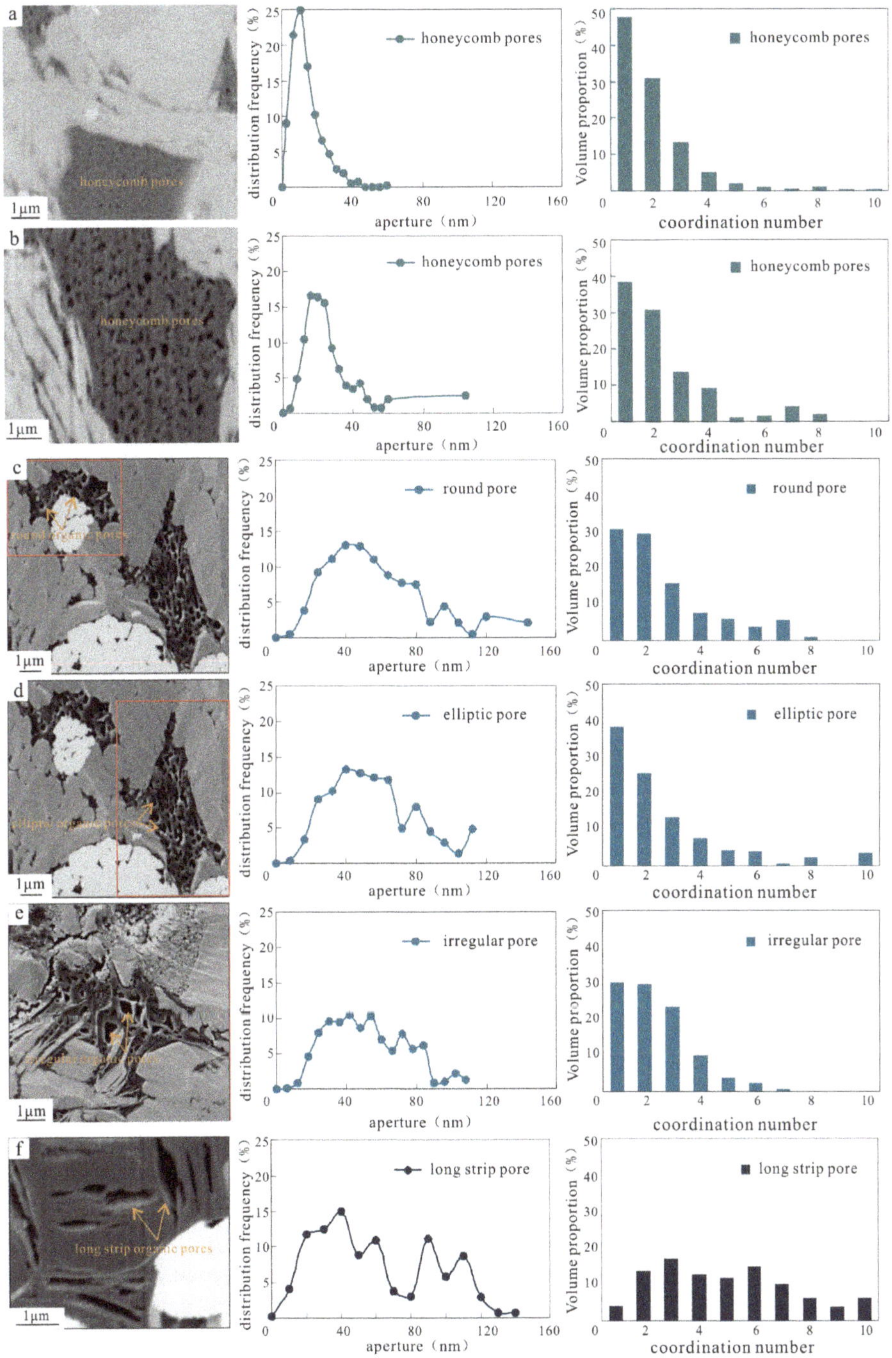

Figure 17. Aperture size distribution and coordination number spread of organic pores with different morphological characteristics. (**a**) Honeycomb pores with small aperture size, W1-1. (**b**) Honeycomb pores with large aperture size, W1-2. (**c**) Round pores in the red box, L203H-7. (**d**) Elliptical pores in the red box, L203H-7. (**e**) Diamond (irregular) pores, L203H-7. (**f**) Long strip organic pores, W1-1.

Suborbicular pores include round pores with good preservation conditions and elliptical and diamond (irregular) pores that are not obvious in later reconstruction. Suborbicular pores are usually formed by the communication and aggregation of honeycomb pores [37,105] and are less affected by the internal stress of the reservoir, and thus, can maintain a good roundness and morphology. They are characterized by varying aperture sizes, distributed in the 20 to 80 nm range, no obvious peaks, and good overall roundness (Figure 17c–e). The fractal dimension of suborbicular pores calculated by the box-counting method based on SEM images is medium, with an average value of 2.949. The data show that the coordination number of round pores is obviously higher than that of honeycomb pores, and the proportions of low and high coordination numbers increase slightly when round pores are transformed into ellipses. When the round pores are transformed into angular and irregular pores, the overall connectivity is slightly reduced. In general, there was little difference between the connectivity of suborbicular pores of different shapes, and the overall connectivity was better than that of honeycomb pores.

The long strip pores are the result of organic pores experiencing large deformations under the influence of strong stress. The aperture size distribution range of long strip organic pores is large, and it is distributed in the 20 to 120 nm range without an obvious peak value. The fractal dimension of long strip pores calculated by the box-counting method based on SEM images is the highest, with an average value of 2.972. The data show that the pore throat coordination number of long strip pores was significantly higher than that of pores of other morphologies, so their connectivity was the best (Figure 17f). It is speculated that with an increase in extrusion, the deformation of organic matter is accompanied by partial shrinkage fractures [106], and, therefore, more organic pores become connected leading to an increase in coordination number.

In conclusion, the honeycomb pore is a type of pore that is in the early stage of thermal evolution or has not undergone significant stress modification and has a small aperture size, low fractal dimension, and poor connectivity and storage-permeability performance. With the evolution process of hydrocarbon generation and tectonic extrusion, the honeycomb pores are gradually connected into suborbicular pores with complex fractal characteristics, and the larger the pore aperture, the better the connectivity and storage-permeability performance. To a certain extent with an increase in aperture size, the fluid pressure decreases and the transformation effect of strong reservoir extrusion on organic pores gradually becomes obvious. In areas with strong local stress, the organic pores with high roundness and large pore diameter are transformed into long strips by stress, and long strip pores have the highest fractal dimension, a complex pore structure, and the best connectivity.

6. Conclusions

In this study, the fractal dimension and deformation characteristics of organic pores under the influence of different stress concentrations were studied in the Weiyuan block in the southwestern Sichuan depression and the Luzhou block in the southern Sichuan fold belt, and the storage and permeability of organic pores with different morphologies were further discussed. The main conclusions of this study are as follows:

(1) In the vertical variation of $Long1_1$ sub-member shale in L203, the brittle mineral framework has a positive impact on the preservation of organic pores. The upper interval of the core samples has a low brittleness index and low organic porosity. The veins or high-angle fractures in the lower interval are more developed, and the local stress extrusion is stronger. The organic pores deformed under the protection of a rigid mineral framework, which manifested as the development of long strip pores and a high fractal dimension. The middle research interval was less affected by the internal stress of the reservoir, resulting in the higher roundness of the organic pores and a lower fractal dimension.

(2) The morphological characteristics of organic pores in the Weiyuan and Luzhou blocks are noticeably different. The N_2 adsorption hysteresis loop of Weiyuan shale is close to type B, indicating that slit-shaped pores are dominant, the proportion of- micropores

is relatively low, the macropores are mostly long strip pores, and the fractal dimension is higher. However, the N_2 adsorption hysteresis loop of Luzhou shale is close to type F, showing the development of ink bottle pores with a relatively high proportion of micropores, and the morphology of macropores is mostly pores with a low fractal dimension.

(3) The factors influencing the fractal and morphology characteristics of the organic pores in the two study areas have been clarified. The formation pressure coefficient of Weiyuan is low, and the tectonic stress has been strong since the Cenozoic. Therefore, the deformation of organic pores is obvious, mostly long strip organic pores have developed and the pore structure is complex. The pressure coefficient of the Luzhou block is high, the compressive stress is weak in the later stage, and suborbicular organic pores have developed.

(4) There are differences in the storage and permeability properties of different forms of organic pores. The morphology and distribution of honeycomb pores are uniform, the fractal dimension is the lowest, and the connectivity is the worst; suborbicular pores have larger pore diameters and medium fractal dimension and connectivity; and long strip pores formed by stress transformation have a large pore aperture size distribution range, strong pore heterogeneity, the highest fractal dimensions, and the best connectivity.

Author Contributions: Conceptualization, D.H. and Y.J.; methodology, J.Z.; software, C.W.; validation, Y.J. and B.M.; formal analysis, Y.W.; investigation, Y.W.; resources, W.L.; data curation, D.H. and Y.J.; writing—original draft preparation, Y.W.; writing—review and editing, Y.W.; visualization, Y.W.; supervision, D.H.; project administration, D.H.; funding acquisition, W.L. All authors have read and agreed to the published version of the manuscript.

Funding: This research was funded by the National Natural Science Foundation of China, grant number 42072121, and 42302175.

Data Availability Statement: The original contributions presented in the study are included in the article, further inquiries can be directed to the corresponding authors.

Conflicts of Interest: The authors declare no conflicts of interest.

References

1. Vengosh, A.; Jackson, R.B.; Warner, N.; Darrah, T.H.; Kondash, A. A Critical Review of the Risks to Water Resources from Unconventional Shale Gas Development and Hydraulic Fracturing in the United States. *Environ. Sci. Technol.* **2014**, *48*, 8334–8348. [CrossRef] [PubMed]
2. Zou, C.N.; Yang, Z.; Cui, J.W.; Zhu, R.K.; Hou, L.H.; Tao, S.Z.; Yuan, X.J.; Wu, S.T.; Lin, S.H.; Wang, L.; et al. Formation mechanism, geological characteristics and development strategy of nonmarine shale oil in China. *Pet. Explor. Dev.* **2013**, *40*, 15–27. [CrossRef]
3. Wang, H.; Chen, L.; Qu, Z.; Yin, Y.; Kang, Q.; Yu, B.; Tao, W.-Q. Modeling of multi-scale transport phenomena in shale gas production—A critical review. *Appl. Energy* **2020**, *262*, 114575. [CrossRef]
4. Hu, Q.; Ewing, R.P.; Rowe, H.D. Low nanopore connectivity limits gas production in Barnett formation. *J. Geophys. Res. Solid Earth* **2015**, *120*, 8073–8087. [CrossRef]
5. Cardott, B.J.; Landis, C.R.; Curtis, M.E. Post-oil solid bitumen network in the Woodford Shale, USA—A potential primary migration pathway. *Int. J. Coal Geol.* **2015**, *139*, 106–113. [CrossRef]
6. Gou, Q.; Xu, S. Quantitative evaluation of free gas and adsorbed gas content of Wufeng-Longmaxi shales in the Jiaoshiba area, Sichuan Basin, China. *Adv. Geo-Energy Res.* **2019**, *3*, 258–267. [CrossRef]
7. Jia, Y.Q.; Han, D.L.; Zhang, J.Z.; Wang, C.C.; Lin, W.; Ren, X.H.; Yang, C.Y.; Chang, L.C. Differences in Pore-forming Efficiency among Organic Macerals and Its Restriction against Reservoir Quality: A Case Study Based on the Marine Shale Reservoir in the Longmaxi Formation, Southern Sichuan Basin, China. *Lithosphere* **2021**, *2021*, 2700912. [CrossRef]
8. Tian, H.; Pan, L.; Zhang, T.; Xiao, X.; Meng, Z.; Huang, B. Pore characterization of organic-rich Lower Cambrian shales in Qiannan Depression of Guizhou Province, Southwestern China. *Mar. Pet. Geol.* **2015**, *62*, 28–43. [CrossRef]
9. Loucks, R.G.; Reed, R.M.; Ruppel, S.C.; Jarvie, D.M. Morphology, Genesis, and Distribution of Nanometer-Scale Pores in Siliceous Mudstones of the Mississippian Barnett Shale. *J. Sediment. Res.* **2009**, *79*, 848–861. [CrossRef]
10. Cao, T.; Song, Z.; Wang, S.; Cao, X.; Li, Y.; Xia, J. Characterizing the pore structure in the Silurian and Permian shales of the Sichuan Basin, China. *Mar. Pet. Geol.* **2015**, *61*, 140–150. [CrossRef]
11. Curtis, M.E.; Cardott, B.J.; Sondergeld, C.H.; Rai, C.S. Development of organic porosity in the Woodford Shale with increasing thermal maturity. *Int. J. Coal Geol.* **2012**, *103*, 26–31. [CrossRef]
12. Loucks, R.G.; Reed, R.M.; Ruppel, S.C.; Hammes, U. Spectrum of pore types and networks in mudrocks and a descriptive classification for matrix-related mudrock pores. *AAPG Bull.* **2012**, *96*, 1071–1098. [CrossRef]

13. Ross, D.J.K.; Marc Bustin, R. The importance of shale composition and pore structure upon gas storage potential of shale gas reservoirs. *Mar. Pet. Geol.* **2009**, *26*, 916–927. [CrossRef]
14. Dewers, T.A.; Heath, J.; Ewy, R.; Duranti, L. Three-dimensional pore networks and transport properties of a shale gas formation determined from focused ion beam serial imaging. *Int. J. Oil Gas Coal Technol.* **2012**, *5*, 229–248. [CrossRef]
15. Jiang, T.; Jin, Z.; Hu, Z.; Du, W.; Liu, Z.; Zhao, J.; Zhang, K. Three-Dimensional Morphology and Connectivity of Organic Pores in Shale from the Wufeng and Longmaxi Formations at the Southeast Sichuan Basin in China. *Geofluids* **2021**, *2021*, 5579169. [CrossRef]
16. Jiang, X.; Deng, S.; Li, H.; Zuo, H. Characterization of 3D pore nanostructure and stress-dependent permeability of organic-rich shales in northern Guizhou Depression, China. *J. Rock Mech. Geotech. Eng.* **2022**, *14*, 407–422. [CrossRef]
17. Mandelbrot, B.B. Stochastic models for the Earth's relief, the shape and the fractal dimension of the coastlines, and the number-area rule for islands. *Proc. Natl. Acad. Sci. USA* **1975**, *72*, 3825–3828. [CrossRef]
18. Sakhaee-Pour, A.; Li, W. Fractal dimensions of shale. *J. Nat. Gas Sci. Eng.* **2016**, *30*, 578–582. [CrossRef]
19. Katz, A.J.; Thompson, A.H. Fractal sandstone pores: Implications for conductivity and pore formation. *Phys. Rev. Lett.* **1985**, *54*, 1325–1328. [CrossRef]
20. Yuan, W.; Pan, Z.; Li, X.; Yang, Y.; Zhao, C.; Connell, L.D.; Li, S.; He, J. Experimental study and modelling of methane adsorption and diffusion in shale. *Fuel* **2014**, *117*, 509–519. [CrossRef]
21. Zhan, H.M.; Li, X.Z.; Hu, Z.M.; Duan, X.G.; Wu, W.; Guo, W.; Lin, W. Fractal Characteristics of Deep Shales in Southern China by Small-Angle Neutron Scattering and Low-Pressure Nitrogen Adsorption. *Fractal Fract.* **2022**, *6*, 484. [CrossRef]
22. Yang, R.; He, S.; Yi, J.; Hu, Q. Nano-scale pore structure and fractal dimension of organic-rich Wufeng-Longmaxi shale from Jiaoshiba area, Sichuan Basin: Investigations using FE-SEM, gas adsorption and helium pycnometry. *Mar. Pet. Geol.* **2016**, *70*, 27–45. [CrossRef]
23. Yao, Y.; Liu, D.; Tang, D.; Tang, S.; Huang, W. Fractal characterization of adsorption-pores of coals from North China: An investigation on CH4 adsorption capacity of coals. *Int. J. Coal Geol.* **2008**, *73*, 27–42. [CrossRef]
24. Shao, X.; Pang, X.; Li, Q.; Wang, P.; Chen, D.; Shen, W.; Zhao, Z. Pore structure and fractal characteristics of organic-rich shales: A case study of the lower Silurian Longmaxi shales in the Sichuan Basin, SW China. *Mar. Pet. Geol.* **2017**, *80*, 192–202. [CrossRef]
25. Tang, X.; Zheng, F.; Liang, G.; Ma, Z.; Zhang, J.; Wang, Y.; Zhang, T. Fractal characterization of pore structure in Cambrian Niutitang shale in northern Guizhou, southwestern China. *Earth Sci. Front.* **2023**, *30*, 110–123. [CrossRef]
26. Zhang, Q.; Dong, Y.H.; Tong, S.Q. Pore-Type-Dependent Fractal Features of Shales and Implications on Permeability. *Fractal Fract.* **2023**, *7*, 803. [CrossRef]
27. Zhang, J.; Li, X.; Xie, Z.; Li, J.; Zhang, X.; Sun, K.; Wang, F. Characterization of microscopic pore types and structures in marine shale: Examples from the Upper Permian Dalong formation, Northern Sichuan Basin, South China. *J. Nat. Gas Sci. Eng.* **2018**, *59*, 326–342. [CrossRef]
28. Guan, Q.Z.; Dong, D.Z.; Wang, S.F.; Huang, J.L.; Wang, Y.M.; Lu, H.; Zhang, C.C. Preliminary study on shale gas microreservoir characteristics of the Lower Silurian Longmaxi Formation in the southern Sichuan Basin, China. *J. Nat. Gas Sci. Eng.* **2016**, *31*, 382–395. [CrossRef]
29. Wang, Y.L.; Jia, Y.Q.; Wang, C.C.; Lin, W.; Zhang, J.Z.; Han, D.L.; Ma, B.Y.; Wang, H.C. Characteristics and Geological Significance of Organic Matter Veins in Shale Reservoir: A Case Study of the Silurian Longmaxi Formation in Luzhou Area, Sichuan Basin. *Minerals* **2023**, *13*, 1080. [CrossRef]
30. Jia, Y.; Liu, Z.; Ren, X.; Zhou, Y.; Zheng, A.; Zhang, J.; Han, D. Organic Matter Type Differentiation Process and Main Control Mechanism: Case study of the Silurian Longmaxi Formation shale reservoir in Weiyuan area. *Acta Sedimentol. Sin.* **2021**, *39*, 341–352. [CrossRef]
31. Teng, G.; Lu, L.; Yu, L.; Zhang, W.; Pan, A.; Shen, B.; Wang, Y.; Yang, Y.; Gao, Z. Formation, preservation and connectivity control of organic pores in shale. *Pet. Explor. Dev.* **2021**, *48*, 687–699. [CrossRef]
32. Chen, J.; Xiao, X. Evolution of nanoporosity in organic-rich shales during thermal maturation. *Fuel* **2014**, *129*, 173–181. [CrossRef]
33. Gai, H.; Tian, H.; Cheng, P.; He, C.; Wu, Z.; Ji, S.; Xiao, X. Characteristics of molecular nitrogen generation from overmature black shales in South China: Preliminary implications from pyrolysis experiments. *Mar. Pet. Geol.* **2020**, *120*, 104527. [CrossRef]
34. Shi, X.; Wu, W.; Xu, L.; Yin, Y.; Yang, Y.; Liu, J.; Yang, X.; Li, Y.; Wu, Q.; Zhong, K.; et al. Thermal Maturity Constraint Effect and Development Model of Shale Pore Structure: A Case Study of Longmaxi Formation Shale in Southern Sichuan Basin, China. *Minerals* **2024**, *14*, 163. [CrossRef]
35. Yuan, G.; Jin, Z.; Cao, Y.; Schulz, H.M.; Gluyas, J.; Liu, K.; He, X.; Wang, Y. Microdroplets initiate organic-inorganic interactions and mass transfer in thermal hydrous geosystems. *Nat. Commun.* **2024**, *15*, 4960. [CrossRef] [PubMed]
36. Gao, Z.; Fan, Y.; Xuan, Q.; Zheng, G. A review of shale pore structure evolution characteristics with increasing thermal maturities. *Adv. Geo-Energy Res.* **2020**, *4*, 247–259. [CrossRef]
37. Liu, B.; Mastalerz, M.; Schieber, J. SEM petrography of dispersed organic matter in black shales: A review. *Earth-Sci. Rev.* **2022**, *224*, 103874. [CrossRef]
38. Shan, C.a.; Shi, Y.; Liang, X.; Zhang, L.; Wang, G.; Jiang, L.; Zou, C.; He, F.; Mei, J. Diagenetic characteristics and microscopic pore evolution of deep shale gas reservoirs in Longmaxi Formation, Southeastern Sichuan basin, China. *Unconv. Resour.* **2024**, *4*, 100090. [CrossRef]

39. Khitab, U.; Umar, M.; Jamil, M. Microfacies, diagenesis and hydrocarbon potential of Eocene carbonate strata in Pakistan. *Carbonates Evaporites* **2020**, *35*, 70. [CrossRef]
40. Wang, R.Y.; Gu, Y.; Ding, W.L.; Gong, D.J.; Yin, S.; Wang, X.H.; Zhou, X.H.; Li, A.; Xiao, Z.K.; Cui, Z.X. Characteristics and dominant controlling factors of organic-rich marine shales with high thermal maturity: A case study of the Lower Cambrian Niutitang Formation in the Cen'gong block, southern China. *J. Nat. Gas Sci. Eng.* **2016**, *33*, 81–96. [CrossRef]
41. Nie, H.; Jin, Z.; Zhang, J. Characteristics of three organic matter pore types in the Wufeng-Longmaxi Shale of the Sichuan Basin, Southwest China. *Sci. Rep.* **2018**, *8*, 7014. [CrossRef] [PubMed]
42. Guo, T. Evaluation of Highly Thermally Mature Shale-Gas Reservoirs in Complex Structural Parts of the Sichuan Basin. *J. Earth Sci.* **2013**, *24*, 863–873. [CrossRef]
43. Guo, X.; Hu, D.; Li, Y.; Wei, Z.; Wei, X.; Liu, Z. Geological factors controlling shale gas enrichment and high production in Fuling shale gas field. *Pet. Explor. Dev.* **2017**, *44*, 513–523. [CrossRef]
44. Li, X.; Zhu, H.; Zhang, K.; Li, Z.; Yu, Y.; Feng, X.; Wang, Z. Pore characteristics and pore structure deformation evolution of ductile deformed shales in the Wufeng-Longmaxi Formation, southern China. *Mar. Pet. Geol.* **2021**, *127*, 104992. [CrossRef]
45. Liang, M.; Wang, Z.; Gao, L.; Li, C.; Li, H. Evolution of pore structure in gas shale related to structural deformation. *Fuel* **2017**, *197*, 310–319. [CrossRef]
46. Ma, Y.; Zhong, N.; Li, D.; Pan, Z.; Cheng, L.; Liu, K. Organic matter/clay mineral intergranular pores in the Lower Cambrian Lujiaping Shale in the north-eastern part of the upper Yangtze area, China: A possible microscopic mechanism for gas preservation. *Int. J. Coal Geol.* **2015**, *137*, 38–54. [CrossRef]
47. Zhu, H.; Ju, Y.; Huang, C.; Han, K.; Qi, Y.; Shi, M.; Yu, K.; Feng, H.; Li, W.; Ju, L.; et al. Pore structure variations across structural deformation of Silurian Longmaxi Shale: An example from the Chuandong Thrust-Fold Belt. *Fuel* **2019**, *241*, 914–932. [CrossRef]
48. Fu, Y.; Jiang, Y.; Dong, D.; Hu, Q.; Lei, Z.; Peng, H.; Gu, Y.; Ma, S.; Wang, Z.; Yin, X.; et al. Microscopic pore-fracture configuration and gas-filled mechanism of shale reservoirs in the western Chongqing area, Sichuan Basin, China. *Pet. Explor. Dev.* **2021**, *48*, 916–927. [CrossRef]
49. Cheng, G.; Jiang, B.; Li, F.; Li, M.; Wu, C. Evolution Mechanism of Pore Structures of Organic-Rich Shale Under Tectonic Deformation: A Comparative Study between Whole Rock and Kerogen Samples. *Nat. Resour. Res.* **2023**, *33*, 263–297. [CrossRef]
50. Shang, F.; Zhu, Y.; Gao, H.; Wang, Y.; Liu, R. Relationship between Tectonism and Composition and Pore Characteristics of Shale Reservoirs. *Geofluids* **2020**, *2020*, 9426586. [CrossRef]
51. Liu, S.G.; Deng, B.; Li, Z.W.; Sun, W. Architecture of basin-mountain systems and their influences on gas distribution: A case study from the Sichuan basin, South China. *J. Asian Earth Sci.* **2012**, *47*, 204–215. [CrossRef]
52. Zhang, G.; Guo, A.; Wang, Y.; Li, S.; Dong, Y.; Liu, S.; He, D.; Cheng, S.; Lu, R.; Yao, A. Tectonics of South China continent and its implications. *Sci. China Earth Sci.* **2013**, *56*, 1804–1828. [CrossRef]
53. Liang, X.; Xu, J.; Wang, Y.; Fang, X.; Zhao, L.; Wu, Y.; Deng, B. The shale gas enrichment factors of Longmaxi Formation under gradient basin-mountain boundary in South Sichuan Basin: Tectono-depositional differentiation and discrepant evolution. *Chin. J. Geol.* **2021**, *56*, 60–81.
54. Yang, R.; He, S.; Wang, X.; Hu, Q.; Hu, D.; Yi, J. Paleo-ocean redox environments of the Upper Ordovician Wufeng and the first member in lower Silurian Longmaxi formations in the Jiaoshiba area, Sichuan Basin. *Can. J. Earth Sci.* **2016**, *53*, 426–440. [CrossRef]
55. Nie, H.K.; He, Z.L.; Wang, R.Y.; Zhang, G.R.; Chen, Q.; Li, D.H.; Lu, Z.Y.; Sun, C.X. Temperature and origin of fluid inclusions in shale veins of Wufeng?Longmaxi Formations, Sichuan Basin, south China: Implications for shale gas preservation and enrichment. *J. Pet. Sci. Eng.* **2020**, *193*, 18. [CrossRef]
56. Wu, J.; Chen, X.; Liu, W.; Wu, W.; Gao, Y.; Luo, C.; Deng, B.; Zhou, Z. Fluid activity and pressure evolution process of Wufeng-Longmaxi shales, southern Sichuan Basin. *Earth Sci. J. China Univ. Geosci.* **2022**, *47*, 518–531. [CrossRef]
57. Dong, M.; Guo, W.; Zhang, L.; Wu, Z.; Ma, L.; Dong, H.; Feng, X.; Yang, Y. Characteristics of paleotectonic stress field and fractures of Wufeng-Longmaxi formations in Luzhou area, southern Sichuan Basin. *Lithol. Reserv.* **2022**, *34*, 43–51. [CrossRef]
58. Huang, H.Y.; He, D.F.; Li, Y.Q.; Li, J.; Zhang, L. Silurian tectonic-sedimentary setting and basin evolution in the Sichuan area, southwest China: Implications for palaeogeographic reconstructions. *Mar. Pet. Geol.* **2018**, *92*, 403–423. [CrossRef]
59. Yan, H.; Peng, X.; Xia, Q.; Xu, W.; Luo, W.; Li, X.; Zhang, L.; Zhu, Q.; Zhu, X.; Liu, X. Distribution features of ancient karst landform in the fourth Member of the Dengying Formation in the Gaoshiti-Moxi region and its guiding significance for gas reservoir development. *Acta Pet. Sin.* **2020**, *41*, 658–670+752. [CrossRef]
60. Ma, X.H.; Xie, J.; Yong, R.; Zhu, Y.Q. Geological characteristics and high production control factors of shale gas reservoirs in Silurian Longmaxi Formation, southern Sichuan Basin, SW China. *Pet. Explor. Dev.* **2020**, *47*, 901–915. [CrossRef]
61. Qiu, Z.; Tao, H.; Lu, B.; Chen, Z.; Wu, S.; Liu, H.; Qiu, J. Controlling Factors on Organic Matter Accumulation of Marine Shale across the Ordovician-Silurian Transition in South China: Constraints from Trace-Element Geochemistry. *J. Earth Sci.* **2021**, *32*, 887–900. [CrossRef]
62. Yang, X.R.; Yan, D.T.; Zhang, B.; Zhang, L.W.; Wei, X.S.; Li, T.; Zhang, J.F.; She, X.H. The impact of volcanic activity on the deposition of organic-rich shales: Evidence from carbon isotope and geochemical compositions. *Mar. Pet. Geol.* **2021**, *128*, 14. [CrossRef]
63. Nie, H.; Bao, S.; Gao, B.; Bian, R.; Zhang, P.; Wu, X.; Ye, X.; Chen, X. A study of shale gas preservation conditiongs for the Lower Paleozoic in Sichuan Basin and its periphery. *Earth Sci. Front.* **2012**, *19*, 280–294.

64. Ma, Y.; Cai, X.; Zhao, P. China's shale gas exploration and development: Understanding and practice. *Pet. Explor. Dev.* **2018**, *45*, 589–603. [CrossRef]
65. Saif, T.; Lin, Q.; Butcher, A.R.; Bijeljic, B.; Blunt, M.J. Multi-scale multi-dimensional microstructure imaging of oil shale pyrolysis using X-ray micro-tomography, automated ultra-high resolution SEM, MAPS Mineralogy and FIB-SEM. *Appl. Energy* **2017**, *202*, 628–647. [CrossRef]
66. *GB/T 20726-2015;* Microbeam Analysis—Selected Instrumental Performance Parameters for the Specification and Checking of Energy Dispersive X-ray Spectrometers for Use in Electron Probe Microanalysis. Standardization Administration of China: Beijing, China, 2015.
67. Ma, B.; Hu, Q.; Yang, S.; Zhang, T.; Qiao, H.; Meng, M.; Zhu, X.; Sun, X. Pore structure typing and fractal characteristics of lacustrine shale from Kongdian Formation in East China. *J. Nat. Gas Sci. Eng.* **2021**, *85*, 103709. [CrossRef]
68. Zhang, J.; Li, X.; Wei, Q.; Sun, K.; Zhang, G.; Wang, F. Characterization of Full-Sized Pore Structure and Fractal Characteristics of Marine–Continental Transitional Longtan Formation Shale of Sichuan Basin, South China. *Energy Fuels* **2017**, *31*, 10490–10504. [CrossRef]
69. Lin, W.; Zhao, X.; Li, M.; Zhuang, Y. Pore Structure Characterization and Fractal Characteristics of Tight Limestone Based on Low-Temperature Nitrogen Adsorption and Nuclear Magnetic Resonance. *Fractal Fract.* **2024**, *8*, 371. [CrossRef]
70. Dathe, A.; Eins, S.; Niemeyer, J.; Gerold, G. The surface fractal dimension of the soil–pore interface as measured by image analysis. *Geoderma* **2001**, *103*, 203–229. [CrossRef]
71. Peng, N.; He, S.; Hu, Q.; Zhang, B.; He, X.; Zhai, G.; He, C.; Yang, R. Organic nanopore structure and fractal characteristics of Wufeng and lower member of Longmaxi shales in southeastern Sichuan, China. *Mar. Pet. Geol.* **2019**, *103*, 456–472. [CrossRef]
72. Zou, X.; Li, X.; Wang, Y.; Zhang, J.; Zhao, P. Reservoir characteristics and gas content of Wufeng-Longmaxi Formations deep shale in southern Sichuan Basin. *Nat. Gas Geosci.* **2022**, *33*, 654–665. [CrossRef]
73. Wang, R.; Hu, Z.; Zhou, T.; Bao, H.; Wu, J.; Du, W.; He, J.; Wang, P.; Chen, Q. Characteristics of fractures and their significance for reservoirs in Wufeng-Longmaxi shale, Sichuan Basin and its periphery. *Oil Gas Geol.* **2021**, *42*, 1295–1306. [CrossRef]
74. Thommes, M.; Kaneko, K.; Neimark, A.V.; Olivier, J.P.; Rodriguez-Reinoso, F.; Rouquerol, J.; Sing, K.S.W. Physisorption of gases, with special reference to the evaluation of surface area and pore size distribution (IUPAC Technical Report). *Pure Appl. Chem.* **2015**, *87*, 1051–1069. [CrossRef]
75. Sing, K.S.W. Reporting physisorption data for gas/solid systems with special reference to the determination of surface area and porosity (Recommendations 1984). *Pure Appl. Chem.* **1985**, *57*, 603–619. [CrossRef]
76. Labani, M.M.; Rezaee, R.; Saeedi, A.; Hinai, A.A. Evaluation of pore size spectrum of gas shale reservoirs using low pressure nitrogen adsorption, gas expansion and mercury porosimetry: A case study from the Perth and Canning Basins, Western Australia. *J. Pet. Sci. Eng.* **2013**, *112*, 7–16. [CrossRef]
77. Barrett, E.P.; Joyner, L.G.; Halenda, P.P. The Determination of Pore Volume and Area Distributions in Porous Substances. I. Computations from Nitrogen Isotherms. *J. Am. Chem. Soc.* **1951**, *73*, 373–380. [CrossRef]
78. Klimakow, M.; Klobes, P.; Rademann, K.; Emmerling, F. Characterization of mechanochemically synthesized MOFs. *Microporous Mesoporous Mater.* **2012**, *154*, 113–118. [CrossRef]
79. Hazra, B.; Wood, D.A.; Vishal, V.; Varma, A.K.; Sakha, D.; Singh, A.K. Porosity controls and fractal disposition of organic-rich Permian shales using low-pressure adsorption techniques. *Fuel* **2018**, *220*, 837–848. [CrossRef]
80. Ning, S.; Xia, P.; Hao, F.; Tian, J.; Fu, Y.; Wang, K. Pore Fractal Characteristics between Marine and Marine–Continental Transitional Black Shales: A Case Study of Niutitang Formation and Longtan Formation. *Fractal Fract.* **2024**, *8*, 288. [CrossRef]
81. Zou, J.; Chen, W.; Yang, D.; Yuan, J.; Jiao, Y.-Y. Fractal characteristics of the anisotropic microstructure and pore distribution of low-rank coal. *AAPG Bull.* **2019**, *103*, 1297–1320. [CrossRef]
82. Milliken, K.L.; Curtis, M.E. Imaging pores in sedimentary rocks: Foundation of porosity prediction. *Mar. Pet. Geol.* **2016**, *73*, 590–608. [CrossRef]
83. Wang, G. Deformation of organic matter and its effect on pores in mud rocks. *AAPG Bull.* **2020**, *104*, 21–36. [CrossRef]
84. Yao, C.; Fu, H.; Ma, Y.; Yan, D.; Wang, H.; Li, Y.; Wang, J. Development characteristics of deep shale fractured veins and vein forming fluid activities in Luzhou block. *Earth Sci.* **2022**, *47*, 1684–1693.
85. Shovkun, I.; Espinoza, D.N. Geomechanical implications of dissolution of mineralized natural fractures in shale formations. *J. Pet. Sci. Eng.* **2018**, *160*, 555–564. [CrossRef]
86. Wu, Q.; Han, D.; Zhang, J.; Wang, C.; Ren, X.; Lin, Z.; Su, M.; Zhu, Y.; Zhang, J. Influence of particle size effect on microfracture development in shale reservoir: A case study of the Silurian Longmaxi Formation shale reservoir in Weiyuan area. *J. Cent. South Univ. (Sci. Technol.)* **2022**, *53*, 3603–3614.
87. Pan, Z.; Liu, D.; Huang, Y.; Zhenxue, J.; Song, Y.; Guo, J.; Li, C. Paleotemperature and paleopressure of methane inclusions in fracture cements from the Wufeng-Longmaxi shales in the Luzhou area, southern Sichuan Basin. *Pet. Sci. Bull.* **2019**, *4*, 242–253.
88. Ju, Y.; Sun, Y.; Tan, J.; Bu, H.; Han, K.; Li, X.; Fang, L. The composition, pore structure characterization and deformation mechanism of coal-bearing shales from tectonically altered coalfields in eastern China. *Fuel* **2018**, *234*, 626–642. [CrossRef]
89. Wang, H.; Shi, Z.; Sun, S.; Zhao, Q.; Zhou, T.; Cheng, F.; Bai, W. Microfacies types and distribution of epicontinental shale: A case study of the Wufeng-Longmaxi shale in southern Sichuan Basin, China. *Pet. Explor. Dev.* **2023**, *50*, 57–71. [CrossRef]
90. Hu, D.; Zhang, H.; Ni, K.; Yu, G. Main controlling factors for gas preservation conditions of marine shales in southeastern margins of the Sichuan Basin. *Nat. Gas Ind.* **2014**, *34*, 17–23.

91. Zhang, K. Preservation Mechanism and Evaluation Method of Marine Shale Gas with Complex Tectonic Background. Ph.D. Dissertation, China University of Petroleum (Beijing), Beijing, China, 2019.
92. Liu, S.; Deng, B.; Zhong, Y.; Ran, B.; Yong, Z.; Sun, W.; Yang, D.; Jiang, L.; Ye, Y. Unique geological featrures of burial and superimposition of the Lower Paleozoic shale gas across the Sichuan Basin and its periphery. *Earth Sci. Front.* **2016**, *23*, 11–28. [CrossRef]
93. Yang, Y. An unrecognized major collision of the Okhotomorsk Block with East Asia during the Late Cretaceous, constraints on the plate reorganization of the Northwest Pacific. *Earth-Sci. Rev.* **2013**, *126*, 96–115. [CrossRef]
94. Zhang, Y.; He, Z.; Lu, S.; Jiang, S.; Xiao, D.; Long, S.; Gao, B.; Du, W.; Zhao, J.; Chen, G.; et al. Characteristics of microorganisms and origin of organic matter in Wufeng Formation and Longmaxi Formation in Sichuan Basin, South China. *Mar. Pet. Geol.* **2020**, *111*, 363–374. [CrossRef]
95. Chang, X.; Huang, Y.; Chen, Z.; Hou, M. The Microscopic Analysis of Pyrite Framboids and Application in Paleo-oceanography. *Acta Sedimentol. Sin.* **2020**, *38*, 150–165. [CrossRef]
96. Shi, Z.; Yuan, Y.; Zhao, Q.; Sun, S.; Zhou, T.; Cheng, F. Paleogeomorphology and oil-bearing shale characteristics of the Wufeng-Longmaxi shale in southern Sichuan Basin, China. *Nat. Gas Geosci.* **2022**, *33*, 1969–1985.
97. Shi, Z.; Zhao, S.; Zhou, T.; Ding, L.; Sun, S.; Cheng, F. Mineralogy and Geochemistry of the Upper Ordovician and Lower Silurian Wufeng-Longmaxi Shale on the Yangtze Platform, South China: Implications for Provenance Analysis and Shale Gas Sweet-Spot Interval. *Minerals* **2022**, *12*, 1190. [CrossRef]
98. Qiu, Z.; Liu, B.; Dong, D.; Lu, B.; Yawar, Z.; Chen, Z.; Schieber, J. Silica diagenesis in the Lower Paleozoic Wufeng and Longmaxi Formations in the Sichuan Basin, South China: Implications for reservoir properties and paleoproductivity. *Mar. Pet. Geol.* **2020**, *121*, 104594. [CrossRef]
99. Xiang-Rong, Y.; De-Tian, Y.; Xiao-Song, W.; Li-Wei, Z.; Bao, Z.; Han-Wen, X.; Yin, G.; Jie, H. Different formation mechanism of quartz in siliceous and argillaceous shales: A case study of Longmaxi Formation in South China. *Mar. Pet. Geol.* **2018**, *94*, 80–94. [CrossRef]
100. Dong, T.; He, Q.; He, S.; Zhai, G.; Zhang, Y.; Wei, S.; Wei, C.; Hou, Y.; Guo, X. Quartz types, origins and organic matter-hosted pore systems in the lower cambrian Niutitang Formation, middle yangtze platform, China. *Mar. Pet. Geol.* **2021**, *123*, 104739. [CrossRef]
101. Cao, T.; Song, Z.; Wang, S.; Xia, J. Characterization of pore structure and fractal dimension of Paleozoic shales from the northeastern Sichuan Basin, China. *J. Nat. Gas Sci. Eng.* **2016**, *35*, 882–895. [CrossRef]
102. Xie, X.; Deng, H.; Li, Y.; Hu, L.; Mao, J.; Li, R. Investigation of the Oriented Structure Characteristics of Shale Using Fractal and Structural Entropy Theory. *Fractal Fract.* **2022**, *6*, 734. [CrossRef]
103. Ju, Y.-W.; Hou, X.-G.; Han, K.; Song, Y.; Xiao, L.; Huang, C.; Zhu, H.-J.; Tao, L.-R. Experimental deformation of shales at elevated temperature and pressure: Pore-crack system evolution and its effects on shale gas reservoirs. *Pet. Sci.* 2024, *in press*. [CrossRef]
104. Ko, L.T.; Ruppel, S.C.; Loucks, R.G.; Hackley, P.C.; Zhang, T.; Shao, D. Pore-types and pore-network evolution in Upper Devonian-Lower Mississippian Woodford and Mississippian Barnett mudstones: Insights from laboratory thermal maturation and organic petrology. *Int. J. Coal Geol.* **2018**, *190*, 3–28. [CrossRef]
105. Zeng, L.; Zhu, H.; Hao, Y.; Shi, E.; Lu, Y.; Qi, Y. Microstructures and Pore Structures of Naturally Deformed Shales: Significance for Shale Gas Exploration in the Complex Tectonic Areas of the Margin of the Sichuan Basin, South China. *Energy Fuels* **2023**, *37*, 10414–10425. [CrossRef]
106. Ko, L.T.; Loucks, R.G.; Zhang, T.; Ruppel, S.C.; Shao, D. Pore and pore network evolution of Upper Cretaceous Boquillas (Eagle Ford-equivalent) mudrocks: Results from gold tube pyrolysis experiments. *AAPG Bull.* **2016**, *100*, 1693–1722. [CrossRef]

fractal and fractional

MDPI

Article

Pore Space Characteristics and Migration Changes in Hydrocarbons in Shale Reservoir

Yiqian Qu [1,2], Siqi Ouyang [1,*], Jianwen Gao [3], Jian Shi [3], Yiying Wu [4], Yuting Cheng [1], Zhen Zhou [1], Zhou Lyu [5], Wei Sun [1] and Hanning Wu [1]

1 State Key Laboratory of Continental Dynamics, Department of Geology, Northwest University, Northern Taibai Street 229, Xi'an 710069, China; quyq@nwu.edu.cn (Y.Q.); 202233554@stumail.nwu.edu.cn (Y.C.); 202221649@stumail.nwu.edu.cn (Z.Z.); sunwei5393901@vip.sohu.com (W.S.); wuhn2506@nwu.edu.cn (H.W.)

2 Xi'an Key Laboratory of Tight Oil (Shale Oil) Development (Xi'an Shiyou University), Xi'an 710065, China

3 Research Institute of Exploration and Development, PetroChina Changqing Oilfield Company, Xi'an 710018, China; gaojianw_cq@petrochina.com.cn (J.G.); shijian2371@163.com (J.S.)

4 College of Urban and Environmental Science, Northwest University, Xuefu Avenue 1, Xi'an 710127, China; 20210148@nwu.edu.cn

5 Research Institute of Petroleum Exploration and Development, China National Petroleum Corporation, Beijing 100083, China; lyuzhou827@petrochina.com.cn

* Correspondence: siqi_oy@163.com

Abstract: The pore structure and mineral characteristics affect the accumulation and migration of hydrocarbons in shale, which determines the production capacity of shale oil. In this study, shale samples from the Chang 7 member of the Ordos Basin in China were selected to investigate the pore space characteristics, the effect of hydrocarbon accumulation on pore heterogeneity, and the hydrocarbon migration changes based on fractal theory, and a series of experiments were conducted involving X-ray diffraction (XRD), total organic carbon (TOC), Soxhlet extraction, and low-temperature nitrogen (N_2) and carbon dioxide (CO_2) adsorption. Then, the factors affecting extraction efficiency in shale pores were discussed. The interparticle pores contributed most to the accumulation of shale oil, and the organic matter (OM) pores contributed positively to the adsorption of hydrocarbons. The accumulation of hydrocarbons in the pore space did not increase the heterogeneity of the shale pore structure. The contents, states, and positions of hydrocarbons changed during the extraction process. Hydrocarbons were redistributed on the pore surface after Soxhlet extraction, and the heterogeneity of hydrocarbon adsorption and pore surface roughness were improved. Some heavy hydrocarbons and adsorbed components were pyrolyzed, resulting in the gradual escape of the adsorbed layer in the large pores. However, the free oil in the small pores diffused to the large pores and reaggregated on the surface, restoring a stable adsorption layer. The extraction rate was closely related to the pore throat structure and the wettability of mineral surfaces. The configuration between pores and throats had a crucial influence on the extraction rate. A high proportion of meso-pores, which effectively connect micro- and macro-pores, had a higher diffusion efficiency and a higher extraction rate. The OM pores with high energy adsorption were located in the micro-pores, and the shale oil existed in a dissolved state with high mobile capacity. The wettability of mineral surfaces affected the adsorption behavior during extraction, and strong oil wetting promoted hydrocarbon re-adsorption in clay minerals, so that the volume of micro-pores was smaller after extraction.

Keywords: pore structure; Soxhlet extraction; shale; fractal theory; Yanchang Formation

1. Introduction

Unconventional resources have received more attention as important alternative energy in recent years [1–3]. Shale reservoirs are known for their low porosity and low permeability, integrated source, storage, and rich organic matter [4–6]. With the development of shale oil and gas exploration and exploitation technology, China has stepped into a rapid development stage [7,8]. China is rich in shale oil resources, characterized by multiple layers, large thickness, high clay mineral contents, strong heterogeneity of pore throat structure, complex mineral composition, hydrocarbon accumulation space, and low exploitation efficiency [3,9,10]. The pore structure and mineral composition affect the accumulation and migration of shale oil, and further determine the productivity. Therefore, the characterization of pore space in shale has received more attention [11–14], which established the geological foundation for the exploration and exploitation of shale oil.

Nano- and micro-pores are the main accumulation space for shale oil, and the volume of nanoscale pores in shale accounts for more than 80% of the total pore volume [15,16]. According to the International Union of Pure and Applied Chemistry (IUPAC) system, pores in shales can be divided into micro- (<2 nm), meso- (2–50 nm), and macro-pores (>50 nm). The pore types of shale mainly include interparticle pores, dissolution pores, intercrystalline pores, organic pores, and micro-fractures [5,17]. The physical and chemical properties of shale reservoirs are different from conventional reservoirs [18], such as capillary force, molecular diffusion, interface effect, etc., which makes permeability and porosity measurements lose accuracy when evaluating reservoir properties. Therefore, quantitative evaluation of pore characteristics in shale from a microscopic perspective is of great significance for understanding shale oil reservoirs.

In recent years, the qualitative–quantitative characterization of the pore structure of shale has gradually become a hot topic in unconventional oil research, including the pore type, size, morphology, structure, and controlling factors [19–23]. Scanning electron microscopy (SEM), transmission electron microscopy (TEM), the mercury intrusion method (MIP), the gas adsorption method (CO_2 and N_2), and nuclear magnetic resonance (NMR) have been used to characterize shale pore space [24–27]. The pore radius range obtained by each method has limitations [28,29]. For instance, the two-dimensional image method can reflect pores from 1 nm to 1 cm, the three-dimensional reconstruction can generally characterize those larger than 0.1 μm, and the quantitative characterization range is between 0.01 μm and 0.01 mm. Low-pressure gas adsorption is considered to be a common method for quantifying shale pore space, which is adaptable and widely accepted for evaluating shale pore space [21,30,31]. Also, the fractal theory has been regarded as a priority method to understand the complexity of pore throat structure [32–36].

The oil exists in organic-rich shale in a variety of states, including free, adsorbed, and dissolved [37–40]. As exploitation proceeds, the state of oil and its location in the pore space are complex and variable, affecting the migration of shale oil, and resulting in extremely low recovery rates from shale oil reservoirs. It is generally believed that shale oil is predominantly free in bedding and cracks, as well as in large interparticle and dissolution pores, and forms continuous aggregates of hydrocarbons, which are predominantly adsorbed in clay mineral intercrystalline pores, OM pores, and pyrite intercrystalline pores [40].

Crude oil in shale pore spaces can be extracted by high-temperature Soxhlet extraction [41]. The comparative analysis of pore distribution characteristics before and after Soxhlet extraction is of great significance for understanding hydrocarbon accumulation and migration in shale reservoirs. The extraction process has indicative significance for clarifying the mobility of shale oil and its migration in the pore space, because it is consistent with the shale oil exploitation process. Previous studies have proved that temperature, wettability, and capillary imbibition have important effects on the extraction process. Under high temperatures, some heavy hydrocarbons (heavy components of free hydrocarbons) and adsorption components are pyrolyzed, the extraction process is accelerated [37,42], and the accumulation state of soluble organic matter in shale is changed dynamically [7]. The imbibition of capillary forces causes the fluid to redistribute in the pore space. Pore throat structure is crucial to the accumulation and migration of shale oil [43]. In the extraction process, oil in nanoscale pores is replaced by water under capillary action. The water is imbibed into the small pores, and then diffused into the meso- and macro-pores [44,45]. Moreover, the movable capacity of shale oil is related to the wettability of mineral surfaces.

Complex physical and chemical reactions occur on different mineral surfaces with different properties of hydrocarbons [7].

Therefore, shale samples from the Chang 7 member of the Yanchang Formation in Ordos Basin, China were selected in this study, with the aim of understanding the characteristics of shale pore space and the hydrocarbon migration in pores before and after extraction. The changes in shale oil content and accumulation state in different sizes of pores before and after extraction were clarified by comparing the N_2 and CO_2 adsorption characteristics. Further, the complexity changes of pore space before and after extraction were compared by combining the analysis with fractal theory. The characteristics of hydrocarbon migration and re-adsorption in shale pores after extraction were studied based on mineral characteristics, TOC content, and pore structure. This study is the basic work to evaluate the mobility and migration of shale oil and has important theoretical significance for enhancing the recovery of shale oil.

2. Samples and Methods

2.1. Geological Background and Samples

The Ordos Basin is a large petroliferous basin in China [46,47], among which the Yanchang Formation of the Triassic system is the main oil reservoir. During the sedimentary period, the Yanchang Formation experienced the process of lake formation, development, and disappearance, and developed the river, delta, and lacustrine sedimentary system [48]. The Yanchang Formation was divided into 10 members from top to bottom (Figure 1). During the sedimentary period of Chang 7, the climate became warmer, and the salinity of the lake decreased. Deep lacustrine and semi-deep lacustrine deposits were formed under the influence of the paleoclimate, and dark lamina-layered shale and massive mudstone were developed. Zhangjiatan shale of Chang 7 is the main source rock in the basin, with a large thickness, high TOC content, and high maturity of organic matter [1,49], and is also the main shale oil reservoir in the basin.

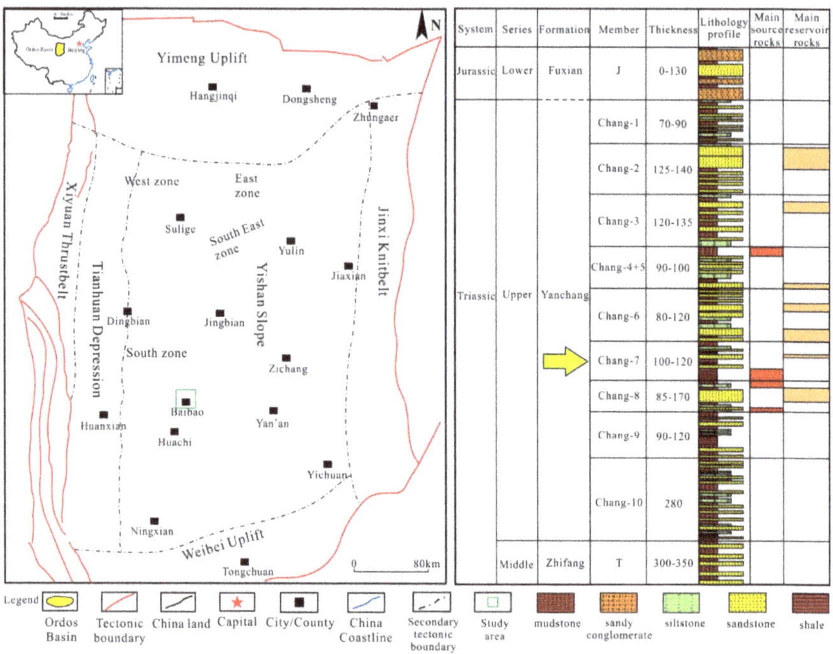

Figure 1. Tectonic units and study area and sampling points in the Ordos Basin.

Six typical shale samples were collected from the Chang 7 member in the Huaqing area in this study. The lithology of the six samples is mainly dark, massive, and lamina-layered shale, and the depth ranged from 2064.3 m to 2205.7 m.

2.2. *Experiments*

The mineral composition and geochemical characteristics of six samples were determined by petrological and geochemical analysis. Soxhlet extraction was used to remove hydrocarbons. Low-temperature N_2 and CO_2 adsorption was performed to study the pore structure of shale samples before and after extraction, respectively.

2.2.1. XRD and TOC

XRD was performed in the State Key Laboratory of Continental Dynamics at Northwest University, P. R. China. Before tests, samples were crushed into powder of more than 200 mesh (74 μm) in an agate mortar. Then, the powder was placed on a glass slide to test.

TOC is an important indicator of the enrichment of organic matter in shale. The TOC content was tested by the LECO-CS230 carbon sulfur analyzer (MI, USA) in the State Key Laboratory of Continental Dynamics at Northwest University, P. R., China. Firstly, about 0.2 g of powder samples of more than 200 mesh were weighed and hydrolyzed with excess hydrogen chloride to remove carbonates. The reaction temperature was maintained at 60~80 °C for about 3 h. Then, the reaction solution was washed to neutral with distilled water, and the TOC content was determined after drying the samples. The unit of total organic carbon content is wt%.

2.2.2. Low-Temperature N_2 and CO_2 Adsorption

The experiments were conducted by a Quanta-chrome Autosorb automatic specific surface and pore analyzer (FL, USA) at the Petroleum Laboratory at Xi'an Shiyou University, P. R. China. Firstly, the samples were crushed and screened by a standard mesh sieve to the particle size of 80~100 mesh and dried for 48 h in a drying oven. N_2 and CO_2 adsorption were performed.

The pore size, measured by nitrogen adsorption, is between 2 and 100 nm [50]. Powder samples of 2–3 g were selected and then loaded into the tube that connected with the adsorption instrument. The powder samples were then dehydrated and degassed at 110 °C for 12 h to remove internal water and volatile materials, and vacuumed to remove other gas components. Then, the samples were backfilled with nitrogen, and the partial pressure of gas was gradually increased and decreased at a constant temperature to measure the adsorption and desorption amount of the sample under the corresponding pressure. According to the IUPAC report of 2015, when the specific surface area of plate pores with a size less than 10 nm is calculated based on the Kelvin equation (such as when using the BJH method), the results will be underestimated by 20~30%. The molecular model and density functional theory (DFT) method can effectively avoid the error caused by the calculation of the Kelvin equation. The accuracy of the calculation result is higher than that of the classical BET method [21]. Therefore, the DFT method was used to obtain parameters such as the pore volume and specific surface area of meso- and macro-pores by N_2 adsorption. The experimental temperature was set at 77 K, and 76 pressure spot recordings were set during adsorption and desorption, with the relative pressure (P/P_0) ranging from 0.05 to 0.99.

The pore size obtained by CO_2 adsorption is mainly less than 2 nm, and 2–3 g powder samples were also selected for testing. The DFT method was also used to interpret the experimental data of CO_2 adsorption and to calculate the parameters such as the volume and specific surface area of micro-pores. The experimental temperature was set at 273.15 K, and 50 pressure spot recordings were set during adsorption.

After the CO_2 and N_2 adsorption, the powder samples were extracted with benzene and chloroform (volume ratio of 1:3) by the Soxhlet extraction method; dissolved residual organic compounds were removed after 60 days and vacuum-dried at 100 °C for 48 h. Then,

N_2 and CO_2 adsorption experiments and pore volume and specific surface area analyses were carried out again.

2.3. Methods

The fractal theory has been widely used to study the complexity and irregularity of pore throat structure since Mandelbrot put forward the fractal theory in 1967 [32,51–53]. The application of fractal theory in the pore throat structure of shale realizes the quantitative characterization of the complexity of pore structure and provides a powerful means for the comparison of the complexity of different samples, which covers the shortage of the uniformity of pore throat structure parameters measured by conventional experiments. The fractal theory is usually combined with experimental parameters from HPMI, CRMI, N_2 adsorption, and NMR to study the complexity of tight reservoirs. Based on the combination of N_2 adsorption and fractal theory, this study compared the complexity of shale pore structure before and after Soxhlet extraction, quantitatively studies the complexity change in shale pore structure, and evaluates the influence of shale oil on pore structure, which is helpful to understand the migration of shale oil.

The Frenkel–Halsey–Hill (FHH) fractal model was used to calculate the fractal dimensions from N_2 adsorption, which was considered to be effective in studying the complexity of pores with different radii [54,55]. The fractal dimension of the shale reservoir is between 2 and 3, and the higher the value, the higher the complexity and irregularity of the pore structure [53]. The FHH model can be expressed as

$$lnV = C + (D - 3)ln\ (ln\frac{P_0}{P}) \tag{1}$$

where V and P represent the equilibrium volume (cm^3/g) and the equilibrium pressure (MPa) of N_2 adsorption, respectively. P_0 stands for saturation pressure (MPa). C is a constant, and dimensionless. $D-3$ denotes the linear correction coefficient, where D is the fractal dimension. Using $\ln\ln P_0/P$ as the horizontal axis and lnV as the vertical axis to make a scatter diagram, the fractal dimension of different samples can be found by fitting the regressed slopes.

3. Results

3.1. TOC and Mineral Composition

The TOC content of the six samples ranged from 0.34 wt% to 13.26 wt%, with an average of 3.84 wt%, representing a high potential for oil production. The XRD results show that clay, quartz, and feldspar are the main minerals (Table 1). The content of clay minerals ranged from 37.3% to 54.5%, averaging 44.6%; the quartz content ranged from 15.5% to 30.6%, averaging 22.9%; and the feldspar content ranged from 21.5% to 34.2%, averaging 25.9%. In addition, carbonatite and pyrite were also developed, with small average contents of 3.7% and 2.5%, respectively. A small amount of siderite existed in samples C4 and C6, with an average content of 1.2%. Illite is the main clay mineral, with an average content of 60.4%, followed by an I/S mixed layer with an average of 28.2%. The chlorite content averaged 11.3%, and smectite and kaolinite were not developed.

Table 1. TOC and mineral composition of samples.

Sample ID	Depth (m)	TOC (wt%)	Detrital Component Content from XRD (%)						Relative Content of Clay Minerals (%)		
			Quartz	Feldspar	Carbonate	Clay	Siderite	Pyrite	Chlorite	Illite	I/S Mixed Layer
C1	2201.2	0.34	30.5	21.5	3.6	40.9	/	3.6	5.4	58.9	35.6
C2	2205.7	1.09	27.4	27.7	6.1	37.3	/	1.5	2.5	65.3	32.2
C3	2132.4	2.41	17.2	34.2	4.2	42.1	/	2.3	23.4	50.8	25.8
C4	2125.2	1.44	16.2	23.5	3.8	54.5	0.2	1.9	9.7	61.5	28.8
C5	2064.3	4.51	30.6	23.9	2.9	41.0	/	1.6	21.5	58.7	19.8
C6	2123.0	13.26	15.5	24.6	1.4	52.2	2.2	4.1	5.6	67.3	27.1

3.2. CO_2 and N_2 Adsorption before and after Soxhlet Extraction

Figure 2 shows the comparison of CO_2 adsorption isotherms of shale samples before and after Soxhlet extraction. The volume of CO_2 adsorption gradually increased with the increase in relative pressure. According to the updated classification of the IUPAC [21], the CO_2 adsorption curve exhibits the type I isotherm, which represents CO_2 adsorption in the micro-pores as a monomolecular layer or micro-pore-filled state. The CO_2 adsorption volume of the initial samples ranged from 0.3169 to 1.2886, and that of samples after extraction ranged from 0.4508 cm^3/g to 1.7221 cm^3/g, up to 17% to 78% from the initial state (Table 2).

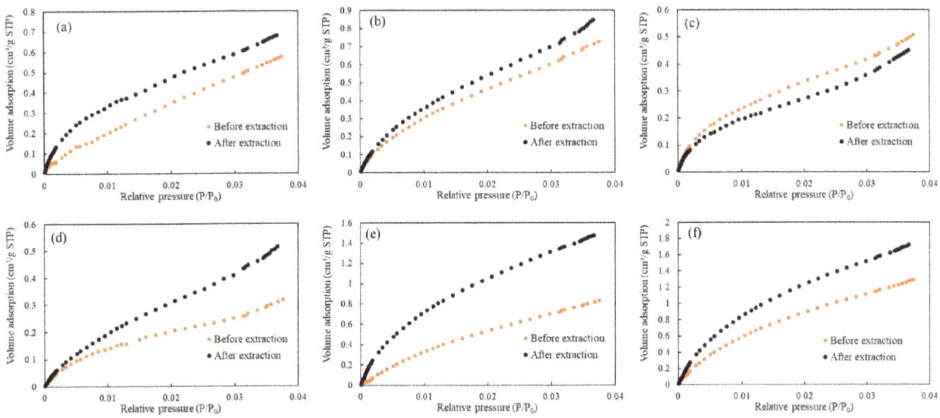

Figure 2. Characteristics of CO_2 adsorption isotherms. (**a**) Sample C1. (**b**) Sample C2. (**c**) Sample C3. (**d**) Sample C4. (**e**) Sample C5. (**f**) Sample C6.

Table 2. Parameters of CO_2 adsorption in initial and Soxhlet extraction samples.

Sample ID	Initial Sample			After Soxhlet Extraction		
	CO_2 Quantity Adsorbed (cm^3/g STP)	Micro-Pore Volume (cm^3/100 g)	Micro-Pore Surface Area (m^2/g)	CO_2 Quantity Adsorbed (cm^3/g)	Micro-Pore Volume (cm^3/100 g)	Micro-Pore Surface (m^2/g)
C1	0.5767	0.0257	0.8791	0.6804	0.0601	2.1142
C2	0.7259	0.0512	1.7128	0.8460	0.0624	2.1171
C3	0.5068	0.0422	1.4828	0.4508	0.0327	1.1978
C4	0.3169	0.0249	0.8550	0.5157	0.0295	1.0123
C5	0.8308	0.0577	1.7889	1.4770	0.1448	4.8371
C6	1.2886	0.1162	3.7192	1.7221	0.1764	5.7932

The N_2 isothermal adsorption curves of the six samples before and after extraction are shown in Figure 3. With the increase in relative pressure, the volume of N_2 adsorption gradually increased, and the nitrogen adsorption curve exhibits an inverse "S" pattern. In the low-pressure stage ($P/P_0 < 0.8$), adsorption gradually changed from micro-pore-filling and monomolecular layer adsorption to multilayer adsorption, where the rapid increase in adsorption volume with the relative pressure was less obvious and was followed by a slow increase in adsorption volume. The amount of adsorbed nitrogen was increased significantly but did not reach saturation at higher relative pressures ($P/P_0 > 0.8$), indicating that capillary condensation of nitrogen occurred. The adsorption and desorption curves clearly show a hysteresis loop, and they did not overlap at relative pressures of 0.45 to 0.9, resulting in an adsorption hysteresis that was mostly caused by capillary condensation in meso-and macro-pores [56,57].

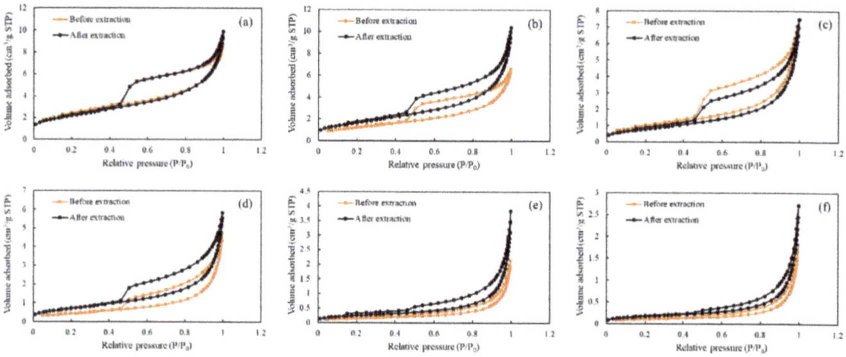

Figure 3. N_2 adsorption and desorption isotherms. (**a**) Sample C1. (**b**) Sample C2. (**c**) Sample C3. (**d**) Sample C4. (**e**) Sample C5. (**f**) Sample C6.

The adsorption and desorption curves show forced closure after the relative pressure dropped to 0.45, which was caused by nitrogen evaporation in the capillary [58]. This showed that the pores of shale were primarily micro- and meso-pores, and capillary condensation occurred in the meso-pores [11,56]. According to the IUPAC, the hysteresis loop types are mainly H3 and H4 [21], indicating that shale pores are dominated by narrow slit-like parallel plate pores [59] and non-rigid aggregates of lamellar particles (like clay minerals). The nitrogen adsorption volume of the samples after extraction was higher than that in the initial state (Figure 3, Table 3).

Table 3. Parameters of N_2 adsorption isotherms in initial and Soxhlet extraction samples.

Sample ID	Initial Sample						
	N_2 Quantity Adsorbed (cm^3/g)	**BET Surface Area (m^2/g)**	**DFT Pore Volume ($cm^3/100$ g)**	**Meso-Pore Volume ($cm^3/100$ g)**	**Macro-Pore Volume ($cm^3/100$ g)**	**Meso-Pore Surface Area (m^2/g)**	**Macro-Pore Surface Area (m^2/g)**
C1	8.6798	8.2261	0.8904	0.6840	0.0748	1.9398	0.0202
C2	6.5852	4.3957	0.7366	0.5638	0.1268	0.3466	0.0278
C3	7.4363	3.4300	1.0803	0.7307	0.3193	1.4640	0.0604
C4	4.7107	1.5939	0.5013	0.3261	0.1753	0.4727	0.0360
C5	2.1385	0.5469	0.3265	0.1665	0.1533	0.2285	0.0268
C6	1.8203	0.3816	0.2787	0.1314	0.1420	0.1518	0.0242
Sample ID	After Soxhlet extraction						
	N_2 quantity adsorbed (cm^3/g)	BET surface area (m^2/g)	DFT pore volume ($cm^3/100$ g)	Meso-pore volume ($cm^3/100$ g)	Macro-pore volume ($cm^3/100$ g)	Meso-pore surface area (m^2/g)	Macro-pore surface area (m^2/g)
C1	9.8757	7.7749	1.4548	0.9679	0.3509	2.4518	0.0641
C2	10.3936	6.0003	1.5187	1.0000	0.4383	2.3071	0.0801
C3	7.4936	2.8556	1.0992	0.7013	0.3731	1.3219	0.0688
C4	5.8187	2.5680	0.5213	0.3612	0.1516	0.6011	0.0327
C5	3.8456	0.9292	0.5760	0.2787	0.2874	0.3914	0.0458
C6	2.7105	0.6351	0.4123	0.2041	0.2011	0.2635	0.0348

3.3. Pore Volumes and Specific Surface Areas

Tables 2 and 3 show, using the DFT model, the pore volumes and specific surface areas of the micro-pores, meso-pores, and macro-pores of shale samples before and after extraction. The specific surface areas of micro-pores ranged from 0.8550 to 3.7492 m^2/g, with an average of 1.7446 m^2/g, while the pore volumes before and after extraction ranged from 0.0249 to 0.1162 $cm^3/100$ g. The meso-pore volumes ranged from 0.1314 to 0.7307 $cm^3/100$ g, with an average of 0.4338 $cm^3/100$ g, and the specific surface areas ranged from 0.1518 to 1.93 m^2/g, with an average of 0.7672 m^2/g. The pore volumes of macro-pores ranged from 0.0748 to 0.3793 $cm^3/100$ g, with an average of

0.1653 cm^3/100 g, and the specific surface areas ranged from 0.0202 to 0.0604 m^2/g, with an average of 0.0326 m^2/g. The average pore volume and specific surface area of micro-pores after extraction were 0.0843 cm^3/100 g and 2.8453 m^2/g, respectively. The average pore volume of meso-pores was 0.5855 cm^3/100 g, and 0.3004 m^2/g was the average for specific surface area. The average pore volume of macro-pores was 1.2228 cm^3/100 g, and the average specific surface area was 0.0544 m^2/g.

Hydrocarbons mainly existed in meso-pores and macro-pores (Figure 4a,c), and the proportion of meso-pores volume in the total volume after extraction ranged from 35.1 to 70.2%, with an average of 56.8%, followed by macro-pores, with an average of 31.9%. The specific surface area of extracted samples was mainly contributed by micro-pores (Figure 4b,d), which accounted for an average of 68.0% of the specific surface area of the extracted samples.

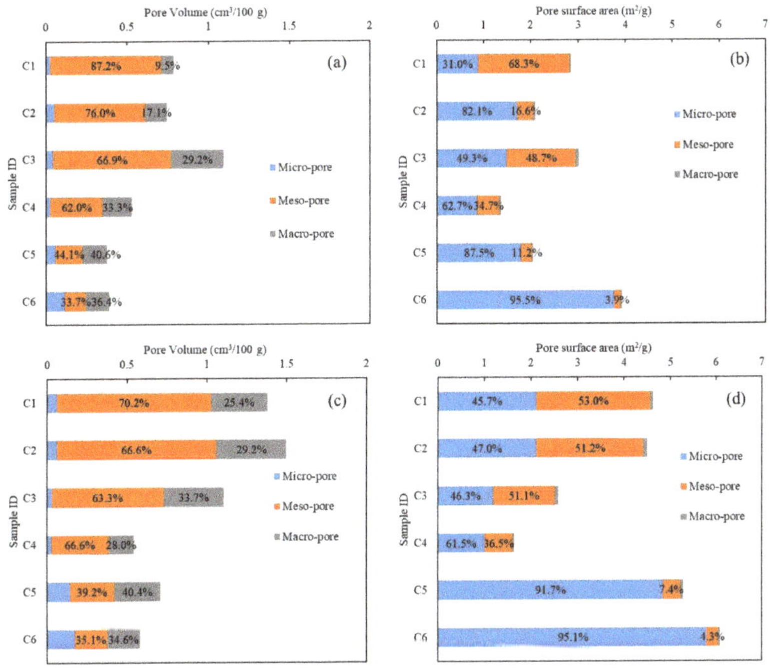

Figure 4. Pore volume and specific surface area distribution of initial and extracted samples. (**a**) Pore volume of initial samples. (**b**) Pore surface area of initial samples. (**c**) Pore volume of Soxhlet extraction samples. (**d**) Pore surface area of Soxhlet extraction samples.

3.4. Fractal Dimensions before and after Soxhlet Extraction

The complexity of pore size distribution, pore structure, and specific surface area is an important index for evaluating shale oil. Based on N_2 adsorption, the fractal theory was used to study the complexity of the pore structure of shale. Under different pressures, N_2 has different adsorption forms and fractal characteristics in the pore throat structure [60]. With the principle of maximizing the fitting coefficient, the isothermal adsorption curves were divided into two intervals, the low-pressure section and the high-pressure section. The least squares principle was used to fit the trend lines of the two sections, respectively, to obtain the slopes and the correlation coefficients R^2 of the different intervals of the samples before and after extraction (Figure 5), and the fractal dimensions D1, D2 of the initial samples and the D1s and D2s of the extraction samples (Table 4) were calculated by using Equation (1). D1 is the fractal dimension under low pressure, which is controlled by van der Waals forces and can be used as an indicator of pore surface roughness. Whereas

N_2 adsorption at relatively high pressure is affected by multilayer adsorption, D2 can be used as an indicator of the irregularity of the pore structure, and D2 becomes larger as the shale pore structure becomes complex. The fractal dimensions D1 and D1s ranged from 2.4909 to 2.6708 and 2.4896 to 2.6386, respectively. The fractal dimensions D2 and D2s ranged from 2.6356 to 2.8770 and 2.6678 to 2.8537, respectively (Table 4), with high values, which confirms the complex pore structure of shale.

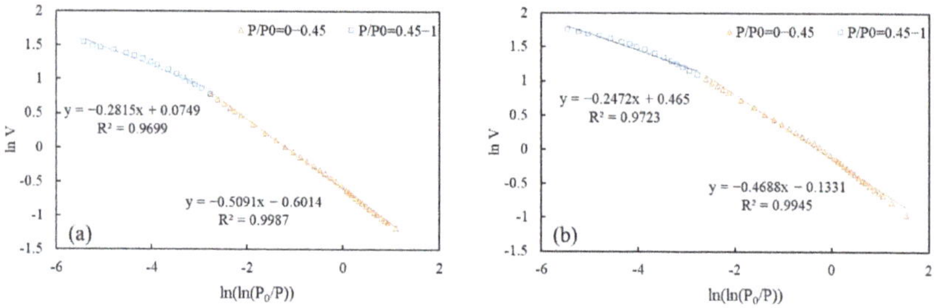

Figure 5. ln V versus ln(ln(P_0/P)) of Sample C4. (**a**) Initial sample. (**b**) After Soxhlet extraction.

Table 4. Pore volume occupied by hydrocarbons and fractal dimensions of initial and Soxhlet extraction samples.

Sample ID	Pore Volume Occupied by Hydrocarbons (cm^3/100 g)			Initial Sample		After Soxhlet Extraction	
	Micro-Pore	Meso-Pore	Macro-Pore	D1	D2	D1 s	D2 s
C1	0.0344	0.2839	0.2761	2.6708	2.8770	2.6386	2.8537
C2	0.0112	0.4362	0.3115	2.5889	2.8459	2.5755	2.8319
C3	−0.0095	−0.0294	0.0538	2.5299	2.8059	2.5038	2.8109
C4	0.0046	0.0351	−0.0237	2.4909	2.7185	2.5312	2.7528
C5	0.0871	0.1122	0.1341	2.5089	2.6855	2.4935	2.6861
C6	0.0602	0.0727	0.0591	2.5166	2.6356	2.4896	2.6678

4. Discussions

4.1. Discussion on Pore Space Occupied by Hydrocarbons

The adsorption volume after extraction was higher than that before extraction except for that of sample C3 (Table 2), based on CO_2 adsorption. This finding indicated that hydrocarbons are present in the micro-pores of shale and that these pores help to accumulate hydrocarbons. The adsorption volume of sample C3 after extraction was lower than before extraction. A decrease in the volume of adsorption occurred as a result of the effect of temperature and capillary pressure during extraction, which probably caused the pyrolysis of heavy hydrocarbons and adsorbed components and re-enrichment by adsorption in organic matter pores. This effect will be further examined in the study that follows. The nitrogen adsorption volume of the samples after extraction was higher than that in the initial state, indicating that the meso-pores and macro-pores are also the pore spaces occupied by hydrocarbons (Figure 3, Table 3).

Hydrocarbons mainly existed in meso-pores and macro-pores (Figure 4a,c). By comprehensive comparison, the specific surface area of extracted samples was mainly contributed by micro-pores (Figure 4b,d), indicating that although the pore volume was small, the specific surface area of micro-pores was large [57,58]. This was mainly related to the large number of organic pores in the micro-pores, which provided a large amount of surface area for hydrocarbon adsorption [5]. It is an important occurrence mode of shale oil.

The volume of micro-pores is positively correlated with TOC (Figure 6a). With higher TOC content, organic pores in shale are more developed and micro-pores are larger. The DFT pore volume calculated based on nitrogen adsorption shows a negative correlation with TOC content (Figure 6b), indicating that organic pores were less developed in meso-pores and macro-pores. The DFT pore volume is positively correlated with the carbonate content (Figure 6d), and negatively correlated with clay mineral content (Figure 6c).

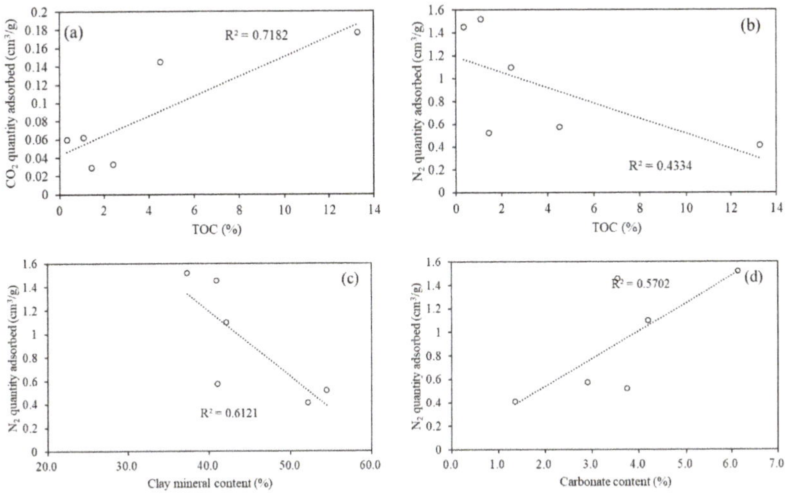

Figure 6. TOC content and mineral contents vs. CO_2 and N_2 quantity adsorbed. (**a**) CO_2 quantity adsorbed vs. TOC. (**b**) N_2 quantity adsorbed vs. TOC. (**c**) N_2 quantity adsorbed vs. clay mineral content. (**d**) N_2 quantity adsorbed vs. carbonate content.

After extraction, the hydrocarbon volume of samples C3 and C4 increased in a certain pore radius, suggesting that these samples have a low residual hydrocarbon content and that the majority of them are in an adsorption state. When hydrocarbons are extracted at high temperatures, their condition changes, and capillary force causes them to reaccumulate in the pores. Figure 4 shows that hydrocarbons were mainly present in meso- and macro-pores, and shale oil was most commonly found with pore radii higher than 5 nm (Figures 7 and 8). The pore distribution curves show that there are obvious peaks between 50 and 100 nm and over 200 nm. The hydrocarbon content in the meso- and macro-pores of samples C1 and C2 was high, and the hydrocarbons in samples C5 and C6 mainly existed in the meso-pores (Figure 8), indicating that inorganic pores are more contributing to the production of shale oil.

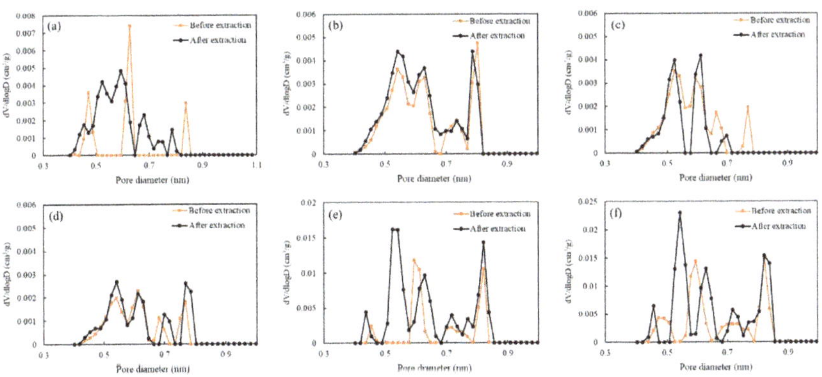

Figure 7. Micro-pore distribution of initial and Soxhlet extraction samples. (**a**) Sample C1. (**b**) Sample C2. (**c**) Sample C3. (**d**) Sample C4. (**e**) Sample C5. (**f**) Sample C6.

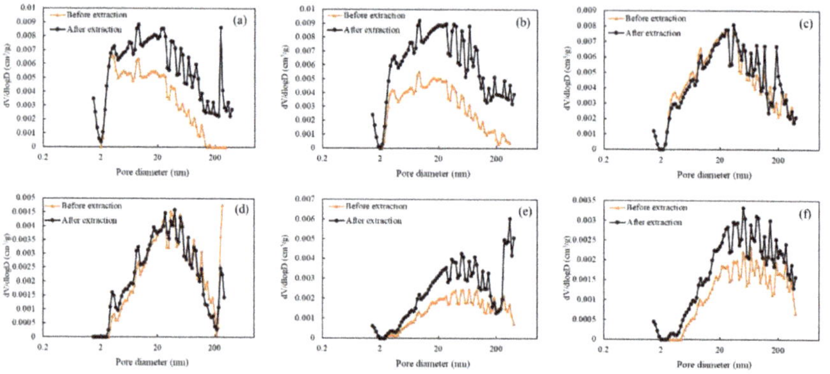

Figure 8. Meso- and macro-pore distribution of initial and Soxhlet extraction samples. (**a**) Sample C1. (**b**) Sample C2. (**c**) Sample C3. (**d**) Sample C4. (**e**) Sample C5. (**f**) Sample C6.

The pore distribution of samples before and after extraction shows obvious differences, especially in micro-pores, suggesting a complex distribution of shale oil in the pores. After extraction, obvious single or multiple peaks appeared in the pore interval of 0.5–0.6 nm and at 0.8 nm (Figure 7). The N_2 adsorption hysteresis loops before and after extraction of meso-pores and macro-pores did not change much (Figure 3), i.e., the hysteresis loop types were all dominated by parallel plates, implying that hydrocarbons existed in the shale pores mainly in an adsorbed state.

4.2. Complexity Change in Pores before and after Extraction

D1s was generally smaller than D1 (Table 4), and the pore surface roughness of the samples after extraction was smaller, indicating that the oil adsorption in the initial samples was heterogeneous, resulting in a high fractal dimension. During the high-temperature extraction process, the residual hydrocarbons in the pore space changed; the heavy components were cracked into light hydrocarbons, most of the light components were extracted, and the residual light components were redistributed on the pore surface, resulting in a smooth pore surface with a more homogeneous roughness. The variation between D2 and D2s was irregular (Table 4), and the accumulation of hydrocarbons in pores did not aggravate the complexity of the pore structure.

According to the relationship between D2, D2s, and pore volume, the pore volume was large and D2 and D2s were large (Figure 9a). For example, in sample C2, the pores were mainly composed of meso-pores and macro-pores, with a high content of brittle minerals, a large proportion of inorganic pores, and a complex pore network formed by intergranular pores, dissolution pores, and organic pores. In both the initial and Soxhlet extraction sample, the complexity of pores was stronger, and the fractal dimension was large. Meanwhile, the high proportion of inorganic pores also implies a large degree of heterogeneity of pore surface, and because of the large differences in roughness on the surfaces of quartz, feldspar, and clay minerals, D1 and D1s of sample C2 were also large (Figure 9a).

The contents and existence forms of minerals directly affect the complexity of the pore throat structure and the roughness of the pore throat surface [61–63]. Therefore, the difference in mineral composition among different samples is the basic geological factor affecting the accumulation of shale oil. The content of brittle minerals is positively correlated with the pore volume (Figure 9d), and the high content of brittle minerals is conducive to the development of primary intergranular pores. In the middle and late diagenesis, part or all of the intergranular pores were filled and cut by clay minerals, forming parallel plate-like pores and providing a large amount of hydrocarbon accumulation space. The dissolution of carbonate minerals increased the reservoir space (Figure 9d). There is no correlation between quartz and pore volume. We found that the addition of I/S greatly increased the complexity of the shale pore space (Figure 9c), and I/S filled or semi-filled intergranular pores and developed curved sheet pores with diverse morphologies, which exacerbated the complexity of pore throat structure.

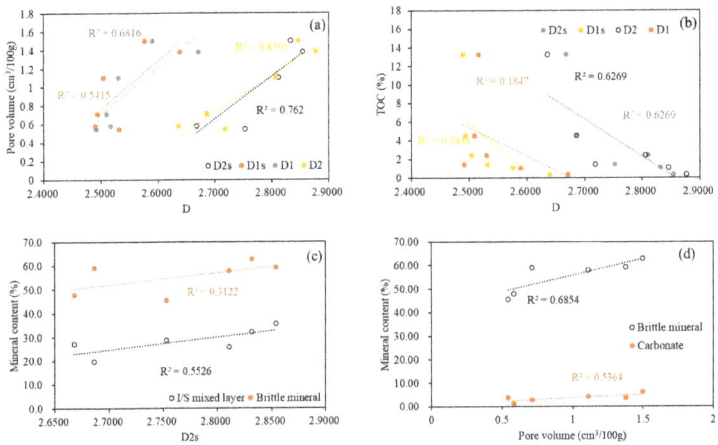

Figure 9. Fractal dimensions vs. pore characteristics and TOC. (**a**) Pore volume vs. D. (**b**) TOC vs. D. (**c**) Mineral content vs. D. (**d**) Mineral content vs. pore volume.

However, the hydrocarbon accumulation space in shale is not only controlled by minerals but also has a strong relationship with the occurrence of kerogen. D2 and D2s are negatively correlated with TOC content (Figure 9b). The higher the TOC content, the higher the proportion of organic matter pores, and the pores are alveolate-connected, resulting in a lower complexity and fractal dimension of pore space. Meanwhile, we found that D1 and D1s are weakly negatively correlated with TOC (Figure 9b), and the surface of the organic matter pores was smooth. The higher the content of organic matter, the lower the roughness of the pore surface of the sample, that is, the smaller the D1 and D1s. This shows that in shale, the occurrence of organic matter has a constructive effect on the homogeneity of the pore throat system and pore surface smoothness.

4.3. *The Change in Hydrocarbon Accumulation before and after Extraction*

Figures 7 and 8 show the volume change of micro-pores, meso-pores, and macro-pores before and after extraction. The pore volume increased after extraction, but not all pore throat size segments of the shale were enlarged. As shown in Figure 8, hydrocarbons in the meso- and macro-pores of samples C1, C2, C5, and C6 were extracted, and it can be seen that the pore volume increased obviously, but the pore volume changes in C3 and C4 were smaller. The volume changes in micro-pores before and after extraction were more complicated, implying that the content, state, and location of hydrocarbons changed during extraction, which also proves that Soxhlet extraction cannot completely remove hydrocarbons in the pore space. Therefore, in the follow-up study, the gas displacement and extraction method can be further used to clarify the characteristics of shale oil accumulation in the pore throat.

4.3.1. Migration and Re-Adsorption of Light Hydrocarbons

Shale oil is characterized by multilayer adsorption in the pore space, showing a solid or solid-like form near the surface of mineral particles, and is mainly free liquid in the center of the pores [7]. However, different from shale gas, it is difficult to define the adsorption state and free state in a single pore, and there is no obvious boundary between the two states [16]. High temperature and high pressure can improve the extraction efficiency of shale oil; a long-term high-temperature action, especially, will cause the pyrolysis of some heavy hydrocarbons and adsorbed components, and the higher the content of heavy components, the greater the content of adsorbed oil. A high-temperature extraction at 150 °C is enough to transform the adsorption components into the free state, and the viscosity of the oil is reduced, so that the extraction efficiency is higher. Also, molecular dynamics simulation confirmed that temperature has an effect on oil fluidity [42]. Previous studies, through adsorption experiments and molecular dynamics simulations, found that the thickness of the oil film showed an obvious pore size effect in micro-pores and meso-pores. The thickness of the oil film increased with the increase in the pore size, and the number of adsorbed layers changed from 1 to 3. However, in macro-pores, the relationship between adsorbed oil film thickness and pore size is weak, and the number of adsorbed layers remains stable at 3–4 layers [64,65].

During the extraction process, the adsorption layer in the macro-pore decreases step by step, and the free oil extracted from the micro-pore can reaggregate in the macro-pore and adsorb again to restore a stable adsorption layer, resulting in extraction difficulties (Figure 10(a-1–a-3)). By sequential solvent extraction, it was believed that free oil mainly existed in pores larger than 5 nm. The maturity of the samples was low, and the residual hydrocarbons were mainly macromolecular heavy hydrocarbons, which mainly existed in meso- and macro-pores. The content of light hydrocarbons was low, and the micro-pores were mainly composed of adsorbed light hydrocarbons. Due to the influence of pore throat structure, the adsorbed hydrocarbons in the micro-pores of samples C3 and C4 were difficult to utilize, resulting in the phenomenon that the hydrocarbons in the micro-pores and meso-pores were almost not extracted. Meso-pores were the main storage space, and the high proportion of meso-pores in samples C1, C2, C5, and C6 facilitated the migration of free oil during the extraction process, resulting in a high extraction rate.

4.3.2. Effect of Pore Throat Structure on Extraction Process

Shale oil diffuses and redistributes along the particle surface. In a well-connected pore structure, shale oil in the adsorption layer of small pores will migrate into large pores as the oil in large pores decreases. The oil in the pores with a large size and good connectivity is easy to extract, and the oil in the nanoscale pores is displaced by capillary force during extraction and then enters into meso- and macro-pores under the action of diffusion. If the pore structure is well connected and the diffusion ability is strong, the shale oil extraction rate is high. Organic pores in samples C5 and C6 were developed, and the proportion of micro-pores, meso-pores, and macro-pores accounted for a small difference, and the pore throat structure was homogeneous (Table 4). The diffusion capacity was strong, and the residual hydrocarbons in organic pores easily entered into meso- and macro-pores, so the extraction rates of micro-pores, meso-pores, and macro-pores were higher (Figure 10(b-1–b-3)). The proportion of meso-pores in samples C3 and C4 was the lowest, and the connectivity between macro- and micro-pores was limited. During extraction, although the capillary force was strong, the diffusion capacity was poor, and shale oil continued to be adsorbed on the particle surfaces where the distribution of pore throats was heterogeneous, resulting in pore blockage and the lowest amount of shale oil extraction (Figure 10(c-1–c-3)).

Samples C1 and C2, with high brittle mineral content (Table 1), had large intergranular pores and a high proportion of meso-pores. Although the extraction rate of residual hydrocarbons in micro-pores was low, the proportion of shale oil in meso- and macro-pores was high, and it was easy to extract, so the extracted oil amount was high. On the other hand, the extraction solvent had fully reacted with the adsorbed oil as well as free heavy oil in the macro-pores during the gradual entry into the small pores, and a long extraction time can also lead to a high extraction rate. Pore throat configuration has a crucial impact on the extraction rate of shale oil. The pore structure of shale is complex, and organic matter pores develop; if meso-pores can effectively communicate with organic matter pores, clay mineral intercrystalline pores, and intergranular pores, higher recovery efficiency can be obtained even if the pore size is small.

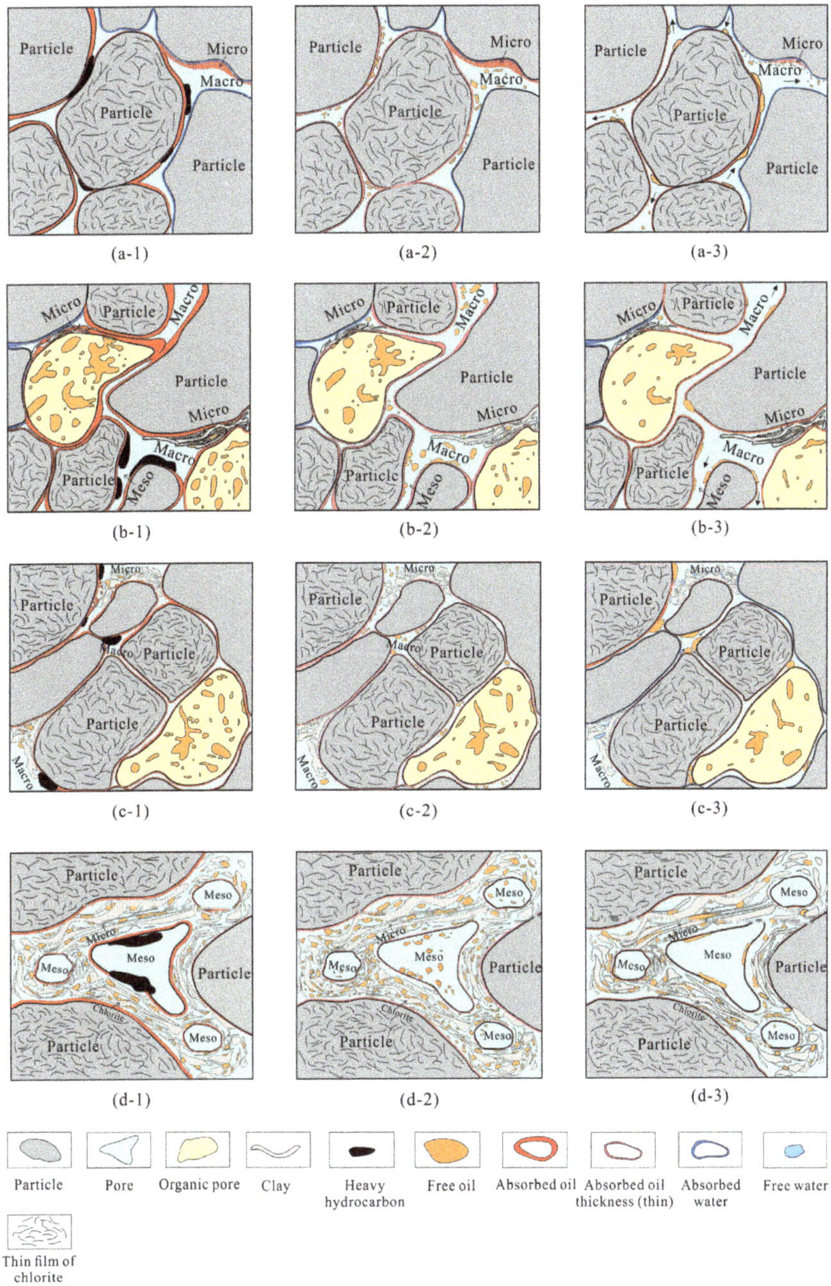

Figure 10. Schematic of hydrocarbon migration in different pore throat structures. (**a-1**) Hydrocarbon distribution of sample with large pores and narrow throat in initial state. (**a-2**) Hydrocarbon distribution of sample with large pores and narrow throat during extraction. (**a-3**) Hydrocarbon distribution of sample with large pores and narrow throat after extraction. (**b-1**) Hydrocarbon distribution of sample

with large pores and large throat (homogeneous) in initial state. (**b-2**) Hydrocarbon distribution of sample with large pores and large throat (homogeneous) during extraction. (**b-3**) Hydrocarbon distribution of sample with large pores and large throat (homogeneous) after extraction. (**c-1**) Hydrocarbon distribution of sample with small pores and narrow throat in initial state. (**c-2**) Hydrocarbon distribution of sample with small pores and narrow throat during extraction. (**c-3**) Hydrocarbon distribution of sample with small pores and narrow throat after extraction. (**d-1**) Hydrocarbon distribution of sample with clay mineral pores in initial state. (**d-2**) Hydrocarbon distribution of sample with clay mineral pores during extraction. (**d-3**) Hydrocarbon distribution of sample with clay mineral pores after extraction.

4.3.3. Effect of Mineral Surface Wettability on the Extraction Process

The interfacial properties between fluids and solids affect oil accumulation in shale, and hydrocarbon compounds with different properties undergo complex physical–chemical reactions. In general, the more oil-wetting characteristics the mineral surface has, the greater the thickness of the adsorption layer on it, and the smaller the movable amount of free oil. Mineral surfaces can adsorb both light hydrocarbon components and large molecules such as non-hydrocarbons and asphaltene with larger polarities. The adsorption of light hydrocarbon components on the mineral surface mainly depends on the physical adsorption from the van der Waals force, which is small, and it is easy to extract the hydrocarbons. Asphaltenes and non-hydrocarbon components contain a large number of polar functional groups with strong polarity, which are more easily adsorbed on the surface of minerals and kerogen through ionic or hydrogen bonds than hydrocarbons [66].

Sample C3 had a high content of chlorite (Table 1), which was mainly a lining filled in in intergranular pores [67], and clay minerals such as chlorite show strong oil-wetting characteristics during diagenetic evolution, which is favorable for the adsorption of shale oil [68,69]. Clay mineral (chlorite) intercrystalline pores are oil-wetting and contain a large number of lipophilic groups. During the extraction process, the residual hydrocarbons in the intergranular pores were further adsorbed by the oil-wetting intergranular pores under the action of capillary force. Secondly, the high fractal dimension of sample C3 (Table 4) and the poor diffusion of oil in it also accelerated the re-aggregation of hydrocarbons in the micro-pores and meso-pores (Figure 10(d-1–d-3)).

A large number of lipophilic groups exist on the surface of organic matter, which can form high-energy adsorption sites with strong adsorption capacity, and shale oil in organic matter pores will exist in a dissolved state with strong fluidity under this strong adsorption. Samples C5 and C6 had a high TOC content, and the micro-pores were dominated by organic matter pores, so the movable oil content was high (Figure 7e,f). Secondly, organic matter pores provided a large number of adsorption sites; the specific surface area of organic matter pores in samples C5 and C6 increased significantly after extraction (Figure 4b,d). If bound water exists on the mineral surface, the amount of adsorbed oil is greatly reduced. In water-bearing samples, the surface of clay, calcite, and quartz can show hydrophilicity [70,71], at which time there is no regularity in the adsorption characteristics of mineral surfaces, and the adsorption amount of oil in shale may be related only to TOC. However, the water saturation of the shale samples from the Yanchang Formation in the study area was less than 10%, so the influence of bound water on extraction is not discussed in this study.

It should be noted that pores provide space for shale oil enrichment, but not all pores contained shale oil, and the properties of the mineral surface determine the state, content, and mobility of the oil in the pore. The exploitation of unconventional reservoirs should pay attention not only to the configuration of pore structure, but also to the wettability of the mineral surface.

5. Conclusions

The hydrocarbon accumulation space characteristics, the effect of hydrocarbon accumulation on pore heterogeneity, the hydrocarbon migration changes, and the factors affecting extraction efficiency in shale pores were investigated in this paper.

Hydrocarbons are mainly found in meso- and macro-pores. The hydrocarbons in the pore space of shale do not aggravate the complexity of the pore structure.

The content, state, and location of hydrocarbons change in a complex process during extraction. After extraction, the hydrocarbons are redistributed on the pore surface, resulting in a decrease in pore surface roughness. Some heavy hydrocarbons and adsorbed components are pyrolyzed by long-term high-temperature extraction, and the adsorption layer in the macro-pores is gradually reduced. The free oil in the small pores diffuses into the large pores and accumulates again, and then is adsorbed on the surface of the large pores again, restoring the stable adsorption layer, resulting in difficult extraction.

The extraction rate was closely related to the pore throat structure and the wettability of the mineral surfaces. (1) Pore throat configuration has a crucial influence on extraction efficiency. The diffusion rate and extraction efficiency are higher when the proportion of meso-pores is high and they can effectively communicate with organic matter pores, clay mineral intercrystalline pores, and intergranular pores. A very large number of organic matter pores is conducive to extraction efficiency due to the strong adsorption force, and shale oil exists in the dissolved state with a high content of mobile oil.

(2) The wettability of clay minerals, such as chlorite, exhibit strong oil wetting, and hydrocarbons are further adsorbed in the intercrystalline pores of clay minerals under the action of strong capillary forces formed by a complex pore throat structure, resulting in the volume of micro-pores after extraction being smaller than that before extraction.

Author Contributions: Conceptualization, Y.Q.; Methodology, S.O.; Software, Y.C. and Z.Z.; Validation, J.G. and J.S.; Writing—original draft, Y.Q.; Writing—review & editing, Y.W. and Z.L.; Supervision, W.S. and H.W.; Funding acquisition, Y.Q. All authors have read and agreed to the published version of the manuscript.

Funding: This study was supported by the National Natural Science Foundation of China (grant No: 42202143), the Open Fund (XSTS-202101) of Xi'an Key Laboratory of Tight oil (Shale oil) Development, and the Natural Science Basic Research Program of Shaanxi (Program No. 2023-JC-QN-0374).

Data Availability Statement: All our data have been displayed in the pictures in this paper.

Acknowledgments: The authors sincerely thank the Department of Geology of Northwest University and PetroChina Changqing Oilfield Company for providing the drill cores used in this study.

Conflicts of Interest: Authors Jianwen Gao and Jian Shi were employed by the PetroChina Changqing Oilfield Company. Author Zhou Lyu was employed by the China National Petroleum Corporation. The remaining authors declare that the research was conducted in the absence of any commercial or financial relationships that could be construed as a potential conflict of interest.

References

1. Jia, C.; Zou, C.; Li, J.; Li, D.; Zheng, M. Assessment criteria, main types, basic features and resource prospects of the tight oil in China. *Acta Pet. Sin.* **2012**, *33*, 343–350.
2. Hao, F.; Zou, H.; Lu, Y. Mechanisms of shale gas storage: Implications for shale gas exploration in China. *AAPG Bull.* **2013**, *97*, 1325–1346. [CrossRef]
3. Zou, C.; Zhu, R.; Chen, Z.; Ogg, J.; Wu, S.; Dong, D.; Qiu, Z.; Wang, Y.; Wang, L.; Lin, S.; et al. Organic-matter-rich shales of China. *Earth-Sci. Rev.* **2019**, *189*, 51–78. [CrossRef]
4. Katsube, T.; Williamson, M. Effects of diagenesis on shale nanopore structure and implications for sealing capacity. *Clay Miner.* **1994**, *29*, 451–461. [CrossRef]
5. Loucks, R.; Reed, R.; Ruppel, S.; Hammes, U. Spectrum of pore types and networks in mudrocks and a descriptive classification for matrixrelated mudrock pores. *AAPG Bull.* **2012**, *96*, 1071–1098. [CrossRef]
6. Merkel, A.; Fink, R.; Littke, R. High pressure methane sorption characteristics of lacustrine shale from the Middle Valley Basin, Scotland. *Fuel* **2016**, *182*, 361–372. [CrossRef]

7. Jin, Z.; Wang, G.; Liu, G.; Gao, B.; Liu, Q.; Wang, H.; Liang, X.; Wang, R. Research progress and key scientific issues of continental shale oil in China. *Acta Pet. Sin.* **2021**, *42*, 821–835. (In Chinese) [CrossRef]
8. Dong, D.; Gao, S.; Huang, J.; Guan, Q.; Wang, S.; Wang, Y. Discussion on the exploration & development prospect of shale gas in the Sichuan Basin. *Nat. Gas Ind. B* **2015**, *2*, 9–23. [CrossRef]
9. Hu, Q.; Liu, X.; Gao, Z.; Liu, S.; Zhou, W.; Hu, W. Pore structure and tracer migration behavior of typical American and Chinese shales. *Pet. Sci.* **2015**, *12*, 651–663. [CrossRef]
10. Zhang, J.; Li, X.; Wei, Q.; Gao, W.; Liang, W.; Wang, Z.; Wang, F. Quantitative characterization of pore-fracture system of organic-rich marine-continental shale reservoirs: A case study of the Upper Permian Longtan formation, Southern Sichuan Basin, China. *Fuel* **2017**, *200*, 272–281. [CrossRef]
11. Mastalerz, M.; Schimmelmann, A.; Drobniak, A.; Chen, Y. Porosity of Devonian and Mississippian New Albany Shale across a maturation gradient: Insights from organic petrology, gas adsorption, and mercury intrusion. *AAPG Bull.* **2013**, *97*, 1621–1643. [CrossRef]
12. Ji, W.; Hao, F.; Schulz, H.; Song, Y.; Tian, J. The architecture of organic matter and its pores in highly mature gas shales of the Lower Silurian Longmaxi Formation in the Upper Yangtze Platform, south China. *AAPG Bull.* **2019**, *102*, 2909–2942. [CrossRef]
13. Huang, H.; Li, R.; Xiong, F.; Hu, H.; Sun, W.; Jiang, Z.; Chen, L.; Wu, L. A method to probe the pore-throat structure of tight reservoirs based on low-field NMR: Insights from a cylindrical pore model. *Mar. Pet. Geol.* **2020**, *117*, 104344. [CrossRef]
14. Zhang, K.; Song, Y.; Jia, C.; Jiang, Z.; Han, F.; Wang, P.; Yuan, X.; Yang, Y.; Zeng, Y.; Li, Z.; et al. Formation mechanism of the sealing capacity of the roof and floor strata of marine organic-rich shale and shale itself, and its influence on the characteristics of shale gas and organic matter pore development. *Mar. Pet. Geol.* **2022**, *140*, 105647. [CrossRef]
15. Bernard, S.; Horsfield, B. Reply to comment on "Formation of nanoporous pyrobitumen residues during maturation of the Barnett Shale (Fort Worth Basin)". *Int. J. Coal Geol.* **2014**, *127*, 114–115. [CrossRef]
16. Lu, S.; Liu, W.; Wang, M.; Zhang, L.; Wang, Z.; Chen, G.; Xiao, D.; Li, Z.; Hu, H. Lacustrine shale oil resource potential of Es3 L sub-member of Bonan Sag, Bohai Bay Basin, Eastern China. *J. Earth Sci.* **2017**, *28*, 996–1005. [CrossRef]
17. Tian, H.; Pan, L.; Zhang, T.; Xiao, X.; Meng, Z.; Huang, B. Pore characterization of organic-rich lower Cambrian shales in Qiannan depression of Guizhou province, southwestern China. *Mar. Pet. Geol.* **2015**, *62*, 28–43. [CrossRef]
18. Lu, H.; Li, Q.; Yue, D.; Wu, Y.; Gao, J.; Wu, S.; Wang, W.; Li, M.; An, K. Quantitative characterization and formation mechanism of the pore system heterogeneity: Examples from organic-rich laminated and organic-poor layered shales of the upper triassic chang 7 member in the southern Ordos Basin, China. *Mar. Pet. Geol.* **2023**, *147*, 105999. [CrossRef]
19. Tang, X.; Jiang, Z.; Jiang, S.; Li, Z.; Peng, Y.; Xiao, D. Effects of organic matter and mineral compositions on pore structures of shales: A comparative study of lacustrine shale in Ordos Basin and Marine Shale in Sichuan Basin, China. *Energy Explor. Exploit.* **2018**, *36*, 28–42. [CrossRef]
20. Zhao, J.; Jin, Z.; Jin, Z.; Hu, Q.; Hu, Z.; Du, W.; Yan, C.; Geng, Y. Mineral types and organic matters of the Ordovician-Silurian Wufeng and Longmaxi shale in the Sichuan Basin, China: Implications for pore systems, diagenetic pathways, and reservoir quality in fine-grained sedimentary rocks. *Mar. Pet. Geol.* **2017**, *86*, 55–674. [CrossRef]
21. Thommes, M.; Kaneko, K.; Neimark, A.; Olivier, J.; Rodriguez-Reinoso, F.; Rouquerol, J.; Sing, K. Physisorption of gases, with special reference to the evaluation of surface area and pore size distribution (IUPAC Technical Report). *Pure Appl. Chem.* **2015**, *87*, 1051–1069. [CrossRef]
22. Wang, M.; Guan, Y.; Li, C.; Liu, Y.; Lu, S. Qualitative description and full-pore-size quantitative evaluation of pores in lacustrine shale reservoir of Shahejie Formation, Jiyang Depression. *Oil Gas Geol.* **2018**, *39*, 1107–1119. [CrossRef]
23. Vivek, K.; Vinod, A.; Alka, D.; Shashanka, P.; Vikram, P.; Priyanka, S. Significance of organo-mineralogical constituents on pore distribution, fractals and gas sorption mechanism of Permian shale beds in Korba sub-basin of the Son-Mahanadi Valley, India. *Geoenergy Sci. Eng.* **2023**, *231*, 212334. [CrossRef]
24. Chalmers, G.; Bustin, R.; Power, I. Characterization of gas shale pore systems by porosimetry, pycnometry, surface area, and field emission scanning electron microscopy/transmission electron microscopy image analyses: Examples from the Barnett, Woodford, Haynesville, Marcellus, and Doig units. *AAPG Bull.* **2012**, *96*, 1099–1119. [CrossRef]
25. Curtis, M.; Cardott, B.; Sondergeld, C.; Rai, C. Development of organic porosity in the Woodford Shale with increasing thermal maturity. *Int. J. Coal Geol.* **2012**, *103*, 26–31. [CrossRef]
26. Zhou, S.; Liu, D.; Cai, Y.; Karpyn, Z.; Yao, Y. Comparative analysis of nanopore structure and its effect on methane adsorption capacity of Southern Junggar coalfield coals by gas adsorption and FIB-SEM tomography. *Microporous Mesoporous Mater.* **2018**, *272*, 117–128. [CrossRef]
27. Liu, Y.; Yao, Y.; Liu, D.; Zheng, S.; Sun, G.; Chang, Y. Shale pore size classification: An NMR fluid typing method. *Mar. Pet. Geol.* **2018**, *96*, 591–601. [CrossRef]
28. Tian, X.; Duan, X.; Sun, M.; Erfan, M.; Hu, Q.; Mehdi, O.; Liu, B.; Ke, Y.; Pan, Z. Evolution of fractal characteristics in shales with increasing thermal maturity: Evidence from neutron scattering, N2 physisorption, and FE-SEM imaging. *Energy* **2024**, *298*, 131342. [CrossRef]
29. Wang, F.; Wang, L. Pore structure analysis and permeability prediction of shale oil reservoirs with HPMI and NMR: A case study of the Permian Lucaogou Formation in the Jimsar Sag, Junggar Basin, NW China. *J. Pet. Sci. Eng.* **2022**, *214*, 110503. [CrossRef]
30. Liu, K.; Ostadhassan, M.; Zou, J.; Gentzis, T.; Rezaee, R.; Bubach, B.; Carvajal-Ortiz, H. Multifractal analysis of gas adsorption isotherms for pore structure characterization of the Bakken Shale. *Fuel* **2018**, *219*, 296–311. [CrossRef]

31. Turlapati, V.; Prusty, B.; Yarlagadda, D.; Pal, S.; Raja, E. Analysis of micropore size distribution using Dubinin's theory of volume filling—Effect of particle size on pore characterization of organic-rich Indian shales. *J. Nat. Gas Sci. Eng.* **2022**, *106*, 104746. [CrossRef]
32. Mandelbrot, B.; Passoja, D.; Paullay, A. Fractal character of fracture surfaces of metals. *Nature* **1984**, *308*, 721–722. [CrossRef]
33. Lai, J.; Wang, G. Fractal analysis of tight gas sandstones using high-pressure mercury intrusion techniques. *J. Nat. Gas Sci. Eng.* **2015**, *24*, 185–196. [CrossRef]
34. Huang, H.; Chen, L.; Sun, W.; Xiong, F.; Ji, W.; Jia, J.; Tang, X.; Zhang, S.; Gao, J.; Luo, B. Pore-throat structure and fractal characteristics of Shihezi formation tight gas sandstones in the Ordos Basin, China. *Fractals* **2018**, *26*, 1840005. [CrossRef]
35. Li, X.; Wei, W.; Wang, L.; Ding, P.; Zhu, L.; Cai, J. A new method for evaluating the pore structure complexity of digital rocks based on the relative value of fractal dimension. *Mar. Pet. Geol.* **2022**, *141*, 105694. [CrossRef]
36. Guan, Q.; Dong, D.; Sun, S.; Hu, L.; Qi, L.; Li, C.; Chen, C.; Chen, X. Fractal characteristics of organic-rich shale pore structure and its geological implications: A case study of the Lower Silurian Longmaxi Formation in the Weiyuan block, Sichuan Basin. *Nat. Gas. Ind.* **2024**, *44*, 108–118.
37. Jarvie, D. Shale Resource Systems for Oil and Gas Part 2: Shale-oil Resource Systems. In *Shale Reservoirs_Giant Resources for the 21st Century: AAPG Memoir*; American Association of Petroleum Geologists; GeoScienceWorld: McLean, VA, USA, 2012; Volume 97. [CrossRef]
38. Wang, M.; Ma, R.; Li, J.; Lu, S.; Li, C.; Guo, Z.; Li, Z. Occurrence mechanism of lacustrine shale oil in the Paleogene Shahejie Formationof Jiyang Depression, Bohai Bay Basin, China. *Pet. Explor. Dev.* **2019**, *46*, 789–802. [CrossRef]
39. Han, Y.; Horsfield, B.; Wirth, R.; Mahlstedt, N.; Bernard, S. Oil retention and porosity evolution in organic rich shales. *AAPG Bull.* **2017**, *101*, 807–827. [CrossRef]
40. Hu, T.; Pang, X.; Jiang, F.; Wang, Q.; Liu, X.; Wang, Z.; Jiang, S.; Wu, G.; Li, C.; Xu, T.; et al. Movable oil content evaluation of lacustrine organic-rich shales: Methods and a novel quantitative evaluation model. *Earth-Sci. Rev.* **2021**, *214*, 103545. [CrossRef]
41. Zhang, H.; Huang, H.; Li, Z.; Liu, M. Oil physical status in lacustrine shale reservoirs—A case study on Eocene Shahejie Formation shales, Dongying Depression, East China. *Fuel* **2019**, *257*, 116027. [CrossRef]
42. Pernyeszi, t.; Patzkó, Á.; Berkesi, O.; Dékány, I. Asphaltene adsorption on clays and crude oil reservoir rocks. *Colloids Surf. A* **1998**, *137*, 373–384. [CrossRef]
43. Liu, B.; Liu, Y.; Liu, Y.; He, J.; Gao, Y.; Wang, H.; Fan, J.; Fu, X. Prediction of low-maturity shale oil produced by in situ conversion: A case study of the first and second members of Nenjiang Formation in the Central Depression, southern Songliao Basin, Northeast China. *Pet. Geol. Exp.* **2020**, *42*, 533–544. [CrossRef]
44. Guo, J.; Li, M.; Chen, C.; Tao, L.; Liu, Z. Experimental investigation of spontaneous imbibition in tight sandstone reservoirs. *J. Pet. Sci. Eng.* **2020**, *193*, 107395. [CrossRef]
45. Gao, H.; Wang, Y.; Xie, Y.; Ni, J.; Li, T.; Wang, C.; Xue, J. Imbibition and oil recovery mechanism of fracturing fluids in tight sandstone reservoirs. *ACS Omega* **2021**, *6*, 1991–2000. [CrossRef] [PubMed]
46. Yang, H.; Fu, J.; He, H.; Liu, X.; Zhang, Z.; Deng, X. Formation and distribution of large low-permeability lithologic oil regions in Huaqing, Ordos Basin. *Pet. Explor. Dev.* **2012**, *39*, 683–691. [CrossRef]
47. Li, P.; Jia, C.; Jin, Z.; Liu, Q.; Zheng, M.; Huang, Z. The characteristics of movable fluid in the Triassic lacustrine tight oil reservoir: A case study of the Chang 7 member of Xin'anbian Block, Ordos Basin, China. *Mar. Pet. Geol.* **2019**, *102*, 126–137. [CrossRef]
48. He, C.; Ji, L.; Su, A.; Wu, Y.; Zhang, M.; Zhou, S.; Li, J.; Hao, L.; Ma, Y. Source-rock evaluation and depositional environment of black shales in the Triassic Yanchang Formation, southern Ordos Basin, north-central China. *J. Pet. Sci. Eng.* **2019**, *173*, 899–911. [CrossRef]
49. Fu, J.; Li, S.; Niu, X.; Deng, X.; Zhou, X. Geological characteristics and exploration of shale oil in Chang 7 Member of Triassic Yanchang Formation, Ordos Basin, NW China. *Pet. Explor. Dev.* **2020**, *45*, 931–945. [CrossRef]
50. Lin, W.; Maria, M.; Arndt, S.; Chen, Y. Influence of Soxhlet extractable bitumen and oil on porosity in thermally maturing organic-rich shales. *Int. J. Coal Geol.* **2014**, *132*, 38–50. [CrossRef]
51. Yang, F.; Ning, Z.; Liu, H. Fractal characteristics of shales from a shale gas reservoir in the Sichuan Basin, China. *Fuel* **2014**, *115*, 378–384. [CrossRef]
52. Anitas, E.; Slyamov, A.; Todoran, R.; Szakacs, Z. Small-angle scattering from nanoscale fat fractals. *Nanoscale Res. Lett.* **2017**, *12*, 389. [CrossRef] [PubMed]
53. Li, T.; Song, Y.; Jiang, Z.; Jiang, S.; Li, Q. Pore structure and fractal characteristics of distinct thermally mature shales. *Energy Fuels* **2019**, *33*, 5116–5128. [CrossRef]
54. Mishra, S.; Mendhe, V.; Varma, A.; Kamble, A.; Sharma, S.; Bannerjee, M.; Kalpana, M. Influence of organic and inorganic content on fractal dimensions of Barakar and Barren Measures shale gas reservoirs of Raniganj Basin, India. *J. Nat. Gas Sci. Eng.* **2018**, *49*, 393–409. [CrossRef]
55. Wang, Y.; Liu, L.; Cheng, H. Gas Adsorption Characterization of Pore Structure of Organic-rich Shale: Insights into Contribution of Organic Matter to Shale Pore Network. *Nat. Resour. Res.* **2021**, *30*, 2377–2395. [CrossRef]
56. Gregg, S.; Sing, K. Adsorption, Surface Area and Porosity. 2. Auflage, academic press, London. *Berichte Der. Bunsenges. Für. Phys. Chem.* **1982**, *86*, 877–962. [CrossRef]
57. Sing, K. Reporting physisorption data for gas/solid systems with special reference to the determination of surface area and porosity (Recommendations 1984). *Pure Appl. Chem.* **1985**, *57*, 603–619. [CrossRef]

58. Groen, J.; Peffer, L.; Pérez-Ramírez, J. Pore size determination in modified micro-and mesoporous materials. Pitfalls and limitations in gas adsorption data analysis. *Microporous Mesoporous Mater.* **2003**, *60*, 1–17. [CrossRef]
59. Tang, X.; Jiang, Z.; Li, Z.; Gao, Z.; Bai, Y.; Zhao, S.; Feng, J. The effect of the variation in material composition on the heterogeneous pore structure of high-maturity shale of the Silurian Longmaxi formation in the southeastern Sichuan Basin, China. *J. Nat. Gas Sci. Eng.* **2015**, *23*, 464–473. [CrossRef]
60. Wang, Y.; Zhu, Y.; Liu, S.; Zhang, R. Pore characterization and its impact on methane adsorption capacity for organic-rich marine shales. *Fuel* **2016**, *181*, 227–237. [CrossRef]
61. Sakhaee-Pour, A.; Bryant, S. Effect of pore structure on the producibility of tight-gas sandstones. *AAPG Bull.* **2014**, *98*, 663–694. [CrossRef]
62. Dutton, S.; Loucks, R. Diagenetic controls on evolution of porosity and permeability in lower Tertiary Wilcox sandstones from shallow to ultradeep (200–6700 m) burial, Gulf of Mexico Basin. U. S. A. *Mar. Pet. Geol.* **2010**, *27*, 69–81. [CrossRef]
63. Chang, J.; Fan, X.; Jiang, Z.; Wang, X.; Chen, L.; Li, J.; Zhu, L.; Wan, C.; Chen, Z. Differential impact of clay minerals and organic matter on pore structure and its fractal characteristics of marine and continental shales in China. *Appl. Clay Sci.* **2022**, *216*, 106334. [CrossRef]
64. Wang, S.; Feng, Q.; Javadpour, F.; Xia, T.; Li, Z. Oil adsorption in shale nanopores and its effect on recoverable oil-in-place. *Int. J. Coal Geol.* **2015**, *147*, 9–24. [CrossRef]
65. Burhan, M.; Shahzad, M.; Ng, K. A universal theoretical framework in material characterization for tailored porous surface design. *Sci Rep.* **2019**, *9*, 8773. [CrossRef]
66. Pan, Y.; Liao, Y.; Shi, Q.; Hsu, C. Acidic and neutral polar NSO compounds in heavily biodegraded oils characterized by negative-ion ESI FT-ICR MS. *Energry Fuel* **2013**, *27*, 2960–2973. [CrossRef]
67. Qu, Y.; Sun, W.; Tao, R.; Luo, B.; Chen, L.; Ren, D. Pore-throat structure and fractal characteristics of tight sandstones in Yanchang Formation, Ordos Basin. *Mar. Pet. Geol.* **2020**, *120*, 104573. [CrossRef]
68. Robin, M.; Rosenberg, E.; Fassi-Fihri, O. Wettability studies at the pore level: A new approach by use of Cryo-SEM. *SPE Form. Eval.* **1995**, *10*, 11–19. [CrossRef]
69. Barclay, S.; Worden, R. Effects of reservoir wettability on quartz cementation in oil fields. *Quartz Cem. Sandstones* **2000**, *29*, 103–117. [CrossRef]
70. Chang, X.; Xue, Q.; Li, X.; Zhang, J.; Zhu, L.; He, D.; Zheng, H.; Lu, S.; Liu, Z. Inherent wettability of different rock surfaces at nanoscale: A theoretical study. *Appl. Surf. Sci.* **2018**, *434*, 73–81. [CrossRef]
71. Tian, S.; Erastova, V.; Lu, S.; Greenwell, H.; Underwood, T.; Xue, H.; Zeng, F.; Chen, G.; Wu, C.; Zhao, R. Understanding model crude oil component interactions on kaolinite silicate and aluminol surfaces: Toward improved understanding of shale oil recovery. *Energy Fuels* **2018**, *32*, 1155–1165. [CrossRef]

fractal and fractional

MDPI

Article

Fractal Characteristics of Pore Throat and Throat of Tight Sandstone Sweet Spot: A Case Study in the East China Sea Basin

Wenguang Wang [1,2,3,*], Chengyan Lin [3] and Xianguo Zhang [3]

[1] Institute of Unconventional Oil & Gas, Northeast Petroleum University, Daqing 163318, China
[2] Key Laboratory of Continental Shale Hydrocarbon Accumulation and Efficient Development, Ministry of Education, Northeast Petroleum University, Daqing 163318, China
[3] School of Geosciences, China University of Petroleum (East China), Qingdao 266580, China; lincy@upc.edu.cn (C.L.); zhangxianguo@upc.edu.cn (X.Z.)
* Correspondence: wangwenguang.0@163.com or wangwg@nepu.edu.cn

Abstract: The study of the fractal characteristics of the pore throat radius (PTR) and throat radius of sweet spots is crucial for the exploration and development of tight gas sandstone. This study used conventional core analysis, X-ray diffraction analysis, scanning electron microscopy (SEM), and constant-rate mercury injection experiment (CRMI), high-pressure mercury injection experiment (HPMI), and nuclear magnetic resonance (NMR) techniques to investigate the fractal characteristics of the PTR and throat radius of the tight sandstone sweet spots of the Huagang Formation in the central uplift belt of the East China Sea Basin. Based on conventional core analysis and SEM, the main pore types of the tight sandstone samples in the Huagang Formation were determined to be intergranular dissolved pore, intragranular dissolved pore, intergranular pore, and moldic pore. HPMI and NMR techniques were used to evaluate the full-size PTR distribution of type I (TI), type II (TII), and type III (TIII) sweet spots. Based on fractal theory, CRMI was used to calculate the fractal dimension of the PTR and throat radius of three types of sweet spots, and the relationship between the fractal dimensions and pore throat structure parameters and mineral composition were investigated. The results showed that the full-size PTR distribution curve exhibited bimodal or unimodal characteristics. The peak values of the PTR distribution of the TI, TII, and TIII sweet spots were mainly concentrated at 0.002–22.5 μm, 0.001–2.5 μm, and 0.0004–0.9 μm, respectively. The fractal dimensions of the PTR and throat radius were calculated. The average throat radius fractal dimensions of the TI, TIII, and TIII sweet spots were 2.925, 2.875, and 2.786, respectively. The average PTR fractal dimensions of the TI, TII, and TIII sweet spots were 2.677, 2.684, and 2.702, respectively. The throat radius fractal dimension of the TI, TII, and TIII sweet spots was positively correlated with mercury saturation, average throat radius, feldspar content, and clay mineral content and negatively correlated with displacement pressure, quartz content, and carbonate cement content. The PTR fractal dimension of the TI, TII, and TIII sweet spots was positively correlated with displacement pressure, quartz content, and carbonate cement content and negatively correlated with feldspar content. The throat size of the TI sweet spot was large, and the heterogeneity of the throat was strong. The PTR heterogeneity of the TI sweet spot was lower than that of the TII and TIII sweet spots. The findings of this study can provide important guidance for the exploration and development of tight gas sandstone.

Keywords: throat radius; fractal dimension; sweet spot; pore throat radius; tight sandstone

1. Introduction

Pore throat structure is a key factor affecting the storage capacity and fluid flow capacity of tight sandstone reservoirs, and it controls the productivity distribution and exploitation efficiency of hydrocarbon reservoirs [1–3]. Tight sandstone reservoirs have poor physical properties, complex pore structures, and strong heterogeneity [4,5]. The main factors determining the reservoir quality are pore throat radius (PTR) and throat radius distribution, while the flow characteristics of fluids in tight reservoirs largely depend on the difference of microscopic PTR and throat radius. Therefore, a correct understanding of the PTR and throat radius of the tight sandstone sweet spot is of great significance for revealing the storage capacity and seepage capacity of tight gas sandstone reservoirs [2,6–8].

Many techniques and methods have been applied to study the microstructure of tight sandstone, including microscopic observation techniques, fluid injection techniques, and spectroscopic analysis techniques [9–12]. The microscopic observation technique is a qualitative research method for pore throat morphology and size; it can intuitively characterize the geometric characteristics of the pore throat and includes SEM and energy spectrum analysis, secondary electron (SE)/backscattered electron (BSE) imaging, and micro/nano CT scanning [13–15]. Fluid injection techniques, including the constant-rate mercury injection experiment (CRMI), the high-pressure mercury injection experiment (HPMI), nuclear magnetic resonance (NMR), low-temperature nitrogen adsorption, and low-temperature CO_2 adsorption, can quantitatively obtain pore throat size and distribution characteristics [4,16,17]. The spectroscopic analysis techniques mainly include focused ion beam scanning electron microscopy (FIB-SEM), neutron small-angle scattering (SAXS), and X-ray small-angle scattering (SAXS) [11]. However, different experimental techniques differ in the measurement range of the PTR, and these testing techniques all have their advantages and limitations in studying pore throat structures [8,18].

Tight sandstone reservoirs often develop nano-sized and micro-sized pore throats with poor connectivity, and the geometric shapes of pore throats are complex and irregular; so, it is difficult to effectively describe and characterize the microscopic pore throat structure with a single experimental method. Therefore, using various experimental techniques to accurately quantify PTR and throat radius is of great significance for the exploration and development of tight gas sandstone. HPMI can characterize the pore throat structure of a reservoir by testing the relationship between mercury saturation and pressure during the mercury injection process. The T_2 (transverse relaxation time) spectra distribution in low-magnetic-field nuclear magnetic resonance has a direct relationship with the pore structure and can to some extent reflect the full-size PTR distribution of the sample. NMR results are related to the T_2 relaxation time; so, T_2 spectra can only be converted into PTR distribution by combining NMR with other experimental techniques. CRMI can obtain the characteristic parameters of pore throats and throats, which have great advantages in the study of the microscopic pore structure characteristics of tight sandstone. PTR, throat radius, and their fractal characteristics in tight sandstone reservoirs are a naturally complex system. Fractal theory can quantitatively characterize the strong heterogeneity and complexity of pore throat structures in tight sandstone reservoirs [6,19]. Fractal dimension (D) is an important and effective index for elucidating pore structure parameters [10,20]. D can quantify the surface roughness of materials, and the D value of sandstone is usually between 2.0 and 3.0 [21]. The size and structure of the throat affect the microscopic heterogeneity of tight sandstone reservoirs, directly affecting the exploration and development of tight gas sandstone [22].

The tight sandstone reservoir of the Huagang Formation in the East China Sea Basin has undergone complex sedimentary evolution and diagenesis alteration, with a complex pore structure. Wang et al. (2024) used various experimental methods, such as thin sections, SEM, X-ray diffraction analysis, scanning electron microscope mineral quantitative evaluation, and HPMI, and studied the fractal dimension of pore throat structure and the influence of diagenesis on the fractal dimension of the tight sandstone sweet spots of the Huagang Formation in the Jiaxing area of the East China Sea Basin [23]. However, the data tested by HPMI show the distribution of pore throat sizes and fail to distinguish between pores and throats. The combination of HPMI and NMR can characterize the full-size PTR distribution of tight sandstone, and CRMI can obtain the PTR and throat radius of tight sandstone. This study is an inheritance and development of previous research and is definitely not a simple repetition. In this study, the full-size PTR distribution of tight sandstone sweet spots is characterized by HPMI and NMR, and the fractal characteristics of the PTR and throat radius of the tight sandstone sweet spot were investigated by combining fractal theory and the CRMI technique. The main purpose of this study is to: (1) characterize the pore throat morphology and size distribution of tight sandstone sweet spots using SEM, CRMI, HPMI, and NMR; (2) quantitatively evaluate the full-size PTR distribution of the tight sandstone sweet spot by combining HPMI and NMR; (3) calculate the fractal dimension of the PTR and throat radius of the tight sandstone sweet spots using the CRMI results; and (4) investigate the relationship between the fractal characteristics, pore throat structure parameters, and mineral composition. The results of this study can provide valuable guidance for the exploration and development of tight gas sandstone.

2. Samples and Methods

2.1. Geological Background and Samples

The East China Sea Basin is a petroliferous basin located in eastern China (Figure 1a) [24]. The basin is composed of three first-order tectonic units: the eastern depression group, the central uplift group, and the western depression group [25,26]. The Xihu Depression is located in the central part of the eastern depression group of the East China Sea Basin, with an area of approximately 59,000 km^2. The western side of the Xihu Depression is bounded by the Haijiao uplift and the Yushan uplift; the eastern side is bounded by the Diaoyudao uplift; the northern side is adjacent to the Fujiang Depression; and the southern side is connected to the Diaobei Depression (Figure 1b) [27]. The Xihu Depression is divided from east to west into an eastern fault terrace belt, a central uplift belt, and a western slope belt [28,29]. A gas field is located in the north-central region of the central uplift belt of the Xihu Depression (Figure 1c). The exploration area of A gas field is approximately 229 km^2, with 4 drilling wells (Figure 1d). The main gas producing layers in the basin are the Huagang Formation and Pinghu Formation. The Huagang Formation can be further divided into 12 subunits from top to bottom, namely the first member—twelfth member (H1–H12). The upper and lower parts of the Huagang Formation are divided into H1–H5 and H6–H12, respectively. According to previous research, the rocks in the upper part of the Huagang Formation belong to braided delta sediments and are the main natural gas reservoirs. The target interval of this study is the tight sandstone of the H4 and H5 members. The burial depth of the tight sandstone of the fourth (H4) and fifth (H5) members in A gas field is approximately 3800–4150 m.

The sampling wells were located in A gas field of the East China Sea Basin, and the samples were taken from the tight sandstone reservoirs of the H4 and H5 members of the Huagang Formation. In this study, 1142 conventional core analysis data items and 108 X-ray diffraction analysis data items were collected from the H4 and H5 tight sandstones of the Huagang Formation in the A gas field of the East China Sea Basin. The total length of the core section for the target interval of the four sampling wells was 129.9 m, including 27.1 m for well A1, 48.3 m for well A2, 45.5 m for well A3, and 9.0 m for well A4. The typical sandstone samples of the tight sandstone of the Huagang Formation from wells A1, A2, A3, and A4 in A gas field were selected, and 158 SEM, 6 CRMI, 13 HPMI, and 13 NMR tests were conducted. Six typical cylindrical samples with a diameter of 2.54 cm and a length of 0.5 cm were selected for CRMI. Thirteen typical cylindrical samples with a diameter of 2.54 cm and a length of 2.5 cm were selected for HPMI. The sample information for the 13 HPMI and 13 NMR tests was the same.

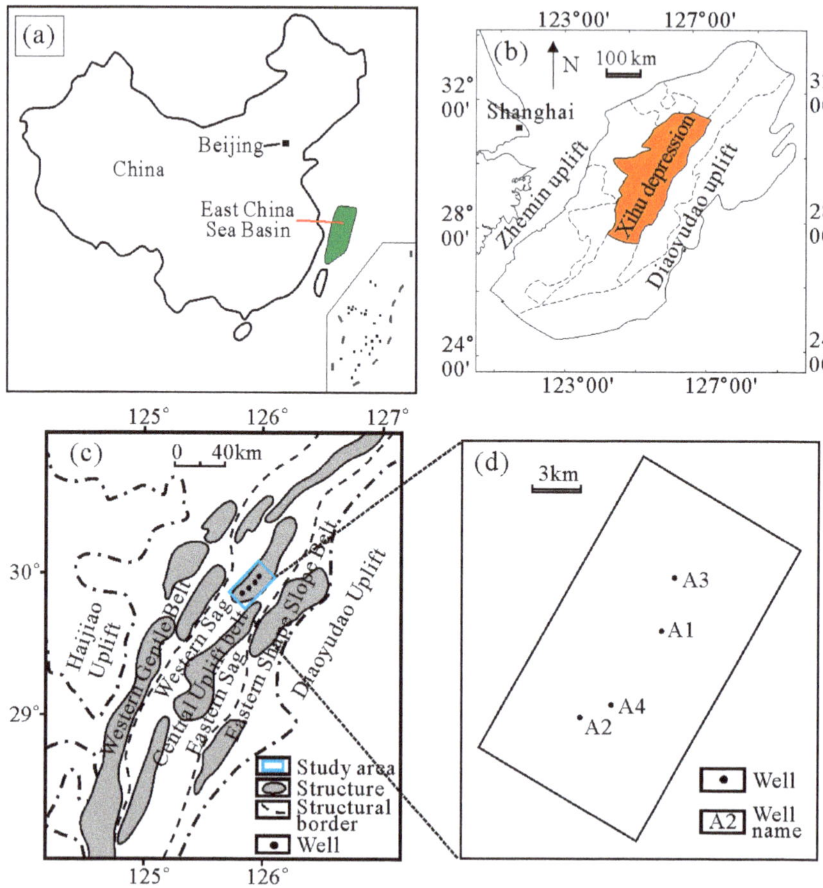

Figure 1. (**a**) Location of the East China Sea Basin [30]. (**b**) Location of Xihu Depression. (**c**) Tectonic unit division of the Xihu Depression and the location of A gas field (modified from [31]). (**d**) A gas field.

2.2. Experimental Measurements

2.2.1. SEM

A total of 158 gold-plated sandstone bulk samples were selected for SEM analysis. The sandstone sample processing procedure included oil washing, sample selection, piling, drying, dust removal, and gold plating. The manufacturer of Quanta FEG 450 scanning electron microscopy is Financial Executives International company (Hillsborough, OR, USA). The type and morphology of the pore throat and mineral distribution of the tight sandstone were studied using Quanta FEG 450 scanning electron microscopy [32,33].

2.2.2. CRMI

Based on a large number of SEM image observations, a total of six typical tight sandstone sweet spot samples were selected for a constant-rate mercury injection experiment (CRMI). The six samples were composed of three type I (TI) sweet spots, one type II (TII) sweet spot, and two type III (TIII) sweet spots. The manufacturer of the ASPE-730 constant-rate mercury injection instrument is American Core Experimental System Company (Houston, TX, USA). A cylindrical sample with a diameter of 2.54 cm and a length of 0.5 cm was drilled from 6 core samples for CRMI. The experiment was carried out using the ASPE-730 constant-rate mercury injection instrument, which injected mercury into a core sample at a quasi-static constant rate of 0.0001 mL/min. The temperature was set to 25 °C, the contact angle was 140°, and the interfacial tension was 485 dyn/cm. The maximum mercury injection pressure of the CRMI was 6.201 MPa, and the corresponding throat radius was about 119 nm.

A constant low speed causes the mercury injection process to be approximated as a quasi-static process. Mercury, as a non-wetting phase, is injected into rock pores at an extremely low and constant rate to ensure a constant contact angle and interfacial tension during the experimental process. The capillary resistance reaches the maximum at the minimum radius of the throat. When mercury breaks through the throat and enters the pores, the pressure drops and fills the pores, causing the pressure to rise again. Every change in pore shape experienced by the mercury front leads to a change in the shape of the meniscus, resulting in a change in the capillary pressure of the system. Under the quasi-static mercury injection condition, the microstructure of the pores is determined according to the natural pressure fluctuation of the meniscus at the mercury injection when it passes through different microscopic pore shapes. By using high-precision pressure sensors to record the changes in pressure with the amount of mercury injected, the throat was distinguished from the pores, and the frequency distribution data of the throat radius, PTR, and pore to throat radius ratio were obtained.

2.2.3. HPMI

Based on a large number of SEM image observations, a total of 13 typical tight sandstone sweet spot samples were selected for high-pressure mercury injection experiments (HPMI). The 13 samples were composed of three TI sweet spot, three TII sweet spots, and seven TIII sweet spots. The manufacturer of the AutoPore IV-9520 mercury injection instrument is Micromeritics Instrument Corporation (Norcross, GA, USA). Cylindrical samples with a diameter of approximately 2.54 cm and a length of approximately 4.5 cm were drilled from 13 core samples for HPMI. Before conducting HPMI, oil and gas should be removed from the sample, and the parameters, such as porosity, permeability, and density, should be measured in advance. HPMI was conducted using the AutoPore IV-9520 mercury injection instrument at a temperature of 25 °C and a relative humidity of 35–50%. The sandstone sample underwent oil and gas washing and drying treatment, with a maximum mercury injection pressure of 116.67 MPa and a corresponding PTR of 0.006 μm. During the experiment, the volume accuracy of the mercury entry and exit was greater than 0.1 μL. HPMI can obtain PTR distribution through a capillary pressure curve. The PTR curve (Equation (1)) was obtained by the relationship between capillary pressure and PTR, and the pore structure of the tight sandstone was evaluated.

$$P_c = \frac{2\sigma\cos\theta}{r} \tag{1}$$

where P_c is the capillary pressure, MPa. σ is the surface tension, N/m. In general, $\sigma = 0.48$ N/m. θ is the wetting angle, °. In general, $\theta = 140°$. r is the pore throat radius, μm.

2.2.4. NMR

Based on a large number of SEM image observations, a total of 13 typical tight sandstone sweet spot samples were selected for the NMR experiment. The 13 samples were composed of three TI sweet spots, three TII sweet spots, and seven TIII sweet spots. The manufacturer of the Magnet2000 nuclear magnetic resonance core analyzer is the Institute of Seepage Mechanics, Chinese Academy of Sciences (Beijing, China). Before conducting the NMR experiment, the oil and gas in the sample need to be removed, and porosity and permeability need to be measured in advance. The NMR experiments were carried out on the core samples using the Magnet2000 nuclear magnetic resonance core analyzer. The NMR frequency was 4.99 MHz, the number of echoes was 2048, the waiting time was 4 s, and the echo interval was 0.4 ms. Core samples saturated with water were placed in a low magnetic-field nuclear magnetic resonance core analyzer for NMR measurement, and the T_2 relaxation time spectra and T_2 cumulative distribution were calculated by inversion. The core samples after centrifugation (centrifugal force 300 psi = 2.07 MPa) were measured by NMR. The measured parameters were the same as those before centrifugation, and the T_2 relaxation time spectra and T_2 cumulative distribution were calculated by inversion. NMR measurements of the core samples after centrifugation (centrifugal force 300 psi = 2.07 MPa) were performed; the measured parameters were the same as those before centrifugation, and the T_2 relaxation time spectra and T_2 cumulative distribution were calculated by inversion. Then, the $T_{2\text{cut-off}}$ value for each sample was determined. The T_2 spectra of a saturated single-phase fluid can reflect the pore throat structure. Therefore, the T_2 relaxation time can be expressed as Equation (2).

$$\frac{1}{T_2} = \rho_2 \frac{S}{V} \tag{2}$$

where T_2 is the surface relaxation time, ms. ρ_2 is the surface relaxation rate (μ m/ms). S/V is the specific surface area, $\mu m^2/\mu m^3$.

2.3. Methodology

2.3.1. NMR Movable Fluid Spectrum Conversion

The PTR volume obtained from HPMI can only characterize a PTR with a large radius. NMR signals can reflect information about the microstructure. In this study, the combination of HPMI data and NMR data can quantitatively characterize the full-size PTR distribution of tight sandstone. By establishing a mathematical relationship between T_2 relaxation time and throat radius r, the NMR T_2 distribution curve can be converted into a PTR distribution curve [15,34].

For simplified spherical and cylindrical pore structures, the relationship between specific surface area and pore radius can be expressed as follows:

$$\frac{S}{V} = \frac{F_S}{r_c} \tag{3}$$

By combining Equations (2) and (3), Equation (4) is obtained.

$$T_2 = \frac{r_c}{\rho_2 F_S} \tag{4}$$

where F_s is the pore shape factor ($F_s = 3$ for spherical pores and $F_s = 2$ for tubular pores), dimensionless. r_c is the pore radius, μm. Most of the experiment results confirm that T_2 distribution is a power function relationship of the pore radius [35], as shown below.

$$T_2 = \frac{r_c^n}{\rho_2 F_S} \tag{5}$$

where n is a power exponent, dimensionless.

The pore radius r_c is equal to the product of the throat radius and the pore to throat radius ratio, and the relationship between T_2 relaxation time and throat radius can be obtained.

$$T_2 = \frac{(C_1 r_t)^n}{\rho_2 F_S} \tag{6}$$

where C_1 is the average pore to throat radius ratio, dimensionless. r_t is the throat radius, μm.

By defining the following equation:

$$C = \frac{(\rho_2 F_S)^{\frac{1}{n}}}{C_1} \tag{7}$$

we can obtain Equation (8).

$$r_t = CT_2^{\frac{1}{n}} \tag{8}$$

By taking the logarithm on both sides of Equation (8), Equation (9) can be obtained.

$$\ln(r_t) = \ln(C) + \frac{1}{n}\ln(T_2) \tag{9}$$

By fitting the values of C and n, the T_2 spectra of a fully saturated core sample can be converted into a PTR distribution curve.

HPMI was used to convert the NMR T_2 spectra, and the full-size PTR distribution of the tight sandstone was finely characterized. The analytical procedure is as follows. (1) The cumulative amplitude percentage from the large to the small values of the NMR T_2 spectra was calculated. (2) The cumulative mercury saturation percentage of the interpolated HPMI data was calculated from the large to the small PTR. (3) Based on the interpolation relationship between the NMR T_2 spectra data S (i) and HPMI PTR data, the interpolated PTR distribution was obtained (Figure 2a). (4) The left part of black dashed line in Figure 2b was selected as the analysis area. The values of C and n were obtained by fitting the relationship between HPMI interpolated r (i) and NMR T_2 (i) in the analysis area by using least square method. The transformation from T_2 relaxation time to PTR can be achieved by substituting C and n into Equation (9).

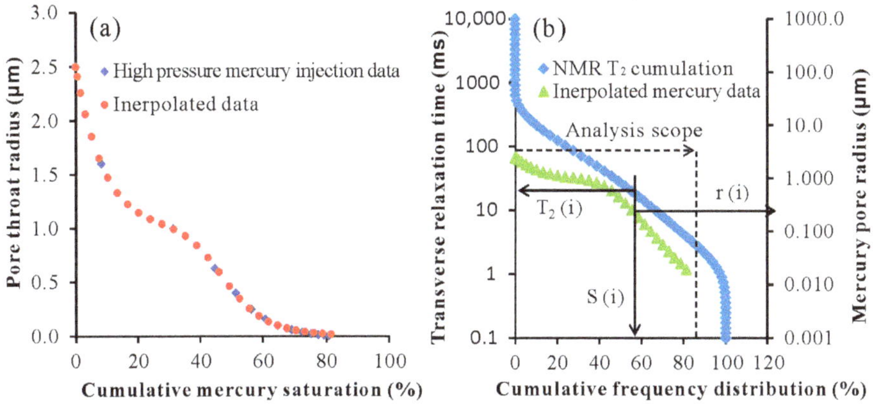

Figure 2. (**a**) Distribution of PTR and cumulative mercury saturation before and after HPMI interpolation. (**b**) Conversion from T_2 NMR spectrum distribution curve to PTR distribution curve.

2.3.2. Fractal Method

In a constant-rate mercury injection experiment, capillary pressure can reflect the difficulty of mercury entering the PTR and throat radius. In this study, the fractal dimension of the PTR and throat radius was calculated based on the CRMI data, and the fractal dimension of the PTR and throat radius of tight sandstone were quantitatively characterized. The analytical procedure for calculating the fractal dimension is similar to that reported by [3,23]. The published mercury intrusion capillary pressure model equation can be used to calculate the fractal dimension [3].

$$\lg[1 - S_{Hg}(> r)] = (3 - D)\lg(r) - (3 - D)\lg r_{max} \tag{10}$$

where D is the "D" of MIP. S_{Hg} is mercury saturation, %. r_{max} is the maximum pore throat radius, measured in μm. r is the pore throat radius, μm.

The fractal dimension D is calculated based on the linear slope λ between lg[1 − S_{Hg} (>r)] and lg(r). The fractal dimension can be derived from the slope of a line (λ = 3 − D), as shown below.

$$D = 3 - \lambda \tag{11}$$

According to fractal theory, pore throat structure fractals can be divided into total fractals and segmented fractals [36]. When the segmented fractal curve of the lg(1 − S_{Hg}) − lg(r) curve is not a straight line and has obvious inflection points, the curve can be divided into small and large PTR, and the fractal dimensions of the small and large PTR can be calculated. These two types of pores have obvious inflection points on the lg(1 − S_{Hg}) and lg(r) curves; then, the fractal dimensions of the small PTR and large PTR are calculated by linear fitting according to the abovementioned fractal dimension calculation procedure. Finally, based on the fractal dimensions and porosity of the small and large PTR, the weighted average method can be used to obtain the fractal dimension (D) of the entire PTR (Equation (12)).

$$D = D_1 \times \frac{\varphi_1}{\varphi_1 + \varphi_2} + D_2 \times \frac{\varphi_2}{\varphi_1 + \varphi_2} \tag{12}$$

where D is the total fractal dimension. D_1 is the fractal dimension of the small pore throat. D_2 is the fractal dimension of the large pore throat. φ_1 is the porosity of the small pore throat, %. φ_2 is the porosity of the large pore throat, %.

Based on the CRMI data, the fractal characteristics of the PTR and throat radius of the sandstone samples were analyzed, and the corresponding fractal dimensions were calculated.

3. Results

3.1. Petrographic Characteristics and Pore Types

The sandstone of the H4 and H5 members in the central uplift belt of the East China Sea Basin is mainly composed of feldspar lithic quartzose, accompanied by a small amount of feldspathic litharenite and litharenite (Figure 3a). The main minerals of these sandstones are quartz, feldspar, and rock fragments, with content ranges of 53.0–75.0%, 8.0–20.0%, and 14.0–37.0%, respectively. The sorting degree of the clastic grains is moderate or good, and the roundness is sub-angular and sub-rounded. Based on bulk rock and clay X-ray diffraction analysis, the content of clay minerals ranges from 2.1% to 35.2%, with an average value of 7.62%. The clay minerals are primarily illite, chlorite, and illite–smectite mixed layers, with a relatively small volume fraction of kaolinite (Figure 3b). Illite has the highest volume fraction, ranging from 0.231% to 15.84%, with an average value of 2.55%. The volume fraction of chlorite is second, ranging from 0.595% to 11.14%, with an average value of 2.44%. The volume fraction of the illite–smectite mixed layer ranges from 0.217% to 15.488%, with an average value of 2.34%. The porosity range of tight sandstone in the H4 and H5 members is 1.2–16.4%, with an average value of 8.27%, and the permeability range is 0.0232–156 mD, with an average value of 1.5 mD.

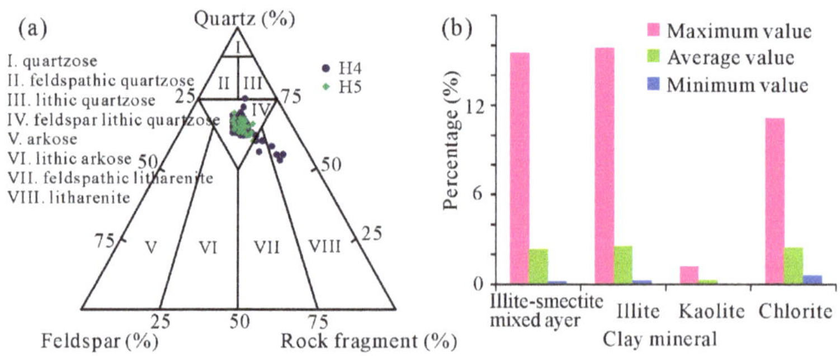

Figure 3. (**a**) Ternary plot of rock types in the H4 and H5 members of A gas field in East China Sea Basin [37]. (**b**) Histogram of clay mineral composition.

Under the thin section and SEM, four types of pores were mainly developed in the tight sandstone reservoirs of the H4 and H5 members in the A gas field, including intergranular pores, intergranular dissolved pores, intragranular dissolved pores, and moldic pores (Figure 4). The petrographic data showed that the average percentage contents of the intergranular pores, intergranular dissolved pores, intragranular dissolved pores, and moldic pores were 0.41%, 1.57%, 0.8%, and 0.74%, respectively.

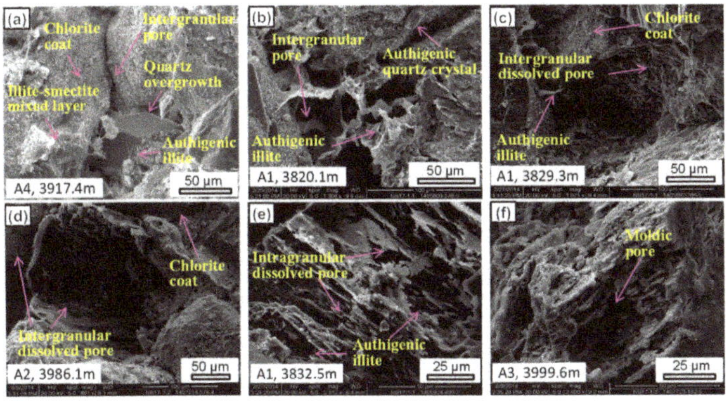

Figure 4. Typical pore types of tight sandstone of the H4 and H5 members in the A gas field. (**a**) SEM image showing chlorite coat, quartz overgrowth, authigenic illite, and intergranular pores. (**b**) SEM image showing authigenic illite, authigenic quartz crystals, and intergranular pores. (**c**) SEM image showing dissolution of debris, intergranular dissolved pores, chlorite coat, and authigenic illite. (**d**) SEM image showing dissolution of debris, intergranular dissolved pores, chlorite coat, and authigenic illite. (**e**) SEM image showing dissolution of feldspar, intragranular dissolved pores, and a small amount of authigenic illite. (**f**) SEM image showing moldic pores and a small amount of authigenic illite.

Intergranular pores are pores between mineral grains after burial diagenesis, typically in the shape of triangles or polygons with smooth straight sides (Figure 4a,b). Most intergranular pores are influenced by diagenesis, such as mechanical compaction and clay mineral cementation filling, forming micro-sized and nano-sized intergranular pores (Figure 4a,b). Intergranular dissolved pores contain some primary intergranular pores, which are the result of further dissolution on the basis of primary intergranular pores (Figure 4c,d). Intergranular dissolved pores are mainly dissolution pores between mineral grains or dissolution pores at mineral edges, accompanied by illite, chlorite, and authigenic quartz. Pores have irregular pore surfaces, ranging in diameter in the order of several hundred micrometers, which provide the main storage space for tight gas sandstones. Intragranular dissolved pores are mainly dissolution pores within the clastic grains, which are further subdivided into feldspar dissolved pores and debris dissolved pores based on the composition of dissolved debris (Figure 4e). A small amount of illite and authigenic quartz is distributed within intragranular dissolved pores, resulting in various types of pores and throats. Moldic pores are dissolution pores where the clastic grains are completely dissolved, leaving only the outer shape (Figure 4f). Illite and authigenic quartz can be seen in the moldic pores.

3.2. Pore Throat Size Distribution Characteristics

3.2.1. Constant-Rate Mercury Injection Experiment Result

CRMI can directly measure the throat radius and the volume distribution controlled by the throat. The pore radius and throat radius distributions were calculated according to the capillary pressure curve of the constant-rate mercury injection experiment (Figure 5 and Table 1). The average pore radius of six typical sandstone samples was mainly distributed in the range of 124.58–154.63 μm, and the average throat radius was mainly distributed in the range of 0.489–2.669 μm. Figure 5 shows the capillary pressure curves of six typical samples. In the pore area, the trend of the total mercury saturation curve was consistent with the pore mercury saturation curve (Figure 5a). In the pore throat area, mercury saturation was controlled by both the pore and the throat. In the throat area, pore mercury saturation stopped increasing as pressure increased, and the throat contributed to total mercury saturation (Figure 5a).

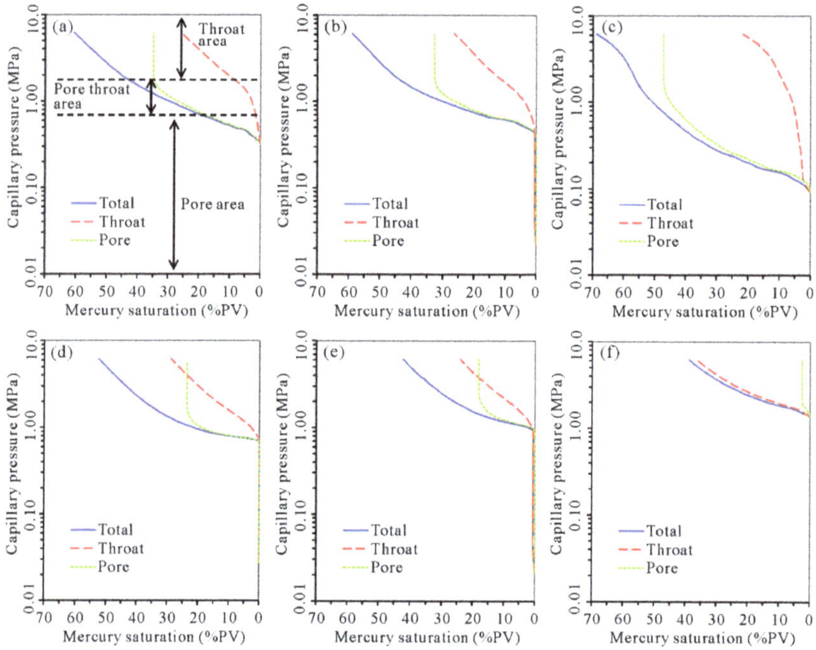

Figure 5. Typical CRMI curve. (**a**–**c**) TI sandstone sweet spot samples. (**d**) TII sandstone sweet spot sample. (**e**,**f**) TIII sandstone sweet spot samples.

Table 1. Parameters of constant-rate mercury injection experiment.

Well Name	Depth (m)	Sweet Spot Type	Displacement Pressure (MPa)	Average Throat Radius (μm)	Average Pore Radius (μm)	Average Pore to Throat Radius Ratio	Average Pore Volume (cm^3/g)	Mercury Saturation (%)
A4	3916.5	Type I	0.379	1.007	124.58	155.3	15.82	60.42
A4	3917.9	Type I	0.479	0.937	147.84	190.1	15.82	58.94
A4	3942.5	Type I	0.112	2.669	134.45	99.8	28.82	68.9
A4	3910.4	Type II	0.689	0.855	148.22	195.1	24.79	52.44
A4	3911.7	Type III	1.017	0.611	154.63	294	24.71	42.2
A4	3915.4	Type III	0.689	0.489	129.23	288.3	10.62	38.9

According to the reservoir properties, the classification evaluation criteria of the reservoir "sweet-spot" in the Huangang Formation were established, namely the TI sweet spot (Φ > 10%, K > 1 mD), TII sweet spot (8% < Φ < 10%, 0.4 mD < K < 1 mD), and TIII sweet spot (Φ < 8%, K < 0.4 mD). Based on the conventional core analysis data and CRMI data, the experimental samples were classified into TI, TII, and TIII (Table 1 and Figure 5). The average displacement pressures of the TI, TII, and TIII sweet spots were 0.323 MPa, 0.689 MPa, and 0.853 MPa, respectively, and the corresponding throat radii were 1.54 μm, 0.855 μm, and 0.55 μm, respectively. The average mercury saturation values of the TI, TII, and TIII sweet spots were 62.75%, 52.44%, and 40.55%, respectively (Figure 5). Due to the limitation of the experimental principle, the average pore radius calculated from the pore to throat radius ratio was between 100 μm and 300 μm, which is inconsistent with the actual situation.

The throat radius distribution range measured by CRMI was 0.118–6.560 μm (Figure 6a). The average throat radius was positively correlated with permeability (Figure 6b), indicating that the throats affected the permeability of the tight sandstone reservoirs.

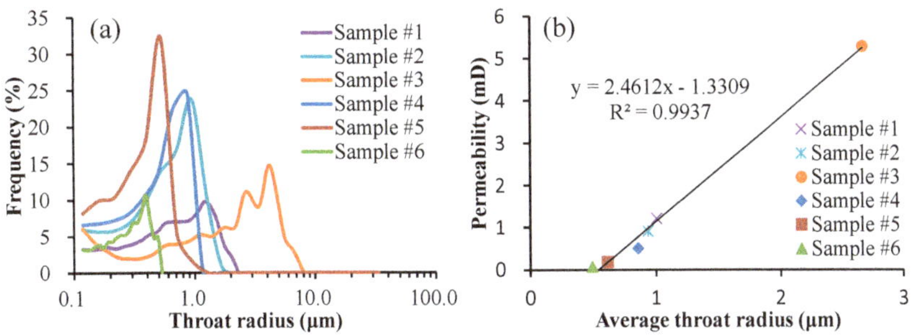

Figure 6. (**a**) Distribution of throat radius. (**b**) Relationship between average throat radius and permeability.

3.2.2. High-Pressure Mercury Injection Experiment Result

HPMI can reflect the pore throat structure and pore radius distribution characteristics of tight sandstone. The average displacement pressure and average PTR of the 13 typical sandstone samples were 1.16 MPa and 0.89 μm, respectively (Table 2). The average values of the maximum mercury saturation, relative sorting coefficient, and skewness of the sample were 77.1%, 0.36, and 1.49, respectively (Table 2).

Table 2. Parameters of high-pressure mercury injection experiment.

Well Name	Depth (m)	Porosity (%)	Permeability (mD)	Sweet Spot Type	Maximum Mercury Saturation (%)	Relative Sorting Coefficient	Skewness	Average Pore Throat Radius (μm)	Displacement Pressure (MPa)	Efficiency of Mercury Withdrawal (%)
A2	3971.6	11.3	2.04	Type I	80.66	0.31	2	0.79	0.3	25.98
A4	3915.9	13.1	16	Type I	95.22	0.34	0.96	7.97	0.01	24.21
A4	3918	12.8	2.79	Type I	96.47	0.22	1.09	0.81	0.2	28.46
A2	3976	9.1	0.728	Type II	96.45	0.2	1.53	0.62	0.3	34.72
A2	3983.6	9.2	0.234	Type II	81.61	0.26	1.66	0.17	1.5	38.50
A4	3910.1	9.3	0.441	Type II	81.78	0.28	1.84	0.38	0.3	37.3
A3	3994	5.7	0.131	Type III	69	0.41	1.49	0.14	1.5	44.69
A3	3999.5	7	0.15	Type III	73.67	0.34	1.5	0.13	1.5	42.32
A3	4012.5	5.3	0.108	Type III	69.46	0.4	1.51	0.16	1.5	40.3
A3	4111.2	4.7	0.0598	Type III	57.99	0.57	1.42	0.06	3	45.36
A3	4112.7	4.7	0.076	Type III	62.58	0.5	1.42	0.08	1.5	43.66
A3	4117.2	5.9	0.0804	Type III	70.66	0.38	1.45	0.11	2	43.76
A3	4120.2	6.3	0.146	Type III	66.25	0.45	1.48	0.13	1.5	43.55

The capillary pressure curve can reflect the development and connectivity of connected pore throats and their controlled pore volumes. The mercury injection curve of HPMI was converted into a PTR distribution curve (Figure 7a,b). The PTR and displacement pressure of different types of sweet spots are different to some extent. The average displacement pressures of the TI, TII, and TIII sweet spots were 0.17 MPa, 0.7 MPa, and 1.79 MPa, respectively (Table 2 and Figure 7a). The TI sandstone sweet spot sample had high porosity and permeability, and the capillary curve showed low displacement pressure and a small inclination degree. The TIII sandstone sweet spot samples had low porosity and permeability, and the capillary curves showed high displacement pressure and a large inclination degree (Table 2 and Figure 7a). The PTR ranges of the TI, TII, and TIII sweet spots were 0.0063–63 μm, 0.0063–1.6 μm, and 0.0063–0.4 μm, respectively, with corresponding average values of 3.19 μm, 0.39 μm, and 0.12 μm (Table 2 and Figure 7b).

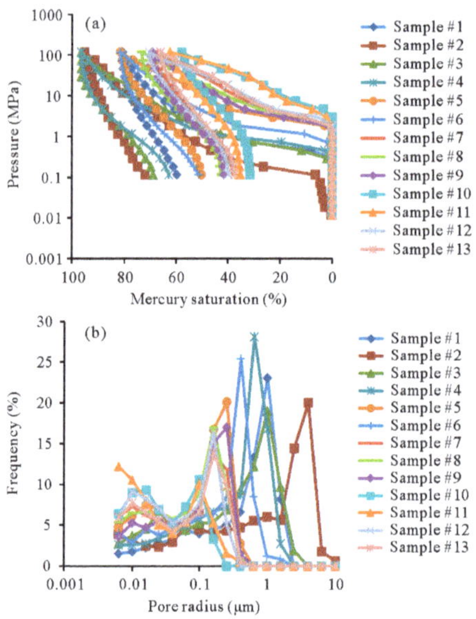

Figure 7. (**a**) Typical high-pressure mercury injection experiment curves. (**b**) The characteristics of pore radius of high-pressure mercury injection experiment.

3.2.3. Nuclear Magnetic Resonance Result

The NMR T_2 spectra of fully saturated water samples can reflect the pore throat structure. The T_2 spectra distribution of the 13 samples under fully saturated water conditions revealed bimodal or unimodal characteristics (Figure 8). The T_2 relaxation time was distributed between 0.1 ms and 10,000 ms. Sample # 1, sample # 2, and sample # 3 are sandstone samples of the TI sweet spot. The T_2 spectra distribution of the TI sweet spot showed a bimodal distribution with an obvious right peak, indicating a large pore throat (Figure 8). Sample # 4, sample # 5, and sample # 6 are sandstone samples of the TII sweet spot. The T_2 spectra distribution of the TII sweet spot showed a bimodal distribution with an obvious right peak. However, the right peak amplitude of the TII sweet spot was lower than that of the TI sweet spot (Figure 8). Sample # 7, sample # 8, sample # 9, sample # 10, sample # 11, sample # 12, and sample # 13 are sandstone samples of the TIII sweet spot. The T_2 spectra distribution of the TIII sweet spot showed bimodal characteristics, and the samples with these bimodal characteristics had a wide range of pore throat sizes (Figure 8).

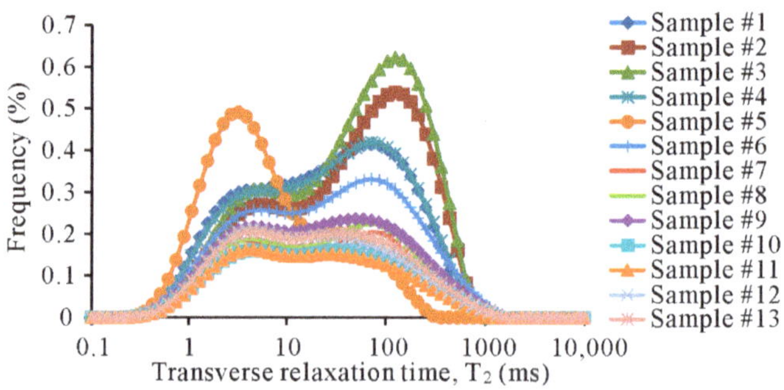

Figure 8. NMR T_2 spectrum of completely saturated water samples.

By placing the porosity cumulative values after centrifugation into the T_2 distribution curve in a fully saturated state, the $T_{2cut\text{-}off}$ value was obtained (Figure 9). This cut-off value can be used to separate the movable and irreducible parts of the fluid. In the fully saturated water state, the part of the T_2 distribution curve where T_2 was less than the $T_{2cut\text{-}off}$ value can be considered as the irreducible area (indicated by the black diagonal line in Figure 9).

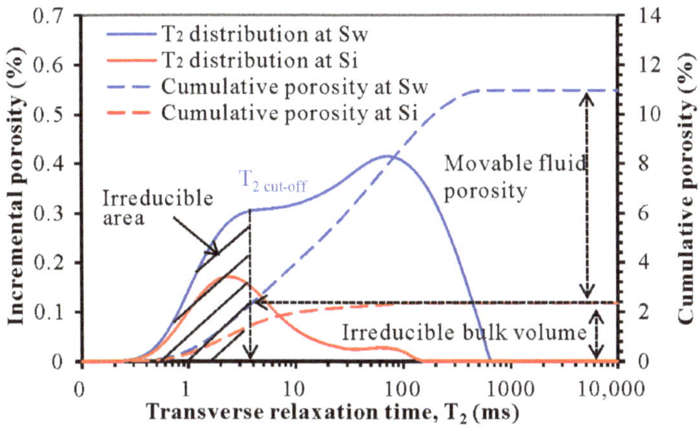

Figure 9. NMR T_2 spectrum of sample # 1 before and after centrifugation.

The $T_{2cut\text{-}off}$ values of the 13 samples were distributed between 3.87 ms and 11.57 ms, with an average value of 8.02 ms (Table 3). The average $T_{2cut\text{-}off}$ values of the TI, TII, and TIII sweet spots were 7.18 ms, 7.3 ms, and 8.7 ms, respectively (Table 3). The range of the movable fluid saturation of the 13 samples was 42.46–72.06%, with an average value of 57.67%. The average movable fluid saturation values of the TI, TII, and TIII sweet spots were 70.19%, 56.21%, and 52.93%, respectively (Table 3).

Table 3. Parameters of nuclear magnetic resonance experiment.

Sample Number	Well Name	Depth (m)	Sweet Spot Type	$T_{2cut\text{-}off}$ (ms)	Movable Fluid Saturation (%)	Movable Fluid Porosity (%)
Sample # 1	A2	3971.6	Type I	3.87	72.06	7.91
Sample # 2	A4	3915.9	Type I	8.03	70.49	8.23
Sample # 3	A4	3918	Type I	9.64	68.02	9.08
Sample # 4	A2	3976	Type II	6.69	67.11	7.43
Sample # 5	A2	3983.6	Type II	5.57	42.46	3.61
Sample # 6	A4	3910.1	Type II	9.64	59.07	5.36
Sample # 7	A3	3994	Type III	6.69	55.85	3.12
Sample # 8	A3	3999.5	Type III	6.69	56.83	3.46
Sample # 9	A3	4012.5	Type III	8.03	57.24	3.98
Sample # 10	A3	4111.2	Type III	11.57	49.24	2.41
Sample # 11	A3	4112.7	Type III	11.57	45.81	2.11
Sample # 12	A3	4117.2	Type III	9.64	49.21	2.9
Sample # 13	A3	4120.2	Type III	6.69	56.34	3.43

3.3. Full-Size Pore Throat Radius Distribution

The full-size PTR distribution was characterized by the combination of HPMI and NMR. In the cross-plot of the T_2 relaxation time and PTR r at the same cumulative distribution frequency, the T_2 and r of the 13 samples showed a distinct two-segment linear relationship and an obvious inflection point (Figure 10a). The least square method was used to fit the C and n values into two segments of the 13 samples, and the results are shown in Table 4. Then, the NMR T_2 spectra could be converted into PTR.

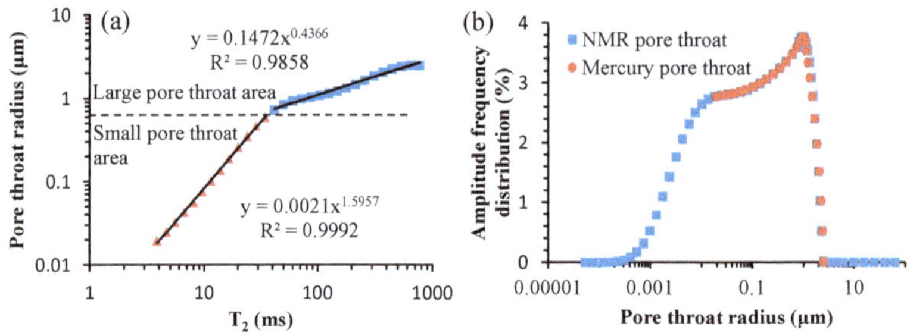

Figure 10. (**a**) Fitting parameters of NMR T_2 spectrum and PTR for sample # 1 sandstone sample. (**b**) Comparison between PTR distribution converted from NMR T_2 spectrum of sample # 1 sandstone sample and the HPMI measurement value.

Table 4. Conversion factors between the NMR T_2 spectrum and PTR and the corresponding correlation index.

Sample Number	Well Name	Depth (m)	Sweet Spot Type	Small Pore Throat			Large Pore Throat			Inflection Point (μm)
				C	n	R^2	C	n	R^2	
Sample # 1	A2	3971.6	Type I	0.0021	0.627	0.9992	0.1472	2.29	0.9858	0.65
Sample # 2	A4	3915.9	Type I	0.1795	2.704	0.9803	0.0039	0.726	0.9312	0.87
Sample # 3	A4	3918	Type I	0.0179	1.277	0.9699	0.0046	0.996	0.9091	1.57
Sample # 4	A2	3976	Type II	0.0178	1.024	0.9939	0.1444	2.573	0.9262	0.95
Sample # 5	A2	3983.6	Type II	0.0012	0.43	0.9925	0.0855	3.125	0.8716	0.2
Sample # 6	A4	3910.1	Type II	0.0042	0.971	0.9768	0.0042	1.169	0.9382	0.62
Sample # 7	A3	3994	Type III	0.0009	0.923	0.9976	0.2413	7.764	0.8789	0.49
Sample # 8	A3	3999.5	Type III	0.0013	0.996	0.9979	0.1397	4.983	0.9123	0.42
Sample # 9	A3	4012.5	Type III	0.0012	0.937	0.9994	0.0536	2.723	0.8822	0.27
Sample # 10	A3	4111.2	Type III	0.0004	0.923	0.9996	0.0244	3.238	0.8791	0.1
Sample # 11	A3	4112.7	Type III	0.0003	0.811	0.9993	0.0287	2.866	0.8774	0.13
Sample # 12	A3	4117.2	Type III	0.0011	0.961	0.9995	0.046	3.147	0.8789	0.18
Sample # 13	A3	4120.2	Type III	0.0006	0.86	0.9996	0.0589	3.112	0.8783	0.26

Figure 10b compares the PTR distribution of sample # 1 sandstone sample based on the NMR and HPMI data. The shape and distribution range of the two curves were similar, and the correlation coefficient was very good. The interpolated HPMI data can accurately characterize a part of the PTR distribution, and the PTR distribution characteristics obtained by NMR had a wider distribution range than those obtained by HPMI (Figure 10b). Therefore, by combining NMR and HPMI, the full-size PTR distribution of the entire sample can be effectively characterized.

The full-size PTR distribution of three types of sweet spots was investigated (Figure 11). The distribution range of the PTR was mainly concentrated between 0.0001 μm and 100 μm, and the distribution curve showed bimodal or unimodal characteristics. The PTR distribution curve of the TI sweet spot showed bimodal characteristics, which were mainly distributed in the right peak area, indicating that the PTR was large. The peak value of the PTR distribution of the TI sweet spot was mainly concentrated between 0.002 μm and 22.5 μm (Figure 11). According to the high-pressure mercury injection experiment, the displacement pressures of the sample # 1, sample # 2, and sample # 3 sandstone samples were 0.3 MPa, 0.01 MPa, and 0.2 MPa, respectively, indicating small displacement pressures and large PTR. The permeability of the sample # 1, sample # 2, and sample # 3 sandstone samples was relatively high, which was consistent with the PTR distribution obtained from HPMI and NMR.

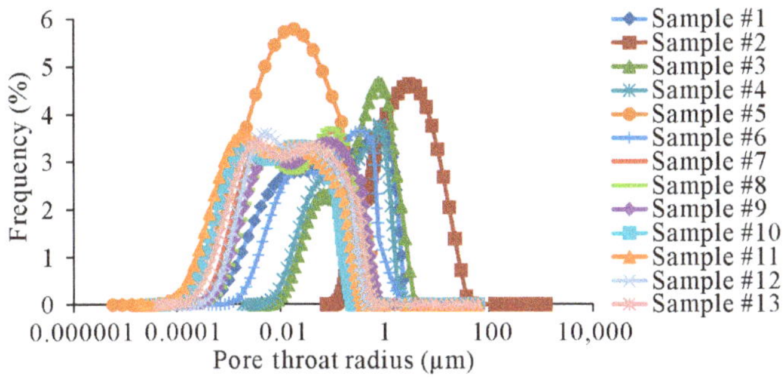

Figure 11. Full-size PTR distribution of tight sandstone in the A gas field of the East China Sea Basin.

Sample # 4, sample # 5, and sample # 6 are sandstone samples of the TII sweet spot. The PTR distribution curves of sample # 4 and sample # 6 had obvious bimodal characteristics, whereas the PTR distribution curve of sample # 5 showed unimodal characteristics. According to the full-size PTR distribution of HPMI and NMR, the peak value of the PTR distribution of the TII sweet spot was mainly concentrated in the range of 0.001–2.5 μm. The PTR distribution curve of the TIII sweet spot showed bimodal characteristics, and the left and right peak areas were similar (Figure 11). According to the full-size PTR distribution of HPMI and NMR, the peak value of the PTR distribution of the TIII sweet spot was primarily concentrated in 0.0004–0.9 μm.

3.4. Fractal Dimension

CRMI data of six sandstone samples were processed; cross-plots of $\lg(1 - S_{Hg}\ (>r))$ and lg(r) were constructed, and the slope of the line was fitted. By substituting the slope into Equation (10), the fractal dimensions of the PTR and throat radius of tight sandstone were obtained. The PTR and throat radius of the tight sandstone of the H4 and H5 members in the central uplift belt of the East China Sea Basin are fractal, while the pores of these samples are not fractal (Figure 12). This may be because large pore distribution is not a determinant factor of pore throat structure heterogeneity, and throat distribution is an important factor in pore throat structure heterogeneity.

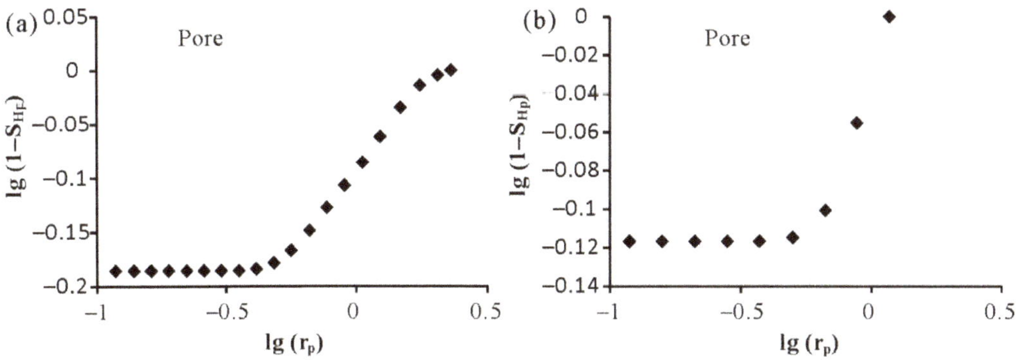

Figure 12. (**a**) Relationship between $\lg(1 - S_{Hg})$ and lg(r) of the pores of TI sweet spot. (**b**) Relationship between $\lg(1 - S_{Hg})$ and lg(r) of the pores of TII sweet spot.

The fractal characteristics of the PTR and throat radius were investigated (Figure 13). The $\lg(1 - S_{Hg}\ (>r))$ and lg(r) curve of the throat radius was clearly divided into two segments at a certain radius, indicating that the tight sandstone sample had a two-segment fractal structure, and the correlation coefficients of the two were greater than 0.9. Figure 13a shows the fractal curve of the throat radius of the TI sweet spot. The correlation coefficient of the left segment of the curve was 0.9335, corresponding to a small throat. The fractal dimension of the small throat was represented by Dt_1. The correlation coefficient of the right segment was 0.9998, corresponding to a large throat. The fractal dimension of the large throat was represented by Dt_2. The $\lg(1-S_{Hg}(>r))$ and $\lg(r)$ curves of the PTR of the TI sweet spot also showed the fractal characteristics of the two segments (Figure 13b). The correlation coefficient of the left segment of the PTR curve for the TI sweet spot was 0.9871, corresponding to a small pore throat. The fractal dimension of the small pore throat was represented by Dp_1. The correlation coefficient of the right segment was 0.9994, corresponding to a large pore throat. The fractal dimension of the large pore throat was represented by Dp_2.

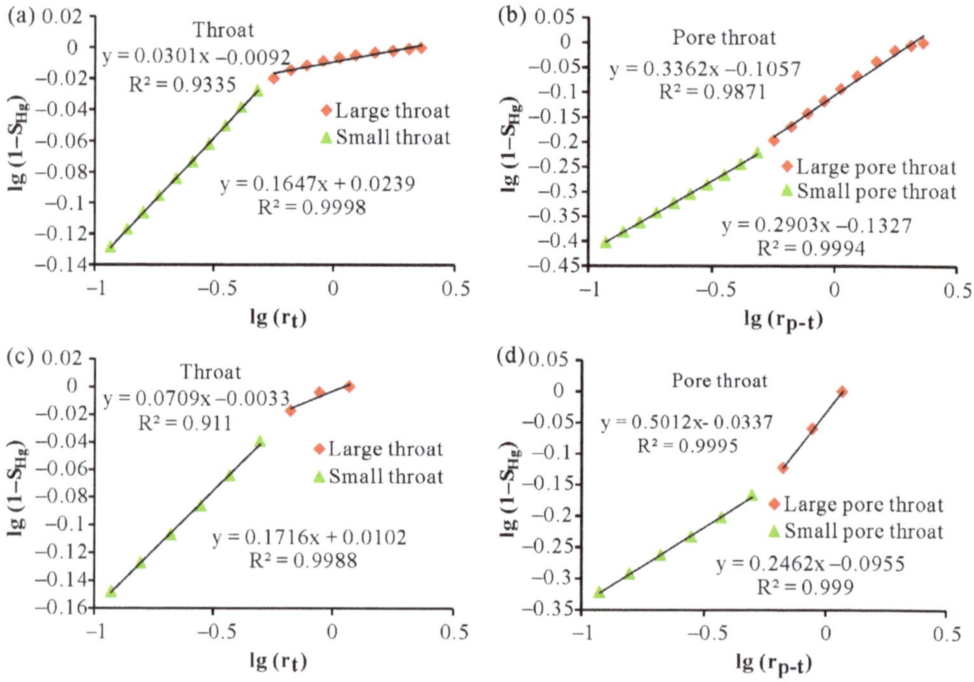

Figure 13. Relationship between $\lg(1-S_{Hg})$ and $\lg(r)$. (**a**) TI sweet spot throat radius. (**b**) TI sweet spot PTR. (**c**) TII sweet spot throat radius. (**d**) TII sweet spot PTR.

In most of the samples, the slope of the large throat segment was lower than that of the small throat segment, and the fractal dimension of the large throat segment was larger than that of the small throat segment (Figure 13a,c). The slope of the large pore throat segment was higher than that of the small pore throat segment, and the fractal dimension of the large pore throat segment was smaller than that of the small pore throat segment (Figure 13b,d). This indicated that the heterogeneity of the small throat was lower than that of the large throat, and the heterogeneity of the small pore throat was higher than that of the large pore throat.

The fractal dimensions of the entire PTR and throat radius of six typical sandstone samples were obtained using the weighted average method. The ranges of the fractal dimensions Dt and Dp of the throat radius and PTR were 2.691–2.954 and 2.648–2.749, respectively, and the corresponding average values were 2.87 and 2.69, respectively. The average fractal dimensions of the throat radius of the TI, TII, and TIII sweet spots were 2.925, 2.875, and 2.786, respectively (Table 5), indicating that the throat structure complexity of the TI sweet spot was higher than that of the TII and TIII sweet spots. The average fractal dimensions of the PTR of the TI, TII, and TIII sweet spots were 2.677, 2.684, and 2.702, respectively (Table 5), and the pore throat structure complexity of the TI sweet spot was lower than that of the TII and TIII sweet spots. The fractal dimension can quantitatively reflect the complexity and heterogeneity of pore throat structure.

Table 5. Fractal dimension calculation of PTR and throat radius.

Well Name	Depth (m)	Sweet Spot Type	Dt_1	R^2	Dt_2	R^2	Dt	Dp_1	R^2	Dp_2	R^2	Dp
A4	3916.5	Type I	2.835	0.9998	2.97	0.9335	2.916	2.71	0.9994	2.664	0.9871	2.682
A4	3917.9	Type I	2.848	0.9994	2.958	0.927	2.905	2.738	0.9982	2.566	0.9754	2.648
A4	3942.5	Type I	2.904	0.9344	2.98	0.9094	2.954	2.762	0.9566	2.671	0.992	2.702
A4	3910.4	Type II	2.828	0.9988	2.929	0.911	2.875	2.754	0.999	2.499	0.9995	2.684
A4	3911.7	Type III	2.833	0.999	2.953	0.8459	2.88	2.777	0.9972	2.706	0.8423	2.749
A4	3915.4	Type III	2.732	0.9969	2.649	0.9828	2.691	2.722	0.997	2.587	0.9885	2.655

4. Discussion

4.1. Full-Size Pore Throat Radius Distribution Characteristics

There are various methods for characterizing pore throat structure, including field emission scanning electron microscopy [38], the gas adsorption method [39], the mercury injection method [40], and nuclear magnetic resonance [34]. Each method has its unique advantages and disadvantages, and it is necessary to combine multiple methods to reveal the full-size pore size distribution of tight reservoirs [41,42]. Previous scholars carried out a large number of studies on full-size pore size characterization, including (1) splicing the pore radius or PTR distributions obtained by low-temperature CO_2 adsorption, low-temperature nitrogen adsorption, and HPMI [39,41]; (2) the characterization of full-size pore size distribution by combining low-temperature nitrogen adsorption and NMR results [43,44]; (3) the study of full-size PTR distribution by combining HPMI and NMR [3]; and (4) quantitatively revealing pore size distribution based on large-field scanning electron microscopy and image recognition techniques [38].

The PTR distribution range of the H4 and H5 tight sandstone in the central uplift belt of the East China Sea Basin is 0.0001–100 μm, and the pore throat structure is complex, which makes it difficult to completely characterize the large-scale PTR distribution in any single experiment. Therefore, combining the results of NMR, low-temperature nitrogen adsorption, HPMI, and CRMI is an effective method for characterizing the full-size PTR distribution of tight sandstone. This study comprehensively applied SEM, HPMI, and NMR techniques, and established a characterization method for the full-size PTR distribution of sweet spots in tight sandstone. By comparing the PTR distribution obtained from HPMI with the cumulative distribution curve of NMR T_2 spectra, HPMI can only characterize large pore throats, and it was difficult for it to reveal many small pore throats. The main peak of the HPMI PTR distribution corresponded to the right peak of the NMR T_2 spectra, and the T_2 spectra were converted into the full-size PTR distribution. The basis for determining the full-size PTR distribution of tight sandstone is as follows. The splicing point between HPMI and NMR was selected at a PTR of 0.1–1.57 μm (Table 4). The distribution of PTR larger than the splicing point was obtained by HPMI. Based on the PTR distribution obtained by NMR and the PTR distribution obtained by HPMI, the distribution of PTR smaller than the splicing point was obtained.

This study combined the results of HPMI and NMR, and quantitatively characterized the full-size PTR distribution of tight sandstone sweet spots. The full-size PTR distribution results of the 13 typical sandstone samples showed that the full-size PTR distribution ranges of the TI, TII, and TIII sweet spots were 0.0004–1265.5 μm, 0.000006–63 μm, and 0.000015–63 μm, respectively (Figure 11), with corresponding the PTR distribution peaks mainly concentrated in 0.002–22.5 μm, 0.001–2.5 μm, and 0.0004–0.9 μm.

4.2. Relationship Between Fractal Dimension and Pore Throat Structure Parameters

Fractal theory is an effective tool to quantitatively characterize the strong heterogeneity and complexity of pore throat structures in tight sandstone reservoirs [19]. All the methods for characterizing the pore throat structure of tight sandstone reservoirs can be used to obtain the fractal dimension of porous rock by establishing fractal models [45]. Previous scholars have carried out a large number of studies on the pore throat structure of tight sandstone, including (1) the fractal dimension derived from complex pore networks obtained from two-dimensional scaled thin sections, SEM, and three-dimensional scaled micro- and nano-CT images [46]; (2) the fractal dimension obtained based on the gas adsorption method [47,48]; (3) the fractal dimension derived from mercury injection capillary pressure data [3,6,15,49]; and (4) the fractal dimension obtained based on nuclear magnetic resonance [9,15]. Wu et al. (2022) studied the pore throat structure and fractal characteristics of Permian Shihezi tight sandstone in Sulige Area, Ordos Basin, by using various experiments involving thin section, SEM, a mercury injection porosimeter, and nuclear magnetic resonance [15]. Based on fractal theory, the fractal dimension of sandstone pore throats was evaluated using HPMI, CRMI, and NMR. Qu et al. (2020) used CRMI results to study the fractal characteristics of pore throats and throats and the relationship between the fractal characteristics and pore throat structure parameters, fluid permeability characteristics, and mineral composition [3]. However, the fractal characteristics of PTR and throat radius in tight sandstone desserts are rarely reported in current studies. Based on the fractal theory, this study used the CRMI results to study the fractal characteristics of the PTR and throat radius of three types of sweet spots in tight sandstone.

With the increase in fractal dimension, the heterogeneity of tight sandstone increases, and the pore throat structure becomes more complex. In order to investigate the relationship between the fractal dimension and pore throat structure parameters of tight sandstone, the intersection analysis of the fractal dimensions Dt and Dp were carried out with mercury saturation, displacement pressure, and average throat radius as parameters (Figure 14). The fractal dimension Dt of the TI, TII, and TIII sweet spots was positively correlated with the mercury saturation (Figure 14a). The fractal dimension Dt and mercury saturation of the TI sweet spot were larger than that of the TII and TIII sweet spots, indicating that the mercury saturation gradually decreased from the TI to the TIII sweet spots (Figure 14a). The fractal dimension Dp of the TI, TII, and TIII sweet spots showed a weak negative correlation with the mercury saturation (Figure 14b).

The fractal dimension Dt of the TI, TII, and TIII sweet spots showed a weak negative correlation with the displacement pressure, whereas the fractal dimension Dp showed a weak positive correlation with the displacement pressure (Figure 14c,d). The displacement pressure of the TI sweet spot was lower than that of the TII and TIII sweet spots. The fractal dimension Dt of the TI, TII, and TIII sweet spots was positively correlated with the average throat radius, whereas the fractal dimension Dp was negatively correlated with the average throat radius (Figure 14e,f). The correlation between the fractal dimension of the throat radius and the average throat radius was higher than that between the fractal dimension of the PTR and the average throat radius. The average throat radius and fractal dimension Dt of the TI sweet spot were high, and the average throat radius and fractal dimension Dt of the TIII sweet spot were low, which indicated that the throat size of the TI sweet spot was large and the heterogeneity of the throat was strong and that there are multiple pore throat shapes.

The fractal dimension Dt of the throat gradually decreased from the TI and TII sweet spots to the TIII sweet spot, indicating that the throat heterogeneity of the TI sweet spot was larger than that of the TII and TIII sweet spots. In contrast, the fractal dimension Dp of the PTR gradually increased or was not significant from the TI and TII sweet spots to the TIII sweet spot, indicating that the heterogeneity of PTR increased or was not significant from the TI and TII sweet spots to the TIII sweet spot.

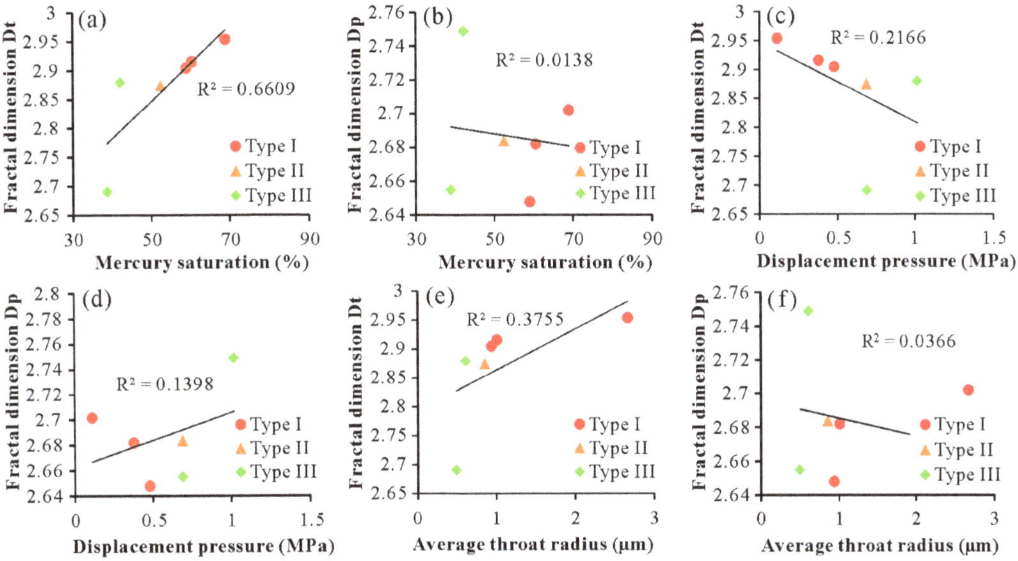

Figure 14. Relationship between throat radius fractal dimension (Dt), PTR fractal dimension (Dp), and pore throat structure parameters. (**a**) Relationship between fractal dimension Dt and mercury saturation. (**b**) Relationship between fractal dimension Dp and mercury saturation. (**c**) Relationship between fractal dimension Dt and displacement pressure. (**d**) Relationship between fractal dimension Dp and displacement pressure. (**e**) Relationship between fractal dimension Dt and average throat radius. (**f**) Relationship between fractal dimension Dp and average throat radius.

4.3. Effect of Mineral Composition on Fractal Dimension

Correlation analysis of mineral composition and fractal dimension was performed (Figures 15 and 16), and the influence of mineral composition and content on the fractal dimension of pore throat structure was investigated. The fractal dimension Dt of the TI, TII, and TIII sweet spots was negatively correlated with quartz content (Figure 15a), whereas the fractal dimension Dp was weakly positively correlated with quartz content (Figure 15b). The negative correlation between the fractal dimension Dt of the throat and the quartz content indicated that the quartz mineral content can improve the throat heterogeneity of tight sandstone to some extent. Quartz minerals had strong anti-compaction and anti-dissolution properties, making the pore throat surface smooth. With the increase in quartz content, the throat heterogeneity of the TI sweet spot was larger than that of the TII and TIII sweet spots. The weak positive correlation between fractal dimension Dp and quartz content can be attributed to the fact that the fractal dimension increased with the increase in quartz content.

The fractal dimension Dt of the TI, TII, and TIII sweet spots was positively correlated with the feldspar content (Figure 15c), whereas the fractal dimension Dp was weakly negatively correlated with the feldspar content (Figure 15d). The positive correlation between fractal dimension Dt and the feldspar content can be attributed to the dissolution of the feldspar, which produced many pores and throats, increasing the throat heterogeneity. If dissolution occurred in the cleavage direction of the feldspar minerals, this created a series of pore throats and increased the heterogeneity of the pore throats. The weak negative correlation between fractal dimension Dp and the feldspar content indicated that the PTR heterogeneity of the TI sweet spots was lower than that of the TII and TIII sweet spots with the increase in feldspar content.

The fractal dimension Dt was negatively correlated with the content of carbonate cement, whereas the fractal dimension Dp was weakly positively correlated with the content of carbonate cement (Figure 15e,f). The carbonate cement mainly occupied pore space rather than throats, and the carbonate cement increased the heterogeneity of the pore throat structure. The fractal dimension Dt was positively correlated with clay mineral content (Figure 16a), whereas the fractal dimension Dp was negatively correlated with clay mineral content (Figure 16b). The increase in clay mineral content reduced intergranular pores, increased throat radius heterogeneity, and reduced PTR heterogeneity.

In summary, from the TI and TII sweet spots to the TIII sweet spot, the throat radius fractal dimension decreased with increasing quartz content, whereas the PTR fractal dimension increased with increasing quartz content. Due to the dissolution of feldspar, the increase in throat radius heterogeneity, and the decrease in PTR heterogeneity, the fractal dimension of the throat radius increased with the increase in feldspar content, whereas the fractal dimension of the PTR decreased with the increase in feldspar content. Considering that carbonate cement occupied pore space, the fractal dimension of the PTR increased with the increase in carbonate cement content. The weak positive correlation between throat fractal dimension Dt and clay mineral content can be attributed to the complex contact between grains and fillings.

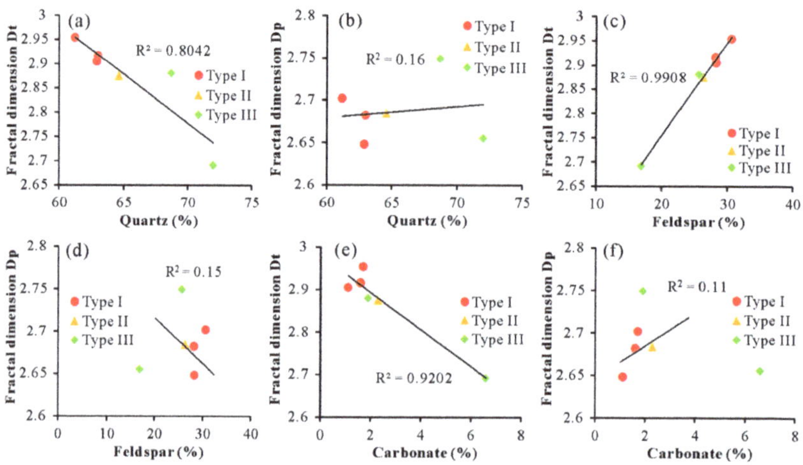

Figure 15. (**a**) Relationship between throat radius fractal dimensions (Dt) and quartz. (**b**) Relationship between PTR fractal dimensions (Dp) and quartz. (**c**) Relationship between fractal dimensions Dt and feldspar. (**d**) Relationship between fractal dimensions Dp and feldspar. (**e**) Relationship between fractal dimensions Dt and carbonate cement. (**f**) Relationship between fractal dimensions Dp and carbonate cement.

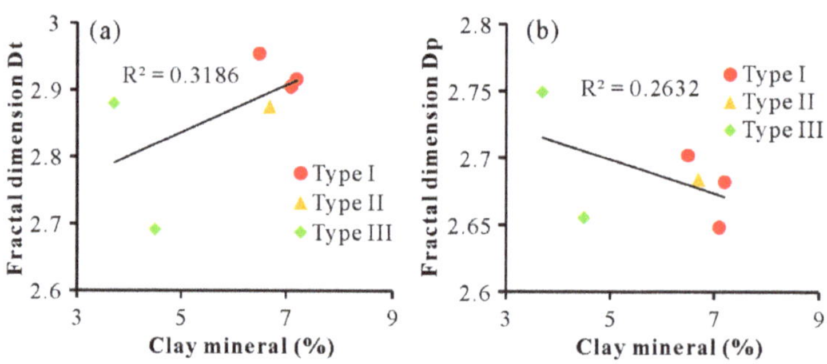

Figure 16. (**a**) Relationship between fractal dimension Dt and clay minerals. (**b**) Relationship between fractal dimension Dp and clay minerals.

5. Conclusions

This study used HPMI to convert NMR T_2 spectra and accurately characterized the full-size PTR distribution of tight sandstone sweet spots. In addition, combined with fractal theory, the fractal characteristics of the PTR and throat radius of tight sandstone sweet spots were quantitatively characterized. The conclusions are summarized as follows.

(1) The tight sandstones of the H4 and H5 members were mainly composed of feldspathic quartzose and a small amount of feldspathic litharenite and litharenite. The porosity range of these tight sandstones was 1.2–16.4%, with an average value of 8.27%. The permeability range was 0.0232–156 mD, and the average value was 1.5 mD. The most common pore types in the research area were feldspar dissolved pores and debris dissolved pores, followed by intragranular dissolved pores, intergranular pores, and moldic pores.

(2) The average throat radius of the CRMI was mainly distributed between 0.489 μm and 2.669 μm, and the average throat radii of the type I (TI), type II (TII), and type III (TIII) sweet spots were 1.54 μm, 0.855 μm, and 0.55 μm, respectively. The average PTR of the HPMI was mainly distributed between 0.06 μm and 7.97 μm, and the average PTRs of the TI, TII, and TIII sweet spots were 3.19 μm, 0.39 μm, and 0.12 μm, respectively. The average $T_{2cut\text{-}off}$ values of the TI, TII, and TIII sweet spots were 7.18 ms, 7.3 ms, and 8.7 ms, respectively. A characterization method for the full-size PTR distribution of the tight sandstone sweet spots was established by integrating HPMI and NMR techniques, and the full-size PTR distribution of the three types of sweet spots was determined. The full-size PTR distribution curve exhibited bimodal or unimodal characteristics, and the peak values of the PTR distribution of the TI, TII, and TIII sweet spots were mainly concentrated in 0.002–22.5 μm, 0.001–2.5 μm, and 0.0004–0.9 μm.

(3) Based on fractal theory, the fractal dimensions of the entire PTR and throat radius were calculated. The throat can be divided into two segments, and the average fractal dimensions Dt_1 and Dt_2 of the small throat and large throat were 2.83 and 2.907, respectively. Similarly, the pore throats can be divided into two segments, and the average fractal dimensions Dp_1 and Dp_2 of the small and large pore throats were 2.74 and 2.65, respectively. The ranges of the fractal dimensions Dt and Dp of the throat radius and PTR were 2.691–2.954 and 2.648–2.749, respectively. The average throat radius fractal dimensions of the TI, TII, and TIII sweet spots were 2.925, 2.875, and 2.786, respectively. The average PTR fractal dimensions of the TI, TII, and TIII sweet spots were 2.677, 2.684, and 2.702, respectively.

(4) The throat radius fractal dimension of the TI, TII, and TIII sweet spots was positively correlated with mercury saturation and average throat radius, and negatively correlated with displacement pressure. The PTR fractal dimension of the TI, TII, and TIII sweet spots was positively correlated with displacement pressure. The throat radius fractal dimension, mercury saturation, and average throat radius of the TI sweet spot were larger than that of the TII and TIII sweet spots, indicating that the throat radius heterogeneity of the TI sweet spot was greater than that of the TII and TIII sweet spots. The PTR fractal dimension of the TI sweet spot was smaller than that of the TII and TIII sweet spots, and the heterogeneity of the PTR increased or was not significant from the TI and TII sweet spots to the TIII sweet spot.

(5) The effect of mineral composition on the fractal dimension was elaborated. The throat radius fractal dimension of the TI, TII, and TIII sweet spots was negatively correlated with quartz content and carbonate cement content, and positively correlated with feldspar content and clay mineral content. The PTR fractal dimension of the TI, TII, and TIII sweet spots was positively correlated with quartz content and carbonate cement content, and negatively correlated with feldspar content and clay mineral content. With the increase in quartz content, the throat radius heterogeneity of the TI sweet spot was greater than that of the TII and TIII sweet spots. The PTR fractal dimension increased with the increase in quartz content. The dissolution of feldspar

produced many pores and throats, which increased the throat radius heterogeneity. With the increase in feldspar content, the PTR heterogeneity from the TIII and TII sweet spots to the TI sweet spot gradually decreased. The carbonate cement mainly occupied the pore space, and the heterogeneity of the PTR increased with the increase in carbonate cement content. The increase in clay minerals complicated the throat radius heterogeneity.

Author Contributions: Conceptualization, W.W.; Methodology, W.W.; Software, W.W.; Validation, W.W.; Formal analysis, W.W.; Investigation, W.W., C.L. and X.Z.; Resources, W.W., C.L. and X.Z.; Data curation, W.W.; Writing—original draft, W.W., C.L. and X.Z.; Writing—review & editing, W.W., C.L. and X.Z.; Visualization, W.W.; Supervision, W.W.; Project administration, W.W., C.L. and X.Z.; Funding acquisition, W.W. All authors have read and agreed to the published version of the manuscript.

Funding: This research was co-funded by the National Natural Science Foundation of China (grant number: 42202144), the Innovative Research Groups of the National Natural Science Foundation of China (grant number: U19B2006), and the Natural Science Foundation of Shandong Province (grant number: ZR2022QD048).

Data Availability Statement: Data is contained within the article.

Conflicts of Interest: The authors declare no conflict of interest.

References

1. Golab, A.N.; Knackstedt, M.A.; Averdunk, H.; Senden, T.; Butcher, A.R.; Jaime, P. 3D porosity and mineralogy characterization in tight gas sandstones. *Lead Edge* **2010**, *29*, 1476–1483. [CrossRef]
2. Yao, Y.B.; Liu, D.M. Comparison of low-field NMR and mercury intrusion porosimetry in characterizing pore size distributions of coals. *Fuel* **2012**, *95*, 152–158. [CrossRef]
3. Qu, Y.Q.; Sun, W.; Tao, R.D.; Luo, B.; Chen, L.; Ren, D.Z. Pore-throat structure and fractal characteristics of tight sandstones in Yanchang Formation, Ordos Basin. *Mar. Pet. Geol.* **2020**, *120*, 104573. [CrossRef]
4. Clarkson, C.R.; Freeman, M.; He, L.; Agamalian, M.; Melnichenko, Y.B.; Mastalerz, M.; Bustin, R.M.; Radlinski, A.P.; Blach, T.P. Characterization of tight gas reservoir pore structure using USANS/SANS and gas adsorption analysis. *Fuel* **2012**, *95*, 371–385. [CrossRef]
5. Chen, L.; Jiang, Z.X.; Liu, K.Y.; Tan, J.Q.; Gao, F.L.; Wang, P.F. Pore structure characterization for organic-rich lower silurian shale in the upper yangtze platform, south China: A possible mechanism for pore development. *J. Nat. Gas Sci. Eng.* **2017**, *46*, 1–15. [CrossRef]
6. Li, P.; Zheng, M.; Bi, H.; Wu, S.T.; Wang, X.R. Pore throat str ucture and fractal characteristics of tight oil sandstone: A case study in the Ordos Basin, China. *J. Pet. Sci. Eng.* **2017**, *149*, 665–674. [CrossRef]
7. Huang, H.X.; Sun, W.; Ji, W.M.; Zhang, R.H.; Du, K.; Zhang, S.H.; Ren, D.Z.; Wang, Y.W.; Chen, L.; Zhang, X. Effects of pore-throat structure on gas permeability in the tight sandstone reservoirs of the Upper Triassic Yanchang formation in the Western Ordos Basin, China. *J. Pet. Sci. Eng.* **2018**, *162*, 602–616. [CrossRef]
8. Nie, R.S.; Zhou, J.; Chen, Z.X.; Liu, J.C.; Pan, Y. Pore structure characterization of tight sandstones via a novel integrated method: A case study of the Sulige gas field, Ordos Basin (Northern China). *J. Asian Earth Sci.* **2021**, *213*, 104739. [CrossRef]
9. Guo, X.B.; Huang, Z.L.; Zhao, L.B.; Han, W.; Ding, C.; Sun, X.W.; Yan, R.T.; Zhang, T.H.; Yang, X.J.; Wang, R.M. Pore structure and multi-fractal analysis of tight sandstone using MIP, NMR and NMRC methods: A case study from the Kuqa depression, China. *J. Pet. Sci. Eng.* **2019**, *178*, 544–558. [CrossRef]
10. Li, Y.; Wang, Z.S.; Pan, Z.J.; Niu, X.L.; Yu, Y.; Meng, S.Z. Pore structure and its fractal dimensions of transitional shale: A cross-section from east margin of the Ordos Basin, China. *Fuel* **2019**, *241*, 417–431. [CrossRef]
11. Wang, W.; Zhao, X.X. Study on characteristics parameters of coal core structure on fractal theory. *J. Henan Univ. (Nat. Sci.)* **2023**, *53*, 729–737.
12. Zhang, C.; Guan, P.; Zhang, J.H.; Liang, X.W.; Ding, X.N.; You, Y. A review of the progress on fractal theory to characterize the pore structure of unconventional oil and gas reservoirs. *Acta Sci. Nat. Universiatis Pekin.* **2023**, *59*, 897–908.
13. Mayo, S.; Josh, M.; Nesterets, Y.; Esteban, L.; Pervukhina, M.; Clennell, M.B.; Maksimenko, A.; Hall, C. Quantitative micro-porosity characterization using synchrotron micro-CT and xenon K-edge subtraction in sandstones, carbonates, shales and coal. *Fuel* **2015**, *154*, 167–173. [CrossRef]
14. Ren, D.Z.; Zhou, D.S.; Liu, D.K.; Dong, F.J.; Ma, S.W.; Huang, H. Formation mechanism of the Upper Triassic Yanchang Formation tight sandstone reservoir in Ordos Basin—Take Chang 6 reservoir in Jiyuan oil field as an example. *J. Pet. Sci. Eng.* **2019**, *178*, 497–505. [CrossRef]

15. Wu, Y.P.; Liu, C.L.; Ouyang, S.Q.; Luo, B.; Zhao, D.D.; Sun, W.; Sarwar Awan, R.; Lu, Z.D.; Li, G.X.; Zang, Q.B. Investigation of pore-throat structure and fractal characteristics of tight sandstones using HPMI, CRMI, and NMR methods: A case study of the lower Shihezi Formation in the Sulige area, Ordos Basin. *J. Pet. Sci. Eng.* **2022**, *210*, 110053. [CrossRef]
16. Nooruddin, H.A.; Hossain, M.E.; Al-Yousef, H.; Okasha, T. Comparison of permeability models using mercury injection capillary pressure data on carbonate rock samples. *J. Pet. Sci. Eng.* **2014**, *121*, 9–22. [CrossRef]
17. Huang, H.X.; Chen, L.; Sun, W.; Xiong, F.Y.; Ji, W.; Jia, J.K.; Tang, X.L.; Zhang, S.H.; Gao, J.W.; Luo, B. Pore-throat structure and fractal characteristics of Shihezi formation tight gas sandstones in the Ordos Basin, China. *Fractals* **2018**, *26*, 1–22. [CrossRef]
18. Liu, W.; Lu, S.F.; Wang, M.; Zhang, S.X.; Liu, Y.; Zhou, N.W.; Guan, Y.; Wang, H.L. Overall pore characterization of tight sandstone reservoir space and its significance: A case study of the outer expansion area of Long 26 in Longhupao oilfield, Songliao Basin. *J. Northeast. Pet. Univ.* **2018**, *42*, 41–51.
19. Mandelbrot, B.B. *The Fractal Geometry of Nature*; Freeman: San Francisco, CA, USA, 1983.
20. Zhang, Z.; Weller, A. Fractal dimension of pore-space geometry of an Eocene sandstone formation Fractal dimension of pore-space geometry. *Geophysics* **2014**, *79*, 377–387. [CrossRef]
21. Li, K.; Horne, R.N. Fractal modeling of capillary pressure curves for the Geysers rocks. *Geothermics* **2006**, *35*, 198–207. [CrossRef]
22. Liu, D.K.; Sun, W.; Ren, D.Z.; Li, C.Z. Quartz cement origins and impact on storage performance in Permian Upper Shihezi Formation tight sandstone reservoirs in the northern Ordos Basin, China. *J. Pet. Sci. Eng.* **2019**, *178*, 485–496. [CrossRef]
23. Wang, W.G.; Lin, C.Y.; Zhang, X.G. Fractal dimension analysis of pore throat structure in tight sandstone reservoirs of Huagang Formation: Jiaxing area of East China Sea Basin. *Fractal Fract.* **2024**, *8*, 374. [CrossRef]
24. Deng, Y.H. Analysis on differences of petroleum type and geological conditions between two depression belts in China offshore. *Acta Pet. Sin.* **2009**, *30*, 1–8.
25. Dai, L.M.; Li, S.Z.; Lou, D.; Liu, X.; Suo, Y.H.; Yu, S. Numerical modeling of late Miocene tectonic inversion in the Xihu Sag, East China Sea Shelf Basin, China. *J. Asian Earth Sci.* **2014**, *86*, 25–37. [CrossRef]
26. Suo, Y.H.; Li, S.Z.; Yu, S.; Somerville, I.D.; Liu, X.; Zhao, S.J.; Dai, L.M. Cenozoic tectonic jumping and implications for hydrocarbon accumulation in basins in the East Asia Continental Margin. *J. Asian Earth Sci.* **2014**, *88*, 28–40. [CrossRef]
27. Liu, J.S.; Zhang, S.P. Natural gas migration and accumulation patterns in the central-north Xihu Sag, East China Sea Basin. *Nat. Gas Geosci.* **2021**, *32*, 1163–1176.
28. Zhang, J.P.; Zhang, T.; Tang, X.J. Basin type and dynamic environment in the East China Sea Shelf Basin. *Acta Geol. Sin.* **2014**, *88*, 2033–2043, (In Chinese with English abstract)
29. Zhang, S.L.; Zhang, J.P.; Tang, X.J.; Zhang, T. Geometry characteristic of the fault system in Xihu sag in East China Sea and its formation mechanism. *Mar. Geol. Quat. Geol.* **2014**, *34*, 87–94, (In Chinese with English abstract) [CrossRef]
30. Wang, W.G.; Lin, C.Y.; Zhang, X.G.; Dong, C.M.; Ren, L.H.; Lin, J.L. Discussion of seismic diagenetic facies of deep reservoir in the East China Sea Basin. *J. Pet. Sci. Eng.* **2022**, *208*, 109352. [CrossRef]
31. Zhu, Y.M.; Li, Y.; Zhou, J.; Gu, S.X. Geochemical characteristics of tertiary coal-bearing source rocks in Xihu depression, East China Sea Basin. *Mar. Pet. Geol.* **2012**, *35*, 154–165. [CrossRef]
32. Wang, W.G.; Lin, C.Y.; Zhang, X.G.; Dong, C.M.; Ren, L.H.; Lin, J.L. Provenance, clastic composition and their impact on diagenesis: A case study of the Oligocene sandstone in the Xihu sag, East China Sea Basin. *Mar. Pet. Geol.* **2021**, *126*, 104890. [CrossRef]
33. Zhang, Z.; He, X.; Tang, X.; Zhu, H. Structural trap characteristics and reservoir types in Xihu Sag, East China Sea Basin. *Mar. Geol. Front.* **2022**, *38*, 27–35.
34. Zhang, F.; Jiang, Z.X.; Sun, W.; Li, Y.H.; Zhang, X.; Zhu, L.; Wen, M. A multiscale comprehensive study on pore structure of tight sandstone reservoir realized by nuclear magnetic resonance, high pressure mercury injection and constant-rate mercury injection penetration test. *Mar. Pet. Geol.* **2019**, *109*, 208–222. [CrossRef]
35. Li, A.F.; Ren, X.X.; Wang, G.J.; Wang, Y.Z.; Jiang, K.L. Characterization of pore structure of low permeability reservoirs using a nuclear magnetic resonance method. *J. China Univ. Pet. (Ed. Nat. Sci.)* **2015**, *39*, 92–98.
36. Zhu, W.B.; Zhang, X.H.; Zhou, D.R.; Fang, C.G.; Li, J.Q.; Huang, Z.Q. New cognition on pore structure characteristics of Permian marine shale in the Lower Yangtze Region and its implications for shale gas exploration. *Nat. Gas Ind.* **2021**, *41*, 41–55. [CrossRef]
37. Folk, R.L.; Andrews, P.B.; Lewis, D.W. Detrital sedimentary rock classification and nomenclature for use in New Zealand. *N. Z. J. Geol. Geophys.* **1970**, *3*, 937–968. [CrossRef]
38. Xie, X.H.; Deng, H.C.; Fu, M.Y.; Hu, L.X.; He, J.H. Evaluation of pore structure characteristics of four types of continental shales with the aid of low-pressure nitrogen adsorption and an improved FE-SEM technique in Ordos Basin, China. *J. Pet. Sci. Eng.* **2021**, *197*, 108018. [CrossRef]
39. Xiang, J.; Zhu, Y.M.; Wang, Y.; Chen, S.B.; Jiang, Z.F. Structural deformation and its pore-fracture system response of the Wufeng-Longmaxi shale in the Northeast Chongqing area, using FESEM, gas adsorption, and SAXS. *J. Pet. Sci. Eng.* **2022**, *209*, 109877. [CrossRef]
40. Peng, S.; Zhang, T.W.; Loucks, R.G.; James, S. Application of mercury injection capillary pressure to mudrocks: Conformance and compression corrections. *Mar. Pet. Geol.* **2017**, *88*, 30–40. [CrossRef]
41. Mayka, S.; Fernandes, C.P.; da Cunha Neto, J.A.B.; Wolf, F.G.; dos Santos, V.S.S. Characterization of pore systems in seal rocks using nitrogen gas adsorption combined with mercury injection capillary pressure techniques. *Mar. Pet. Geol.* **2013**, *39*, 138–149.

42. Xiao, D.S.; Zheng, L.H.; Xing, J.L.; Wang, M.; Wang, R.; Guan, X.D.; Guo, X.Y. Coupling control of organic and inorganic rock components on porosity and pore structure of lacustrine shale with medium maturity: A case study of the Qingshankou Formation in the southern Songliao Basin. *Mar. Pet. Geol.* **2024**, *164*, 106844. [CrossRef]
43. Gao, Y.; Wang, M.; Li, Y.Y.; Jiang, Z.X.; Deng, Y.; Qin, J.H. Multi-scale pore structure characterization of lacustrine fine-grained mixed sedimentary rocks and its controlling factors: A case study of Lucaogou Formation in Jimusar Sag. *Energy Fuels* **2023**, *37*, 977–992. [CrossRef]
44. Teng, J.B.; Qiu, L.W.; Zhang, S.P.; Ma, C.F. Origin and diagenetic evolution of dolomites in paleogene Shahejie Formation lacustrine organic shale of Jiyang depression, Bohai Bay Basin, east China. *Pet. Explor. Dev.* **2022**, *49*, 1251–1265. [CrossRef]
45. Zhang, Q.P. Study on Microscopic Pore Throat Structure and Grading Evaluation of Chang 7 Tight Sandstone Reservoirs in Jiyuan Area, Ordos Basin. Ph.D. Thesis, Northwest University, Xi'an, China, 2022; pp. 1–15.
46. Hansen, J.P.; Skjeltorp, A.T. Fractal pore space and rock permeability implications. *Phys. Rev. B* **1988**, *38*, 2635–2638. [CrossRef] [PubMed]
47. Ma, B.Y.; Hu, Q.H.; Yang, S.Y.; Zhang, T.; Qiao, H.G.; Meng, M.M.; Zhu, X.C.; Sun, X.H. Pore structure typing and fractal characteristics of lacustrine shale from Kongdian Formation in East China. *J. Nat. Gas Sci. Eng.* **2021**, *85*, 1–16. [CrossRef]
48. Chang, J.Q.; Fan, X.D.; Jiang, Z.X.; Wang, X.M.; Chen, L.; Li, J.T.; Zhu, L.; Wan, C.X.; Chen, Z.X. Differential impact of clay minerals and organic matter on pore structure and its fractal characteristics of marine and continental shales in China. *Appl. Clay Sci.* **2022**, *216*, 106334. [CrossRef]
49. Li, K.W. Analytical derivation of Brooks-Corey type capillary pressure models using fractal geometry and evaluation of rock heterogeneity. *J. Pet. Sci. Eng.* **2010**, *73*, 20–26. [CrossRef]

fractal and fractional

MDPI

Article

Introducing an Enhanced Cumulative Size Distribution Model to Unearth the Origins of Macropore Heterogeneity of the Chang-7 Shale Oil Reservoirs in the Ordos Basin

Ziyi Wang [1], Lin Dong [1,*], Zhijun Jin [1,2], Xubin Wang [1], Jinhua Fu [3], Xianyang Liu [3] and Rukai Zhu [4]

1 Key Laboratory of Orogenic Belts and Crustal Evolution, School of Earth and Space Sciences, Peking University, Beijing 100871, China; ziyi-wang@pku.edu.cn (Z.W.); jinzj1957@pku.edu.cn (Z.J.); xubinwang@pku.edu.cn (X.W.)
2 Institute of Energy, Peking University, Beijing 100871, China
3 PetroChina Changqing Oilfield Company, Xi'an 710018, China; fjh_cq@petrochina.com.cn (J.F.); lxy3_cq@petrochina.com.cn (X.L.)
4 Research Institute of Petroleum Exploration & Development, PetroChina, Beijing 100083, China; zrk@petrochina.com.cn
* Correspondence: lin.dong@pku.edu.cn

Abstract: Characterizing the fractal nature of shale pores is critical for understanding their heterogeneity, which significantly impacts hydrocarbon mobility. A critical technique in this analysis involves the digital evaluation of scanning electron microscopy (SEM) images. While the cumulative size distribution model is a viable method in SEM fractal analysis, its accurate application necessitates exhaustive microimaging of entire sample cross-sections, rendering it impractical. To address this challenge, we propose an innovative approach that develops a universal equation for this model, along with numerical solving methods. This approach expands the model's applicability, enabling accurate fractal analysis of entire samples using only randomly selected microscopic fields. It enhances the existing framework for fractal modeling and potentially serves as a valuable tool for studying fractal structures. Using this method, we investigated the fractal characteristics of various macropore types across different lithological units in the Chang-7 shale of the Ordos Basin. Our findings indicate that macropore heterogeneity is influenced by variations in pore type composition, their response under compaction, and organic acid dissolution, all of which are driven by differences in lithology. These findings provide new insights into the mechanisms driving macropore heterogeneity from the perspective of lithological and pore-type variations, potentially yielding valuable implications for the exploration and development of Chang-7 shale oil.

Keywords: shale; fractal analysis; cumulative size distribution model; universal equation; macropore heterogeneity origin

1. Introduction

Shale oil and gas resources have profoundly impacted the global energy landscape. As unconventional reservoirs, shale formations are known for their intricate pore systems exhibiting pronounced heterogeneity, which poses substantial challenges to the efficient mobilization and extraction of hydrocarbons [1–3]. On the other hand, this pore complexity has also been identified as potentially advantageous for CO_2 adsorption, presenting an opportunity to leverage shale formations as long-term carbon storage reservoirs [4]. Fractal theory provides a mathematical framework for capturing the heterogeneity and self-similarity observed in complex structures [5]. Researchers have observed clear negative correlations between fractal dimension and permeability in many tight formations [6–8], highlighting the potential impact of pore complexity on fluid mobility. Consequently, the fractal natures of pores have become a pivotal parameter in reservoir evaluation, leading to a notable uptick in research endeavors focused on the in-depth examination of pore heterogeneity within shale formations [9–12].

A variety of analytical techniques have been implemented in fractal studies, such as low-pressure gas adsorption (LPGA) [11], mercury injection capillary pressure (MICP) [13], and nuclear magnetic resonance (NMR) [14]. Despite providing a general view of pore structures at a whole-rock scale, these methodologies falter in discerning between different pore categories or examining specific pore attributes within individual lithological units of a core sample, such as laminae less than one millimeter in thickness. Such limitations may result in the oversight of critical pore details and ambiguities in explaining certain phenomena. For instance, the heterogeneity of shale pores can originate from variations in lithological composition and/or the dominance of particular pore types. However, the inherent constraints of whole-rock analysis techniques hinder the investigation of these potential influencing factors, thereby limiting a deeper understanding of the origin of pore heterogeneity.

To address these constraints, scanning electron microscope (SEM) digital technology offers a promising avenue, facilitating the precise identification of fractal dimensions associated with various pore types within targeted regions of cross-sectional samples. For this technology, previous research has explored various fractal models for pore analysis on two-dimensional planes, such as the box-counting method [15], the perimeter-area model [16], and the long axis-area model [17]. These approaches require complex geometric iterative calculations or regression analyses focusing on various parameters such as area, perimeter, size, or long axis of each pore unit. Other than that, the cumulative size distribution model [18] utilizes a simpler data framework that relies solely on pore count-size distribution, thereby simplifying pore parameter extraction and fractal dimension calculation processes. This simplicity is especially advantageous when dealing with highly heterogeneous shale reservoirs, where obtaining a substantial amount of data with sufficient statistical coverage is crucial to ensure that the extracted fractal parameters accurately represent the overall characteristics of the study object. Furthermore, this model facilitates the efficient utilization of existing pore count-size distributions of shale samples, enabling the calculation of fractal dimensions based solely on these data in some studies.

Despite its significant advantages, the wide adoption of the cumulative size distribution model in pore characterization studies has been hindered. Its optimal utilization is contingent upon the exhaustive analysis of the complete cross-sectional area of a sample via microscopy—an approach often impractical in real-world scenarios. Estimations derived from the analysis of partial cross-sectional areas introduce considerable inaccuracies, as this approach inherently restricts the analysis to pores smaller than the field of view, resulting in measurement uncertainties due to the absence of data on larger pores (detailed mathematical proofs are provided in Section 4.2.1). Thus, the utilization of the cumulative size distribution model remains underexplored, with its current application primarily intertwined with other formulas to construct numerical models [18–20]. This study aims to address these challenges by formulating a universal equation for the cumulative size distribution model and devising a method to accurately compute numerical solutions for fractal dimensions. This advancement extends the applicability of the cumulative size distribution model, broadening its feasibility from the analysis of entire profiles to the evaluation of partial areas accessible through microscopic examination. Our research offers a fresh perspective on computing fractal dimensions, with implications for advancing related research endeavors.

The Chang-7 Member of the Triassic Yanchang Formation, situated in the Ordos Basin, represents a significant shale oil reservoir in China, holding substantial promise for exploration and development [21]. It is predominantly composed of clay shale, massive siltstone, and silty laminae (Figure 1) [22]. Previous studies have quantitatively analyzed the surface macroporosity of these lithological units using microscopic techniques. The results indicate that silty laminae (averaging 1.37%) and massive siltstone (averaging 2.24%) exhibit more developed macropores compared to clay shale (averaging 0.65%) [23]. However, a gap remains in the fractal analysis of different pore types within these units, which limits our comprehensive understanding of the factors controlling the heterogeneity of pore structures in shale formations. Notably, current efforts to advance shale oil extraction focus largely on recovering free oil primarily housed in macropores exceeding 50 nm [24,25], underscoring the critical need for a thorough comprehension of the macropore structures within shale formations [23]. Therefore, in this study, we examined the macropore heterogeneity of the Chang-7 shale through a systematic analysis of whole-rock samples and their individual lithologies. The fractal dimensions of the whole-rock macropores were ascertained using MICP measurements, revealing potential relationships between the fractal dimensions and lithological composition. This initial analysis highlighted the necessity of employing SEM for further detailed examination. Subsequently, the fractal nature across diverse pore types and lithologies was examined using SEM digital techniques in tandem with an innovative analytical approach. Furthermore, we validated the representativeness of the computed fractal dimension values, bolstering the reliability of our findings. Through this comprehensive inquiry, this research shed new light on the mechanisms underlying macropore heterogeneity, potentially offering valuable implications for the exploration and development of Chang-7 shale oil.

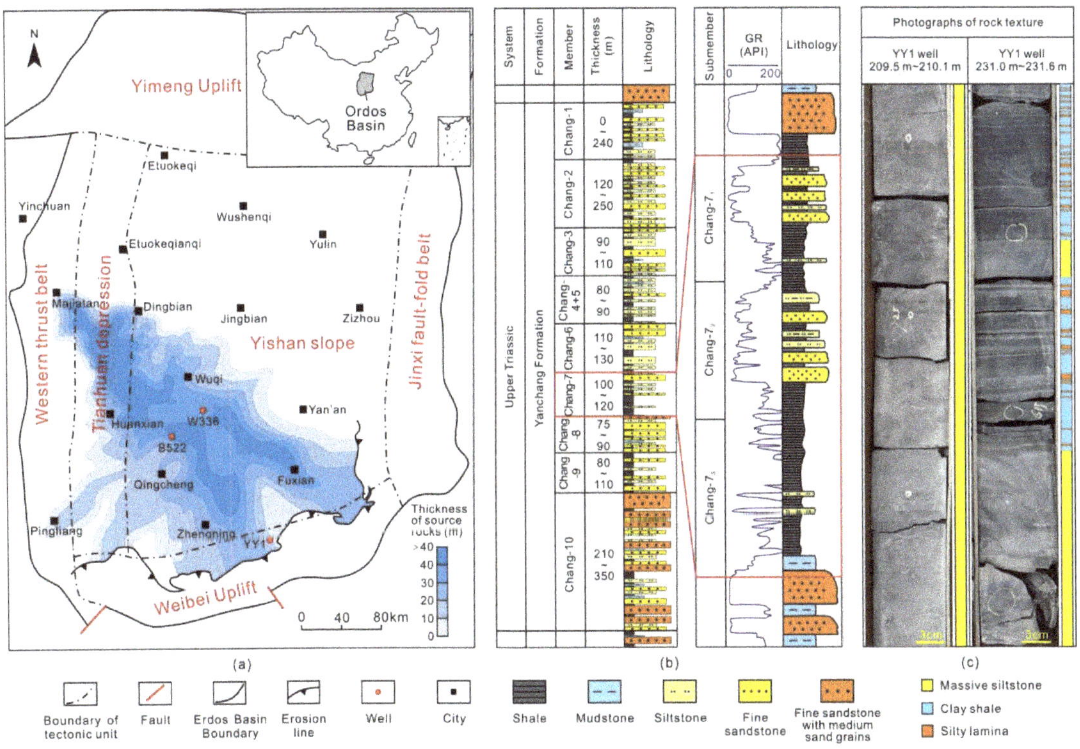

Figure 1. (**a**) Geological map of the Ordos Basin showing the thickness of the source rocks within the Chang-7 Member and the positions of the sampling wells. (**b**) Stratigraphic column illustrating the Upper Triassic Yanchang Formation. (**c**) Photos of rock texture displaying the siltstone layer and the clay shale interbedded with silty laminae. The figure is adapted from work by Wang et al. [23].

2. Geological Setting

The Ordos Basin is a Mesozoic intracratonic basin located in Central China [26,27], spanning approximately 32×10^4 km^2 in area (Figure 1a). During the Late Triassic, the basin underwent a notable phase of lacustrine expansion, resulting in the deposition of a series of fluvial and lacustrine sediments. The Yanchang Formation was deposited during this period [27,28] and consists of ten members, named Chang-1 to Chang-10, from top to bottom, based on marker beds, sedimentary cycles, and lithological association (Figure 1b) [29,30]. Of these members, the Chang-7 Member was deposited during the most significant lake transgression of the Mesozoic, leading to the formation of organic-rich black shale and dark-colored mudstone [27,29]. The Chang-7 Member is recognized as the most significant shale oil reservoir in the Ordos Basin, commonly referred to as Chang-7 shale. The Chang-7 shale mainly contains types I and II kerogen and has a total organic carbon (TOC) of up to 38% [31,32]. The vitrinite reflectance (R_o) value in the shale varies between 0.6% and 1.2%, with shale having R_o values below 1.0%, covering approximately 90% of the total shale distribution area [33,34]. Due to the varying degrees of burial and uplift processes experienced by Chang-7 shale in different regions within the basin, the current burial depths of the shale exhibit significant variability, generally ranging from several hundred meters to around 3000 m. The Chang-7 Member is stratigraphically subdivided into three submembers in ascending order: the Chang-7_3 at the base, followed by Chang-7_2, and lastly, Chang-7_1 at the top (Figure 1b). The Chang-7_1 and Chang-7_2 submembers are predominantly sandstone and siltstone, interspersed with layers of dark shale. In contrast, the Chang-7_3 is notably characterized by its high concentration of organic-rich black shale, with interspersed silty laminae (Figure 1c) [29]. Currently, commercial extraction efforts in the Chang-7 Member primarily focus on the Chang-7_1 and Chang-7_2 submembers [34].

3. Samples and Experiments

3.1. Sample Preparation

We collected sixteen core samples from the Upper Triassic Chang-7 Member of the Ordos Basin, with the locations of three sampling wells depicted in Figure 1a. The specific depths of the samples are detailed in Table S1 of the Supplementary Material. These samples, which are block-shaped with sides ranging from approximately 2 cm to 4 cm, encompass one or several lithological types, including clay shale, massive siltstone, and silty lamina. To accurately assess the lithological composition of each sample, we sectioned the cores into rectangular blocks and quantified the area proportion of the various lithologies within the profile perpendicular to the bedding planes. We then split each block along the profile, with each half (weighing over 20 g) earmarked for distinct analytical procedures. For the first set of samples, we prepared them by crushing, sieving, and using the fraction that passed through a 100-mesh screen for X-ray diffraction (XRD) analysis while selecting particles in the 20 to 35 mesh range for MICP measurement. Prior to MICP testing, we subjected the samples to a ten-day Soxhlet extraction with dichloromethane to remove pore-occluding bitumen. Following the extraction, we dried the samples at 80 °C for a minimum of 72 h until a constant weight was achieved. This temperature was chosen because it does not exceed the maximum temperatures that any of the collected samples experienced throughout their geological history (approximately 90 °C to 140 °C) [35,36], thereby preventing potential alterations to organic matter at higher temperatures. In this study, we opted for particulate rather than bulk samples for MICP analysis because particulate samples can be more thoroughly cleared of pore-occluding oil and attain higher levels of mercury intrusion saturation [13,37]. This makes them particularly suitable for studying shale oil reservoirs. The second set of samples was prepared into thin and thick sections for examination under optical and electron microscopes, respectively. The thick sections underwent the same oil removal process as those prepared for MICP before being cut into 1.5 cm squares for SEM analysis. We polished the surface of each sample to a fine 0.5 μm finish using diamond paste and then employed argon ion polishing to achieve an ultra-smooth finish appropriate for SEM observation. To prevent charging during SEM analysis, we coated the sample surfaces with a chromium layer approximately 5 nm thick.

3.2. Experimental Procedures

X-ray diffraction analysis was conducted on approximately 1 g of each ground sample using a Rigaku D/max-2500PC X-ray diffractometer (Tokyo, Japan), followed by computer diffractogram analysis to identify and semi-quantitatively assess the relative abundances (wt%) of different mineral phases. Mercury intrusion capillary pressure measurements were performed with a Micromeritics PoreMaster GT60 (Orlando, FL. Initially, a blank correction was applied for enhanced accuracy. Subsequently, approximately 0.5 g of each sample was weighed, loaded into the apparatus, and degassed under a vacuum of 50 μm Hg for 30 min. The mercury intrusion process monitored the infused mercury volume into the sample across a pressure range from 1 psi to 60,000 psi. Pore volume (cm^3/g) and pore size distribution data were derived using the Washburn equation [38], with detailed experimental procedures outlined in Yu et al. [39]. Quantitative analysis of different types of macropores across various lithologies was performed on a FEI Quanta 650 FEG field emission scanning electron microscope. High-resolution imaging was conducted at magnifications of 15,000 times (view field size = 17 μm × 20 μm), targeting stochastic regions set by random coordinates within the predefined areas for analysis across sample surfaces. Areas analyzed within each lithological sample exceeded 10,000 μm^2, with image resolution surpassing 0.1032 pixels/nm. Diverse types of macropores within these images were quantitatively extracted using Fiji-ImageJ 2.1.0 software [40], employing minimum Feret diameter as the metric for assessing pore size diameters. For a comprehensive understanding of the processes involved, a study by Wang et al. [23] provides an extensive description of these experimental procedures.

4. Development of Macropore Fractal Models

4.1. Fractal Model Based on MICP Analysis

Expanding upon the contributions of Pfeifer and Avnir [41], Friesen and Mikula established a fractal model that links the differential of mercury intrusion volume to the fractal dimension D_{Hg}, derived from mercury injection capillary pressure measurements [42]:

$$\lg\left(\frac{\mathrm{d}V_P}{\mathrm{d}P}\right) \propto (D_{\mathrm{Hg}} - 4)\lg P \tag{1}$$

where V_P represents the cumulative volume of mercury (cm^3/g) intruded at pressure P (MPa). Thus, a scatter plot of $\lg(\mathrm{d}V_P/\mathrm{d}P)$ against $\lg P$ can be employed to determine the fractal dimension.

When assessing bulk samples, the fractal dimension reflecting the pore structure typically spans from 2 to 3 [43]. This range is, however, not consistently observed in particle samples subject to MICP experiments. It is crucial to recognize that MICP analyses of particulate samples capture both the intrinsic pores of the reservoirs and the spaces between rock particles, consequently distorting the interpretive outcomes. This highlights the necessity of applying a conformance correction [13] to improve the reliability of the results. The plot of $\lg(\mathrm{d}V_P/\mathrm{d}P)$ versus $\lg P$ for particulate samples reveals distinct slope variations across four pore size intervals, signifying sequential Stages A to D during the mercury intrusion process. In Stage A, mercury merely occupies interparticle spaces, leading to a fractal dimension D_1 that ranges between 1 and 2 [42]. In Stage B, mercury intrusion progresses beyond the interparticle voids and infiltrates the internal pore spaces of the samples. During this phase, the fractal dimension D_2 closely aligns with the natural fractal dimension of pore structures, which typically falls in the range between 2 and 3. During Stage C, marked by a fractal dimension D_3 between 3 and 4, there is significant pore compression due to increased mercury pressure [42,44]. As mercury pressure escalates, observable damage to the pore structure occurs, rendering fractal analysis impracticable, signifying the onset of Stage D [45].

The dataset from Stage B integrates information from both the interparticle voids and the inherent pores within the reservoirs. An innovative method introduced by Yu et al. [13]

was developed specifically to address and eliminate interference from interparticle voids. During Stage A, the cumulative intruded mercury volume $V_{P,\mathrm{stageA}}$ measured at pressure P is solely ascribed to interparticle voids and is governed by the following equation:

$$\lg\left(\frac{\mathrm{d}V_{P,\mathrm{stageA}}}{\mathrm{d}P}\right) = (D_1 - 4)\lg P + a \quad (2)$$

where the intercept a is constant. Both a and D_1 can be derived from Stage A data by implementing Equation (2). Given that interparticle voids exhibit consistent fractal traits [42], the pressure data from Stage B can be substituted into Equation (2) to estimate the differential of the mercury volume intruded into the interparticle voids $\mathrm{d}V_{P,\mathrm{stageB\text{-}iv}}$. Subsequently, the differential of the mercury volume that has intruded into the shale pores $\mathrm{d}V_{P,\mathrm{stageB\text{-}sp}}$ is given by

$$\mathrm{d}V_{P,\mathrm{stageB\text{-}sp}} = \mathrm{d}V_{P,\mathrm{stageB}} - \mathrm{d}V_{P,\mathrm{stageB\text{-}iv}} \quad (3)$$

Substituting $\mathrm{d}V_{P,\mathrm{stageB\text{-}sp}}$ for $\mathrm{d}V_P$ in Equation (1), D_{Hg} can be determined via linear regression analysis of $\lg(\mathrm{d}V_{P,\mathrm{stageB\text{-}sp}}/\mathrm{d}P)$ against $\lg P$. Typically, the pore sizes analyzed in Stage B predominately occupy the macroporous scale, and consequently, D_{Hg} ultimately reflects the fractal character of the macropores found within shale reservoirs.

4.2. Fractal Models Based on SEM Analysis

4.2.1. Traditional Cumulative-Size Distribution Model and Its Limitations

The fractal characteristics of pore structures can also be evaluated using SEM digital analysis. In two-dimensional profiles, the fractal cumulative-size distribution model is given as follows [18,19]:

$$N(L \geq \lambda) = \left(\frac{\lambda_{\max}}{\lambda}\right)^{D_f} \quad (4)$$

where $N(L \geq \lambda)$ represents the number of pores with a size not less than λ (nm), $\lambda_{\max}$ denotes the largest pore size (nm), and D_f is the fractal dimension of pores, which varies between 1 and 2 within a two-dimensional framework. Therefore, there is the following relationship:

$$\lg N(L \geq \lambda) = -D_f \cdot \lg\lambda + b \quad (5)$$

where, for a certain sample, b is a constant that depends on the values of $\lambda_{\max}$ and D_f. By conducting a linear regression between $\lg N(L \geq \lambda)$ and $\lg\lambda$, the negative of the slope yields the fractal dimension.

It is critical to acknowledge that conducting a quantitative pore analysis across the entirety of a core sample's cross-sectional area, indicated by A_t (μm^2), is not feasible in SEM experiments. Instead, a widely accepted method entails the examination of randomly selected regions within the microscopic field of view, encompassing a total area denoted as A_{SEM} (μm^2). Due to the limited size of the SEM field of view, the maximum pore size observed, denoted as $\lambda_{\mathrm{SEM,max}}$, is typically smaller than the true maximum pore size of the entire sample, represented by $\lambda_{\max}$. As a result, the quantity of pores identified through SEM, represented by N_{SEM}, is lower than the expected value for the analyzed cross-sectional area, computed as $(A_{\mathrm{SEM}}N)/A_t$. Consequently, using the ratio $(A_t N_{\mathrm{SEM}})/A_{\mathrm{SEM}}$ as a direct replacement for N in Equation (5) could introduce inaccuracies in estimating the fractal dimension, owing to this discrepancy.

4.2.2. An Enhanced Cumulative-Size Distribution Model

To address the limitations of traditional methods, we endeavored to enhance the existing cumulative size distribution model. In fact, in Equation (4), $N(L \geq \lambda)$ can be expressed with the following equations:

$$N(L \geq \lambda) = N(L \geq \lambda_{\mathrm{SEM,max}}) + N(\lambda \leq L < \lambda_{\mathrm{SEM,max}}) \quad (6)$$

Moreover, $N_{SEM}(\lambda \le L < \lambda_{SEM,max})$ represents the expected value of $N(\lambda \le L < \lambda_{SEM,max})$:

$$N_{SEM}(\lambda \le L < \lambda_{SEM,max}) = \frac{A_{SEM} N(\lambda \le L < \lambda_{SEM,max})}{A_t} \tag{7}$$

Upon integrating Equation (6) into Equation (4) and reorganizing, the resultant expression is

$$\lg N(\lambda \le L < \lambda_{SEM,max}) = D_f \cdot \lg\left(\frac{\lambda_{max}}{\lambda}\right) - \lg\left(1 + \frac{N(L \ge \lambda_{SEM,max})}{N(\lambda \le L < \lambda_{SEM,max})}\right) \tag{8}$$

Incorporating Equation (7) into the left side and Equation (4) into the right side of Equation (8), we derive

$$\lg N_{SEM}(\lambda \le L < \lambda_{SEM,max}) + \lg\frac{A_t}{A_{SEM}} = D_f \cdot \lg\left(\frac{\lambda_{max}}{\lambda}\right) - \lg\left(1 + \frac{\left(\frac{\lambda_{max}}{\lambda_{SEM,max}}\right)^{D_f}}{\left(\frac{\lambda_{max}}{\lambda}\right)^{D_f} - \left(\frac{\lambda_{max}}{\lambda_{SEM,max}}\right)^{D_f}}\right) \tag{9}$$

Subsequent simplification leads to

$$\lg N_{SEM}(\lambda \le L < \lambda_{SEM,max}) + \lg\frac{\lambda_{SEM,max}^{D_f}}{\lambda_{SEM,max}^{D_f} - \lambda^{D_f}} = -D_f \cdot \lg\lambda + D_f \cdot \lg\lambda_{max} - \lg A_t + \lg A_{SEM} \tag{10}$$

Assuming SEM analysis covers the whole shale profile area, it follows that $N(L \ge \lambda) = N_{SEM}(\lambda \le L < \lambda_{SEM,max}) + 1$, $A_t = A_{SEM}$, and $\lambda_{max} = \lambda_{SEM,max}$. By incorporating these conditions into Equation (10) and subsequently reorganizing, we derive Equation (4). Hence, the traditional cumulative size distribution model represents a specific scenario within the broader framework of Equation (10). As such, we propose Equation (10) as the general equation for the cumulative-size distribution method used in determining the pore fractal dimension through microscopic evaluations. For a designated sample and analytical procedure, the parameters D_f, λ_{max}, $\lambda_{SEM,max}$, A_t and A_s remain constant. Thus, Equation (10) can be simplified to

$$\lg N_{SEM}(\lambda \le L < \lambda_{SEM,max}) + F_a = -D_f \cdot \lg\lambda + c \tag{11}$$

where c is a constant, and F_a is a corrective function related to λ and D_f, expressed as

$$F_a = \lg\frac{\lambda_{SEM,max}^{D_f}}{\lambda_{SEM,max}^{D_f} - \lambda^{D_f}} \tag{12}$$

Evidently, as $\lambda_{SEM,max}$ greatly exceeds λ, F_a approaches zero. Performing a linear regression of $\lg N_{SEM}(\lambda \le L < \lambda_{SEM,max})$ against $\lg\lambda$, the negative slope provides a close estimate of the pore fractal dimension. However, as λ increases, neglecting F_a leads to a significant undervaluation of the terms on the left side of Equation (11), consequently resulting in an overestimated fractal dimension value.

To address this, we present a numerical method for determining the fractal dimension. By substituting a hypothetical value D_i for D_f in Equation (12) and integrating it with Equation (11), we derive the following relationship:

$$\lg N_{SEM}(\lambda \le L < \lambda_{SEM,max}) + \lg\frac{\lambda_{SEM,max}^{D_i}}{\lambda_{SEM,max}^{D_i} - \lambda^{D_i}} = -D_o \cdot \lg\lambda + c \tag{13}$$

where D_o signifies the observed values of fractal dimension when assuming D_i values. It is evident that Equation (13) is valid when $D_o = D_i = D_f$. Differentiating Equation (13) with respect to λ and rearranging, we find

$$D_o = -\frac{D_i \cdot \lambda^{D_i}}{\lambda_{SEM,max}^{D_i} - \lambda^{D_i}} - \frac{\lambda \cdot dN_{SEM}(\lambda \le L < \lambda_{SEM,max})}{d\lambda \cdot N_{SEM}(\lambda \le L < \lambda_{SEM,max})} \tag{14}$$

When plotting D_o against D_i, Equation (14) is expected to intersect the line $D_o = D_i$ within the interval $1 < D_i < 2$. If a unique intersection point exists within this range, it suggests that D_i at this point accurately represents the fractal dimension. While directly graphing Equation (14) may be impractical, we can determine its coordinates accurately by selecting a spectrum of D_i values and conducting regression analysis on the left side of Equation (13) relative to lgλ. This approach yields a set of D_o values corresponding to each D_i.

When a unique intersection within the definition domain ($1 < D_i < 2$) is confirmed on the plot of D_i versus D_o, determining the numerical solution for the fractal dimension transforms into an optimization problem:

$$D_{SEM} = \text{argmin}_{D_i} |D_o - D_i| \text{ s.t. } 1 < D_i < 2 \tag{15}$$

as required, by specifying D_i to a specific decimal precision during the optimization, the estimated fractal dimension D_{SEM} can be obtained with the desired precision.

To enhance computational precision and ensure comprehensive statistical representation in SEM analysis, it is crucial to examine extensive view fields to capture a sufficient number of pores. In light of fractal geometry theory, the cumulative pore count diminishes exponentially with enlarging pore size, as described by Equation (4). As λ approaches $\lambda_{SEM,max}$, the number of extracted pores $N_{SEM}(\lambda \le L < \lambda_{SEM,max})$ substantially decreases, potentially to single digits. This reduction yields diminished accuracy in fractal dimension estimations, as it is measured by changes across orders of magnitude. Therefore, for precise fractal dimension calculations, it is critical to select an interval for λ that includes well-developed pores. Considering that nanoscale pores are predominant in shale oil reservoirs, this study focuses on the λ range from 50 nm to 1000 nm within $N_{SEM}(\lambda \le L < \lambda_{SEM,max})$ for macropore fractal dimension analysis. To validate our fractal analysis within this pore size range, we assess the goodness-of-fits of the regression models. Furthermore, to verify the statistical robustness of the calculated fractal dimension D_{SEM}, we plot its relationship with the SEM-investigated area. As the examined area enlarges, D_{SEM} is expected to stabilize, showing minimal variation and maintaining consistency within a specific range. This pattern indicates that the derived D_{SEM} faithfully represents the comprehensive fractal properties of the target sample.

5. Results and Discussion

5.1. Petrological Characteristics

X-ray diffraction analysis reveals that the Chang-7 shale oil reservoir samples are primarily composed of clay minerals (35.7% on average), feldspar (23.5% on average), and quartz (23.2% on average). Additionally, there is a notable amount of pyrite, with an average of 11.0%, and a smaller quantity of carbonate, averaging 6.6%. These findings are depicted in Figure 2. Optical and electron microscopy analyses have delineated the lithologic and structural attributes of the Chang-7 shale oil reservoirs. Based on grain size and bedding thickness, the reservoirs are classified into clay shale, massive siltstone, and silty laminae. Clay shale consists of alternating organic-rich dark-colored laminae and mineral-rich light-colored laminae (Figure 3a,b), with over 50% of the mineral grains classified as clay-sized (<3.9 μm). This high concentration of organic matter renders the clay shale a primary source rock throughout the Ordos Basin. Siltstones with individual layer thicknesses exceeding 1 cm are categorized as massive siltstones. These siltstones mainly consist of silt-sized terrigenous and volcanic particles ranging from 0.39 μm to 62.5 μm, with a significant amount of clay matrix inhabiting the inter-clast spaces (Figure 3c,d). Silty laminae, which display grain size characteristics akin to massive siltstones, are distinguished by a single layer thickness of less than 1 cm and are typically found between clay shale layers (Figure 3e,f). Silty laminae are spatially more proximal to the source rock compared to massive siltstone with higher thickness. By investigating the cross-sectional area of different lithologies in the core samples, we quantified the percentage composition of each lithology, as represented in Figure 2.

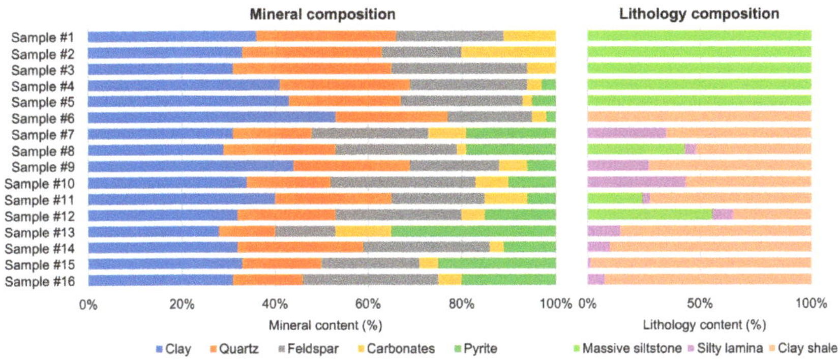

Figure 2. Percent stacked bar charts showing mineral and lithology compositions of the core samples from the Chang-7 shale oil reservoirs. The mineral content data are sourced from a study by Wang et al. [23], and lithology composition was quantified by assessing the area ratio of different lithologies in the profile perpendicular to the bedding planes. The data are comprehensively listed in Table S1.

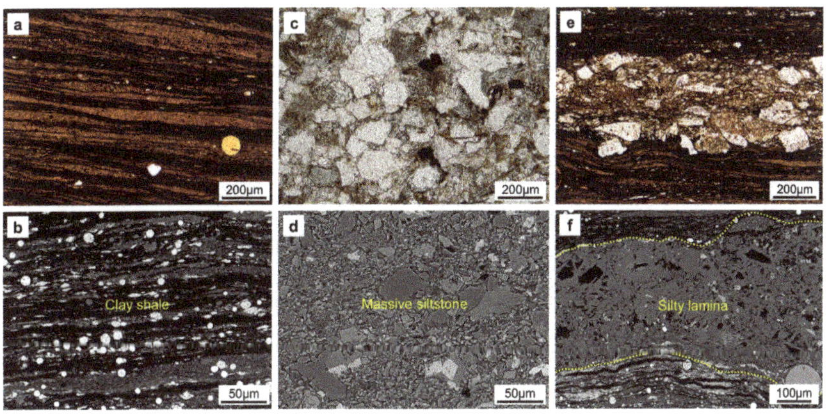

Figure 3. Micrographs of varied lithologies within the Chang-7 shale oil reservoirs. (**a**) Transmitted light and (**b**) backscattered electron images of clay shale; (**c**) transmitted light and (**d**) backscattered electron images of massive siltstone; and (**e**) transmitted light and (**f**) backscattered electron images of silty lamina between clay shale layers.

5.2. Macropore Fractal Analysis for Whole Rocks

5.2.1. Determination of Fractal Dimensions Using MICP

Building upon the approach outlined in Section 4.1, we analyzed the fractal dimensions of macropore structures based on MICP results. Following Equation (1), we generated scatter plots portraying the relationship between $\lg(dV_P/dP)$ and $\lg P$. It can be observed in Figure 4a,b that the scatter plot can be divided into four intervals based on the varying correlation between the dependent and independent variables across different pore size ranges. Each segment denotes a separate mercury intrusion phase. The initial three phases exhibit fractal behavior, enabling us to ascertain the respective fractal dimensions D_1, D_2, and D_3. The dataset from Stage A primarily reflects the mercury injection progress as it fills the spaces between sample particles. In contrast, data from Stage B indicate the simultaneous filling of both interparticle spaces and intrinsic pores within the reservoir samples. Utilizing Equation (2), we extrapolated the $\lg(dV_P/dP)$ for mercury intruded within interparticle voids from those determined during Stage A to predict the corresponding values in Stage B. The simulated dataset is represented by the black circles, depicted in Figure 4a,b. Subsequently, we applied Equation (3) to determine the change in mercury volume that intruded into the intrinsic pores of reservoir samples, denoted as $dV_{P,\text{stageB-sp}}$. By conducting linear regression analysis of $\lg(dV_{P,\text{stageB-sp}}/dP)$ against $\lg P$, we calculated the adjusted fractal dimension D_{Hg} (Figure 4c,d). The regression equations and calculated fractal dimension values for all of the whole-rock samples are summarized in Table S2.

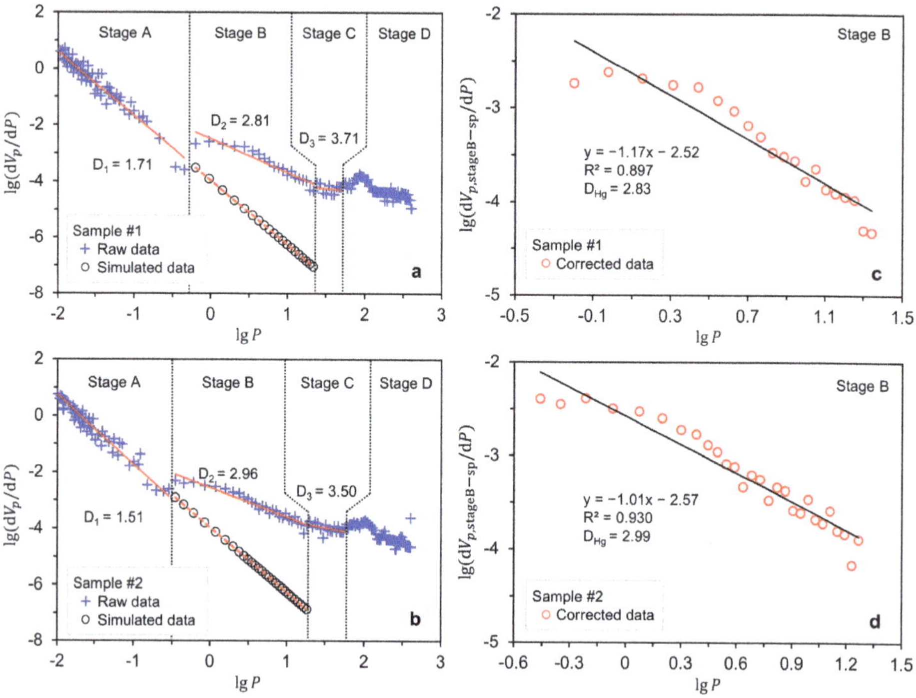

Figure 4. Mercury intrusion process stages and fractal dimension analysis. Plots of $\lg(dV_P/dP)$ versus $\lg P$ illustrates sequential four stages during the mercury intrusion process for samples (**a**) #1 and (**b**) #2. Plots of $\lg(dV_{P,\text{stageB-sp}}/dP)$ versus $\lg P$ shows the calculation of fractal dimension values for samples (**c**) #1 and (**d**) #2. Details on regression equations and fractal dimensions for additional samples are listed in Table S2.

5.2.2. Relationship Between Fractal Dimensions and Petrological Properties

The exploration of the relationship between fractal dimensions and petrological characteristics provides valuable insights into the factors influencing pore heterogeneity. Figure 5a,b presents the mineralogical ternary plots for the analyzed Chang-7 shale oil reservoir samples, with varying color gradients indicating the range of macropore fractal dimensions. The absence of a regular distribution pattern for samples with different D_{Hg} suggests that mineral composition may not be the predominant factor in determining the fractal dimensions of macropores at the whole-rock level. Conversely, Figure 5c,d illustrates the correlation between lithological composition and macropore fractal dimensions. Notably, samples primarily composed of clay shale and silty lamina exhibit a tendency of decreasing D_{Hg} as the proportion of silty lamina increases. This trend suggests a potential link between lithological composition and fractal dimensions, underscoring the need for a comprehensive investigation to elucidate the mechanisms underlying macropore heterogeneity. While the MICP technique is effective for investigating and characterizing pore heterogeneity, its scope is limited to whole-rock samples rather than individual lithologies, particularly the thin silty laminae in the Chang-7 shale (Figure 3f). In order to capture variations in different pore types across diverse lithologies, we conducted an SEM digital analysis.

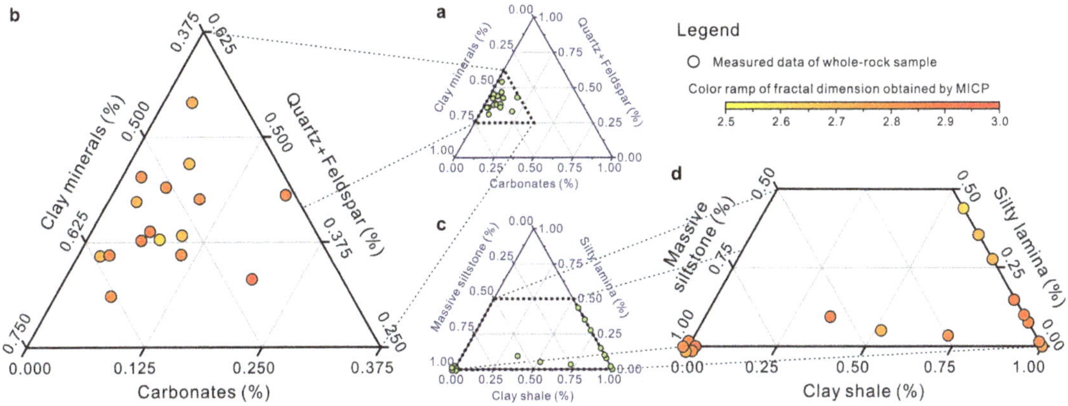

Figure 5. Relationships between petrological properties and macropore fractal dimensions tested by MICP. (**a**,**b**) Ternary plots illustrating mineral contents and their relationship with macropore fractal dimensions. (**c**,**d**) Ternary plots displaying lithology composition and its association with macropore fractal dimensions.

5.3. Fractal Nature of Multi-Type Macropores Across Various Lithologies

5.3.1. Macropore Types and Count-Size Distribution

Using SEM techniques, we identified a diverse range of pore types within the Chang-7 shale oil reservoirs, predominantly comprising pores among clay minerals, intergranular pores, and dissolution pores in carbonates and feldspar (Figure 6a,b). Additionally, a smaller subset of pore types, including intercrystalline pores, microfractures and organic pores, were also observed (Figure 6c–e). To perform a quantitative analysis, we evaluated the relative abundance and count-size distribution of inter-clay mineral, intergranular, and dissolution pores across different lithologies, with a particular focus on those smaller than one micrometer (Figure 7). Secondary pore types, due to their negligible proportions, were combined into a single category to facilitate streamlined analysis and enhance the clarity of the presentation of results. Intriguingly, despite clay shale boasting higher contents of clay minerals compared to massive siltstone and silty lamina, it paradoxically contained a smaller proportion of inter-clay mineral pores (Figure 7). This seemingly contradictory observation can be elucidated through detailed microscopic examination. As the primary source rock in the Ordos Basin, clay shale is rich in organic matter, which frequently occludes the original inter-clay mineral pores (Figure 6f), resulting in a tangible decrease in their prevalence. In contrast, massive siltstone and silty lamina possess relatively lower contents of clay minerals but a significant presence of rigid detrital minerals with plenty of clay-rich matrix, which obstructs intergranular space. Such a structure aids in the preservation of inter-clay mineral pores during compaction, thus enhancing their visibility in these lithologies. A comparative analysis of different pore types across various lithologies lays the foundation for discerning the macropore heterogeneous nature and origin within the Chang-7 shale oil reservoirs.

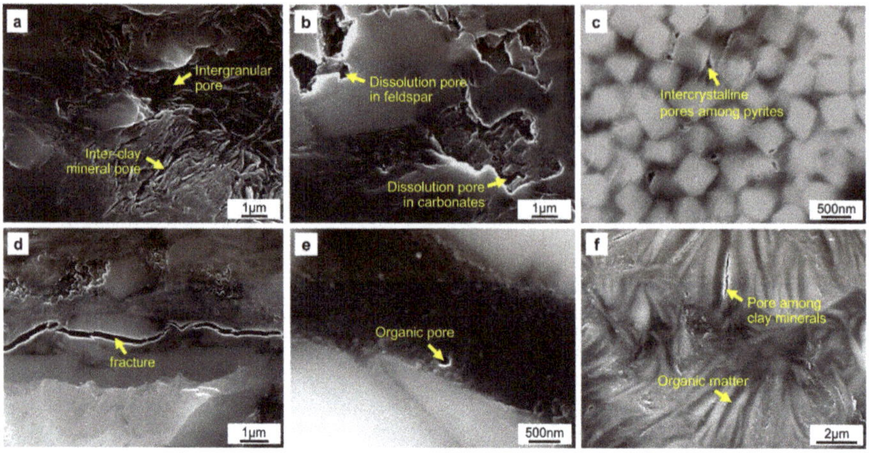

Figure 6. Secondary electron images showing the pore types observed in the Chang-7 shale oil reservoirs. (**a**) Intergranular pores and inter-clay mineral pores. (**b**) Dissolution pores in feldspar and carbonates. (**c**) Intercrystalline pores among pyrites. (**d**) Fracture. (**e**) Organic pore. (**f**) Pores among clay minerals and pore-occluding organic matters.

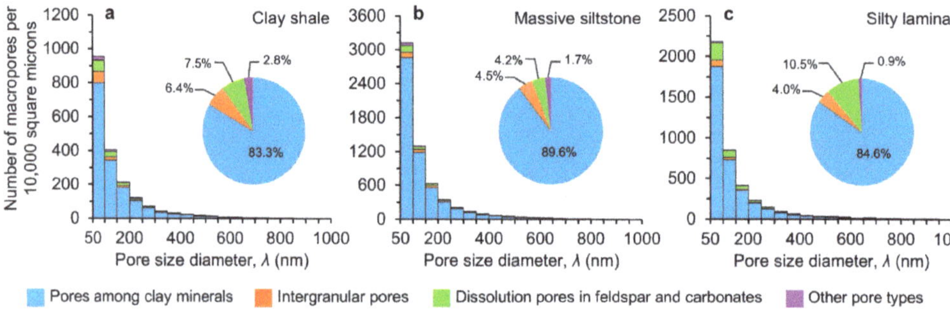

Figure 7. Quantity percentages and count-size distributions of various pore types in (**a**) clay shale, (**b**) massive siltstone, and (**c**) silty lamina, focusing on macropores smaller than one micrometer. The depicted pore counts represent the average number for each lithology. Before calculation, pore numbers across all lithological samples were normalized based on frequency in the same area (10,000 μm^2). Only samples passing the fractal dimension representativeness tests, as detailed in Section 5.3.2, are included in this dataset.

5.3.2. Fractal Dimension Assessment by SEM Analysis

Following the methodology outlined in Section 4.2.2, we first calculated an array of D_o values for a range of assumed D_i values and plotted the calculations onto scatter graphs to illustrate the relationship between D_i and D_o. The resulting graphs show that each curve intersects the line where $D_o = D_i$ at only one point within the parameter range of $1 < D_i < 2$ (Figure 8). Utilizing Equation (15), we calculated the fractal dimensions of all lithologic types and summarized the results in Table S3.

As mentioned in Section 4.2.2, the validity of the model relies heavily on the statistical coverage of the pore data used for calculations. Therefore, we evaluated the impact of variations in input parameters on the model's performance to ensure that our results are both reliable and comprehensive. The results of the linear regression of $\lg N_{SEM}(\lambda \leq L < \lambda_{SEM,max}) + F_a$ against $\lg\lambda$ for lithological samples are displayed in Figure 9a–c and Table S3. The regression models exhibit high goodness-of-fit, with determination coefficients R^2 ranging between 0.940 and 0.997. These results indicate that the regression models maintain relatively stable slopes when λ ranges from 50 nm to 1000 nm, suggesting that the corresponding data can reliably characterize the fractal nature of the pore structures within this scale interval. Figure 9d–f illustrates the variation in fractal dimension calculations corresponding to the increasing SEM-investigated areas. A trend toward stabilization of fractal dimensions with larger SEM examination areas becomes apparent, with deviations under 0.1 from the final value serving as a threshold for achieving representativeness. When the derived fractal dimension maintains a deviation below this threshold, it is considered emblematic of the macropore fractal nature for a sample (Figure 9d,e, Table S3). Conversely, fractal dimensions failing to maintain a deviation below 0.1 as the SEM-studied area increases are deemed non-representative of the sample's overall characteristics (Figure 9f, Table S3). In this research, only data that passed this validation criterion were utilized for characterizing macropore heterogeneity and analyzing its origins.

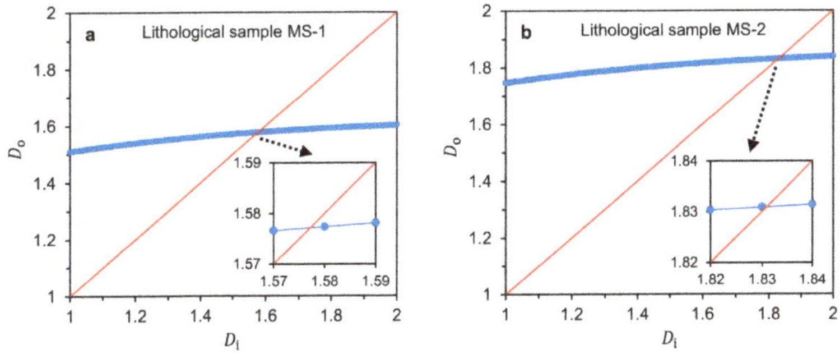

Figure 8. Cross-plot of the assumed fractal dimension D_i against the observed fractal dimension D_o for the lithological samples (**a**) MS-1 and (**b**) MS-2, following the method delineated in Section 4.2.2. Each resulting curve (illustrated by the blue line) intersects the line where $D_o = D_i$ (illustrated by the red line) at only one point within the $1 < D_i < 2$ range. Other lithological samples also exhibit similar unique intersection points within this range. The D_i values at these intersection points represent the fractal dimension, and their numerical solutions are detailed in Table S3.

Ourmethodology facilitated the refinement of the existing cumulative size distribution model by accurately calculating the correction function F_a in Equation (11), thereby minimizing the errors stemming from the lack of comprehensive microscopic analysis over the full cross-sectional areas of a sample. This refinement ensures the acquisition of unbiased fractal dimensions. Conversely, forcing the application of the traditional cumulative size distribution model without accounting for the inherent limitations of SEM technology, i.e., neglecting the correction function F_a, leads to an overestimation of the fractal dimension. The biased fractal dimensions can even exceed the theoretical maximum value of 2, as evidenced in Figure 10.

As illustrated in Figure 11, the macropore fractal dimensions vary noticeably among different lithologies within the Chang-7 shale oil reservoirs. In particular, clay shales display the highest D_{SEM} values, ranging from 1.62 to 1.91, with an average value of 1.80. This is followed closely by massive siltstone, which shows D_{SEM} values ranging from 1.44 to 1.91, with an average of 1.72. In contrast, the macropores within silty lamina present the lowest D_{SEM}, extending from 1.31 to 1.87 and averaging a mere 1.58. These observations corroborate the whole-rock analyses previously demonstrated in Figure 5, confirming the considerable impact of lithology on the heterogeneity of macropores within the Chang-7 shale oil reservoirs. Based on the pore quantity-size distribution data, we further calculated the fractal dimensions associated with three primary types of macropores (inter-clay mineral pores, intergranular pores, and dissolution pores in feldspar and carbonates) across various lithologies (Figure 12). The outcomes are visually depicted in Figure 13.

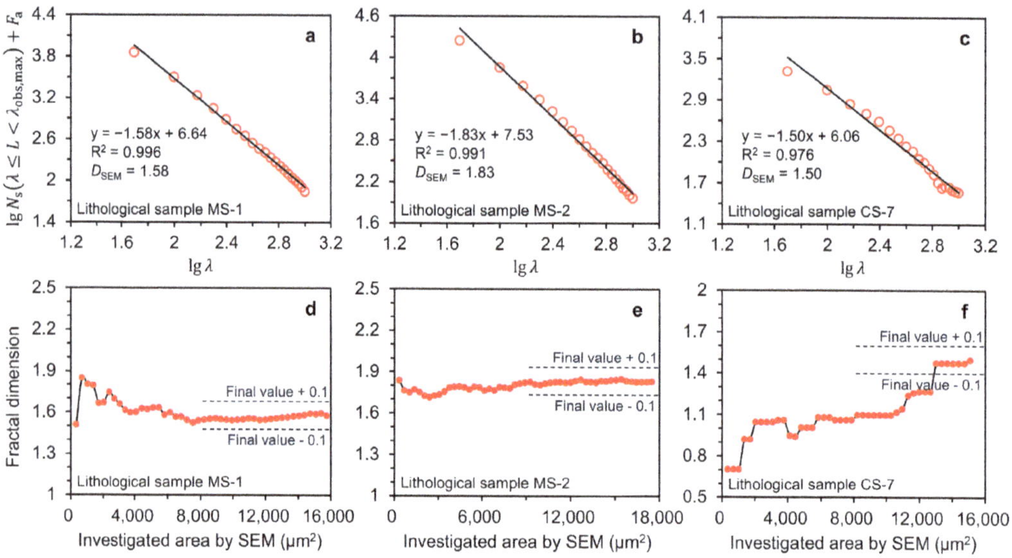

Figure 9. Plots of $\lg N_{\text{SEM}}(\lambda \leq L < \lambda_{\text{SEM,max}}) + F_a$ versus $\lg\lambda$ for lithological samples (**a**) MS-1, (**b**) MS-2, and (**c**) SL-11(1), as derived from SEM data. The representativeness of the fractal data was assessed by examining the variations in fractal dimensions with an increasing SEM-investigated area. Lithological samples (**d**) CS-6 and (**e**) MS-1 show fractal dimension stability within 0.1 of the final value, confirming their representativeness, while fluctuations in (**f**) CS-7 exceed this threshold, indicating a failure in the representativeness criteria. Representative validation results for all lithological samples are detailed in Table S3.

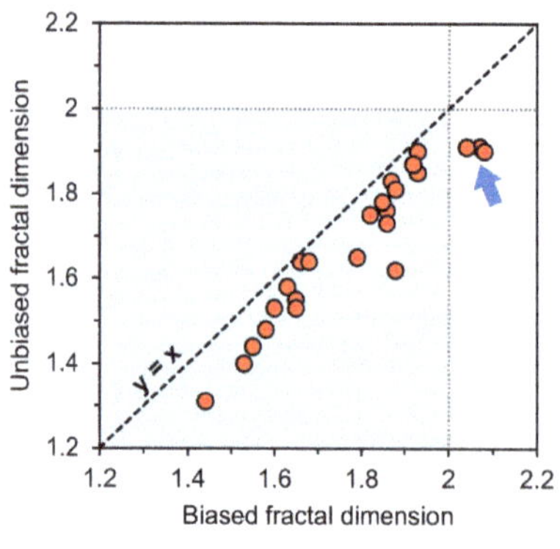

Figure 10. Comparison of biased and unbiased fractal dimensions: biased values exceed unbiased values and sometimes surpass the theoretical maximum of 2 (see points indicated by the arrow). Unbiased fractal dimensions were derived using the improved cumulative-size distribution model, whereas biased fractal dimensions were obtained by applying the traditional cumulative-size distribution model without accounting for the inherent limitations of SEM analysis. This dataset includes only samples that passed the fractal dimension representativeness test.

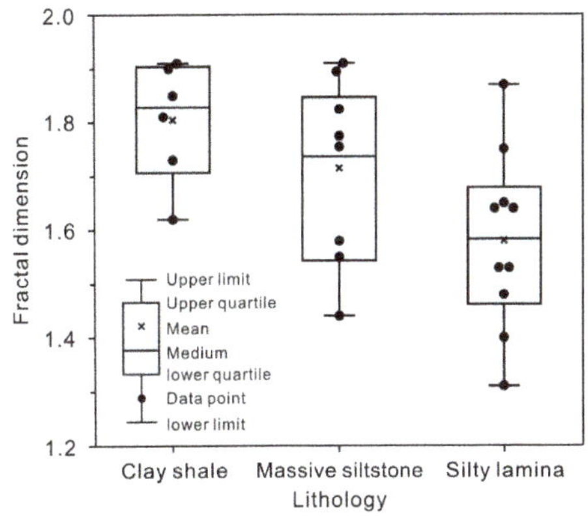

Figure 11. Box-plot of the macropore fractal dimension within the clay shale, massive siltstone, and silty lamina of the Chang-7 shale oil reservoirs. The data included are restricted to lithological samples that fulfill the fractal dimension representativeness verification criteria.

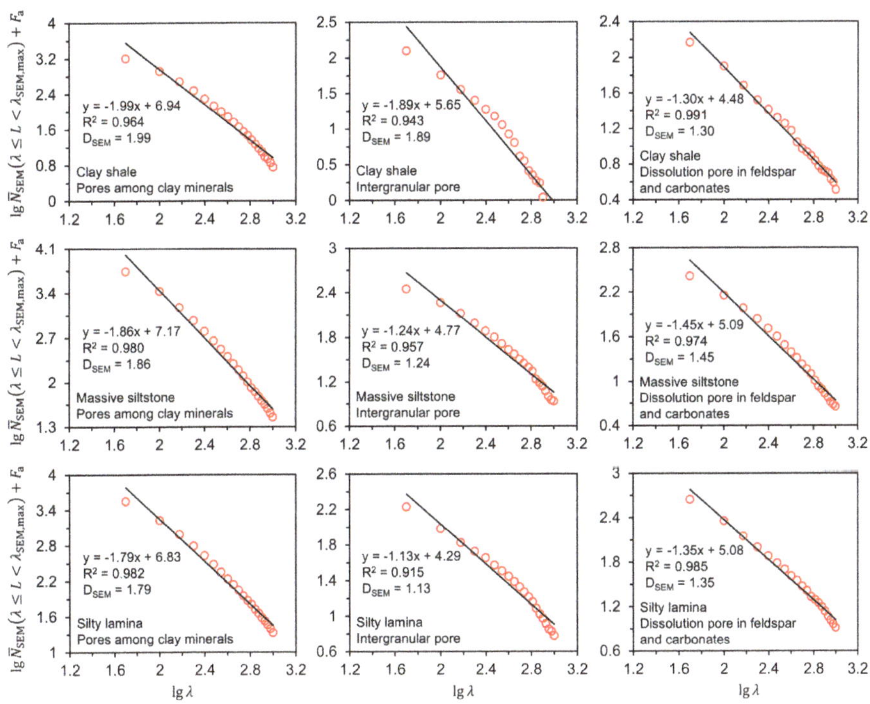

Figure 12. Plots of $\lg\overline{N}_{\mathrm{SEM}}(\lambda \leq L < \lambda_{\mathrm{SEM,max}}) + F_a$ versus $\lg\lambda$ for different pore type across various lithologies. $\overline{N}_{\mathrm{SEM}}$ represents the average number of specific pore types observed in the SEM-studied areas of samples with same lithology, normalized to a frequency per 10,000 μm^2.

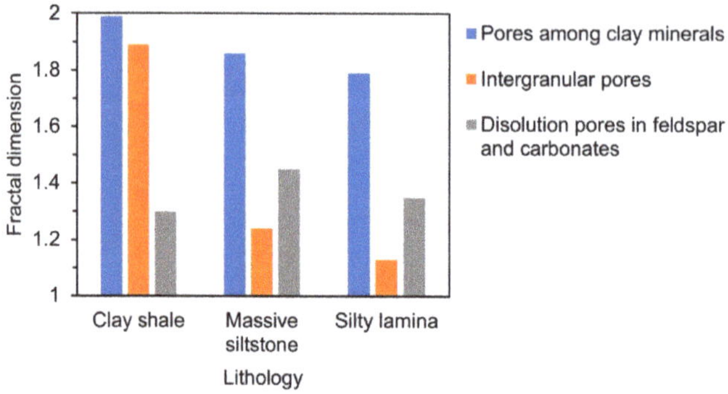

Figure 13. Bar chart showing the fractal dimension values of different macropore types across various lithologies within the Chang-7 shale oil reservoirs.

5.4. Mechanistic Insights into Macropores Heterogeneity in the Chang-7 Shale Oil Reservoirs

The fractal dimension data of different pore types within various lithologies (Figure 13) are crucial for unraveling the underlying origins of macropore heterogeneity in the Chang-7 shale oil reservoirs. Notably, inter-clay mineral pores consistently exhibit the highest fractal dimensions across all examined lithologies, indicating considerable heterogeneity in their macropore networks. This suggests that the structure of clay-associated pores is intrinsically more complex compared to intergranular and dissolution pores. This complexity likely has a negative impact on hydrocarbon flow permeability.

Previous studies have identified distinct fractal characteristics associated with different types of pores in shale reservoirs [15,46]. Moreover, our research further demonstrates that even the same type of pores can show significant fractal dimension variations across different lithologies. For instance, the intergranular pores within clay shale exhibit a higher fractal dimension than those in massive siltstone or silty laminae (Figure 13). This could be due to a lack of detrital grains in clay shale, which would otherwise provide a rigid framework, resulting in reduced resistance to compaction. As a result, pore sizes are smaller, and pore structures are more complex. These findings are also consistent with the largest fractal dimensions observed for inter-clay mineral pores in clay shale relative to the other lithologies. Consequently, clay shale displays the most complex macropore system, primarily caused by substantial compaction effects.

Compared to massive siltstone, the silty laminae exhibit lower fractal dimensions in both inter-clay mineral and intergranular pores. This difference may stem from multiple factors, and among them, a prominent cause could be the more frequent presence of volcanic fragments in the silty laminae of our samples, as opposed to the primarily terrigenous detritus found in massive siltstone (Figure 3). Volcanic fragments are more angular compared to regular terrigenous particles, which usually undergo relatively long-distance transport [47]. Thus, under similar compaction, these angular volcanic fragments are more prone to forming point contacts, facilitating the development of a rigid framework that is more conducive to maintaining larger intergranular and inter-clay mineral pores.

On the other hand, the observed lower fractal dimensions within silty laminae, as contrasted with massive siltstone, might result from the difference in dissolution effect. As illustrated in Figure 13, the fractal dimension reflects the disparity in feldspar and carbonate dissolution macropores across distinct lithologies. The highest values are found in massive siltstone, followed by silty laminae, with the lowest values in clay shale. Such variation may be tied to the different degrees of organic acid dissolution, which differentially promotes the enlargement of dissolution macropores and simplification of pore networks across each lithology. Being the primary source rock, clay shale generates considerable organic acids that trigger pronounced dissolution. As silty laminae are situated closer to the source rock than massive siltstone, they endure more intense organic acid dissolution and thus exhibit lower heterogeneity in dissolution pores. The dissolution effect is thus pivotal in reducing the complexity of macropore structures within the silty laminae when compared to massive siltstone. Furthermore, the significance of the pore-type composition in influencing heterogeneity also warrants careful consideration. Figures 7 and 13 reveal that, in comparison to massive siltstone, silty laminae exhibit a higher proportion of dissolution pores characterized by a relatively low fractal dimension, in addition to a lower percentage of inter-clay mineral pores, which possess a relatively high fractal dimension. This differential pore-type composition likely constitutes an additional pivotal factor that imparts silty laminae with a lower heterogeneity in their macropore system.

Overall, within the Chang-7 shale oil reservoir, silty laminae exhibit the least macropore heterogeneity, followed by moderate heterogeneity in massive siltstone, while the macropores in clay shale exhibit the highest level of heterogeneity. The heterogeneity attributes have been proven to arise from a combination of factors, such as the pore type composition, compaction effects, and the influence of organic acid dissolution.

Previous research has demonstrated that the macroporosity of the silty lamina, ranging from 0.23% to 3.26%, is generally lower than that of massive siltstone, which varies from 1.31% to 3.76% [23]. Despite the lower degree of pore development in silty laminae, their relatively simpler macropore structure, primarily attributed to enhanced organic acid dissolution, may improve hydrocarbon migration to some extent. Additionally, there is a significant overlap in the macroporosity range between silty lamina and massive siltstone. When their macroporosity levels are similar, the impact of pore structure complexity on fluid flow may be further amplified, underscoring the significant potential of intervals rich in silty laminae for the future exploration and development of Chang-7 shale resources.

5.5. Implications for Methodology: Applicability, Advantages, and Limitations

In this study, we utilized SEM digital analysis techniques and a modified cumulative-size distribution model to investigate the heterogeneity of macropores in Chang-7 shale. This approach effectively overcomes the inherent limitations of the traditional cumulative-size distribution model, which often impede the accurate calculation of fractal dimensions. Our modifications were mathematically derived within the framework of fractal theory, based on statistical and geometric principles, without introducing additional assumptions. Consequently, the improved model has the potential to be applied to any medium that conforms to fractal theory, such as the pores in other types of formations beyond shale. This enhances the existing fractal modeling framework and may serve as a valuable tool for the study of various fractal structures.

The SEM technique, integral to our method, facilitates quantitative research on specific objects within defined sample regions. This capability allows us to precisely obtain the fractal dimensions of different pore types across various lithological units, providing robust and detailed evidence for revealing the origin of macropore heterogeneity in Chang-7 shale. In contrast, techniques such as MICP, LPNA, and NMR, which rely on the analysis of whole-rock samples, do not offer the same level of specificity and detail.

However, it is crucial to recognize that the current automation level of SEM digital analysis is lower compared to whole-rock analysis techniques, resulting in a higher workload. Especially when analyzing pores, additional representativeness tests are essential to ensure that the extracted pore data have adequate statistical coverage and accurately reflect the comprehensive pore characteristics of the sample. Therefore, we recommend selecting appropriate analytical techniques based on the specific research objectives. For studies where whole-rock fractal analysis is sufficient, more convenient whole-rock methods should be employed. When detailed fractal analysis of specific regions and pore types is required, the SEM quantitative technique, which provides more comprehensive and precise information, should be used.

6. Conclusions

In this research, we carried out a comprehensive analysis of the macropores within Chang-7 shale oil reservoirs, leveraging MICP analysis and SEM digital techniques to examine both whole-rock samples and various individual lithological units. The MICP results indicated a weak correlation between the fractal dimensions of macropores and the mineral composition. However, a trend was observed where fractal dimensions decreased as the proportion of silty lamina increased, particularly in samples predominantly made up of clay shale and silty lamina, highlighting the significant role of lithology in influencing macropore heterogeneity. To further investigate this aspect, we utilized SEM analysis to uncover the macroporous heterogeneity present across different lithologies.

The traditional cumulative size distribution model requires pore quantity-size distribution data to calculate the fractal dimension. However, it also necessitates a comprehensive SEM analysis of the entire cross-sectional areas of a sample for precise outcomes. To overcome this limitation, we introduced a new method involving the establishment of a universal equation and a computational methodology for this model. This advancement broadens the model's applicability, allowing for accurate fractal analysis of the entire sample through the examination of only randomly selected cross-sectional areas. Our approach enhances the existing framework for fractal modeling and potentially serves as a valuable tool for studies of fractal structures.

SEM fractal analysis sheds light on the divergences in macropore types across varied lithologies, underscoring its importance for understanding the origins of macropore heterogeneity. Among different lithologies, clay shale displayed the highest level of macroporous heterogeneity, followed by massive siltstone and then silty lamina. These distinctions arise from multiple factors, including the diversity in pore type composition, the ability to preserve pore space during compaction, and the degree of organic acid dissolution, varying across lithologies.

Supplementary Materials: The following supporting information can be downloaded at: https://www.mdpi.com/article/10.3390/fractalfract8120732/s1.

Author Contributions: Conceptualization, Z.W. and L.D.; methodology, Z.W. and L.D.; formal analysis, Z.W., L.D., and X.W.; investigation, Z.W. and X.W.; resources, J.F., X.L., and R.Z.; writing—original draft, Z.W. and L.D.; writing—review and editing, Z.W. and L.D.; supervision, Z.J.; project administration, Z.J.; funding acquisition, Z.J. All authors have read and agreed to the published version of the manuscript.

Funding: This research was funded by the National Natural Science Foundation of China, grant numbers 42090021 and 42373050.

Data Availability Statement: The original contributions presented in the study are included in the article and Supplementary Materials; further inquiries can be directed to the corresponding authors.

Conflicts of Interest: Authors Jinhua Fu and Xianyang Liu are employed by the PetroChina Changqing Oilfield Company. The remaining authors declare that the research was conducted in the absence of any commercial or financial relationships that could be construed as a potential conflict of interest.

References

1. Yang, J.; Wang, M.; Wu, L.; Liu, Y.; Qiu, S.; Xu, P. A Novel Monte Carlo Simulation on Gas Flow in Fractal Shale Reservoir. *Energy* **2021**, *236*, 121513. [CrossRef]
2. Song, W.; Yao, J.; Zhang, K.; Yang, Y.; Sun, H. Understanding Gas Transport Mechanisms in Shale Gas Reservoir: Pore Network Modelling Approach. *Adv. Geo-Energy Res.* **2022**, *6*, 359–360. [CrossRef]
3. Yang, Y.; Xiao, W.; Zheng, L.; Lei, Q.; Qin, C.; He, Y.; Liu, S.; Li, M.; Li, Y.; Zhao, J.; et al. Pore Throat Structure Heterogeneity and Its Effect on Gas-Phase Seepage Capacity in Tight Sandstone Reservoirs: A Case Study from the Triassic Yanchang Formation, Ordos Basin. *Pet. Sci.* **2023**, *20*, 2892–2907. [CrossRef]
4. Ozotta, O.; Ostadhassan, M.; Liu, K.; Lee, H.; Pu, H.; Kolawole, O.; Malki, M.L. Time-Dependent Impact of CO_2-Shale Interaction on CO_2 Storage. In Proceedings of the 15th International Conference on Greenhouse Gas Control Technologies, Abu Dhabi, United Arab Emirates, 15–18 March 2021; pp. 15–18.

5. Mandelbrot, B.B. Stochastic Models for the Earth's Relief, the Shape and the Fractal Dimension of the Coastlines, and the Number Area Rule for Islands. *Proc. Natl. Acad. Sci. USA* **1975**, *72*, 3825–3828. [CrossRef]
6. Peng, C.; Zou, C.; Yang, Y.; Zhang, G.; Wang, W. Fractal Analysis of High Rank Coal from Southeast Qinshui Basin by Using Gas Adsorption and Mercury Porosimetry. *J. Pet. Sci. Eng.* **2017**, *156*, 235–249. [CrossRef]
7. Yang, F.; Ning, Z.; Liu, H. Fractal Characteristics of Shales from a Shale Gas Reservoir in the Sichuan Basin, China. *Fuel* **2014**, *115*, 378–384. [CrossRef]
8. Yao, Y.; Liu, D.; Tang, D.; Tang, S.; Huang, W.; Liu, Z.; Che, Y. Fractal Characterization of Seepage-Pores of Coals from China: An Investigation on Permeability of Coals. *Comput. Geosci.* **2009**, *35*, 1159–1166. [CrossRef]
9. Chandra, D.; Bakshi, T.; Bahadur, J.; Hazra, B.; Vishal, V.; Kumar, S.; Sen, D.; Singh, T.N. Pore Morphology in Thermally-Treated Shales and Its Implication on CO_2 Storage Applications: A Gas Sorption, SEM, and Small-Angle Scattering Study. *Fuel* **2023**, *331*, 125877. [CrossRef]
10. Zhang, J.; Tang, Y.; Ge, D.; Sun, P.; Zou, X. Full-Scale Nanopore System and Fractal Characteristics of Clay-Rich Lacustrine Shale Combining FE-SEM, Nano-CT, Gas Adsorption and Mercury Intrusion Porosimetry. *Appl. Clay Sci.* **2020**, *196*, 105758. [CrossRef]
11. Liu, Y.; Zhang, H.; Zhang, Z.; Jing, L.; Liu, K. Quantifying the Pore Heterogeneity of Alkaline Lake Shale during Hydrous Pyrolysis by Using the Multifractal Method. *Fractal Fract.* **2024**, *8*, 335. [CrossRef]
12. Zhang, Q.; Dong, Y.; Tong, S. Pore-Type-Dependent Fractal Features of Shales and Implications on Permeability. *Fractal Fract.* **2023**, *7*, 803. [CrossRef]
13. Yu, Y.; Luo, X.; Wang, Z.; Cheng, M.; Lei, Y.; Zhang, L.; Yin, J. A New Correction Method for Mercury Injection Capillary Pressure (MICP)to Characterize the Pore Structure of Shale. *J. Nat. Gas Sci. Eng.* **2019**, *68*, 102896. [CrossRef]
14. Zhang, N.; Wang, X.; Wang, S.; Wang, R.; Wu, J.; Li, Z.; Song, Y. Multifractal Characteristics on Pore Structure of Longmaxi Shale Using Nuclear Magnetic Resonance (NMR). *Geoenergy Sci. Eng.* **2024**, *241*, 213176. [CrossRef]
15. Lu, H.; Li, Q.; Yue, D.; Wu, Y.; Gao, J.; Wu, S.; Wang, W.; Li, M.; An, K. Quantitative Characterization and Formation Mechanism of the Pore System Heterogeneity: Examples from Organic-Rich Laminated and Organic-Poor Layered Shales of the Upper Triassic Chang 7 Member in the Southern Ordos Basin, China. *Mar. Pet. Geol.* **2023**, *147*, 105999. [CrossRef]
16. Jiao, K.; Yao, S.; Liu, C.; Gao, Y.; Wu, H.; Li, M.; Tang, Z. The Characterization and Quantitative Analysis of Nanopores in Unconventional Gas Reservoirs Utilizing FESEM-FIB and Image Processing: An Example from the Lower Silurian Longmaxi Shale, Upper Yangtze Region, China. *Int. J. Coal Geol.* **2014**, *128–129*, 1–11. [CrossRef]
17. Pan, J.; Wang, K.; Hou, Q.; Niu, Q.; Wang, H.; Ji, Z. Micro-Pores and Fractures of Coals Analysed by Field Emission Scanning Electron Microscopy and Fractal Theory. *Fuel* **2016**, *164*, 277–285. [CrossRef]
18. Yu, B.; Li, J.; Li, Z.; Zou, M. Permeabilities of Unsaturated Fractal Porous Media. *Int. J. Multiph. Flow* **2003**, *29*, 1625–1642. [CrossRef]
19. Yu, B.; Cheng, P. A Fractal Permeability Model for Bi-Dispersed Porous Media. *Int. J. Heat Mass Transf.* **2002**, *45*, 2983–2993. [CrossRef]
20. Zuo, M.; Chen, H.; Liu, X.; Liu, H.; Wu, Y.; Qi, X. Fractal Model of Spontaneous Imbibition in Low-Permeability Reservoirs Coupled with Heterogeneity of Pore Seepage Channels and Threshold Pressure. *Pet. Sci.* **2024**, *21*, 1002–1017. [CrossRef]
21. Hu, T.; Pang, X.; Jiang, F. Advances in Whole Petroleum System Theory and New Directions for Petroleum Geology Development. *Adv. Geo-Energy Res.* **2024**, *11*, 1–5. [CrossRef]
22. Gao, C.; Chen, Y.; Yin, J.; Liang, Q.; Hao, S.; Zhang, L.; Zhao, Q.; Sun, J.; Xu, J. Experimental Study on Reservoir Characteristics and Oil-Bearing Properties of Chang 7 Lacustrine Oil Shale in Yan'an Area, China. *Open Geosci.* **2022**, *14*, 234–251. [CrossRef]
23. Wang, Z.; Dong, L.; Jin, Z.; Zou, S.; Fu, J.; Zhu, R. Efforts to Untie the Multicollinearity Knot and Identify Factors Controlling Macropore Structures in Shale Oil Reservoirs. *Adv. Geo-Energy Res.* **2024**, *11*, 194–207. [CrossRef]
24. Sing, K.S.W.; Everett, D.H.; Haul, R.A.W.; Moscou, L.; Pierotti, R.A.; Rouquerol, J.; Siemieniewska, T. Reporting Physisorption Data for Gas/Solid Systems with Special Reference to the Determination of Surface Area and Porosity (Recommendations 1984). *Pure Appl. Chem.* **1985**, *57*, 603–619. [CrossRef]
25. Li, J.; Lu, S.; Cai, J.; Zhang, P.; Xue, H.; Zhao, X. Adsorbed and Free Oil in Lacustrine Nanoporous Shale: A Theoretical Model and a Case Study. *Energy Fuels* **2018**, *32*, 12247–12258. [CrossRef]
26. Guo, H.; Jia, W.; Peng, P.; Lei, Y.; Luo, X.; Cheng, M.; Wang, X.; Zhang, L.; Jiang, C. The Composition and Its Impact on the Methane Sorption of Lacustrine Shales from the Upper Triassic Yanchang Formation, Ordos Basin, China. *Mar. Pet. Geol.* **2014**, *57*, 509–520. [CrossRef]
27. Lin, S.; Hou, L.; Luo, X.; Wu, Y. A Millimeter-Scale Insight into Formation Mechanism of Lacustrine Black Shale in Tephra Deposition Background. *Sci. Rep.* **2022**, *12*, 11511. [CrossRef]
28. Liu, C.; Zhao, H.; Gui, X.; Yue, L.; Zhao, J.; Wang, J. Space-Time Coordinate of the Evolution and Reformation and Mineralization Response in Ordos Basin. *Acta Geol. Sin.* **2006**, *80*, 617–638. [CrossRef]
29. Fu, J.; Li, S.; Guo, Q.; Guo, W.; Zhou, X.; Liu, J. Enrichment Condition and Favorable Area Optimization of Continental Shale Oil in Ordos Basin. *Acta Pet. Sin.* **2022**, *43*, 1702–1716. [CrossRef]
30. Cao, Z.; Liu, G.; Zhan, H.; Li, C.; You, Y.; Yang, C.; Jiang, H. Pore Structure Characterization of Chang-7 Tight Sandstone Using MICP Combined with N2GA Techniques and Its Geological Control Factors. *Sci. Rep.* **2016**, *6*, 36919. [CrossRef]
31. Hou, L.; He, K.; Zhai, J.; Mi, J.; Weng, N. Compositional Kinetics for Hydrocarbon Evolution in the Pyrolysis of the Chang 7 Organic Matter: Implications for in-Situ Conversion of Oil Shale. *J. Anal. Appl. Pyrolysis* **2022**, *162*, 105434. [CrossRef]

32. Zhao, W.; Hu, S.; Hou, L. Connotation and Strategic Role of In-Situ Conversion Processing of Shale Oil Underground in the Onshore China. *Pet. Explor. Dev.* **2018**, *45*, 563–572. [CrossRef]
33. Zhao, W.; Zhu, R.; Hu, S.; Hou, L.; Wu, S. Accumulation Contribution Differences between Lacustrine Organic-Rich Shales and Mudstones and Their Significance in Shale Oil Evaluation. *Pet. Explor. Dev.* **2020**, *47*, 1079–1089. [CrossRef]
34. Cui, J.; Zhu, R.; Li, S.; Qi, Y.; Shi, X.; Mao, Z. Development Patterns of Source Rocks in the Depression Lake Basin and Its Influence on Oil Accumulation: Case Study of the Chang 7 Member of the Triassic Yanchang Formation, Ordos Basin, China. *J. Nat. Gas Geosci.* **2019**, *4*, 191–204. [CrossRef]
35. Wang, Z.; Fu, J.; Liu, X.; Li, S.; Zhang, C.; Liang, X.; Dong, L. The Influence of Hydrothermal Activities on Shale Oil Reservoirs during the Burial Period of the Upper Triassic Chang 7 Member, Ordos Basin. *Oil Gas Geol.* **2023**, *44*, 899–909. [CrossRef]
36. Ren, Z.; Zhang, S.; Gao, S.; Cui, J.; Liu, X. Research on Region of Maturation Anomaly and Formation Time in Ordos Basin. *Acta Geol. Sin.* **2006**, *80*, 675–684. [CrossRef]
37. Comisky, J.T.; Santiago, M.; McCollom, B.; Buddhala, A.; Newsham, K.E. Sample Size Effects on the Application of Mercury Injection Capillary Pressure for Determining the Storage Capacity of Tight Gas and Oil Shales. In Proceedings of the Society of Petroleum Engineers—Canadian Unconventional Resources Conference 2011, CURC 2011, Calgary, AB, Canada, 19–21 October 2011; p. 149432.
38. Washburn, E.W. The Dynamics of Capillary Flow. *Phys. Rev.* **1921**, *18*, 206–209. [CrossRef]
39. Yu, Y.; Luo, X.; Cheng, M.; Lei, Y.; Wang, X.; Zhang, L.; Jiang, C.; Zhang, L. Study on the Distribution of Extractable Organic Matter in Pores of Lacustrine Shale: An Example of Zhangjiatan Shale from the Upper Triassic Yanchang Formation, Ordos Basin, China. *Interpretation* **2016**, *5*, SF109–SF126. [CrossRef]
40. Abràmoff, M.D.; Magalhães, P.J.; Ram, S.J. Image Processing with ImageJ. *Biophotonics Int.* **2004**, *11*, 36–41. [CrossRef]
41. Pfeifer, P.; Avnir, D.; Farin, D. Ideally Irregular Surfaces, of Dimension Greater than Two, in Theory and Practice. *Surf. Sci.* **1983**, *126*, 569–572. [CrossRef]
42. Friesen, W.I.; Mikula, R.J. Fractal Dimensions of Coal Particles. *J. Colloid Interface Sci.* **1987**, *120*, 263–271. [CrossRef]
43. Zhang, B.; Li, S. Determination of the Surface Fractal Dimension for Porous Media by Mercury Porosimetry. *Ind. Eng. Chem. Res.* **1995**, *34*, 1383–1386. [CrossRef]
44. Li, Y.; Lu, G.; Rudolph, V. Compressibility and Fractal Dimension of Fine Coal Particles in Relation to Pore Structure Characterization Using Mercury Porosimetry. *Part. Part. Syst. Charact.* **1999**, *16*, 25–31. [CrossRef]
45. Bustin, R.M.; Bustin, A.M.M.; Cui, X.; Ross, D.J.K.; Pathi, V.S.M. Impact of Shale Properties on Pore Structure and Storage Characteristics. In Proceedings of the Society of Petroleum Engineers—Shale Gas Production Conference 2008, Fort Worth, TX, USA, 16–18 November 2008; pp. 32–59.
46. Wang, P.; Jiang, Z.; Ji, W.; Zhang, C.; Yuan, Y.; Chen, L.; Yin, L. Heterogeneity of Intergranular, Intraparticle and Organic Pores in Longmaxi Shale in Sichuan Basin, South China: Evidence from SEM Digital Images and Fractal and Multifractal Geometries. *Mar. Pet. Geol.* **2016**, *72*, 122–138. [CrossRef]
47. Xi, K.; Li, K.; Cao, Y.; Liu, M.; Niu, X.; Zhu, R.; Wei, X.; You, Y.; Liang, X.; Feng, S. Laminae Combination and Shale Oil Enrichment Patterns of Chang 73 Sub-Member Organic-Rich Shales in the Triassic Yanchang Formation, Ordos Basin, NW China. *Pet. Explor. Dev.* **2020**, *47*, 1342–1353. [CrossRef]

fractal and fractional

MDPI

Article

Fractal Characteristics of Pore Structure of Longmaxi Shales with Different Burial Depths in Southern Sichuan and Its Geological Significance

Manping Yang [1,2], Yanyan Pan [1,2], Hongye Feng [2,*], Qiang Yan [2], Yanjun Lu [1], Wanxin Wang [1], Yu Qi [1] and Hongjian Zhu [1,*]

1 School of Vehicle and Energy, Yanshan University, Qinhuangdao 066000, China
2 Institute of Mineral Resources, Chinese Academy of Geological Sciences, Beijing 100037, China
* Correspondence: fenghongye18@mails.ucas.ac.cn (H.F.); zhuhongjian@ysu.edu.cn (H.Z.)

Abstract: Burial depth can significantly impact the pore structure characteristics of shale. The Lower Silurian Longmaxi Shale in the Weiyuan block of the Sichuan Basin is a marine formation that we studied for deep shale gas exploration. We used two groups of Longmaxi samples, outcrop shale and middle-deep shale, to investigate the pore structure fractal features at varying burial depths using a combination of mineralogy, organic geochemistry, scanning electron microscopy (SEM), and low-temperature gas (CO_2, N_2) adsorption. The V-S fractal model was used to determine the fractal dimension (Dc) of micropores, and the FHH fractal model was used to determine the fractal dimension (D_N) of mesopores. The findings indicate that the pore morphology of organic matter becomes irregular and more broken as the burial depth increases, as does the content and maturity of organic matter. The pore size of organic matter gradually decreases, the SSA (BET, DR) and PV (BJH, DA) of shale pores increase, the pore structure becomes more complex, and the average shale pore size decreases. According to this study, the organic matter content and its maturity show an increasing trend as burial depth increases. Meanwhile, the organic matter's pore morphology tends to be irregular, and fracture rates rise, which causes the organic matter's pore size to gradually decrease. In addition, the SSA (comprising the values assessed by BET and DR techniques) and PV (evaluated by BJH and DA methods) of shale pores grew, suggesting that the pore structure became more complex. Correspondingly, the average pore size of the shale decreased. The fractal dimensions of the micropores (D_C), mesoporous surface (D_{N1}), and mesoporous structure (D_{N2}) of outcrop shale are 2.6728~2.7245, 2.5612~2.5838, and 2.7911~2.8042, respectively. The mean values are 2.6987, 2.5725, and 2.7977, respectively. The D_C, D_{N1}, and D_{N2} of middle-deep shale are 2.6221~2.7510, 2.6085~2.6390, and 2.8140~2.8357, respectively, and the mean values are 2.7050, 2.6243, and 2.8277, respectively. As the fractal dimension grows, the shale's pore structure becomes more intricate, and the heterogeneity increases as the buried depth increases. The fractal dimension has a positive association with the pore structure parameters (SSA, PV), TOC, and R_o and a negative association with the mineral component (quartz, feldspar, clay mineral) contents. Minerals like quartz, feldspar, and clay will slow down the expansion of pores, but when SSA and PV increase, the pore heterogeneity will be greater and the pore structure more complex.

Keywords: Sichuan Basin; Longmaxi shale; burial depth; pore structure; fractal characteristics

1. Introduction

Shale gas is a special kind of natural gas that can produce and store energy on its own [1]. Thanks to developments in geological theory and shale gas exploration technology, China's shale gas production has increased dramatically in recent years. The Lower Paleozoic Wufeng-Longmaxi Formation in the Sichuan Basin and its surrounding areas is the primary target for shale gas development in China. Shale gas demonstration zones have been constructed at the state level, including Jiaoshiba, Weiyuan, and Changning-Zhaotong [2]. Reservoirs of deep shale have been found that are more than 3500 m below the surface. With their substantial oil and gas potential, deep shale reservoirs will progressively become more relevant in the subsequent exploration phase. The deep shale reservoirs differ from the shallow ones in that they are buried deeper, are exposed to more extreme temperatures and pressures, and have a higher degree of densification [3]. Comparing the pore structure characteristics of shale reservoirs with different burial depths is essential for promoting the development of deep shale gas.

A variety of pore categorization methods have been established based on pore size, pore matrix, pore morphology, and pore connectivity [4–7]. The International Union of Pure and Applied Chemistry (IUPAC) proposed a pore classification standard, which is now extensively used in the quantitative description of shale pore. IUPAC separates pores into the following three categories: Micropores (less than 2 nm), mesopores (between 2 nm and 50 nm), and macropores (more than 50 nm). According to Bu et al., mesopores or micropores offer more surface area and adsorption energy for adsorbed gas, whereas macropores offer room for free gas [8]. The storage capacity, desorption difficulties, and seepage characteristics of shale gas are directly determined by the pore structure characteristics of shales. These factors also impact the gas content, reserve estimate, and development appraisal of shales [9].

The two methods currently being studied regarding the nanopore structure of shale are high-resolution imaging technology and fluid/non-fluid injection technology. The development of field emission scanning electron microscopy (FESEM), focused ion beam scanning electron microscopy (FIBSEM), focused ion beam helium ion microscopy (FIB-HIM), and transmission electron microscopy (TEM) has greatly promoted the intuitive and qualitative observation of shale's organic, mineral structure and pore fissure structure, and has continuously deepened the study of ultra-structural morphology and the genesis of oil and gas reservoirs [10,11]. Furthermore, pore size distribution (PSD), specific surface area (SSA), pore volume (PV), porosity, and other parameters are frequently obtained through the use of nuclear magnetic resonance (NMR), mercury intrusion porosimetry (MIP), gas (N_2 and CO_2) adsorption (GA), and other experiments for the quantitative analysis of shale pore structure [12–14]. Therefore, the investigation of shale pore structure using a variety of experimental techniques has grown significantly in recent years.

Pore geometry with uneven surfaces and shapes may be well described by fractal theory. The pore structure of the marine shale of Changning, southern Sichuan's Longmaxi Formation, was examined by Li et al. using the FHH model [15]. Also using the FHH model, Li et al. performed a more thorough fractal and pore structure investigation of the deep shale of the Wufeng Formation-Longmaxi Formation [16]. Using a variety of fractal techniques, Ma et al. conducted a fractal analysis and investigated the pore structure of lacustrine shale in China's Kongdian Formation [17]. Three types of study are now being conducted on the fractal dimension of shale: the FHH model, the BET model, and the thermodynamic model. The most popular approach and the best fit for shale reservoirs is the fractal FHH model [18].

The outcrop and the core (middle-deep) shale of Wei X well of Longmaxi Formation in southern Sichuan were selected as the research materials, and the shale petrology, mineralogy, organic geochemistry, and other aspects were studied. The pore structure of shale with varying burial depths was thoroughly described from both qualitative and quantitative perspectives using SEM, GA, and other testing methods. Based on this, an appropriate fractal model was chosen to examine the shale pores at various scales. Ultimately, a relationship was found between the fractal dimension, mineral composition, organic matter content, organic matter maturity, and shale pore structure characteristics. By comparing the pore structure features of shales with varying burial depths, this study aimed to investigate the resource potential of deep shale reservoirs using the shale in the Weiyuan block as an example.

2. Geological Setting

The Weiyuan block, with an area of around 6500 km^2, is situated south of the Sichuan Basin. The tectonic position in the southwestern Sichuan paleo-central slope of the low-folded zone and the development of the Weiyuan dorsal tectonics, Triassic System, and Jurassic System are exposed, with a longitudinal lack of the Devonian and Carboniferous Systems. In the northwestern part of the dorsal tectonics, rock layers of the Zhiru System are lacking to varying degrees [19]. The Chuanzhong Palaeouplift, Qianzhong Palaeouplift, and Yichang Underwater Uplift surround the research region, which was in a stagnant semi-deepwater-deepwater shelf depositional environment throughout the Late Ordovician to Early Silurian. Black shales that were heavily bedded (10~30 m thick), high-maturity (R_o > 2%), rich in organic matter (TOC > 2 wt.%), and very brittle were deposited. The depth of the Wufeng-Longmaxi Formation's shales in the Weiyuan area ranges from 1500 to 4000 m, and they progressively deepen from north to south in an easterly direction due to the influence of several tectonic movements in the later stages, including the Caledonian, Hercynian, Indo-Chinese, Yanshanian, and Himalayan periods [20]. The most advantageous shale gas exploration and production system in the Sichuan Basin is the Longmaxi Formation, which produces siliceous, carbonaceous, calcareous, and clayey shale from bottom to top, locally intermingled with black siltstone shale [21].

3. Samples and Experiments

3.1. Samples

The 20 shale samples utilized came from the Lower Silurian Longmaxi Formation in the Weiyuan block, southern Sichuan. Three samples were taken from the measured profile of the field outcrops; six samples were taken from the middle shale formation, with sampling depths ranging from 2651.09 to 2943.8 m; and 11 samples were taken from the deep shale formation, with sampling depths ranging from 3552.58 to 3574.54 m. Seventeen middle and deep shale samples came from the same locations and wells as the samples that were characterized extensively in previous work (data from Hao et al., 2023) [22]. The 20 collected samples were examined for TOC, R_o, XRD analysis, and SEM observation. Six of them were tested for GA.

3.2. Experiments

The LECO CS-230 carbon-sulfur analyzer was used to determine TOC. The materials were crushed to 200 mesh prior to the experiment, and the inorganic carbon was eliminated by adding a diluted hydrochloric acid. The samples were then neutralized with distilled water and dried. TOC was ascertained in the experiment after 1.0 g of iron flux and 1.0 g of tungsten flux were applied to the dried samples. A Japanese Rigaku TTR III multifunctional X-ray diffractometer was used for XRD investigations. A suitable powder sample was evaluated during the test and analysis using the following parameters: tube voltage of 35 kV, tube current of 30 mA, scanning range of $2\theta = 5\text{~}80^\circ$, and scanning speed of 2°/min. By examining the XRD pattern features and diffraction peak strength of distinct mineral crystals, the content of different minerals in the sample was ascertained. The pore morphology of the shale samples was characterized using a Zeiss Sigma 500 SEM. Before the test, the samples were processed into cubes about 1 cm in size. Subsequently, the samples were dried in an oven. A flat and smooth surface was selected and tested after gold plating on the smooth surface. During the test, the secondary electron resolution reached 1 nm, the magnification was 200,000 to 800,000 times, and the acceleration voltage adjustment range was 0.02–30 kV. The ASAP 2460 surface area and porosity analyzer was used for GA experiments. Prior to the experiment, the samples were crushed to 200 meshes. In order to eliminate moisture and volatile materials from the sample, it was initially placed in a degassing station and allowed to degas at 110 °C for 10 h. The degassed samples were then transferred to the analysis station, where high quality N_2 (purity > 99.9%) was used as the adsorbate for the adsorption tests, which were conducted at 77.35 K. The sample pretreatments for the CO_2 adsorption experiments were similar to those of the N_2 adsorption experiments. The CO_2 adsorption curves were obtained at 273 K after the powder samples were degassed for 12 h. The CO_2 was used to characterize the pore information of micropores and N_2 was used to identify the pore information of mesopores. The DR and BET models were used to calculate the SSA of shale pores. The DA and BJH models were used to determine PV and PSD.

3.3. Fractal Theory

The fractal dimension of the micropores was calculated using the fractal V-S model based on CO_2 adsorption measurements, and the formula was [23,24]:

$$\ln(V) = \frac{3}{D_C}\ln S + C \tag{1}$$

In the equation: V—gas adsorption volume cm^3/g; S—pore cumulative specific surface area m^2/g; D_c—microporous fractal dimension; C—constant;

Shale fractal dimension was calculated using the fractal FHH model based on N_2 adsorption tests [18,25] and the formula was [26]:

$$\ln(V) = (D_N - 3) \times \ln\left(\ln\frac{P_0}{P}\right) + C \tag{2}$$

In the equation: V—Adsorption gas volume at varying relative pressures, cm^3/g; P—equilibrium pressure, MPa; P_0—Adsorbed gas saturation vapor pressure, MPa; C—constant; D_N—fractal dimension of the porous materials.

The range of the fractal dimension (D) is two to three. Three denotes very complicated surfaces or uneven pore structure, and two denotes flawlessly regular smooth solid surfaces or homogenous pore structure.

4. Results

4.1. Organic Geochemistry and Mineralogy

The TOC values range from 1.67 wt.% to 6.36 wt.%, with an average of 3.471 wt.%, indicating that the shales have favorable conditions for gas production, demonstrating good potential for hydrocarbon production. The TOC values of outcrop and middle-deep shale are 1.67 wt.%~1.94 wt.% and 2.24 wt.%~6.36 wt.%, respectively, and the average values are 1.79 wt.% and 3.77 wt.%, respectively [22]. It was discovered that middle-deep shale has a greater TOC value than outcrop. This shows that as burial depth increases, the TOC values progressively increase as well.

The degree of organic matter conversion to hydrocarbons is directly reflected in the maturity of organic matter, making it a crucial indicator of shale hydrocarbon generation potential. The maturity of organic matter is generally reflected by the R_o. However, there is a lack of vitrinite in Longmaxi shales, so the asphaltene reflectance (R_b) is measured in this experiment. The empirical formula $R_o = 0.618 \times R_b + 0.4$, presented by Jacob et al. in 1985, was used to obtain the R_o value [27]. The outcrop shales' R_o fell within the high maturation stage, ranging from 1.78% to 1.80%, with an average of 1.795%. Meanwhile, the R_o of the middle-deep shales ranges from 2.032% to 2.26%, with an average of 2.143%, which belongs to the over-mature stage. The middle-deep shales, on the other hand, have an R_o that falls into the over-mature stage and varies from 2.032% to 2.26%, with an average of 2.143%. The high to over-mature state that the shales are now in provides a solid basis upon which the shale reservoir's pore structure can develop. As a consequence, it is also found that as burial depth grows, so does the shale's organic matter maturity.

Silica, carbonate, and clay minerals constitute the majority of the mineral makeup of the shales (Figure 1A). Siliceous minerals are dominated by quartz, with contents ranging from 23.5% to 52%, with an average of 35.82%. The feldspar contents range from 1.3% to 10.4%, with an average of 5.755%. Calcite and dolomite make up the majority of carbonate minerals. No carbonate minerals are detected in the outcrop shales. The carbonate contents of the middle-deep shales range from 10.9% to 52.1%, with an average of 30.94%. The clay mineral contents range from 15.6% to 48.3%, with an average of 29.29%. Illite predominates among the clay minerals. The contents of illite, I/S mixed layer, and chlorite are 49~83%, 4~37%, and 5~22%, respectively, and the average values are 66.6%, 19.7%, and 13.7%, respectively. (Figure 1B). The clay composition is dominated by illite, suggesting that the shales are at a high late diagenetic stage [28].

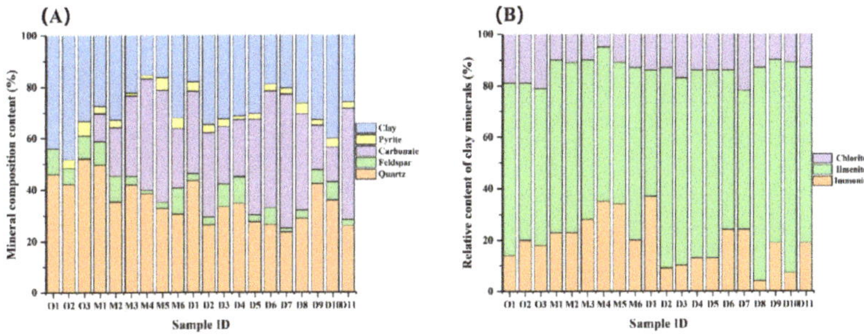

Figure 1. Mineral composition (**A**) and the relative contents of clay mineral (**B**). Data from Hao et al. [22].

4.2. Pore Types in Shale at Different Burial Depths

Organic pores, inorganic pores (intercrystalline pores, intracrystalline-dissolved pores, clay mineral interlayer pores), and microfractures represent several categories of pores that can be distinguished by the SEM (Figure 2). The middle-deep shale has well-developed organic pores. In the deep shale, the morphology of organic pores is distributed heterogeneously. Most of them are honeycomb, elliptical, and irregular. Some of the morphologies are more ruptured. The pore size is below 500 nm (Figure 2A). This could have something to do with the high gas production during the thermal development of maturation and the collapse and carbonation of the organic pore skeleton following the over-maturing of the deep shale [28]. The morphology of organic pores in the middle shale is comparatively homogeneous. They are mainly circular organic pores and have large pore sizes, mainly concentrated below 2 μm (Figure 2B). As burial depth increases, the organic pore size is increasingly smaller. The formation of organic pores is not observed in outcrop shales.

It could be because the outcrop shales have experienced a range of geologic processes over the course of their geologic history, such as compaction, diagenesis, and weathering. The growth and production of organic matter pores may be impacted by these processes. Furthermore, one of the main reasons limiting the formation of organic matter pores is exposure to the environment. The outcrop shale is exposed to the surface and is affected by weathering, erosion, and other environmental elements, which may also destroy the original pore structure and make it difficult to preserve organic pores.

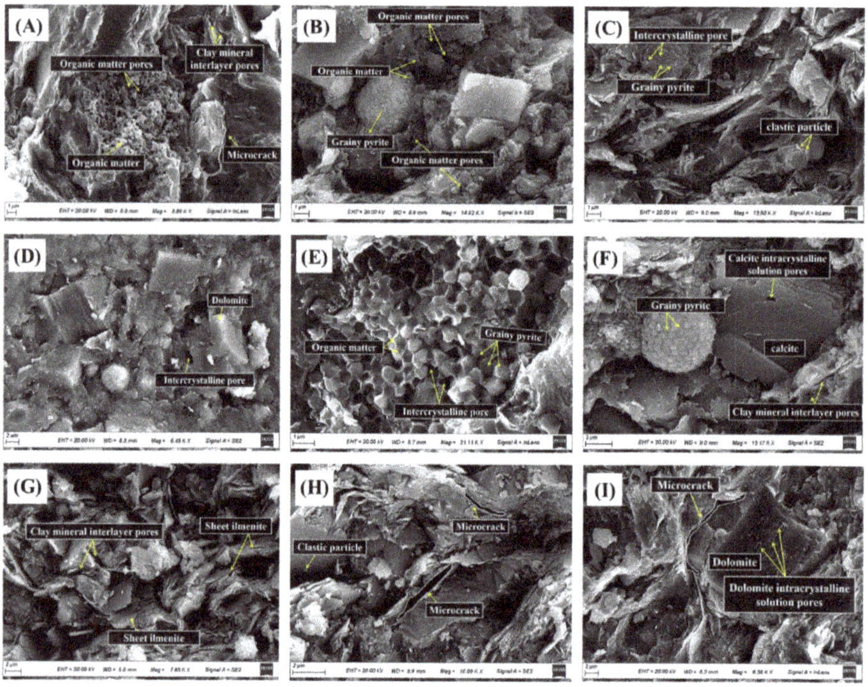

Figure 2. SEM images of Longmaxi shales at different burial depths. (**A**): Sample D10, organic pores, clay mineral interlayer pores, microcracks; (**B**): Sample M1, organic pores; (**C**): Sample O1, intergranular pores; (**D**): Sample D5, intergranular pores; (**E**): Sample D7, intergranular pores; (**F**): Sample M1, clay mineral interlaminar pores, calcite intergranular dissolution pores; (**G**): Sample O1, clay mineral interlaminar pores; (**H**): Sample O1, microcracks; (**I**): Sample M2, dolomite intergranular dissolution pores, microcracks.

The bulk of crystalline pyrite (Figure 2B,C,E,F), which is produced in anoxic environments, is found in middle-deep shales and outcrops. Crystalline pyrite comprises a high number of pyrite crystals with uniform cubic grains. The shape of intergranular pores developed between pyrite is mostly irregular. The pore size ranges from tens of nanometers to hundreds of nanometers. The type of pores is usually isolated in the shale matrix. Gas seepage is hampered by the inadequate connection between these pores [4]. Crystalline pyrite is mostly seen in deep shales in association with organic matter, forming organic pores within the organic matter (Figure 2E). These organic pores adsorbed on pyrite particles have some ability to adsorb natural gas and are a very important pore type in shales [29].

Intracrystalline solution pores form in the middle-deep shales, especially in carbonate minerals like calcite and dolomite. Most of the pores have irregular shapes, such as oval, elongated, and triangular (Figure 2F,I). This phenomenon may occur because of the release of organic acids during the decomposition of organic matter. These organic acids can dissolve carbonate minerals in large quantities, resulting in the formation of dissolution pores, and the pore size of these pores is often less than 1 μm. Most of these pores are relatively isolated with poor connectivity, making little contribution to shale gas seepage [28].

The primary constituents of clay minerals are illite and I/S mixed layer. Numerous interlayer pores of clay minerals develop in both the outcrop and the middle-deep shales. Illite is in the form of thin lamellae with distinct slit-shaped pores developed between the lamellae (Figure 2A,F,G) and pore sizes below 500 nm. During sedimentary burial, montmorillonite transforms into I/S mixed layer and illite, resulting in pores created by volume reduction. These interlayer pores can serve as tiny transport channels for shale gas seepage since they are somewhat connected to one another.

As microfractures grow, they can store free gas and aid in the desorption of adsorbed gas, which is a crucial channel for gas seepage in shales [30]. Microfractures mainly develop at the edges of clay minerals in outcrops and middle-deep shales. They are jagged and curved, with large extension lengths. The width of microfractures reaches the nanometer scale (Figure 2A,H,I). The generation of these microfractures may be related to post-generation modification effects such as tectonic movements, shale reservoir fracturing effects, and differential horizontal compaction [29]. Microfractures may provide important gas storage space and promote the extension of fracture-induced fractures in the later stages of fracturing.

4.3. Pore Structure

4.3.1. CO_2 Isothermal Adsorption Curve

The CO_2 adsorption curve type aligns with the IUPAC-defined type I isotherm (Figure 3A) [31]. The shale exhibits micropore-filling properties, suggesting that it has a certain number of micropores [32]. The CO_2 adsorption curves for shallow–middle–deep samples are similar. The adsorption volume also exhibited a rising trend when the relative pressure increased in the region of $P/P_0 < 0.035$. However, there are differences in the maximum adsorption volume. The outcrop samples have the smallest adsorption volume., which is 1.2158 cm^3/g~1.2320 cm^3/g. The middle–deep samples have a higher adsorption volume than the outcrop samples, which is 1.8041 cm^3/g~2.4985 cm^3/g. This demonstrates that the microporous pore structure of shale samples varies depending on their depth.

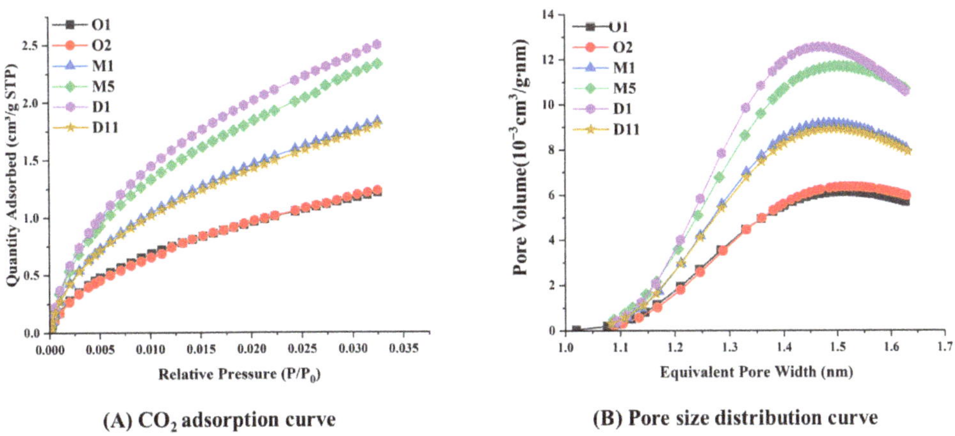

Figure 3. CO_2 adsorption curves (**A**) and PSD (**B**) of samples based on CO_2 adsorption experiments.

DR and DA models were used to determine the SSA and PV of shale micropores at different depths (Table 1). The SSA of outcrop and middle–deep shale micropores ranges from 9.4408 to 9.8153 m^2/g and 14.124 to 20.7664 m^2/g, with mean values of 9.6281 m^2/g and 16.9684 m^2/g, respectively. The PV ranges from 0.0051 to 0.0052 cm^3/g and 0.0068 to 0.0094 cm^3/g, with mean values of 0.0052 cm^3/g and 0.0079 cm^3/g, respectively. The SSA and PV of shale micropores increase with increasing burial depth, which further indicates that the number of micropores rises and the pore structure becomes increasingly complicated.

Table 1. SSA and PV of pores were obtained based on gas adsorption experiments. Four middle and deep shale samples came from the same locations and wells as the samples that were characterized extensively in the previous work (data of BET surface area from Hao et al., 2023) [22].

Sample Number	Low Temperature CO_2 Adsorption			Low Temperature N_2 Adsorption		
	DR Micropore Surface Area (m^2/g)	DA Micropore Volume (cm^3/g)	Mean Equivalent Pore Width (nm)	BET Surface Area (m^2/g)	BJH Pore Volume (cm^3/g)	BJH Average Pore Diameter (nm)
O1	9.4408	0.0051	1.7056	8.8087	0.0117	7.6146
O2	9.8153	0.0052	1.7149	11.3879	0.0092	5.6448
M1	14.7504	0.0068	1.6589	21.2077	0.0185	5.0852
M5	18.2329	0.0094	1.6903	23.2946	0.0211	5.6191
D1	20.7664	0.0087	1.6306	29.3799	0.0260	5.3489
D11	14.1240	0.0069	1.6713	21.7908	0.0222	5.6811

The PSD of shale micropores is described using the DA model. The PSD peaks of O1, O2, M1, M5, D1, and D11 are 1.5125 nm, 1.5227 nm, 1.4857 nm, 1.5045 nm, 1.4662 nm, and 1.4948 nm, respectively (Figure 3B). The pore sizes of outcrop and middle–deep shale are concentrated at 1.5125~1.5227 nm and 1.5125~1.5227 nm, respectively. According to Table 1, the average pore diameters of outcrops and middle–deep samples are 1.7056~1.7149 nm and 1.6306~1.6903 nm, respectively, and the average values are 1.7103 nm and 1.6628 nm, respectively. Regardless of the PSD or average pore size, the outcrop shale has a larger pore size than the middle–deep shale. This might be because the deep shale has a higher formation pressure.

4.3.2. N_2 Isothermal Adsorption Curve

Overall, the isothermal adsorption curves of the shallow–middle–deep shale samples primarily exhibit an inverse S shape, according to the results of the N_2 adsorption–desorption curves (Figure 4). The isothermal adsorption curves are classified as type IV in accordance with the IUPAC classification scheme for isothermal adsorption curves [31], and the hysteresis loops are comparable to type H3, indicating that the pores in the shales are mesopores originating from slits.

Three phases may be distinguished in the adsorption process based on the trend of the N_2 adsorption curve: (1) The adsorption curve climbs gradually and has a short slope when the relative pressure (P/P_0) is less than 0.45. The presence of micropores is shown by the adsorption volume not being 0 when the relative pressure (P/P_0) is 0; (2) The capillary condensation phenomenon causes a hysteresis loop to emerge between the adsorption-desorption curves when the relative pressure (P/P_0) is between 0.45 and 0.9. The hysteresis loop of the middle–deep sample is larger than that of the outcrop. The larger the hysteresis loop, the greater the number of mesopores, meaning that there are more mesopores in the middle–deep sample; (3) The adsorption curve steepens quickly when the relative pressure (P/P_0) is between 0.9 and 1, and it does not attain adsorption saturation when the relative pressure is approximately equal to 1. The cause is gas condensation, which shows that there are some macropores or microcracks in the shale. The adsorption volume of middle–deep shales is greater than that of outcrop shales when the adsorption volumes of shales with varying burial depths are compared. This suggests that the middle–deep shales contain more pores and more intricate pore structures.

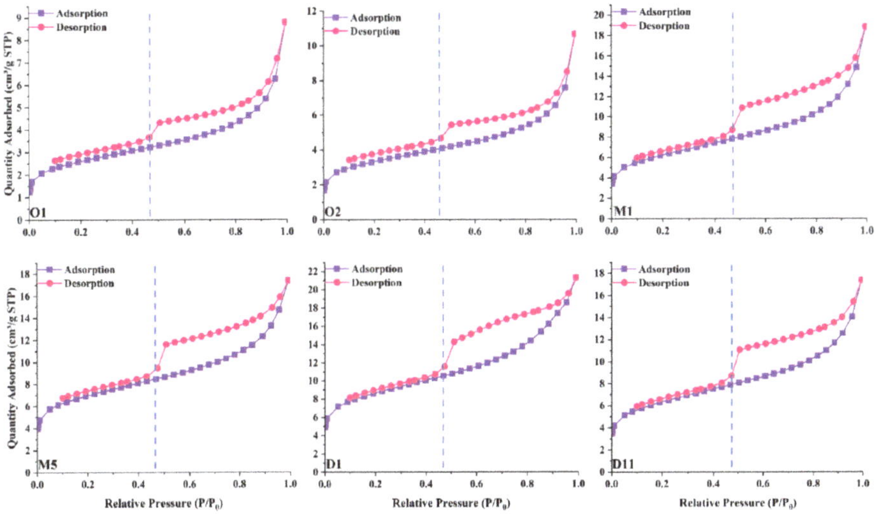

Figure 4. N_2 adsorption curves of shale with different burial depths. Note: The blue dotted line represents a relative pressure of 0.45.

The SSA and PV of mesopores are calculated using the BET and BJH models, respectively. Table 1 displays the specifics. The SSA of the outcrops and middle–deep shale mesopores are 8.8087~11.3879 m^2/g and 21.2077~29.3799 m^2/g, with mean values of 10.0983 m^2/g and 23.9183 m^2/g, respectively. The PV is 0.0092~0.0117 cm^3/g and 0.0185~0.0260 cm^3/g, with mean values of 0.0104 cm^3/g and 0.0220 cm^3/g, respectively. According to the aforementioned investigation, the middle–deep shale's SSA and PV are greater than those of the outcrop shale. This indicates that the middle–deep shale's pores are more developed and have a more complex pore structure.

The PSDs of shale mesoporous were analyzed in conjunction with BJH theory. As shown in Figure 5, the samples O1, O2, M1, M5, D1, and D11 all have two peaks, which are 2.46 nm and 29.82 nm, 2.86 nm and 7.84 nm, 2.65 nm and 20.87 nm, 2.66 nm and 21.58 nm, 2.86 nm and 12.54 nm, and 2.65 nm and 15.62 nm, respectively. This shows that the probability of aperture appearing at this peak is the highest. The average pore sizes of the outcrop and middle–deep samples are 5.6448~7.6146 nm and 5.0852~5.6811 nm, respectively, and the mean values are 6.6297 nm and 5.4336 nm, respectively (Table 1). The outcrop shale has greater average pore sizes than the middle and deep shale. This is because, after the shale is deposited, the pressure of the layers above it rises as the buried depth increases, improving compaction. This compaction will compress the pores and decrease their size in the middle–deep shale. The compaction of outcrop shale is relatively weak, so the pores can maintain a relatively large pore size to a certain extent. Stronger heterogeneity and a more complicated pore structure are found in shale with lower pore sizes.

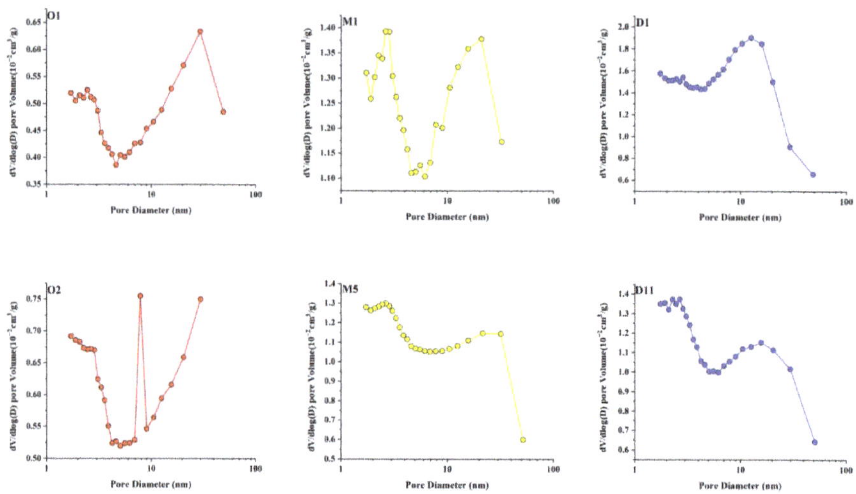

Figure 5. Pore size distribution curve of shale based on N_2 adsorption experiments.

4.3.3. Fractal Characteristics

The fractal dimension (D_C) of the micropores was determined using the fractal V-S model based on CO_2 adsorption tests. Figure 6 displays the fractal fitting findings. With an average value of 2.7029 and a correlation coefficient (R^2) more than 0.99, the D_C of the shale micropores at various depths varies from 2.6221 to 2.7510. The highly complex and diverse pore structure of micropores is indicated by the fractal dimension being near 3. The D_C of outcrop and middle–deep shale are 2.6728~2.7245 and 2.6221~2.7510, respectively, and the average values are 2.6987 and 2.7050, respectively. As shale burial depth increases, the pore structure becomes more complex and varied. A higher percentage of adsorbed gas content is seen in shale with larger fractal dimensions because the more micropores develop, the more gas adsorption sites and adsorption potentials are available.

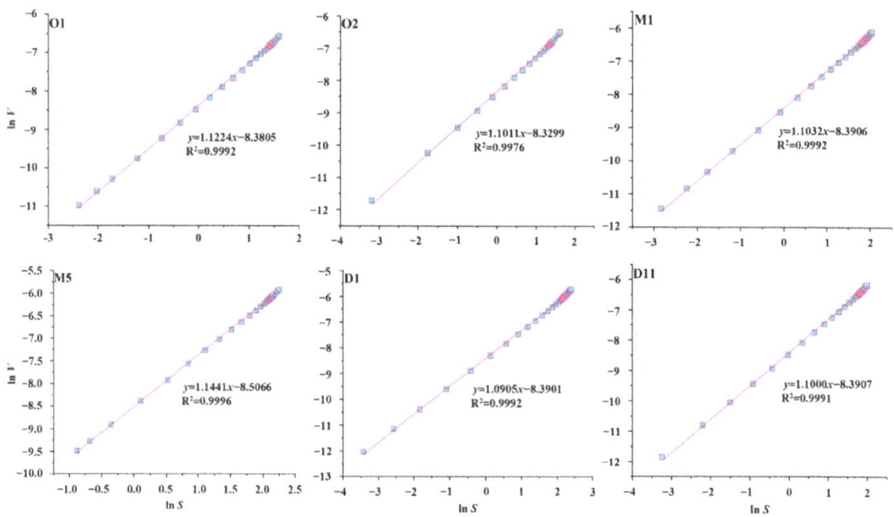

Figure 6. Shale pore fractal fitting curves based on CO_2 adsorption experiments.

A fractal FHH model derived from N_2 adsorption studies was used to determine the fractal dimension (D_N) of the mesopores. Because of the hysteresis phenomenon, the adsorption and desorption curves do not overlap at a relative pressure of 0.45, indicating that the pore structure is significantly changed before and after this relative pressure. The fractal dimension of shale is determined by linearly fitting the adsorption curves with $P/P_0 = 0.45$ as the cut-off line, as seen in Figure 7. Double fractals are discovered to be a characteristic of shale pores. The D_{N1} ($P/P_0 < 0.45$) mainly reflects the monomolecular-multimolecular adsorption and filling occurring within the micropores and can characterize the irregularity and complexity of the pore surface. The D_{N2} ($P/P_0 > 0.45$) usually characterizes the complexity and irregularity of the interior space of the pores, mainly reflecting the capillary cohesion.

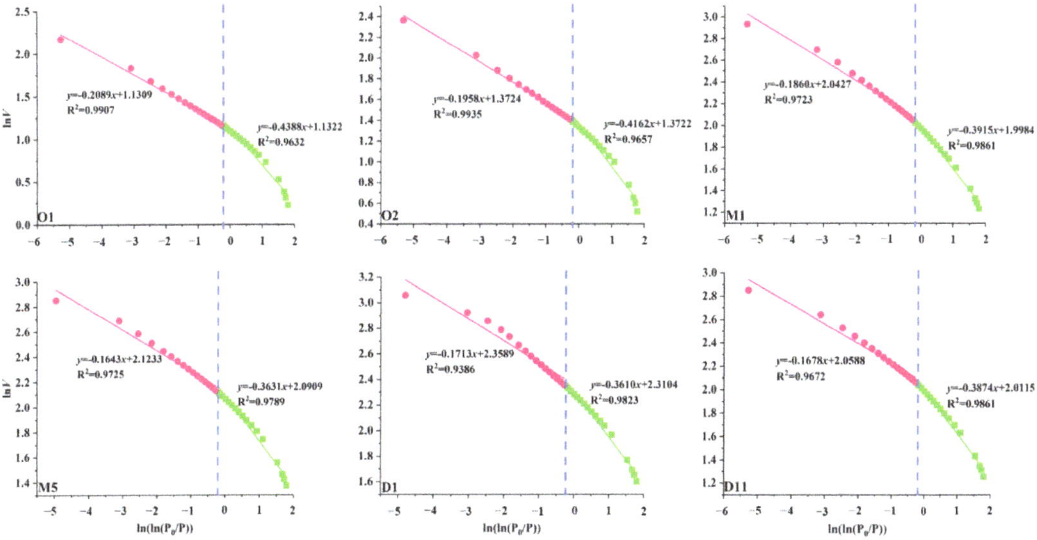

Figure 7. Shale pore fractal fitting curves based on N_2 adsorption experiments. Note: Green represents the fractal dimension D_{N1} of relative pressure < 0.45; red represents the fractal dimension D_{N2} with relative pressure > 0.45; the blue dotted line represents relative pressure = 0.45.

Fractal fitting results are shown in Figure 7. D_{N1} and D_{N2} are 2.5612~2.6390 and 2.7911~2.8357, respectively, and the mean values are 2.6070 and 2.8177, respectively. Both D_{N1} and D_{N2} are close to 3, but D_{N2} is larger than D_{N1}, demonstrating that the inside structure of the pore is more complicated than the surface. The D_{N1} of outcrop and middle–deep shale are 2.5612~2.5838 and 2.6085~2.6390, respectively, and the average values are 2.5725 and 2.6243, respectively. D_{N2} are 2.7911~2.8042 and 2.8140~2.8357, respectively, and the mean values are 2.7977 and 2.8277, respectively. When shale is exposed to the surface, its fractal dimension is at a low level, and the pore structure is relatively simple. However, with the deepening of burial depth, the fractal dimension progressively grows, the pore structure of the shale becomes more intricate, and the heterogeneity is significantly enhanced.

5. Discussion

5.1. Effect of Pore Structure on Fractal Dimension

The intricacy of shale pore structure may be quantitatively described by the fractal dimension, which also reveals the shale pores' potential for gas adsorption. In general, the more complex the pores and the greater the gas adsorption capacity, the bigger the fractal dimension [33]. It was demonstrated that the adsorbed gas content of shale is controlled by SSA, whereas the free gas content is controlled by the PV. The majority of PV is contributed by micropores and mesopores. Micropores contribute significantly more to shale's SSA. Typically, micropores account for less than 20% of PV. Nonetheless, over 80% of SSA may be supplied by micropores [34].

The fractal dimension has a strong positive link with both SSA and PV (Figure 8). This finding is in line with Zhao et al.'s research, which found a favorable correlation between fractal dimension and shale's pore structure [35]. The correlation coefficient between D_C and (DR) SSA is 0.5976, and the correlation coefficients between D_{N1} and D_{N2} and (BET) SSA are 0.9159 and 0.7638, respectively (Figure 8A). The correlation coefficient between D_C and (DA) PV is 0.6063, and the correlation coefficients between D_{N1} and D_{N2} and (BJH) PV are 0.7833 and 0.7364, respectively (Figure 8B). The fractal dimension rises as SSA and PV grow, according to the research above. This demonstrates that the more developed shale micropores and mesopores are, the greater the heterogeneity, the more complex the pores, and the higher the PV and SSA.

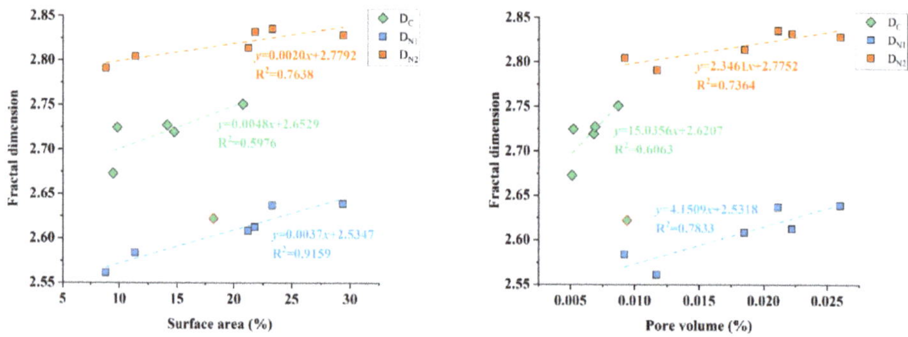

(A) Correlation between fractal dimension and specific surface area

(B) Correlation between fractal dimension and pore volume

Figure 8. Correlation between fractal dimension and SSA (**A**) and PV (**B**) of shale samples.

5.2. Effect of TOC Content and R_o on Fractal Dimension

The content and thermal evolution degree of organic matter, a crucial carrier for the production of organic pores, influence not only the pore shape of organic matter but also the distribution of pore sizes, which in turn influences the fractal dimension. The analysis focuses on the relationships between D_C, D_{N1}, D_{N2}, and TOC. The findings demonstrate a positive relationship between TOC and D_C, D_{N1}, and D_{N2} (Figure 9A), with correlation coefficients of 0.4932, 0.7646, and 0.4871, respectively. This suggests that as TOC rises, so does the complexity of the shale pore structure and its adsorption capacity. Thus, it can be concluded that more organic micropores and mesopores form during the pyrolysis process as TOC rises. Additionally, the higher SSA and rougher surfaces of these pores contribute to their heterogeneity and complex pore structure. As shale's pore structure becomes more complex, its fractal dimension grows, giving gas molecules more adsorption sites and enhancing its storage capacity.

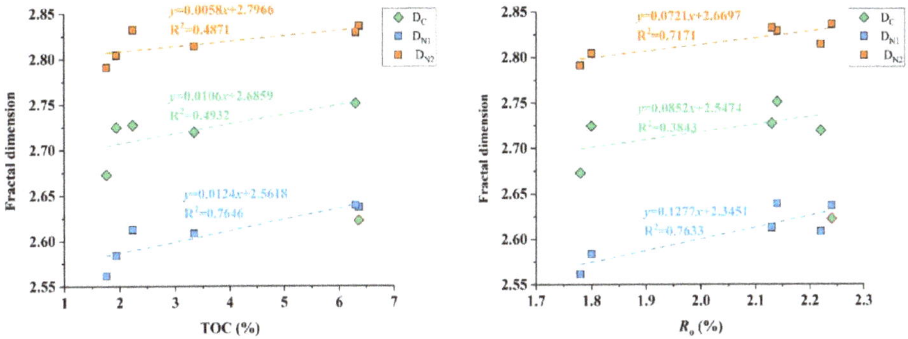

(A) Correlation between fractal dimension and TOC content

(B) Correlation between fractal dimension and R_o

Figure 9. Correlation between fractal dimension and TOC content (**A**) and thermal maturity of organic matter (**B**) in Longmaxi shales.

The morphology and level of development of organic pores in shales are directly linked to the stage of hydrocarbon evolution. The morphology, pore size, and development degree of the organic pores have obvious differences at different stages of thermal evolution. Therefore, the degree of thermal development experienced by organic matter will also become an essential component impacting the change of the fractal dimension of shale. The degree of thermal evolution and D_C, D_{N1}, and D_{N2} are shown to be positively correlated, according to an analysis of the relationship between R_o and fractal dimension (Figure 9B), with correlation coefficients of 0.3843, 0.7633, and 0.7171, respectively. As organic matter reaches the high maturity stage, its ability to produce hydrocarbons increases. Additionally, the organic pores have a simple pore structure and are primarily round and oval in shape. The aromatization of organic matter intensifies in the overmature stage as maturity increases. Additionally, the pore's inner wall gets rougher, and it also reduces the support capacity of the pore. The pore becomes distorted, its connection decreases, its structure becomes complex, and its fractal dimension increases when subjected to the pressure of the rock layer above it [28].

5.3. Effect of Mineral Composition on Fractal Dimension

The degree of pores development is directly linked to the magnitude of the fractal dimension. Thus, the fractal dimension is affected by a number of factors governing shale pores. Shale mineral composition, TOC, degree of thermal evolution, burial depth, and stage of diagenetic development are the primary contributing factors. The diagenetic minerals, as the direct carriers of inorganic pore development, control the development of inorganic pores. This paper mainly analyzes the correlation between fractal dimension and mineral compositions such as quartz, feldspar, and clay minerals (Figure 10). The findings indicate that there is no relationship between quartz content and D_C. There is a weak negative correlation between D_{N1} and D_{N2} and the amount of quartz; their respective correlation coefficients are 0.1342 and 0.4385 (Figure 10A). There is a strong negative correlation between D_C, D_{N1}, and D_{N2} and the amount of feldspar; their respective correlation coefficients are 0.6112, 0.6102, and 0.7818 (Figure 10B). Additionally, D_C, D_{N1}, and D_{N2} exhibit a strong negative correlation with clay mineral content; their respective correlation coefficients are 0.4334, 0.8763, and 0.7952 (Figure 10C).

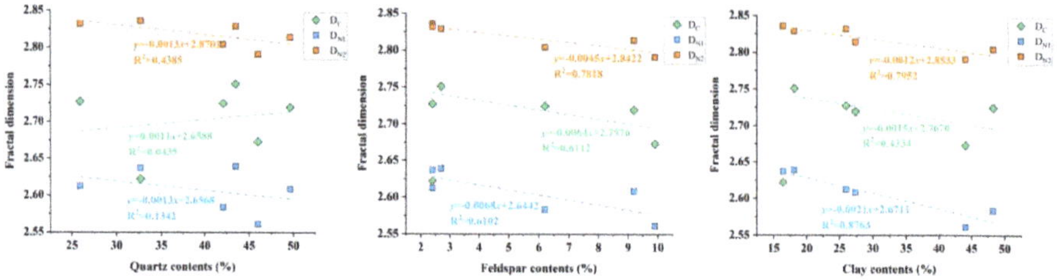

(A) Correlation between fractal dimension and quartz content **(B) Correlation between fractal dimension and feldspar content** **(C) Correlation between fractal dimension and clay content**

Figure 10. Correlation between fractal dimension and mineral composition of Longmaxi shales.

D_{N1} and D_{N2} have a weakly negative correlation with quartz content but no correlation with D_C (Figure 10A). The fractal dimension of mesopores decreases with increasing quartz content, indicating a more homogeneous pore structure with less heterogeneity. The crystalline structure of quartz is often uniform. The fractal dimension of the pore structure is reduced because the pore scale is primarily macropores, and the pores are comparatively more homogeneous [36]. Some researchers have discovered a negative correlation between fractal dimension and quartz content when it comes to the impact of quartz on shale pore development [36,37], which is in line with the findings of this investigation.

As aluminosilicate, feldspar is easily dissolved by fluids under buried conditions. In a buried environment, feldspar, an aluminosilicate mineral, is vulnerable to fluid erosion. As a result, feldspar readily develops into mesopores with enormous pore sizes and macropores when formation fluid dissolution is present. Wang et al.'s research revealed that with the increase of feldspar content, the fractal dimension gradually rose, which reflected the shale pore structure's growing complexity. They hypothesized that this may be the result of many dissolving pores in feldspar forming later on [38]. On the opposite hand, this study's D_C, $D_{N1,}$ and D_{N2} exhibit a strong negative association with its feldspar content (Figure 10B). It is hypothesized that this may be due to the strong brittleness of feldspar. Under the formation conditions, cleavage fractures are developed. The feldspar content and D_C, $D_{N1,}$ and D_{N2} are negatively correlated as a result of this portion of the cleavage fractures increasing the equivalent pore width [35].

The formation of shale pores is significantly influenced by clay minerals, one of the major inorganic minerals [39,40]. According to the correlation study, there is a negative link between the clay mineral content and D_C, $D_{N1,}$ and D_{N2} (Figure 10C). The high amount of clay minerals, their small mineral particles, and their poor compaction resistance are thought to be the possible causes. The clay mineral particles tend to align during long-term compaction diagenesis, which reduces pore space and results in a simple pore shape [16,35].

In clay minerals, illite, I/S mixed layer, and chlorite are formed. As a result, the relationship between fractal dimension and illite, I/S mixed layer, and chlorite content will be analyzed (Figure 11). These findings indicate a negative association between illite content and D_C, $D_{N1,}$ and D_{N2}, with corresponding correlation values of 0.4352, 0.4522, and 0.2187 (Figure 11A). The content of the I/S mixed layer has a positive association with D_C, $D_{N1,}$ and D_{N2}, with corresponding correlation values of 0.6759, 0.8136, and 0.5168 (Figure 11B). D_C, $D_{N1,}$ and D_{N2} have a negative association with the chlorite content; the correlation coefficients are 0.2244, 0.5764, and 0.5499 (Figure 11C). As the amount of illite and chlorite in shale increases, its fractal dimension drops and its pore structure becomes more uniform, consistent with the correlation between fractal dimension and clay mineral content. As a result, clay minerals are influenced by illite and chlorite. The I/S mixed layer, on the other hand, has a positive association with the fractal dimension. This might be due to the incomplete conversion of montmorillonite into illite, and some montmorillonite influences the shale's pore structure. When Nie et al. studied the marine-continental transitional coal-bearing shale, they also looked at the connection between the fractal dimension and the amount of the three clay minerals mentioned above [35]. The findings indicate that there is no association between chlorite and the fractal dimension, but illite and the I/S mixed layer have a positive correlation with the fractal dimension. Different from the relationship between illite, chlorite, and fractal dimension in this study, it is speculated that it may be due to the different research objects. Marine shale is used in this study, and marine-continental transitional shale is the research object.

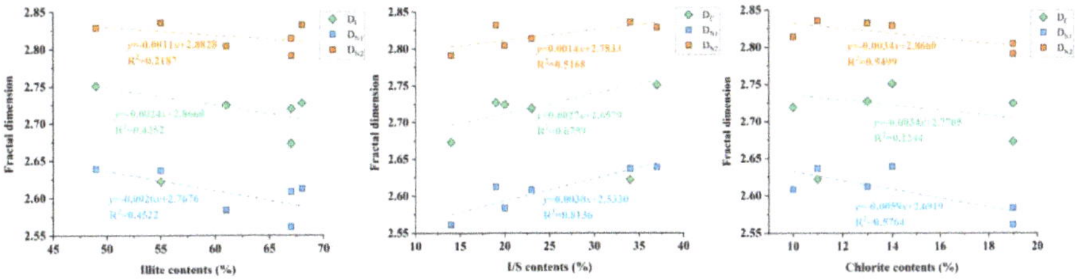

(A) Correlation between fractal dimension and illite content

(B) Correlation between fractal dimension and I/S content

(C) Correlation between fractal dimension and chlorite content

Figure 11. Relationship between clay mineral composition and fractal dimension of Longmaxi shales.

5.4. Effect of Burial Depth on Fractal Dimension and Geologic Significance

The fractal dimension of shale is significantly influenced by burial depth. The shale in the outcrop has a small fractal dimension and a very simple pore structure. The fractal dimension and the complexity of the pore structure increase as the burial depth increases because the middle–deep shale experiences stronger compaction, the pore diameter decreases, the heterogeneity is improved, and the pore surface becomes rougher.

The temperature and overburden pressure of shale will rise in proportion to the burial depth. There is a positive correlation between burial depth and shale's fractal dimension. On the one hand, the shale organic matter moves into the high-over mature stage as the burial depth increases and the earth's temperature rises. The number of irregular organic pores is significantly increased by the high hydrocarbon production of organic matter. However, as compaction is increased, the shale's pore structure is crushed and distorted, increasing the pore structure's complexity and fractal dimension.

With a high degree of thermal evolution and a high organic matter content, the Longmaxi shales in the Weiyuan Block in south Sichuan provide a solid basis for shale gas development. Furthermore, the middle–deep Longmaxi shales in southern Sichuan include well-developed organic pores, including micropores and mesopores, which have the capacity to produce source-generation and source-reservoirs on a small scale.

6. Conclusions

(1) Longmaxi shales have an average TOC of 3.471 wt.%, with values ranging from 1.67 wt.% to 6.36 wt.%. The phase of organic matter maturity is high to over-mature. The deeper the burial, the higher the organic matter content and maturity. Silica, carbonate, and clay minerals make up the majority of the minerals found in shale.

(2) The morphology of organic matter pores gradually becomes irregular and tends to break as the burial depth increases, and the pore size likewise decreases accordingly. In the shale samples, it can be observed that intergranular pores of pyrite, interlayer pores of clay minerals, and tiny cracks have developed. Furthermore, with the increase in burial depth, both SSA (measured by BET and DR methods) and PV (measured by BJH and DA

methods) show an increasing trend, making the pore structure more complex. At the same time, as burial depth increases, the average shale pore size steadily decreases.

(3) In comparison with the outcrop shale, the middle–deep shale has greater fractal dimensions D_C, D_{N1}, and D_{N2}, suggesting a more complex pore structure and more heterogeneity as the buried depth increases. D_C, D_{N1}, and D_{N2} have a positive correlation with the pore structure parameters, TOC and R_o, and a negative correlation with the mineral composition. The more complicated the pore structure, the stronger the pore heterogeneity becomes as SSA and PV rise. Minerals like quartz, feldspar, and clay will slow down the formation of nanopores.

(4) The middle–deep shales are subjected to stronger compaction as the burial depth increases. The pores become smaller, and the heterogeneity is enhanced. Consequently, the fractal dimension and the pore structure's complexity both rise. A reference for estimating the resource potential of deep shale reservoirs in the Weiyuan block is supplied through a comparison of the pore structure and fractal features of shale at various depths, which reveals that the burial depth considerably impacts the fractal dimension of shale.

Author Contributions: Conceptualization, M.Y. and Y.P.; methodology, H.F.; software, Y.P.; validation, Y.L., W.W. and Y.Q.; formal analysis, Y.P.; investigation, H.F.; resources, H.Z.; data curation, Y.P.; writing—original draft preparation, M.Y. and Y.P.; writing—review and editing, Y.P., Q.Y. and H.Z.; visualization, H.F.; supervision, Y.P.; project administration, H.F.; funding acquisition, H.Z. All authors have read and agreed to the published version of the manuscript.

Funding: This research was financially supported by National Natural Science Foundation of China (grant numbers: 42102186, 71991482), Hebei Natural Science Foundation (grant number: D2023203008), Science Research Project of Hebei Education Department (BJK2022018), the Geological Survey Project (grant number: DD20221795), and the Opening Foundation of State Key Laboratory of Continental Dynamics, Northwest University (grant number: 21LCD10).

Data Availability Statement: The original contributions presented in the study are included in the article, further inquiries can be directed to the corresponding authors.

Conflicts of Interest: The authors declare that they have no known competing financial interests or personal relationships that could have appeared to influence the work reported in this paper.

References

1. Zhu, H.J.; Lu, Y.J.; Pan, Y.Y.; Qiao, P.; Raza, A.; Liu, W. Nanoscale mineralogy and organic structure characterization of shales: Insights via AFM-IR spectroscopy. *Adv. Geo-Energy Res.* **2024**, *13*, 231–236. [CrossRef]
2. Zou, C.N.; Yang, Z.; Dai, J.X.; Dong, D.Z.; Zhang, B.M.; Wang, Y.M.; Deng, S.H.; Huang, J.L.; Liu, K.Y.; Yang, C.; et al. The characteristics and significance of conventional and unconventional Sinian-Silurian gas systems in the Sichuan Basin, central China. *Mar. Pet. Geol.* **2015**, *64*, 386–402. [CrossRef]
3. Li, Y.R.; Liu, X.G.; Cai, C.H.; Hu, Z.M.; Wu, B.; Mu, Y.; Duan, X.G.; Zhang, Q.X.; Zeng, S.T.; Guo, J.S.; et al. Pore structure characteristics and their controlling factors of deep shale: A case study of the Lower Silurian Longmaxi Formation in the Luzhou area, southern Sichuan Basin. *ACS Omega* **2022**, *7*, 14591–14610. [CrossRef] [PubMed]
4. Loucks, R.G.; Reed, R.M.; Ruppel, S.C.; Jarvie, D.M. Morphology, genesis, and distribution of nanometer-scale pores in siliceous mudstones of the Mississippian Barnett Shale. *J. Sediment. Res.* **2009**, *79*, 848–861. [CrossRef]
5. Slatt, R.M.; O'Brien, N.R. Pore types in the Barnett and Woodford gas shales: Contribution to understanding gas storage and migration pathways in fine-grained rocks. *AAPG Bull.* **2011**, *95*, 2017–2030. [CrossRef]
6. Rouquerol, J.; Avnir, D.; Fairbridge, C.W.; Everett, D.H.; Haynes, J.M.; Pernicone, N.; Ramsay, J.D.F.; Sing, K.S.W.; Unger, K.K. Recommendations for the characterization of porous solids (Technical Report). *Pure Appl. Chem.* **1994**, *66*, 1739–1758. [CrossRef]
7. Desbois, G.; Urai, J.L.; Kukla, P.A. Morphology of the pore space in claystones–Evidence from BIB/FIB ion beam sectioning and cryo-SEM observations. *eEarth* **2009**, *4*, 1–19. [CrossRef]
8. Bu, H.L.; Ju, Y.W.; Tan, J.Q.; Wang, G.C.; Li, X.S. Fractal characteristics of pores in non-marine shales from the Huainan coalfield, eastern China. *J. Nat. Gas Sci. Eng.* **2015**, *24*, 166–177. [CrossRef]
9. Liu, S.G.; Ye, Y.H.; Ran, B.; Jiang, L.; Li, Z.; Li, J.; Song, J.; Jiao, K.; Li, Z.; Li, Y. Evolution and implications of shale pore structure characteristics under different preservation conditions. *Reser. Eval. Dev.* **2020**, *10*, 1–11.

10. Jiao, K.; Yao, S.P.; Liu, C.; Gao, Y.Q.; Wu, H.; Li, M.C.; Tang, Z.Y. The characterization and quantitative analysis of nanopores in unconventional gas reservoirs utilizing FESEM–FIB and image processing: An example from the lower Silurian Longmaxi Shale, upper Yangtze region, China. *Int. J. Coal Geol.* **2014**, *128*, 1–11. [CrossRef]
11. Zhang, J.Z.; Li, X.Q.; Xie, Z.Y.; Li, J.; Zhang, X.Q.; Sun, K.X.; Wang, F.Y. Characterization of microscopic pore types and structures in marine shale: Examples from the Upper Permian Dalong formation, Northern Sichuan Basin, South China. *J. Nat. Gas Sci. Eng.* **2018**, *59*, 326–342. [CrossRef]
12. Wang, Y.; Han, D.; Lin, W.; Jia, Y.; Zhang, J.; Wang, C.; Ma, B. Factors Controlling Differences in Morphology and Fractal Characteristics of Organic Pores of Longmaxi Shale in Southern Sichuan Basin, China. *Fractal Fract.* **2024**, *8*, 555. [CrossRef]
13. Zhang, J.Z.; Li, X.Q.; Zou, X.Y.; Zhao, G.J.; Zhou, B.G.; Li, J.; Xie, Z.Y.; Wang, F.Y. Characterization of the full-sized pore structure of coal-bearing shales and its effect on shale gas content. *Energy Fuels* **2019**, *33*, 1969–1982. [CrossRef]
14. Gou, Q.Y.; Xu, S.; Hao, F.; Yang, F.; Shu, Z.G.; Liu, R. The Effect of Tectonic Deformation and Preservation Condition on the Shale Pore Structure Using Adsorption-Based Textural Quantification and 3D Image Observation. *Energy* **2021**, *219*, 119579. [CrossRef]
15. Li, H.; Zhou, J.L.; Mou, X.Y.; Guo, H.X.; Wang, X.X.; An, H.Y.; Mo, Q.W.; Long, H.Y.; Dang, C.X.; Wu, J.F.; et al. Pore structure and fractal characteristics of the marine shale of the Longmaxi Formation in the Changning Area, Southern Sichuan Basin, China. *Front. Earth Sci.* **2022**, *10*, 1018274. [CrossRef]
16. Li, X.M.; Wang, Y.R.; Lin, W.; Ma, L.H.; Liu, D.X.; Liu, J.R.; Zhang, Y. Micro-pore structure and fractal characteristics of deep shale from Wufeng Formation to Longmaxi Formation in Jingmen exploration area, Hubei Province, China. *J. Nat. Gas Geosci.* **2022**, *7*, 121–132. [CrossRef]
17. Ma, B.Y.; Hu, Q.H.; Yang, S.Y.; Zhang, T.; Qiao, H.G.; Meng, M.M.; Zhu, X.C.; Sun, X.H. Pore structure typing and fractal characteristics of lacustrine shale from Kongdian Formation in East China. *J. Nat. Gas Sci. Eng.* **2021**, *85*, 103709. [CrossRef]
18. Jia, A.Q.; Hu, D.F.; He, S.; Guo, X.W.; Hou, Y.G.; Wang, T.; Yang, R. Variations of pore structure in organic-rich shales with different lithofacies from the Jiangdong block, fuling shale gas field, SW China: Insights into gas storage and pore evolution. *Energy Fuels* **2020**, *34*, 12457–12475. [CrossRef]
19. Zeng, Q.C.; Chen, S.; He, P.; Yang, Q.; Guo, X.L.; Chen, P.; Dai, C.M.; Li, X.; Gai, S.H.; Deng, Y.; et al. Quantitative prediction of shale gas sweet spots based on seismic data in Lower Silurian Longmaxi Formation, Weiyuan area, Sichuan Basin, SW China. *Pet. Explor. Dev.* **2018**, *45*, 422–430. [CrossRef]
20. Hao, Y.X.; Wu, L.; Jiang, W.; Qian, C.; Zhou, X.; Wang, Y.L. Development characteristics and controlling mechanism of different microfracture combinations in shale reservoir: A case study of silurian longmaxi formation in Weiyuan area. *Pet. Res.* 2024, *in press*. [CrossRef]
21. Zhang, J.R.; Xu, J.T.; Jiang, S.M.; Tang, S. Quantitative identification and distribution of quartz genetic types based on QemScan: A case study of Silurian Longmaxi Formation in Weiyuan area, Sichuan Basin. *Pet. Res.* **2021**, *6*, 423–430. [CrossRef]
22. Hao, Y.X.; Li, Y.Z.; Zhang, Q.; Jiang, W.; Liang, X.S.; Wang, B.Q.; Lu, Y.J.; Zhu, H.J. Reservoir characterization of the shallow to deep Longmaxi formation in the Weiyuan Block, southwestern Sichuan Basin. *Energy Explor. Exploit.* **2023**, *41*, 2058–2077. [CrossRef]
23. Mandelbrot, B.B. Fractal Geometry of Nature, WH Freeman, New York, 1982. *Math. Gaz.* **1984**, *68*, 71–72.
24. Wang, X.Q.; Zhu, Y.M.; Wang, Y. Fractal Characteristics of Micro- and Mesopores in the Longmaxi Shale. *Energies* **2020**, *13*, 1349. [CrossRef]
25. Li, Z.; Liang, Z.K.; Jiang, Z.X.; Gao, F.L.; Zhang, Y.H.; Yu, H.L.; Xiao, L.; Yang, Y.D. The Impacts of Matrix Compositions on Nanopore Structure and Fractal Characteristics of Lacustrine Shales from the Changling Fault Depression, Songliao Basin, China. *Minerals* **2019**, *9*, 127. [CrossRef]
26. Avnir, D.; Jaroniec, M. An isotherm equation for adsorption on fractal surfaces of heterogeneous porous materials. *Langmuir* **1989**, *5*, 1431–1433. [CrossRef]
27. Jacob, H. Disperse solid bitumens as an indicator for migration and maturity in prospecting for oil and gas. *Erdol Kohle Erdgas Petrochem.* **1985**, *38*, 365.
28. Shan, C.A.; Shi, Y.K.; Liang, X.; Zhang, L.; Wang, G.C.; Jiang, L.W.; Zou, C.; He, F.Y.; Mei, J. Diagenetic characteristics and microscopic pore evolution of deep shale gas reservoirs in Longmaxi Formation, Southeastern Sichuan basin, China. *Unconv. Resour.* **2024**, *4*, 100090. [CrossRef]
29. Zhang, Z.N.; Lin, M.R.; Xi, K.L.; Li, J.M. Pore development characteristics and controlling factors of Lower Cambrian Qiongzhusi shale gas reservoir in Sichuan Basin, China. *Unconv. Resour.* **2024**, *4*, 100100. [CrossRef]
30. Shao, X.H.; Pang, X.Q.; Li, Q.W.; Wang, P.W.; Chen, D.; Shen, W.B.; Zhao, Z.F. Pore structure and fractal characteristics of organic-rich shales: A case study of the lower Silurian Longmaxi shales in the Sichuan Basin, SW China. *Mar. Pet. Geol.* **2017**, *80*, 192–202. [CrossRef]
31. Everett, D.H.; Koopal, L.K. International union of pure and applied chemistry. *Polymer* **2001**, *31*, 1598.

32. Thommes, M.; Kaneko, K.; Neimark, A.V.; Olivier, J.P.; Rodriguez-Reinoso, F.; Rouquerol, J.; Sing, K.S.W. Physisorption of gases, with special reference to the evaluation of surface area and pore size distribution (IUPAC Technical Report). *Pure Appl. Chem.* **2015**, *87*, 1051–1069. [CrossRef]
33. Yang, Y.Y.; Zhang, J.C.; Xu, L.F.; Li, P.; Liu, Y.; Dang, W. Pore structure and fractal characteristics of deep shale: A case study from Permian Shanxi Formation shale, from the Ordos Basin. *Acs Omega* **2022**, *7*, 9229–9243. [CrossRef] [PubMed]
34. Nie, W.C.; Zhang, T.S.; Wang, M.W.; Wu, W.; Tan, X.C. Fractal Characteristics and Interfering Factors of Microscopic pores in Marinecontinental Transitional Coal Shale: A case study of the Taiyuan Formation in the northern Qinshui Basin. *Acta Sedimentol. Sin.* **2024**, *42*, 1047–1057. (In Chinese)
35. Zhao, Y.; Li, L.; Si, Y.H.; Wang, H.M. Fractal Characteristics and Controlling Factors of Pores in Shallow Shale Gas Reservoirs: A Case Study of Longmaxi Formation in Zhaotong Area, Yunnan Province. *J. Jilin Univ. (Earth Sci. Ed.)* **2022**, *52*, 1813–1829. (In Chinese)
36. Zhang, J.Z.; Xiao, X.; Wang, J.G.; Lin, W.; Han, D.L.; Wang, C.C.; Li, Y.; Xiong, Y.; Zhang, X.C. Pore structure and fractal characteristics of coal-bearing Cretaceous Nenjiang shales from Songliao Basin, Northeast China. *J. Nat. Gas Geosci.* **2024**, *9*, 197–208. [CrossRef]
37. Li, A.; Ding, W.L.; He, J.H.; Dai, P.; Yin, S.; Xie, F. Investigation of pore structure and fractal characteristics of organic-rich shale reservoirs: A case study of lower Cambrian Qiongzhusi formation in Malong block of Eastern Yunnan province, south China. *Mar. Pet. Geol.* **2016**, *70*, 46–57. [CrossRef]
38. Wang, D.X.; Zeng, Z.P.; Hu, H.Y.; Wang, T.; Zhu, G.G.; Li, S.T.; Li, L. Fractal characteristics of pore structure of deep continental shale of Lower Wuerhe Formation in central Junggar Basin and its geological significance. *Pet. Geol. Recovery Effic.* **2024**, *31*, 23–35. (In Chinese)
39. Gao, M.; Yang, M.; Lu, Y.; Levin, V.A.; He, P.; Zhu, H. Mechanical characterization of uniaxial compression associated with lamination angles in shale. *Adv. Geo-Energy Res.* **2024**, *13*, 56–68. [CrossRef]
40. Zhu, H.; Huang, C.; Ju, Y.; Bu, H.; Li, X.; Yang, M.; Chu, Q.; Feng, H.; Qiao, P.; Qi, Y.; et al. Multi-scale multidimensional characterization of clay-hosted pore networks of shale using FIBSEM, TEM, and X-ray micro-tomography: Implications for methane storage and migration. *Appl. Clay Sci.* **2021**, *213*, 106239. [CrossRef]

fractal and fractional

MDPI

Article

Pore Structure and Fractal Characteristics of Inter-Layer Sandstone in Marine–Continental Transitional Shale: A Case Study of the Upper Permian Longtan Formation in Southern Sichuan Basin, South China

Jianguo Wang [1,2], Jizhen Zhang [1,2,*], Xiao Xiao [1,2], Ya'na Chen [3] and Denglin Han [4]

1 School of Resource and Environment, Yangtze University, Wuhan 430100, China
2 Hubei Key Laboratory of Complex Shale Oil and Gas Geology and Development in Southern China, Yangtze University, Wuhan 430100, China
3 PetroChina Research Institute of Petroleum Exploration & Development, Beijing 100083, China
4 Laboratory of Reservoir Microstructure Evolution and Digital Characterization, Yangtze University, Wuhan 430100, China
* Correspondence: zhangjz@yangtezu.edu.cn or zhangjz1991@126.com

Abstract: With the evolution of unconventional oil and gas exploration concepts from source rocks and reservoirs to carrier beds, the inter-layer sandstone carrier bed within marine–continental transitional shale strata has emerged as a significant target for oil and gas exploration. The inter-layer sandstone is closely associated with the source rock and differs from conventional tight sandstone in terms of sedimentary environment, matrix composition, and the characteristics of reservoir microscopic pore development. Preliminary exploration achievements display that the inter-layer sandstone is plentiful in gas content and holds promising prospects for exploration and development. Consequently, it is essential to investigate the gas-rich accumulation theory specific to the inter-layer sandstone reservoir in transitional facies. Pore development characteristics and heterogeneity are crucial aspects of oil and gas accumulation research, as they influence reservoir seepage performance and capacity. This paper employs total organic carbon analysis, X-ray diffraction, rock thin section examination, field emission scanning electron microscopy, physical analysis, high-pressure mercury intrusion analysis, gas adsorption experiments, and fractal theory to explore the reservoir development characteristics of the sandstone samples from the Upper Permian marine–continental transitional facies Longtan Formation in the southern Sichuan Basin. It also attempts to combine high-pressure mercury intrusion analysis and gas adsorption experiments to describe the structural and fractal characteristics of pores at different scales in a segmented manner. The findings reveal that the sandstone type of the Longtan Formation is mainly lithic sandstone. The pore size distribution of the sandstone primarily falls below 30 nm and above 1000 nm, with the main pore types being inter-granular pores and micro-fractures in clay minerals. The pore volume and specific surface area are largely attributed to the micropores and mesopores of clay minerals. The pore morphology is complex, exhibiting strong heterogeneity, predominantly characterized by slit-like and ink bottle-like features. Notably, there are discernible differences in reservoir structural characteristics and homogeneity between muddy sandstone and non-muddy sandstone. The pore morphology is complex, exhibiting strong heterogeneity, predominantly characterized by slit-like and ink bottle-like features. Notably, there are discernible differences in reservoir structural characteristics and homogeneity between muddy sandstone and non-muddy sandstone.

Keywords: pore structure; fractal; marine–continental transitional; Longtan formation; shale inter-layer sandstone

1. Introduction

As an important alternative field for natural gas resources, marine–continental transitional shale formations have received increasing attention in China. Marine–terrestrial transitional strata in Carboniferous–Permian strata are widely distributed across major basins in China [1–3]. China has substantial marine–terrestrial transitional shale gas resources, with the proven reserves amounting to 19.8×10^{12} m^3 as of 2021, accounting for a quarter of the total shale gas resources in China, highlighting its significant exploration potential [4–6].

Due to the complex sedimentary environment, the transition phase formation often forms a natural gas superimposed system with multiple types of shale gas, coalbed methane, and sandstone gas. However, previous studies on marine–terrestrial transition mostly focused on shale strata, with insufficient emphasis placed on the inter-layer sandstone [7–10].

Guo et al., 2023, combined with previous experience, achieved a major breakthrough of natural gas in the sandstone carrier bed of Qiongzhusi Formation in the Sichuan Basin based on the theory of hydrocarbon accumulation in the carrier bed, which promoted the evolution of unconventional gas exploration target from hydrocarbon source rock and reservoir to the carrier bed, and the carrier bed can also be used as reservoir [11]. This has guided the exploration process of natural gas in marine–continental transitional shale formation, with carrier beds in shale strata, such as sandstone, emerging as a novel field for natural gas exploration and development.

Inter-layer sandstone of transitional facies is different from conventional tight sandstone. On the macroscopy, sedimentary environment is the most basic factor controlling reservoir performance [12–14]. The transitional strata are seriously affected by the change of sedimentary facies in the vertical direction, which often presents the phenomenon of frequent superposition of shale, sandstone, coal, and other lithology [9,15,16]; it can even be seen that the inversion of sedimentary microfacies occurs under the frequent transgression and regression (frequent inter-bedding of shale and sandstone) [17]; the heterogeneity of the transitional strata on the plane is very strong, and adjacent wells often show completely different lithology combinations [1]. The conventional tight sandstone is mostly developed in the river delta and channel microfacies of the continental strata. The multi-stage sandstone are superimposed on each other in the vertical direction, and are constantly distributed in a large range on the plane [13,18–20]. On the microscopy, the content and spatial distribution characteristics of matrix components affect the microscopic pore development characteristics of the reservoir [21–24]. Due to the thin inter-bedded with source rocks or surrounded by source rocks, the organic debris residues are commonly distributed in the components of the transitional sandstone, which makes the sandstone have certain hydrocarbon generation potential as well as the source rocks [6,9]. The conventional tight sandstone is thickly inter-bedded with the source rock or deposited on the source rock, which is less affected by the source rock, and there is no or rare organic debris residue in the component [25]. Whether from the macroscopic or microscopic point, the marine–continental transitional shale inter-layer sandstone is quite different from the conventional tight sandstone, and has certain particularity. Therefore, it is necessary to research the transitional inter-layer sandstone reservoir to clarify its microscopic pore development characteristics. This will contribute to the gas-rich accumulation mechanism of sandstone and adjacent shale and coal seams, and to offer a number of theoretical experiences for the exploration and development of unconventional natural gas.

Reservoir development characteristics (physical properties, pore development characteristics, pore structure characteristics and pore heterogeneity, etc.) affect the reservoir and seepage performance of the reservoir, which have always been the most basic content of reservoir research [2,18,22,26,27]. Tight reservoirs typically exhibit intricate pore structures (nano to micron pores), along with distinctive physical characteristics [27–29]. To effectively articulate the development features of reservoirs, galore quantitative and qualitative technical methodologies have been invented. A range of scanning electron microscopy (SEM) techniques and nanoelectron computed tomography (CT) scans are frequently employed to qualitatively assess the characteristics of pore development, including type, morphology, and size. Additionally, high-pressure mercury intrusion analysis (HPMI) and nitrogen (N_2) and carbon dioxide (CO_2) adsorption experiments are utilized to quantitatively evaluate various parameters related to pore structure characteristics, such as pore size distribution (PSD), pore volume (PV), and specific surface area (SSA), among others [19,21,30]. The pore heterogeneity present in marine–continental transitional shale reservoirs is significantly pronounced, which will inevitably influence the microscopic properties of the reservoir [31,32]. In order to study its mechanism of action, it is urgent to be able to quantitatively characterize the pore heterogeneity. With the development of fractal geometry theory, scholars have gradually applied it to the oil and gas exploration. Studies indicate that the rock pore throat system exhibits a complex geometric structure, which can be effectively analyzed through the fractal theory [33–37]. The fractal dimension (D) can be used as a quantitative index to characterize the pore heterogeneity. The D between 2 and 3 can effectively characterize the heterogeneity of rock pores. The closer the fractal dimension value is to 2, the more regular the pore morphology is, the stronger the rock homogeneity is, and vice versa [18,34]. The fractal dimension of pore can be computed using experimental methods such as SEM, nuclear magnetic resonance (NMR), nuclear magnetic resonance cryoporometry (NMR-C), gas adsorption (GA), and HPMI [2,3,19,38–41].

The Upper Permian Longtan Formation in the central and southern Sichuan Basin is a famous set of marine–continental transitional strata in China [42,43]. Past research on it has mostly focused on the shale strata. The Longtan formation has developed rich inter-layer sandstone [44], and recent exploration well results have found that the sandstone also contains abundant natural gas resources, showing great exploration potential. However, there has not yet been any clear exploration progress. To clearly understand the pore structure characteristics and heterogeneity of the inter-layer sandstone in the Longtan Formation, this paper analyzes the collected sandstone samples using methods such as total organic carbon (TOC) analysis, rock thin sections, X-ray diffraction (XRD), rock porosity and permeability analysis, field emission scanning electron microscopy (FE-SEM), HPMI, GA experiments, and fractal theory. It also attempts to combine HPMI and gas adsorption experiments to describe the structural and fractal characteristics of pores at different scales in a segmented manner. This approach has not been reported in recent articles related to sandstone reservoirs, demonstrating innovation of research.

2. Materials and Methods

2.1. Geological Setting and Samples

The Sichuan Basin, located in the western region of the Yangtze Platform, represents one of the crucial petroliferous basins in China [42,43]. Many complex tectonic movements in the geological history period have resulted in the occurrence of six structure areas in the present Sichuan Basin. The research area is mainly located in the low-steep fold belt of southern Sichuan in the southwestern Sichuan Basin (Figure 1a) [2,44,45].

The Dongwu movement, occurring during the transition from the Middle to Late Permian period, resulted in the Sichuan Basin developing a paleogeographic configuration characterized by high southwest and low northeast (West Land, East Sea). Notably, there was a pronounced sedimentary differentiation observed within the strata of the late Permian. This differentiation is specifically evidenced by the shift from marine strata in the northeastern region to continental strata in the southwestern region. The Longtan Formation marine–continental transitional coal-bearing shale strata has been extensively deposited in the central to southern parts of the Sichuan Basin [2,46,47].

In the process of sequence stratigraphy, it was determined that the Upper Permian Longtan Formation (P_3l) in the Sichuan Basin has experienced two transgression–regression cycles, which divided the stratigraphy of the Longtan Formation into three segments came from below. The lithology of the Longtan Formation in southern Sichuan is dominated by dark gray to gray black mud shale, mud shale inter-bedded with siltstone, thin-bedded sandstone, and coal seams [44].

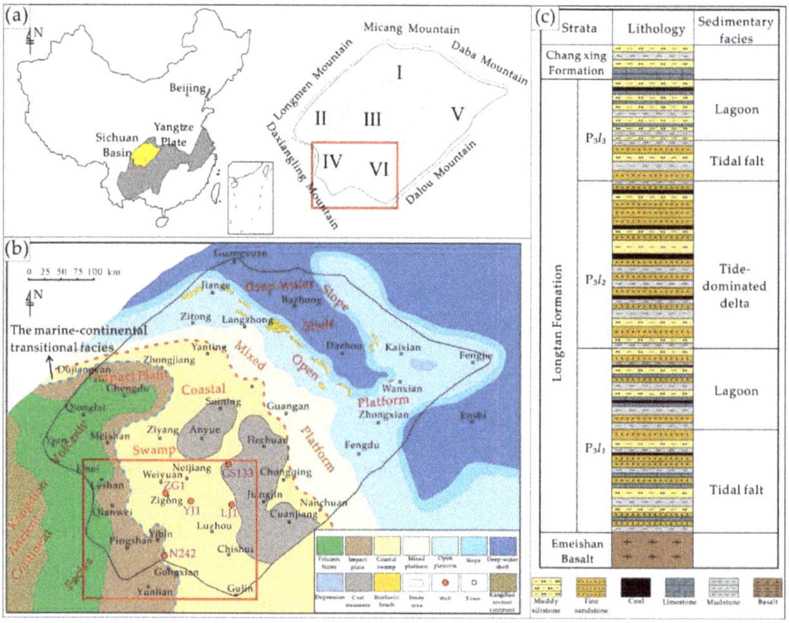

Figure 1. (**a**) Structural background map of Sichuan Basin [2]; (**b**) paleogeographic map and Sampling Well Locations of the Longtan Formation in the Study Area [47]; and (**c**) stratigraphic Column of Longtan Formation [47].

The paleogeographic map of the Longtan Formation reveals the distribution of sedimentary facies in the Permian era (Figure 1b) [2]. The transitional strata of Longtan Formation were mainly in the coastal swamp facies. The continual changes in sedimentary microfacies have brought about obvious lithological alteration in the Longtan Formation, and the southwest direction of the Sichuan Basin was greatly affected by the input of terrigenous materials, and the sandstone inter-layer was richly developed, and the grain size of the sand was coarse [44].

Twelve fresh sandstone samples were selected from the cores of the Longtan Formation 1–3 sections (depth of 2948.5–3907.39 m) of five wells (ZG1, YJ1, LJ1, GS133, and

N242) in the study area (southern Sichuan Basin). After sealing and packaging, they were promptly transported to the laboratory for subsequent experimental analysis. All experiments were conducted in the laboratory of Yangtze University. According to the categorization scheme of sandstone grain size [48], the sandstone samples of Longtan Formation are classified. There are five types of Longtan Formation sandstone: medium sand (grain size 0.25–0.5 mm), fine sand (grain size 0.125–0.25 mm), very fine sand (grain size 0.0625–0.125 mm), coarse silt (grain size 0.0315–0.0625 mm), and fine silt (grain size 0.0039–0.0078 mm). The sandstone of the Longtan Formation generally contains muddy component, so the sandstone can be further divided into muddy sandstone and non-muddy sandstone according to the amount of muddy content.

2.2. Total Organic Carbon (TOC) Analysis

Firstly, the selected samples were smashed into particles below 200 mesh (80 μm) with the Chinese national standard T/CNIA 0137-2022, the "Relationship between the mesh size of the test sieve and the number of mesh openings" [1], part of which was used for TOC analysis and the other part for mineralogical analysis.

The appropriate amount of particle samples was selected and washed several times with 10% hydrochloric acid to take out the inorganic carbon and water in the sample, and then dried at 80 °C. Finally, the TOC was measured by LECO-230 carbon sulfur analyzer that produced by the American company LECO at the Yangtze University, and the instrument has an analysis error of $\leq$0.4% for TOC content [3]. The experiment was conducted in accordance with the Chinese national standard (GB/T 19145-2022), the "Determination of Total Organic Carbon in Sedimentary Rock" [1,29].

2.3. X-Ray Diffraction (XRD) and Thin Sections

The crushed samples were blended with ethanol, ground, and smeared on a glass slide to create a stochastically oriented powder preparation. The Bragg–Brentano geometric reflection mode analysis was performed using a Brook D8 Advanced diffractometer from the German company Bruker at the Yangtze University, and the X-ray diffraction pattern of the sample was recorded. The standard for quantitative analysis of mineral content came from the standard powder diffraction analysis database provided by the Joint Committee on Power Diffraction-International Center for Diffraction Data (JCPDS–ICDD) [1]. The Cu Kα-radiation was operated at 35 mA and 45 kV, and the step-by-step scanning measurement was performed from 2° to 70° at a speed of 4°/min and the measurement environment were humidity of 40% and temperature of 24 °C. In order to assure the accuracy of the collected rock composition data, at least 50 points were counted in each slice [13,49]. The experiment followed the Chinese oil and gas industry standard (SY/T 5163-2018), the "Analysis method for clay minerals and ordinary non-clay minerals in sedimentary rocks by the x-ray diffraction" [1,29].

Rock thin sections can be used to study mineralogical characteristics, reservoir space types, and the types and idiosyncrasy of mineral growth. The thin section photos used in this article were taken by the BX51 metallographic microscope produced by the Japanese company Olympus at the Yangtze University. The light source was a 12 V/100 W halogen lamp, and the optical acquisition system was provided by Olympus Soft Imaging Solutions GmbH's Olympus Stream [18].

2.4. Rock Property Analysis

Porosity and permeability were measured by KXD-II porosity instrument produced by the Chinese company Nantong Huaxing Petroleum Instrument at the Yangtze University, and the measurement inaccuracy of the instrument is $\leq$0.5%. The principle is to place the rock sample in a smaller container, apply high pressure at one end of the sample, and record the time of pressure dissipation. The permeability and porosity of the samples were calculated by recording the process of pressure depletion with time. The experiment was performed under the requirements of pressure of 9.68 MPa, temperature of 25 °C, and dry nitrogen as the carrier gas. Porosity and permeability were measured according to the Chinese national standard of the GB/T 29172–2012, the "Practices for core analysis" [3,18].

2.5. Field Emission Scanning Electron Microscopy (FE-SEM) Analysis

FE-SEM analysis is often used to elucidate the micro pore characteristics in tight reservoirs. Before observation, the samples were first ground and polished with Ar ions using the IB09010CP ion section polisher from the Japanese company JEOL at the Yangtze University, and then a 10 nm thick gold film was deposited on the surface to improve the electric conductivity, in order to observed the pore space more clearly. Then, the FE-SEM pore images of the samples were generated by using a S3400 scanning electron microscope produced by the Chinese company Tianmei Yituo Laboratory Equipment (Shanghai, China) with a maximum resolution of 4 nm under the conditions of an acceleration voltage of 20 keV and a working distance of 8–9 mm. The experimental analysis was performed according to the Chinese oil and gas industry standard SY/T 5162–2021, the "Analytical method of rock sample by scanning electron microscope" [1,2].

2.6. High-Pressure Mercury Injection (HPMI) and GA Experiments

HPMI as a direct and effective pore structure characterization method, and has been widely used in the evaluation of tight reservoirs. Its effective measurement range is 1.8–100,000 nm, which is often used to determine the pore structure characteristics of macropores (>50 nm). The HPMI test was carried out using a AutoPore V 9600 Mercury injection apparatus from the American company Micro Instruments at the Yangtze University, and the measurement error of the apparatus is $\leq$0.05%. The measurement was performed under the condition of interfacial tension of 485 dyn/cm and wet contact angle of 130°, and the minimum scale of measurement was 3 nm. The experiment followed the Chinese national standard GB/T 21650.3-2011, the "Pore size distribution and porosity of solid materials by mercury porosimetry and gas adsorption. Part 3:Analysis of micropores by gas adsorption" [30,35].

Gas adsorption experiments were performed with respect to CO_2 adsorption and N_2 adsorption. N_2 adsorption experiments are commonly used to determine the pore structure characteristics of mesopores (2–50 nm), while CO_2 adsorption experiments are used to survey the characteristics of micropores (<2 nm). Firstly, the sample was crushed below 250 mm, and 120 g of the sample was weighed and dried at 60 °C for 48 h to eliminate bound water and capillary water. Then, the low-temperature N_2 and CO_2 adsorption–desorption experiments were carried out using BSD-660M A6B6M Surface Area Analyzer and Pore Size Analyzer from the Chinese company Beishide Instrument Technology (Bei jing, China), and the measurement inaccuracy of the analyzer is $\leq$1%. The experimental pressure range was 0–101.3 kPa [2]. The analysis of mesoporous PV and SSA are based on the nonlocal density functional theory (NLDFT) and Barrett–Joyner–Halenda (BJH) models, while the analysis of microporous PV and SSA is based on the Dubinin–Radushkevich (DR) theory [50–52].

2.7. Fractal Analysis

The purpose of this study is to explore the heterogeneity characteristics of various scales of inter-layer sandstone in detail. Therefore, the fractal dimension of macropores and ultra-large macropores were determined by high-pressure mercury intrusion method, and the fractal dimension of mesopores and micropores were calculated by N_2 gas adsorption method.

Predecessors have carried out a lot of research on the method of calculating the fractal dimension of HPMI capillary pressure curve [33,35,39]. The double logarithmic curve relationship between S_{Hg} and P_c can be used to calculation the fractal dimension of pores. The method is laid out in the following equations:

$$\lg(S_{Hg}) = (D-3)\lg(P_0) - (D-3)\lg(P_c) \tag{1}$$

$$D = 3 + k \tag{2}$$

Among them, S_{Hg} is the mercury saturation corresponding to capillary pressure P_c, and P_0 is the displacement pressure, k is the slope of the double logarithmic curve, and D is the fractal dimension.

Many methods have been built to calculate the fractal dimension of porous media by gas adsorption experiments, such as thermal thermodynamics, fractal Brunauer–Emmett–Teller (BET), and fractal Frenkel–Halsey–Hill (FHH). The FHH model is considered to be the most effective method to calculate the fractal dimension based on gas adsorption experiments [3,26,41,53–55]. The method is laid out in the following equations:

$$\ln(V/V_0) = A[\ln(\ln(P_0/P))] + C \tag{3}$$

$$D = 3 + A \tag{4}$$

Among them, V is the volume of adsorbed gas molecules at balance pressure P, V_0 is the volume of monolayer molecular coverage, P_0 is the saturation pressure of the gas, A is the slope of the line, C is a constant, and the fractal dimension D can be acquired by A.

3. Results

3.1. TOC, and Petrological and Physical Characteristics

The TOC results of the shale inter-layer sandstone in the Longtan Formation are shown in the Figure 2. The TOC is distributed between 0.53% and 2.3% (average of 1.15%). Although the TOC content of sandstone is not as high as that of shale in Longtan Formation, it also has certain hydrocarbon generation potential [2,17,46].

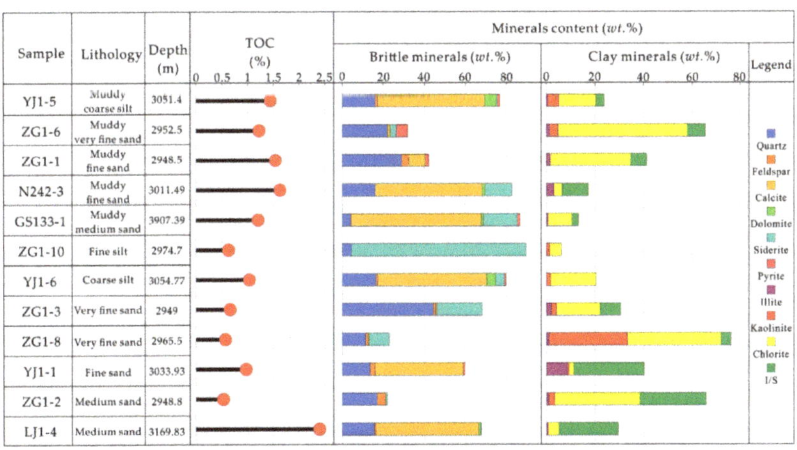

Figure 2. Comparison diagram of sandstone component content of Longtan Formation.

Combined with XRD and rock thin section analysis, it can be seen that the Longtan Formation sandstone in southern Sichuan is different from the conventional tight sandstone, which is mainly lithic sandstone (Figures 2 and 3b). The X-ray diffraction pattern of the sample is shown in Figure 3a. The content of quartz ranges from 4.16 to 44.43 wt% (average of 17.24 wt%), the average content of feldspar is only 1.47%, and the rock fragments value is more than 50% (average of 81.41 wt%). The type of rock debris is mainly sedimentary rock, followed by metamorphic rock, and the content of magmatic rock is the least. The content of clay minerals in the sandstone of the Longtan Formation ranges from 5.96 to 76.08% (average of 37.19 wt%). Chlorite is rich in development, with an average of 20.92 wt%. The average value of the illite/smectite mixed layer (I/S) is 10.61 wt%. The content of illite and kaolinite are both below 5 wt%. The distribution range of carbonate minerals dominated by calcite is between 0.32 and 85.26 wt% (average 39.4 wt%). A small amount of pyrite and other minerals are also developed in the sandstone of Longtan Formation, with an average content of less than 5 wt%.

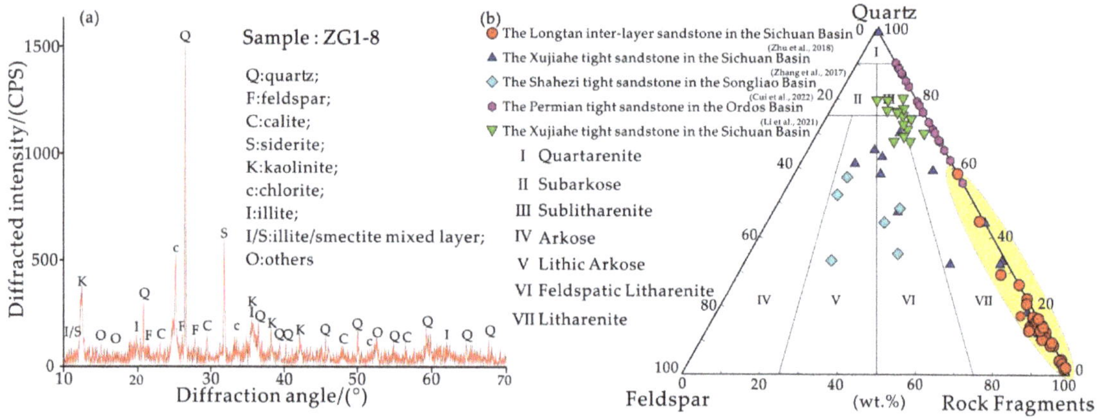

Figure 3. (**a**) XRD pattern; and (**b**) sandstone type triangle diagram of Longtan Formation [18,27,41,55].

In the identify of thin sections, the sandstone samples of Longtan Formation show the directional arrangement of minerals (Figure 4b), the linear contact relationship between minerals (Figure 4a), and the rigid fracture of brittle minerals (Figure 4a), indicating that the sandstone has experienced strong compaction. In the muddy sandstone samples, the widely distributed muddy components (mainly clay minerals and TOC) are usually banded and agglomerated (Figure 4b,c). The distribution of quartz particles is scattered, mostly silt grade, with smooth surface and irregular shape (Figure 4d). Feldspar is mostly corroded and irregular in shape (Figure 4c). Calcite, dolomite, and other inter-particle cements can be seen locally in the thin section (Figure 4b). A large amount of siderite is developed in some siltstone samples, and siderite is distributed in lumps, spheres, or ooids, or filled in inter-granular in the form of cement (Figure 4e). In addition, abundant banded and agglomerated organic carbonaceous fragments can be seen in the muddy sandstone thin sections, which are often filled with different kinds of minerals or other components (Figure 4f).

Porosity and permeability are important characteristics that affect the seepage capacity of reservoirs and represent the quality of pore connectivity. The results of porosity and permeability test are as follows (Figure 5): The porosity of sandstone in Longtan Formation is distributed between 1.08 and 8.02% (average of 3.52%). The distribution range of permeability is 0.014–3.099 mD (average of 0.497 mD). The muddy sandstone of Longtan Formation generally shows abnormally high permeability, especially the muddy fine sandstone, which is up to 3.099 mD. With the increase or decrease in grain size, the abnormal degree of permeability gradually decreases.

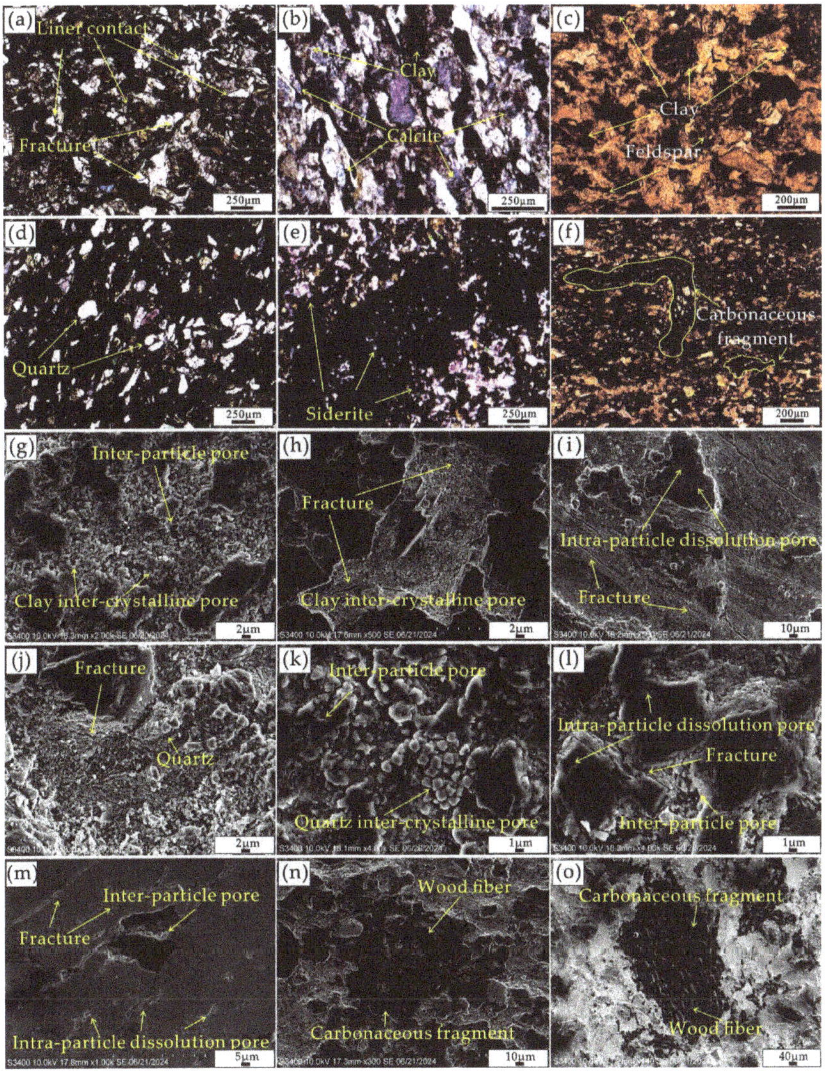

Figure 4. (**a**–**f**) are sandstone thin section identification of Longtan Formation: (**a**) the linear contact relationship of minerals and brittle minerals micro-fracture, sample ZG1-8, orthogonal light; (**b**) directional arrangement of minerals and cementation of carbonate minerals, sample ZG1-6, orthogonal light; (**c**) muddy components and feldspar dissolution, sample N242-3; (**d**) authigenic quartz, sample ZG1-1, orthogonal light; (**e**) siderite enrichment, sample ZG1-3, orthogonal light; and (**f**) carbonaceous fragment filled with minerals components, sample N242-3. (**g**–**o**) are pore types of sandstone reservoir in Longtan Formation: (**g**) inter-particle pores and inter-crystalline pores of clay minerals, sample GS133-1; (**h**) clay minerals micro-fracture, sample ZG1-1; (**i**) brittle minerals intra-particle dissolution pores, sample ZG1-2; (**j**) quartz distributed inside clay minerals, sample N242-3; (**k**) inter-crystalline pores in brittle mineral aggregates, sample YJ1-6; (**l**) brittle minerals inter-particle fracture, sample GS133-1; (**m**) brittle minerals micro-fracture, sample ZG1-1; and (**n**,**o**), carbonaceous fragment and wood fiber filled in it, sample ZG1-1 and ZG1-8.

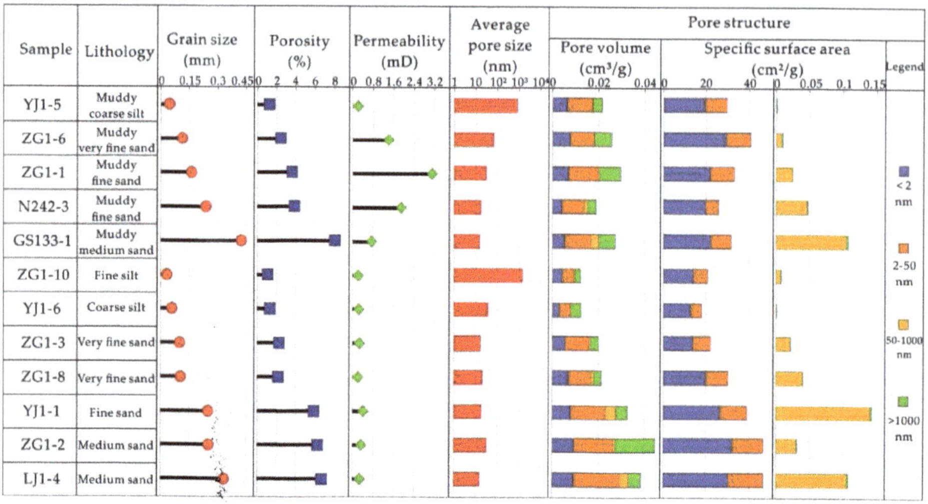

Figure 5. Comparison diagram of pore structure parameters of sandstone in Longtan Formation.

3.2. Pore Types and SEM Results

Based on the field scanning electron microscope of argon ion polishing, the development characteristics of pore type, morphology, and scale of Longtan Formation sandstone were analyzed. It is found that the Longtan Formation sandstone mainly develops pores and fractures of inorganic minerals, such as inter-crystalline pores of clay minerals, inter-particle pores of minerals, dissolution pores of brittle minerals, and micro-fractures.

The inter-crystalline pores of clay minerals are mainly developed in the authigenic clay mineral aggregates filled in the original inter-particle pores. Such pores in muddy sandstone are rich in development, and the pores are distributed from the nanometer to micrometer scale, and the shapes are mostly irregular multilateral shapes (Figure 4g,h). Residual authigenic quartz and other minerals can be seen in clay minerals (Figure 4j).

There are abundant inter-particle pores developed between brittle minerals (quartz, feldspar, carbonate minerals, etc.) and clay minerals. These pores have a wide range of distribution and have different shapes, and most of them are residual pores formed by the compaction of primary inter-particle pores and filling of other minerals (Figure 4k,m). When the inter-particle pores diameter is greater than 5 μm, they are classified as micro-fractures. The aggregates of quartz and other minerals can be seen under the microscope, and the inter-crystalline pores are developed in the aggregate. These pores are mostly below the micron level in size, and most of them are triangular and irregular in shape (Figure 4k).

Intra-particle dissolution pores are developed in brittle minerals such as feldspar. The pore size of dissolution pores is relatively small, mostly on a nanoscale level, and the morphology is relatively regular, round or oval (Figure 4l).

Micro-fractures are great percolation channels in tight reservoirs, and micro-fractures are abundant in Longtan Formation sandstone. With regard to the micro-fractures in clay minerals, some are inter-layer structural fractures formed under tectonic action, while some are shrinkage fractures formed due to volume shrinkage during the transformation of illite to smectite (Figure 4h–j). There are micro-fractures produced by mineral break under compaction in brittle minerals. The overall length of these micro-fractures is not as long as that of clay minerals, but their morphology is more regular (Figure 4m).

Carbonaceous fragments can also be observed in some muddy sandstone samples, and the remains of plant wood fiber with relatively stable structure and various minerals can be seen in carbonaceous fragments. The distribution area of carbonaceous fragments is large, the diameter is tens or even hundreds of microns, and the shapes are various (Figure 4n); the darker carbonaceous fragments indicate a more developed coastal swamp environment from higher plants (Figure 4o).

3.3. Pore Structure

Due to the influence of the surrounding shale formation, the change in pore structure characteristics of sandstone reservoir will be very complicated. The previous pore size division scheme of tight sandstone and shale cannot accurately characterize each pore. Therefore, combined with the previous pore size division scheme, on the basis of International Union of Pure and Applied Chemistry (IUPAC) pore size division [3,35], the pore size of Longtan Formation sandstone is divided into the following four types: micropore (<2 nm), mesopore (2–50 nm), macropore (50–1000 nm), and ultra-large macropore (>1000 nm).

3.3.1. Characteristic of HPMI Curves

The HPMI curves of sandstone in Longtan Formation have various forms, and four common forms are shown in Figure 6.

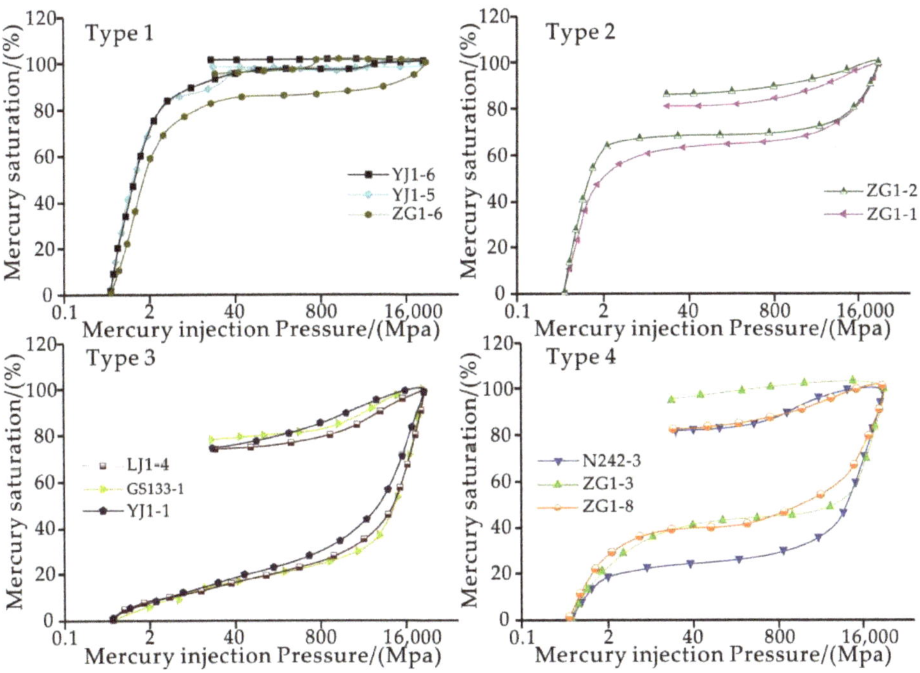

Figure 6. Mercury saturation curves of different types of Longtan formation sandstone.

Type 1: Under low pressure < 2 Mpa, the mercury saturation (S_{Hg}) increases rapidly with the growth of pressure. When S_{Hg} is about 80%, the mercury entry curve tends to be gentle, indicating that micron-sized pores are abundant. After that, as the pressure sustains to increase, the mercury inlet curve changes slowly. The mercury removal curve shows that this type of mercury removal efficiency is extremely low, indicating that the pore connectivity of the sample is poor [2,56].

Type 2: Similar to the type 1, S_{Hg} increased rapidly under low pressure, but the mercury entry curve tended to be gentle at about 50% of S_{Hg}, and then increased rapidly under high pressure > 10,000 Mpa, indicating that the pores above 1 μm and below 30 nm are abundant. The mercury removal shows that the efficiency is better than Type 1.

Type 3: S_{Hg} increased slowly with the increase of pressure until it increased rapidly under high pressure, indicating that the pores below 30 nm are developed in large quantities, and the mercury removal efficiency is similar to that of Type 2.

Type 4: Under low pressure, S_{Hg} increases rapidly, but the rate is far less than that of Type 1 and Type 2. The growth rate of S_{Hg} decreases at about 20%, and then increases rapidly under high pressure. The pores are mainly below 30 nm in size, and micron-sized macropores.

3.3.2. Characteristic of N_2 and CO_2 Adsorption Curves

The N_2 adsorption and desorption isotherms of the sandstone samples of the Longtan Formation is as follows (Figure 7): The shape of the adsorption curve can characterize the distribution of the pore system. The curved part of the adsorption curve is called the hysteresis loop, and the hysteresis loop can be used to judge the shape of the pore. According to the classification criteria of IUPAC, all adsorption curves are divided into six types (I–VI), and the hysteresis loop have two classification methods, (H_1–H_4) and (A–E); the latter makes up for some shortcomings of the former. Nowadays, the type of hysteresis curve is often judged by combining the two methods [3,57,58]. The adsorption isotherms for all sandstone samples from the Longtan Formation exhibit a consistent trend of extension prior to a P/P_0 value of 0.45, which signifies the adsorption of nitrogen gas (N_2) onto the pore surfaces of the sandstone. The adsorption isotherms are classified as type IV, suggesting the presence of a continuous open pore network. The observed hysteresis loops predominantly fall into type B and type H_4 categories, indicative of slit-shaped and ink-bottle shaped pores, respectively.

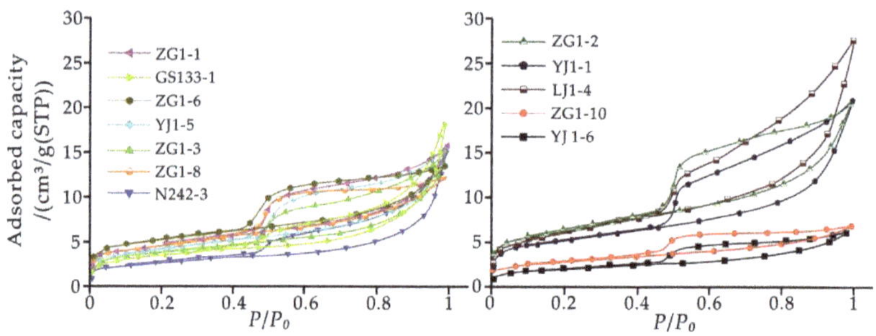

Figure 7. N_2 adsorption–desorption isotherms of Longtan formation sandstone.

The type of CO_2 adsorption–desorption isotherms of Longtan Formation sandstone is close to the Type I isotherm (Figure 8), indicating that it has an open micropore system. The CO_2 adsorption curve generally grows rapidly when $P/P_0 < 0.005$, indicating that the CO_2 gas tends to adsorb on the pore surface with larger pore size under low pressure. As the pressure increases, the adsorption curve gradually flattens, representing the adsorption of CO_2 gas in small-pore-size pores.

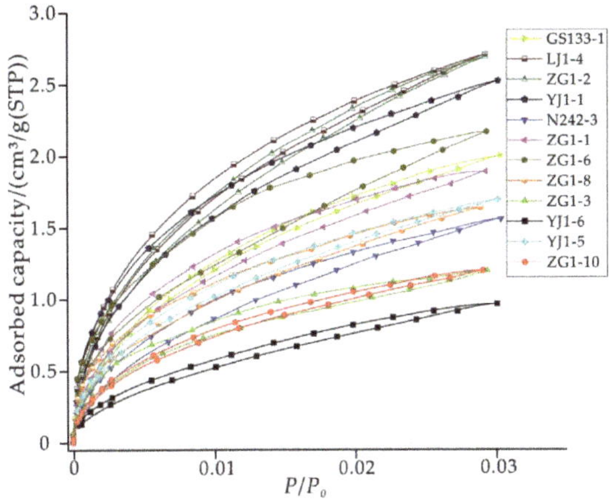

Figure 8. CO_2 adsorption–desorption isotherms of Longtan formation sandstone.

3.3.3. Pore Structure Parameter

To effectively delineate the pore structure attributes of sandstone within the Longtan Formation, the pore size, PV, and SSA of sandstone pores were analyzed through HPMI, and N_2 and CO_2 adsorption experiments (Figure 5).

The total PV of the sandstone samples of the Longtan Formation ranges between 0.012–0.045 m^3/g (average of 0.025 m^3/g). The distribution range of ultra-large macropores' PV based on HPMI data is 0.0026–0.0174 m^3/g (average of 0.0062 m^3/g); the macropores' PV is 0–0.0034 m^3/g (average of 0.0013 m^3/g). The N_2 adsorption data based on NLDFT show that the PV of mesopores is distributed in the ranges of 0.0047–0.02 m^3/g (average of 0.011 m^3/g). The CO_2 adsorption data based on DR show that the PV of micropores ranges of 0.0031–0.0092 m^3/g (average of 0.0064 m^3/g). In summary, mesopores are the main contributors to the PV of the Longtan Formation sandstone, accounting for 45.07%, followed by micropores (26.24%) and ultra-large macropores (23.87%), and the contribution of macropores is the lowest (4.8%).

The total SSA values of sandstone in Longtan Formation range from 18.169 to 46.696 m^2/g (average of 32.145 m^2/g). The contribution of SSA decreases with the increase in pore size. The CO_2 adsorption data based on DR method show that micropores is the most favorable contributors to SSA, which is distributed between 12.776 and 32.332 m^2/g (average of 21.882 m^2/g), with a contribution ratio of 68.07%. The N_2 adsorption data based on the BJH method indicate that mesopores are secondary contributors to the SSA. The distribution range of mesopores is 5.392–16.199 m^2/g (average of 10.217 m^2/g), and the contribution is 31.81%. HPMI data show that the contribution of macropores SSA is very low, only 0.12%, distributed in 0–0.141 m^2/g (average of 0.039 m^2/g); the ultra-large macropores are 0.0003–0.0038 m^2/g (average of 0.0013 m^2/g), and the contribution are negligible (0.004%).

The joint characterization map of pore size based on HPMI and N_2 and CO_2 adsorption could intuitively show the important information of pore structure, such as the range of pore size distribution and the prevalent pore size, and the contribution of pore size to PV and SSA [3]. Due to the differences in different technical methods, the overlapping parts of the test data need to be weighted. The pore size and PV joint characterization distribution map of Longtan Formation sandstone is as follows (Figure 9): Combined with the above, it can be seen that, for sandstone with different particle sizes, as the particle size increases, the PV and SSA also increase. The pore size of Longtan Formation sandstone showed a trimodal distribution, which are 0.35–1 nm, 10–30 nm, and above 10 μm, respectively, which proves the complexity of pore distribution.

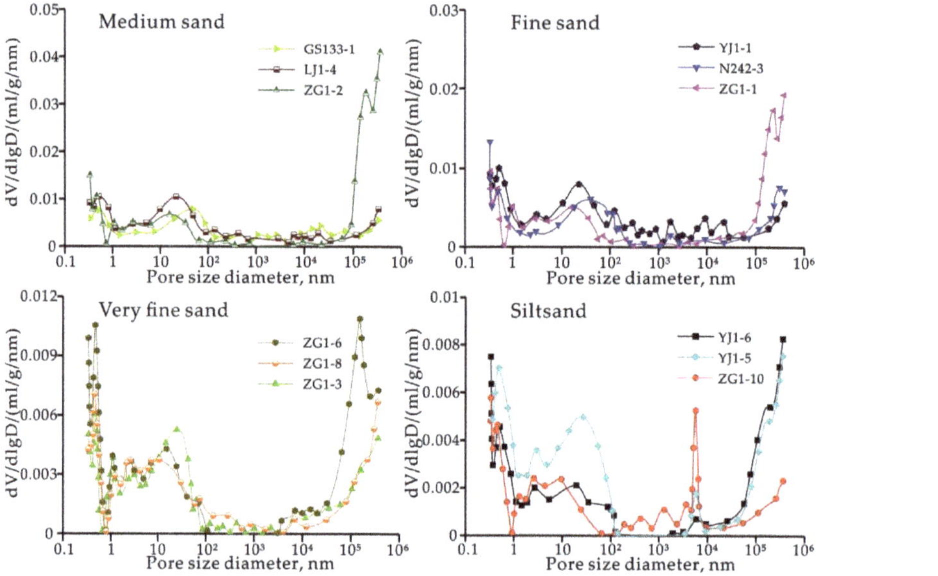

Figure 9. The pore size to PV joint characterization diagram of different grain size sandstone in Longtan Formation.

3.4. Fractal Characteristics

In an effort to explore the heterogeneity characteristics of pores with different sizes in Longtan Formation sandstone, the D_{III} and D_{IV} were calculated based on HPMI data, and the D_I and D_{II} were calculated based on the FHH model with N_2 adsorption data. D_I, D_{II}, D_{III}, and D_{IV} represent the fractal dimensions of micropores, mesopores, macropores, and ultra-large macropores, respectively. The calculation results are show in Table 1.

Table 1. Fractal characteristics of Longtan formation sandstone.

Sample	Fractal Characteristic							
	Micropore <2 nm		Mesopore 2–50 nm		Macropore 50–1000 nm		Ultra Macropore >1000 nm	
	D_I	R^2	D_{II}	R^2	D_{III}	R^2	D_{IV}	R^2
YJ1-5	2.490	0.9574	2.705	0.9954	/	/	2.942	0.8874
ZG1-6	2.639	0.9971	2.756	0.9951	2.991	0.7493	2.929	0.7803
ZG1-1	2.644	0.9985	2.724	0.9963	2.974	0.9199	2.954	0.8708
N242-3	2.526	0.9827	2.671	0.9952	2.907	0.8993	2.914	0.9777
GS133-1	2.467	0.9890	2.684	0.9974	2.880	0.9903	2.731	0.9206
ZG1-10	2.616	0.9958	2.739	0.9888	2.945	0.8983	2.733	0.9177
YJ1-6	2.545	0.9945	2.716	0.9876	/	/	2.945	0.8174
ZG1-3	2.612	0.9971	2.696	0.9987	2.954	0.9215	2.889	0.8354
ZG1-8	2.629	0.9983	2.746	0.9892	2.893	0.9774	2.929	0.8334
YJ1-1	2.529	0.999	2.690	0.999	2.802	0.9811	2.726	0.9727
ZG1-2	2.627	0.9995	2.719	0.9973	2.981	0.9572	2.983	0.7777
LJ1-4	2.532	0.9927	2.661	0.9990	2.836	0.9980	2.801	0.9767

NOTE: "/" indicates that the data have not been measured.

The lg (S_{Hg}) to lg (P_c) diagram (Figure 10) according to HPMI data shows the fractal particularities of macropores of 50–1000 nm and ultra-large macropores of 1000–100,000 nm. The fractal dimension of pores larger than 100,000 nm is greater than 3, which does not have fractal meaning. There are some samples (YJ1-5 and YJ1-6) whose pore fractal characteristics of 50–1000 nm is not obvious, so they are not recorded. The $\ln(V/V_0)$ to $\ln(\ln(P_0/P))$ diagram drawn from N_2 adsorption data shows the fractal characteristics of mesopores and micropores (Figure 11). The correlation coefficients achieved by the two methods are pass 0.7; the data of nitrogen adsorption, especially, are greater than 0.98, which show that the sandstone of Longtan Formation have obvious fractal characteristics.

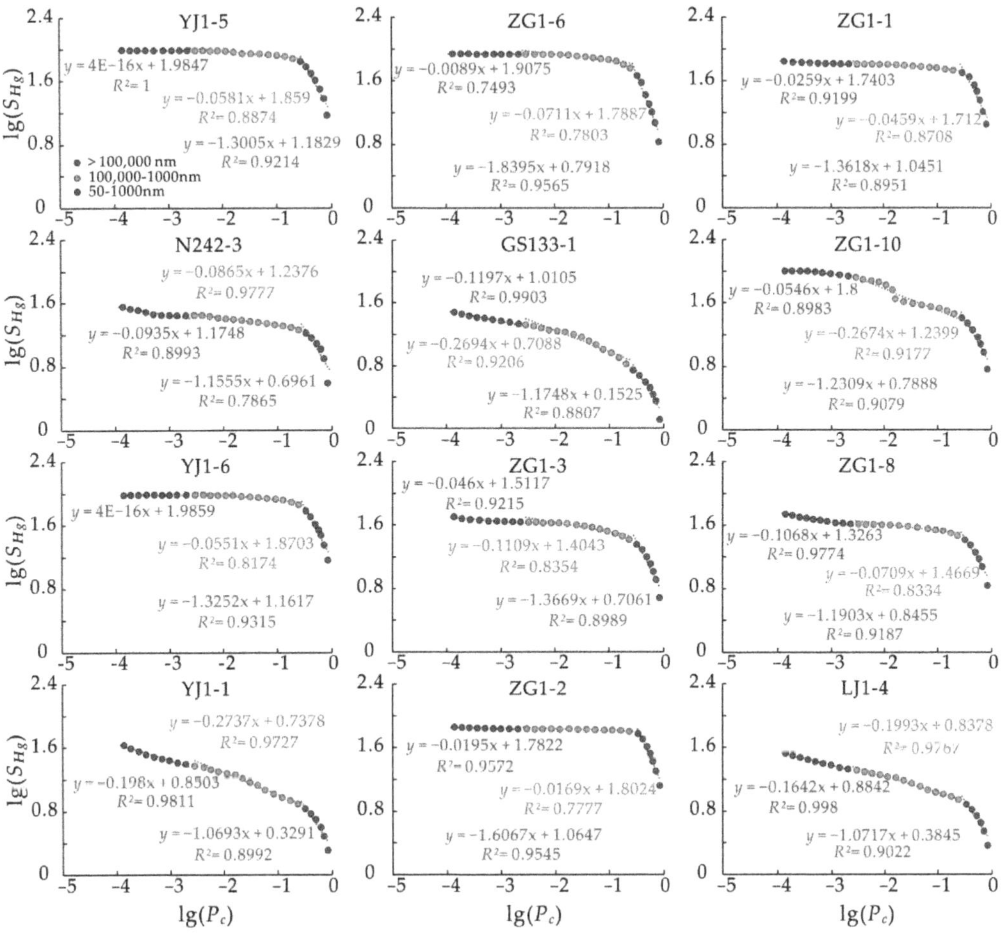

Figure 10. Relationship diagram between lg (S_{Hg}) and lg (P_c) based on HPMI data.

The D_I distribution of all samples of sandstone in Longtan Formation is between 2.467 and 2.644 (average of 2.571), the D_{II} distribution range is 2.661–2.756 (average of 2.709), the D_{III} distribution is from 2.802 to 2.981 (average of 2.916), and the D_{IV} is 2.731–2.983 (average of 2.873). The fractal dimension of macropores is larger than micropores, which suggests that the heterogeneity of large-pore-size pores is stronger.

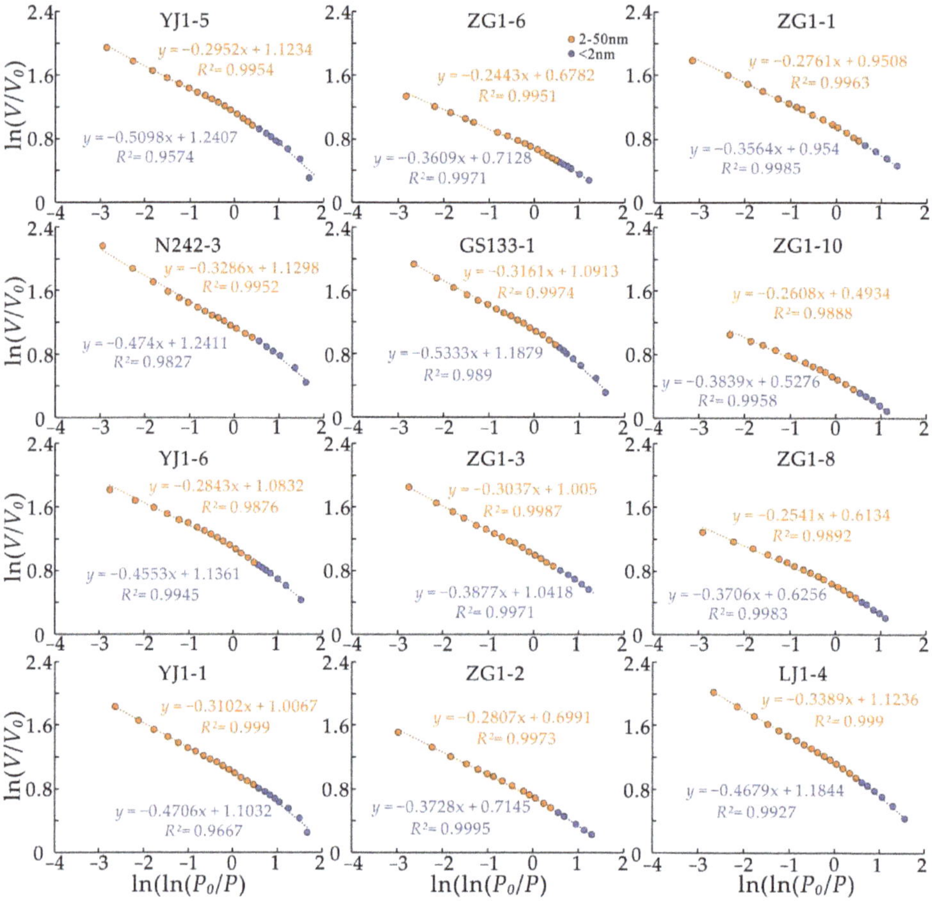

Figure 11. Relationship curve of ln(V) to ln (ln(P_0/P)) based on N_2 adsorption experimental data.

4. Discussion

4.1. Control Factors of Pore Structure Characteristics

The characteristics of the pore structure are influenced by multiple factors. A significant positive correlation exists between the total PV and the total SSA (R^2 is 0.93) in the inter-layer sandstone of the Longtan Formation. The PV of macropores, mesopores, and micropores are strongly positively correlated with SSA (the R^2 is 1, 0.92, and 0.89, respectively); only the PV of ultra-large macropores has no correlation with SSA (Figure 12a). These are consistent with the study of shale [2,3,59,60]. The grain size of sandstone is a crucial factor affecting the characteristics of pore structure. Generally, with the increase in grain size, the reservoir space will be better [13]. The grain size of sandstone within the Longtan Formation exhibits a strong positive correlation with PV and SSA, but their correlation is slightly different in different PSD ranges (Figure 12a). The grain size mainly affects the PV of mesopores and macropores and the SSA of micropores and pores larger than 50 nm. The average pore size of sandstone is extensively distributed, which is one of the pieces of evidence of the strong microscopic heterogeneity of sandstone. There is a strong negative correlation between the grain size and average pore size (R^2 is −0.56) (Figure 12a). The average pore size generally has a weak negative relationship with PV and

SSA (Figure 12a), and this suggests that small pores play a significant role in contributing to PV and SSA.

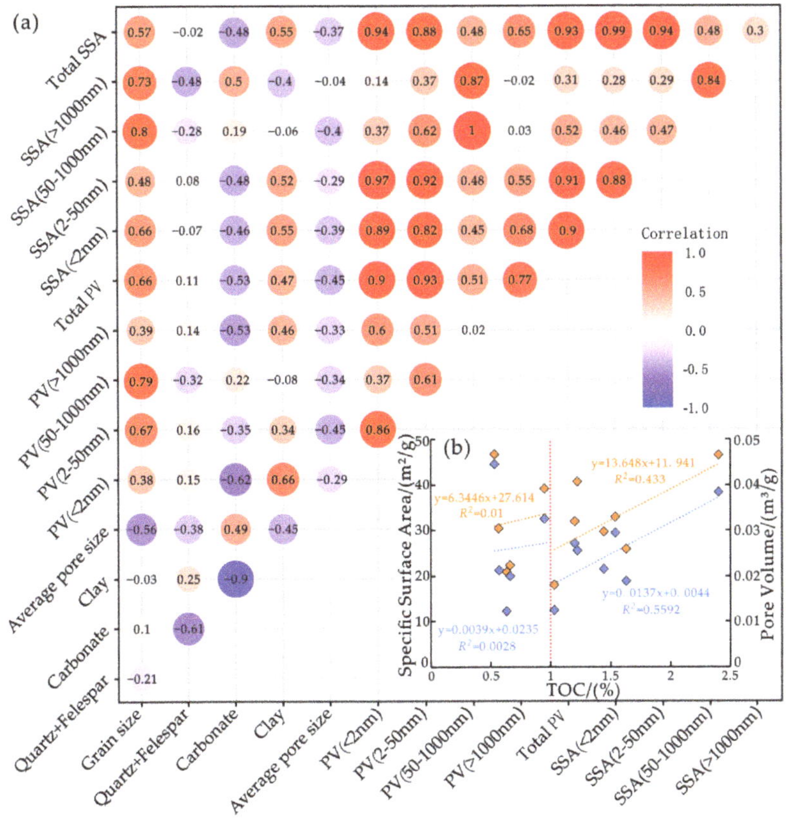

Figure 12. (**a**) Correlation heat map between different parameters of Longtan Formation sandstone; and (**b**) interrelation between the PV and SSA and TOC.

Prior research concerning the Longtan Formation shale have shown that clay minerals and TOC are significant determinants influencing the pore structure of shale [2,59]. The high content of TOC (>1%) in sandstone still promotes the growth of PV and SSA (Figure 12b). Clay minerals have a good correlation with total PV and total SSA (the R^2 is 0.47 and 0.55, respectively); it also shows a positive relation with the PV and SSA of micropores and mesopores, as well as the PV of ultra-large macropores (especially mesopores) (Figure 12a). These indicate that clay minerals are the major contributors to the micro- and mesopores' space, and also provide some micron-sized micro-fractures. The graphics of pores and fractures of clay minerals in FE-SEM can also support these results (Figure 4h–j). The quartz and feldspar content have almost no correlation with PV and SSA (Figure 12a), indicating that they have little effect on pore space. Carbonate minerals have a significant negative correlation with PV and SSA (Figure 12a); this is mainly due to the presence of carbonate minerals in inter-particle pores in the form of cements (Figure 4b,e); micropores, especially, are most affected by carbonate minerals (R^2 is -0.62).

The physical properties of reservoirs are intricately linked to their pore structure. Reservoir physical properties are closely related to pore structure. The porosity of sandstone in the Longtan Formation has a good positive correlation with PV and SSA (the R^2 is 0.5901 and 0.4417, respectively), but permeability has no obvious correlation with PV and SSA in all sandstone (Figure 13). However, after the sandstone is divided into muddy and non-muddy according to the content of argillaceous components, the results have changed. The porosity of non-muddy sandstone exhibits a stronger correlation with PV and SSA (the R^2 is 0.9287 and 0.926, respectively) (Figure 13a,b). Permeability also has a good correlation in non-muddy sandstone (the R^2 is 0.4365 and 0.453, respectively) (Figure 13c,d). Nevertheless, the muddy sandstone does not show the same correlation as the above, which indicates that the content of muddy components affects the physical properties of the Longtan Formation sandstone, especially the permeability.

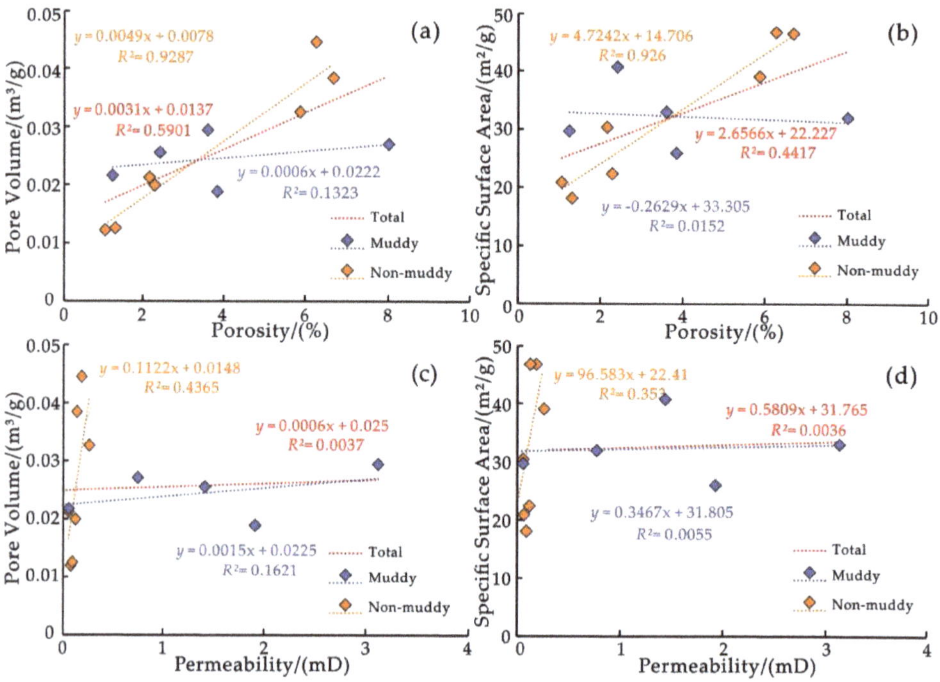

Figure 13. Relationship between pore structure parameters and porosity (**a**,**b**) and permeability (**c**,**d**) in different sandstone types.

4.2. The Association Between Fractal Properties and Pore Structure

Fractal dimension can serve as a quantitative measure for assessing the complexity and heterogeneity of pore structures [2,60]. Combining with Table 1 and Figure 14a, the fractal dimension of the sandstone in the Longtan formation shows a tendency of decreasing with an increasing grain size, and the fractal dimensions of pores at different scales also exhibit a weak negative correlation with grain size, which is similar to studies in shale and tight sandstone [2,3,26], but not as significant, indicating that the fractal dimension of the Longtan Formation sandstone is certainly influenced by other factors. The PV and SSA of micro- and mesopores have no significant correlation with the corresponding fractal dimensions D_I and D_{II}. The PV and SSA of macropores have an excellent negative correlation with the corresponding fractal dimension D_{III} (the R^2 are 0.841 and 0.8396, respectively). Only SSA has a strong negative correlation with the corresponding D_{IV} (R^2 is 0.7631) (Figure 14b–d). The findings indicate that the expansion of micro- and mesopores does not significantly influence the heterogeneity of the pore structure, whereas the formation of macropores tends to diminish pore heterogeneity. Nevertheless, as the pore size continues to increase, the SSA of ultra-large macropores progressively diminishes its influence on the heterogeneity of pores. The results of macropores are consistent with the study of Longtan Formation shale [2,47]. However, the PV and SSA of micro- and mesopores of Longtan Formation shale promote the fractal dimension, but there is no such result in sandstone. In conjunction with the preceding analysis of the factors influencing pore structure characteristics, it is evident that the micro- and mesopores, which significantly contribute to the pore structure in sandstone, are influenced by a variety of factors, and their properties have changed greatly compared with shale.

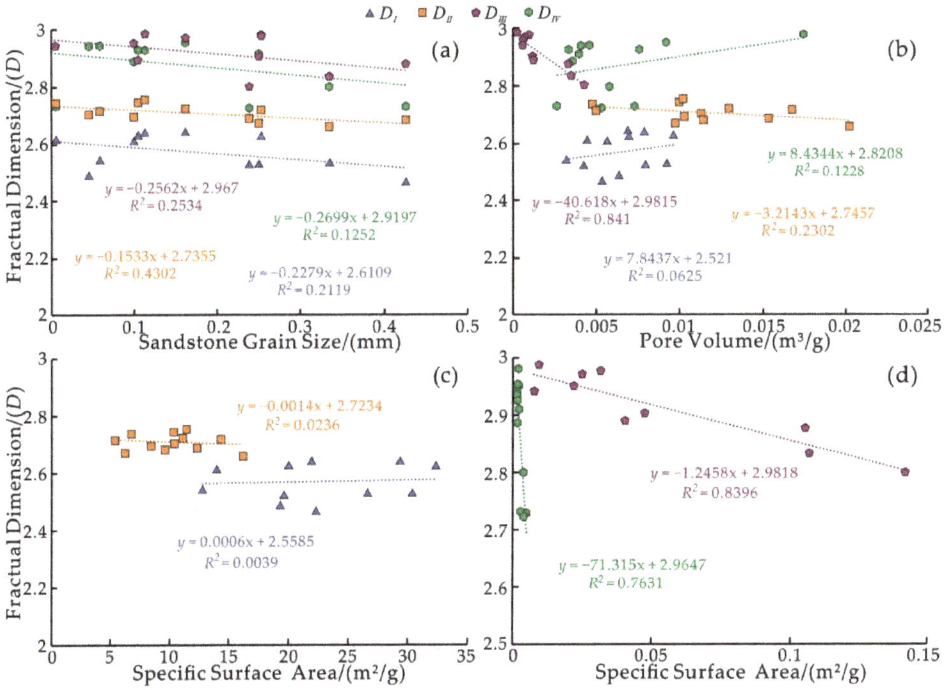

Figure 14. Correlation between pore fractal dimensions of different scale pores and grain size (**a**) and pore structure parameters (**b**–**d**).

A weak negative correlation has been observed between the porosity and fractal dimension of sandstone within the Longtan Formation (Figure 15a). Among them, D_{II} has the highest correlation (R^2 is 0.3968), indicating that an increase in porosity is associated with a more uniform pore morphology, and the mesopores are most affected. There is no significant relevance between the permeability and fractal dimension of sandstone in Longtan Formation (Figure 15b). However, in muddy sandstone, permeability exhibits a strong positive correlation with D_I, D_{II}, D_{III}, and D_{IV} (the R^2 is 0.7312, 0.4776, 0.5911, and 0.7652, respectively) (Figure 15c); it shows that, with the increase in permeability in muddy sandstone, the pore heterogeneity is stronger. In non-muddy sandstone, the permeability has adverse correlation with the fractal dimension (Figure 15d); that is, the increase in permeability will inhibit the heterogeneity of pores. This is similar to the study of tight sandstone, but its degree of inhibition is less than that of tight sandstone [33,35]. This phenomenon can be primarily attributed to the significant influence exerted by the neighboring shale on the sandstone of the Longtan Formation.

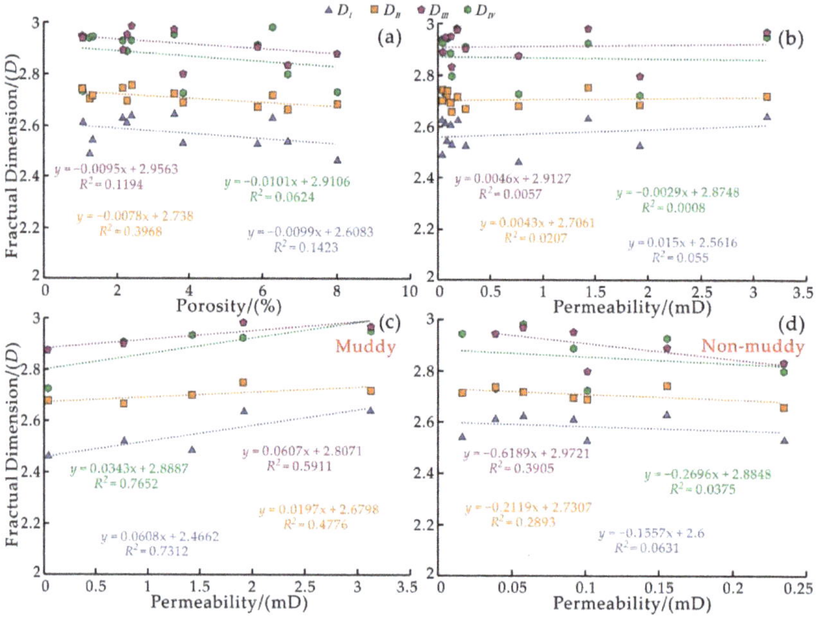

Figure 15. Correlation between pore fractal dimensions of different scale pores and porosity (**a**) and permeability (**b–d**).

4.3. The Relationship Between Fractal Characteristics and Sandstone Composition

Inorganic minerals affect the heterogeneity of sandstone pores. Clay minerals have a promoting effect on the fractal dimension, in which micropores and macropores have the best correlation with the corresponding D_I and D_{IV} (the R^2 is 0.3623 and 0.4616, respectively) (Figure 16a). It indicates that, the higher the clay mineral content, the more complex the pore morphology of micropores and macropores, and the stronger the heterogeneity. Quartz and feldspar minerals also have a weak promoting effect on the fractal dimension, and their correlation are smaller than that of clay minerals (Figure 16c). The carbonate minerals have an inhibitory effect on the fractal dimension, especially on the micropores and ultra-large macropores (the R^2 is 0.519 and 0.5721, respectively) (Figure 16e). The above results show that clay and quartz minerals promote the growth of pore fractal dimension, while carbonate minerals play a completely opposite role. The reasons are as follows: In the process of the diagenetic compaction of sandstone in the Longtan Formation, clay minerals are easily compacted and plastically deformed, so micro- and mesopores with complex pore morphology are mainly developed. In addition, clay minerals also produce some volumetric shrinkage fractures formed by the conversion of illite to smectite [23,24]. These micro-fractures have large pore sizes but irregular shapes (Figure 4h–j). Brittle minerals will produce micro-fractures with regular morphology during compaction and fracturing (Figure 4m). The content of carbonate minerals in the sandstone of Longtan Formation is rich, and its inhibitory effect on fractal dimension is less affected by other factors. However, quartz and feldspar minerals are less and dispersed in sandstone, often symbiotic with clay minerals, restricted by clay minerals, and could not exert their own characteristics. The cementation of carbonate minerals in the sandstone of Longtan Formation makes the primary inter-particle pores smaller, so the micropores are often controlled by carbonate minerals.

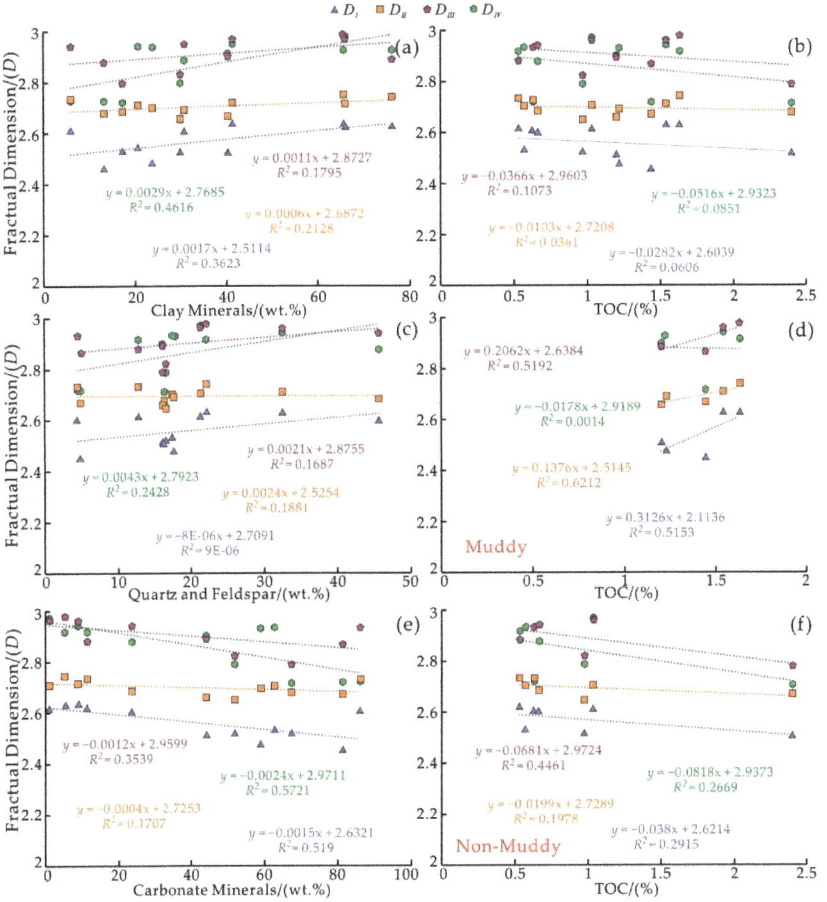

Figure 16. Interrelation between pore fractal dimensions of different scale pores and clay minerals, quartz, carbonate minerals, and TOC (**a**–**c**,**e**); and correlation between TOC and fractal dimensions of different scale pores in muddy and non-muddy sandstone (**d**,**f**).

The influence of TOC content on pore heterogeneity in sandstone of the Longtan Formation is different in muddy sandstone and non-muddy sandstone. The fractal dimension of TOC content in non-muddy sandstone has a slight negative correlation (Figure 16f), indicating that TOC enrichment in non-muddy sandstone will inhibit pore heterogeneity. In muddy sandstone, except for D_{IV}, the TOC have a good positive correlation with D_I, D_{II}, and D_{III} (the R^2 is 0.5153, 0.6212, and 0.5192, respectively) (Figure 16d), indicating that the enrichment of organic matter will make the pore morphology more complex, which is consistent with the research results in shale [2,3].

5. Conclusions

(1) The sandstone type of the Longtan Formation is mainly lithic sandstone; the grain size is mainly siltstone to medium sandstone. The minerals are mainly clay minerals and carbonate minerals (average of 37.19 and 39.4 wt%, respectively), and the content of quartz feldspar is low (average of 18.58 wt%). Low-abundance organic matters are also developed in sandstone. Longtan sandstone reservoir has lower porosity (mean of 3.52%) and permeability (mean of 0.497 mD).

(2) The pore size of Longtan Formation sandstone is divided into four types: micropore (<2 nm), mesopore (2–50 nm), macropore (50–1000 nm), and ultra-large macropore (>1000 nm). The distribution of pore sizes predominantly occurs below 30 nm and above 1000 nm. The primary types of pores include inter-crystalline pores associated with clay minerals and micro-fractures in inorganic minerals. Additionally, remnants of organic material are observable in samples of muddy sandstone. The pore morphology typically exhibits characteristics of ink-bottle-shaped and slit-like pores. The average total PV of sandstone is 0.025 m^3/g, and the average total SSA is 32.145 m^2/g. Micropores and mesopores are the main contributors to PV and SSA. The pore heterogeneity of the sandstone in the Longtan Formation is strong, the average fractal dimensions of the four types of pores are as follows: D_I is 2.571, D_{II} is 2.709, D_{III} is 2.916, and D_{IV} is 2.873, respectively.

(3) The PV of the sandstone of the Longtan Formation is proportional to the SSA. The clay minerals provide the main micropores' and mesopores' space; quartz and feldspar show similar characteristics to clay minerals due to the limitation of clay minerals. The cementation of carbonate minerals is the most crucial factor to reduce the pore space of sandstone. TOC only promotes the growth of pore space at high content (>1%).

(4) The development of macropores will reduce the factual dimension of pores. However, with the further increase in the pore size, the SSA of ultra-large macropores progressively loses the control impact on the heterogeneity of pores. Clay minerals promote the pore fractal dimensions of micropores and ultra-large macropores, while carbonate minerals inhibit the fractal dimensions of micropores and ultra-large macropores. The enrichment of organic matter in muddy sandstone generally increases the fractal dimension of pores.

(5) Combined with all the analysis, it can be seen that the pore structure characteristics and heterogeneity characteristics of muddy sandstone are closer to those of shale. The non-muddy sandstone is similar to the conventional tight sandstone in heterogeneity, but it is controlled by the adjacent shale, and the similarity is low.

Author Contributions: Conceptualization, J.Z.; methodology, J.Z.; software, J.W. and Y.C.; validation, J.W. and X.X.; formal analysis, J.W. and X.X.; investigation, J.W. and J.Z.; resources, J.Z.; data curation, J.Z. and J.W.; writing—original draft preparation, J.W. and J.Z.; writing—review and editing, J.W. and J.Z.; visualization, D.H.; supervision, J.Z.; project administration, J.Z.; funding acquisition, J.Z. All authors have read and agreed to the published version of the manuscript.

Funding: This work is funded by the National Natural Science Foundation of China (Grant No. 42202141), the Guiding Project of Scientific Research Plan of Hubei Provincial Department of Education (Grant No. B2023022), and the Natural Science Foundation of Hubei Province, China (Grant No. 2021CFB370).

Data Availability Statement: The data that support the findings of this study are available from the corresponding author upon reasonable request. All the data can be found in the paper.

Conflicts of Interest: The authors declare that they have no conflicts of interest.

References

1. Zhang, J.Z.; Li, X.Q.; Zhang, X.Q.; Zhang, M.; Cong, G.S.; Zhang, G.W.; Wang, F.Y. Geochemical and Geological Characterization of Marine–Continental Transitional Shales from Longtan Formation in Yangtze Area, South China. *Mar. Pet. Geol.* **2018**, *96*, 1–15. [CrossRef]
2. Zhang, J.Z.; Li, X.Q.; Wei, Q.; Sun, K.X.; Zhang, G.W.; Wang, F.Y. Characterization of Full-Sized Pore Structure and Fractal Characteristics of Marine–Continental Transitional Longtan Formation Shale of Sichuan Basin, South China. *Energy Fuels* **2017**, *31*, 10490–10504. [CrossRef]
3. Zhang, J.Z.; Tang, Y.J.; He, D.X.; Sun, P.; Zou, X.Y. Full-Scale Nanopore System and Fractal Characteristics of Clay-Rich Lacustrine Shale Combining FE-SEM, Nano-CT, Gas Adsorption and Mercury Intrusion Porosimetry. *Appl. Clay Sci.* **2020**, *196*, 105758. [CrossRef]

4. Wu, J.; Wang, H.Y.; Shi, Z.S.; Wang, Q.; Zhao, Q.; Dong, D.Z.; Li, S.X.; Liu, D.X.; Sun, S.S.; Qiu, Z. Favorable Lithofacies Types and Genesis of Marine-Continental Transitional Black Shale: A Case Study of Permian Shanxi Formation in the Eastern Margin of Ordos Basin, NW China. *Pet. Explor. Dev.* **2021**, *48*, 1315–1328. [CrossRef]
5. Dong, D.Z.; Wang, Y.M.; Li, X.J.; Zou, C.N.; Guan, Q.Z.; Zhang, C.C.; Huang, J.L.; Wang, S.F.; Wang, H.Y.; Liu, H.L.; et al. Breakthrough and Prospect of Shale Gas Exploration and Development in China. *Nat. Gas Ind.* **2016**, *3*, 12–26. [CrossRef]
6. Zhao, B.S.; Li, R.X.; Qin, X.L.; Wang, N.; Zhou, W.; Khaled, A.; Zhao, D.; Zhang, Y.N.; Wu, X.L.; Liu, Q. Geochemical Characteristics and Mechanism of Organic Matter Accumulation of Marine-Continental Transitional Shale of the Lower Permian Shanxi Formation, Southeastern Ordos Basin, North China. *J. Pet. Sci. Eng.* **2021**, *205*, 108815. [CrossRef]
7. Xi, Z.D.; Tang, S.H.; Zhang, S.H.; Sun, K. Pore Structure Characteristics of Marine–Continental Transitional Shale: A Case Study in the Qinshui Basin, China. *Energy Fuels* **2017**, *31*, 7854–7866. [CrossRef]
8. Jiang, S.; Tang, X.L.; Cai, D.S.; Xue, G.; He, Z.L.; Long, S.X.; Peng, Y.M.; Gao, B.; Xu, Z.Y.; Dahdah, N. Comparison of Marine, Transitional, and Lacustrine Shales: A Case Study from the Sichuan Basin in China. *J. Pet. Sci. Eng.* **2017**, *150*, 334–347. [CrossRef]
9. Yang, C.; Xiong, Y.Q.; Zhang, J.C.; Liu, Y.K.; Chen, C. Comprehensive Understanding of OM-Hosted Pores in Transitional Shale: A Case Study of Permian Longtan Shale in South China Based on Organic Petrographic Analysis, Gas Adsorption, and X-Ray Diffraction Measurements. *Energy Fuels* **2019**, *33*, 8055–8064. [CrossRef]
10. Peng, Y.X.; Guo, S.B.; Zhai, G.Y.; Shi, D.S.; Chen, R. Determination of Critical Parameters for Evaluating Coal Measure Shale Gas in China. *Mar. Pet. Geol.* **2019**, *109*, 732–739. [CrossRef]
11. Guo, T.L.; Xiong, L.; Ye, S.J.; Dong, X.X.; Wei, L.M.; Yang, Y.T. Theory and Practice of Unconventional Gas Exploration in Carrier Beds: Insight from the Breakthrough of New Type of Shale Gas and Tight Gas in Sichuan Basin, SW China. *Pet. Explor. Dev.* **2023**, *50*, 27–42. [CrossRef]
12. Yang, X.G.; Guo, S.B. Reservoirs Characteristics and Environments Evolution of Lower Permian Transitional Shale in the Southern North China Basin: Implications for Shale Gas Exploration. *J. Nat. Gas Sci. Eng.* **2021**, *96*, 104282. [CrossRef]
13. Yang, W.; Wang, Q.Y.; Wang, Y.H.; Jiang, Z.X.; Song, Y.; Li, Y.H.; Liu, D.; Zuo, R.S.; Gu, X.M.; Zhang, F. Pore Characteristic Responses to Categories of Depositional Microfacies of Delta-Lacustrine Tight Reservoirs in the Upper Triassic Yanchang Formation, Ordos Basin, NW China. *Mar. Pet. Geol.* **2020**, *118*, 104423. [CrossRef]
14. Chen, Y.H.; Wang, Y.B.; Guo, M.Q.; Wu, H.Y.; Li, J.; Wu, W.T.; Zhao, J.Z. Differential Enrichment Mechanism of Organic Matters in the Marine-Continental Transitional Shale in Northeastern Ordos Basin, China: Control of Sedimentary Environments. *J. Nat. Gas Sci. Eng.* **2020**, *83*, 103625. [CrossRef]
15. Zhang, L.F.; Dong, D.Z.; Qiu, Z.; Wu, C.J.; Zhang, Q.; Wang, Y.M.; Liu, D.X.; Deng, Z.; Zhou, S.W.; Pan, S.Q. Sedimentology and Geochemistry of Carboniferous-Permian Marine- Continental Transitional Shales in the Eastern Ordos Basin, North China. *Palaeogeogr. Palaeoclimatol. Palaeoecol.* **2021**, *571*, 110389. [CrossRef]
16. Li, G.; Yang, Z.B.; Gao, W.; Zhang, Z.G.; Jiang, B.R.; Lu, B.J. Characteristics of Coal-Measure Gas Reservoirs in Thin Interbedded Marine–Continental Transitional Facies and Optimization of Combined Production: Examples from the Tucheng Syncline in Western Guizhou. *Nat. Resour. Res.* **2022**, *31*, 1503–1522.
17. Luo, W.; Hou, M.C.; Liu, X.C.; Huang, S.G.; Chao, H.; Zhang, R.; Deng, X. Geological and Geochemical Characteristics of Marine-Continental Transitional Shale from the Upper Permian Longtan Formation, Northwestern Guizhou, China. *Mar. Pet. Geol.* **2018**, *89*, 58–67. [CrossRef]
18. Cui, H.; Zhu, S.F.; Wang, J.P.; Gao, Y.; Wan, C.F.; Tong, H. Physical Properties, Pore-Throat Structure, Fractal Characteristics and Their Effects on the Gas-Bearing Capacity of Tight Sandstone: A Case Study from the Northern Tianhuan Depression, Ordos Basin, China. *Nat. Resour. Res.* **2022**, *31*, 1559–1584. [CrossRef]
19. Huang, W.B.; Lu, S.F.; Hersi, O.S.; Wang, M.; Deng, S.W.; Lu, R.J. Reservoir Spaces in Tight Sandstones: Classification, Fractal Characters, and Heterogeneity. *J. Nat. Gas Sci. Eng.* **2017**, *46*, 80–92. [CrossRef]
20. Zou, C.N.; Yang, Z.; Tao, S.Z.; Yuan, X.J.; Zhu, R.K.; Hou, L.H.; Wu, S.T.; Sun, L.; Zhang, G.S.; Bai, B.; et al. Continuous Hydrocarbon Accumulation over a Large Area as a Distinguishing Characteristic of Unconventional Petroleum: The Ordos Basin, North-Central China. *Earth-Sci. Rev.* **2013**, *126*, 358–369. [CrossRef]
21. Lai, J.; Wang, G.W.; Wang, Z.Y.; Chen, J.; Pang, X.J.; Wang, S.C.; Zhou, Z.L.; He, Z.B.; Qin, Z.Q.; Fan, X.Q. A Review on Pore Structure Characterization in Tight Sandstones. *Earth-Sci. Rev.* **2018**, *177*, 436–457. [CrossRef]
22. Ross, D.J.K.; Marc Bustin, R. The Importance of Shale Composition and Pore Structure upon Gas Storage Potential of Shale Gas Reservoirs. *Mar. Pet. Geol.* **2009**, *26*, 916–927. [CrossRef]
23. Wang, R.; Shi, W.Z.; Xie, X.Y.; Zhang, W.; Qin, S.; Liu, K.; Busbey, A.B. Clay Mineral Content, Type, and Their Effects on Pore Throat Structure and Reservoir Properties: Insight from the Permian Tight Sandstones in the Hangjinqi Area, North Ordos Basin, China. *Mar. Pet. Geol.* **2020**, *115*, 104281. [CrossRef]
24. Jiu, B.; Huang, W.H.; Li, Y.; He, M.Q. Influence of Clay Minerals and Cementation on Pore Throat of Tight Sandstone Gas Reservoir in the Eastern Ordos Basin, China. *J. Nat. Gas Sci. Eng.* **2021**, *87*, 103762. [CrossRef]

25. Zou, C.N.; Zhu, R.K.; Liu, K.Y.; Su, L.; Bai, B.; Zhang, X.X.; Yuan, X.J.; Wang, J.H. Tight Gas Sandstone Reservoirs in China: Characteristics and Recognition Criteria. *J. Pet. Sci. Eng.* **2012**, *88–89*, 82–91. [CrossRef]
26. Li, A.; Ding, W.L.; He, J.H.; Dai, P.; Yin, S.; Xie, F. Investigation of Pore Structure and Fractal Characteristics of Organic-Rich Shale Reservoirs: A Case Study of Lower Cambrian Qiongzhusi Formation in Malong Block of Eastern Yunnan Province, South China. *Mar. Pet. Geol.* **2016**, *70*, 46–57. [CrossRef]
27. Zhang, L.C.; Lu, S.F.; Xiao, D.S.; Li, B. Pore Structure Characteristics of Tight Sandstones in the Northern Songliao Basin, China. *Mar. Pet. Geol.* **2017**, *88*, 170–180. [CrossRef]
28. Li, F.L.; Wang, M.Z.; Liu, S.B.; Hao, Y.W. Pore Characteristics and Influencing Factors of Different Types of Shales. *Mar. Pet. Geol.* **2019**, *102*, 391–401. [CrossRef]
29. Zhang, J.Z.; Lin, W.; Li, M.T.; Wang, J.G.; Xiao, X.; Li, Y.; Zhang, X.C. Evolution Mechanism of Microscopic Pore System in Coal-Bearing Marine–Continental Transitional Shale with Increasing Maturation. *Minerals* **2023**, *13*, 1482. [CrossRef]
30. Gao, H.; Cao, J.; Wang, C.; He, M.Q.; Dou, L.B.; Huang, X.; Li, T.T. Comprehensive Characterization of Pore and Throat System for Tight Sandstone Reservoirs and Associated Permeability Determination Method Using SEM, Rate-Controlled Mercury and High Pressure Mercury. *J. Pet. Sci. Eng.* **2019**, *174*, 514–524. [CrossRef]
31. Feng, G.; Li, W.; Shang, Z.F. Scale-dependent fractal properties and geological factors for the pore structure in shale: Insights from field emission scanning electron microscopy and fluid intrusion. *Energy Fuels* **2023**, *37*, 16519–16535. [CrossRef]
32. Han, Y.H.; Jiang, Z.X.; Liang, Z.K.; Lai, Z.Y.; Liang, H.Y.; Wu, Y.H.; Guo, J.; Wang, B. Full pore size fractal characteristics of Longmaxi Formation shale in Luzhou-Changning area and its influence on seepage. *Energy Fuels* **2023**, *37*, 11083–11096. [CrossRef]
33. He, T.P.; Zhou, Y.Q.; Chen, Z.B.; Zhang, Z.W.; Xie, H.Y.; Shang, Y.H.; Cui, G.X. Fractal characterization of the pore-throat structure in tight sandstone based on low-temperature nitrogen gas adsorption and high-pressure mercury injection. *Fractal Fract.* **2024**, *8*, 356. [CrossRef]
34. Ning, S.T.; Xia, P.; Hao, F.; Tian, J.Q.; Fu, Y.; Wang, K. Pore fractal characteristics between marine and marine–continental transitional black shales: A case study of Niutitang Formation and Longtan Formation. *Fractal Fract.* **2024**, *8*, 288. [CrossRef]
35. Liu, S.N.; Qiao, J.C.; Zeng, J.H.; Li, C.M.; Liu, Y.Z.; Kong, Z.; Liu, X.L. Heterogeneity of Micro- and Nanopore Structure of Lacustrine Shales with Complex Lamina Structure. *Fractal Fract.* **2024**, *8*, 245. [CrossRef]
36. Mandelbrot, B.B.; Good, I.J. Fractals: Form, Chance, and Dimension. *J. Am. Stat. Assoc.* **1978**, *73*, 438. [CrossRef]
37. Giri, A.; Tarafdar, S.; Gouze, P.; Dutta, T. Fractal Pore Structure of Sedimentary Rocks: Simulation in 2-d Using a Relaxed Bidisperse Ballistic Deposition Model. *J. Appl. Geophy.* **2012**, *87*, 40–45. [CrossRef]
38. Zhang, Z.Y.; Weller, A. Fractal Dimension of Pore-Space Geometry of an Eocene Sandstone Formation. *Geophysics* **2014**, *79*, 377–387. [CrossRef]
39. Wu, Y.P.; Liu, C.L.; Ouyang, S.Q.; Luo, B.; Zhao, D.D.; Sun, W.; Awan, R.S.; Lu, Z.D.; Li, G.X.; Zang, Q.B. Investigation of Pore-Throat Structure and Fractal Characteristics of Tight Sandstones Using HPMI, CRMI, and NMR Methods: A Case Study of the Lower Shihezi Formation in the Sulige Area, Ordos Basin. *J. Pet. Sci. Eng.* **2022**, *210*, 110053. [CrossRef]
40. Xu, S.; Yang, Z.M.; Wu, S.T.; Wang, L.; Wei, W.; Yang, F.; Cai, J.C. Fractal Analysis of Pore Structure Differences Between Shale and Sandstone Based on the Nitrogen Adsorption Method. *Nat. Resour. Res.* **2022**, *31*, 1759–1773. [CrossRef]
41. Zhu, F.; Hu, W.X.; Cao, J.; Sun, F.N.; Liu, Y.F.; Sun, Z.M. Micro/Nanoscale Pore Structure and Fractal Characteristics of Tight Gas Sandstone: A Case Study from the Yuanba Area, Northeast Sichuan Basin, China. *Mar. Pet. Geol.* **2018**, *98*, 116–132. [CrossRef]
42. Potter, C.J. Paleozoic Shale Gas Resources in the Sichuan Basin, China. *Sci. Bull.* **2018**, *102*, 987–1009. [CrossRef]
43. Dai, J.X.; Ni, Y.Y.; Liu, Q.Y.; Wu, X.Q.; Gong, D.Y.; Hong, F.; Zhang, Y.L.; Liao, F.R.; Yan, Z.M.; Li, H.W. Sichuan Super Gas Basin in Southwest China. *Pet. Explor. Dev.* **2021**, *48*, 1251–1259. [CrossRef]
44. Lin, L.B.; Yu, Y.; Zhai, C.B.; Li, Y.H.; Wang, Y.N.; Liu, G.X.; Guo, Y.; Gao, J. Paleogeography and Shale Development Characteristics of the Late Permian Longtan Formation in Southeastern Sichuan Basin, China. *Mar. Pet. Geol.* **2018**, *95*, 67–81. [CrossRef]
45. Miao, Z.S.; Pei, Y.W.; Su, N.; Sheng, S.Z.; Feng, B.; Jiang, H.; Liang, H.; Hong, H.T. Spatial and Temporal Evolution of the Sinian and Its Implications on Petroleum Exploration in the Sichuan Basin, China. *J. Pet. Sci. Eng.* **2022**, *210*, 110036. [CrossRef]
46. Yang, K.; Zhang, B.; Yao, Y.J.; Yang, H.; Zhang, H.Q.; Xiao, W.; Wang, Y. Organic Matter Accumulation Mechanism and Characteristics in Marine-Continental Transitional Shale: A Case Study of the Upper Permian Longtan Formation from the Well F5 in Sichuan Basin, China. *J. Pet. Sci. Eng.* **2022**, *208*, 109604. [CrossRef]
47. Zhang, J.Z.; Lin, W.; Li, M.T.; Wang, J.G.; Xiao, X.; Chen, Y.C. Pore Structure and Heterogeneity Characteristics of Coal-Bearing Marine-Continental Transitional Shales from the Longtan Formation in the South Sichuan Basin, China. *Minerals* **2024**, *14*, 588. [CrossRef]
48. Blair, T.C.; McPherson, J.G. Grain-Size and Textural Classification of Coarse Sedimentary Particles. *J. Sediment. Res.* **1999**, *69*, 6–19. [CrossRef]
49. Guo, S.; Lyu, X.X.; Zhang, Y. Relationship between Tight Sandstone Reservoir Formation and Hydrocarbon Charging: A Case Study of a Jurassic Reservoir in the Eastern Kuqa Depression, Tarim Basin, NW China. *J. Nat. Gas Sci. Eng.* **2018**, *52*, 304–316. [CrossRef]

50. Kupgan, G.; Liyana-Arachchi, T.P.; Colina, C.M. NLDFT Pore Size Distribution in Amorphous Microporous Materials. *Langmuir* **2017**, *33*, 11138–11145. [CrossRef] [PubMed]
51. Barrett, E.P.; Joyner, L.G.; Halenda, P.P. The Determination of Pore Volume and Area Distributions in Porous Substances. I. Computations from Nitrogen Isotherms. *J. Am. Chem. Soc.* **1951**, *73*, 373–380. [CrossRef]
52. Dubinin, A.M. A Study of the Porous Structure of Active Carbons Using a Variety of Methods. *Q. Rev. Chem. Soc.* **1955**, 101–114. [CrossRef]
53. Li, Z.T.; Liu, D.M.; Cai, Y.D.; Wang, Y.P.; Teng, J. Adsorption Pore Structure and Its Fractal Characteristics of Coals by N_2 Adsorption/Desorption and FESEM Image Analyses. *Fuel* **2019**, *257*, 116031. [CrossRef]
54. Brunauer, S.; Emmett, P.H.; Teller, E. Adsorption of Gases in Multimolecular Layers. *J. Am. Chem. Soc.* **1938**, *60*, 309–319. [CrossRef]
55. Li, Y.; Fan, A.P.; Yang, R.C.; Sun, Y.P.; Lenhardt, N. Sedimentary Facies Control on Sandstone Reservoir Properties: A Case Study from the Permian Shanxi Formation in the Southern Ordos Basin, Central China. *Mar. Pet. Geol.* **2021**, *129*, 105083. [CrossRef]
56. Klaver, J.; Hemes, S.; Houben, M.; Desbois, G.; Radi, Z.; Urai, J.L. The Connectivity of Pore Space in Mudstones: Insights from High-pressure Wood's Metal Injection, BIB-SEM Imaging, and Mercury Intrusion Porosimetry. *Geofluids* **2015**, *15*, 577–591. [CrossRef]
57. Wang, Q.C.; Chen, D.X.; Gao, X.Z.; Wang, F.W.; Li, J.H.; Liao, W.H.; Wang, Z.Y.; Xie, G.J. Microscopic Pore Structures of Tight Sandstone Reservoirs and Their Diagenetic Controls: A Case Study of the Upper Triassic Xujiahe Formation of the Western Sichuan Depression, China. *Mar. Pet. Geol.* **2020**, *113*, 104119. [CrossRef]
58. Lykiema, J.; Sing, K.S.W.; Haber, J.; Kerker, M.; Wolfram, E.; Block, J.H.; Churaev, N.V.; Everett, D.H.; Hansen, R.S.; Haul, R.A.W.; et al. Reporting Physisorption Data for Gas/Solid Systems with Special Reference to the Determination of Surface Area and Porosity. *Pure Appl. Chem.* **1984**, *57*, 603–619.
59. Zhang, Q.; Liu, R.H.; Pang, Z.L.; Lin, W.; Bai, W.H.; Wang, H.Y. Characterization of Microscopic Pore Structures in Lower Silurian Black Shale(S_1l), Southeastern Chongqing, China. *Mar. Pet. Geol.* **2016**, *71*, 250–259. [CrossRef]
60. Cao, T.T.; Deng, M.; Song, Z.G.; Luo, H.Y.; Hursthouse, A.S. Characteristics and Controlling Factors of Pore Structure of the Permian Shale in Southern Anhui Province, East China. *J. Nat. Gas Sci. Eng.* **2018**, *60*, 228–245. [CrossRef]

fractal and fractional

MDPI

Article

Fractal Analysis of Volcanic Rock Image Based on Difference Box-Counting Dimension and Gray-Level Co-Occurrence Matrix: A Case Study in the Liaohe Basin, China

Sijia Li [1], Zhuwen Wang [2] and Dan Mou [1,*]

[1] School of Mathematics and Statistics, Beihua University, Jilin 132000, China; lisijia@beihua.edu.cn
[2] College of Geo-Exploration Science and Technology, Jilin University, No. 938, Ximinzhu Street, Changchun 130026, China; wangzw@jlu.edu.cn
* Correspondence: moudan@beihua.edu.cn

Abstract: Volcanic rocks, as a widely distributed rock type on the earth, are mostly buried deep within basins, and their internal structures possess characteristics by irregularity and self-similarity. In the study of volcanic rocks, accurately identifying the lithology of volcanic rocks is significant for reservoir description and reservoir evaluation. The accuracy of lithology identification can improve the success rate of petroleum exploration and development as well as the safety of engineering construction. In this study, we took the electron microscope images of four types of volcanic rocks in the Liaohe Basin as the research objects and comprehensively used the differential box-counting dimension (DBC) and the gray-level co-occurrence matrix (GLCM) to identify the lithology of volcanic rocks. Obtain the images of volcanic rocks in the research area and conduct preprocessing so that the images can meet the requirements of calculations. Firstly, calculate the different box-counting dimension. Divide the grayscale image into boxes of different scales and determine the differential box-counting dimension based on the variation of grayscale values within each box. The differential box-counting dimension of basalt ranges from 1.7 to 1.75, that of trachyte ranges from 1.82 to 1.87, that of gabbro ranges from 1.76 to 1.79, and that of diabase ranges from 1.78 to 1.82. Then, the gray-level co-occurrence matrix is utilized to extract four image texture features of volcanic rock images, namely contrast, energy, entropy, and variance. The recognition of four types of volcanic rock images is achieved by combining the different box-counting dimension and the gray-level co-occurrence matrix. This method has been experimentally verified by volcanic rock image samples. It has a relatively high accuracy in identifying the lithology of volcanic rocks and can effectively distinguish four different types of volcanic rocks. Compared with single-feature recognition methods, this approach significantly improves recognition accuracy, offers reliable technical support and a data basis for volcanic rock-related geological analyses, and drives the further development of volcanic rock research.

Keywords: fractal dimension; different box-counting dimension; volcanic rock; gray-level co-occurrence matrix; fractal characteristics

1. Introduction

1.1. The Theoretical Basis of Volcanic Rock Lithology Identification

Against the background of the increasing demand for oil and gas resources and the growing difficulty in oil and gas exploration, identifying the lithology of volcanic rocks constitutes an important part in the reservoir evaluation of volcanic rock formations and oil and gas exploration [1–4].

The formation of volcanic rocks is a complex geological process involving the eruption, cooling, and solidification of high-temperature magma and the subsequent geological transformation. During this process, differences in physical properties, such as mineral composition, degree of crystallization, grain size, and arrangement, directly determine the lithological characteristics of volcanic rocks.

During the formation of volcanic rocks, the crystallization and aggregation of minerals exhibit complex self-organizing phenomena, endowing them with fractal structures at different scales. For example, during volcanic eruptions, the rapid or slow cooling of magma leads to different degrees of mineral crystallization, thereby forming fractal structures with different degrees of complexity. As a quantitative index, the fractal dimension can effectively reflect the complexity of this structure and provides an important basis for distinguishing different lithologies [5,6].

Regarding texture features, parameters extracted from the Gray-Level Co-Occurrence Matrix (GLCM), such as contrast, correlation, energy, and entropy, are closely related to the arrangement and distribution of minerals inside the rock. The size, shape of mineral grains, and the spatial relationships among them determine the texture features of the image.

1.2. Comparison of Advantages Between Traditional Methods and Fractal Dimension

In recent years, with the renewal and development of logging equipment, logging data have become more abundant. As a result, a variety of logging lithology interpretation methods based on logging response characteristics have emerged, ranging from cross-plot interpretation methods to multivariate statistical analysis methods, and then to machine learning algorithms that have emerged with the application of big data, gradually improving the timeliness of logging lithology interpretation [7–10].

Compared with the various logging lithology interpretation methods developed in recent years based on logging response characteristics, using the fractal dimension of volcanic rock core images to identify volcanic rock lithology has unique advantages [11–13].

From the perspective of cost, the renewal and maintenance costs of logging equipment are high, while the cost of obtaining core images is relatively low, which makes our method more feasible in research and application scenarios with limited resources. In terms of intuitiveness, methods based on logging data require complex data processing and analysis, while the fractal dimension analysis of core images can more intuitively reflect lithological characteristics, facilitating rapid understanding and judgment by researchers. In addition, under some complex geological conditions, logging data are easily interfered with, leading to deviations in interpretation results. However, the fractal dimension identification of volcanic rock core images is relatively less sensitive to geological conditions and has stronger stability.

1.3. Fractal Geometry

In view of the complexity and irregularity of volcanic rocks, traditional Euclidean geometric theories are unable to accurately describe these objects [14]. Therefore, the introduction of fractal characteristics serves as a new approach to the study of volcanic rocks. Fractal dimension (FD) is an important parameter for describing fractal characteristics, which can reflect the surface roughness and structural irregularity of fractal objects [15–17]. Generally speaking, a larger fractal dimension value represents a more complex structure. Research is conducted by calculating the fractal dimension of well logging curves and the distribution of volcanic lithology in the formation, and the corresponding lithological profile is established. The larger the fractal dimension of the well logging curve is, the more complex the internal structure and composition of the corresponding volcanic rocks will be [18,19].

With the continuous development of digitalization, the application of digital image processing technology in the geological field has opened up new avenues and infused new vitality into the identification of volcanic rock lithology. In two-dimensional images, the pore structure is studied by calculating the fractal dimension of the grayscale CT images of rocks [20]. The fractal dimension of the rock pore structure increases with the increase in the porosity ratio. Even with the same porosity, the fractal dimensions may vary. In the three-dimensional CT images of sandstone reservoir rocks, their fractal dimensions, porosity, and sphericity are estimated to analyze the microstructure differences among sandstone reservoir rocks with different permeabilities [21].

1.4. Gray Release Co-Occurrence Matrix

In the 1970s, Haralick proposed the gray-level co-occurrence matrix (GLCM), which was used to describe and analyze the texture features of image [22,23]. Image texture features are one of the important features used to identify the objects or regions of interest in an image, which describe the surface characteristics of the image or a certain region of the image. Due to the fact that different rock structures possess distinct image texture characteristics, the classification of rocks can be achieved by utilizing image texture features [24]. The gray-level co-occurrence matrix was applied to the analysis of computed tomography images of different rock types. The results demonstrated that each rock type had a unique GLCM, based on which the rocks could be classified [25–27].

1.5. DBC-GLCM

The method proposed in this paper, which comprehensively utilizes the Differential Box-counting Dimension (DBC) and the Gray-Level Co-occurrence Matrix (GLCM), has unique advantages.

The DBC method can effectively capture the complexity and self-similarity characteristics of rock textures by calculating the fractal dimension of volcanic rock images. Due to the differences in the microstructures of volcanic rocks with different lithologies, these differences will be reflected in the fractal dimension, thus providing a quantitative basis for lithology identification. Compared with traditional methods, the DBC method does not rely on the collection of a large number of rock samples, and it can rapidly process a large amount of image data, improving the identification efficiency.

The GLCM method starts from the spatial dependence of image gray levels and extracts texture features such as energy, contrast, and entropy. These features can reflect information such as the arrangement of internal mineral particles and the distribution uniformity of volcanic rocks. They are of great significance for differentiating lithologies with similar appearances but different microstructures.

Combining the DBC and GLCM methods gives rise to a comprehensive lithology identification system. This system can not only fully leverage the advantages of the two methods and describe the lithological characteristics of volcanic rocks from multiple dimensions, but it can also improve the accuracy and stability of lithology identification. Compared with traditional methods, this DBC-GLCM-based approach can better address the challenges of volcanic rock lithology identification under complex geological conditions, providing more reliable technical support for petroleum exploration and engineering construction.

2. Samples and Methods

2.1. Geological Background and Samples

The Liaohe Basin is located in the plain area of central and southern Liaoning Province, China, and it is part of the Bohai Bay Basin, with a total area of approximately 25,500 square kilometers [28,29]. Among them, the eastern sag of the Liaohe Basin is the research area of this paper [30,31]. The research area is shown in Figure 1 [32]. The eastern sag of the Liaohe Basin is a NE-trending half-graben sag. A large number of volcanic rocks have developed in the strata. These volcanic rocks have not only completely recorded the process of Paleogene volcanism in this area but also constitute the most important oil and gas reservoirs in this area. Conducting comprehensive, in-depth, and systematic research on it holds significant importance for understanding the Paleogene tectonic evolution and dynamic processes in the Liaohe and Bohai Bay basins, as well as for practical hydrocarbon exploration and application. Obtain 2000 typical volcanic rock samples in the eastern sag area of the Liaohe Basin. In this paper, taking the microscopic images of four types of volcanic rock cores, namely basalt, trachyte, gabbro, and diabase, as the research objects, combined with the lithofacies characteristics of Cenozoic intermediate–basic volcanic rocks in the Liaohe Basin, the lithology of these four types of volcanic rocks was identified by using DBC and GLCM.

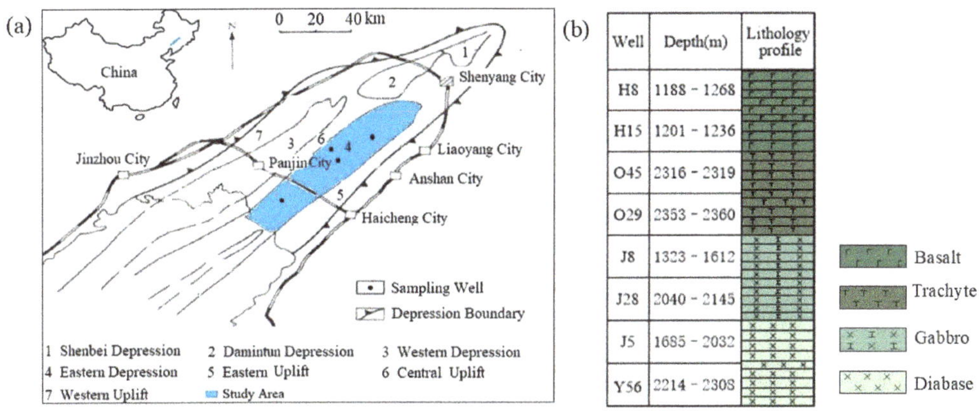

Figure 1. (**a**) The location of the Liaohe Basin; (**b**) four types of volcanic rocks studied.

2.2. Experimental Data Preprocessing

2.2.1. Image Gray Release

A color image contains multiple color channels (such as the three RGB channels). When processing it, the data volume is relatively large and complex, and gray scaling can reduce the computational difficulty. When R = G = B, the color represents a grayscale color, where the value of R = G = B is called the grayscale value. Therefore, each pixel in a grayscale image only requires one byte to store the grayscale value (also known as the intensity value or brightness value), and the grayscale range is from 0 to 255. Commonly used gray scaling methods include the maximum-value method, the average-value method, and the weighted-average method, as shown in Figure 2. To ensure the consistency of the analysis, all original images adopt a common resolution with a pixel size of 519 × 392. In Matlab, the resolution of an image can be adjusted using the imresize function.

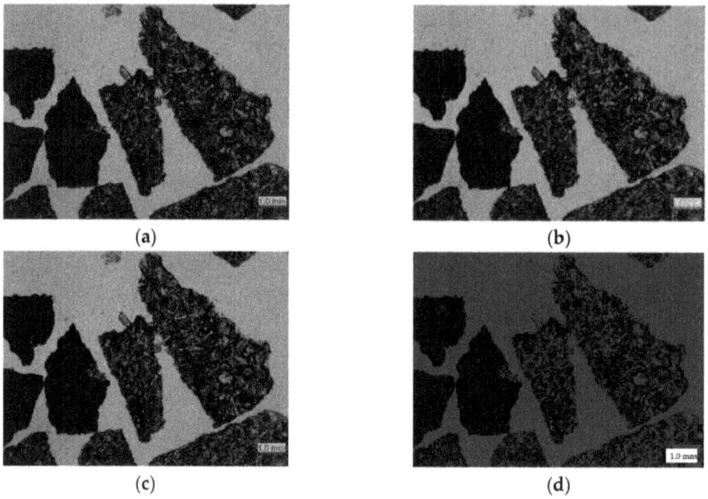

Figure 2. Basalt image gray release processing: (**a**) microscopic image, (**b**) maximum method gray image, (**c**) average method gray image, (**d**) weighted average method gray image.

In practical applications, the weighted-average method is widely recognized. It can generally meet the gray scaling requirements in most cases. The grayscale image obtained through this method is visually closer to the human eye's perception of a color image. It can preserve the details and contrast of the image well, facilitating subsequent processing and analysis based on the grayscale image [33].

Grayscale Scaling Range: Scaling the grayscale image and altering its grayscale range can affect the image's contrast and detail representation. If the grayscale scaling range is too large, some subtle texture features in the image may be overmagnified. As a result, during the calculation of the fractal dimension, these details, which might not be dominant originally, can have a significant impact on the result, making the calculated fractal dimension unstable or inaccurate. Conversely, if the grayscale scaling range is too small, the image contrast decreases, and some important texture features may be weakened. This will also affect the accuracy of the fractal dimension calculation and prevent it from truly reflecting the fractal characteristics of the volcanic rock image.

2.2.2. Filtering Processing

Before analyzing an image, one of the most crucial steps is to remove unwanted noisy pixels from the image. This task is referred to as filtering or denoising. Image filtering is classified into linear filtering and non-linear filtering. Commonly used filtering methods include mean filtering, median filtering, Gaussian filtering, etc. In this study, a Gaussian filter with a kernel size of 5×5 and a mean squared deviation of 1.5 was first applied to reduce Gaussian noise. Subsequently, a median filter with a kernel size of 3×3 was used to remove salt-and-pepper noise. The following are the results of the filtering process applied to the microscopic images of basalt [34,35].

(1) Gaussian filtering: It belongs to linear smoothing filtering. The principle is to perform a weighted average on the image. The value of each pixel is obtained by a weighted average of its own value and those of other pixels in the neighborhood. Figure 3 shows the result of Gaussian filtering.

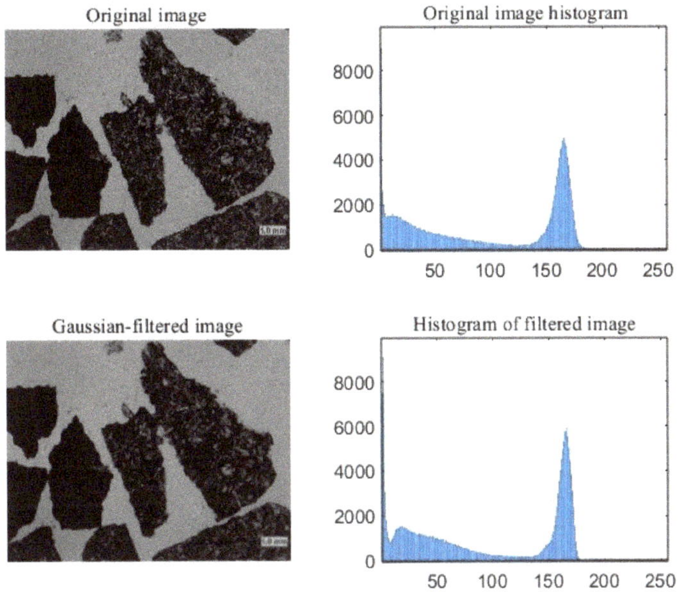

Figure 3. Gaussian filter image.

(2) Median filtering: It is a type of non-linear filtering. Its principle is to replace the grayscale value of a pixel with the median value of the grayscale values in the neighborhood of that pixel. The advantage of this method is that the obtained pixel values are mostly real values, and isolated noise points can be easily removed. Figure 4 shows the result of median filtering.

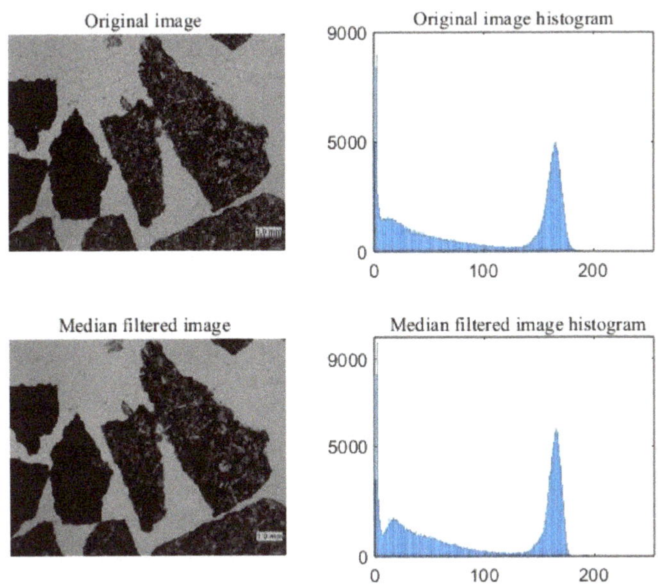

Figure 4. Median filter image.

2.2.3. Binarization

Convert the obtained image into grayscale, with the brightness range from 0 to 255. During the binarization process, the size of each pixel in the image is compared with the threshold value. If the grayscale value of a pixel is greater than or equal to the threshold T, the grayscale value of this pixel is set to a fixed value (usually 255, representing white); if the grayscale value of a pixel is less than the threshold T, the grayscale value of this pixel is set to another fixed value (usually 0, representing black). In this way, a grayscale image can be converted into a binary image with only two colors, black and white.

The threshold can be determined using global and local thresholding methods. However, different thresholding methods will have a certain impact on the results. When extremely high or low thresholds are used during the binarization process, the generated binary image may lose some important texture information, resulting in inaccurate fractal dimension values. Global thresholding methods, such as the Otsu method, also known as the maximum between-class variance method, is an adaptive threshold-selection approach. The core idea of this method is to analyze the grayscale histogram of the image to find an optimal threshold that maximizes the between-class variance between the foreground and the background. In the calculation of the fractal dimension of volcanic rock images, the Otsu method can binarize the images reasonably according to the grayscale distribution characteristics of the images themselves, effectively avoiding the subjectivity of artificially setting thresholds. Generally speaking, in the calculation of the fractal dimension, the binary image obtained by this method can effectively reflect the macroscopic outline and main structural features of volcanic rocks. The calculated fractal dimension is relatively stable and highly accurate and can be used for the preliminary classification and feature extraction of volcanic rock images. Local thresholding methods calculate a unique local threshold using the brightness of detected pixels in the neighborhood around a pixel (i.e., the window). This local approach can effectively identify the local features of pixels in images with varying backgrounds, including illumination or background noise. However, it requires more computational effort. For complex volcanic rock images, due to its inability to adapt to the local variations of the image, the binarization effect may be poor, which in turn can lead to significant deviations in the calculated results of the fractal dimension [36–38].

In this study, we used the Otsu method to determine the threshold. Through Matlab calculation, the threshold of Figure 5a was obtained as 92. If the grayscale value of a pixel is greater than or equal to the threshold of 92, the grayscale value of this pixel is set to 255 (i.e., white); if the grayscale value of a pixel is less than the threshold of 92, the grayscale value of this pixel is set to 0 (i.e., black). A binary image with only two colors, black and white, was obtained as shown in the Figure 5c.

(a) (b) (c)

Figure 5. Basalt image preprocessing: (**a**) microscopic image, (**b**) gray image, (**c**) binary image.

2.3. Methods

2.3.1. DBC

Its main idea is as follows as shown in Figure 6: An image with a pixel size of $M \times M$ is regarded as a three-dimensional (3-D) spatial surface, where (x, y) represent the positions of pixels on the two-dimensional spatial plane of the image and (z) represents the gray level of pixels. The (x, y) plane is divided into non-overlapping grids of size $s \times s$. The proportion of each block is $r = s/M$, where $\frac{M}{2} \geq s > 1$ and s are integers. In this case, a non-square grid introduces 0 to form a complete grid with a size of $s \times s$ pixels. There is a column of $s \times s \times h$ on each grid, where h represents the height of each box. n_r is the total number of boxes required to fill each grid at a ratio r, and the subscript r indicates the result obtained using the ratio r. If the total number of gray levels is G, then $G/h = M/s$, $h = sG/M$. Let the minimum and maximum gray levels in the (i, j)th block fall into the kth and lth bins, respectively. Then

$$n_r(i, j) = l - k + 1 \tag{1}$$

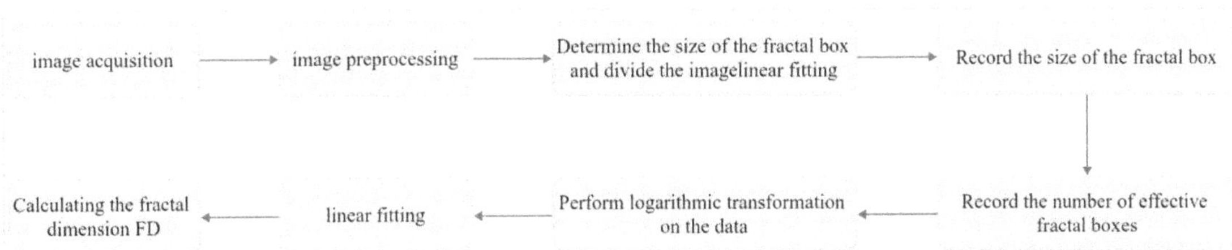

Figure 6. DBC Flowchart.

For example, when $s = h = 3$, assign numbers 1, 2, 3..., and the minimum and maximum gray levels of the image in the grid fall into box numbers 3 and 1, respectively. Then the number of boxes covering this block is calculated as follows:

$n_r = 3 - 1 + 1 = 3$, as shown in Figure 7 [39].

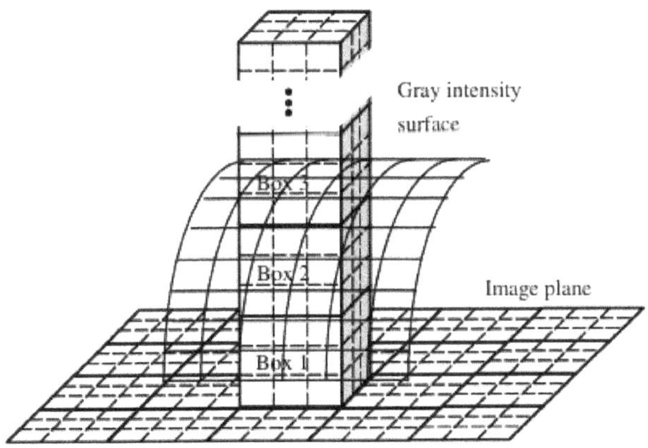

Figure 7. Sketch of determination of the number of boxes by the DBC method.

N_r is the number of boxes required to fill the entire image at a proportion r.

$$N_r = \sum_{i,j} n_r(i, j) \tag{2}$$

Perform the least squares linear fitting estimation on the slope of the curve within the range of $\log(N_r)$-$\log(1/r)$ to obtain the FD, namely the fractal dimension [39].

$$FD = \lim_{r\to 0}\frac{\log(N_r)}{\log(1/r)} \tag{3}$$

2.3.2. GLCM

As shown in Figure 8. Firstly, grayscale processing is carried out on the volcanic rock images. Then, the GLCM can be defined as the probability of the occurrence of two pixels with gray values of i and j, whose adjacent distance is d and directions are θ (0°, 45°, 90°, 135°) within the image domain $N_x \times N_y$.

$$P(i,j|d,\theta) = \#\{(x_1,y_1),(x_2,y_2) \in N_x \times N_y | f(x_1,y_1) = i, f(x_2,y_2) = j\} \tag{4}$$

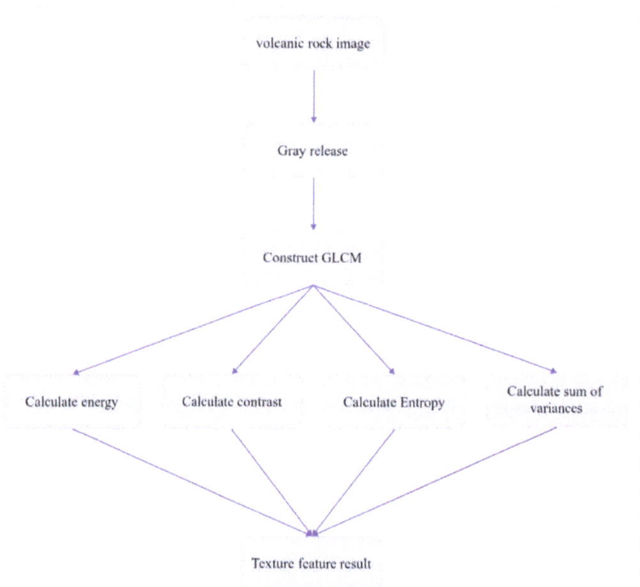

Figure 8. GLCM flowchart.

In the formula, #(x) represents the number of elements in set x that satisfy certain conditions. d is the distance between (x_1,y_1) and (x_2,y_2), and θ is the included angle between the coordinates of (x_1,y_1) and (x_2,y_2), as shown in Figure 9.

The GLCM can extract texture features in four directions (0°, 45°, 90°, 135°). Here, contrast, energy, entropy, variance, and mean sum are listed, and their formulas are as follows [40–42]:

- Contrast W_1: Contrast is manifested by the difference in brightness of pixel points in image textures. It is related to the gray values and positions of pixels and reflects the amplitude of local grayscale changes in an image.

$$W_1 = \sum\sum p(i,j)(i-j)^2 \tag{5}$$

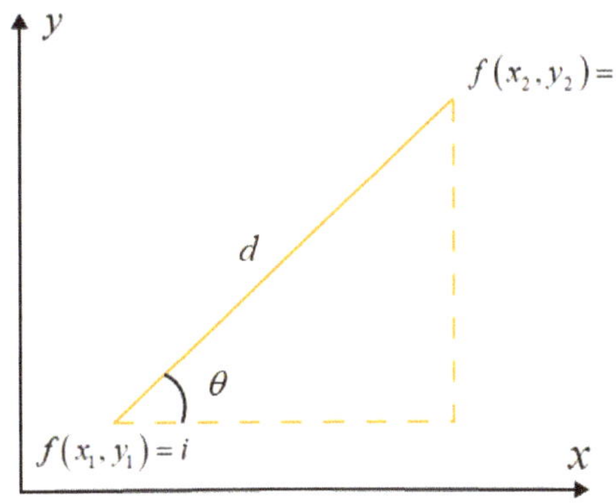

Figure 9. GLCM sketch map of space position.

- Energy W_2: Energy, also known as angular second moment, is the sum of the squares of all elements in the gray-level co-occurrence matrix. It can reflect the uniformity of the gray distribution and the coarseness or fineness of the texture in an image.

$$W_2 = \sum\sum p(i,j)^2 \tag{6}$$

- Entropy W_3: Entropy represents the amount of information in an image and can characterize the complexity of textures. It reflects the randomness or complexity of textures in an image. The greater the entropy value is, the more complex the texture will be.

$$W_3 = -\sum\sum p(x,j)\log_{10} p(i,j) \tag{7}$$

- Sum of variances W_4: Among them, m is the mean value of p.

$$W_4 = -\sum\sum (i-m)^2 p(i,j) \tag{8}$$

- Sum of averages W_5: The mean value sum is a measure of the average gray value of the pixel points within the image area, reflecting the brightness and darkness of the image and is applicable to grayscale images.

$$P_X(k) = \sum\sum p(i,j), k = 2,3\cdots$$

$$W_5 = \sum k \times P_X(k) \tag{9}$$

3. Results

The dataset is divided into a training set, a test set, and a validation set. The training set accounts for 70%, the test set makes up 15%, and the remaining 15% is used for validation. This stratification is based on the distribution of different lithologies in the dataset to ensure that each subset contains a representative proportion of each rock type.

In this paper, the different box-counting dimension and gray-level co-occurrence matrix are utilized to comprehensively discriminate the lithology of volcanic rocks. The process of this discrimination method is illustrated in Figure 10 [43,44].

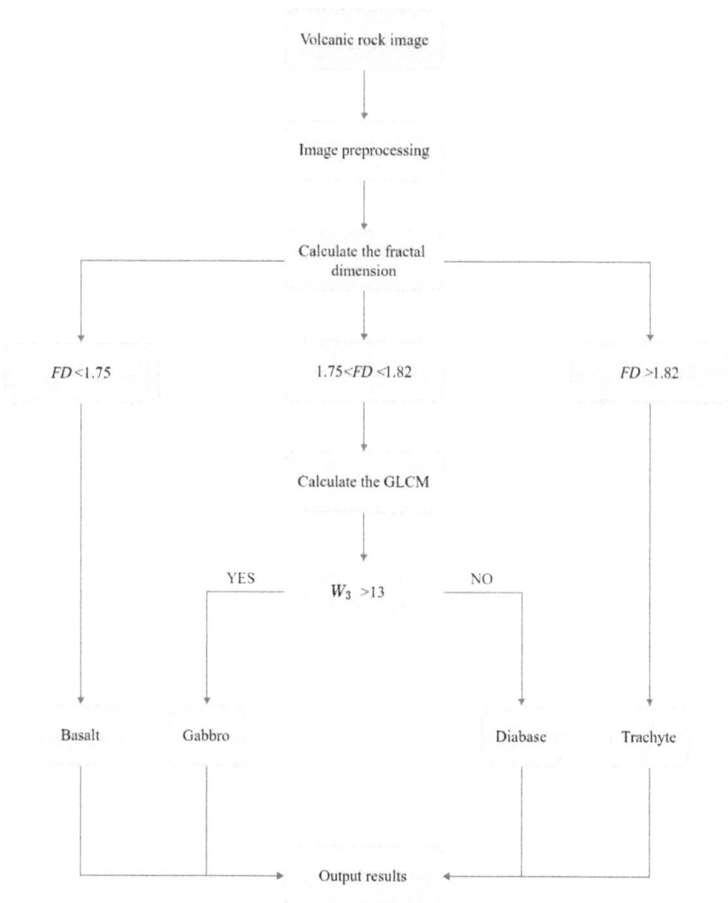

Figure 10. Flowchart for comprehensively discriminating volcanic lithology based on differential box-counting dimension and gray-level co-occurrence matrix.

3.1. DBC Experiment Results

The fractional box-counting dimension of volcanic rocks was calculated using the Fraclab toolbox in Matlab. Nine points were selected within each box, and a least squares linear fitting was performed. The results are shown in Figure 11. The linear equation $y = kx + b$ was obtained, and the absolute value of the slope k of this straight line is the fractional box dimension; that is, $|k| = FD$.

1. As can be seen from Figure 12, the difference box-counting dimension of volcanic rocks in this area ranges from 1.6 to 1.9. The surface texture of volcanic rocks is relatively complex, and its difference box-counting dimension is between 1 and 2 (the difference box-counting dimension of a straight line is 1, and that of a planar figure is 2).

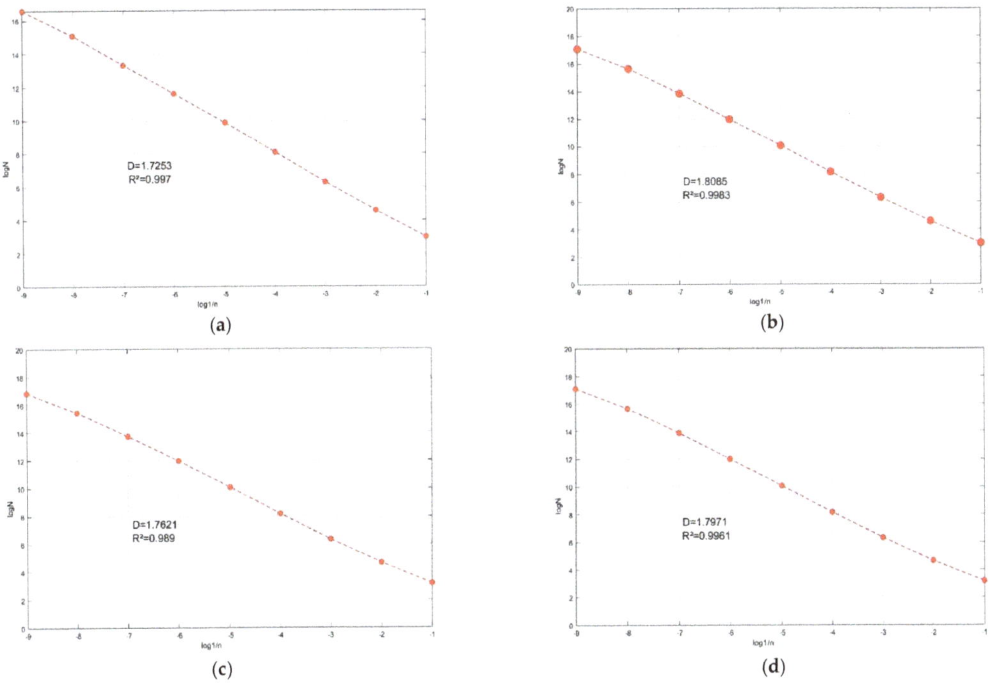

Figure 11. Plot the relationship between ln(*N*) and ln(*1/n*) for volcanic rock images using the DBC dimension method. (**a**) Fractal dimension of the basalt image, in Well D15–2106 m. (**b**) Fractal dimension of the trachyte image, in Well O20–2260 m. (**c**) Fractal dimension of the gabbro image, in Well J28–1332 m. (**d**) Fractal dimension of the diabase image, in Well H28–1135 m.

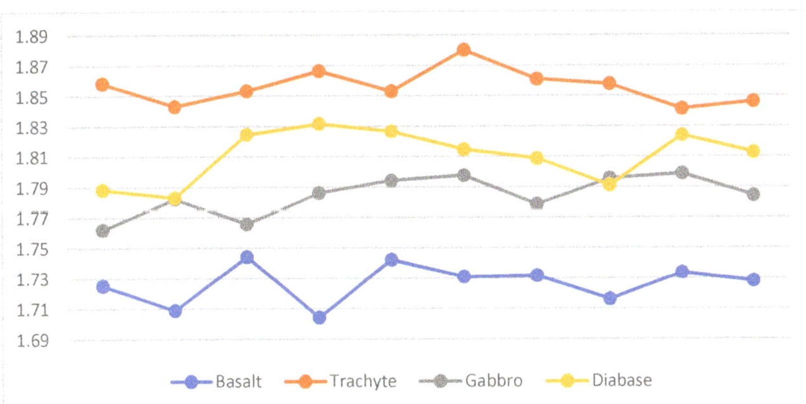

Figure 12. Range of fractal dimension of four kinds of volcanic rocks.

2. The difference box-counting dimensions of trachyte, diabase, and gabbro are all larger than that of basalt, indicating that the surface structures of trachyte, diabase, and gabbro are more complex than that of basalt. As can be seen from the obtained data, the differential box-counting dimension can, to a certain extent, distinguish basalt and trachyte from the other two types of volcanic rocks. When the fractal dimension ranges of diabase and gabbro overlap (1.76–1.84), the FD-GLCM method is comprehensively utilized to distinguish the lithologies of these three types of volcanic rocks.

3.2. GLCM Experiment Results

In this study, diabase and gabbro samples were selected as the test objects. The gray-level co-occurrence matrices were calculated when $d = 1, 2, 3$, and 4, respectively. For the distances ($d = 1$ to $d = 4$), we chose these values because they can capture texture information at different scales. A smaller distance ($d = 1$) focuses on the local texture details, while larger distances (up to $d = 4$) can reveal more global and coarser-scale texture features. Analyze the influence of different values of d on the texture parameter values. To neglect the impact of directional differences and make the characteristic parameter values independent of image rotation, the characteristic parameter values calculated here are taken as the average of the values in four directions: 0°, 45°, 90°, 135°. Generally, contrast, energy (angular second moment), entropy, and variance are used as texture feature parameters. Although the sum of means is not a traditional texture feature parameter, it reflects the overall grayscale characteristics of an image and can serve as an indicator of color shade and intensity. It holds significant importance for enhancing texture expression and classification. In Tables 1 and 2, W_1, W_2, W_3, W_4, and W_5 represent contrast, energy, entropy, variance, and mean value, respectively.

Table 1. Analysis of gabbro samples.

d	W_1	W_2 (1×10^{-4})	W_3	W_4	W_5
$d = 1$	18,575.288	4.115	13.231	123.162	3406.308
$d = 2$	18,550.121	3.279	13.615	123.061	3406.229
$d = 3$	18,526.246	3.006	13.759	122.965	3405.892
$d = 4$	18,505.105	2.821	13.847	122.871	3405.526

Table 2. Analysis of diabase samples.

d	W_1	W_2 (1×10^{-4})	W_3	W_4	W_5
$d = 1$	32,059.146	2.461	13.657	169.219	3423.735
$d = 2$	32,035.065	1.892	13.962	169.144	3425.269
$d = 3$	32,014.594	1.787	14.057	169.078	3427.228
$d = 4$	31,994.295	1.702	14.119	169.013	3428.788

As can be seen from Tables 1 and 2, when the distance between pixel pairs d varies from 1 to 4, the values of W_1, W_2, W_3, W_4, and W_5 are all affected to a certain extent. W_3 shows an upward trend as the value of d increases; while W_1, W_2, and W_4 exhibit a downward trend with the increase in d. Corresponding to the change in d value in the texture characteristics of volcanic rocks, it can be indicated that when d is within the range of 3–5, it represents the periodic change in the volcanic rock texture corresponding to the co-occurrence matrix. Therefore, it is considered that when $d = 4$, the calculated texture feature parameter values have better representativeness for the texture.

For gabbro and diabase images, when $d = 4$, their gray-level co-occurrence matrices are calculated and shown in Tables 3 and 4, respectively.

The contrast W_1 mainly reflects the degree of local grayscale differences in the image. Although there are certain differences in mineral grain size between gabbro and diabase, the manifestation of this grain-size difference in the image grayscale contrast is not sufficient to form distinct features that can accurately distinguish the two. The energy W_2 is vulnerable to interference from factors such as local noise in the image and small-scale mineral impurities. In actual rock images, these local interfering factors may cause fluctuations in the energy values, thus affecting its reliability as a means to distinguish between gabbro and diabase. Entropy W_3 has unique advantages in differentiating between gabbro and diabase.

It can better capture the differences between the two types of rocks from the perspectives of texture complexity and randomness. Entropy is sensitive to the uniformity of the grayscale distribution in an image. The differences in the size, shape, and distribution uniformity of mineral grains in gabbro and diabase lead to variations in the grayscale distribution of the images, thus resulting in different entropy values. For example, the texture formed by the interweaving of mineral grains in diabase may result in a relatively non-uniform grayscale distribution, leading to a larger entropy value. In contrast, the equigranular texture of gabbro may make the grayscale distribution more uniform, with a relatively smaller entropy value. This threshold is obtained through calculation and comparison. By comparing the magnitude relationship between the entropy value and the threshold, these two types of rocks can be distinguished to a certain extent.

Table 3. Gray-level co-occurrence matrix of gabbro images.

Sample	W_1	W_2 (1×10^{-4})	W_3	W_4	W_5
1	18,597.376	4.959	13.092	3406.319	123.252
2	21,326.265	5.118	12.873	2851.409	135.922
3	18,911.328	1.212	13.948	2696.592	127.337
4	18,453.301	1.553	13.612	2978.959	124.396
5	17,703.911	6.291	12.773	4230.387	116.075
6	13,253.303	1.706	13.778	3667.949	97.905
7	23,534.587	4.499	12.618	2156.852	146.211
8	27,924.793	9.424	11.761	1865.669	161.428
9	18,904.292	6.849	12.361	4359.485	120.601
10	25,812.256	4.467	12.925	2061.523	154.114

Table 4. Gray-level co-occurrence matrix of diabase images.

Sample	W_1	W_2 (1×10^{-4})	W_3	W_4	W_5
1	32,045.486	2.865	13.627	3424.492	169.1774
2	29,986.944	0.827	14.209	2795.745	164.898
3	20,582.479	0.949	13.836	2206.793	135.557
4	19,231.518	1.119	13.601	1848.355	131.845
5	26,672.479	1.586	14.188	4477.683	148.979
6	25,713.986	3.138	13.634	2599.351	152.035
7	22,705.355	1.158	13.653	2586.818	141.8398
8	20,343.634	1.849	13.509	1571.144	137.013
9	19,205.051	6.55	12.634	1972.271	131.274
10	22,749.615	5.87	12.083	5925.378	132.083

Evidently, for the images of gabbro and diabase, the lithology of volcanic rocks can be discriminated through a comprehensive analysis of the W_3 value. By integrating two different sets of experimental data from the differential box-counting dimension and the gray-level co-occurrence matrix, basalt, trachyte, diabase, and gabbro can be distinguished.

4. Discussion

4.1. Model Evaluation

In this study, accuracy, precision, recall, and F_1-score were used as evaluation indicators to conduct a comparative analysis of the results of the random forest model and FD-GLCM. The results are shown in Table 5.

Table 5. Comparison of lithology identification results between the random forest model and FD-GLCM method.

Lithology	Random Forest Model			FD-GLCM		
	Precision/1	Recall/1	F_1	Precision/1	Recall/1	F_1
Basalt	0.81	0.91	0.85	0.91	0.92	0.915
Trachyte	0.87	0.92	0.894	0.88	0.90	0.89
Gabbro	0.82	0.67	0.738	0.67	0.67	0.67
Diabase	0.76	0.67	0.712	0.87	0.78	0.823
Accuracy/1		0.88			0.86	

As can be seen from the table, the recognition effects of random forest and FD-GLCM on gabbro and diabase are not very good. The three evaluation indicators, namely precision, recall, and F_1-score, are not high, resulting in a relatively large number of misjudgments for this type of lithology. Firstly, the fractal dimension values and the texture features derived from GLCM of a small number of diabase samples are extremely close to those of gabbro. This similarity makes it difficult for our classification method to accurately distinguish between them. In addition, during the sampling and imaging processes, some samples may be affected by local rock heterogeneity or small-scale impurities, which also leads to misjudgments. However, both methods have relatively good identification results for basalt and trachyte. Overall, the identification accuracy of the random forest model is approximately 88%, and that of the FD-GLCM (Feature-Distribution Gray-Level Co-occurrence Matrix, if this is what it stands for; replace it with the accurate full name if otherwise) is approximately 86%.

4.2. Analysis of Application Effect

Take Well H24 in the study area as an example, and compare the prediction results with the lithologic section. In the well section with a depth ranging from 2870 to 2980 m, as shown in Figure 13, the lithological profile of the volcanic rock is basalt and diabase. The interval from 2942 to 2951 m is diabase. The random forest method misclassified diabase as gabbro at 2946 m and 2929 m. The FD-GLCM method misidentified diabase as gabbro at 2942 m and 2945 m. Both of the above-mentioned methods wrongly judged diabase as gabbro. Overall, except for individual thin layers, the prediction results are basically consistent with the lithologic section, indicating a relatively good prediction effect.

In lithology identification, there is an overlapping range in the differential box-counting dimensions of gabbro and diabase. Therefore, the gray-level co-occurrence matrix (GLCM) is further utilized to discriminate the lithology. Due to the fact that the texture features derived from GLCM of some gabbro and diabase are highly similar, misjudgments between gabbro and diabase are likely to occur.

4.3. Computational Complexity of DBC and GLCM Methods

The time complexity of the DBC method is proportional to the image size and the number of scales taken into account, where the main computational cost stems from grid partitioning and grayscale change analysis. The spatial complexity is primarily determined by the image size.

The time complexity of the GLCM method is related to the image size, the number of gray levels, and the number of distances and angles considered and is significantly influenced by the number of gray levels. The spatial complexity is mainly determined by the number of gray levels (for storing the GLCM matrix) but is also related to the image size. In practical applications, it is necessary to select an appropriate method or optimize the algorithm according to the specific image data and computing resources so as to improve the computational efficiency and reduce resource consumption.

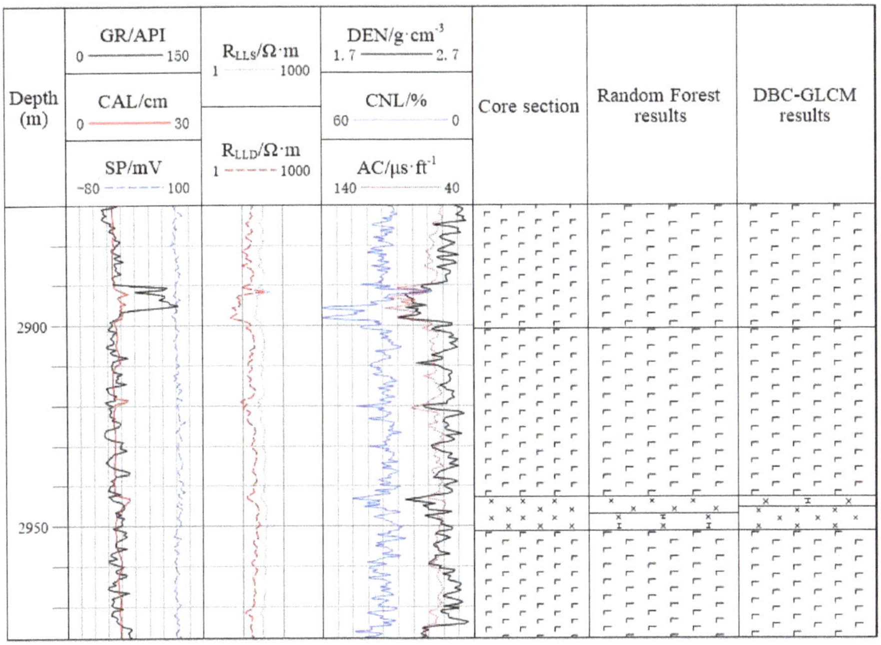

Figure 13. H24well, depth 2850–2950 m predict the lithology results of volcanic rocks.

4.4. Limitation

Regarding the sensitivity to image resolution, we acknowledge that when the image resolution is low, detailed information in the image will be lost. There will be spatial blurring between pixels, making it impossible to accurately calculate the differential box-counting dimension and affecting the construction of the gray-level co-occurrence matrix. When the image resolution is high, some unnecessary noise will be introduced. This noise can interfere with the calculation of the differential box-counting dimension. The calculation of the gray-level co-occurrence matrix (GLCM) increases significantly, and at high resolutions, the GLCM matrix becomes extremely large. The DBC and GLCM methods have certain limitations when dealing with high-resolution images. In practical applications, appropriate adjustments and optimizations are required according to specific circumstances to ensure the effectiveness and accuracy of these methods.

4.5. Practical Application and Scalability

In practical geological exploration, our method can be integrated into the existing logging and core analysis workflows. For instance, during logging operations, our method can be used to process the acquired images in real time to identify the lithology of volcanic rocks. Subsequently, this information can be utilized to guide subsequent exploration and production decisions, such as determining the location of potential oil-bearing reservoirs. For the discovered oil and gas reservoirs, using this method to conduct a detailed analysis of the lithology of the surrounding volcanic rocks helps to describe the reservoir characteristics more comprehensively. This provides a scientific basis for formulating reservoir exploitation plans and improving the oil and gas recovery rate.

Regarding scalability, we analyzed the computational complexity of our method. Although the current implementation incurs a certain computational cost, we have proposed several strategies for handling larger datasets. For instance, we can utilize parallel computing techniques to accelerate the processing of a large number of images. Additionally, we can adopt distributed computing techniques, distributing data-processing tasks to multiple computing nodes for parallel processing to improve computational efficiency and reduce the overall computing time. For example, by using distributed computing frameworks such as Apache Spark, it is possible to rapidly process large-scale image data and compute features such as the Gray-Level Co-occurrence Matrix (GLCM).

5. Conclusions

(1) The images of volcanic rocks indicate that the microscopic structures of volcanic rocks are fractal structures with self-similarity, the complexity of which can be described by the magnitude of the fractal dimension. The value of the differential box-counting dimension of volcanic rocks is in direct proportion to the complexity of the rock surface. For rocks with more complex surface textures, the calculated values of the differential box-counting dimension will be larger.

(2) The differential box-counting dimension of basalt ranges from 1.7 to 1.75, that of trachyte ranges from 1.82 to 1.87, that of gabbro ranges from 1.76 to 1.79, and that of diabase ranges from 1.78 to 1.82. The application of the differential box-counting dimension can distinguish basalt from the other three types of volcanic rocks.

(3) Since the differential box-counting dimensions of trachyte, diabase, and gabbro overlap to some extent, we further comprehensively utilize the differential box-counting dimension and the gray-level co-occurrence matrix method to distinguish the lithology of volcanic rocks. In terms of the W_3 value, it is gabbro when $W_3 > 14$ and diabase when $W_3 < 14$. Compared with the method of identifying volcanic rocks based on a single feature, the method that combines the differential box-counting dimension and the gray-level co-occurrence matrix is more accurate in identifying the lithology of volcanic rocks.

Author Contributions: Conceptualization, S.L.; software, Z.W.; validation, Z.W.; formal analysis, S.L.; investigation, D.M.; re-sources, Z.W.; data curation, S.L.; writing—original draft preparation, S.L. and D.M.; visu-alization, Z.W.; supervision, Z.W.; project administration, D.M.; funding acquisi-tion, D.M. All authors have read and agreed to the published version of the manuscript.

Funding: This research was funded by Natural Science Foundation of Jilin Province with grant number YDZJ202501ZYTS661, the Education Department of Jilin Province under grant number No. JJKH20250805KJ, Chunhui Project Foundation of the Education Department of China with grant number HZKY20220367 and the Postgraduate Innovation Program Project of Beihua University with grant number 2024039.

Data Availability Statement: The data used in this study are restricted due to privacy and confidentiality agreements with the data providers. Interested researchers can submit a data access request to moudan@beihua.edu.cn.

Conflicts of Interest: The authors declare no conflict of interest.

References

1. Zou, C.-N.; Zhao, W.-Z.; Jia, C.-Z.; Zhu, R.-K.; Zhang, G.-Y.; Zhao, X.; Yuan, X.-J. Formation and distribution of volcanic hydrocarbon reservoirs in sedimentary basins of China. *Pet. Explor. Dev.* **2008**, *3*, 257–271. [CrossRef]
2. Liu, B.; Zhang, B.; Guo, Q.; Tan, N.; Wang, J.; Chen, X. Discovery and exploration inspiration of deep volcanic gas reservoirs in eastern sag of Liaohe depression. *China Pet. Explor.* **2020**, *25*, 33–43.

3. He, H.; Li, S.; Liu, C.; Kong, C.; Jiang, Q.; Chang, T. Characteristics and quantitative evaluation of volcanic effective reservoirs: A case study from Junggar Basin, China. *J. Pet. Sci. Eng.* **2020**, *195*, 107723. [CrossRef]
4. Hu, C.; Deng, Q.; Lin, L.; Hu, M.; Hou, X.; Zong, L.; Song, P.; Kane, O.I.; Cai, Q.; Hu, Z. Lower Cretaceous volcanic-sedimentary successions of the continental rift basin in the Songliao Basin, northeast China: Implication in high-quality reservoir prediction and hydrocarbon potential. *Mar. Pet. Geol.* **2023**, *158*, 106540. [CrossRef]
5. Fan, C.; Li, H.; Qin, Q.; Shang, L.; Yuan, Y.; Li, Z. Formation Mechanisms and Distribution of Weathered Volcanic Reservoirs: A Case Study of the Carboniferous Volcanic Rocks in Northwest Junggar Basin, China. *Energy Sci. Eng.* **2020**, *8*, 8. [CrossRef]
6. Kousehlar, M.; Weisenberger, T.; Tutti, F.; Mirnejad, H. Fluid Control on Low-temperature Mineral Formation in Volcanic Rocks of Kahrizak, Iran. *Geofluids* **2012**, *12*, 295–311. [CrossRef]
7. Dong, S.; Wang, Z.; Zeng, L. Lithology identification using Kernel Fisher discriminant analysis with well logs. *J. Pet. Sci. Eng.* **2016**, *143*, 95–102. [CrossRef]
8. Xu, T.; Chang, J.; Feng, D.; Lv, W.; Kang, Y.; Liu, H.; Li, J.; Li, Z. Evaluation of active learning algorithms for formation lithology identification. *J. Pet. Sci. Eng.* **2021**, *206*, 108999. [CrossRef]
9. Duan, Y.; Xie, J.; Su, Y.; Liang, H.; Hu, X.; Wang, Q.; Pan, Z. Application of the decision three method to lithology identification of volcanic rocks-taking the Mesozoic in the Laizhouwan Sag as an example. *Sci. Rep.* **2020**, *11*, 19209.
10. Lyu, Q.; Luo, S.; Guan, Y.; Fu, J.; Niu, X.; Xu, L.; Feng, S.; Li, S. A New Method of Lithologic Identification and Distribution Characteristics of Fine—Grained Sediments: A Case Study in Southwest of Ordos Basin, China. *Open Geosci.* **2019**, *26*, 17–28. [CrossRef]
11. Xu, Z.; Shi, H.; Lin, P.; Liu, T. Integrated lithology identification based on images and elemental data from rocks. *J. Pet. Sci. Eng.* **2021**, *205*, 108853. [CrossRef]
12. Lai, J.; Fan, X.; Liu, B.; Pang, X.; Zhu, S.; Xie, W.; Wang, G. Qualitative and quantitative prediction of diagenetic facies via well logs. *Mar. Pet. Geol.* **2020**, *120*, 104486. [CrossRef]
13. Shehata, A.A.; Osman, O.A.; Nabawy, B.S. Neural network application to petrophysical and lithofacies analysis based on multi-scale data: An integrated study using conventional well log, core and borehole image data. *J. Nat. Gas Sci. Eng.* **2021**, *93*, 104015. [CrossRef]
14. Mandelbrot, B.; Passoja, D.; Paullay, A. Fractal character of fracture surfaces of metal. *Nature* **1984**, *308*, 721–722. [CrossRef]
15. Pentland, A.P. Fractal-based description of natural scenes. *IEEE Trans. Pattern Anal. Mach. Intell.* **1984**, *6*, 661–674. [CrossRef]
16. Peleg, S.; Naor, J.; Hartley, R.; Avnir, D. Multiple resolution texture analysis and classification. *IEEE Trans. Pattern Anal. Mach. Intell.* **1984**, *4*, 518–523. [CrossRef]
17. Sarkar, N.; Chaudhuri, B.B. An efficient differential box-counting approach to compute fractal dimension of image. *IEEE Trans. Syst. Man Cybern.* **1994**, *24*, 115–120. [CrossRef]
18. Leary Peter, C. Deep borehole log evidence for fractal distribution of fractures in crystalline rock. *Geophys. J. Int.* **1991**, *107*, 615. [CrossRef]
19. Jorge, L.; Luis, O.; Jerson, G. Identification of Natural Fractures using Resistive Image Logs, Fractal Dimension and Support Vector Machines. *Ing. E Investig.* **2016**, *36*, 125.
20. Peng, R.; Yang, Y.; Ju, Y.; Mao, L.; Yang, Y. Computation of Fractal Dimension of Rock Pores Based on Gray CT Images. *Chin. Sci. Bull.* **2011**, *56*, 2256–2266. [CrossRef]
21. Xia, Y.; Cai, J.; Perfect, E.; Wei, W.; Zhang, Q.; Meng, Q. Fractal Dimension, Lacunarity and Succolarity Analyses on CT Images of Reservoir Rocks for Permeability Prediction. *J. Hydrol.* **2019**, *579*, 124198. [CrossRef]
22. Haralick, R.M.; Shanmugam, K.; Dinstein, I. Textural features for image classification. *IEEE Trans. Syst. Man Cybern.* **1973**, *SMC-3*, 610–621. [CrossRef]
23. Haralick, R.M. Statistical and structural approaches to texture. *Proc. IEEE* **1979**, *67*, 786–804. [CrossRef]
24. Garg, M.; Dhiman, G. A novel content-based image retrieval approach for classification using GLCM features and texture fused LBP variants. *Neural Comput. Appl.* **2020**, *33*, 1311–1328. [CrossRef]
25. Singh, A.; Armstrong, R.T.; Regenauer-Lieb, K.; Mostaghimi, P. Rock characterization using gray-level co-occurrence matrix: An objective perspective of digital rock statistics. *Water Resour. Res.* **2019**, *55*, 1912–1927. [CrossRef]
26. Xie, Y.; Wang, J. Study on the identification of the wood surface defects based on texture features. *Optik* **2015**, *126*, 2231–2235.
27. Saptarshi, C.; Debangshu, D.; Sugata, M. Integration of morphological preprocessing and fractal based feature extraction with recursive feature elimination for skin lesion types classification. *Comput. Methods Programs Biomed.* **2019**, *178*, 201–218.
28. Peng, W.; Hu, G.; Feng, Z.; Liu, D.; Wang, Y.; Lv, Y.; Zhao, R. Origin of Paleogene Natural gases and discussion of abnormal carbon isotopic composition of heavy alkanes in the Liaohe Basin, NE China. *Mar. Pet. Geol.* **2018**, *92*, 670–684. [CrossRef]
29. Huang, S.; Feng, Z.; Gu, T.; Gong, D.; Peng, W.; Yuan, M. Multiple origins of the Paleogene natural gases and effects of secondary alteration in Liaohe Basin northeast China: Insights from the molecular and stable isotopic compositions. *Int. J. Coal Geol.* **2017**, *172*, 134–148. [CrossRef]

30. Li, S.; Huang, Y.; Feng, Y.; Guo, Y.; Zhang, B.; Wang, P. Cenozoic extension and strike-slip tectonics and fault structures in the eastern sag, Liaohe depression. *Chin. J. Geophys.-Chin. Ed.* **2020**, *63*, 612–626.
31. Han, R.; Wang, Z.; Wang, W.; Xu, F.; Qi, X.; Cui, Y.; Zhang, Z. Igneous rocks lithology identification with deep forest: Case study from eastern sag, Liaohe Basin. *J. Appl. Geophys.* **2023**, *208*, 104892. [CrossRef]
32. Wang, W.; Lin, C.; Zhang, X. Fractal dimension analysis of pore throat structure in tight sandstone reservoirs of Huagang formation: Jiaxing area of east China sea basin. *Fractal Fract.* **2024**, *8*, 374. [CrossRef]
33. Asawari, K.; Aparna, J. Gray-scale image compression technique: A review. *Int. J. Comput. Appl.* **2015**, *131*, 22–25.
34. Saha, S.K.; Pradhan, S.; Barai, S.V. Use of machine learning based technique to X-ray microtomographic images of concrete for phase segmentation at meso-scale. *Constr. Build. Mater.* **2020**, *249*, 118744. [CrossRef]
35. Sherzer, G.L.; Alghalandis, Y.F.; Peterson, K.; Shah, S. Comparative study of scale effect in concrete fracturing via Lattice Discrete Particle and Finite Discrete Element Models. *Eng. Fail. Anal.* **2022**, *135*, 106062. [CrossRef]
36. Gencel, O.; Nodehi, M.; Bozkurt, A.; Sarı, A.; Ozbakkaloglu, T. The use of computerized tomography (CT) and image processing for evaluation of the properties of foam concrete produced with different content of foaming agent and aggregate. *Constr. Build. Mater.* **2023**, *399*, 132433. [CrossRef]
37. Yoon, J.; Kim, H.; Sim, S.-H.; Pyo, S. Characterization of porous cementitious materials using microscopic image processing and X-ray CT analysis. *Materials* **2020**, *13*, 3105. [CrossRef]
38. Sherzer, G.L.; Alghalandis, Y.F.; Peterson, K. Introducing fracturing through aggregates in LDPM. *Eng. Fract. Mech.* **2022**, *261*, 108228. [CrossRef]
39. Li, J.; Du, Q.; Sun, C. An improved box-counting method for image fractal dimension estimation. *Pattern Recognit.* **2009**, *42*, 2460–2469. [CrossRef]
40. Chinmaya, P.; Ayan, S.; Nihar, K.M.; Debotosh, B. Differential box counting methods for estimating fractal dimension of gray-scale images: A survey. *Chaos Solitons Fractals* **2019**, *126*, 178–202.
41. Cristina, M.; Laura, F.; Riccardo, G.; Ernestina, C.; Riccardo, R. Glcm, an image analysis technique for early detection of biofilm. *J. Eng.* **2016**, *185*, 48–55.
42. Gabriel, R.G.; Gabriel, P.; Alejandro, P.; Alejandro, C.; Cristina, W.; Samanta, S.S.; Emilia, L.; Nieves, P. Assessment of multi-data sentinel-1 polarizations and GLCM texture features capacity for onion and sunflower classification in an irrigated valley: An object level approach. *Agronomy* **2020**, *10*, 845.
43. Paladugu, R.; Veera, M.R.; Bhima, P.R. Optimal GLCM combined FCM segmentation algorithm for detection of kidney cysts and tumor. *Multimed. Tools Appl.* **2019**, *78*, 18419–18441.
44. Redouan, K.; Sidi, M.F.; Abderrahmane, S.; Rajaa, T.; Mustapha, T. A combined method of fractal and GLCM features for MRI and CT scan images classification. *Signal Image Process.* **2014**, *5*, 4.

fractal and fractional

Article

Analysis of Multifractal Characteristics and Detrended Cross-Correlation of Conventional Logging Data Regarding Igneous Rocks

Shiyao Wang [1], Dan Mou [1,*], Xinghua Qi [2,3] and Zhuwen Wang [2]

[1] School of Mathematics and Statistics, Beihua University, Jilin 132000, China; wsy@beihua.edu.cn
[2] College of Geo-Exploration Science and Technology, Jilin University, Changchun 130026, China; qixh20@mails.jlu.edu.cn (X.Q.); wangzw@jlu.edu.cn (Z.W.)
[3] School of Mining Engineering and Geology, Xinjiang Institute of Engineering, Urumqi 830023, China
* Correspondence: moudan@beihua.edu.cn

Abstract: In the current context of the global energy landscape, China is facing a growing challenge in oil and gas exploration and development. It is difficult to evaluate the log data because of the lithological composition of igneous rocks, which displays an unparalleled degree of complexity and unpredictability. Against this backdrop, this study deploys advanced multifractal detrended fluctuation analysis (MF-DFA) to comprehensively analyze key parameters within igneous rock logging data, including natural gamma-ray logging, resistivity logging, compensated neutron logging, and acoustic logging. The results unequivocally demonstrate that these logging data possess distinct multifractal characteristics. This multifractality serves as a powerful tool to elucidate the inherent complexity, heterogeneity, and structural and property variations in igneous rocks caused by diverse geological processes and environmental changes during their formation and evolution, which is crucial for understanding the subsurface reservoir behavior. Subsequently, through a series of rearrangement sequences and the replacement sequence on the original logging data, we identify that the probability density function and long-range correlation are the fundamental sources of the observed multifractality. These findings contribute to a deeper theoretical understanding of the data-generating mechanisms within igneous rock formations. Finally, multifractal detrended cross-correlation analysis (MF-DCCA) is employed to explore the cross-correlations among different types of igneous rock logging data. We uncover correlations among different igneous rocks' logging data. These parameters exhibit different properties. There are negative long-range correlations between natural gamma-ray logging and resistivity logging, natural gamma-ray logging and compensated neutron logging in basalt, and resistivity logging and compensated neutron logging in diabase. The logging data on other igneous rocks have long-range correlations. These correlation results are of great significance as they provide solid data support for the formulation of oil and gas exploration and development plans.

Keywords: igneous rock; logging data; MF-DFA; MF-DCCA; multifractality; long-range correlation

1. Introduction

In the modern energy exploration field, exploring for oil and gas resources in complex geological structures has become the core focus. Igneous rock is one of the three major types of rock. Its strata are rich in oil and gas resources, which are of great significance for ensuring energy supply. Meanwhile, igneous rock is also a prospective reservoir for hydrocarbons [1]. Its mineral composition, rock structure, and other properties are crucial for oil and gas storage. When there are more soluble minerals in the mineral composition of igneous rock, dissolution pores may form under later geological processes, increasing the reservoir performance. Also, if well preserved in the later tectonic movements, the primary pores formed during magmatic activities can provide effective space for oil and gas storage [2–4]. However, the lithological composition of igneous rock is complex and uncertain, which brings great challenges to oil and gas exploration. Accurately grasping the characteristics of igneous rocks is the key to efficient oil and gas exploration. Among the numerous methods and data available for studying the characteristics of igneous rocks, conventional logging data play a pivotal role due to their unique advantages [5,6]. Therefore, discussing the logging data regarding igneous rocks can provide valuable references for evaluating the value of their potential hydrocarbon reservoirs and guide more targeted exploration and development in the hydrocarbon field.

Igneous rocks, formed through the cooling and solidification of magma or lava, possess unique lithological and physical characteristics [6,7]. The internal structures of different types of igneous rocks have significant differences [8,9]. This is mainly the result of the combined effects of various geological processes such as magmatic crystallization, volcanic eruption, and subsequent tectonic activities. At the same time, mineral phase transitions and element migrations also play a crucial role in the evolution of these structures. In terms of mineral phase transitions, during the complex geological history of igneous rocks, changes in temperature, pressure, and chemical environment can trigger mineral phase transitions. These transitions not only change the physical properties of the minerals themselves but also directly affect the overall internal structure of the igneous rocks. As for element migrations, when elements move within igneous rocks, they can change the chemical composition and distribution of minerals. This, in turn, affects the bonding between minerals and the overall stability of the rock structure. In addition, the diversity of magma composition and crystallization kinetic conditions also play a crucial role in controlling the rock structure. Different volatile contents can affect the viscosity and eruption mode of magma and further influence the rock structure; the composition of trace elements may participate in the formation of minerals during the crystallization process, changing the types and properties of minerals. Changes in the degree of supercooling and the cooling rate gradient can greatly affect the growth rate and morphology of crystals. Rapid cooling often results in a fine-grained texture, while slow cooling may lead to a coarse-grained texture. All the above factors are intertwined, making the internal structures of different types of igneous rocks highly heterogeneous. For example, the rapid cooling of magma during a volcanic eruption leads to the formation of fine-grained textures and numerous vesicles in basalt, while gabbro undergoes a slow cooling process and finally forms a coarser-grained and more evenly distributed structure. However, it cannot be simply assumed that slow cooling necessarily results in a single-textured rock with a simple mineral assemblage. Because gabbro, during the cooling process, is affected by various factors such as its magma composition and the physical and chemical conditions of the surrounding environment, the types and arrangements of minerals are also relatively complex. The differences in these formation mechanisms lead to changes in porosity, permeability, and fluid-carrying capacity, which are key factors determining the existence and productivity of hydrocarbon reservoirs [8,10].

This paper focuses on the research of the eastern sag of the Liaohe Basin. The eastern sag, which has long been recognized as the magmatic activity center of the Liaohe Basin, is a secondary tectonic unit within the Cenozoic pull-apart basin in the northern section of the Tan-Lu Fault Zone. In this area, igneous rocks are extensively developed [11]. In this paper, basalt, diabase, gabbro, tuff, and magmatic breccia, among other igneous rocks, are selected for research according to the tertiary classification principle of igneous rock lithology [12]. Basalt, as a basic extrusive rock, is of great significance in plate tectonics research. In the field of resources and environment, its porous structure can be used as water and oil reservoirs, and its weathered soil also has certain functions [13]. As a hypabyssal basic intrusive rock, diabase is a sub-volcanic rock with the typical features of diabase structure. Its mineral composition is composed of plagioclase and pyroxene, and it often exhibits a massive structure feature macroscopically. Diabase shows significant differences from extrusive rocks [14]. Gabbro is a basic plutonic intrusive rock, which can reveal the magma crystallization process in the deep earth and can be used to compare the evolution of basic magma in different environments [15]. Tuff, as a pyroclastic product of volcanic eruption, records volcanic activity information, and its chemical composition is valuable for the study of paleoenvironment and paleoclimate [16]. Magmatic breccia is a rock formed during the ascent of magma. The formation reasons mainly include the explosion of gases within the magma due to pressure release during magma ascent or the mixing of magma with surrounding volcanic debris followed by cooling and solidification [17]. Magmatic breccia's clasts are mainly homologous volcanic fragments, and it is cemented by a volcanic ash matrix. In contrast, tectonic breccia is formed by the fracturing of surrounding rocks due to tectonic forces. It is often cemented by substances like calcite, quartz, or clay minerals precipitated from groundwater during the post-tectonic movement. These differences highlight the uniqueness and iconic features of magmatic breccia in the genetic classification system of pyroclastic rocks.

Logging data, which are rich in geological information and feature high precision, continuity, and multi-parameter characteristics, can record the physical property changes of underground rocks in detail. In the research field of igneous rocks, these characteristics enable logging data to play a crucial role in analyzing the formation processes and mineral compositions of igneous rocks [18]. For example, changes in natural gamma-ray logging (GR) data can reflect variations in the content of radioactive minerals in igneous rocks. Different types of igneous rocks have different radioactive mineral contents. Establishing a radioactive parameter database and studying the correlation between potassium content and feldspar abundance help identify the mineral composition and content of igneous rocks through logging data. This provides a basis for the classification and study of igneous rocks [19–21]. Resistivity logging (RT) is crucial in studying the electrical conductivity of igneous rocks. The electrical conductivity of igneous rocks is affected by their mineral composition, pore structure, and fluid contained. The dual water model is introduced to consider the influence of free water and bound water in rock pores on resistivity. This can effectively explain the characteristics of hydrocarbon-bearing igneous rock reservoirs and judge their hydrocarbon-bearing conditions. It is of great significance for evaluating the hydrocarbon-bearing properties of igneous rock reservoirs [22,23]. Due to the differences in hydrogen content among different minerals and pore fluids in igneous rocks, compensated neutron log (CNL) data can obtain the hydrogen index in igneous rocks [24]. The mineral response matrix (olivine-pyroxene-feldspar) is established based on the known neutron-scattering cross-sections of these minerals. By measuring the compensated neutron log (CNL) data and comparing them with the values predicted by the matrix, the content and properties of different minerals in igneous rocks can be analyzed more accurately, and then the performance of igneous rock reservoirs can be evaluated. Acoustic logging (AC) data also play an important role in igneous rock exploration. They can effectively distinguish the lithology of igneous rocks based on the differences in the acoustic properties of different igneous rocks. The time-average equation method is used to estimate the effective porosity of igneous rocks. This is very crucial for evaluating the reservoir performance of igneous rock reservoirs [25,26]. By comprehensively using logging data such as GR, RT, CNL, and AC to conduct a systematic analysis of igneous rocks, various information regarding igneous rock strata can be comprehensively and accurately revealed. This provides a solid decision-making basis for geological exploration and oil/gas development in igneous rock areas and significantly improves the exploration efficiency and success rate [27].

Igneous rocks, as complex geological bodies, have physical and chemical connections among internal mineral particles. Phase changes of minerals or migration of trace elements in local areas transmit changes through lattice diffusion, fluid migration, etc., increasing the complexity and self-similarity of the measured data set, thus leading to the emergence of multifractal characteristics [28]. This makes it extremely challenging to find underlying patterns and correlations in igneous rock logging data, as traditional statistical methods have difficulty capturing the complex relationships and hidden patterns in the data [29]. In recent years, multifractal analysis has emerged as a powerful tool for studying complex systems. Multifractal theory can effectively analyze the non-uniformity of data on multiple spatial scales, providing a more comprehensive understanding of the underlying processes. For example, in the study of time-series data, multifractal analysis has been used to reveal complex volatility patterns and market inefficiencies [30–35]. In the context of geosciences, multifractal analysis has been applied to study the fractal characteristics of river networks, soil particle size distributions, and rock fracture patterns [36–38].

In addition to the multifractal characteristics of individual logging data, the correlations among different logging data are also of great significance. Detrended cross-correlation analysis (DCCA) and its multifractal extension, multifractal detrended cross-correlation analysis (MF-DCCA), can be used to study the cross-correlations between different time-series data [39,40]. These methods can reveal the hidden relationships between different physical properties of igneous rocks, such as the relationship between porosity and permeability or the correlation between fluid saturation and electrical conductivity. Understanding these correlations can help in more accurate reservoir characterization and the prediction of reservoir performance.

This study requires complex mathematical methods such as MF-DFA and MF-DCCA. These methods are highly technical and complex, and there are certain difficulties in their practical application and theoretical explanation. Nevertheless, they are powerful tools for revealing the multifractal characteristics and interrelationships of igneous rock data. Therefore, we aim to comprehensively analyze the multifractal characteristics of conventional logging data on igneous rocks using MF-DFA and investigate the detrended cross-correlations among different logging data using MF-DCCA. The results of this study are expected to provide new insights into the complex nature of igneous rocks and contribute to more efficient oil and gas exploration in igneous rock formations. It can effectively reveal the hidden correlation patterns between different logging parameters. Based on the analysis of igneous rock logging data and the correlation analysis among them, we can more accurately evaluate key parameters of the reservoir, such as its oil-bearing properties and permeability. This provides a more reliable data foundation for reserve calculation, helps to formulate more efficient oil and gas exploration and development plans, and greatly promotes the research of igneous petrology and the progress of oil and gas exploration and development technologies. The flowchart of the multifractal analysis in this paper is shown in Figure 1. The materials and methods are introduced in Section 2. Section 3 briefly introduces the results. The conclusions are drawn in Section 4.

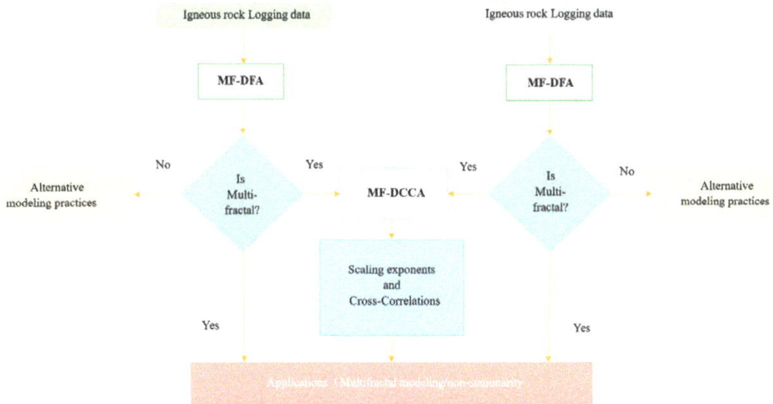

Figure 1. The flowchart of the multifractal analysis.

2. Materials and Methods

2.1. Materials

This paper chooses to analyze the conventional GR, RT, Density logging (DEN), CNL, and AC data on basalt, diabase, gabbro, tuff, and magmatic breccia in the igneous rocks of the eastern sag of the Liaohe Basin. The tuff studied in this paper is mainly composed of volcanic ash and contains minerals such as quartz and feldspar. Magmatic breccia is composed of volcanic clasts and lava fragments and contains minerals such as olivine and pyroxene.

Given the complex lithology of igneous rocks within a well, when selecting data, it is crucial to ensure that, at a specific depth, the igneous rocks exhibit only a single lithology [41]. Additionally, the quality and accessibility of logging data can impose limitations on the research conclusions, thereby affecting the accuracy and reliability of the findings. In geological research, the local changes in igneous rock logging data have a non-negligible impact. Small-scale local stress changes cause rock fractures, and the cracks continuously extend, altering the rock structure. Local fluid surges carry minerals to diffuse, changing the composition of the surrounding rocks. These changes make the logging data exhibit more variable characteristics. The relationships between the data become more complex, and noise interference increases, greatly enhancing the complexity of the data set. At the same time, fractal structures are generated within the rock due to local changes, and the physical processes show similarities at different scales. This makes the logging data present similar patterns at different scales, increasing the self-similarity. Therefore, we need to carry out a lot of data preprocessing work. For example, in this study, we need to use ResForm software (ResForm3.0) to extract the igneous data from many types of wells, and the igneous rocks contain many rock types, which requires the use of Excel and other tools to screen the data, and then we selected basalt from 1827 to 2213 m in Well Jie 13, diabase from 3628 to 3752 m in Well Jia 26, gabbro from 1323 to 1612 m in Well Jia 28, tuff from 3578 to 3771 m in Well Yu 68, and magmatic breccia from 3456 to 3613 m in Well Hong 22 as the research objects. Subsequently, Python software (Phython3.12) is utilized to detect and process outliers, and then the data are corrected according to requirements to provide data that meet the requirements for multifractal analysis. The logging deployment map of the middle and southern sections in the eastern sag of the Liaohe Basin is shown in Figure 2.

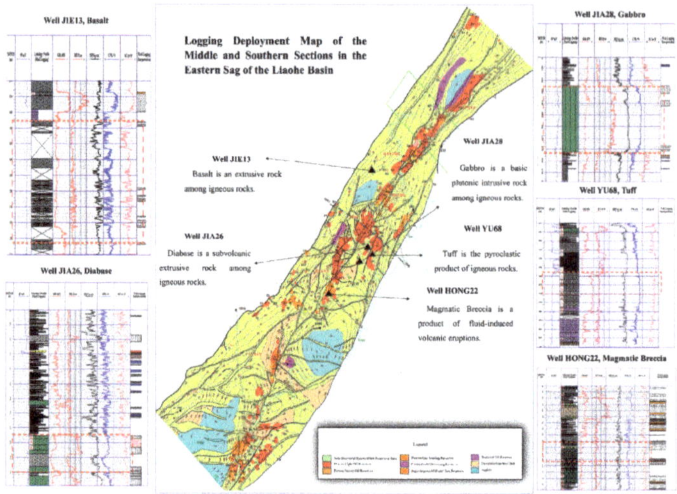

Figure 2. Logging deployment map of the middle and southern sections in the eastern sag of the Liaohe Basin.

Table 1 shows the descriptive statistical analysis of the corresponding logging data for the above-mentioned various rocks. As can be seen from Table 1, the mean and variance of gabbro's RT data are relatively high, mainly because the RT values are influenced by various metal minerals. The metal minerals that have a greater impact on the resistivity of gabbro are mainly sulfides and oxides. For sulfides, pyrrhotite affects gabbro's resistivity due to its conductivity and structural changes under high temperature and high-pressure conditions. Pyrite, common in gabbro, significantly changes the resistivity with its conductivity, pore distribution, and the rock-structure transformation caused by high temperature and high pressure. Chalcopyrite, often seen in magmatic-hydrothermal igneous rocks, impacts the resistivity through its conductivity and symbiotic relationships. Regarding oxides, magnetite, with good conductivity and magnetism, notably alters the rock's electrical properties when abundant, and its interaction with other minerals under high temperature and high pressure further affects the resistivity. Ilmenite, with a relatively high content, also influences the resistivity due to its structural changes and reactions with surrounding minerals under such conditions. Thus, under the influence of metal minerals, high temperature, and high pressure, the internal structure of gabbro changes, resulting in a substantial increase in resistivity. Subsequently, we found that the variation range of the DEN data on all rocks does not exceed 1, and the variance is nearly 0, indicating that the fluctuations in the DEN data on rocks are relatively small. However, in the multifractal analysis, data fluctuation is a key factor. If the data fluctuation is small, then the data do not meet the data conditions required by the multifractal analysis. Therefore, later on, we will conduct a separate multifractal discussion on the DEN data on rocks. From the perspectives of skewness and kurtosis, the skewness of the logging data corresponding to each rock is not 0, and the kurtosis is not 3, indicating that the logging data do not follow a normal distribution. Moreover, the Jarque–Bera (J–B) statistic shows that the logging data regarding each rock rejects the null hypothesis of the Gaussian distribution at a 1% confidence level, indicating that the distribution characteristics of the logging data are relatively complex. Therefore, the assumption of a normal distribution cannot accurately describe the fluctuation characteristics of the logging data. In summary, this paper focuses on the study of igneous rocks. We selected MF-DFA for the analysis. The goal was to analyze the multifractality of the logging data on basalt, diabase, gabbro, tuff, and magmatic breccia. After that, we chose MF-DCCA. This method was used to analyze the cross-correlations among the logging data. By using these two methods, we aim to gain a deeper understanding of the characteristics and relationships within the logging data regarding different igneous rocks.

Table 1. Descriptive statistical analysis of well logging data.

Rock Type	Logging Parameter	Min/m	Max/m	Mean/m	S.D./m	Skew	Kurt	J-B
Basalt	GR	12.11	139.86	38.83	498.09	2.05	3.83	2259.53 ***
	RT	0.73	49.64	4.50	20.30	3.48	17.71	34,119.20 ***
	DEN	1.85	2.82	2.37	0.03	−0.22	−0.14	1293.12 ***
	CNL	20.11	108.34	43.46	195.20	1.10	1.24	1017.19 ***
	AC	55.52	150.02	93.32	369.59	0.78	0.27	1272.20 ***
Diabase	GR	15.95	121.03	47.07	527.98	0.68	−0.13	486.47 ***
	RT	1.61	2724.01	209.41	139,607.61	3.27	12.23	5337.32 ***
	DEN	2.09	2.87	2.66	0.02	−1.27	1.27	393.34 ***
	CNL	6.04	47.77	25.42	83.02	0.64	−0.66	627.66 ***
	AC	53.43	149.47	67.88	90.21	4.54	35.15	46,512.70 ***
Gabbro	GR	49.73	133.05	68.63	153.30	1.71	2.87	1128.42 ***
	RT	3.73	14,590.20	3229.36	7,727,472.39	1.05	0.60	975.59 ***
	DEN	2.04	2.81	2.71	0.01	−2.36	8.21	4756.56 ***
	CNL	10.07	49.78	15.90	28.75	2.01	4.98	1934.15 ***
	AC	46.86	112.78	50.55	16.70	9.19	120.91	1,371,764.03 ***
Tuff	GR	30.86	210.20	135.83	223.02	−1.61	17.00	13,346.48 ***
	RT	3.90	218.27	12.73	171.09	10.20	129.66	1,064,387.02 ***
	DEN	2.39	2.83	2.59	0.00	−0.28	2.37	46.05 ***
	CNL	7.75	26.91	11.24	4.56	2.40	8.98	3799.86 ***
	AC	55.74	91.13	67.97	22.20	1.47	3.03	558.91 ***
Magmatic breccia	GR	24.08	99.41	38.06	100.54	1.54	3.15	997.85 ***
	RT	2.09	37.91	10.35	53.43	1.41	1.53	944.40 ***
	DEN	2.23	2.81	2.60	0.01	−0.49	0.19	516.32 ***
	CNL	16.69	40.45	26.69	26.78	0.18	−0.59	689.39 ***
	AC	54.66	100.16	70.02	60.58	0.81	0.55	595.26 ***

Note: *** indicates that the Jarque–Bera (J–B) statistic rejects the null hypothesis with a significance of 1% (9.21).

2.2. Methods

In the field of geological data processing, principal component analysis (PCA) and the random forest algorithm in machine learning are commonly used methods. PCA can reduce the dimensionality of data and extract main features, and its calculation is relatively simple, with intuitive results. However, it has obvious deficiencies in mining complex features and correlation relationships in geological data. Although the random forest algorithm has the ability to classify, predict, and process high-dimensional data, it is difficult to reveal the complex internal characteristics of the data. These limitations restrict their application in some geological research with high requirements for in-depth data analysis.

In sharp contrast, the multifractal technique has unique advantages. The multifractal technique can deeply mine the hidden complex features and correlation relationships in geological data, providing more detailed information for geological research. Especially in analyzing the multifractal properties of igneous rock logging data and the long-term correlations between different parameters, it plays an irreplaceable role. Although the multifractal technique has some drawbacks, such as a relatively complex calculation process, high requirements for computing resources and time, and the need for professional knowledge and experience for result interpretation, which is somewhat subjective, overall, its advantages in mining deep-level data and information are obvious. Compared with traditional statistical methods and the random forest algorithm, the multifractal technique can provide a more balanced perspective in complex geological data processing, opening up broader perspectives and more in-depth analysis approaches for geological research.

Although MF-DCCA is often used for time series, its application to depth sequences is still feasible. A time series is an arrangement of data in the time dimension. A depth sequence, on the other hand, is an arrangement of data in the depth dimension, such as underground depth. Both of them represent an ordered data set and may contain complex structures and variations. In multifractal analysis, the focus is on the self-similarity and irregularity of data at different scales. Whether it is time or depth, as long as the data set has such complex variation characteristics at different scales, multifractal analysis can be applied. Here, we mainly introduce MF-DCCA. This is because an MF-DFA can be deduced from an MF-DCCA. Therefore, the specific steps of MF-DCCA in this paper are as follows.

Suppose there are two depth series, namely, depth series $\{x(t)\}$ and $\{y(t)\}$, where $t = 1, 2, \ldots, N$ represents two depth series with a length of N.

The first step is to calculate the "profile" of each depth series:

$$\begin{aligned} X(t) &= \sum_{k=1}^{t} (x(k) - \overline{x}), \\ \overline{x} &= \sum_{k=1}^{N} x(k)/N, t = 1, 2, \ldots, N; \\ Y(t) &= \sum_{k=1}^{t} (y(k) - \overline{y}), \\ \overline{y} &= \sum_{k=1}^{N} y(k)/N, t = 1, 2, \ldots, N. \end{aligned} \tag{1}$$

where $\overline{x}$ and $\overline{y}$ denote the mean values of depth series $\{x(t)\}$ and $\{y(t)\}$, respectively.

The second step is to partition the sequence $X(t)$ and $Y(t)$ into non-overlapping sub-intervals of length s. The number of sub-intervals totals $N_s = \text{int}(N/s)$. In view of the fact that the length of the sequence N is usually not an integer multiple of s, in order to avoid the loss of information and to make all the data of the sequence $X(t)$ and $Y(t)$ participate in the computation, the partitioning process is usually repeated from the end of the sequence, so a total of $2N_s$ sub-intervals are obtained.

The range of the scale s is taken as $[s_{\min} = \max(20, N/100), s_{\max} = \min(20s_{\min}, N/10)]$, which yields

$$\begin{aligned} &X_v(k) = \{X(t), (v-1)s + 1 \le t \le vs\}, \\ &v = 1, 2, \ldots, 2N_s, k = 1, 2, \ldots, s; \\ &Y_v(k) = \{Y(t), (v-1)s + 1 \le t \le vs\}, \\ &v = 1, 2, \ldots, 2N_s, k = 1, 2, \ldots, s. \end{aligned} \tag{2}$$

In the third step, the least-squares method is applied to each sub-sequence. The local trend is fitted through polynomial fitting to obtain the fitting curves $\widetilde{X}_v(k)$ and $\widetilde{Y}_v(k)$. Then, the local trend on each sub-interval is eliminated with them. Then, the detrended covariance of each sub-sequence is

$$F_v(s) = \frac{1}{s}\sum_{k=1}^{s}\left|X_v(k) - \widetilde{X}_v(k)\right| \cdot \left|Y_v(k) - \widetilde{Y}_v(k)\right|. \tag{3}$$

In the fourth step, the order q descending convergence covariance fluctuation function is computed:

$$\begin{aligned} F_{xy}(q,s) &= \{\frac{1}{2N_s}\sum_{v=1}^{2N_s}[F_v(s)]^{q/2}\}^{1/q}, q \neq 0; \\ F_{xy}(q,s) &= \exp\{\frac{1}{4N_s}\sum_{v=1}^{2N_s}\ln F_v(s)\}, q = 0. \end{aligned} \tag{4}$$

In the fifth step, if there is a long-range cross-correlation between the two depth series, then the volatility function $F_{xy}(q,s)$ has a power law relationship with the depth scale s:

$$F_{xy}(q,s) \sim s^{h_{xy}(q)}, \tag{5}$$

In the formula, the scaling index $h_{xy}(q)$ is the generalized cross-Hurst index. When $h_{xy}(q)$ changes with the change in q, it is said that the cross-correlation of the two depth series has multifractal characteristics; otherwise, it has monofractal characteristics. When $h_{xy}(q) > 0.5$, the two depth series have long-range correlations. When $h_{xy}(q) = 0.5$, their cross-relationship can be described by a random walk. When $h_{xy}(q) < 0.5$, there is a negative long-range correlation, that is, anti-persistence.

In the sixth step, the Multifractal Mass Exponent is further obtained $\tau_{xy}(q)$:

$$\tau_{xy}(q) = qh_{xy}(q) - 1. \quad (6)$$

The Singularity Strength α_{xy} and the Multifractal Spectrum Function $f_{xy}(\alpha)$ are obtained by the Legendre transform:

$$\begin{aligned} \alpha_{xy}(q) &= \tau'_{xy}(q) = h_{xy}(q) + qh'_{xy}(q); \\ f_{xy}(\alpha) &= q(\alpha_{xy} - h_{xy}(q)) - 1. \end{aligned} \quad (7)$$

The above is the MF-DCCA, and when the depth series $\{x(t)\}$ and $\{y(t)\}$ are the same, the MF-DCCA transforms into the MF-DFA.

In addition, both $\Delta h_{xy} = \max(h_{xy}) - \min(h_{xy})$ and $\Delta\alpha_{xy} = \max(\alpha_{xy}) - \min(\alpha_{xy})$ can be used to measure the intensity of the multifractal, with Δh_{xy} being larger when the degree of the multifractal and the intensity of data fluctuation are richer and stronger; $\Delta\alpha_{xy}$ is the width of the multifractal spectrum, with $\Delta\alpha_{xy}$ being larger when there is more variation in the singularity and more unevenness in the distribution of the depth series and the absolute degree of data fluctuation is larger. The multifractal spectrum $f_{xy}(\alpha)$ portrays the local scalability of different fractal subsets; if $f_{xy}(\alpha)$ presents as a single-peaked bell shape, it indicates a multifractal feature; while for a single fractal, $f_{xy}(\alpha)$ is a point.

For the choice of order q, the fluctuation function $F_{xy}(q, s)$ will tend to stabilize when the absolute value of order q is larger than a certain range. The order q is often restricted to a specified range. Based on the practice of various references, this paper sets the range of q to $\{-10, -9, \ldots, 9, 10\}$. In addition, when fitting polynomials in the detrending step, this paper adopts second-order polynomials for the calculation, tests first-order and third-order polynomials, and finds that there is no big difference in the conclusions.

3. Results

3.1. Results of Multifractal Detrended Fluctuation Analysis

3.1.1. Multifractality Analysis

First, we perform multifractal analysis on all igneous DEN data, and the results are shown in Figure 3. It can be seen from Figure 3 that the generalized Hurst index does not change with the change in order. Combined with the descriptive statistical analysis results of all igneous rock DEN data, it shows that all igneous rock DEN data are not multifractal.

Figures 4–8 show the analysis of multifractals for basalt, diabase, gabbro, tuff, and magmatic breccia logging data (GR, RT, CNL, and AC), respectively. From Figures 4a, 5a, 6a, 7a and 8a, it can be seen that the generalized Hurst index of each rock's logging data decreases with the increase in q, showing a nonlinear relationship. This nonlinear change feature reflects the multifractal of the depth sequence, which means that the logging data have complex changes under different conditions. As can be seen from Figures 4b, 5b, 6b, 7b and 8b, each curve in each scale index graph is nonlinear, which further demonstrates the multifractal characteristics of the depth sequence, indicating that the related logging indicators of rocks, such as physical properties, will vary in a complex and uneven manner with different depth scales. It can be seen from Figures 4c, 5c, 6c, 7c and 8c that the curves in each multifractal spectrum satisfy the inverted parabola, which also shows that the depth sequence has multiple fractals. We find that the generalized Hurst index, scaling index, and multifractal spectrum confirm the multifractality of logging data from different angles when multifractal analysis is carried out on logging data. Therefore, it can be concluded that the logging data regarding basalt, diabase, gabbro, tuff, and magmatic breccia all have multifractality.

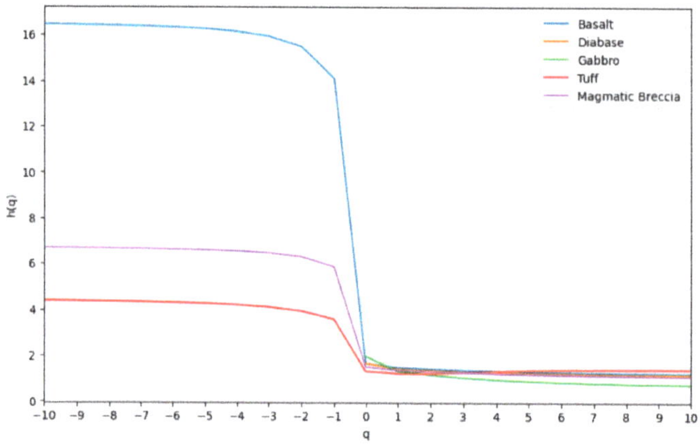

Figure 3. Multifractality of DEN data regarding all igneous rocks.

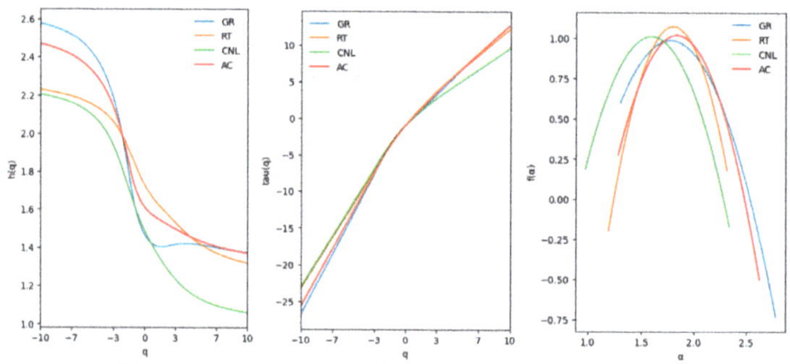

Figure 4. Multifractal analysis of basalt logging data. (**a**) Generalized Hurst index; (**b**) scale index; (**c**) multifractal spectrum.

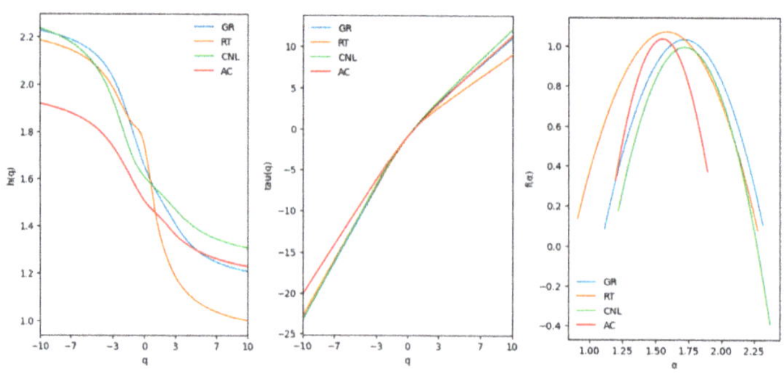

Figure 5. Multifractal analysis of diabase logging data. (**a**) Generalized Hurst index; (**b**) scale index; (**c**) multifractal spectrum.

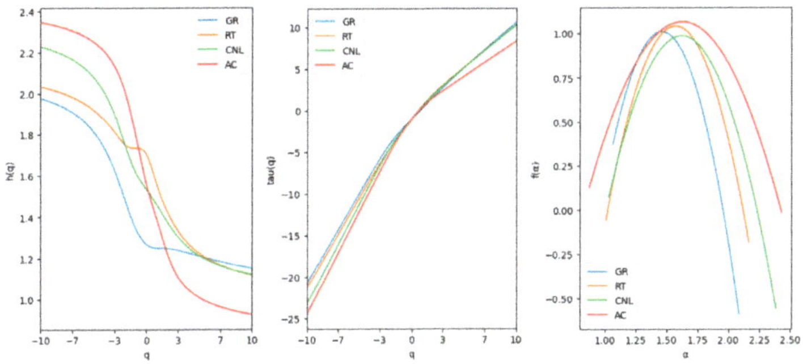

(**a**) (**b**) (**c**)

Figure 6. Multifractal analysis of gabbro logging data. (**a**) Generalized Hurst index; (**b**) scale index; (**c**) multifractal spectrum.

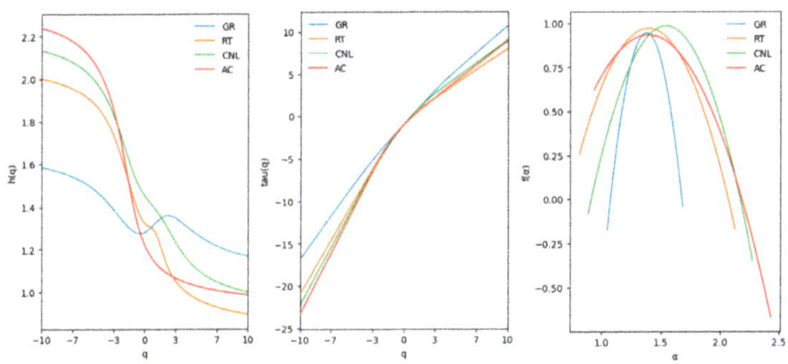

(**a**) (**b**) (**c**)

Figure 7. Multifractal analysis of tuff logging data. (**a**) Generalized Hurst index; (**b**) scale index; (**c**) multifractal spectrum.

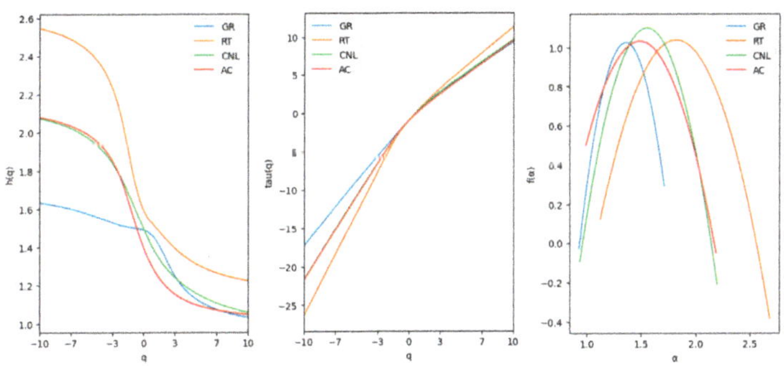

(**a**) (**b**) (**c**)

Figure 8. Multifractal analysis of magmatic breccia logging data. (**a**) Generalized Hurst index; (**b**) scale index; (**c**) multifractal spectrum.

Table 2 shows the statistics of the generalized Hurst exponent ($\Delta h = \max h(q) - \min h(q)$) and the spectral width ($\Delta\alpha = \max(\alpha) - \min(\alpha)$) of basalt logging data. It can be seen from Table 2 that, in basalt, the spectral width of GR data ($\Delta\alpha = 1.477$) is the widest, indicating that its multifractal strength is the strongest. In the depth range of 1827–2213 m, the maximum spectral width of basalt GR is mainly due to mineral and pore factors. The basalt mineral composition at this depth is complex, and the distribution of radioactive minerals, such as potassium feldspar, is uneven, which makes the GR data fluctuate greatly.

Concurrently, the pore size, distribution, and pore fluid properties are unstable, resulting in variable states of radioactive substances in the pores, which greatly increases the complexity and variability of GR data. CNL data, AC data, and RT data are relatively less affected by these factors, but the spectral width of GR data is prominent.

Table 2. Δh and $\Delta\alpha$ of basalt logging data.

Logging Parameter	Δh	$\Delta\alpha$
GR	1.206	1.477
RT	0.912	1.121
CNL	1.145	1.361
AC	1.100	1.343

Table 3 shows the statistics of Δh and $\Delta\alpha$ of diabase logging data. It can be seen from Table 3 that, in diabase, the spectral width of RT data is the widest, indicating that its multifractal strength is the strongest. In the depth section of 3628–3752 m, the maximum spectral width of RT ($\Delta\alpha = 1.372$) in diabase is mainly due to the mineral characteristics of rocks and fluid factors of pore structure. In terms of mineralogy, the types, contents, and distribution of conductive minerals in diabase vary significantly. For example, the local enrichment or dispersion of conductive minerals such as pyrite makes RT data fluctuate greatly, resulting in the largest spectral width. However, the data regarding GR, CNL, and AC are not as significantly affected by these factors as RT, but the RT spectral width is prominent.

Table 3. Δh and $\Delta\alpha$ of diabase logging data.

Logging Parameter	Δh	$\Delta\alpha$
GR	1.020	1.205
RT	1.186	1.372
CNL	0.929	1.157
AC	0.561	0.699

Table 4 shows the statistics of Δh and $\Delta\alpha$ of gabbro logging data. It can be seen from Table 4 that, in gabbro, the spectral width of AC data ($\Delta\alpha = 1.556$) is the widest, indicating that its multifractal strength is the strongest. In the depth range of 1323–1612 m, the maximum AC spectral width of gabbro is mainly due to the pore structure of the rock. The pore size, shape, and distribution of gabbro are extremely complex and uneven, which significantly impacts the propagation of sound waves. The pores of different scales and connectivity cause complex changes such as scattering, refraction, and reflection of sound waves, resulting in large fluctuations in AC data and the largest spectral width. However, the distribution of hydrogen content in gabbro at this depth is relatively stable, the distribution of conductive minerals is relatively uniform, and the content of radioactive minerals is low, but the AC spectral width of gabbro is prominent.

Table 4. Δh and $\Delta\alpha$ of gabbro logging data.

Logging Parameter	Δh	$\Delta\alpha$
GR	0.781	1.020
RT	0.923	1.154
CNL	1.092	1.352
AC	1.355	1.556

Table 5 shows the statistics of Δh and $\Delta\alpha$ of tuff logging data. It can be seen from Table 5 that, in tuff, the spectral width of AC data ($\Delta\alpha = 1.492$) is the widest, indicating that its multifractal strength is the strongest. In the depth section of 3578–3771 m, tuff has a wide variety of pore types, including primary and secondary pores, and the pore size is significantly different, the shape is irregular, and the distribution is extremely uneven. When propagating in it, these pore characteristics cause sound waves to encounter complex phenomena such as refraction, reflection, and scattering, resulting in larger fluctuations in AC data and wider spectral width.

Table 5. Δh and $\Delta\alpha$ of tuff logging data.

Logging Data	Δh	$\Delta\alpha$
GR	0.417	0.638
RT	1.105	1.311
CNL	1.132	1.385
AC	1.251	1.492

Similar to gabbro, the characteristics of different types of igneous rocks also have a significant impact on logging data. At the same time, tuff's mineral composition is complex. Mineral interfaces strongly interfere with sound waves during propagation, exacerbating AC data complexity. At 3578–3771 m, tuff's radioactive minerals, pore fluid, and conductive minerals are stable, with little impact on GR data, CNL data, and RT data. However, the AC spectral width is most prominent in tuff logging data.

Table 6 shows the statistics of Δh and $\Delta\alpha$ of the log data regarding the magmatic breccia. It can be seen from Table 6 that, among the magmatic breccia, the spectral width of the RT data ($\Delta\alpha = 1.562$) is the widest, indicating that its multifractal strength is the strongest. In the depth range of 3456–3613 m, the maximum spectral width of RT is mainly due to its unique rock structure and mineral composition characteristics. Magmatic breccia is cemented with rock debris of different sizes and shapes. This structure results in extremely complex pore spaces, wide pore sizes, and variable connectivity. The distribution of pore fluids in it is extremely uneven, and the distribution of conductive minerals in magmatic breccia also shows a high degree of dispersion. There may be high concentrations of conductive minerals in local areas, while other areas are relatively scarce. All these factors make the resistivity data regarding magmatic breccia fluctuate greatly, and the spectrum width of RT data is the largest.

Table 6. Δh and $\Delta\alpha$ of magmatic breccia logging data.

Logging Parameter	Δh	$\Delta\alpha$
GR	0.602	0.783
RT	1.322	1.562
CNL	1.015	1.259
AC	1.031	1.194

Overall, the logging data regarding different igneous rocks (GR, RT, CNL, and AC) all exhibit multifractal properties, and due to the uniqueness of mineralogical characteristics (mineral species, content, distribution, etc.) and pore structure characteristics (pore size, shape, distribution, connectivity, fluid properties, etc.), exploring the internal relationship between multifractal natures and reservoirs is crucial for understanding the features of igneous rock reservoirs.

Reservoirs, as crucial carriers for oil and gas storage and migration, directly impact the efficiency and benefits of oil and gas extraction [42]. The multifractal features of igneous rock logging data play a pivotal role in reservoir evaluation. Multifractal features can reflect reservoir heterogeneity. A more complex multifractal pattern indicates stronger heterogeneity, which is unfavorable for oil and gas extraction. This is because magma composition and cooling rate influence the rock pore structure, affecting the heterogeneity reflected by multifractal features. For instance, rocks from basic and acidic magma, as well as those with different cooling rates, have distinct pore characteristics, ultimately influencing reservoir exploitation difficulty. Furthermore, multifractal features are closely related to reservoir pore structure, permeability, and oil-bearing properties, assisting in reservoir quality assessment. Fracture development and mineral alteration impact permeability, which is reflected in multifractal features and affects reservoir seepage capacity evaluation. Porosity and rock wettability, which determine the distribution of oil and gas in the reservoir, are also mirrored in multifractal features, helping to judge reservoir oil-bearing quality. These geological variables shape the multifractal features of igneous rock logging data and comprehensively influence reservoir evaluation.

3.1.2. Analysis of the Causes of Multifractals

A large amount of research literature [43–45] shows that there are two main causes of the formation of multifractal features: they are caused by the correlation between different ranges of large fluctuations and small fluctuations, that is, the correlation multifractal; they are caused by the probability distribution of data fluctuations with "sharp peaks and thick tails", that is, the distribution multifractal. We explore the contribution of the above two reasons to the formation of multifractal features through random rearrangement and phase randomization. Random rearrangement can destroy the long-memory characteristics of the sequence and retain its probability distribution characteristics; phase randomization can weaken the non-Gaussian distribution characteristics of the sequence and retain its long-range correlation characteristics. In order to analyze the causes of multifractal features, the rearrangement sequence generated by the original sequence after random rearrangement and the replacement sequence generated by phase randomization are carried out in multifractal analysis, respectively. Further, the index to measure the multifractal feature strength of the rearrangement sequence and the replacement sequence is obtained by the calculation of Δh, in turn, Δh_{per} and Δh_{sur}, and the original sequence is recorded as Δh_{ori}. Then, we calculate the size of $\Delta\Delta h_{per} = \Delta h_{ori} - \Delta h_{per}$ and $\Delta\Delta h_{sur} = \Delta h_{ori} - \Delta h_{sur}$, respectively. The magnitudes of $\Delta\Delta\widetilde{h}_{per} = 100\% \times \Delta\Delta h_{per}/(\Delta\Delta h_{sur} + \Delta\Delta h_{per})$ and $\Delta\Delta\widetilde{h}_{sur} = 100\% \times \Delta\Delta h_{sur}/(\Delta\Delta h_{sur} + \Delta\Delta h_{per})$ are compared to determine the long-range correlation of the data and the primary and secondary contributions of the probability distribution of the data to the causes of multifractality [46].

First, we randomly rearranged and randomized the logging data for 1827–2213 m regarding basalt in 13 wells in the eastern sag of the Liaohe Basin, 3628–3752 m regarding diabase in 26 wells, 1323–1612 m regarding gabbro in 28 wells, 3578–3771 m regarding tuff in 68 wells, and 3456–3613 m regarding magmatic breccia in Red 22 wells, respectively, and then we used MF-DFA for analysis. Finally, the $\Delta\Delta\widetilde{h}_{per}$ and $\Delta\Delta\widetilde{h}_{sur}$ of each logging data were calculated. The results are shown in Table 7.

It can be seen from Table 7 that the GR data regarding basalt, gabbro, and amphibolite, the CNL and AC data regarding diabase, gabbro, and tuff, and the RT data regarding tuff are much greater than $\Delta\Delta\widetilde{h}_{sur}$ for $\Delta\Delta\widetilde{h}_{sur}$. This is because, during the cooling process of magma, due to the different crystallization orders and conditions of different minerals, the mineral composition content is concentrated within a certain range, but due to the small differences in local temperature, pressure, and other conditions and the late geological transformation (such as weathering, cracks caused by tectonic movements, etc., which affect the physical properties of rocks), some areas with abnormal compositions are generated, which leads to the complexity of the formation process of igneous rocks and produces the probability distribution of a "peak thick tail", which makes it play a major role in the genesis of multifractal characteristics. The $\Delta\Delta\widetilde{h}_{sur}$ and $\Delta\Delta\widetilde{h}_{per}$ of basalt RT and AC data, diabase GR and RT data, and amphibolite RT data and CNL data are close to 50%, indicating that the long-range correlation of these depth sequences plays an almost consistent role with the probability distribution of the sequence; while the CNL data regarding basalt, RT data regarding gabbro, GR data regarding tuff, and AC data regarding amphibolite are much larger than $\Delta\Delta\tilde{h}_{per}$ for $\Delta\Delta\tilde{h}_{sur}$, which means that these logging data have similar structures and laws at different depth scales, so the long-range correlation of these depth sequences plays a decisive role in the genesis of multifractal features.

Table 7. Genesis analysis of multifractal characteristics of rock logging data.

Rock Type	Logging Parameter	$\Delta\Delta\tilde{h}_{per}$/%	$\Delta\Delta\tilde{h}_{sur}$/%
Basalt	GR	15.81	84.19
	RT	44.43	55.57
	CNL	83.71	16.29
	AC	57.99	42.01
Diabase	GR	55.67	44.33
	RT	48.84	51.16
	CNL	41.96	58.04
	AC	13.03	86.97
Gabbro	GR	29.07	70.93
	RT	97.47	2.53
	CNL	31.08	68.92
	AC	28.27	71.73
Tuff	GR	75.72	24.28
	RT	23.91	76.09
	CNL	29.71	70.29
	AC	43.90	56.10
Magmatic breccia	GR	29.94	70.06
	RT	48.45	51.55
	CNL	51.20	48.80
	AC	72.13	27.87

3.2. Results of Multifractal Detrended Cross-Correlation Analysis

3.2.1. Cross-Correlation Test

Podobnik et al. [47] proposed a new cross-correlation statistic, used to determine the existence of the cross-correlation relationship between two time series. The method is as follows: Suppose there are two depth series $\{x_i\}_{i=1}^{N}$ and $\{y_i\}_{i=1}^{N}$ with a length of N. Firstly, calculate the cross-correlation coefficients of the two depth series.

$$C_i = \frac{\sum\limits_{k=i+1}^{N} x_k y_{k-1}}{\sqrt{\sum\limits_{k=1}^{N} x_k^2 \sum\limits_{k=1}^{N} y_k^2}}. \tag{8}$$

If the two sequences $\{x_i\}_{i=1}^{N}$ and $\{y_i\}_{i=1}^{N}$ are independent of each other, then the cross-correlation coefficients C_i and C_k are uncorrelated.

$$E(C_i C_k) \propto \sum_{j=i+1}^{N} \sum_{h=k+1}^{N} E(x_j y_{j-i} x_h y_{h-k}) = \sum_{j=1}^{N} \sum_{h=1}^{N} E(x_j y_h) E(x_{j-i} y_{h-k}). \tag{9}$$

So, when $i \neq k$, the variance of the cross-correlation coefficient C_i is as follows:

$$V(C_i) = E(C_i^2) = \frac{\sum\limits_{j=i+1}^{N} \sum\limits_{h=k+1}^{N} E(x_j y_{j-i} x_h y_{h-k})}{\sum\limits_{j=1}^{N} \sum\limits_{h=1}^{N} E(x_j^2 x_h^2)} = \frac{\sum\limits_{j=i+1}^{N} \sum\limits_{h=k+1}^{N} E(x_j y_{j-i} x_h y_{h-k})}{\sigma^2 \sigma'^2 N^2}, \tag{10}$$

where $E(x_i x_h) = \sigma^2 \delta_{j,k}$ and $E(y_i y_h) = \sigma'^2 \delta_{j,k}$; thus,

$$V(C_i) = \frac{\sum_{j=i+1}^{N} \sum_{h=k+1}^{N} \delta_{j,h}\delta_{j-i,h-i}}{N^2} = \frac{\sum_{j=i+1}^{N} \delta_{i,j}}{N^2} = \frac{N-i}{N^2}. \quad (11)$$

When N is relatively large, the cross-correlation coefficient C_i asymptotically follows a normal distribution, so $C_i/\sqrt{(N-i)/N^2}$ asymptotically follows a normal distribution with an expectation of 0 and a variance of 1. According to the definition of the chi-squared distribution χ^2, the cross-correlation statistic $Q_{CC}(m) = N^2 \sum_{i=1}^{m} \frac{C_i^2}{N-i}$.

The cross-correlation statistic $Q_{CC}(m)$ approximately follows the χ^2 distribution, with the degree of freedom being m. If the degree of freedom m is given and $Q_{CC}(m) \geq \chi^2(m)$, it indicates that there is a cross-correlation relationship between the two depth sequences.

The correlation test between different igneous rock logging data is shown in Figure 9. The results show that, at the significance level of 5%, with the increase in degrees of freedom, the cross-correlation statistics $Q_{CC}(m)$ of each pair of depth series in each period are always higher than the critical value of the $\chi^2(m)$ distribution. Therefore, we can conclude that there is a significant cross-correlation between depth series.

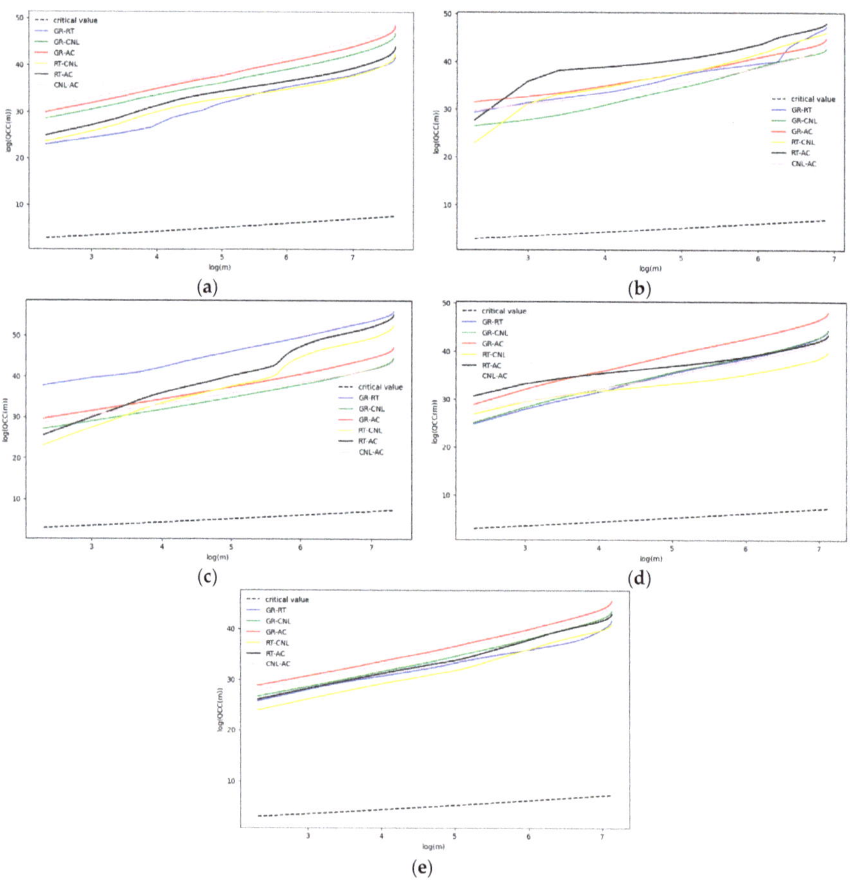

Figure 9. Results of multifractal detrended cross-correlation analysis. (**a**) Basalt; (**b**) diabase; (**c**) gabbro; (**d**) tuff; (**e**) magmatic breccia.

3.2.2. Cross-Correlation Analysis

The MF-DCCA was used to analyze the cross-correlation between the logging data regarding basalt, diabase, gabbro, tuff, and magmatic breccia in igneous rocks. The cross-correlation results between different igneous rock logging data are shown in Table 8.

Table 8. Δh_{xy} and $\Delta \alpha_{xy}$ between igneous rock logging data.

Rock Type	Logging Parameter	Δh_{xy}	$\Delta \alpha_{xy}$
Basalt	GR-RT	0.32	0.42
	GR-CNL	0.39	0.50
	GR-AC	0.65	0.90
	RT-CNL	0.40	0.48
	RT-AC	0.57	0.76
	CNL-AC	0.73	0.90
Diabase	GR-RT	0.61	0.77
	GR-CNL	0.76	0.97
	GR-AC	0.68	0.90
	RT-CNL	0.34	0.49
	RT-AC	0.71	0.87
	CNL-AC	0.96	1.21
Gabbro	GR-RT	0.74	0.93
	GR-CNL	0.96	1.20
	GR-AC	0.83	1.05
	RT-CNL	0.65	0.82
	RT-AC	0.89	1.10
	CNL-AC	0.73	0.92
Tuff	GR-RT	0.75	0.97
	GR-CNL	0.56	0.74
	GR-AC	0.64	0.79
	RT-CNL	0.80	1.03
	RT-AC	0.55	0.73
	CNL-AC	0.47	0.64
Magmatic breccia	GR-RT	0.79	1.03
	GR-CNL	1.01	1.26
	GR-AC	0.88	1.08
	RT-CNL	0.88	1.09
	RT-AC	0.75	0.93
	CNL-AC	1.19	1.43

As can be seen from Table 8, the results of Δh_{xy} and $\Delta \alpha_{xy}$ among the logging data regarding various igneous rocks all show correlations between them. However, different igneous rocks exhibit different properties, which contain rich geological information. The inherent connections revealed by these correlations offer crucial clues for our in-depth understanding of the geological structure and physical properties of igneous rocks. These clues are essential for subsequent logging interpretation and reservoir evaluation. There are negative long-range correlations between GR and RT, GR and CNL in basalt, and RT and CNL in diabase. In basalt, this means that the distribution of radioactive minerals that determine the natural gamma value and the minerals or pore structures affecting resistivity and neutron porosity show inverse change relationships. In diabase, the factors affecting resistivity and those affecting neutron porosity in the rock develop in opposite directions. There are long-range correlations among the logging data regarding other igneous rocks, which indicates that the minerals and pore structures corresponding to various physical properties change synergistically on a large scale and have closer internal connections.

For logging interpretation and reservoir evaluation work, these correlation results are of great significance. When evaluating reservoir characteristics, different correlation features can help judge the continuity, heterogeneity, and fluid distribution of the reservoir. Thus, they provide solid data support for the formulation of oil and gas exploration and development plans.

Specifically, the long-term correlations among various logging parameters have multiple important impacts on oil and gas exploration strategies. In well-location deployment, in areas where there is a negative long-term correlation between natural gamma-ray logging and resistivity logging, as well as between natural gamma-ray logging and compensated neutron logging in basalt, due to the complex underground geological conditions and large variations in reservoir properties, well-location deployment needs to be carried out with caution. The subsurface structure can be explored by densifying well locations or using advanced seismic exploration methods to prevent missing reservoirs or drilling into non-reservoir areas. In areas with long-term positive-correlated igneous rock logging data, the geological parameters are relatively stable, and the reservoir properties are somewhat predictable. Well-location deployment can be optimized by combining other geological data to improve the success rate of drilling. In terms of reservoir evaluation, the long-term correlations of different logging parameters are beneficial for comprehensively judging reservoir properties. In a positively correlated parameter combination, when one parameter shows favorable characteristics for oil and gas storage, other parameters may also have a similar trend, strengthening the basis for judging high-quality reservoirs. For negatively correlated parameters, in-depth analysis is required to avoid misjudgment. Meanwhile, long-term correlations can also help identify complex reservoir features such as fracture and pore-structure complexity, providing accurate basic data for reservoir modeling and numerical simulation. In reservoir simulation and the formulation of extraction plans, understanding long-term correlations helps to set the boundary and initial conditions of reservoir simulation. For areas with strong positive correlations, conventional extraction methods can be adopted, and the flow laws of oil and gas can be predicted based on the coordinated changes in parameters. For negatively correlated areas, flexible extraction plans need to be formulated to prevent problems such as water-channeling and gas-channeling. Moreover, according to the correlations between logging parameters and key extraction parameters, the strategies for improving recovery efficiency can be adjusted pertinently, and appropriate stimulation measures can be selected to enhance oil and gas recovery.

4. Discussion

First, we use MF-DFA to study the multifractality of igneous rock logging data. This effectively solves the problem of understanding the internal complexity and heterogeneity of igneous rocks. It can clearly show the differences in the structure and properties of igneous rocks during their long geological history. These differences are often overlooked in traditional data interpretation, which are caused by different geological processes and environmental changes. Second, we analyze the causes of the multifractality of igneous rock logging data. One reason is that the concentration and anomaly of mineral components in igneous rocks lead to the non-uniformity of component content data. The other reason is that, by maintaining the correlation of data on a large scale, local changes on a small scale can have an impact on a larger range. This enhances the complexity and self-similarity of the data, further resulting in the emergence of multifractal characteristics. Finally, we use MF-DCCA to discuss the correlations among igneous rock logging data. The correlations among different igneous rock logging data reveal the differences in rock physical properties, reflecting the internal relationships of characteristics such as the internal mineral composition and pore structure of rocks. These properties are closely related to the storage and migration of oil and gas, which help to understand the distribution laws of oil and gas in these igneous rock reservoirs. Meanwhile, the rock-characteristic information provided by the results obtained through the MF-DCCA method can help us more accurately map the underground rock distribution, thus better judging the possible locations of oil and gas reservoirs, as well as the storage and flow laws of oil and gas within them. This provides a key basis for oil and gas exploration and development, improves exploration efficiency and development success rate, and helps identify potential oil and gas reservoir areas. In the process of geological modeling, based on the correlation results of igneous rock logging data obtained from MF-DCCA analysis, incorporating these correlations into geological modeling can provide more accurate constraints for geological modeling, more accurately depict the spatial distribution and physical property changes of underground rocks, and construct a three-dimensional model that is more in line with the actual underground geological conditions. For example, when establishing an igneous rock reservoir model, fully considering the positive and negative correlations among different logging data can more reasonably allocate reservoir parameters and improve the accuracy of the model in predicting the location and scale of oil and gas reservoirs. In addition, by continuously and deeply analyzing newly acquired logging data using MF-DCCA, the geological model can be continuously updated and optimized to better adapt to the complex and changeable underground geological conditions, providing a more reliable basis for subsequent oil and gas exploration and development.

Combined with the research results of this paper, the following steps can be taken to incorporate the multifractal results of igneous rock logging data into the actual exploration and drilling decision-making process. First, we collect GR, RT, AC, and CNL logging data on igneous rocks, use the multifractal detrended fluctuation analysis method to clarify the commonalities and characteristics of the data, and integrate them to establish a database. Second, the multifractal detrended cross-correlation analysis method will be used to analyze the correlations among different igneous rock logging data and evaluate the impact of parameter combinations on geological judgment based on the results. Third, we combine the multifractal characteristics of igneous rock logging data with the correlation analysis results, construct an accurate geological model that can reflect the underground distribution, characteristics, and relationship with the surrounding geological environment of igneous rocks, and optimize the model using multifractal characteristics. Then, based on the geological model, we determine the key exploration areas and directions according to the characteristics and distribution of igneous rocks, use the characteristics of logging data to predict the location and quality of reservoirs, and reasonably arrange exploration resources. Finally, during drilling, we collect and analyze new logging data in real time, compare it with the previous multifractal analysis results, and adjust the exploration and drilling decisions according to the actual situation. If there are deviations, we re-evaluate and correct the geological model.

Although certain achievements have been made in the current research, there is still room for improvement. In terms of lithology classification, supervised learning algorithms such as the support vector machine (SVM) can be used. By training on a large number of known lithology igneous rock logging data, a lithology classification model can be established. This model can be used to classify new data during data collation, improving the accuracy and convenience of data collation. When the amount of igneous rock logging data is large, parallel-computing technologies such as the Python multiprocessing module can be adopted. Decomposing the data-processing tasks into multiple subtasks and running them simultaneously on multiple processor cores can significantly shorten the time for data collation and cleaning. The application of these technologies and the deepening of research are expected to further improve the method of integrating the multifractal results of igneous rock logging data into the exploration and drilling decision-making process and promote the development of the earth science field in oil and gas exploration and related research.

Author Contributions: Conceptualization, S.W.; software, Z.W.; validation, X.Q.; formal analysis, S.W.; investigation, D.M.; resources, Z.W.; data curation, S.W.; writing—original draft preparation, S.W. and D.M.; visualization, X.Q.; supervision, Z.W.; project administration, D.M.; funding acquisition, D.M. All authors have read and agreed to the published version of the manuscript.

Funding: This research was funded by Natural Science Foundation of Jilin Province with grant number YDZJ202501ZYTS661, the Education Department of Jilin Province under grant number No. JJKH20250805KJ; 20224BR98740058 and the Postgraduate Innovation Program Project of Beihua University with grant number 2024037.

Data Availability Statement: The data used in this study are restricted due to privacy and confidentiality agreements with the data providers. Interested researchers can submit a data access request to moudan@beihua.edu.cn.

Conflicts of Interest: The authors declare no conflict of interest.

References

1. Schutter, S. Hydrocarbon Occurrence and Exploration in and Around Igneous Rocks. *Geol. Soc. Lond. Spec. Publ.* **2003**, *214*, 7–33. [CrossRef]
2. Tang, H.; Tian, Z.; Gao, Y.; Dai, X. Review of Volcanic Reservoir Geology in China. *Acta Pet. Sin.* **2022**, *232*, 104158. [CrossRef]
3. Yao, W.J.; Chen, Z.H.; Hu, T.T.; Liang, Z.L.; Jia, C.M.; Wu, K.Y.; Pan, T.; Yu, H.T.; Dang, Y.F. Storage Space, Pore Structure, and Primary Control of Igneous Rock Reservoirs in Chepaizi Bulge, Junggar Basin, Western China: Significance for Oil Accumulation. *J. Pet. Sci. Eng.* **2020**, *195*, 107836. [CrossRef]
4. Chen, Z.H.; Wang, X.Y.; Wang, X.L.; Zhang, Y.G.; Yang, D.S.; Tang, Y. Characteristics and Petroleum Origin of the Carboniferous Volcanic Rock Reservoirs in the Shixi Bulge of Junggar Basin, Western China. *Mar. Petrol. Geol.* **2017**, *80*, 517–537. [CrossRef]
5. Li, N.; Qiao, D.X.; Li, Q.F.; Wu, H.L.; Fu, Y.S.; Dong, L.X.; Feng, Q.F.; Wang, K.W. Theory on logging interpretation of igneous rocks and its application. *Pet. Explor. Dev.* **2009**, *36*, 683–692.
6. Han, R.Y.; Wang, Z.W.; Wang, W.H.; Xu, F.H.; Qi, X.H.; Cui, Y.T.; Zhang, Z.T. Igneous rocks lithology identification with deep forest: Case study from eastern sag, Liaohe basin. *J. Appl. Geophys.* **2023**, *208*, 104892. [CrossRef]
7. Liu, Z.L.; Wang, Z.W.; Zhou, D.P.; Zhao, S.Q.; Xiang, M. Pore distribution characteristics of the igneous reservoirs in the Eastern Sag of the Liaohe depression. *Open Geosci.* **2017**, *9*, 161–173.
8. Annen, C.; Blundy, J.D.; Leuthold, J.; Sparks, R.S.J. Construction and evolution of igneous bodies: Towards an integrated perspective of crustal magmatism. *Lithos* **2015**, *230*, 206–221. [CrossRef]
9. Smith, J.A.; Johnson, K.R.; Brown, L.D. Petrology and geochemistry of igneous rocks: Implications for hydrocarbon exploration. *J. Geochem. Explor.* **2018**, *190*, 107–120.
10. Fisher, A.T. Permeability within basaltic oceanic crust. *Rev. Geophys.* **1998**, *36*, 143–182. [CrossRef]
11. Wang, W.H.; Wang, Z.W.; Han, R.Y.; Xu, F.H.; Qi, X.H.; Cui, Y.T. Lithology classification of volcanic rocks based on conventional logging data of machine learning: A case study of the eastern depression of Liaohe oil field. *Open Geosci.* **2021**, *13*, 1245–1258.
12. Wang, P.J.; Zheng, C.Q.; Shu, P.; Liu, W.Z.; Huang, Y.L.; Tang, H.F.; Cheng, R.H. Classification of Deep Volcanic Rocks in Songliao Basin. *Pet. Geol. Oilfield Dev. Daqing* **2007**, *26*, 17–22.
13. Ma, X.H.; Li, G.H.; Ying, D.L.; Zhang, B.J.; Li, Y.; Dai, X.; Fan, Y.; Zeng, Y.X. Distribution and Gas-Bearing Properties of Permian Igneous Rocks in Sichuan Basin, SW China. *Pet. Explor. Dev.* **2019**, *46*, 228–237. [CrossRef]
14. Sun, A.; Huang, Y.L.; Li, J.; Feng, Y.H.; Wang, J.F.; Wang, P.J. Oligocene Diabase of Eastern Sag in Liaohe Basin, NE China: Characteristics, Identification and Hydrocarbon Accumulation. *Oil Gas Geol.* **2016**, *37*, 372–380.
15. Jiao, S.T.; Zhou, Y.Z.; Zhang, Q.; Jin, W.J.; Liu, Y.P.; Wang, J. Study on intelligent discrimination of tectonic settings based on global gabbro data from GEOROC. *Acta Petrol. Sin.* **2018**, *34*, 3189–3194.
16. Kanakiya, S.; Adam, L.; Rowe, M.C.; Lindsay, J.M.; Esteban, L. The role of tuffs in sealing volcanic conduits. *Geophys. Res. Lett.* **2021**, *48*, e2021GL095175. [CrossRef]
17. Sillitoe, R.H. Ore-related breccias in volcanoplutonic arcs. *Econ. Geol.* **1985**, *80*, 1467–1514. [CrossRef]
18. Han, R.Y.; Wang, Z.W.; Wang, W.H.; Xu, F.H.; Qi, X.H.; Cui, Y.T. Lithology identification of igneous rocks based on XGboost and conventional logging curves, a case study of the eastern depression of Liaohe Basin. *J. Appl. Geophys.* **2021**, *195*, 104480.
19. Wang, Z.H.; Shu, X.M.; Sun, Z.C.; Luo, X.J.; Dai, X.J.; Dai, Y. Igneous lithology identification and lithofacies classification in the basin using logging data: Taking Junggar Basin as an example. *Earth Sci. Front.* **2015**, *22*, 254.
20. Moskalewicz, D. Spektrometryczne profilowanie gamma w odsłonięciach geologicznych: Metody i przykłady zastosowania. *Przegl. Geol.* **2022**, *70*, 806–815. [CrossRef]

21. Huang, B.Z.; Pan, B.Z. Characteristics of log responses and lithology determination of igneous rock reservoirs. *J. Geophys. Eng.* **2004**, *1*, 51–55. [CrossRef]
22. Guo, Y.H.; Pan, B.Z.; Zhang, L.H.; Lei, J.; Fan, Y.F.; Ruhan, A.; Yan, D.D.; Zhao, Y.Q. A Study on Water Saturation Predictions in Igneous Reservoirs Based on the Relationship Between the Transverse Relaxation Time and the Resistivity Index. *J. Pet. Sci. Eng.* **2021**, *208*, 109519. [CrossRef]
23. İnce, İ. Predicting index-mechanical properties of igneous rock using electrical resistivity method. *Mining Metall. Explor.* **2022**, *39*, 531–539. [CrossRef]
24. Pechnig, R.; Delius, H.; Bartetzko, A. Effect of compositional variations on log responses of igneous and metamorphic rocks. II: Acid and intermediate rocks. *Geol. Soc. Lond. Spec. Publ.* **2005**, *240*, 279–300. [CrossRef]
25. Yari, M.; Bagherpour, R. Implementing acoustic frequency analysis for development the novel model of determining geomechanical features of igneous rocks using rotary drilling device. *Geotech. Geol. Eng.* **2018**, *36*, 1805–1816. [CrossRef]
26. Wang, W.H.; Wang, P.J.; Wang, Z.W.; Xu, F.; Qi, X.H.; Han, R.Y.; Cui, Y.T. Acoustic time–frequency characteristics of igneous reservoir with different fluid properties. *Acta Geophys.* **2021**, *69*, 1753–1768. [CrossRef]
27. Qiu, X.L.; Fu, Y.X.; Yan, Z.D.; Zheng, X.M.; Wang, M.X.; Lan, Z. Effective reservoir identification and sweet spot prediction in Chang 8 Member tight oil reservoirs in Huanjiang area, Ordos Basin. *Open Geosci.* **2024**, *16*, 20220584. [CrossRef]
28. Wang, X.R.; Pan, B.Z.; Guo, Y.H.; Wang, Q.H.; Guan, Y. Multi-scale data joint inversion of minerals and porosity in altered igneous reservoirs—A case study in the South China Sea. *Pet. Sci.* **2024**, *21*, 206–220. [CrossRef]
29. Dentith, M.; Enkin, R.J.; Morris, W.; Adams, C.; Bourne, B. Petrophysics and mineral exploration: A workflow for data analysis and a new interpretation framework. *Geophys. Prospect.* **2020**, *68*, 178–199. [CrossRef]
30. Tzanis, C.G.; Koutsogiannis, I.; Philippopoulos, K.; Kalamaras, N. Multifractal Detrended Cross-Correlation Analysis of Global Methane and Temperature. *Remote Sens.* **2020**, *12*, 557. [CrossRef]
31. Delbianco, F.; Tohmé, F.; Stosic, T.; Stosic, B. Multifractal behavior of commodity markets: Fuel versus non-fuel products. *Phys. A* **2016**, *457*, 573–580. [CrossRef]
32. Wang, F.; Ye, X.; Wu, C.X. Multifractal characteristics analysis of crude oil futures prices fluctuation in China. *Phys. A* **2019**, *533*, 122021. [CrossRef]
33. Mensi, W.; Sensoy, A.; Vo, X.V.; Kang, S.H. Impact of COVID-19 outbreak on asymmetric multifractality of gold and oil prices. *Resour. Policy* **2020**, *69*, 101829. [CrossRef]
34. Wang, F.; Xu, J.; Fan, Q.J. Statistical properties of the detrended multiple cross-correlation coefficient. *Commun. Nonlinear Sci.* **2021**, *99*, 105781. [CrossRef]
35. Xiang, J.; Xu, Y.P.; Yuan, J.; Wang, Q.; Wang, J.; Deng, X.J. Multifractal Analysis of River Networks in an Urban Catchment on the Taihu Plain, China. *Water* **2019**, *11*, 2283. [CrossRef]
36. Gao, Z.Y.; Niu, F.J.; Lin, Z.J.; Luo, J. Fractal and Multifractal Analysis of Soil Particle-Size Distribution and Correlation with Soil Hydrological Properties in Active Layer of Qinghai–Tibet Plateau, China. *Catena* **2021**, *203*, 105373. [CrossRef]
37. Boulassel, A.; Zaourar, N.; Gaci, S.; Boudella, A. A new multifractal analysis-based for identifying the reservoir fluid nature. *J. Appl. Geophys.* **2021**, *185*, 104185. [CrossRef]
38. Sui, L.L.; Wang, H.Y.; Wu, J.S.; Zhang, J.W.; Yu, J.; Ma, X.Y.; Sun, Q.J. Fractal description of rock fracture networks based on the space syntax metric. *Fract. Fractal.* **2022**, *6*, 353. [CrossRef]
39. Podobnik, B.; Stanley, H.E. Detrended Cross-Correlation Analysis: A New Method for Analyzing Two Nonstationary Time Series. *Phys. Rev. Lett.* **2008**, *100*, 084102. [CrossRef]
40. Zhou, W.X. Multifractal detrended cross-correlation analysis for two nonstationary signals. *Phys. Rev. E* **2008**, *77*, 066211. [CrossRef]
41. Ge, X.M.; Fan, Y.R.; Zhu, X.J.; Deng, S.G.; Wang, Y. A method to differentiate degree of volcanic reservoir fracture development using conventional well logging data—An application of kernel principal component analysis (KPCA) and multifractal detrended fluctuation analysis (MFDFA). *IEEE J.-STARS* **2014**, *7*, 4972–4978. [CrossRef]
42. Guo, S.M.; Yang, C.B.; Han, L.G.; Feng, Y.Z.; Zhao, J.M. Lithology classification of igneous rocks using C-band and L-band dual-polarization SAR data. *Open Geosci.* **2023**, *15*, 20220465. [CrossRef]
43. Lin, J.S.; Chen, Q. Fault Diagnosis of Rolling Bearings Based on Multifractal Detrended Fluctuation Analysis and Mahalanobis Distance Criterion. *Mech. Syst. Signal Process.* **2013**, *38*, 515–533. [CrossRef]
44. Lu, X.S.; Tian, J.; Zhou, Y.; Li, Z.H. Multifractal Detrended Fluctuation Analysis of the Chinese Stock Index Futures Market. *Phys. A* **2012**, *392*, 1452–1458. [CrossRef]
45. Wu, Y.; He, Y.; Wu, M.W.; Lu, C.; Gao, S.Y.; Xu, Y.W. Multifractality and Cross-Correlation Analysis of Streamflow and Sediment Fluctuation at the Apex of the Pearl River Delta. *Sci. Rep.* **2018**, *8*, 16553. [CrossRef]

46. Yang, J.; Feng, Y.; Yang, H. Research on the Four Agricultural Commodity Futures Markets Based on National Food Security under the Impact of Extreme Events. *Chin. J. Manag.* **2024**, *21*, 96.
47. Podobnik, B.; Horvatic, D.; Petersen, A.M.; Stanley, H.E. Cross-correlations Between Volume Change and Price Change. *Proc. Natl. Acad. Sci. USA* **2009**, *106*, 22079–22084. [CrossRef]

fractal and fractional

Article

Fractal Analysis of Organic Matter Nanopore Structure in Tectonically Deformed Shales

Mingliang Liang [1,2,*], Min Dong [1,2], Zongxiu Wang [1,2], Kaixun Zhang [1,2], Xiaoshi Li [1,2] and Xingqiang Feng [1,2]

1 Key Laboratory of Petroleum Geomechanics Chinese Geological Survey, Institute of Geomechanics, Chinese Academy of Geological Sciences, Beijing 100081, China
2 Key Laboratory of Paleomagnetism and Tectonic Reconstruction, Ministry of Natural Resources, Beijing 100081, China
* Correspondence: liangmingl09@mails.ucas.ac.cn

Abstract: Fractal analysis was used to characterize the organic matter nanopore structure in tectonically deformed shales, providing insights into the heterogeneity and complexity of the pore network. Shale samples from different tectonic deformation styles (undeformed, brittle deformed, and ductile deformed) in the Lower Cambrian Niutitang Formation in western Hunan, South China, were collected. By comprehensively applying techniques such as low-temperature gaseous (CO_2 and N_2) adsorption (LTGA), scanning electron microscopy (SEM), and ImageJ analysis, we accurately obtained key parameters of the pore structure. The results show ductile deformation reduces fractal dimension (D_M) by ~0.2 compared to brittle deformed shale, reflecting the homogenization of organic nanopore structures. Brittle deformation leads to a more complex pore network, while ductile deformation reduces the complexity of the organic nanopore structure. The fractal dimensions are affected by various factors, with micropore development being crucial for undeformed shale, clay and pore length–width ratio dominating in brittle deformed shale, and all-scale pores being key for ductile deformed shale. This study provides the first comparative analysis of fractal dimensions across undeformed, brittle deformed, and ductile deformed shales, revealing distinct pore structure modifications linked to deformation styles. These findings not only enhance our understanding of the influence mechanism of tectonic deformation on shale pore structure and fractal characteristics but also provide a theoretical basis for optimizing shale gas exploration and production strategies. These findings offer a framework for predicting gas storage and flow dynamics in tectonically complex shale reservoirs. For instance, in areas with different tectonic deformation styles, we can better evaluate the gas storage capacity and production potential of shale reservoirs according to the obtained fractal characteristics, which is of great significance for efficient shale gas development.

Keywords: organic matter pore; fractal characteristics; gas shale; brittle deformation; ductile deformation

1. Introduction

In recent years, shale gas, an important unconventional natural gas resource, has attracted significant global attention [1–5], and the transformation of shale from traditional source and cap rocks into unconventional reservoirs. Unlike traditional reservoirs, shale is considered a tight reservoir due to its abundance of nano pores [4,6–8]. These pore characteristics play an important role in shale gas exploration and production, as they directly influence the storage and migration of shale gas within the reservoir [6–10]. The pores in shale are highly complex, with a wide range of pore sizes, shapes, and connectivity patterns [7,11]. These pores can be classified into micropores (pore diameter < 2 nm), mesopores (2–50 nm), and macropores (>50 nm) according to the International Union of Pure and Applied Chemistry (IUPAC) classification [12]. Based on their genesis, pores can be classified as intergranular pores, intragranular pores (intergranular pores), organic matter (OM) pores, and microcracks [8,10,13]. OM-host pores make a significant contribution to the pore network of shale reservoirs [14] and are critical in controlling the gas content and methane adsorption capacity of black shales [15,16]. Therefore, the analysis of organic pores is significant for the assessment of reservoir quality and the exploration of shale gas resources.

The presence of organic pores is a key differentiating factor between shale reservoirs and conventional oil and gas reservoirs. It has been demonstrated [14–16] that organic matter OM-hosted pores are significant constituents of the pore system of black shales. Understanding the OM pore structure of shale is essential for accurately assessing shale gas reservoirs. Previous studies have shown that the adsorbed gas can account for up to 20–85% of the total gas content [17], and organic matter nanopores are the main contributors to this adsorbed gas storage. Numerous studies have validated the pivotal regulatory impact of organic pores on the enrichment and productivity of shale gas [3,18–21]. The formation of organic pores is primarily initiated during the process of hydrocarbon generation and expulsion, which is concomitant with the thermal evolution of hydrocarbon source rocks [20]. Analyses of shale samples with different maturity levels indicate that the development of organic pores is controlled by the maturity and type of organic matter in combination [20,22]. The OM pore volume, geometry, pore size distribution, and specific surface area are key parameters for estimating the amount of gas that can be stored in the shale. By analyzing these parameters, geologists can predict the potential productivity of a shale gas reservoir. When examining the detailed organic matter nanopore network within shale reservoirs, using SEM and other high-resolution microscope image analysis techniques can help observe the geometry and pore size in shale [14]. The pore size distribution, pore volume, and specific surface area can be quantitatively characterized using low-pressure N_2/CO_2 adsorption experiments [23–26]. Moreover, SEM images can be analyzed using image processing software such as ImageJ to extract quantitative information on pore geometry, roundness, and aspect ratio [27,28].

Fractal theory offers a new scientific method for exploring the complex pore architecture of shales [22,29]. The fractal dimension (D) quantifies the roughness of pore surfaces and structural irregularities, with higher values indicating more complex pore structures. Studies have shown that the pore structure of shales exhibits similarities within a certain scale range, and its spatial spreading pattern is between 2D and 3D. Researchers are examining the fractal properties of shale through image techniques and low-pressure nitrogen adsorption, aiming to connect these properties to OM pore structure. The fractal dimension tends to be greater with smaller pore sizes and higher porosities, whereas the influence of shale composition is unclear, sometimes leading to entirely different findings in different studies. Thus, recognizing the factors that affect OM pore fractal dimension within the geological environment is crucial, making it a primary focus of this research.

The pore structure was controlled by the composition of shales, while the fractal dimension was controlled by the pore specific surface area, pore shape, and pore size [30]. Certainly, the generation and development of shale organic matter pores are mainly controlled by the original sedimentary environment and the basic characteristics of shale organic matter, including total organic content (TOC), types of organic matter, thermal maturity, mineral composition, and rigid framework [6–8,10,11]. On the other hand, in the late evolutionary stage of organic matter pores, tectonic stress and structural deformation may also affect the shape and size of organic matter pores, which in turn affects the methane adsorption capacity of shale and shale gas potential [29,31–33]. For shales in tectonically complex areas in southern China, researchers have carried out a lot of work to study the effects of tectonic deformation on the pore structure of shale reservoirs, including the organic matter pores. Liang et al. point out that the pore surface area and adsorption capacity of shale are affected by tectonic stress and structural deformation, while the total pore volume and micropores (<2 nm and mainly organic matter pores) content were not significantly affected by tectonic deformation [31]. Fractal analysis by Wang et al. suggests that increased tectonic compression causes organic pores to be predominantly long and slit-shaped, exhibiting high fractal dimensions [29]. Other researchers have argued that organic matter pores are poorly developed in deformed shale samples, but nanopores in organic matter and clay mixtures are well developed in both brittle deformed and ductile deformed shales [32]. Overall, previous studies show evidence that organic matter's origin dictates the potential for organic pore formation, with tectonic deformation processes having a secondary impact by modifying nanopore structures. These studies indicate the importance of organic matter pore structure and its fractal dimension characteristics as affected by tectonic deformation. However, as different organic matter types have different effects on the primary development of organic matter pores, different tectonic deformation types have different modifications and effects on organic matter pores, but little is known about the interrelationships between different tectonic deformation styles and the fractal characteristics of organic matter pores in shale.

In this study, fractal analysis of organic pore feature investigations is performed using three sets of shale samples (undeformed shale, brittle deformed shale, and ductile deformed shale) collected from the Lower Cambrian Niutitang Formation with different tectonic deformation styles in western Hunan, Middle Yangtze, South China, were selected. The relationship between fractal dimension and geological conditions, including TOC, mineralogical, and structure styles, was investigated. The influence of structural deformation on fractal patterns and pore structure was studied. These results provide new evidence for the evolution of pore structures of shale affected by different tectonic deformations.

2. Samples and Methods

2.1. Sample Collection

The Lower Paleozoic Cambrian strata in the west of Hunan and Hubei have good petroleum geological conditions and huge resource potential for shale gas exploration. It has experienced complex multi stage tectonic movements, and the differences in shale reservoirs in the area are closely related to the tectonic deformation mechanism [27,31,33,34]. Shale samples were carefully collected from three outcrops with well-documented deformation (undeformation, brittle and ductile deformation) (Figure 1). Samples are sitting at western Hunan, South China, and the detachment structural deformation control across the Niutitang formations shale in this area [27,34]. As the main detachment structure belt [27], the Niutitang formations shale layer developed multi-layer subdivided slip structural deformation. The samples subset is based on detailed geological stratigraphy specimens and thin section characteristics (Figures 1 and 2). A suite of samples was subset to three sets of brittle deformed shale, ductile deformed shale, and undeformed shale for this study due to their variability of texture, fabric, and structure properties (Figures 1 and 2). Undeformed shale (UDS) samples were collected from an outcrop that maintains the original sedimentary stratigraphy (Figure 1a), and samples maintain the original parallel laminations with poorly developed fractures (Figure 2a). Brittle deformed shale (BDS) samples were collected from an outcrop where a folded structure was developed (Figure 1b). BDS development with single-phase cracks was filled with secondary mineral veins (Figure 2b). Ductile deformation shale (DDS) samples were collected from a slip zone where multiple phases of deformation had developed (Figure 1c). The DDS samples have developed multi-angle fault mirrors, multi-phase fracture cross-development, the development of mylonite structures (Figure 2c), and the phenomenon of fragmentation and rearrangement of minerals and organic matter by intense stress (Figures 1 and 2). A total of 12 samples were collected; the sample numbers were abbreviated as UDS1, UDS2, UDS3, and UDS4 (undeformed shale); BDS1, BDS2, BDS3, and BDS4 (brittle deformed shale); and DDS1, DDS2, DDS3, and DDS4 (ductile deformed shale). The pore structure development characteristics of similar fold-development-related brittle deformed shales [33] and slip-related mylonitized ductile deformed shales [35] in different regions and horizons have been analyzed by previous studies. Similar tectonic styles of brittle and ductile deformation are widely developed in shale formations in the outer regions of the Sichuan Basin in South China, which is a typical tectonic deformation style. In this study, two typical brittle and ductile deformation styles are selected for comparative study, which has certain regional significance. The basic features of the experimental samples are shown in Table 1.

Figure 1. Outcrops and deformation characteristics of shale samples. (**a**) Undeformed shale (UDS) outcrop; (**b**) Brittle deformed shale (BDS) outcrop with folded structure; (**c**) Ductile deformed shale (DDS) outcrop in the slip zone. The length of the hammer is 38 cm, and the diameter of the coin is 20.5 mm. See Figure 2 and Table 1 for texture and structure properties of these samples.

Table 1. The basic features of the experimental samples.

Sample ID	TOC (%)	VRb (%)	δC_{13} ‰ (PDB)	Quartz (%)	Clay (%)	Other Minerals (%)	Texture and Fabric Feature of Macroscopic Hand Specimens
UDS1	6.38	3.12	−34.2	80	16	4	Shale primary structure can be observed. Shale has original parallel bedding.
UDS2	8.44	3.15	−33.7	77	14	9	
UDS3	8.12	3.15	−33.6	66	27	7	
UDS4	3.57	3.13	−34.2	66	33	1	
BDS1	8.95	3.12	−34.3	57	36	7	Shale has original parallel bedding. Shale shows cleavage structure and cleavage surface is smooth with fine-grained powder coatings. Fractures and mineral filling development.
BDS2	11.5	3.19	−33.5	85	10	5	
BDS3	18	3.28	−33.7	85	15	-	
BDS4	9.11	3.30	−32.5	61	31	8	
DDS1	11.9	3.89	−32.8	80	15	5	The primary structure of shale is damaged and the parallel bedding has disappeared due to mylonitization. The plastic deformation of shale is obvious. Shale structure is loose and fragile. Shale shows cleavage structure and cleavage surface is smooth with fine-grained powder coatings.
DDS2	20.7	3.48	−32.7	79	17	4	
DDS3	14.3	3.39	−32.8	72	25	3	
DDS4	15.8	3.37	−32.8	65	28	7	

TOC—total organic carbon (%); VRb—vitrinite reflectance values of solid bitumen (%). Other minerals (%): feldspar, carbonate, pyrite, and siderite.

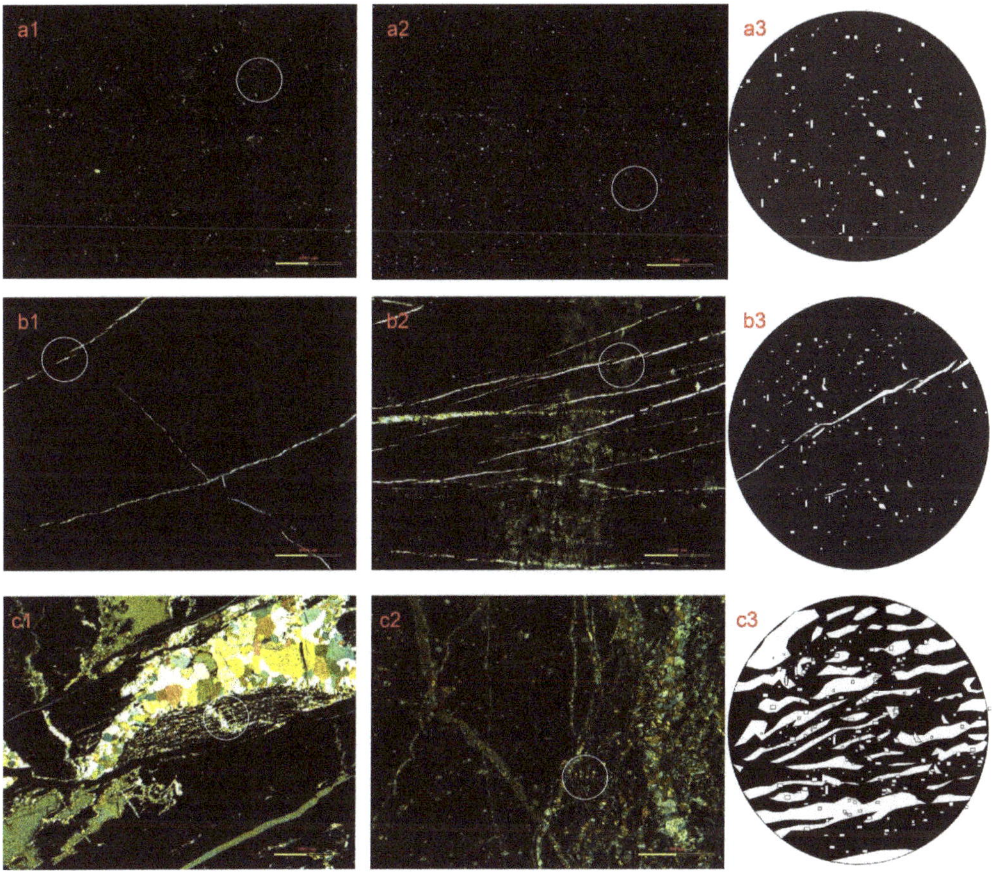

Figure 2. Microscopic characteristics of shale samples. (**a**) Undeformed shale with original parallel laminations and poorly-developed fractures; (**b**) Brittle deformed shale with single-phase cracks and secondary mineral veins; (**c**) Ductile deformed shale with multi-angle fault mirrors, multi-phase fracture cross-development, mylonite structures, and mineral-organic matter fragmentation and rearrangement.

2.2. Experimental and Methods

To evaluate the influence of different tectonic deformations on organic matter pore characteristics by comparing the brittle deformed shale, ductile deformed shale, and undeformed sample subsets, 12 shale samples were collected in this study (Table 1). Among these samples, all samples were analyzed to determine the organic geochemistry and mineralogy, refs. [23,26] low-temperature gas (CO_2 and N_2) adsorption (LTGA) to quantify and determine the pore structure, including the pore-size distribution (PSD), surface area (SA), and pore volume (PV), and scanning electron microscopy (SEM), used for directly observing the microstructure of pore geometry, including the pore morphology, pore size, and the spatial distribution of pores. In addition, Frenkel–Halsey–Hill (FHH) model [29, 30,36,37] was used to analyze low-temperature nitrogen adsorption data to obtain fractal dimension characteristics of organic nanopores (0–200 nm). ImageJ software was used [27, 28] to analyze the shape, length–width ratio (LWR), and roundness characteristics of organic pores from SEM images. Comprehensively analyzing the above experimental results, the fractal features and pore structure characteristics of organic matter pores in different tectonically deformed shales can be obtained.

2.2.1. Geochemical and Mineralogical

Samples were ground to 180–200 mesh before conducting geochemical and mineralogical analyses. The TOC content was obtained using a Leco C/S-344 Carbon/Sulfur analyzer (Leco, St. Joseph, MI, USA), and the stable carbon isotope was measured with a Finnigan MAT 252 mass spectrometer (Thermo Fisher Scientific, Bremen, Germany). To determine the bitumen reflectance values (VRb) in the samples, a 3Y-Leica DMR XP microscope with a microphotometer was employed, ensuring at least 30 measurements were taken on bitumen particles for each sample. For X-ray diffraction (XRD) analysis, the crushed samples were mixed with ethanol, manually ground with a mortar and pestle, and then mounted on glass slides, using a D/Max-III analyzer (Rigaku, Tokyo, Japan) at 40 kV and 30 mA. A semi-quantitative estimation of mineral contents was achieved by measuring the area under the curve for the peaks.

2.2.2. Low-Temperature Gas Adsorption (LTGA)

Low-temperature gas adsorption (LTGA) is a powerful method for characterizing nanopores. To fully evaluate the organic matter nanopore structure, samples were crushed (2–5 mm) and dried at 110 °C and then both CO_2 and N_2 gas adsorption analyses were performed. Low-temperature gas adsorption (LTGA) analyses have been used to measure the nanopores of pore size distribution from 0.33 nm to 200 nm using both nitrogen adsorption at −196 °C and carbon dioxide adsorption at 0 °C by a Micromeritics ASAP 2020 HD88 analyzer (Micromeritics, Norcross, GA, USA). The PSD, PV, and SA analyses combined the N_2 and CO_2 gas adsorption by the same calculation models of density function theory (DFT) [31,38–40]. The development of DFT models has led to a better understanding of adsorption processes in well-ordered systems compared to the more conventional models [41], used in the present study to quantify the calculation of micropores (0.3–2 nm), mesopores (2–50 nm), and macropores (50–200 nm) parameters by N_2 and CO_2 gas adsorption data.

2.2.3. Scanning Electron Microscopy (SEM)

Scanning electron microscopy (SEM) is a valuable tool for directly observing the microstructure of shale, including the pore morphology, pore size, and the spatial distribution of pores. In the SEM analysis of shale samples, the samples were first prepared. The shale samples were cut into small pieces and mounted on aluminum stubs using conductive carbon tape. To ensure good electrical conductivity and prevent charging effects during the electron-beam scanning, the samples were coated with a thin layer of gold using a sputter coater. The thickness of the coating was typically around 10–20 nm. After sample preparation, the samples were placed in the SEM chamber. The SEM was operated at an accelerating voltage of 10–20 kV, depending on the sample characteristics and the required resolution. The electron beam was scanned across the sample surface, and the secondary-electron images were collected. The SEM images were then analyzed using ImageJ—analysis software to measure the organic nanopore size and shape.

2.2.4. ImageJ Analysis

SEM images provide a detailed view of the organic pore structure in shale. We mainly use the ImageJ software (v1.52a) to extract and quantitatively count the morphological features (circularity and length/width ratio) of organic matter pores in SEM images in this study. We employed manual adjustment by experts to evaluate the image quality under different thresholds and noise reduction parameters. Multiple evaluation indicators, such as image entropy and edge preservation index, were used to comprehensively assess the quality of the processed images. The optimal settings were selected based on the combination of parameters that maximized the image entropy while maintaining good edge preservation. To clearly distinguish organic matter pores from other features (such as mineral pores), we established a set of strict criteria. First, we considered the morphological characteristics of pores, such as roundness and aspect ratio. Organic matter pores generally have more irregular shapes compared to mineral pores. Second, we analyzed the spatial distribution of pores. Organic matter pores are often associated with organic matter aggregates. In addition, for each sample, we selected multiple SEM images ($n \geq 3$) for comparative experiments and calculated the average values of pore morphology characteristic parameters. Considering that the main purpose of this part of the research is to compare the pore morphology characteristics under different deformation styles, all images and threshold selections were carried out under the same experimental conditions and with the same level of manual experience. We believe that this can yield quality data that meets the research objectives. The first stage entailed converting the SEM image of the shale into an 8 bit grayscale image (Figure 3a). Then the threshold is adjusted in the Threshold tool, and then the Median filter is used to reduce the noise of the image; the organic matter region (Figure 3b) and the organic matter pore region (Figure 3c) can be extracted, respectively. The black area in the picture is obtained as organic matter pores (Figure 3d). The quantitative parameters for each pore were derived from ImageJ software, which used the Kraver method [42] to calculate the circularity and length/width ratio (LWR).

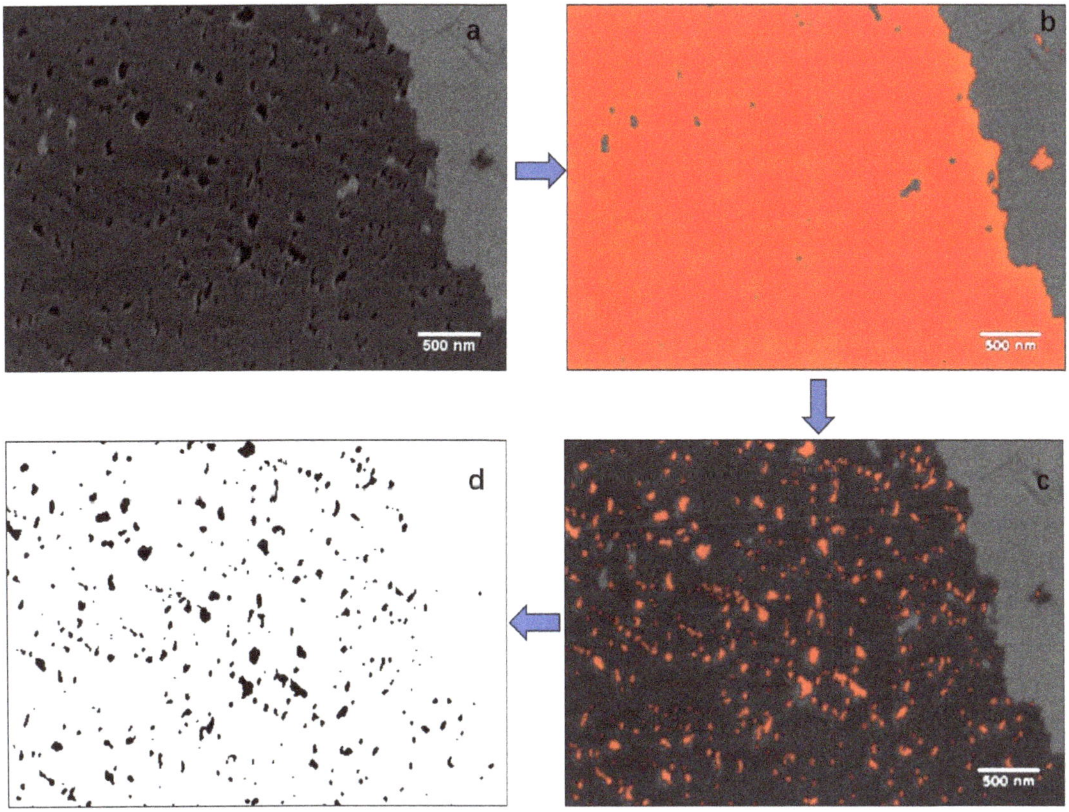

Figure 3. Process of organic matter pore extraction and analysis using ImageJ software. (**a**) SEM image of shale converted to 8 bit grayscale image; (**b**) Extracted organic matter region; (**c**) Extracted organic matter pore region; (**d**) Organic matter pores presented as black areas.

2.2.5. Fractal Dimension

The fractal dimension of the pore structure is a key parameter to quantify the complexity of the pore system. Fractal dimension D is a quantitative tool for describing the surface roughness or structural irregularity of shale pore structures. A D value between 2 and 3 can indicate the heterogeneity of the pore structure or surface. D = 2 indicates a completely smooth surface or uniform pore structure, while D = 3 signifies a completely irregular or rough surface or non-uniform pore structure. Frenkel–Halsey–Hill (FHH) method was used to calculate the fractal dimension from low-temperature nitrogen adsorption (LTNA) data [29,43,44]. The FHH equation is as follows:

$$\mathrm{Ln}\,(V/V_0) = K \times \ln(\ln(P_0/P)) + C$$

where V is the volume of N_2 adsorbed under equilibrium pressure, V_0 is the volume of monomolecular adsorption gas, P is the equilibrium pressure, P_0 is the saturation pressure, K is related to the fractal dimension (D = 3 + K), and C is a constant. By fitting the LTNA data to the FHH equation, the fractal dimension D can be calculated.

3. Results

3.1. Geochemical and Mineralogical Characteristics

The samples' organic geochemistry and mineralogy data are detailed in Table 1. Shale's composition, comprising organic matter and minerals, is thought to significantly affect organic pore structure [45]. The TOC provides a clear indication of the ability to generate gas hydrocarbons and influences the state and quantity of hydrocarbon gas potential. The TOC of the Niutitang Formation shale ranges from 3.57 wt% to 20.7 wt% in this study. All of the undeformed shales, brittle deformed shales, and ductile deformed shales show high-quality organic matter richness features (TOC > 2%) and are within the gas generation window (VRb values of 3.12% to 3.89%), indicating that the shale samples possess a high richness of organic matter suitable for generating shale gas over geological periods. Tectonically deformed shales appear to have a higher TOC content than undeformed shales, probably because shales with a high TOC content are relatively softer in terms of rock mechanics and are more susceptible to deformation under tectonic stress. Ductile deformed shale samples have VRb values (3.37% to 3.89%) that are higher than those of brittle deformed shale samples (3.12% to 3.30%) and undeformed samples (3.12% to 3.15%), indicating that tectonic stress and deformation affect the maturation process of organic matter [46]. All samples have similar carbon isotope values of −34.3‰ to −32.5‰. These carbon isotopic compositions allow good insights into the provenance of the organic matter, the isotope analysis often used for paleoenvironmental reconstructions [47]. Based on the range of carbon isotope values, we assume that the organic matter in the Niutitang shales is 100% marine organic matter [47]. Consistent with earlier findings, structural deformation slightly raises the maturity level of organic matter in shale. Quartz and clay were the predominant components in all samples (Table 1), with their combined totals exceeding 90%. These were categorized as biogenic siliceous shale because of the quartz content surpassing 60% and a TOC content greater than 3.7%. While the brittle deformed shales contain higher quartz contents (85%), which may be related to fracture development with quartz vein-filling contributions.

3.2. Pore Morphology by SEM

Using ImageJ software for graphic processing (Figure 4), the organic pore in the SEM image consists of a series of pixels, and the pore area is defined as the number of pixels in its block. The organic matter content (OM%) and the organic pore content (OP%) were extracted by the threshold function of ImageJ software, respectively, and then the organic matter pore porosity (OMP%) in this study represents the amount of organic pore space in the organic matter rather than the amount of organic pore space in the whole rock, which was calculated and obtained as follows:

$$\mathrm{OMP} = \mathrm{OP}/\mathrm{OM}$$

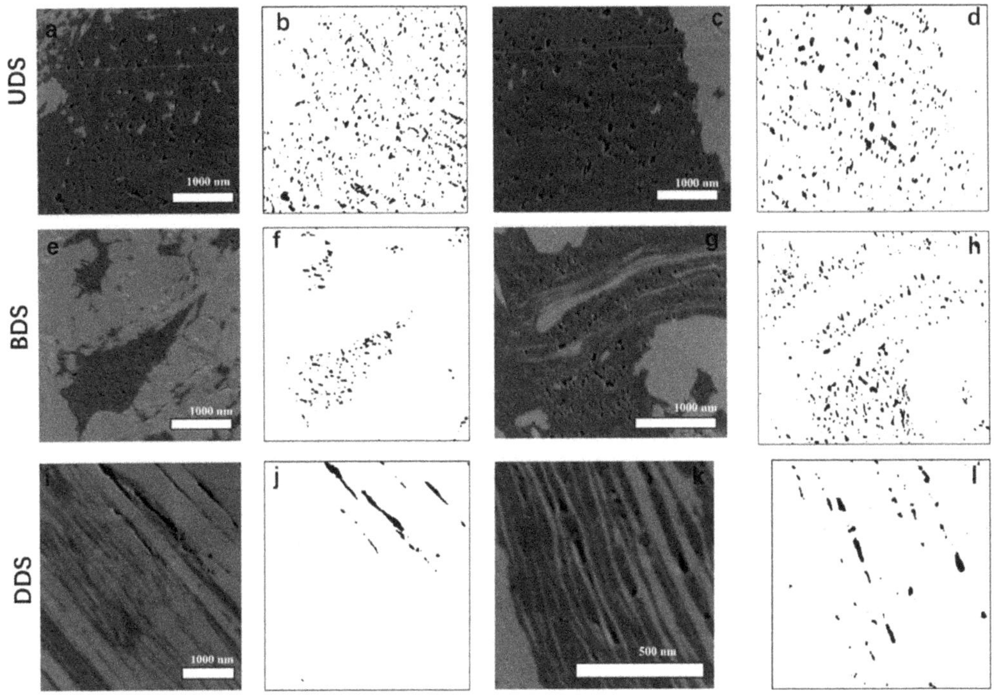

Figure 4. Schematic diagram of parameters measurement for organic pores in SEM image processed by ImageJ software for different types of deformation. Undeformed shale in (**a**–**d**); Brittle deformed shale in (**e**–**h**); Ductile deformed shale in (**i**–**l**).

The perimeter can be obtained by accumulating the distances of the block boundary pixels. The length and width of the pores can be defined by the Feret diameter. With the pore area, perimeter, length, and width parameters, the complexity of the pore shape can be described in terms of roundness (Circ.) and aspect ratio (LWR). The roundness parameter (Circ.) of the pore is calculated from the pore area S and circumference C of the pore with the following equation:

$$\mathrm{Circ.} = 4\pi S/C^2$$

The shape factor Circ. characterizes the roundness of the pore boundary and has a value between 0 and 1. Circ. values closer to 1 indicate that the pore shape is closer to a regular circle. The LWR is calculated as the ratio of the pore length (maxFeret diameter) to the pore width (miniFeret diameter). LWR indicates the flatness and directionality of the pore shape, with larger LWR values indicating that the pore tends to be narrower and longer. Thus, both Circ. and LWR values are unitless. The pore morphology by SEM of all samples is shown in Table 2.

The box plots (Figure 5) show that ductile deformed shale has the smallest organic matter porosity (OMP) as deformation intensity increases, only 2.39 to 3.82%. Whereas brittle deformed shales have OMPs between 4.16 and 7.77%, undeformed shales have OMPs between 7.57 and 9.56%. Tectonic deformation has had a limited effect on the roundness of the organic matter pores of the shale, with undeformed shale ranging from 0.64 to 0.72, while brittle deformed shales have roundness between 0.71 and 0.77, and ductile deformed shales have roundness between 0.45 and 0.70. However, the effect of tectonic deformation on the aspect ratios of the organic matter pores in shales is significant, with the LWR of undeformed shales ranging from 1.75 to 1.86, while brittle deformed shales have LWR between 1.88 and 1.99, and ductile deformed shales have LWR between 2.34 and 3.45. It suggests that tectonic stresses act on organic pores, making them narrower and more elongated, which is consistent with the previous understanding that strong tectonic deformation develops organic pores dominated by long strip pores and slit-shaped pores [29].

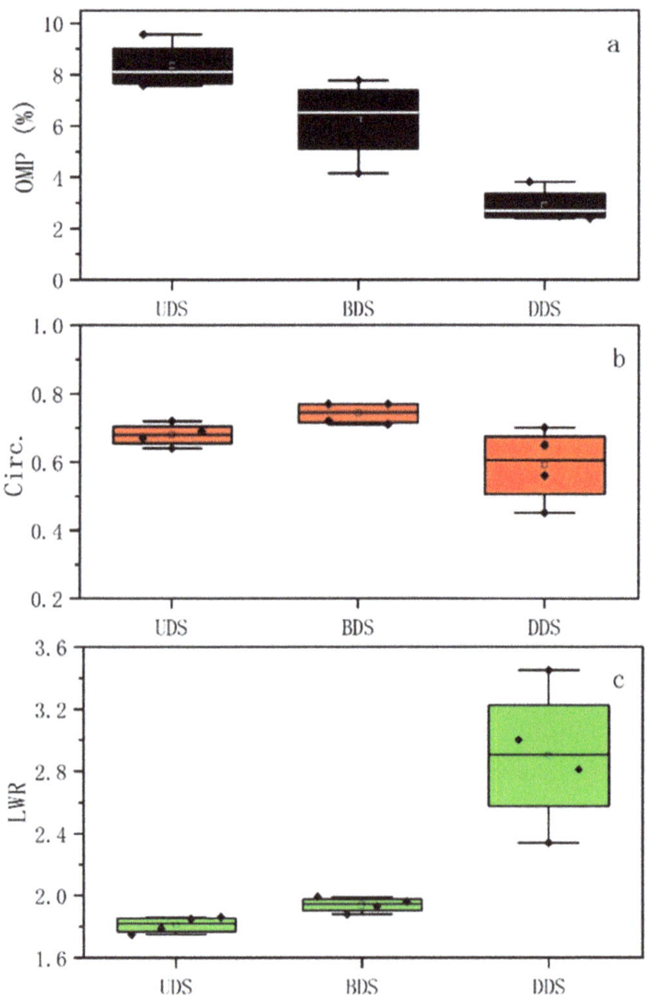

Figure 5. Box plots showing the organic matter porosity (OMP) (**a**), roundness (Circ.) (**b**), and aspect ratio (LWR) (**c**) of shale samples with different deformations.

Table 2. Quantitative pore parameters of the shale samples.

Sample ID	Pore Volume				Adsorption Capacity		Relative Content * (%)			Pore Parameters from SEM		
	Micropore (μL/g)	Mesopore (μL/g)	Macropore (μL/g)	Total Pore (μL/g)	CO_2 (mL/g)	N_2 (mL/g)	Micropore (%)	Mesopore (%)	Macropore (%)	OMP (%)	Circ. (0–1)	LWR
UDS1	7.59	18.18	3.38	29.15	3.43	29.59	26.04	62.37	11.60	8.47	0.72	1.85
UDS2	5.52	19.26	2.17	26.95	2.78	21.05	20.48	71.47	8.05	7.73	0.69	1.79
UDS3	6.76	16.10	3.02	25.88	3.11	20.10	26.12	62.21	11.67	9.56	0.64	1.86
UDS4	2.87	12.90	4.26	20.01	1.40	17.75	14.34	64.47	21.29	7.57	0.67	1.75
BDS1	5.42	12.61	2.00	20.03	2.70	13.95	27.06	62.96	9.99	7.77	0.77	1.93
BDS2	8.45	12.92	4.44	25.81	4.80	15.35	32.74	50.06	17.20	6.99	0.77	1.88
BDS3	9.52	17.31	2.90	29.73	4.92	20.15	32.02	58.22	9.75	4.16	0.71	1.96
BDS4	7.22	12.38	2.05	21.65	3.71	16.20	33.35	57.18	9.47	6.05	0.72	1.99
DDS1	5.93	4.75	3.72	14.4	3.76	6.58	41.18	32.99	25.83	2.47	0.45	2.81
DDS2	1.56	2.56	2.09	6.21	1.09	3.79	25.12	41.22	33.66	3.82	0.56	3.00
DDS3	2.88	3.52	2.45	8.85	1.85	4.98	32.54	39.77	27.68	2.39	0.70	2.34
DDS4	1.79	2.12	2.01	5.92	1.30	3.23	30.24	35.81	33.95	2.90	0.65	3.45

* The relative contents of different pore size distributions with pore volume were calculated by the same model of DFT. OMP: porosity % of pores in organic matters from SEM images using ImageJ.

3.3. Gas Adsorption Isotherms from LTGA

CO_2 Gas Adsorption analyses have been used to measure the adsorption capacity in micropores of pore size distribution from 0.33 nm to 1.5 nm using carbon dioxide adsorption at 0 °C. Figure 6 presents the CO_2 adsorption curves of all shale samples. These curves bear a resemblance to type I adsorption isotherms, suggesting the presence of micropores within the samples. As depicted in Figure 6b, the BDS samples exhibit the highest maximum adsorption capacity, spanning from 2.7 to 4.92 mL/g, with an average value of 4.03 mL/g (Table 2). The DDS samples have a relatively low maximum adsorption capacity, ranging from 1.09 to 3.76 mL/g, with an average of 2.0 mL/g. The UDS samples have medium adsorption capacity, ranging from 1.40 to 3.43 mL/g, with an average of 2.68 mL/g. On the one hand, the CO_2 gas adsorption capacity of shale is affected by the content of TOC, and undeformed shale with lower TOC content (3.57–8.44%) has a moderate adsorption capacity. On the other hand, strong ductile tectonic deformation effects may reduce the adsorption capacity of shale micropores, resulting in UDS with low gas adsorption despite high TOC content (11.9–20.7%). In contrast, the organic matter micropores in brittle deformed shale may be limited by tectonic stress and retain a high CO_2 gas adsorption capacity. Organic matter content can determine the basis of gas adsorption, and shales with low organic matter content also have low adsorption capacity. The tectonic stress background and deformation style can affect the present adsorption capacity of shale, with brittle deformed shale having little or no effect and ductile deformed shale having a significant decrease in the gas adsorption capacity of micropores.

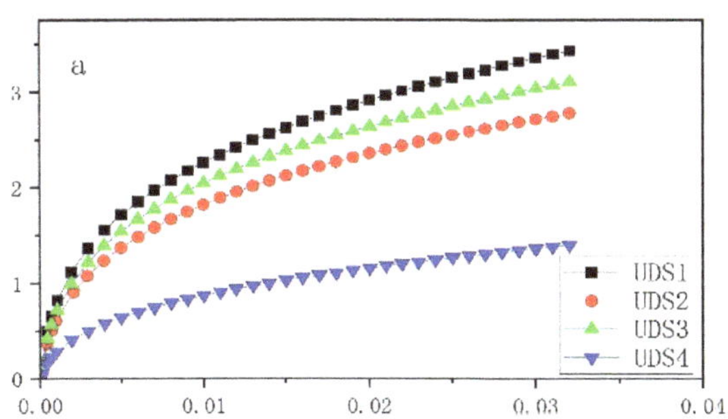

Figure 6. *Cont.*

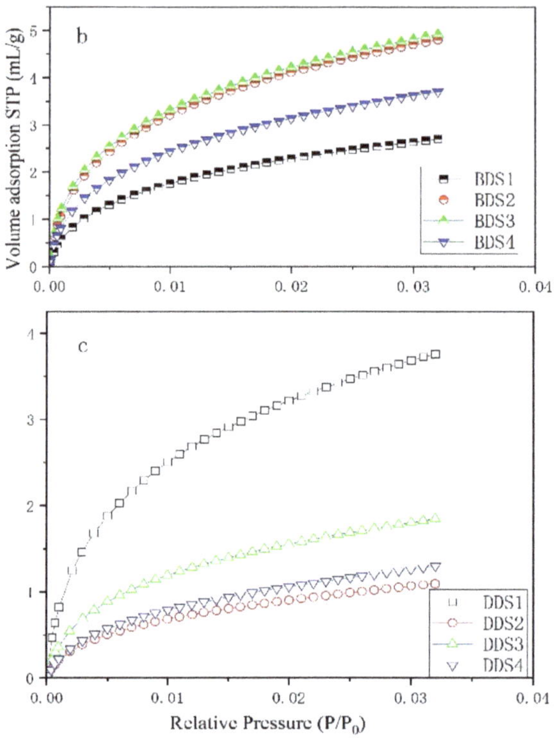

Figure 6. CO_2 adsorption curves of shale samples. (**a**) Undeformed shale; (**b**) Brittle deformed shale; (**c**) Ductile deformed shale.

N_2 gas adsorption analyses have been used to measure the adsorption capacity in nanopores of pore size distribution from 0.37 nm to 200 nm using N2 at −196 °C. Figure 7 displays the N_2 adsorption/desorption curves of three shale groups; the desorption curve is always on top of the adsorption curve, resulting in a loop [12,38]. The N_2 adsorption curves of both UDS and BDS samples are of H2-type. This indicates that the pores are mainly ink-flask-type. In contrast, the N_2 adsorption curves of DDS samples are of H3-type (Figure 7c). It means the pores are mainly fissure-type. As presented in Table 2, undeformed shale has the highest maximum nitrogen adsorption capacity. It ranges from 17.75 to 29.59 mL/g, with an average of 22.12 mL/g. Brittle deformed shale comes next. Its adsorption capacity ranges from 13.95 to 20.15 mL/g, with an average of 16.14 mL/g. Ductile deformed shale has the minimum adsorption capacity. It ranges from 3.23 to 6.58 mL/g, with an average of 4.65 mL/g. In the nanopore range of 0.37–200 nm, the tectonic stress background and structural deformation style significantly affect the nitrogen adsorption capacity of shale, with brittle deformed shale having less effect as the deformation intensity and deformation style vary, while ductile deformed shale significantly reduces the gas adsorption capacity of the nanopores. Combined with the previous changes in CO_2 micropore adsorption capacity, brittle tectonic deformation mainly has a greater effect on the mesopores and macropores in organic nanopores, while ductile tectonic deformation produces a greater weakening effect on the all-scale micropores, mesopores, and macropores in organic nanopores.

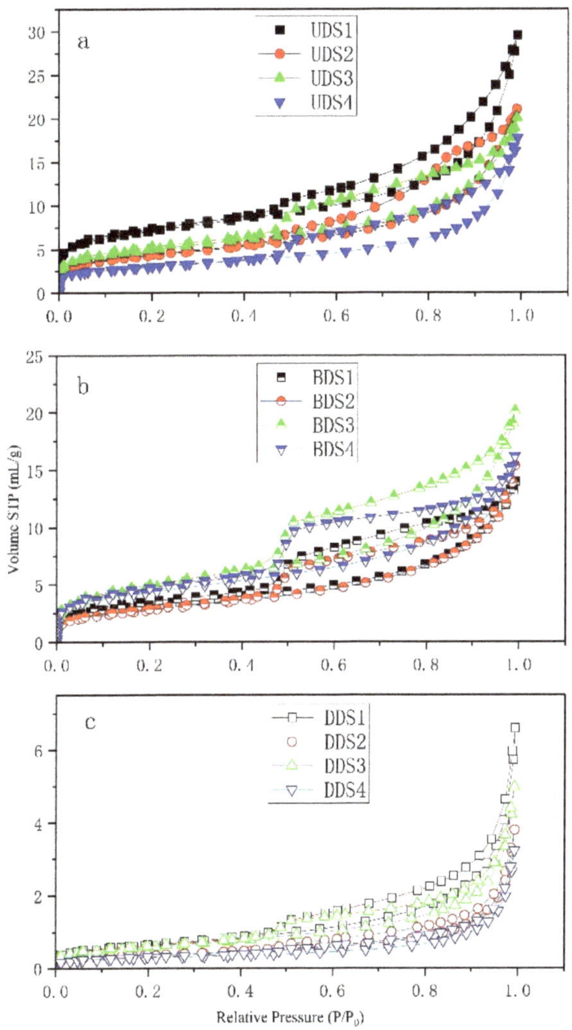

Figure 7. N_2 adsorption/desorption curves of three shale groups. (**a**) Undeformed shale; (**b**) Brittle deformed shale; (**c**) Ductile deformed shale.

3.4. Pore Structure from LTGA

The pore structure was derived by analysis of the combined N_2 and CO_2 gas adsorption data by the same calculation models of density function theory (DFT). The development of density function theory (DFT) models has enhanced our understanding of gas adsorption in well-ordered systems compared to traditional models [12,41]. DFT calculation models offer a more reliable way for pore size analysis across all nanopores [39,40]. According to SEM picture analysis, 0.33–200 nm can represent the pore size of most organic matter pores. Low-temperature gas adsorption analyses measure pore size distributions (PSD) of micropores (0.33–2 nm), mesopores (2–50 nm), and macropores (50–200 nm). Two CO_2 and N_2 gas adsorption methods were combined using incremental PV (Figure 8) and SA plots (Figure 9). In Figures 8 and 9, DFT-model incremental curves against pore diameter accurately reflect pore size distributions.

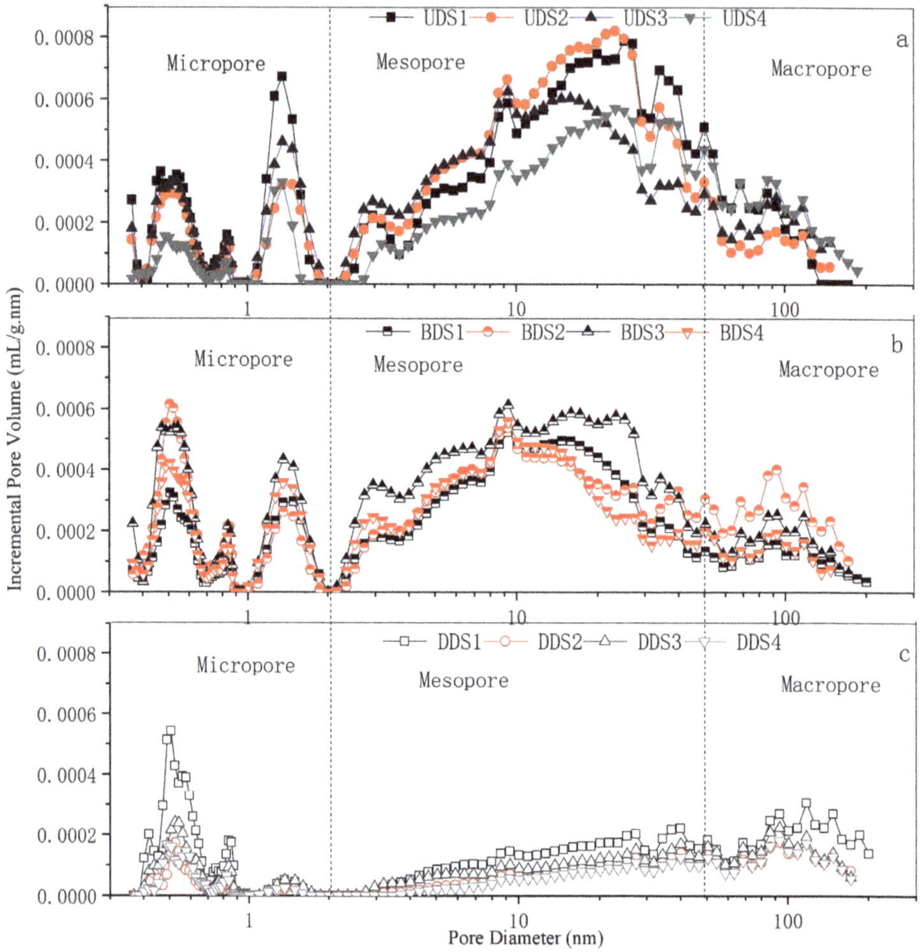

Figure 8. Incremental pore volume (PV) plots of shale samples based on DFT-model analysis of combined N_2 and CO_2 gas adsorption data. (**a**) Undeformed shale; (**b**) Brittle deformed shale; (**c**) Ductile deformed shale.

In terms of total nanopore volume, there is little difference between the organic matter nanopore volume of undeformed shale and brittle deformed shale. The total pore volume of undeformed shale is in the range of 20.01–29.15 μL/g, with an average of 25.5 μL/g (Table 2). The total pore volume of brittle deformed shale is in the range of 20.03–29.73 μL/g, with an average of 24.31 μL/g. The total pore volume of ductile deformed shale is in the range of 5.92–14.4 μL/g, with an average of 8.85 μL/g. In terms of the percentage of the pore volume of different pore structures, the pore volume of undeformed shale and brittle deformed shale is dominated by mesopores, whereas the structure of pore volume of ductile deformed shale is more balanced (Figure 10), with a decrease in the percentage of mesopores and an increase in micropores and macropores.

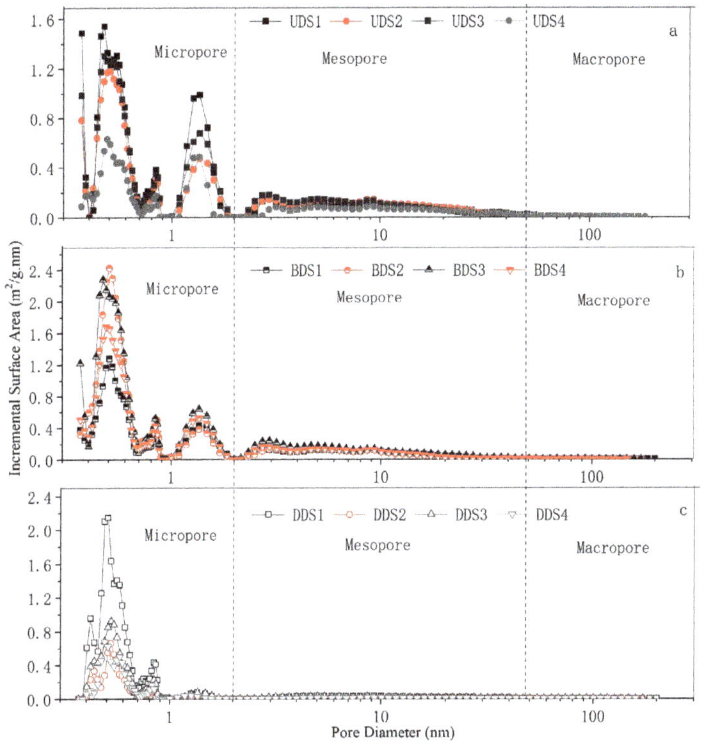

Figure 9. Incremental specific surface area (SA) plots of shale samples based on DFT-model analysis of combined N_2 and CO_2 gas adsorption data. (**a**) Undeformed shale; (**b**) Brittle deformed shale; (**c**) Ductile deformed shale.

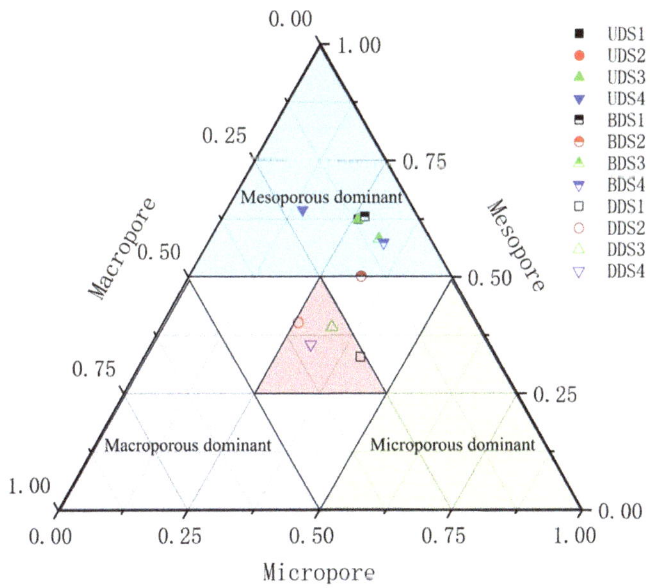

Figure 10. Ternary phase diagram of pore structure composition for different deformation type shales.

The Dubinin–Radushkevich (DR) and Brunauer–Emmett–Teller (BET) models calculate total surface area (SA) using CO_2 and N_2 adsorption data, respectively. Table 3 shows SA values calculated by different models. Brittle deformed shales always have a higher specific surface area, benefiting from a higher TOC and limited influence by tectonic stresses. The structure of the pore volume of all samples had the highest proportion of mesopore volume, but the structure of the pore's specific surface area of all samples had the largest proportion of microporous contribution (>80%) to the specific surface area. Both in terms of pore volume and specific surface area, tectonic deformation has a greater impact on meso- and macropores, while tectonic deformation has less impact on micropores (0.33–2 nm).

Table 3. Pore surface area characteristics of the samples.

Sample ID	Total SA of Gas Absorption by Different Model				SA of Gas Absorption by Same Model (DFT)				Relative Content * (%)		
	CO_2 (DR) (m^2/g)	CO_2(DFT) (m^2/g)	N_2(BET) (m^2/g)	N_2(DFT) (m^2/g)	Micropore (m^2/g)	Mesopore (m^2/g)	Macropore (m^2/g)	Total pore (m^2/g)	Micropore (%)	Mesopore (%)	Macropore (%)
UDS1	25.77	26.74	25.54	18.00	22.10	3.17	0.10	25.37	87.11	12.50	0.39
UDS2	21.20	21.54	15.36	10.23	16.77	3.66	0.06	20.49	81.84	17.86	0.29
UDS3	23.54	24.11	17.94	12.02	19.71	3.64	0.08	23.43	84.12	15.54	0.34
UDS4	10.52	10.94	10.51	6.77	8.27	2.12	0.11	10.5	78.76	20.19	1.05
BDS1	21.44	20.77	11.67	7.81	16.20	2.84	0.05	19.09	84.86	14.88	0.26
BDS2	37.15	36.96	10.03	6.05	27.89	2.92	0.11	30.92	90.20	9.44	0.36
BDS3	36.36	38.25	17.04	10.15	30.31	4.20	0.07	34.58	87.65	12.15	0.20
BDS4	28.06	28.77	15.57	10.30	22.21	3.13	0.05	25.39	87.48	12.33	0.20
DDS1	29.47	28.83	2.26	1.50	20.68	0.78	0.08	21.54	96.01	3.62	0.37
DDS2	8.89	8.53	1.25	0.82	5.37	0.36	0.05	5.78	92.91	6.23	0.87
DDS3	14.70	14.22	2.12	1.36	9.77	0.61	0.06	10.44	93.58	5.84	0.57
DDS4	10.92	10.09	1.00	0.66	6.10	0.28	0.04	6.42	95.02	4.36	0.62

* The relative contents of different pore size distributions with SA were calculated by the same model of DFT.

In general, brittle deformation has a limited effect on the total pore volume and total pore specific surface area of organic pores in shale, while ductile deformation significantly reduces the pore volume and specific surface area of organic pores. In terms of pore structure, the distribution of pore volume and specific surface area of brittle deformed shale is not much different from that of undeformed shale. The pore volume is dominated by the significant dominance of mesopores, and the percentage of mesopore pore volume is greater than 50%. The specific surface area is dominated by micropores, with the percentage of microporous specific surface area greater than 80%, while the percentage of mesoporous specific surface area is greater than 10%. However, the distribution of pore volume of ductile deformed shale is more balanced, and the percentage of mesopore is 37.03%. The distribution of the specific surface area of ductile deformed shale is more extreme; the percentage of microporous area is greater than 90%.

3.5. Fractal Dimensions from the N_2 Adsorption

This study utilized the FHH model to calculate the fractal dimension and analyzed N_2 adsorption/desorption data to evaluate the complexity of shale and kerogen nanopores. The fairly good fitting results (see Figures 11–13 and Table 4) indicate different fractal characteristics at the two intervals. Based on these phenomena, the fractal dimensions Ds and Dm were calculated at the relative pressure (P/P_0) range of 0–0.45 and 0.45–1 using the FHH equation [48]. Based on previous research results [29], at low relative pressures ($P/P_0 < 0.45$), it mainly reflects the characteristics of micropore surfaces, which is related to monolayer adsorption under the action of van der Waals forces. At high relative pressures ($P/P_0 > 0.45$), capillary condensation occurs in mesopores and macropores, affecting the calculation of the fractal dimension and reflecting the complexity of mesopore and macropore structures. The fitting equations, correlation coefficients of R^2, and fractal dimension values (Ds and Dm) are summarized in Table 4. Curve fitting for the N_2 adsorption fractal dimension characteristics is shown in Figures 11–13, yielding an excellent correlation coefficient (R^2) ranging from 0.9361 to 0.9984 (Table 4), showing a strong fit between them. In this research, we used the surface fractal dimension (Ds) at $0 < P/P_0 < 0.45$ to characterize the roughness of shale pore surfaces. This Ds dimension reflects the influence of the van der Waals forces phenomenon of monolayer adsorption in micropores with diameters smaller than 3 nm and dominated by adsorption pores containing adsorbed gas. A larger Ds indicates rougher pore surfaces. Rougher surfaces mean more adsorption sites on the pore surface, resulting in a higher shale adsorption capacity (Table 4). On the other hand, the matrix fractal dimension (Dm) at $0.45 < P/P_0 < 1$ reflects the heterogeneity and complexity within the pore structure, especially indicating the spatial complexity and irregularity of mesopores and macropores with diameters larger than 3 nm, mainly composed of flow-through pores containing free gas [49]. It is affected by the capillary condensation effect. A higher Dm value corresponds to a more intricate pore structure. Such a complex structure makes shale gas desorption, diffusion, and percolation more difficult.

Table 4. Fractal dimensions calculated from N_2 adsorption data using FHH method.

Sample ID	P/P_0 = 0–0.45			P/P_0 = 0.45–1		
	Fitting Equation	R^2	D_S	Fitting Equation	R^2	D_M
UDS1	y = −0.3564x + 2.1274	0.9916	2.6436	y = −0.2764x + 2.1503	0.9851	2.7236
UDS2	y = −0.4377x + 1.6561	0.9942	2.5623	y = −0.3157x + 1.7296	0.9651	2.6843
UDS3	y = −0.4038x + 1.7965	0.9927	2.5962	y = −0.2634x + 1.8728	0.9702	2.7366
UDS4	y = −0.4238x + 1.2816	0.9961	2.5762	y = −0.34x + 1.3339	0.9851	2.66
BDS1	y = −0.4237x + 1.3697	0.9907	2.5763	y = −0.2597x + 1.5008	0.9361	2.7403
BDS2	y = −0.4849x + 1.2479	0.9969	2.5151	y = −0.3063x + 1.3827	0.9678	2.6937
BDS3	y = −0.4657x + 1.7655	0.9959	2.5343	y = −0.2615x + 1.8765	0.9628	2.7385
BDS4	y = −0.409x + 1.6566	0.9938	2.5910	y = −0.2412x + 1.7601	0.9561	2.7588
DDS1	y = −0.4894x − 0.2237	0.9972	2.5106	y = −0.42x − 0.1289	0.9922	2.58
DDS2	y = −0.5028x − 0.8218	0.9937	2.4972	y = −0.4542x − 0.8151	0.9967	2.5458
DDS3	y = −0.5101x − 0.2926	0.9969	2.4899	y = −0.3909x − 0.2273	0.9972	2.6091
DDS4	y = −0.4488x − 1.0709	0.9944	2.5512	y = −0.4809x − 1.1008	0.9984	2.5191

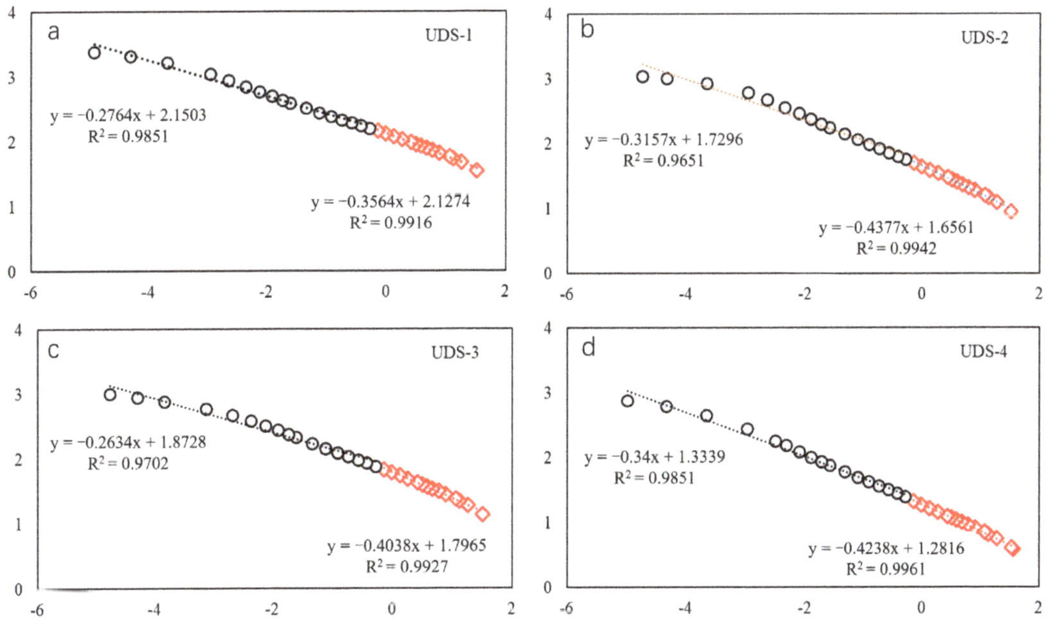

Figure 11. FHH plots for calculating fractal dimensions of undeformed shale samples from N_2 adsorption data.

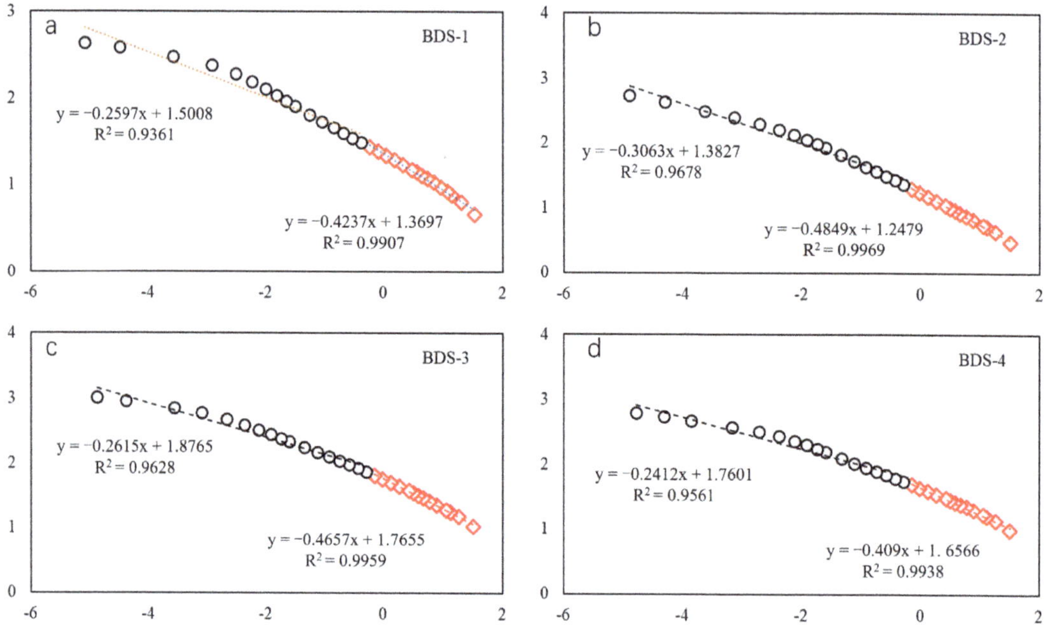

Figure 12. FHH plots for calculating fractal dimensions of brittle deformed shale samples from N_2 adsorption data.

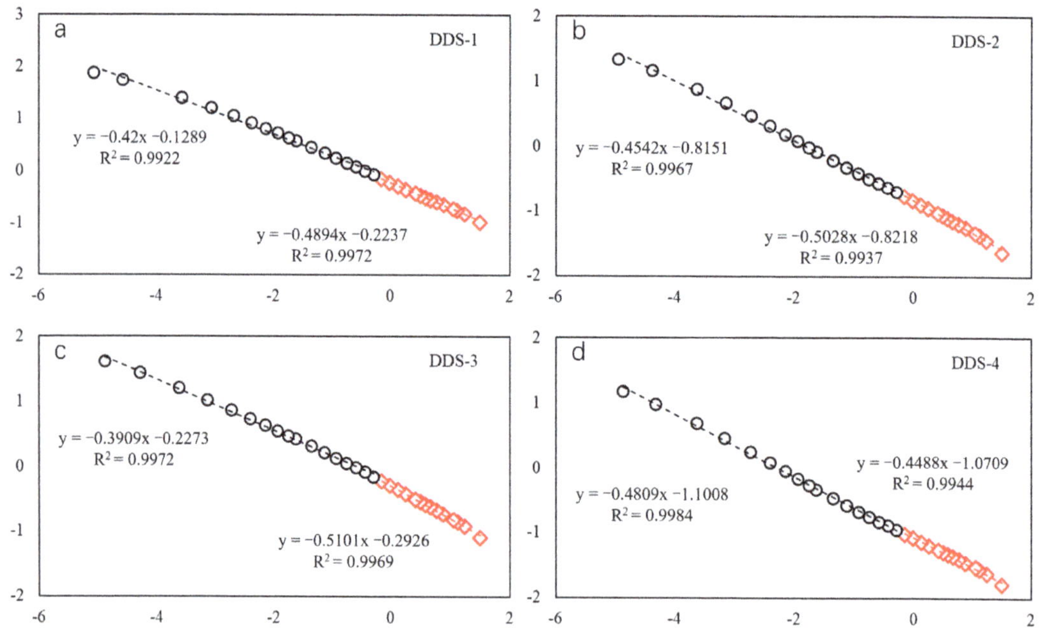

Figure 13. FHH plots for calculating fractal dimensions of ductile deformed shale samples from N_2 adsorption data.

The FHH plots of four undeformed shale samples are illustrated in Figure 11. The UDS fractal dimension Ds values range from 2.5623 to 2.6436, with a mean of 2.5946. The UDS fractal dimension Dm values are from 2.66 to 2.7366, with an average of 2.7011, indicating high pore structure complexity. The fractal dimension of brittle deformed shale is similar to undeformed shale samples (Figure 12). The BDS fractal dimension Ds values range from 2.5151 to 2.591, with a mean of 2.5542. The BDS fractal dimension Dm values are from 2.6937 to 2.7588, with an average of 2.7328. The BDS fractal dimension Ds values are slightly lower than UDS values, while Dm is slightly higher than UDS, indicating a bit of higher complexity of organic pore structure in brittle deformed shale samples. Ductile deformed shale exhibits different fractal dimension characteristics from undeformed shale and brittle deformed shale (Figure 13). The DDS fractal dimension Ds values range from 2.4899 to 2.5512, with a mean of 2.5122. The DDS fractal dimension Dm values are from 2.5191 to 2.6091, with an average of 2.5635. The fractal dimensions Ds and Dm of ductile deformed shale are relatively consistent, and both are about 2.5, indicating that the complexity of all scale organic pore structures is similar and both are relatively low. Compared with UDS and BDS, the fractal dimension Ds and Dm of ductile deformed shale are both very low (Table 4), indicating that the complexity of different pore structures of DDS are both lower than that of UDS and BDS. Especially, the value of Dm decreases from more than 2.7 in UDS and BDS to about 2.5 in DDS, which shows a significant decrease in the fractal dimension of mesopores and macropores in DDS, indicating that the complexity of mesopores and macropores in ductile deformed shale decreases significantly. The following discussion will cover the relationships between pore structures and fractal dimensions and the effects of different tectonic deformations on these organic pore features.

4. Discussion

4.1. Comparison of Pore Structure Fractal Characteristics Between Brittle and Ductile Deformation Shales

4.1.1. Differences in Fractal Dimensions

The fractal dimensions calculated from LTNA data show distinct differences between brittle deformed and ductile deformed shales. As presented in the results section, the average fractal dimension DS and DM for brittle deformed shales were 2.5542 and 2.7328, respectively. While for ductile deformed shales, they were 2.5122 and 2.5635, respectively. Both fractal dimension D_S and D_M of ductile deformed shale are lower than brittle deformed shales (Figure 14).This significant difference indicates that the organic nanopore network structure in the micropore, mesopore, and macropore range of brittle deformed shales is more complex than that of ductile deformed shales.

The development of fractures during brittle deformation is a major factor contributing to the higher D_M in brittle deformed shales. When shale undergoes brittle deformation, tensile and shear fractures are formed [33]. These fractures can create a highly tortuous and branched pore network. For example, in a brittle deformed shale sample, a fracture may intersect with multiple intergranular and intragranular pores, connecting them in a complex way. The presence of these fractures increases the number of pore connection points and the irregularity of the pore network, which is reflected in the higher fractal dimension. In contrast, during ductile deformation, the rock deforms plastically, and the pore structure is more homogenized. The crystal plastic deformation and pressure solution creep processes tend to reduce the complexity of the organic pore network in the mesopore and macropore range. For instance, the alignment of minerals during crystal-plastic deformation can lead to a more regular arrangement of pores, resulting in a lower fractal dimension.

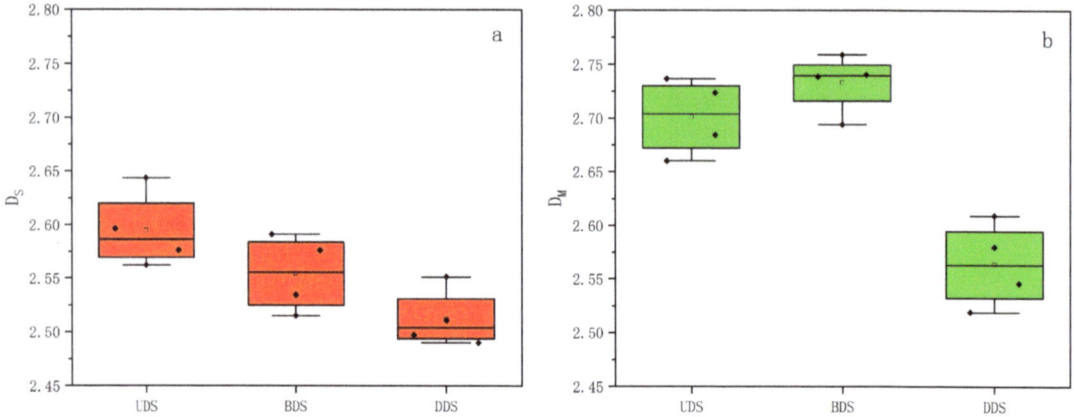

Figure 14. Box plots of fractal dimensions D_S (**a**) and D_M (**b**) for undeformed shales (UDS), brittle deformed shales (BDS), and ductile deformed shales (DDS).

4.1.2. Impact of Deformation Type on Pore Structure Complexity

The different deformation types, brittle and ductile, have distinct impacts on the complexity of the shale pore structure, which can be well understood from the perspective of fractal characteristics. For brittle deformed shales, the formation of fractures is the key factor increasing pore structure complexity. The fractures not only increase the porosity and permeability of the shale but also create a complex network of pore connections. The large-scale fractures can act as main flow channels for gas, while the small-scale fractures can connect organic pores and other different types of pores, such as intergranular and intragranular pores. This complex pore network structure is reflected in the relatively high fractal dimension calculated from LTNA data. This is consistent with previous work on the development of high aspect ratios and high fractal dimensions in the strongly deformed Longmaxi Shale under strong tectonic stress [29]. It suggests that appropriate structure deformation of OM pores under tectonic deformation, especially brittle deformation, is conducive to the development of connectivity, but with the transformation of deformation to ductility and further increase of aspect ratio, the fractal dimension and pore connectivity will decrease.

In ductile deformed shales, the plastic deformation mechanisms, such as crystal plastic deformation and pressure–solution creep, lead to a different kind of pore structure complexity. The development of micro folds and foliations, as well as the preferred orientation of clay minerals and kerogen, can create a complex pore surface and pore size distribution. The low fractal dimension D_M calculated from LTNA data for ductile deformed shales indicates that the macropore and mesopore structure is more homogenized and may have changed from the original ink-bottle type with complex edges to a simpler, narrower pore. This homogenized OM pore structure is beneficial for gas adsorption, as it maintains a larger specific surface area and adsorption sites for shale gas. However, the relatively lower fractal dimension calculated from LTNA data for ductile deformed shales implies that the mesopore and macropore network is less complex, which may result in a relatively lower gas flow rate compared to brittle deformed shales. In summary, brittle deformation mainly affects the fracture development of shale, while ductile deformation has a more significant impact on all-scale pore structures.

4.2. Factors Affecting the Fractal Characteristics of Shale Organic Pore Structure in Different Tectonic Deformation

Correlation analysis was carried out to explore the relationships between the fractal characteristics of the pore structure and other factors. The Pearson correlation coefficient was calculated to measure the linear relationship between variables (Figures 15–17). The color bar on the right-hand side of the heatmap indicates the relationship between colors and correlation coefficients. Red colors signify positive correlations. The deeper the red, the closer the correlation coefficient is to 1, indicating a stronger positive correlation. Blue colors represent negative correlations. The deeper the blue, the closer the correlation coefficient is to −1, indicating a stronger negative correlation. Colors close to white indicate weak correlations, close to 0. Cells in the heatmap are labeled with specific correlation coefficient values. A positive value indicates a positive linear relationship, while a negative value indicates a negative linear relationship. The higher the absolute value of the correlation coefficient (AVOCC), the stronger the correlation with different factors. The close the AVOCC is to 1, the stronger the linear relationship. We used the p-values and Pearson correlation coefficients and presented them in Figures 15–17. In this study, we chose the general criteria for p-values: 0.01, 0.05, and 0.1. In general, if the p-value is less than 0.01, it indicates at least 99% certainty; if the p-value is less than 0.05 (and greater than or equal to 0.01), it indicates at least 95% certainty; and if the p-value is less than 0.1 (and greater than or equal to 0.05), it indicates at least 90% certainty. Some cells are marked with asterisks (*, **, and ***). These asterisks denote the statistical significance levels of the correlations. Typically, * indicates a p-value less than 0.1, ** indicates a p-value less than 0.05, and *** indicates a p-value less than 0.01. Cells with asterisks imply that the correlations between the corresponding variables are statistically significant, meaning they are unlikely to occur randomly.

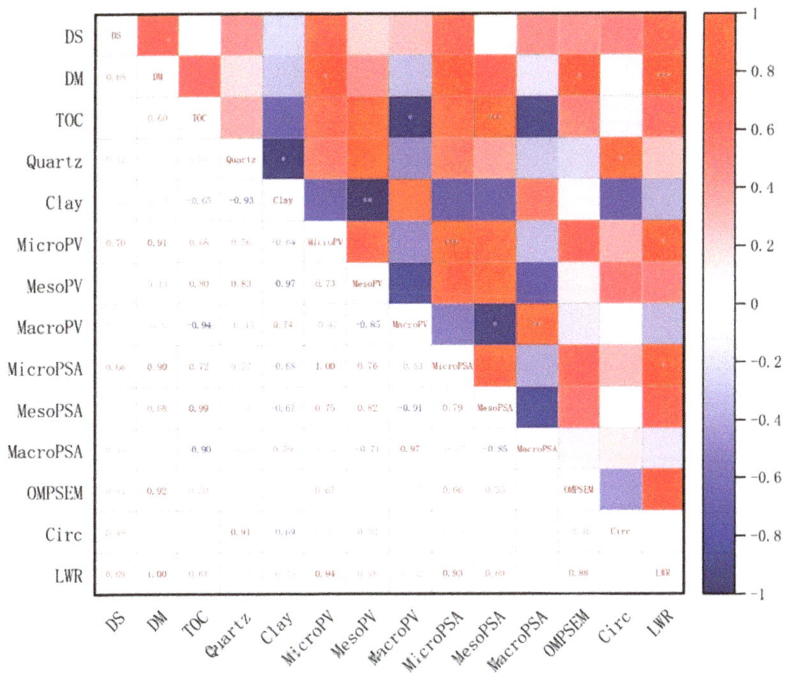

Figure 15. Correlation plot between the fractal dimension and different parameters of undeformed shale (*: $p < 0.1$; **: $p < 0.05$; ***: $p < 0.01$).

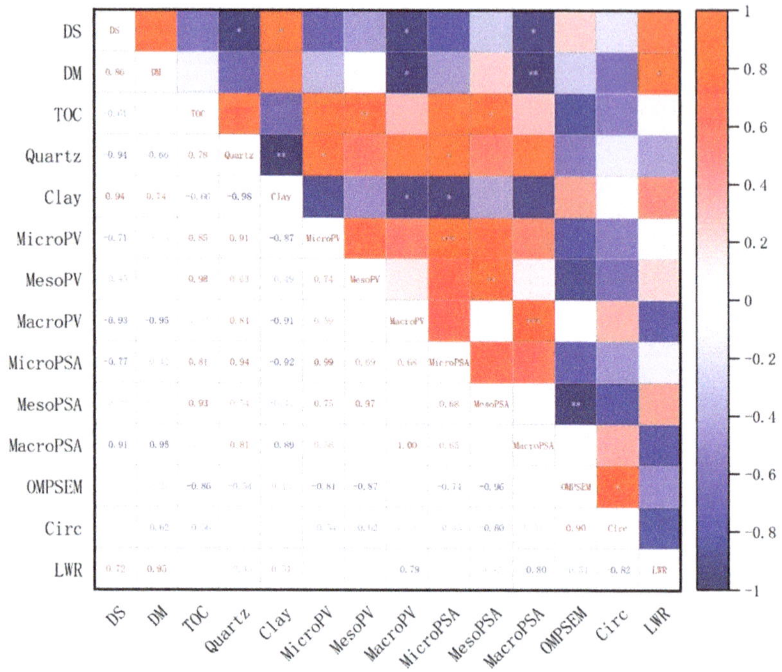

Figure 16. Correlation plot between the fractal dimension and different parameters of brittle deformed shale (*: $p < 0.1$; **: $p < 0.05$; ***: $p < 0.01$).

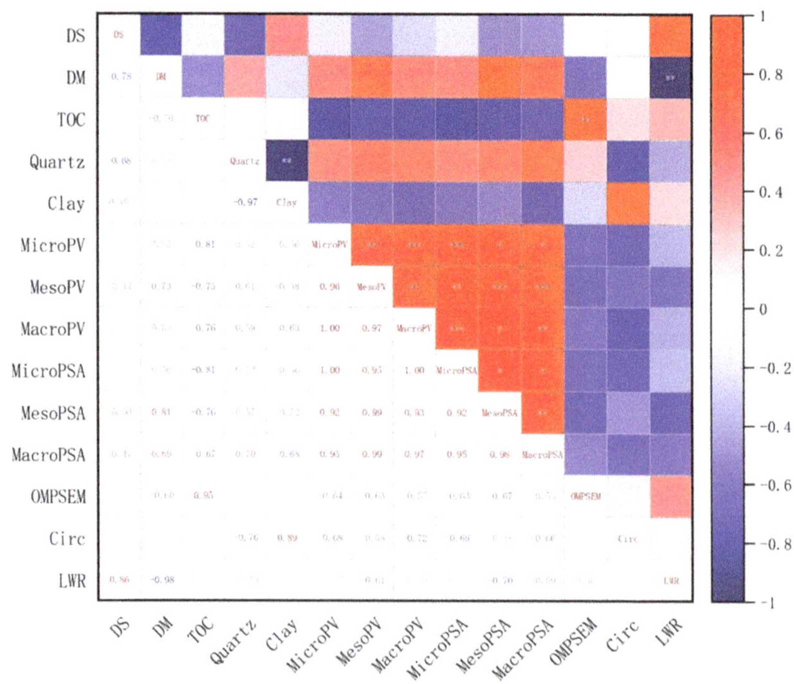

Figure 17. Correlation plot between the fractal dimension and different parameters of ductile deformed shale (*: $p < 0.1$; **: $p < 0.05$; ***: $p < 0.01$).

Figure 15 illustrates the correlation plot between the fractal dimension and UDS different parameters. For undeformed shale, the fractal dimension Ds is positively correlated with all parameters except clay content, with the best correlation coefficient of 0.70 with microPV and 0.69 with LWR. The fractal dimension D_M is positively correlated with most parts of parameters except clay content and macroPV and macroSA. The best correlation coefficients are 0.92 with OMP_{SEM}, 0.91 with microPV, and 0.90 with microPSA. When the significance markers of the correlations are considered, the fractal dimension D_M of the undeformed shale has good determinism with the microporous pore volume, organic matter porosity, and organic matter pore aspect ratio LWR parameters. In particular, we stated that the correlation coefficient between the fractal dimension D_M and LWR is 1, with a p-value less than 0.01 (***), showing a significant positive correlation within the 99% confidence interval. It is shown that the fractal dimension of undeformed shale is mainly affected by the development of micropores and pore shapes. The fractal dimensions D_S and D_M are also well positively correlated (0.65) with each other, indicating a synergistic evolutionary relationship between them in UDS.

Figure 16 illustrates the correlation plot between the fractal dimension and BDS's different parameters. For brittle deformed shale, the fractal dimension Ds is positively correlated with parameters of clay content, LWR, and OMP_{SEM}, with the best correlation coefficient of 0.94 with clay and 0.72 with LWR. The fractal dimension D_M is positively correlated with parameters LWR and clay content mesoPSA. The best correlation coefficient of 0.95 with LWR and 0.74 with clay. It is shown that the fractal dimension of brittle deformed shale is mainly affected by the development of clay and pore length–width ratio. The fractal dimensions D_S and D_M are also well positively correlated with each other in BDS the correlation coefficient is 0.86, indicating a synergistic evolutionary relationship between them in BDS. Considering the significance markers of the correlations, the fractal dimension Ds of brittle deformed shale shows strong determinism with clay content, whereas the fractal dimension DM demonstrates strong determinism with the organic matter pore aspect ratio LWR.

In the case of ductile deformed shale (DDS), Figure 17 reveals distinct relationships. The fractal dimension Ds is negatively correlated with all parameters except LWR and clay content; the positive correlation coefficient is 0.86 with LWR and 0.49 with clay. The fractal dimension D_M is positively correlated with parameters quartz content, all scale organic pore PV, and SA. The best correlation coefficient is 0.81 with mesoPSA and 0.73 with mesoPV. It is shown that the fractal dimension D_M of ductile deformed shale is affected by all scale pores. The fractal dimensions D_S and D_M are very poorly negatively correlated with each other in DDS; the correlation coefficient is −0.78, indicating a decoupling of evolutionary relationships between them in DDS. Meanwhile, when the significance markers of correlation are considered, the fractal dimensions D_S and D_M of ductile deformed shale do not have positive correlation certainty with all the parameters, but rather the fractal dimension D_M has a better negative correlation certainty with the organic matter pore aspect ratio LWR parameter. It shows that the complex types of organic matter mesopores and macropores, have been significantly negatively correlated with the values of pore aspect ratio in ductile deformed shales.

In conclusion, different tectonic deformations lead to diverse influencing factors on the fractal characteristics of shale organic pore structures. The development of micropores is crucial for undeformed shale, clay, and pore length–width ratio dominance in brittle deformed shale, and all-scale pores are key for ductile deformed shale. The pore aspect ratios LWR in undeformed and brittle deformed shales follow the same trend as the fractal dimension D_M of meso- and macropores, with the fractal dimension D_M becoming higher and pore shapes becoming more complex as the LWR increases. However, when the deformation reaches the ductile stage, after a large number of pores are compressed and larger LWR, the fractal dimension D_M starts to decrease and the pore complexity decreases. Additionally, the relationships between D_S and D_M vary greatly among different types of shale, reflecting the complex impacts of tectonic deformation on the pore structure's fractal features. TOC variability might influence the correlations observed. Shales with different TOC contents could have distinct pore structures and fractal characteristics. For example, higher TOC might lead to more organic-rich pores, which could in turn affect the relationships between fractal dimensions and other parameters. In addition, the compression of brittle minerals under stress and the directionality of clay minerals affect the fractal dimension of organic nanopores in shales [29,50]. Future studies could further explore the role of TOC and mineral variability in more detail.

4.3. *Geological Significance of Shale Organic Nanopore Fractal Characters*

Tectonic deformation is a fundamental geological process that has a profound impact on the organic nanopore structure of shale. Shale formations are often subjected to various tectonic forces, such as compression, extension, and shear, during their geological history [27,29,31–33]. These forces can cause the shale to undergo two main types of deformation: brittle deformation and ductile deformation. Brittle deformation in shale occurs when the rock is subjected to stress levels that exceed its strength, leading to the formation of fractures and the fragmentation of the rock mass. The main mechanisms of brittle deformation include tensile fracturing and shear fracturing. Brittle deformation typically occurs under high-stress conditions and low-temperature environments. In brittle deformed shale, fractures and microfractures are commonly developed. These fractures can be created by the tensile stress that exceeds the rock's strength or by the shear stress that causes the rock to break along planes of weakness. In areas with strong tectonic compression, shale may experience faulting and fracturing, resulting in the formation of large-scale fractures that can significantly change the pore structure. These fractures can increase the porosity and permeability of the shale by providing new flow paths for gas. However, if the fractures are not well connected or are filled with secondary minerals during diagenesis, they may not effectively enhance gas migration.

Ductile deformation, on the other hand, occurs under relatively high temperatures and low stress rate conditions. In ductile deformed shale, the rock deforms plastically, resulting in the development of micro folds, foliations, and preferred orientations of pores. These features can also alter the organic nanopore structure of the shale. The development of micro folds can create new pore spaces within the shale matrix of clay and kerogen, while the preferred orientation of organic pores can affect the connectivity of pores. In addition, ductile deformation can cause the original organic pores to be distorted or compressed, changing their size and shape. The influence of tectonic deformation on shale organic pore structure is not only related to the type of deformation but also to the intensity and duration of the tectonic forces. Stronger and longer-lasting tectonic forces can lead to more significant changes in the pore structure. Moreover, the interaction between different types of tectonic deformation can further complicate the organic nanopore structure evolution. Shale may first experience brittle deformation, creating fractures, and then undergo toughness deformation, which can modify the shape and connectivity of these fractures. Given the significant impact of tectonic deformation on shale organic pore structure, due to the complexity of pore structure and strong heterogeneity of shale reservoirs, the relationship between organic pore characteristics and tectonic deformation still needs further study in the future.

Fractal dimension is a key parameter for quantifying the complexity of pore structures. The fractal characteristics of shale organic pore structures are closely related to the gas storage capacity, which is of great significance for accurately assessing shale gas reserves. In undeformed and brittle-deformed shales, the relatively higher fractal dimension D_M (>2.7) in the mesopore and macropore range calculated from LTNA data implies a complex pore-network structure. A high D_M value indicates a complex and irregular mesopore and macropore structure. When gas flows through such a structure, it needs to overcome more resistance, such as narrow pore necks and complex curved paths. Narrow pore necks can limit the gas flow rate, and complex curved paths increase the tortuosity of the gas flow, resulting in a slowdown of gas diffusion and seepage rates. By comparing the fractal dimensions of brittle and ductile deformed shales, we can determine how different types of tectonic deformation affect the complexity of pore structures. A lower fractal dimension D_M (~2.5) in ductile deformed shales may indicate that the pore structure of organic matter gradually becomes homogeneous and stabilized under strong ductile deformation. More homogeneous and stable organic matter pores indicate that the shale still has some gas adsorption capacity even after intense tectonic stress, but the reduced gas adsorption capacity. The possible reduced connectivity of the organic matter pores increases the difficulty of shale gas desorption and fracking development in the ductile deformed shale region. Moreover, understanding the implications of organic pore structure fractal characteristics for shale gas can help in formulating more effective production strategies. For UDS and BDS, efforts should be focused on optimizing the use of natural fractures and enhancing the connectivity of the organic pore network. For ductile deformed shales, strategies should be centered around promoting gas desorption from the organic nanopore structure and improving the gas flow rate in the less complex mesopore and macropore range. And targeted stimulation strategies like creating new, larger fracture networks may be needed to improve gas production efficiency due to the lower gas diffusivity indicated by the reduced fractal dimension. Fractal dimension may not only respond to pore structure, but may also be an indicator of coordinated changes in permeability. Future work could integrate permeability measurements to validate the relationship between fractal dimensions and gas flow.

5. Conclusions

In this study, a comprehensive investigation was carried out on the fractal characteristics of the pore structure in shales subjected to undeformation, brittle and ductile deformation. The sample size may affect the universality of the conclusions; future research could expand the sampling range to verify the reliability of the current findings. However, our experimental methods are reproducible; the entire experimental process strictly follows relevant standards and specifications. Additionally, we compared the geological characteristics and pore structure features of the Niutitang Formation shale with those of other typical shale basins (such as the Longmaxi Formation shale). It is necessary to consider the impacts of geological condition differences on pore structures and fractal characteristics when applying the results to other different shale stratums and basins. The results revealed significant differences in the organic nanopore structure characteristics between the different types of deformed shales.

For undeformed and brittle deformed shales, the pore size distribution was relatively broad, with a dominant proportion of mesopores. The peak of the pore size distribution curve was mainly concentrated in the 2–50 nm range, and the macropores also contributed to the pore structure, providing large-scale flow paths for gas migration. The specific surface area calculated from LTNA data was relatively high, ranging from 13.95 to 29.59 mL/g, while the total pore volume ranged from 20.01 to 29.73 μL/g, and mesopores contributed the largest proportion to the total pore volume. The fractal dimensions D_S calculated from $P/P0 < 0.45$ in LTNA data ranged from 2.5151 to 2.6436, indicating a complex pore network structure in the micropore. The fractal dimensions D_M calculated from P/P0 > 0.45 in LTNA data ranged from 2.66 to 2.7588, indicating a more complex pore network structure in the mesopore and macropore range due to the maintenance of the original organic pore during tender tectonic stress and brittle deformation.

In contrast, ductile deformed shales had a pore size distribution with a peak mainly in the micropore and macropore range. The proportion of mesopores has decreased to be equivalent to that of micropores and macropores. The total pore volume was lower, ranging from 5.92 to 14.40 μL/g. The total pore surface area was ranging from 5.78 to 21.54 m^2/g, with micropores contributing the largest proportion (92.91–96.01%). The D_S for ductile deformed shales ranged from 2.4899 to 2.5512, showing a less complex pore network in the micropore. The D_M for ductile deformed shales ranged from 2.5191 to 2.6091, showing a less complex pore network in the mesopore and macropore range due to the plastic flow and homogenization of the pore structure during ductile deformation.

The comparison of fractal characteristics among undeformed, brittle, and ductile deformed shales showed that the differences in fractal dimensions were significant. The deformation type had a distinct impact on the nanopore structure complexity, with brittle deformation being more limited in modifying the organic nanopore structure of shale. And ductile deformation has a more significant influence on the mesopore. In general, brittle deformation led to a more complex pore network, while ductile deformation reduced the complexity of the organic nanopore structure.

Author Contributions: Conceptualization, M.L. and Z.W.; methodology, M.L., M.D. and X.L.; software, M.L.; validation, K.Z. and X.F.; formal analysis, M.L.; investigation, Z.W. and X.F.; resources, Z.W.; data curation, M.L.; writing—original draft preparation, M.L.; writing—review and editing, Z.W. and X.F.; funding acquisition, M.L., M.D. and X.L. All authors have read and agreed to the published version of the manuscript.

Funding: This research was funded by the Fundamental Research Funds for Chinese Academy of Geological Sciences (grant No. JKYQN202338 and DZLXJK202208), the National Natural Science Foundation of China (NSFC, 41802158), and the China Geological Survey (CGS, DD20242231 and DD20250209006).

Data Availability Statement: The article consolidates the original findings of this study. For any other inquiries, please contact the corresponding author.

Acknowledgments: The LTGA analysis was performed with the support of Tao Zhang from the Beijing Center for Physical & Chemical Analysis, Beijing Academy of Science and Technology. The authors would like to express their sincere appreciation for his assistance.

Conflicts of Interest: The authors declare that they have no known competing financial interests or personal relationships that could have appeared to influence the work reported in this paper.

References

1. Curtis, J.B. Fractured shale-gas systems. *Aapg Bull.* **2002**, *86*, 1921–1938.
2. Zou, C.; Yang, Z.; Dong, D.; Zhao, Q.; Chen, Z.; Feng, Y.; Li, J.; Wang, X. Formation, Distribution and Prospect of Unconventional Hydrocarbons in Source Rock Strata in China. *Earth Sci.* **2022**, *47*, 1517–1533.
3. Dong, D.; Zou, C.; Dai, J.; Huang, S.; Zheng, J.; Wang, Y.; Li, X.; Guan, Q.; Zhang, C.; Huang, J. Suggestions on the development strategy of shale gas in China. *Nat. Gas Geosci.* **2016**, *27*, 397–406. [CrossRef]
4. Sun, C.; Nie, H.; Dang, W.; Chen, Q.; Zhang, G.; Li, W.; Lu, Z. Shale gas exploration and development in China: Current status, geological challenges, and future directions. *Energy Fuels* **2021**, *35*, 6359–6379. [CrossRef]
5. Ross, D.J.K.; Bustin, R.M. Shale gas potential of the Lower Jurassic Gordondale member, Northeastern British Columbia, Canada. *Bull. Can. Pet. Geol.* **2007**, *55*, 51–75. [CrossRef]
6. Hu, Q.; Ewing, R.P.; Rowe, H.D. Low nanopore connectivity limits gas production in Barnett formation. *J. Geophys. Res. Solid Earth* **2015**, *120*, 8073–8087. [CrossRef]

7. Tian, H.; Pan, L.; Zhang, T.; Xiao, X.; Meng, Z.; Huang, B. Pore characterization of organic-rich Lower Cambrian shales in Qiannan Depression of Guizhou Province, Southwestern China. *Mar. Pet. Geol.* **2015**, *62*, 28–43. [CrossRef]
8. Loucks, R.G.; Reed, R.M.; Ruppel, S.C.; Jarvie, D.M. Morphology, Genesis, and Distribution of Nanometer-Scale Pores in Siliceous Mudstones of the Mississippian Barnett Shale. *J. Sediment. Res.* **2009**, *79*, 848–861. [CrossRef]
9. English, J.M.; English, K.L.; Corcoran, D.V.; Toussaint, F. Exhumation charge: The last gasp of a petroleum source rock and implications for unconventional shale resources. *AAPG Bull.* **2016**, *100*, 1–16. [CrossRef]
10. He, S.; Li, H.; Qin, Q.R.; Long, S.X. Influence of mineral compositions on shale pore development of Longmaxi Formation in the Dingshan area, southeastern Sichuan Basin, China. *Energy Fuels* **2021**, *35*, 10551–10561. [CrossRef]
11. Cao, T.; Song, Z.; Wang, S.; Cao, X.; Li, Y.; Xia, J. Characterizing the pore structure in the Silurian and Permian shales of the Sichuan Basin, China. *Mar. Pet. Geol.* **2015**, *61*, 140–150. [CrossRef]
12. Sing, K.S.W. Reporting physisorption data for gas/solid systems with special reference to the determination of surface area and porosity, IUPAC. *Pure Appl. Chem.* **1985**, *57*, 603–619. [CrossRef]
13. Wang, Y.; Cheng, H. Advances in microscopic pore structure characterization of fine-grained mudrocks. *Energy Fuels* **2023**, *37*, 1495–1510. [CrossRef]
14. Liu, B.; Mastalerz, M.; Schieber, J. SEM Petrography of Dispersed Organic Matter in Black Shales: A Review. *Earth Sci. Rev.* **2022**, *224*, 103874. [CrossRef]
15. Bakshi, T.; Vishal, V. A review on the role of organic matter in gas adsorption in shale. *Energy Fuels* **2021**, *35*, 15249–15264. [CrossRef]
16. Yang, W.; Wang, Y.; Yan, F.; Si, G.; Lin, B. Evolution characteristics of coal microstructure and its influence on methane adsorption capacity under high temperature pyrolysis. *Energy* **2022**, *254*, 124262. [CrossRef]
17. Awan, R.S.; Liu, C.; Yasin, Q.; Liu, B.; Wood, D.A.; Feng, D.; Iltaf, K.H. Geologic characterization of the Lower Paleozoic black shale of the Niutitang Formation: Implications for a shale gas potential in western Hunan, China. *Mar. Pet. Geol.* **2024**, *163*, 106756. [CrossRef]
18. Loucks, R.G.; Reed, R.M.; Ruppel, S.C. Spectrum of pore types and networks in mudrocks and a descriptive classification for matrix-related mudrock pores. *AAPG Bull.* **2012**, *96*, 1071–1098. [CrossRef]
19. Ko, L.T.; Loucks, R.G.; Zhang, T.W. Pore and pore network evolution of upper cretaceous boquillas (Eagle Ford-equivalent) mudrocks: Results from gold tube pyrolysis experiments. *AAPG Bull.* **2016**, *100*, 1693–1722. [CrossRef]
20. Chen, J.; Xiao, X. Evolution of nanoporosity in organic-rich shales during thermal maturation. *Fuel* **2014**, *129*, 173–181. [CrossRef]
21. Zhang, Y.; He, Z.; Jiang, S.; Lu, S.; Xiao, D.; Chen, G.; Zhao, J. Factors affecting shale gas accumulation in overmature shales case study from lower Cambrian shale in western Sichuan Basin, South China. *Energy Fuels* **2018**, *32*, 3003–3012. [CrossRef]
22. Ning, S.; Xia, P.; Hao, F.; Tian, J.; Fu, Y.; Wang, K. Pore Fractal Characteristics between Marine and Marine–Continental Transitional Black Shales: A Case Study of Niutitang Formation and Longtan Formation. *Fractal Fract.* **2024**, *8*, 288. [CrossRef]
23. Okolo, G.N.; Everson, R.C.; Neomagus, H.W.J.P.; Roberts, M.J.; Sakurovs, R. Comparing the porosity and surface areas of coal as measured by gas adsorption, mercury intrusion and SAXS techniques. *Fuel* **2015**, *141*, 293–304. [CrossRef]
24. Brunauner, S.; Emmett, S.P.H.; Teller, E.J. Adsorption of gases in multimolecular layers. *J. Am. Chem. Soc.* **1938**, *60*, 309–319. [CrossRef]
25. Klimakow, M.; Klobes, P.; Rademann, K.; Emmerling, F. Characterization of mechanochemically synthesized MOFs. *Microporous Mesoporous Mater.* **2012**, *154*, 113–118. [CrossRef]
26. Han, H.; Cao, Y.; Chen, S.; Lu, J.; Huang, C.; Zhu, H.; Zhan, P.; Gao, Y. Influence of particle size on gas-adsorption experiments of shales: An example from a Longmaxi Shale sample from the Sichuan Basin, China. *Fuel* **2016**, *186*, 750–757. [CrossRef]
27. Yu, Y.; Wang, Z.; Zhang, K.; Feng, X.; Cheng, M. Effects of shear on development characteristics of organic matter pores in shale: A case study of shale in the Niutitang Formation of the well XAD1. *J. Petr. Sci. Eng.* **2022**, *211*, 110166. [CrossRef]
28. Tang, L.; Song, Y.; Jiang, Z.; Jiang, S.; Li, Q. Pore Structure and Fractal Characteristics of Distinct Thermally Mature Shales. *Energy Fuels* **2019**, *33*, 5116–5128. [CrossRef]
29. Wang, Y.; Han, D.; Lin, W.; Jia, Y.; Zhang, J.; Wang, C.; Ma, B. Factors Controlling Differences in Morphology and Fractal Characteristics of Organic Pores of Longmaxi Shale in Southern Sichuan Basin, China. *Fractal Fract.* **2024**, *8*, 555. [CrossRef]
30. Guan, D.Z.; Dong, D.Z.; Sun, S.S.; Hu, L.X.; Qi, L.; Li, C.D.; Chen, C.; Chen, X. Fractal characteristics of organic-rich shale pore structure and its geological implications: A case study of the Lower Silurian Longmaxi Formation in the Weiyuan block, Sichuan Basin. *Nat. Gas Ind.* **2024**, *44*, 108–118.
31. Liang, M.; Wang, Z.; Gao, L.; Li, C.; Li, H. Evolution of pore structure in gas shale related to structural deformation. *Fuel* **2017**, *197*, 310–319. [CrossRef]
32. Zhu, H.; Ju, Y.; Qi, Y.; Huang, C.; Zhang, L. Impact of tectonism on pore type and pore structure evolution in organic-rich shale: Implications for gas storage and migration pathways in naturally deformed rocks. *Fuel* **2018**, *228*, 272–289. [CrossRef]
33. Li, X.; Wang, J.; Li, Z.; Zhang, K.; Feng, X.; Wang, Z. Impact of brittle deformation on pore structure evolution in shale: Samples collected from different fault positions. *Energy Fuels* **2022**, *36*, 1441–1456. [CrossRef]

34. Hu, J.; Tang, S.; Zhang, S. Investigation of pore structure and fractal characteristics of the Lower Silurian Longmaxi shales in western Hunan and Hubei Provinces in China. *J. Nat. Gas. Sci. Eng.* **2016**, *28*, 522–535. [CrossRef]
35. Ma, Y.; Zhong, N.; Han, H. Definition and sturcture characteristics of pores in mylonitized organic-rich. *Shales Earth Sci.* **2014**, *44*, 2202–2210. [CrossRef]
36. Zhan, H.M.; Li, X.Z.; Hu, Z.M.; Duan, X.G.; Wu, W.; Guo, W.; Lin, W. Fractal Characteristics of Deep Shales in Southern China by Small-Angle Neutron Scattering and Low-Pressure Nitrogen Adsorption. *Fractal Fract.* **2022**, *6*, 484. [CrossRef]
37. Yang, R.; He, S.; Yi, J.; Hu, Q. Nano-scale pore structure and fractal dimension of organic-rich Wufeng-Longmaxi shale from Jiaoshiba area, Sichuan Basin: Investigations using FE-SEM, gas adsorption and helium pycnometry. *Mar. Pet. Geol.* **2016**, *70*, 27–45. [CrossRef]
38. Thommes, M.; Kaneko, K.; Neimark, A.V. Physisorption of gases, with special reference to the evaluation of surface area and pore size distribution (IUPAC Technical Report). *Pure Appl. Chem.* **2015**, *87*, 1051–1069. [CrossRef]
39. Thommes, M.; Cychosz, K.A. Physical adsorption characterization of nanoporous materials: Progress and challenges. *Adsorption* **2014**, *20*, 233–250. [CrossRef]
40. Landers, J.; Gor, A.G.A.; Neimark, V. Density functional theory methods for characterization of porous materials. *Chem. Inf.* **2014**, *437*, 3–32. [CrossRef]
41. Groen, J.C.; Peffer, L.A.A.; Perez-Ramírez, J. Pore size determination in modified micro- and mesoporous materials. Pitfalls and limitations in gas adsorption data analysis. *Microporous Mesoporous Mater.* **2003**, *60*, 1–17. [CrossRef]
42. Klaver, J.; Desbois, G.; Urai, J.L.; Littke, R. BIB-SEM study of the pore space morphology in early mature Posidonia Shale from the Hils area, Germany. *Int. J. Coal Geol.* **2012**, *103*, 12–25. [CrossRef]
43. Zhang, J.; Li, X.; Xie, Z.; Li, J.; Zhang, X.; Sun, K.; Wang, F. Characterization of microscopic pore types and structures in marine shale: Examples from the Upper Permian Dalong formation, Northern Sichuan Basin, South China. *J. Nat. Gas Sci. Eng.* **2018**, *59*, 326–342. [CrossRef]
44. Lin, W.; Zhao, X.; Li, M.; Zhuang, Y. Pore Structure Characterization and Fractal Characteristics of Tight Limestone Based on Low-Temperature Nitrogen Adsorption and Nuclear Magnetic Resonance. *Fractal Fract.* **2024**, *8*, 371. [CrossRef]
45. Ross, D.J.K.; Bustin, R.M. The importance of shale composition and pore structure upon gas storage potential of shale gas reservoirs. *Mar. Pet. Geol.* **2009**, *26*, 916–927. [CrossRef]
46. Pan, J.N.; Zhu, H.T.; Hou, Q.L. Macromolecular and pore structures of Chinese tectonically deformed coal studied by atomic force microscopy. *Fuel* **2015**, *139*, 94–101. [CrossRef]
47. Lewan, M.D. Stable carbon isotopes of amorphous kerogens from Phanerozoic sedimentary-rocks. *Geochim. Cosmochim. Acta* **1986**, *50*, 1583–1591. [CrossRef]
48. Liang, Z.; Jiang, Z.; Han, Y.; Wang, B.; Wu, W.; Li, Z.; Li, Y.; Xue, Z. A modified dent-fractal mathematical model to investigate the water vapor adsorption on nanopore structure heterogeneity from the Longmaxi shale, Sichuan Basin, China. *Energy Fuels* **2023**, *37*, 9131–9150. [CrossRef]
49. Zou, J.; Chen, W.; Yang, D.; Yuan, J.; Jiao, Y.Y. Fractal characteristics of the anisotropic microstructure and pore distribution of low-rank coal. *AAPG Bull.* **2019**, *103*, 1297–1319. [CrossRef]
50. Jiang, W.; Zhang, Y.; Ma, T.; Chen, S.; Hu, Y.; Wei, Q.; Zhuang, D. Pore Structure and Its Fractal Dimension: A Case Study of the Marine Shales of the Niutitang Formation in Northwest Hunan, South China. *Fractal Fract.* **2025**, *9*, 49. [CrossRef]

fractal and fractional

MDPI

Article

Pore–Fracture Structure and Fractal Features of Carboniferous Taiyuan Formation Hydrocarbon Source Rocks as Investigated Using MICP, LFNMR, and FESEM

Dun Wu [1,2,*], Liu Zhao [1], Guangqing Hu [3] and Wenyong Zhang [3]

1 AnHui Provincial Key Laboratory of Intelligent Underground Detection, Anhui Province Intelligent Underground Exploration and Environmental Geotechnical Engineering Research Center, College of Civil Engineering, Anhui Jianzhu University, Hefei 230601, China; 19155351239@163.com
2 School of Earth and Space Sciences, University of Science and Technology of China, Heifei 230026, China
3 Exploration Research Institute, Anhui Province Bureau of Coal Geology, Hefei 230088, China; huguangqing117@163.com (G.H.); adam854621@163.com (W.Z.)
* Correspondence: wudun@ahjzu.edu.cn

Abstract: The pore structure of reservoir rocks was a crucial factor affecting hydrocarbon production. Accurately characterized the micropore structure of different types of rock reservoirs was of great significance for unconventional natural gas exploration. In this study, multiple observation methods (field emission scanning electron microscope (FESEM) and low-field nuclear magnetic resonance (LFNMR)) and physical tests (mercury injection capillary pressure (MICP)) were employed, and double logarithmic plots for fractal fitting were illustrated. The fractal dimension of 15 samples was calculated using fractal theory to systematically investigate the pore–fracture structure and fractal characteristics of hydrocarbon source rock (limestone, mudstone, and sandstone) samples from the Late Carboniferous Taiyuan Formation in the Huainan coalfield. MICP experiments revealed that sandstone reservoirs had larger and more uniformly distributed pore throats compared to mudstone and limestone, exhibiting superior connectivity and permeability. The T_2 spectrum characteristic maps obtained using LFNMR were also consistent with the pore distribution patterns derived from MICP experiments, particularly showed that sandstone types exhibited excellent signal intensity across different relaxation time periods and had a broader T_2 spectrum width, which fully indicated that sandstone types possess superior pore structures and higher connectivity. FESEM experiments demonstrated that sandstone pores were highly developed and uniform, with sandstone fractures dominated by large fractures above the micrometer scale. Meanwhile, the FESEM fractal dimension results indicated that sandstone exhibits good fractal characteristics, validating its certain oil storage capacity. Furthermore, the FESEM fractal dimension exhibited a good correlation with the porosity and permeability of the hydrocarbon source rock reservoirs, suggesting that the FESEM fractal dimension can serve as an important parameter for evaluating the physical properties of hydrocarbon source rock reservoirs. This study enriched the basic geological theories for unconventional natural gas exploration in deep coal-bearing strata in the Huainan coalfield.

Keywords: pore structure; fractal dimension; fracture features; MICP; LFNMR; FESEM; carboniferous Taiyuan Formation; hydrocarbon source rock

1. Introduction

In recent years, along with the deepening of unconventional natural gas exploration activities, unconventional natural gas resources have received more and more attention and become the focus of research and practice, and the intensity of the research on them has also shown a trend of continuous increase [1–3]. As the core carrier of oil and gas storage and the key path for oil and gas seepage, the pore–fracture structures, with their complex characteristics, have a profound impact on the physical properties of reservoirs, oil and gas transportation, and storage efficiency, as well as the assessment of tight oil and gas resources reserves and the scale development of economic benefits [4,5]. Therefore, an in-depth investigation of the differences in the interlayer structure among multiple reservoirs is of great academic and practical value for the efficient exploration and utilization of coal system unconventional natural gas resources. The pore–fracture structure of hydrocarbon reservoirs refers to the geometric features, size distribution, connectivity, and configuration relationship of pores and throats in the reservoir, which constitutes the general appearance of pores and connected throats [6]. Currently, the study of pore structure in tight reservoirs constitutes a major hotspot and difficulty in the field of tight oil and gas reservoir research [7,8].

In recent years, many scholars have made many research advances through different pore structure characterization techniques [9]. In the field of observational description, the effects of microwave-assisted cyclic oxidation on the internal and surface structures of coal have been investigated based on spectra obtained from nuclear magnetic resonance (NMR) and surface roughness data acquired from atomic force microscopy (AFM). The results indicated that microwave-assisted cyclic oxidation is most effective in altering the internal and surface morphologies of coal [10]. Furthermore, the genesis of clay minerals in glutenite reservoirs and their impact on reservoir quality have been studied using field emission scanning electron microscopy (FESEM) and X-ray diffraction (XRD) methods. The results showed that the genesis of clay minerals is closely related to rock–fluid interactions, and their impact on reservoir quality depends on the intensity of fluid injection and changes in diagenetic environments [3]. Additionally, X-ray computed tomography (XCT) scanning was employed to analyze the structural characteristics of reservoir formations in the Xihu Sag of the East China Sea Shelf Basin and their impact on permeability by scanning two sets of samples. The results revealed that smaller throats (<5 μm) contributed differently to permeability (0% for Sample 1 and 100% for Sample 5). When these throats significantly contribute to permeability, high-resolution XCT and magnified studies become crucial [11]. In the field of physical testing methods, the pore evolution patterns of tar-rich coal have been unveiled through the utilization of mercury intrusion capillary pressure (MICP), low-temperature N_2 adsorption, and low-temperature CO_2 adsorption techniques. The research identifies three stages of pore development in tar-rich coal, categorized as Stage 1 with R_O, max < 0.9%, Stage 2 with R_O, max = 0.9 to 3%, and Stage 3 with R_O, max > 3%. There is a strong correlation between these pore development stages and the increase in coal maturation [12]. Experimental methods such as optical microscopy, field emission scanning electron microscopy (FESEM), focused ion beam scanning electron microscopy (FIB-SEM), and micro-nano CT scanning can be used to visually and precisely observe the morphology of pore structures [13–15]. However, due to the limitations of high resolution, these instruments can only describe local pore structure patterns, leading to highly uniform and non-representative pore information. Gas adsorption, constant-rate/high-pressure mercury injection, and nuclear magnetic resonance (NMR) testing can characterize a broad range of pore areas, making it easy to quantify the characteristics of storage spaces [16–18]. Based on previous research, LFNMR, which provides extensive pore area characterization, was selected as the primary method. The spectral information generated by LFNMR is combined with images from MICP, which has advantages in characterizing macropores, serving as verification and auxiliary checks. According to the aforementioned characterization methods, this study employed a comprehensive and scientific approach combining MICP, LFNMR, and FESEM to characterize the pore and fracture structure of the source rocks from the HN-1 well in the Late Carboniferous Taiyuan Formation.

Fractal geometry is a mathematical discipline that emerged in the late 1970s, focusing on describing irregular shapes and random phenomena [19]. This theory defines irregular structures with self-similarity as fractals, characterized by a certain degree of similarity between their parts and the whole, which is quantitatively described by the fractal dimension. Currently, fractal theory has been widely applied in the field of geology [20]. In the study of source rocks, the FESEM fractal dimension can effectively distinguish the modification of pore structures by different diagenetic processes. For instance, sandstone with a large pore throat system exhibits a low fractal dimension, while clay mineral alignments or organic matter fractures in mudstones display a high fractal dimension [21]. Furthermore, the integrated analysis of FESEM fractal dimensions with MICP and LFNMR data can reveal the synergistic effects of pores across multiple scales. Additionally, the fractal dimension not only describes the pore size and uniformity of distribution in shale but also reflects the complexity of pore morphology. Therefore, the fractal dimension has become an important parameter for quantitatively describing the pore structure of shale, and it can be used to quantitatively characterize the heterogeneity of shale gas reservoirs [22].

In recent years, the exploration of unconventional natural gas in the deep coal measures of the Huainan coalfield has continued to deepen, yet it also faces geological challenges such as fractured and soft coal seams. These challenges have prompted various scholars to conduct research at different levels [23,24]. Furthermore, the predicted geological resource volume of coalbed methane in coal seams shallower than 2000 m in the Huaian–Huaibei region amounts to 8984.69×10^8 m^3, accounting for approximately 90% of the deep coalbed methane resources in East China. Regardless of coal seam gas content or coalbed methane resource abundance, the Huainan coalfield exhibits higher resource potential [25], and the aforementioned content is a key focus for future exploration of unconventional natural gas in deep coal measures in Anhui. Based on this, this study selects the HN-1 well in the Taiyuan Formation of the Late Carboniferous in the Huainan coalfield as the research object. Building upon traditional geological exploration methods, multiple cutting-edge experimental techniques are introduced, including MICP, LFNMR, FESEM, and polarized optical microscopy (POM). The aim is to deeply reveal the microstructure characteristics of different types of rocks, such as limestone, sandstone, and mudstone, focusing on key parameters like pore size distribution, pore connectivity, and permeability. Moreover, by calculating the fractal dimension using SEM, combined with pore data obtained from MICP and LFNMR, the heterogeneous evolution of micropores and fractures in the source rocks of the Taiyuan Formation was quantitatively analyzed. Relevant reports have been published on studying rock pore–fracture structures using methods such as FESEM [3], LFNMR [10], and MICP [12]. However, this study is the first to systematically combine these three methods to conduct a comprehensive and in-depth analysis of the pore structure and fracture characteristics of the source rocks of the Taiyuan Formation in the Huainan coalfield. Specifically, MICP technology is utilized to conduct a detailed analysis of the pore structure of source rocks in the Taiyuan Formation, revealing differences in pore morphology and distribution among different rock types. Through LFNMR technology, important information regarding the distribution and fluidity of internal fluids within rocks is obtained. Meanwhile, FESEM and POM allow for direct observation of the micromorphology and mineral composition of rock surfaces. Furthermore, this study delves into the differences and complementarity of various experimental methods in characterizing the internal microstructure of rocks. These findings enrich the fundamental geological theories underlying unconventional natural gas exploration in deep coal seams in the Huainan coalfield.

2. Geological Setting

2.1. Stratigraphy of the Study Area

The stratigraphic composition of the Huainan coalfield is primarily dominated by the Upper Paleozoic, with its Lower Paleozoic distributed at the edges of the coalfield, all extensively covered by the Quaternary. Its coal-bearing strata belong to the North China-type Carboniferous–Permian coal measures, encompassing the Benxi Formation and Taiyuan Formation of the Late Carboniferous, as well as the Shanxi Formation, Lower Shihezi Formation, and Upper Shihezi Formation of the Early and Late Permian, respectively [26]. Among them, the Taiyuan Formation overlays the Benxi Formation. The source rocks of the Taiyuan Formation are primarily composed of limestone, sandstone, mudstone, carbonaceous shale, and thin coal seams, which are deposited in a transitional environment from shallow sea to coast [26,27]. Within this, the limestone is stable and relatively thick, with a thickness of 10 to 20 m, accounting for 37% of the total thickness. The sandstone (accounting for approximately 14% of the thickness) is mainly medium- to fine-grained quartz sandstone of marine–terrestrial transitional facies, while the mudstone (accounting for approximately 23% of the thickness) is rich in organic matter [28] (Figure 1). The lithological characteristics primarily reflect the depositional processes of a series of platform–barrier island complex systems at the marine–terrestrial interface. The L1, L3, L4, and L12 limestone horizons within the Taiyuan Formation are stable, especially L4 and L12 limestones, which are not only stable in horizon but also possess significant thicknesses, generally ranging from 10 to 20 m [28]. Based on lithological characteristics, the Taiyuan Formation is further subdivided into three limestone groups: C_3I (L1 to L4 limestone), C_3II (L5 to L10 limestone), and C_3III (L11 to L13 limestone). Additionally, there exists a disconformity contact relationship between the Taiyuan Formation and the underlying Ordovician Majiagou Formation [29].

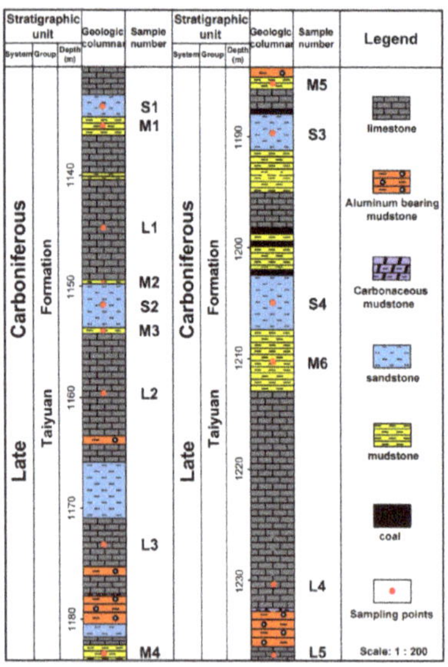

Figure 1. Geological histogram and sampling point distribution map of the Taiyuan Formation of the HN-1 well in the Huainan coalfield.

2.2. Tectonics of the Study Area

The Huainan coalfield is located in the northern part of Anhui Province, on both sides of the Huai River, spanning the counties (districts) of Fengtai, Yingshang, Lixin, and Mengcheng in the cities of Huainan, Fuyang, and Bozhou. The geographic location is at the southern edge of the North China Craton Basin, and the coalfield extends from Chuxian County in the east to the vicinity of Fuyang in the west, spreading in a northwesterly direction in the plane, with an overall long ellipse shape, a length of about 180 km, a width of about 20–30 km, and a geographic area of about 3654 km^2 [27], surrounded by large basins, such as the Ordos Basin and Bohai Bay Basin, and neighboring the Tantanlu Fracture Zone and the Suru Orogenic Belt (Figure 2a). The HN-1 well is located in the alluvial plain of the Huaihe River, with a low relief, generally 22–26 m above sea level, high in the west and low in the east. The stratigraphic strike is from east to west, inclined to north, with a large dip angle of 60° to 90°. Figure 2b shows the geotectonic map of the Panxie mining area of this study, as well as the distribution of faults and folds and the location of the HN-1 well.

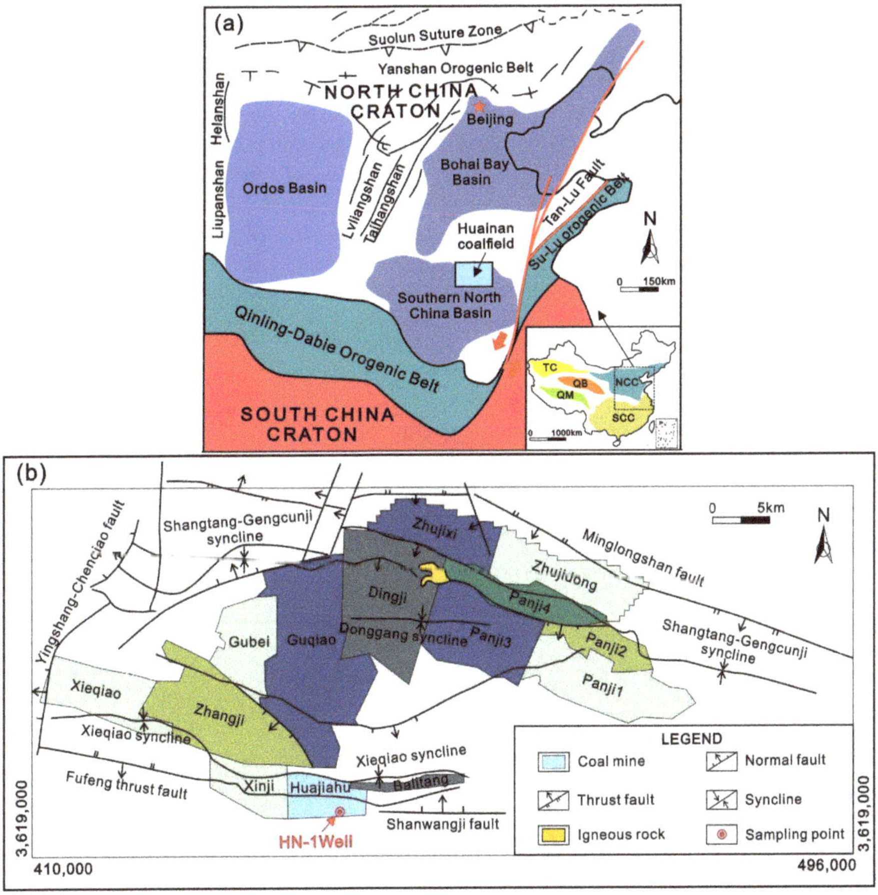

Figure 2. (**a**) Distribution map of major basins in North China (modified from [30]), The red bi-directional arrow in the figure indicates a strike-slip fault in the Tanlu Fault Zone east of the collision zone between the South China Craton and Qinling-Dabie; (**b**) Geological structure map of Panxie mining area, Huainan coalfield. ((**a**) shows TC: Tarim Craton; QB: Qaidam Block; QM. Qiangtang Block).

3. Methodology

The classification of pore sizes in porous media is fundamental for evaluating pore structure characteristics. In this study, a comprehensive pore size classification scheme proposed by both Hodot and IUPAC was adopted to analyze the pore structure of the rocks (Figure 3). Figure 3 illustrates that various techniques are available for characterizing the pore size distribution in porous media, primarily categorized into observational description methods and physical testing methods. By integrating these two types of methods, a full-scale (macro-meso-micro) characterization of the pore structure in porous media can be achieved. Therefore, in this study, we chose to combine three techniques: MICP, LFNMR, and FESEM. Additionally, we calculated the fractal dimension of images obtained through FESEM to comprehensively characterize the pore–fracture structure and fractal features of the hydrocarbon source rocks from the Late Carboniferous Taiyuan Formation. Firstly, MICP technology was employed to quantitatively assess the pore structure of the rock samples. This method is crucial for understanding the size and morphology of pores in sandstone, mudstone, and limestone, particularly as it reveals that sandstone has larger and more uniformly distributed pore throats compared to other lithologies. Subsequently, the LFNMR technology was used to further verify and refine the MICP results. LFNMR indirectly reflects the pore structure by measuring the relaxation time of fluids within the rocks. The T_2 spectrum characteristic plot displays signal intensities across different relaxation time intervals, thereby reflecting the distribution of pore sizes. FESEM offers high-resolution imaging capabilities, enabling the revelation of pore and fracture structures within rock samples. Furthermore, through the calculation of fractal dimensions, it is possible to quantitatively describe the complexity of source rock reservoir structures. Notably, FESEM images also exhibit microstructural differences between various types of rocks, which further underscores the importance of combining multiple techniques for a comprehensive characterization of the internal microstructure of rocks.

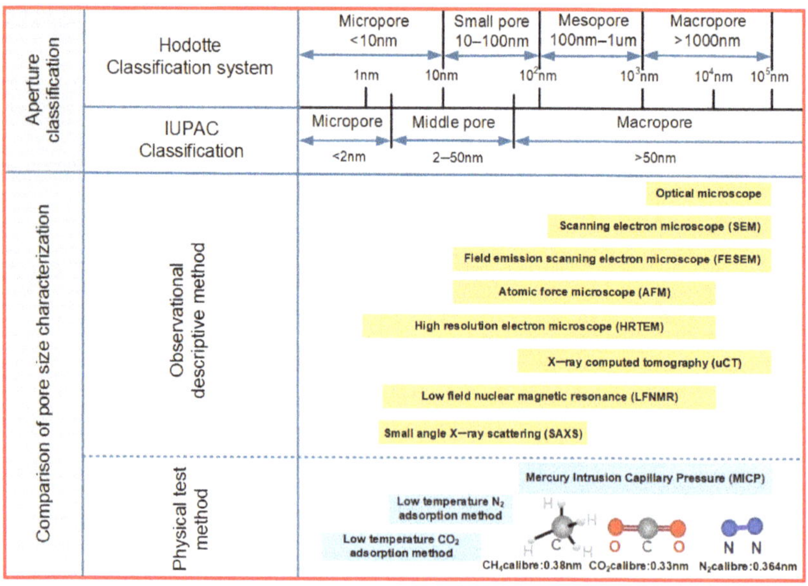

Figure 3. Information summary of different representation methods for multi-scale rock physical structure characterization.

4. Samples and Experiments

4.1. Hydrocarbon Rock Sample Collection

The HN-1 well is the first investigation well for the Taiyuan Formation in the Huainan coalfield, and the vast majority of the rock samples were used for physical and chemical analyses and resolved adsorption tests. Therefore, a representative portion of the retained rock samples was utilized in this study for special pore structure testing and analysis. According to the current regulations of the "Coal Rock Sample Taking Program" (GB/T 19222-2003) [31], samples of hydrocarbon source rocks (limestone, mudstone, and sandstone) were collected from the Late Carboniferous Taiyuan Formation strata in the Huainan coalfield. Due to the limited thickness of thin coal seams and carbonaceous shales, which contribute minimally to reservoir capacity and permeability [27], and their low compressive strength and sampling integrity, with a tendency to be fragile, this study focuses on sampling the more critical core lithologies within the source–reservoir system. Specifically, five limestone samples, four sandstone samples, and six mudstone samples were collected (Figure 1).

4.2. Experimental Methods

During a systematic investigation of the pore structure and fracture characteristics of the hydrocarbon source rocks (including limestone, mudstone, and sandstone) from the Late Carboniferous Taiyuan Formation in the Huainan coalfield, an experimental sequence of MICP-LFNMR-FESEM was followed, thereby establishing an in-depth understanding of the microstructure of different types of rocks. Figure 4 presents the cores sampled from the Taiyuan Formation at depths ranging from 1150 to 1156 m, which serve as representative samples for testing the rocks.

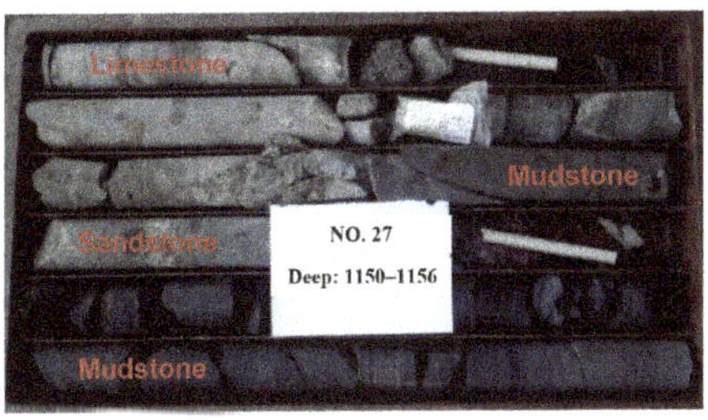

Figure 4. Physical sample image of source rock from the 1150–1156 m interval of the Taiyuan Formation at the HN-1 well.

Firstly, quantitative analysis of pore structure characteristics was conducted using MICP experiments, with particular attention paid to the size distribution and connectivity of pore throats. Simultaneously, the results from MICP provided crucial clues for uncovering the relationship between pore size distribution and the width of the LFNMR T_2 spectrum. Subsequently, the application of LFNMR technology further enhanced the understanding of pore structures. The T_2 spectrum characteristic map of LFNMR clearly demonstrated strong signal intensity and a broad T_2 spectrum width for sandstone types within different relaxation time periods, which directly reflected the favorable pore structure and high connectivity of the sandstone. The results from LFNMR and MICP can mutually validate each other. Finally, through FESEM observations, high-resolution morphological information (including micro- and nanoscale pore fractures) of the rock sample surfaces was obtained, and through the calculation of fractal dimension, the complexity of the source rock reservoir structure was quantitatively described. This morphological evidence provided an intuitive interpretative framework for the results of MICP and LFNMR. Detailed experimental methods for the aforementioned three experiments will be elaborated on in the following sections.

4.2.1. Mercury Intrusion Capillary Pressure (MICP)

The MICP experiment employed an AutoPore IV 9505 porosimeter manufactured by Micromeritics Instrument Corporation, Norcross, Georgia, USA. This instrument is primarily used to measure key parameters in materials with macroporous structures, such as total pore volume, pore size distribution, and porosity. Its low-pressure analysis range spans from 1.38 to 310 kPa and is capable of analyzing pore size distributions from 3.6 mm to 1000 μm in diameter. Meanwhile, high-pressure analysis can reach pressures up to 227,527 kPa, allowing the instrument to accurately measure pore sizes as small as 0.0050 mm (i.e., 5 μm). The entire experimental process was strictly conducted in accordance with the standard "Method for Measurement of Capillary Pressure Curve of Rocks" (GB/T 29171-2012) [32], and carried out in a constant temperature laboratory environment throughout. During the experiment, the integrated temperature control system of the Micromeritics AutoPore IV 9505 instrument was employed to precisely maintain the ambient temperature within the range of 20.1 °C to 20.3 °C. A high-precision thermocouple was used to monitor the temperature of the sample chamber in real time, ensuring that the error was controlled within ±0.1 °C. Additionally, the relative humidity of the experimental environment was maintained between 66% and 70%. Prior to the experiment, samples underwent preprocessing, including cleaning, drying, and weighing, to ensure the accuracy of the experimental results. Subsequently, the pretreated samples were carefully placed in the instrument's sample chamber, and appropriate pressure ranges and measurement parameters were set according to the experimental requirements. During the high-pressure mercury injection process, the instrument automatically recorded and analyzed data at different pressure points, thereby obtaining detailed pore structure parameters, including pore diameter, pore size distribution, porosity, and permeability. Throughout the experiment, the instrument's status and experimental conditions were closely monitored to ensure data accuracy and reliability.

4.2.2. Low-Field Nuclear Magnetic Resonance (LFNMR)

Before conducting the LFNMR experiment, the rock samples must be precisely prepared into a cylindrical plug with a diameter of 25 mm and a length of 4.5 mm, ensuring that its surface is smooth and flawless. Subsequently, a Memmert UN 110 constant temperature drying oven (Schwabach, Bavaria, Germany), equipped with a PID automatic feedback control system, was utilized. Real-time monitoring of temperature fluctuations within the oven was conducted using a high-precision platinum resistance temperature sensor (PT100) manufactured by JUMO GmbH & Co. KG (Fulda, Hesse, Germany). To ensure that the temperature remained stable within the range of 70 ± 0.3 °C. The calibration of the drying oven was strictly performed in accordance with ISO 9001 standards, with periodic verification conducted using an NIST-traceable Type K thermocouple manufactured by JUMO GmbH & Co. KG, Fulda, Hesse, Germany (with an error of ±0.2 °C). Additionally, multipoint temperature distribution tests were conducted to confirm temperature uniformity. After 24 h of drying, the moisture within the samples was thoroughly removed. After drying, the sample was introduced into a vacuum pressure water saturation device and subjected to saturation treatment with distilled water for 72 h under a constant pressure of 20 Mpa. During this process, regular checks are required to maintain stable pressure and water saturation within the device, ensuring thorough saturation and pretreatment of the sample. The instrument used in the experiment was the AVANCE III/WB-400 solid wide-bore superconducting LFNMR spectrometer manufactured by Bruker BioSpin GmbH

(Rheinstetten, Baden-Württemberg, Germany). This instrument boasts high precision and is capable of identifying molecular chemical structures, as well as the microscopic morphology of solids or semi-solids. The superconducting magnet was utilized to maintain a constant low temperature through a liquid helium circulation system, with the temperature controlled at 32 ± 0.01 °C. Prior to the experiment, the temperature of the sample chamber was calibrated using a built-in platinum resistance temperature sensor (PT100). Real-time feedback adjustment was performed through the instrument software to ensure temperature stability during the relaxation time measurement. The magnetic field strength was set to 9.4 Tesla, providing sufficient intensity to excite the LFNMR phenomenon. Additionally, an appropriate rotor needs to be selected based on the sample size and experimental requirements. Among them, the 4 mm rotor can reach a maximum rotational speed of 15 kHz, suitable for small-volume samples or high-resolution testing, whereas the 7 mm rotor has a maximum rotational speed of 8 kHz, suitable for larger samples or routine testing. Prior to the experiment, the pore size test range must also be calibrated to ensure that it accurately covers the range of 0.1 to 1000 mm, meeting the needs of testing different types of samples.

4.2.3. Field Emission Scanning Electron Microscopy (FESEM)

To more accurately describe the morphology of the sample surface, the FESEM experimental procedure is outlined as follows: Initially, the rock samples undergo meticulous treatment using an argon ion polisher, model EM TIC 3X, manufactured by Leica Microsystems GmbH (Wetzlar, Hesse, Germany). During the polishing process, the polishing time is strictly controlled to ensure the smoothness of the sample surface; concurrently, the ion beam energy is set to a specific value to avoid damaging the microstructure of the samples. Upon completion of polishing, the sample surface undergoes rigorous inspection to confirm the absence of scratches or residues. The observation surface is not coated with carbon or gold to prevent unnecessary interference or obscuring of the micro-features on the sample surface. Subsequently, the samples are placed on the sample stage of the FESEM, model Zeiss Sigma 500, produced by Carl Zeiss Microscopy GmbH (Oberkochen, Baden-Württemberg, Germany). The position and angle of the samples are meticulously adjusted to achieve optimal observation results. Prior to activating the microscope, the operating parameters of the microscope are preset based on the characteristics of the samples and the intended observation objectives. These parameters include the accelerating voltage (0.02–30 kV), low vacuum range (2–133 Pa), and beam current (3 pA–20 nA), ensuring the acquisition of high-resolution images. With a resolution exceeding 0.8 nm, the Zeiss Sigma 500 microscope can precisely capture and characterize the micro morphological features of the sample surface. During the experiment, the operating state of the microscope is continuously monitored and adjusted to maintain optimal image quality throughout. This includes periodic checks and calibrations of the microscope's focus, brightness, contrast, and other parameters to guarantee image clarity and accuracy.

4.2.4. Calculation of Fractal Dimension

Fractal theory was established in the 1970s and has since evolved into various types of fractal dimensions, including topological dimension, Hausdorff dimension, self-similarity dimension, correlation dimension, and information dimension [33]. Based on SEM experimental results, this paper utilizes the Image J tool (version 1.54) and adopts the Hausdorff dimension, specifically the box-counting dimension method, to quantitatively analyze the pore distribution characteristics on the surface of coal. The specific steps are as follows: Starting with the original SEM image, an appropriate threshold is selected to convert the grayscale image into a black-and-white binary image. Boxes of different side lengths (δ) are then used to cover the binary image area, where different δ values correspond to grids of different sizes, typically $\delta = 2^{-k}$. The number of small boxes containing pores is counted, and the fractal dimension is then calculated using Equation (1) [34]. In a log–log coordinate system, the least squares method is employed to perform regression analysis on the number of fractures $N(\delta)$ and the grid side length δ. The slope of the regression line represents the fractal dimension of the fracture surface.

$$D = -\lim_{\delta \to 0} \frac{\log(N(\delta))}{\log(\delta)} \tag{1}$$

In the equation, D represents the fractal dimension, δ denotes the side length of the box, and $N(\delta)$ is the number of boxes required to cover the pore area on the coal surface.

5. Results and Analysis

5.1. Results of MICP Test

5.1.1. Pore Throat Size

The pore throat size in rocks refers to the dimensions of pores or channels within rocks, serving as the primary pathways for fluid flow in underground reservoirs. The size, shape, and distribution of these pore throats influence fluid velocity, flow rate, and permeability [35,36], making it necessary to quantitatively characterize the pore throat size. Studies indicate that a lower displacement pressure corresponds to coarse pore throats, good permeability, and excellent pore structure; conversely, a higher displacement pressure indicates fine pore throats, low permeability, and poor pore structure [17].

Additionally, based on the pore throat interval, pore throat sizes can be categorized as follows: micropores with a radius < 0.05 μm; small pores with 0.05 μm $\leq$ radius < 0.5 μm; medium pores with 0.5 μm $\leq$ radius < 5 μm; large pores with 5 μm $\leq$ radius < 50 μm. The experimental test results of 15 hydrocarbon source rock samples from the HN-1 well in the study area (Table 1) indicate that, in terms of displacement pressure, limestone ranges from 13.78 to 20.66 Mpa, averaging at 17.22 Mpa; sandstone ranges from 5.50 to 13.78 Mpa, averaging at 7.57 Mpa; and mudstone ranges from 11.26 to 16.34 Mpa, averaging at 13.78 Mpa. In terms of average pore throat interval, limestone ranges from 0.009 μm to 0.016 μm, averaging at 0.013 μm; sandstone ranges from 0.015 μm to 0.033 μm, averaging at 0.025 μm; and mudstone ranges from 0.012 μm to 0.015 μm, averaging at 0.013 μm. During the research process, it was observed that pore throat size parameters significantly influence rock porosity. To further analyze this correlation, a linear correlation plotting method was employed (Figure 5).

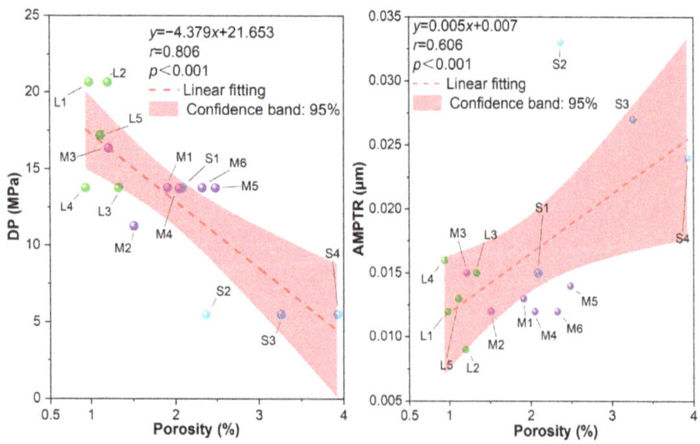

Figure 5. Relationship between roaring parameters and porosity of three rock samples.

Table 1. Mercury intrusion capillary pressure experimental parameters for rock samples.

Sample Number	Porosity (%)	DP (Mpa)	AMPTR (μm)	RSC (D)	UC (α)	PTD (S_{kp})	MMS (S_{max})	RMS (%)	MRE (%)
L1	0.98	20.66	0.012	89.85	0.325	−0.287	76.53	25.72	66.40
L2	1.20	20.66	0.009	98.05	0.266	−0.504	78.46	28.46	63.73
L3	1.33	13.78	0.015	73.32	0.294	−0.131	86.29	31.39	63.62
L4	0.94	13.78	0.016	74.56	0.299	−0.099	85.02	40.46	52.41
L5	1.11	17.22	0.013	83.94	0.310	−0.255	81.57	31.51	61.54
Average	1.11	17.22	0.013	83.94	0.291	−0.255	81.57	31.51	61.54
SD	0.18	3.98	0.003	12.04	0.249	0.185	4.80	6.41	6.22
S1	2.09	13.78	0.015	77.32	0.286	−0.289	90.07	27.28	69.71
S2	2.37	5.50	0.033	52.08	0.250	−0.244	88.25	33.55	61.99
S3	3.26	5.50	0.027	67.21	0.201	−0.422	90.59	26.32	70.95
S4	3.93	5.51	0.024	61.31	0.183	−0.265	95.26	42.73	55.14
Average	2.91	7.57	0.025	64.48	0.230	−0.305	91.04	32.47	64.45
SD	0.84	4.14	0.008	10.58	0.047	0.080	2.98	7.55	7.36
M1	1.91	13.78	0.013	82.62	0.235	−0.345	89.00	13.29	85.07
M2	1.51	11.26	0.012	79.51	0.221	−0.247	79.15	15.41	69.26
M3	1.21	16.34	0.015	80.55	0.261	−0.441	90.13	27.15	84.69
M4	2.05	13.73	0.012	88.46	0.241	−0.353	89.42	12.64	82.46
M5	2.49	13.78	0.014	76.86	0.260	−0.275	88.78	28.54	67.85
M6	2.33	13.78	0.012	89.04	0.228	−0.420	80.92	13.36	83.49
Average	1.83	13.78	0.013	82.84	0.241	−0.347	86.23	18.40	78.80
SD	0.30	0.00	0.001	6.09	0.017	0.073	4.60	8.79	9.52

DP: Discharge pressure; AMPTR: Arithmetic mean pore throat interval; RSC: Relative sorting coefficient; UC: Uniformity coefficient; PTD: Pore throat distortion; MMS: Maximum mercury saturation; RMS: Residual mercury saturation; MRE: Mercury removal efficiency; SD: Standard deviation (1σ).

The results indicate that all three rock types exhibit characteristics of microporosity development. As shown in Figure 5, sandstone has a relatively low entry pressure, reflecting coarser pore throats in the reservoir, which in turn contributes to its better permeability and pore structure. A significant negative correlation exists between the entry pressure and the porosity of source rocks. This is because rocks with low porosity are dominated by small pores, leading to increased capillary resistance. Therefore, higher entry pressure is required to overcome this resistance. Additionally, there is a positive correlation between the average pore throat interval and porosity. In source rocks or reservoirs, the formation of pores is usually accompanied by the expansion of pore throats. Based on this characteristic, entry pressure and average pore throat interval can be utilized to assess porosity.

5.1.2. Pore Throat Sorting Characteristics

The pore throat sorting characteristics refer to the variability and regularity of parameters, such as pore throat size, shape, and distribution [37]. These characteristics reflect the storage and permeability capabilities of rocks, which are of great significance for the formation and development of oil and gas reservoirs. According to related research, parameters such as the relative sorting coefficient (D), uniformity coefficient (α), and pore throat skewness (S_{kp}) have a direct impact on pore throat sorting characteristics. The relative sorting coefficient (D) is a dimensionless coefficient; the smaller its value, the more uniform the pore throat distribution. The uniformity coefficient (α) is also a dimensionless coefficient ranging from 0 to 1; the closer its value is to 0, the more uniform the throat distribution. The S_{kp} indicates a positive skewness if greater than 0, with coarser pore throats; it indicates a symmetrical distribution of coarse and fine pore throats if equal to 0 and a negative skewness if less than 0, with finer pore throats. The experimental test results indicate (Table 1) that for limestone, D ranges from 73.32 to 98.05, with an average of

83.94; α ranges from 0.263 to 0.323, averaging 0.291; and S_{kp} ranges from −0.504 to −0.009, averaging −0.255. For sandstone, D ranges from 52.08 to 77.32, averaging 64.48; α ranges from 0.183 to 0.286, averaging 0.230; and S_{kp} ranges from −0.422 to −0.244, averaging −0.305. For mudstone, D ranges from 76.86 to 89.04, averaging 82.84; α ranges from 0.221 to 0.261, averaging 0.241; and S_{kp} ranges from −0.441 to −0.247, averaging −0.347. During the research, a correlation analysis was also conducted between relevant parameters in pore throat sorting characteristics and porosity (Figure 6).

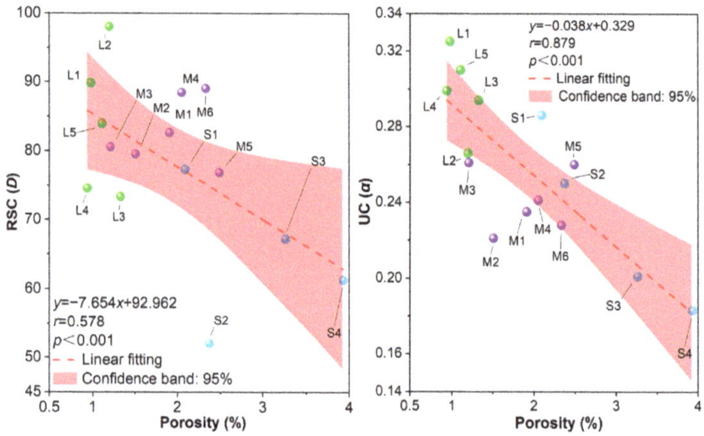

Figure 6. Relationship between pore throat sorting characteristics and porosity of three rock samples.

As can be seen in Figure 6, the pore throat distribution in sandstone is relatively uniform. However, throat distribution in limestone, sandstone, and mudstone exhibits uneven characteristics. The skewness of pore throats in limestone, sandstone, and mudstone is less than 0, indicating that they all belong to the category of negative skewness, with relatively small skewness values. Furthermore, there is a significant negative correlation between the uniformity coefficient and the relative sorting coefficient with porosity. Since rocks with low porosity are dominated by primary pores, which have irregular pore shapes and a wide distribution range of pore throat sizes, the uniformity coefficient and sorting coefficient are relatively high. Based on this characteristic, the uniformity coefficient can be utilized to assess porosity.

5.1.3. Pore–Throat Connectivity and Seepage Performance

Rock pore–throat connectivity refers to the degree of connection between pores and throats within the rock, while permeability describes the ability of fluids to flow through these pores and throats within the rock [35]. Studying rock pore–throat connectivity and permeability aids in determining the reserves and recoverability of oil and gas deposits. Therefore, it is necessary to quantitatively characterize pore–throat connectivity and permeability.

Connectivity and permeability are related to maximum mercury injection saturation, residual mercury saturation, and mercury withdrawal efficiency, with the relevant parameters outlined in Table 1. The maximum mercury saturation (S_{max}) of limestone ranged from 76.53 to 86.29, with an average of 81.57. The residual mercury saturation (%) ranged from 25.72 to 40.46, with an average of 31.51. The mercury withdrawal efficiency ranged from 52.41 to 66.40, with an average of 61.54. The S_{max} of sandstone in the study area ranged from 88.25 to 95.26, with an average of 91.04. The S_{max} of the sandstone in the study area ranged from 88.25 to 95.26, with an average of 91.04. The residual mercury saturation (%) ranged from 27.28 to 42.73, with an average of 32.47, and the mercury withdrawal efficiency ranged from 55.14 to 70.95, with an average of 64.45. The S_{max} of mudstone in the study area ranged from 79.15 to 90.13, with an average of 86.23. The S_{max} of the mudstone in the study area ranged from 79.15 to 90.13, with an average of 86.23. The residual mercury saturation (%) ranged from 13.29 to 28.54, with an average of 18.40, and the mercury withdrawal efficiency ranged from 67.85 to 85.07, with an average of 78.80. Correlation plots were made with the same two points as those mentioned above for the purpose of analyzing the results (Figure 7).

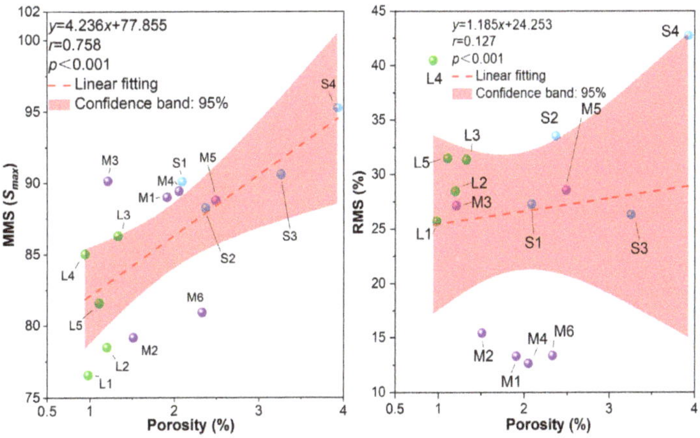

Figure 7. Relationship between connectivity, permeability, and porosity of three rock samples.

As can be seen from Figure 7, in terms of the average maximum mercury saturation and average displacement efficiency, sandstone exhibits characteristics of high maximum mercury saturation and high mercury displacement efficiency. This indicates that the pore structure of sandstone in the study area is more uniform and has better connectivity, followed by mudstone. In terms of pore–throat connectivity and permeability performance, there is a positive correlation between maximum mercury saturation and rock porosity. This is because rocks with low porosity are mainly composed of isolated pores or micropores, with narrow or blocked throat passages, which results in mercury being able to enter only partially connected pores, thereby limiting the maximum mercury saturation. Based on this characteristic, the porosity and connectivity of rocks can be evaluated through maximum mercury saturation.

5.2. Capillary Pressure Curve Characteristics

Figures 8 and 9 show the characteristics of the capillary pressure curves and the distribution of the pore throats in some of the limestones, sandstones, and mudstones in the study area, respectively. The capillary pressure curves show that, overall, the flat section of the capillary pressure curves of limestones, sandstones, and mudstones in the study area is higher, indicating that the pore throats tend to be concentrated in fine pore throats with fine crookedness. The flat section of the mercury feed curve is almost absent and tilted to the lower right, indicating that the pore throat interval is poorly sorted. On the whole, the pore throats of limestone, mudstone, and sandstone in the study area are on the thin side, and the pore structure is poor. The pore throat distribution indicates that the pore throat interval distribution curves of limestone, sandstone, and mudstone in the study area are all of the fine single-peaked type, and the pore throat radii of L1 and L2 are mainly distributed in the range of 0.00404 μm to 0.044 μm. The distribution of pore throat interval of L3 and L4 ranges from 0.00404 μm to 0.0630 μm. The pore throat interval distribution range of sandstone is mainly 0.01 μm to 0.16 μm, and the part larger than 0.1 μm is fine throat. The radius distribution of the pore throat of mudstone mainly ranges from 0.004 μm to 0.063 μm, which is a fine throat. In general, the overall pore throat interval of sandstone is larger.

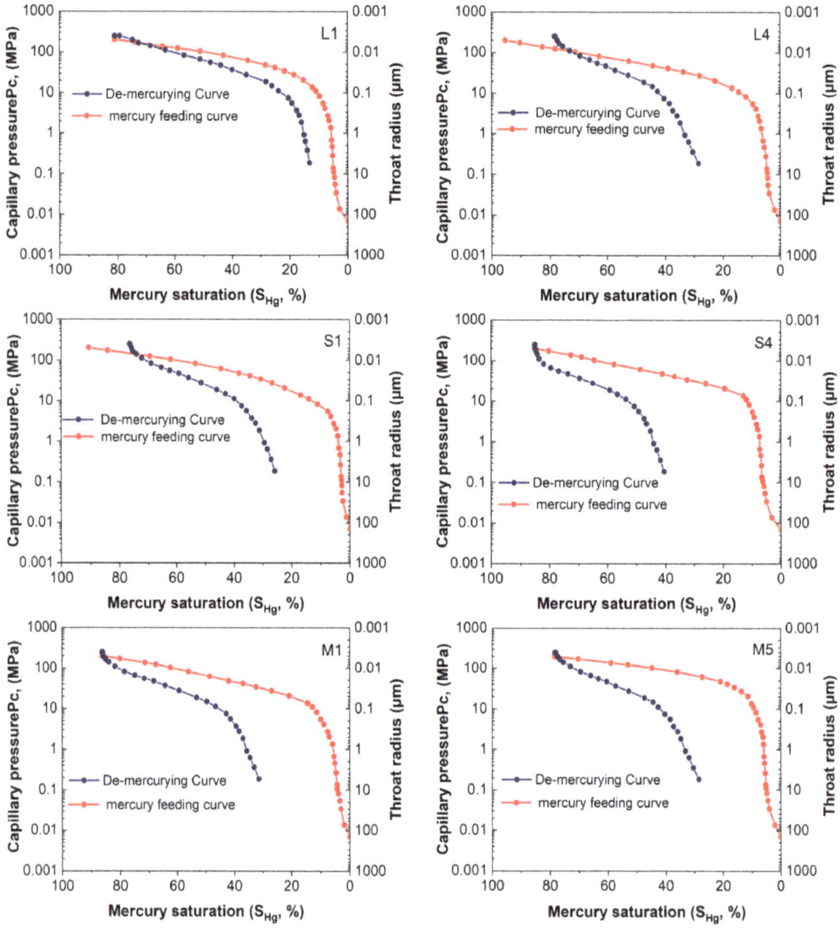

Figure 8. Capillary pressure curves of three rock samples.

In addition, the pore throat interval at the peak of limestone ranges from 0.025 μm to 0.040 μm, and the pore throat distribution frequency (PTDF) at the peak is about 15%, and that of sandstone ranges from 0.0634 μm to 0.100 μm, and the PTDF at the peak fluctuates in a small range above and below 10%. The pore throat interval at the peak of mudstone is 0.025 μm to 0.040 μm, and the PTDF at the peak is about 12%. The pore throat interval at the peak is generally low and has a long tail trailing to the end of the fine roar channel, which is poorly sorted. However, their permeability contributions are all greater than 50%, and in the single-peak roar distribution curve, the permeability contribution is always biased toward the coarser roar section, and the pore contribution of the permeability to the finer roar section is small, generally less than 10%.

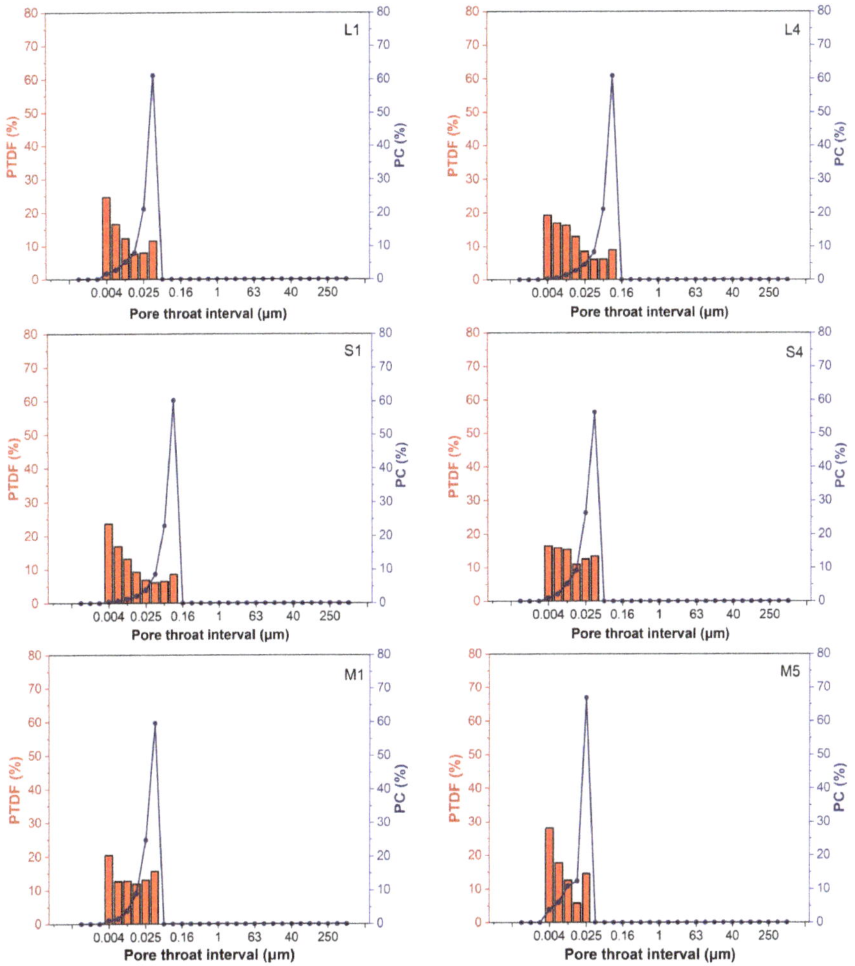

Figure 9. Distribution of pore throat interval for three types of rock samples.

5.3. Results of LFNMR Experiments

The LFNMR T_2 spectrum is based on the mathematical inversion of the signal decay process of a multi-segment echo string. The decay of the spin echo string reflects the number and distribution of hydrogen nuclei in the medium, and the echo amplitude decreases with time. This decay information can be utilized to determine the pore content and the pore size distribution [38]. The initial signal amplitude of the echo string is converted to obtain the porosity, and the decay rate reflects the pore size distribution and fluid type. Therefore, samples with high peaks in the T_2 spectrum have large porosity, and multiple peaks or broad peak features represent diverse pore structures [39]. During low-field LFNMR experiments, according to Coates et al. [40], the lateral relaxation characteristics of fluids in porous media can be expressed by the following mathematical equation:

$$\frac{1}{T} = \frac{1}{T_2} = \rho^2 \frac{S}{V} \tag{2}$$

where ρ^2 is the surface relaxation rate, μm/ms; S is the pore surface area, μm^2; and V is the pore volume, μm^3.

5.3.1. Low-Field Nuclear Magnetic Resonance Porosity

The results of LFNMR porosity experiments conducted on 15 hydrocarbon source rock samples from the HN-1 well in the study area (Table 2) indicate that the porosity (%) of limestone samples ranges from 1.55 to 2.19, averaging at 1.88%; the porosity (%) of sandstone samples ranges from 1.92 to 4.21, averaging at 3.02%; and the porosity (%) of mudstone samples ranges from 1.03 to 2.77, averaging at 2.01%. Combining these findings with the rock porosity analysis obtained through high-pressure mercury injection, which yields an average porosity of 1.11% for limestone, 2.91% for sandstone, and 1.92% for mudstone, it can be clearly deduced from the integrated results of both the LFNMR and high-pressure mercury injection methods that the porosity distribution patterns among different rock types exhibit a high degree of consistency. Specifically, sandstone samples generally have the highest porosity, followed by mudstone samples, while limestone samples have the lowest porosity. This discovery provides critical insights into the reservoir properties and pore structure characteristics of hydrocarbon source rocks in the study area.

Table 2. LFNMR porosity results of 15 rock samples.

Sample Number	Porosity (%)	Average Porosity (%)
L1	1.81	
L2	1.55	
L3	2.19	1.88 (n = 5)
L4	2.11	
L5	1.76	
S1	1.92	
S2	2.61	3.02 (n = 4)
S3	3.32	
S4	4.21	
M1	1.99	
M2	1.03	
M3	1.39	2.01 (n = 6)
M4	2.42	
M5	2.77	
M6	2.50	

5.3.2. Pore Size Distribution

Figure 10a presents the LFNMR analysis results for six representative rock samples (L3, L4, S1, S3, M2, and M3). The T_2 spectrum characteristic maps of these samples were characterized using LFNMR, and the study found that the pore distribution patterns obtained were consistent with those obtained from high-pressure mercury injection tests. The sandstone types exhibited strong signal intensity across different relaxation time periods and had broader T_2 spectra, indicating good pore structure and connectivity in sandstone. For limestone, the spectrum relaxation time of L4 is approximately 0.01 ms, with a right peak roughly located between 1 and 10 ms and a maximum relaxation time of 500 to 1000 ms. The T_2 spectrum peak shape is a bimodal type dominated by the left peak, with distinct wave troughs, inflection points, and other features, suggesting a tendency for uniform pore distribution. Pores of various diameters are developed, with a wide range of pore sizes, primarily micropores and mesopores, along with a small number of macropores [23,41]. This is consistent with the LFNMR characteristics of the L1 sample. L3 has a smaller amplitude and a maximum relaxation time of less than 1000 ms. The T_2 spectrum peak shape can be regarded as a left-converging unimodal type, indicating a narrow range of pore size distribution, primarily micropores, with a small number of mesopores and macropores developed. The signal intensity and T_2

spectrum width of limestone types do not perform well, which is consistent with the LFNMR characteristics of the L2 and L5 samples. For sandstone, the spectra of both S1 and S3 are bimodal types dominated by the left peak, with obvious wave troughs and inflection points, indicating a tendency for uniform pore distribution. Pores of different levels are continuously and excessively developed, primarily micropores and mesopores, with a small number of macropores. This is consistent with the LFNMR characteristics of the S2 and S4 samples. Additionally, the good signal intensity across different relaxation time periods and the broader T_2 spectrum of sandstone types indicate their superior pore structure and connectivity. For mudstone, the spectra of both M2 and M3 are bimodal types dominated by the left peak, sharing features similar to those of the aforementioned sandstone bimodal types. However, in terms of the signal intensity range across different relaxation time periods and the width of the T_2 spectrum, M3 performs most prominently among the 6 samples, while M2 has a lower signal intensity range and T_2 spectrum width, indicating uneven pore structure distribution and connectivity within mudstone types.

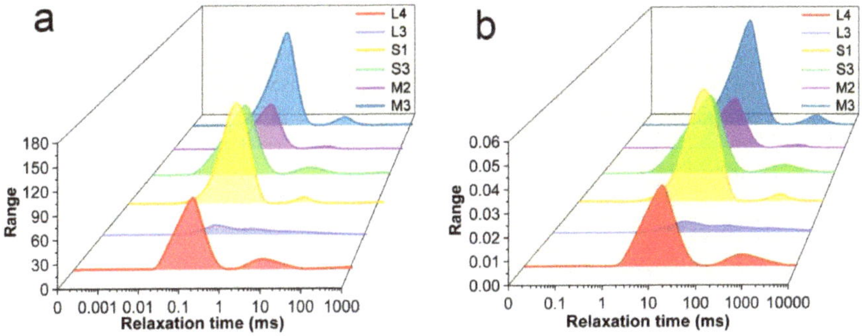

Figure 10. (**a**) LFNMR T_2 spectrum characteristics of representative rock samples; (**b**) LFNMR pore size distribution of representative rock samples.

The characterization of the full pore size distribution of rock samples can be achieved using LFNMR experimental methods [42–44]. From Equation (2), it can be seen that there is a certain correlation between LFNMR T_2 parameters and pore structure parameters, and ρ_2 is an established constant in a single sample; then, the LFNMR T_2 spectra can be transformed into pore size distribution curves after the conversion relationship between T_2 and (S/V) has been obtained using other experimental or mathematical methods. Based on previous research methods [42–44], Equation (2) is deformed as

$$\frac{1}{T_2} = \rho^2 \frac{F_S}{r} \tag{3}$$

where r is the pore radius (ratio of pore surface area to pore volume), μm; and F_S is the pore shape factor. It is usually considered that the pores in shale are mostly cylindrical, and F_S can be taken as 2 [43,44]. According to the study of Sonddergeld et al. [45] for the surface relaxation rate of shale, ρ^2 can be taken as 0.05 μm/ms. Therefore, Equation (4) can be further transformed into

$$r = \frac{T_2}{10} \tag{4}$$

Equation (4) is utilized to calculate the pore size distribution for the six rock samples (Figure 10b). Except for S3, the pore diameters of the remaining samples are mainly distributed within the range of 0 to 300 nm, with peak pore diameters primarily within 1 to 100 nm, indicating that the samples are dominated by nanoscale pores. Based on previous classifications of rock nanoscale pores [46], pores can be categorized as small pores (with diameters less than 10 nm), transitional pores (with diameters between 10 and 100 nm), mesopores (with diameters between 100 and 1000 nm), and macropores (with diameters of 1000 nm or greater). Figure 10b shows that the pore diameters of the rock samples are mainly distributed between 0 and 100 nm, belonging to the joint development of small and transitional pores. Additionally, the pore size distribution curve of S3 exhibits a broader peak shape and a larger peak pore diameter, suggesting that It contains a relatively higher abundance of macropores. However, the pores in S3 are still predominantly contributed by nanopores.

5.4. Pore Types and Their Distribution Characteristics

5.4.1. Organic Matter Type and Development of Micropores Based on POM

The observation conditions (both transmitted light and fluorescence) of the polarized optical microscope (POM) for the samples in this study of deep coal-measure hydrocarbon source rocks are shown in Figure 11. Under transmitted light, L1 exhibits a large amount of calcareous skeletal biodebris, with asphaltic organic matter in the limestone appearing as a brownish-yellow color (Figure 11a). In fluorescence mode, it maintains a brownish-yellow hue, with fractures and pores in the limestone filled with abundant brownish-yellow and brown asphaltic organic matter, along with a significant presence of pyrite (Figure 11b). Under transmitted light, L2 contains numerous fractures and pores filled with asphalt, and biodebris also shows significant development of asphaltic organic matter or organic inclusions (Figure 11c). In fluorescence, the asphaltic organic matter exhibits flow traces, and brighter yellow-green fluorescent organic inclusions are visible in calcite veins or fractures and pores of the limestone (Figure 11d). In transmitted light, fine sandstone S1 has grains with distinct angularities, poor rounding, and poor sorting, with calcite and muddy cementation, and pores filled with abundant asphalt (Figure 11e). The fluorescence grade indicates oil stain level, with pores showing a large amount of yellow fluorescent asphaltic organic matter, and quartz grain fractures containing numerous yellow-green fluorescent organic inclusions (Figure 11f). Under transmitted light, fine sandstone S2, containing argillaceous debris, is filled with solid asphalt and abundant organic matter, dominated by vitrinite, with a small amount of inertinite, and a relatively developed exinite content exceeding 5% (Figure 11g). In fluorescence, exinite is visible in the argillaceous debris, and bright yellow fluorescent organic inclusions are seen in quartz secondary enlargement rims or fractures (Figure 11h). Under transmitted light, dark mudstone M1 is characterized by a large amount of mineral asphalt matrix present in pores and fractures of the mudstone, with vitrinite distributed in banded patterns (Figure 11i). In fluorescence, it exhibits a weak yellow fluorescent matrix asphaltic body, containing numerous sporinite bodies and microspore bodies, along with inertinite macerals (Figure 11j). Under transmitted light, M2 mudstone appears brownish-red, with organic matter dominated by vitrinite and being non-fluorescent. Exinite is relatively well developed, typically around 5%, and is mainly composed of microspore bodies, with primarily sporinite bodies making up the remainder (Figure 11k). In fluorescence, it appears yellow, with a small amount of asphaltene bodies or alginite bodies present in pores and fractures of the mudstone, and pyrite filling on the surface or inside (Figure 11l).

Figure 11. POM images of different types of rock samples: (**a**,**b**) L1 limestone, buried at depths of 1140.45–1149.40 m ((**a**) transmitted light; (**b**) fluorescence); (**c**,**d**) L4 limestone, buried at a depth of 1231.50 m ((**c**) transmitted light; (**d**) fluorescence); (**e**,**f**) S2 sandstone, buried at depths of 1149.75–1153.65 m (**e**) transmitted light; (**f**) fluorescence); (**g**,**h**) S4 sandstone, buried at depths of 1201.80–1206.65 m ((**g**) transmitted light; (**h**) fluorescence); (**i**,**j**) M2 mudstone, buried at a depth of 1164.10 m ((**i**) transmitted light; (**j**) fluorescence); (**k**,**l**) M5 mudstone, buried at depths of 1181.95–1183.20 m ((**k**) transmitted light; (l) fluorescence).

5.4.2. Surface Nanoscale Pore Fracture Characteristics Based on FESEM

The POM results were observed and analyzed, providing preliminary insights into the internal morphology and structural characteristics of various rock samples. However, due to limitations in instrument magnification, resolution, and manual secondary modifications, POM typically only reveals micrometer-scale observations and may also show artifact pores or fractures. Therefore, based on the POM observations, the samples underwent an Ar ion polishing treatment, followed by more detailed observations of typical sections of the rock samples using FESEM, enabling accurate characterization of the internal morphology and structure of different rock samples at the nanoscale. The FESEM experimental results obtained are shown in Figure 12. For the experimental results, Nano measurer software (version 1.2) was utilized to deeply interpret the image content and accurately measure the diameters of some typical pores or fractures depicted. L1 limestone exhibits long pores above 100 nm and dissolved pores below 100 nm (ranging from 339 nm to 37 nm), but the number of developed pores and the average pore diameter are inferior to those in sandstone. Besides quartz mineral filling on the surface, there are also obvious incompletely fractured fissures and a large number of matrix components. The fractures developed in L1 limestone are mainly above the nanoscale (Figure 12a). The pore development in L2 limestone is primarily below 100 nm (ranging from 216 nm to 56 nm) and relatively sparse. Additionally, no obvious fissures were found, but it exhibits a certain mineral composition and layered structure (Figure 12b). In S1 sandstone, no significant dissolved pore development is observed. Although micrometer-scale macropores are present, the pores in S1 are still mainly composed of nanopores. Furthermore, S1 sandstone exhibits two distinct fractures or fissures, possibly caused by deposition or tectonic movements, which are primarily micrometer-scale and serve as important channels for oil and gas migration in the formation, demonstrating a certain oil storage capacity of the sandstone (Figure 12c). The pore development in S2 sandstone is more mature, with highly developed interparticle pores not exceeding 100 nm in size and obviously dissolved pores ranging from nanometers to micrometers. Other pores are above 100 nm, along with a few micrometer-scale macropores (Figure 12d). Additionally, clay minerals within S2 sandstone mostly exhibit granular structures. No fractures develop in S2 sandstone, which may be influenced by the observation location; however, the presence of fractures to some degree in S2 sandstone can be inferred from the observations in S1 sandstone. M1 mudstone, besides quartz mineral filling and the development of nanoscale fractures, also exhibits more nanopore development, with pore sizes ranging from 200 nm to 40 nm (Figure 12e). In M2 mudstone, a large number of dissolved pores are found, with sizes at the micrometer scale and irregular shapes, spreading in a bead-like or floral pattern and exhibiting a layered structure. Besides the development of nanopores (ranging from 160 nm to 60 nm), there are also a large number of pores with even smaller diameters and a few inconspicuous fractures (Figure 12f).

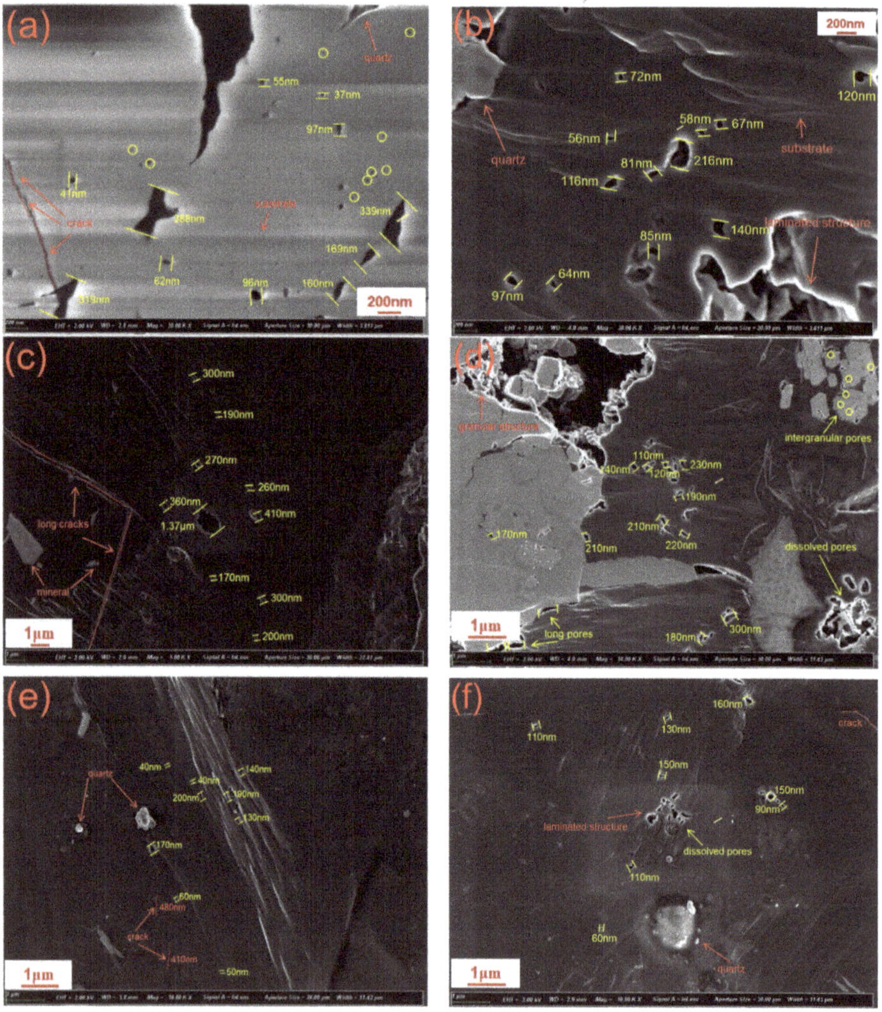

Figure 12. FESEM results of different rock samples: (**a**) L1 limestone, burial depth 1140.45–1149.40 m; (**b**) L4 limestone, burial depth 1231.50 m; (**c**) S2 sandstone, burial depth 1149.75–1153.65 m; (**d**) S4 sandstone, burial depth 1201.80–1206.65 m; (**e**) M2 mudstone, burial depth 1164.10 m; (**f**) M5 mudstone, burial depth 1181.95–1183.20 m.

Based on the observation results from FESEM, all hydrocarbon source rock samples exhibited highly developed nanopores at higher magnifications. Additionally, some rocks also developed fractures. In conjunction with related research, it is known that the degree of pore and fracture development can reflect the occurrence of unconventional natural gas in deep coal measures [47]. Therefore, this indicates that the HN-1 well in the Taiyuan Formation of the Late Carboniferous in the Huainan coalfield has favorable unconventional natural gas reservoir conditions. The FESEM results show that the overall pore development in sandstone is relatively good, along with the presence of micrometer-scale fractures, followed by mudstone. The pore development in limestone is poorer, which aligns with the experimental results from LFNMR and MICP. Consequently, the FESEM results also suggest that the combination of LFNMR and MICP is more conducive to characterizing the internal microstructure of different rocks, thereby providing theoretical guidance for revealing the adsorption, diffusion, and seepage mechanisms of unconventional natural gas in deep coal measures.

5.4.3. Fractal Dimension Calculation Based on FESEM

According to fractal theory, when the FESEM image captures the two-dimensional projection morphology of the material surface, the fractal dimension of the rock typically lies between 1 and 2 [48]. A higher fractal dimension indicates a more complex pore morphology and poorer permeability; conversely, a lower fractal dimension suggests simpler pore morphology, better permeability, and pore shapes closer to circularity [46]. In this study, based on the image results obtained from FESEM, to more clearly observe the nanoscale micropore structure of the samples, images at a resolution of 1 μm were selected for fractal dimension calculation for all 15 samples. Figure 13 illustrates the process using Sample S4 as an example. Based on the original FESEM image (Figure 13a), a suitable threshold Is selected to convert the grayscale Image Into a black-and-white binary image (Figure 13b). In determining the threshold, a method reported in the related literature is adopted, which involves comparing the original FESEM image with the generated pore extraction image to ascertain the most appropriate threshold. Ultimately, it is observed that a relatively ideal pore extraction effect can be achieved when the threshold is between 70 and 80; hence, 75 is selected as the threshold for this experiment. Subsequently, under five fractal scale conditions for surface cracks, $\delta = L_0$, $L_0/2$, $L_0/4$, $L_0/8$, and $L_0/16$, the sample surface is covered with corresponding grids, and the number of cracks $N(\delta)$ within each grid is calculated (Figure 13c). Given that the pore and crack sizes of the sample are nanoscale, combined with the size of the observed sample area, L_0 is set to 4 μm. Formula 1 is then used to calculate the fractal dimension. In a logarithmic–logarithmic coordinate system, the least squares method is employed for regression analysis of the crack count $N(\delta)$ and grid side length δ (Figure 13d). The slope of the regression line represents the fractal dimension of the crack surface.

The fractal dimension calculation results for various types of rock samples are detailed in Table 3. As can be seen from Table 3, the fitting coefficients R^2 are all greater than 0.99, indicating that the box-counting method is reasonable for calculating the pore fractals of coal surfaces. In terms of fractal dimensions, limestone samples range from 1.796 to 1.990, with an average of 1.913; sandstone samples range from 1.341 to 1.769, with an average of 1.467; and mudstone samples range from 1.568 to 1.972, with an average of 1.799. The fractal dimension results suggest that sandstone samples have the smallest fractal dimension, implying a more uniform pore structure distribution, relatively better connectivity and permeability, and thus favorable reservoir properties, followed by mudstone. These fractal dimension results are consistent with previous research conclusions [21] and also align with the results obtained from the MICP and LFNMR experiments conducted in this study.

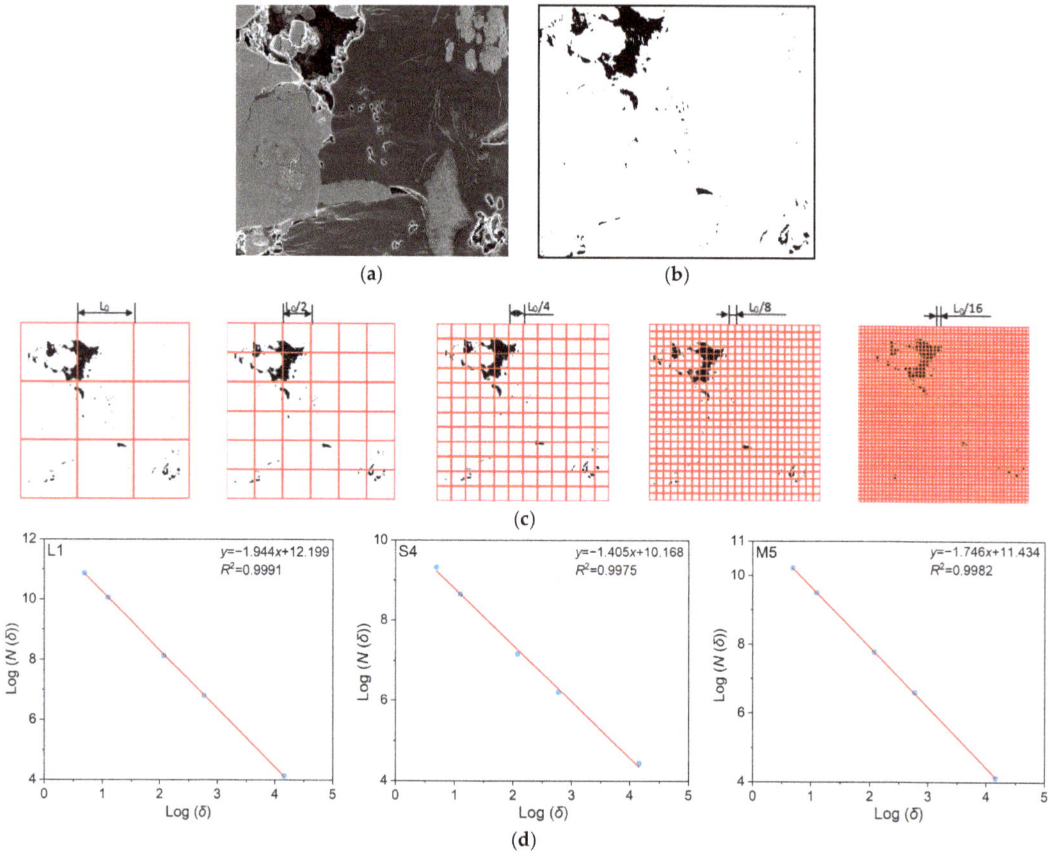

Figure 13. Schematic diagram for the calculation of fractal dimension of S4 sandstone samples ((**a**) original FESEM Image, (**b**) binary black-and-white image, (**c**) box-covering image, and (**d**) fractal dimension fitting image of some samples).

Table 3. Fractal dimensions of various rock samples.

Sample Number	Fitting Formula	R^2	D
L1	y = −1.944x + 12.199	0.9991	1.944
L2	y = −1.985x + 14.178	0.9938	1.985
L3	y = −1.851x + 11.647	0.9971	1.851
L4	y = −1.796x + 15.318	0.9956	1.796
L5	y = −1.990x + 12.211	0.9994	1.990
S1	y = −1.769x + 13.094	0.9934	1.769
S2	y = −1.341x + 11.681	0.9929	1.341
S3	y = −1.351x + 14.163	0.9983	1.351
S4	y = −1.405x + 10.168	0.9975	1.405
M1	y = −1.821x + 10.927	0.9919	1.821
M2	y = −1.568x + 12.981	0.9907	1.568
M3	y = −1.972x + 13.144	0.9964	1.972
M4	y = −1.895x + 12.657	0.9951	1.895
M5	y = −1.746x + 11.434	0.9982	1.746
M6	y = −1.789x + 14.212	0.9937	1.789

5.5. Differences Between MICP and LFNMR Porosity

In this study, two characterization techniques, MICP and LFNMR, were employed to test and analyze the porosity of 15 samples. While both methods can effectively reveal the pore structure characteristics of the materials, the experimental results showed that there were certain discrepancies in the specific porosity values obtained through MICP and LFNMR. To investigate the overall difference between these two methods in assessing rock porosity, further analyses using the *T*-test (with a significance level of 0.05 for the t-value) and correlation tests were conducted to determine whether there were statistically significant differences in porosity between the three different types of rocks tested under MICP and LFNMR conditions (Figure 14).

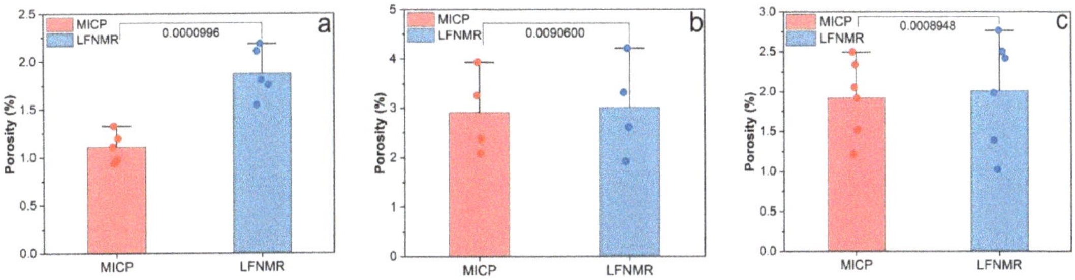

Figure 14. Analysis of significant differences in porosity of different rock samples under MICP and LFNMR test conditions ((**a**) limestone samples, (**b**) sandstone samples, and (**c**) mudstone samples).

According to Figure 14, the porosity of the three types of rocks studied exhibited distinct characteristics when measured using MICP and LFNMR, indicating that the same sample may yield different experimental results when tested using different methods. As shown in Figure 15, the porosity of rocks obtained through high-pressure mercury injection has good applicability to LFNMR. Combining the information from Figure 14, it is necessary to effectively assess the pore diameter size of the samples to enhance the accuracy of the LFNMR analysis results.

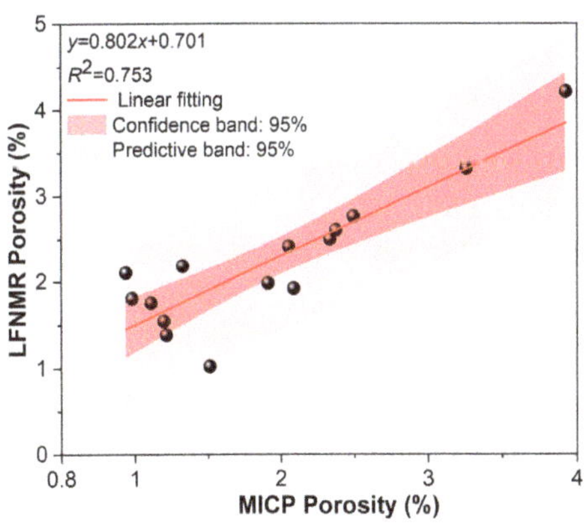

Figure 15. Porosity correlation between the two testing methods.

As shown in Figure 3, MICP primarily characterizes pore sizes larger than 50 nm and may not provide accurate representations of smaller pores [47,49]. In contrast, LFNMR characterization covers a pore size ranging from 2 nm to 1000 nm, offering advantages in representing both larger and smaller pores [26,50]. Therefore, it is expected that the average porosity of the different types of rock samples tested using LFNMR in this study is generally higher than that tested using MICP. Furthermore, the experimental results indicate that sandstone has the largest overall pore radius, followed by mudstone, and limestone has the smallest. This characteristic results in relatively smaller porosity values when MICP is used to characterize limestone samples, leading to the largest difference in the average porosity between MICP and LFNMR among the three types of rock samples for limestone, while the difference is the smallest for sandstone.

LFNMR and MICP cover pore size ranges of 2–1000 nm and >50 nm, respectively, and the data from these two methods essentially represent complementary subsets of the same pore system. The geometric mean method integrates the advantages of LFNMR (sensitive to small pores) and MICP (sensitive to large pores) through the nature of mathematical constraints, thus providing a more comprehensive characterization of pore structure. Furthermore, for samples with large data spans (e.g., S4 with an LFNMR value of 4.21% and an MICP value of 3.93%), the geometric mean method can effectively mitigate the influence of extreme values. In summary, this study adopts the algorithm theory of the geometric mean to obtain reasonable values for rock porosity based on both the MICP and LFNMR methods (see Table 4).

Table 4. Porosity of each rock sample calculated using the geometric mean method.

Sample Number	MICP Porosity (%)	LFNMR Porosity (%)	Geometric Mean Porosity %
L1	0.98	1.81	1.33
L2	1.20	1.55	1.36
L3	1.33	2.19	1.70
L4	0.94	2.11	1.41
L5	1.11	1.76	1.40
S1	2.09	1.92	2.00
S2	2.37	2.61	2.49
S3	3.26	3.32	3.29
S4	3.93	4.21	4.07
M1	1.91	1.99	1.95
M2	1.51	1.03	1.25
M3	1.21	1.39	1.30
M4	2.05	2.42	2.23
M5	2.49	2.77	2.63
M6	2.33	2.50	2.42

5.6. Relationship Between FESEM Fractal Dimension and Reservoir Physical Properties

Porosity and permeability are crucial parameters for evaluating reservoir physical properties [51]. Exploring the correlation between FESEM fractal dimension and reservoir physical properties holds significant geological importance. For rock reservoirs, the development of pore structures can enhance reservoir porosity, with large-pore structures effectively increasing porosity, while small-pore structures have a relatively weaker impact on porosity. A higher fractal dimension of a sample indicates a more complex pore structure and a larger pore-specific surface area [52]. In this study, the increase in fractal dimension D is primarily attributed to the rise in the number of nanoscale pores. Consequently, as shown in Figure 16, a negative correlation is observed between fractal dimension D and porosity. This finding suggests that FESEM fractal characteristics significantly influence the complexity of the micropore structure in source rocks. Lower displacement pressures are indicative of good permeability and excellent pore structure, whereas higher displacement pressures imply poorer permeability and worse pore structure. Figure 16 also reveals a positive correlation between fractal dimension D and displacement pressure. In summary, the FESEM fractal dimension exhibits a good correlation with porosity and permeability in source rock reservoirs, indicating that it can serve as an important parameter for evaluating the physical properties of source rock reservoirs.

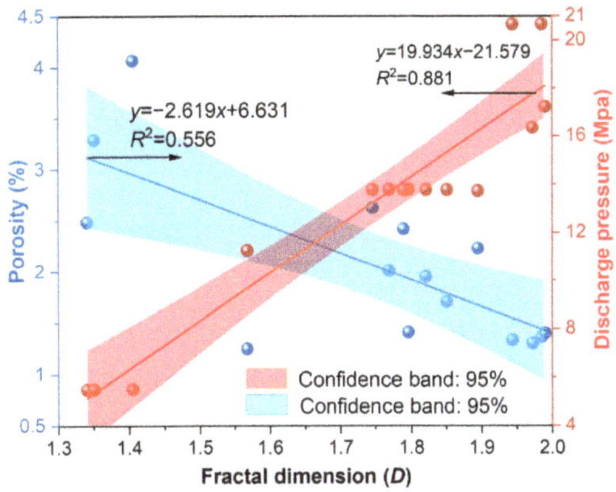

Figure 16. Correlation between fractal dimension, porosity, and displacement pressure.

6. Outlook

This paper systematically combines three techniques (MICP, LFNMR, and FESEM) for the first time to characterize the pore structure and fracture features of source rocks (including limestone, mudstone, and sandstone) from the Late Carboniferous Taiyuan Formation in Huainan on a multi-scale and multi-dimensional basis. This study achieves a full-scale pore structure analysis from macro to micro scales, revealing the differences in pore structures among different lithologies and their impacts on reservoir properties. It provides a new technical pathway for the evaluation of unconventional natural gas reservoirs in coal measures and holds significant theoretical value and scientific importance. From this study, it can be observed that MICP, LFNMR, and FESEM yielded abundant experimental results and new discoveries in the characterization of rock pore and fracture structures. However, it is worth noting, with slight regret, that this research only conducted quantitative characterization and differentiated interpretation of the pore and fracture structures of source rocks from a single well (HN-1 well), making it difficult to comprehensively reflect the regional variability and general laws of the pore structures of source rocks from the Taiyuan Formation of the Late Carboniferous in the Huainan coalfield. Additionally, the study overlooked the importance of establishing a unified multi-scale pore structure model for different types of source rocks. Based on current research progress, the authors believe that the following three directions deserve further exploration:

(1) Deepening the understanding of pore structures and optimizing their practical applications, including expanding the sample size and regional research scope to cover different geological settings to enhance the universality of the data;
(2) Developing high-resolution multi-scale characterization techniques, such as Nano-CT and FIB-SEM, combined with in situ experiments to simulate dynamic changes in formation environments for multi-factor coupling analysis; at the same time, establishing a comprehensive geological–geochemical–pore evolution model by comprehensively considering mineral composition, organic matter types, and fluid–rock interactions;
(3) Utilizing experimental data to construct numerical simulation and prediction models, such as pore network models or machine learning algorithms, combining microscopic pore characteristics with macroscopic reservoir properties to optimize unconventional natural gas development strategies and assess the impact of pore structures on efficient unconventional natural gas development.

7. Conclusions

This study employed multiple technical approaches, including MICP, LFNMR, FESEM, and POM, and constructed double logarithmic scatter plots. By integrating fractal theory, the fractal dimensions of 15 samples were calculated, systematically analyzing the pore–fracture structure and fractal characteristics of different lithologies within the deep Taiyuan Formation of the HN-1 well. The results show that sandstone possesses relatively coarse throat passages, good permeability, as well as excellent pore structure, exhibiting the best fluid flow performance. Limestone displays uneven pore development, whereas mudstone falls between these two extremes. Furthermore, the organic matter in the rocks exhibits varying developmental characteristics, significantly impacting the pore structure. The observation results from FESEM further confirm the highly developed porosity and excellent connectivity within the sandstone. Additionally, the fractal dimension analysis using FESEM indicates that the sandstone exhibits good fractal characteristics, validating its certain oil storage capacity.

(1) The MICP test results indicate that the displacement pressures for limestone, sandstone, and mudstone in the study area are 17.22 MPa, 7.57 MPa, and 13.78 MPa, respectively. This suggests that sandstone has coarser pore throats, better permeability, and a superior pore structure. Sandstone exhibits the lowest mean D value (64.48), while the mean α values for the three rock types are similar (ranging from 0.2 to 0.3), indicating a uniform distribution of pore throats. The skewness of pore throats in all three rock types is less than 0, and the kurtosis is less than 1, presenting a negatively skewed, fine pore throat distribution with a flat kurtosis curve. Sandstone has an average S_{max} of 91.04 and a mercury withdrawal efficiency of 64.45%, indicating excellent connectivity and fluid flow properties, followed by mudstone. DP and α are negatively correlated with porosity, while K_p and S_{max} are positively correlated with porosity.

(2) The T_2 spectrum characteristic maps of six representative samples were characterized using LFNMR, and the study found that the pore distribution patterns were consistent with those obtained from MICP experiments, the signal intensity of sandstone types in different relaxation time segments was better, and the width of the T_2 spectra was wider, which indicated that the sandstone types had better pore structure and connectivity. The T_2 spectra of different rock types have the double-peak type with the left peak dominant, and the wave valley and inflection point are obvious, which indicates that the pore distribution tends to be uniform, and the pores of different diameters are developed with a wide range of pore diameters, among which micropores and mesopores are dominant, and a small number of macropores are developed at the same time. However, the spectra of the limestones do not show significant characteristics, which indicates that the pore development of limestones is not uniform.

(3) The results of POM indicate that the rocks are predominantly characterized by organic matter development. Specifically, limestones are dominated by bituminous matter and organic inclusions, sandstones are primarily composed of vitrinite with a small amount of inertinite, and exhibit relatively developed exinite, while mudstones are mainly composed of vitrinite with well-developed exinite and some inertinite macerals. The FESEM results reveal that the pore development in limestones is mostly below 100 nm, with highly developed nanoscale pores in the matrix. Sandstones exhibit good and uniform pore development, with the emergence of microscale macropores; however, the pores in sandstones are still primarily composed of nanopores. Mudstone samples contain numerous pores dissolved in quartz minerals, ranging from nanoscale to microscale in size, with mostly irregular pore shapes. The degree of fracture development varies significantly among the three rock types, but sandstone

fractures are dominated by macrofractures above the micron scale, demonstrating a certain oil storage capacity of the sandstone.

(4) The MICP method mainly characterizes pore sizes larger than 50 nm and may not be accurate enough for smaller pores. In contrast, the LFNMR method mainly characterizes pore sizes from 2 nm to 1000 nm, which is advantageous for both large and small pores. This characteristic leads to the relatively small value of the MICP method in characterizing the porosity of limestone samples, which results in the largest difference between the pressed mercury method and the LFNMR method in the mean porosity values of limestone and the smallest difference in sandstone among the three types of rock samples.

(5) The results of the fractal dimensions indicate that sandstone samples have the lowest fractal dimension, implying a more uniform pore structure distribution, better connectivity, and relatively higher permeability, thus exhibiting superior reservoir performance, followed by mudstone. These fractal dimension results are consistent with the experimental results obtained from the MICP and LFNMR conducted in this study. Additionally, there is a good correlation between the fractal dimensions derived from field emission scanning electron microscopy (FESEM) and the porosity and permeability of source rock reservoirs, suggesting that FESEM fractal dimensions can serve as an important parameter for evaluating the physical properties of source rock reservoirs.

Author Contributions: Conceptualization, D.W. and G.H.; methodology, D.W.; software, D.W.; validation, D.W. and L.Z.; formal analysis, L.Z.; investigation, G.H.; resources, D.W.; data curation, L.Z.; writing—original draft preparation, L.Z.; writing—review and editing, D.W.; visualization, D.W.; supervision, W.Z.; project administration, D.W.; funding acquisition, D.W. All authors have read and agreed to the published version of the manuscript.

Funding: This research was funded by the National Natural Science Foundation of China (grant number 41972170), the Anhui University of Architecture High-level Talent Introduction Program (grant number 2022QDZ22), and the Anhui Province Public Welfare Geological Work Program (grant number 2016-g-3-33).

Data Availability Statement: Data will be made available on request.

Conflicts of Interest: The authors declare no conflicts of interest.

References

1. Yan, X.; Dai, C.; Wang, R.; Liu, H.; Meng, S.; Jin, X.; Wu, Y. Experimental study on countercurrent imbibition in tight oil reservoirs using nuclear magnetic resonance and AFM: Influence of liquid–liquid/solid interface characteristics. *Fuel* **2024**, *358*, 130026. [CrossRef]
2. Liu, B.; Gao, S.; Mohammadian, E. Comprehensive outlook into critical roles of pressure, volume, and temperature (PVT) and phase behavior on the exploration and development of shale oil. *Energy Fuels* **2022**, *36*, 14534–14553. [CrossRef]
3. Yuan, D.E.; Yan, D.T.; Tang, Y. Genesis of clay minerals in gravel reservoirs and their influence on reservoir quality—An example from the Triassic Baikouquan Formation in the western part of the Mahu Depression, Junggar Basin. *J. Geol.* **2024**, *98*, 3588–3600.
4. Qin, Y.; Shen, J.; Shen, Y.L. Co-mining compatibility of stacked gas-bearing systems—A common geological problem in coal system "three gases" and deep coalbed methane mining. *J. Coal* **2016**, *41*, 14–23.
5. Zhang, X.; Wei, B.; You, J.; Liu, J.; Wang, D.; Lu, J.; Tong, J. Characterizing pore-level oil mobilization processes in unconventional reservoirs assisted by state-of-the-art nuclear magnetic resonance technique. *Energy* **2021**, *236*, 121549. [CrossRef]
6. Zhao, X.; Yang, Z.; Zhou, S.; Luo, Y.; Liu, X.; Zhang, Y.; Niu, Z. Research on characterization and heterogeneity of microscopic pore throat structures in tight oil reservoirs. *ACS Omega* **2021**, *6*, 24672–24682. [CrossRef]
7. Du, S.; Shi, Y. Nanostructures in tight oil reservoirs: Multiple perspectives. *Int. J. Hydrogen Energy* **2024**, *49*, 884–896. [CrossRef]
8. Davoodi Shadfar, S.; Al-Shargabi, M.; Wood, D.A.; Rukavishnikov, V.S.; Minaev, K.M. Experimental and field applications of nanotechnology for enhanced oil recovery purposes: A review. *Fuel* **2022**, *324*, 124669. [CrossRef]

9. Wang, Y.; Liu, L.; Zheng, S.; Luo, Z.; Sheng, Y.; Wang, X. Full-scale pore structure and its controlling factors of the Wufeng-Longmaxi shale, southern Sichuan Basin, China: Implications for pore evolution of highly overmature marine shale. *J. Nat. Gas Sci. Eng.* **2019**, *67*, 134–146. [CrossRef]
10. He, L.; Cao, J.; Lu, J.; Lin, B.; Lu, Y.; Shi, S.; Liu, M. Effect of microwave-assisted cyclic oxidation on the coal internal and surface structure based on NMR and AFM. *Energy* **2024**, *288*, 129872.
11. Zeng, F.; Dong, C.; Lin, C.; Tian, S.; Wu, Y.; Lin, J.; Zhang, X. Pore structure characteristics of reservoirs of Xihu Sag in East China Sea Shelf Basin based on dual resolution X-ray computed tomography and their influence on permeability. *Energy* **2022**, *239*, 122386. [CrossRef]
12. Shi, S.; He, J.; Zhang, X.; Yu, Z.; Wang, J.; Yang, T.; Wang, W. Pore structure evolution of tar-rich coal with temperature-pressure controlled simulation experiments. *Fuel* **2023**, *354*, 129298. [CrossRef]
13. Zou, C.; Zhu, R.; Bai, B.; Yang, Z.; Wu, S.; Su, L.; Dong, D.; Li, X. The first discovery of nanopores in oil and gas reservoirs in China and its scientific value. *Acta Petrol. Sin.* **2011**, *27*, 1857–1864.
14. Kirchofer, A.; Firouzi, M.; Psarras, P.; Wilcox, J. Modeling CO_2 Transport and Sorption in Carbon Slit Pores. *J. Phys. Chem. C* **2017**, *121*, 21018–21028. [CrossRef]
15. Bernard, S.; Wirth, R.; Schreiber, A. Formation of nanoporous pyrobitumen residues during maturation of the Barnett Shale (Fort Worth Basin). *Int. J. Coal Geol.* **2012**, *103*, 3–11. [CrossRef]
16. Li, J.; Hu, L.; Jiang, W.; Cai, M.; He, J.; Wang, Q.; Li, D. Shale pore characteristics and their impact on the gas-bearing properties of the Longmaxi Formation in the Luzhou area. *Sci. Rep.* **2024**, *14*, 16896. [CrossRef]
17. Jiang, Z.; Zhu, Y.; Li, P.; Wang, Y.; Xiang, J. Applicability of Nuclear Magnetic Resonance Experiment in Analyzing Pore and Fluid Distribution Characteristics of Tight Sandstone: A Case Study in the Julu Area, Bohai Bay Basin, China. *ACS Omega* **2023**, *8*, 43903–43919. [CrossRef] [PubMed]
18. Yang, Q.; Xue, J.; Li, W.; Du, X.; Ma, Q.; Zhan, K.; Chen, Z. Comprehensive evaluation and interpretation of mercury intrusion porosimetry data of coals based on factal theory, Tait equation and matrix compressibility. *Fuel* **2021**, *298*, 120823. [CrossRef]
19. Mandelbrot, B.B. *The Fractal Geometry of Nature*; W.H. Freeman: New York, NY, USA, 1982.
20. Zhang, G.; Guo, J.; Xu, B.; Xu, L.; Dai, Z.; Yin, S.; Soltanian, M.R. Quantitative analysis and evaluation of coal mine geological structures based on fractal theory. *Energies* **2021**, *14*, 1925. [CrossRef]
21. Wang, D.; Zeng, Z.P.; Hu, H.Y.; Wang, T.; Zhu, G.G.; Li, S.T.; Li, L. Pore structure fractal characteristics and geological significance of deep terrestrial shales in the Lower Urho Formation, central Junggar Basin. *Oil Gas Geol. Recovery* **2024**, *31*, 23–35.
22. Zhong, B.; Zhu, Y.; Feng, G.; Xiang, J.; Wang, Y. Matrix Compression and Pore Heterogeneity in the Coal-Measure Shale Reservoirs of the Qinshui Basin: A Multifractal Analysis. *Fractal Fract.* **2024**, *8*, 580. [CrossRef]
23. Li, L.L. Evaluation of Coal System Oil and Gas Combined Exploration and Extraction Potential of Taiyuan Formation in Huainan Area. Master's Thesis, Nanjing University, Nanjing, China, 2019.
24. Zhang, W.Y.; Dou, X.Z.; Liu, G.J. Geochemical characterization and hydrocarbon potential of hydrocarbon source rocks in the deep coal system of the Panxie Mining Area, Huainan. *J. Coal Sci.* **2020**, *45*, 731–739.
25. Sang, S.; Han, S.; Zhou, X.; Liu, S.; Wang, Y. Deep coalbed methane resources and prospects for exploration and development in East China. *Reserv. Eval. Dev.* **2023**, *13*, 403–415.
26. Wu, D.; Wu, J.; Hu, G.; Cai, G.; Gao, X. Fine Structure Characterization of Representative Coal-Bearing Rocks of the Late Carboniferous Taiyuan Formation in Huainan Coalfield: Combined SEM, FESEM, and μCT Techniques. *ACS Omega* **2024**, *9*, 2850–2865. [CrossRef]
27. Wu, D.; Zhang, W.Y.; Liu, G.J. Characteristics and geological significance of germanium in Taiyuan coal formation of Huainan Coalfield, Anhui, China. *Int. J. Coal Sci. Technol.* **2020**, *7*, 662–675. [CrossRef]
28. Wu, D.; Zhang, W.Y.; Liu, G.J.; Zhan, R.; Hu, G.Q. Unconventional oil and gas exploration and development technology of the Taiyuan Formation coal system in Huainan Coalfield. *Coalf. Geol. Explor.* **2017**, *45*, 13–18. [CrossRef]
29. Chen, B.Y.; Liu, G.J.; Wu, D.; Sun, R.Y. Elemental geochemical characterization of the Taiyuan Formation aluminous mudstone in Huainan Coalfield. *Coalf. Geol. Explor.* **2018**, *46*, 21–27.
30. Yu, K.; Gan, Y.T.; Ju, Y.W.; Shao, C.J. Influence of sedimentary environment on the brittleness of coal-bearing shale: Evidence from geochemistry and micropetrology. *J. Pet. Sci. Eng.* **2020**, *185*, 106603. [CrossRef]
31. GB/T 19222-2003. Available online: https://codeofchina.com/standard/GBT19222-2003.html (accessed on 16 March 2025).
32. GB/T 29171-2012. Available online: https://codeofchina.com/standard/GBT29171-2012.html (accessed on 16 March 2025).
33. Xie, H.P. *Introduction to Fractal-Rock Mechanics*; Science Publishing House: Beijing, China, 1997; pp. 18–23.
34. Feder, J. *Fractals*; Plenum Press: New York, NY, USA, 1988.
35. Yang, X. Research on Rock Pore Imaging Based on Oblique Incident Light Reflection Difference Technique. Master's Thesis, China University of Petroleum (Beijing), Beijing, China, 2022.
36. Gao, J.D.; Kong, D.B.; Peng, Y.F.; Zhou, Y.Z.; Liu, Y.W.; Zhu, W.Y. Pore-scale mechanisms and hysteresis effect during multi-cycle injection and production process in underground hydrogen storage reservoir. *Energy* **2023**, *283*, 129007. [CrossRef]

37. Fang, H.H.; Wang, Z.F.; Sang, S.X.; Liu, S.Q.; Gu, C.C.; Yang, J.; Huang, Y.H. Correlation Evaluation and Schematic Analysis of Influencing Factors Affecting Pore and Fracture Connectivity on the Microscale and Their Application Discussion in Coal Reservoir Based on X-ray CT Data. *ACS Omega* **2023**, *8*, 11852–11867. [CrossRef]
38. Liu, X.; Rao, L.; Zhuang, T. Characterization of Micro-nano pore structure of tight sandstone based on nuclear magnetic resonance experiments (NMR). In Proceedings of the International Field Exploration and Development Conference, Wuhan, China, 20–22 September 2023; Springer Nature: Singapore, 2023; pp. 208–224.
39. Yao, Y.B.; Liu, D.M.; Che, Y.; Tang, D.Z.; Tang, S.H.; Huang, W.H. Petrophysical characterization of coals by low-field nuclear magnetic resonance (NMR). *Fuel* **2009**, *89*, 1371–1380. [CrossRef]
40. Coates, G.R.; Xiao, L.Z.; Prammer, M.G. *NMR Logging Principles and Applications*; Gulf Publishing Company: Houston, TX, USA, 1999; *in press*.
41. Wen, X.; Chen, Y.; Pu, R. Analysis of the factors influencing movable fluid in shale oil reservoirs: A case study of Chang 7 in the Ordos Basin, China. *Geomech. Geophys. Geo-Energy Geo-Resour.* **2024**, *10*, 177. [CrossRef]
42. Yao, Y.B.; Liu, D.M. Physical properties and fluid characterization of shale reservoirs based on nuclear magnetic resonance relaxation spectroscopy. *J. Coal* **2018**, *43*, 181–189.
43. Li, A.; Ding, W.L.; Wang, R.Y. Petrophysical characterization of shale reservoir based on nuclear magnetic resonance (NMR) experiment: A case study of Lower Cambrian Qiongzhusi Formation in eastern Yunnan Province, South China. *J. Nat. Gas Sci. Eng.* **2016**, *37*, 29–38. [CrossRef]
44. Liu, Y.; Yao, Y.B.; Liu, D.M.; Zheng, S.J.; Sun, G.X.; Chang, Y.H. Shale pore size classification: An NMR fluid typing method. *Mar. Pet. Geol.* **2018**, *96*, 591–601. [CrossRef]
45. Sondergeld, C.H.; Ambrose, R.J.; Rai, C.S. Microstructural studies of gas shales. In Proceedings of the SPE Unconventional Gas Conference, Pittsburgh, PA, USA, 23–25 February 2010. Paper SPE-131771 MS.
46. Curtis, J.B. Fractured shale-gas system. *AAPG Bull.* **2002**, *86*, 1921–1938.
47. Qin, L.; Li, S.G.; Zhai, C.; Lin, H.F.; Zhao, P.X.; Yan, M.; Shi, Y. Joint analysis of pores in low, intermediate, and high rank coals using mercury intrusion, nitrogen adsorption, and nuclear magnetic resonance. *Powder Technol.* **2020**, *362*, 615–627. [CrossRef]
48. Li, K.; Jia, Y.; Han, Y.L.; Lei, F.; Lin, F.H.; Wang, T.; Chen, W. Quantitative characterisation of microstructure of coal carbonate rocks in tectonic regions based on MIP and SEM. *China Coal* **2025**, *51*, 204–214.
49. Wang, J.; Zhang, J.; Xiao, X.; Chen, Y.N.; Han, D. Pore Structure and Fractal Characteristics of Inter-Layer Sandstone in Marine–Continental Transitional Shale: A Case Study of the Upper Permian Longtan Formation in Southern Sichuan Basin, South China. *Fractal Frac.* **2025**, *9*, 11. [CrossRef]
50. Shen, Y.J.; Wang, X.; Zhao, X.; Zhao, C.H.; Wang, S.Q.; Guo, C.; Shi, Q.M.; Ma, W. Characteristics of multiscale pore structure of oil-rich coal in Yushenfu mining area. *Coalf. Geol. Explor.* **2021**, *49*, 33–41.
51. Li, J.; Liu, D.; Yao, Y.; Cai, Y.; Qiu, Y. Evaluation of the reservoir permeability of anthracite coals by geophysical logging data. *Int. J. Coal Geol.* **2011**, *87*, 121–127. [CrossRef]
52. Liu, L.H.; Hu, H.Y.; Zhan, X.F.; Wang, T.; Tang, T.; Lei, L. Pore structure and NMR fractal characteristics of deep shale in Longmaxi Formation, Luzhou Block. *Chin. Sci. Technol. Pap.* **2023**, *18*, 501–511.

fractal and fractional

MDPI

Article

Characterization of Pore Heterogeneity in Lacustrine Shale Based on MIP, LTNA, NMR, and Multifractal Characteristics: A Case Study of the Jurassic Dongyuemiao Member, China

Xu Wu [1,2], Yifan Gu [1,2,*], Yuqiang Jiang [1,2], Zhanlei Wang [1,2] and Yonghong Fu [1]

1 School of Geoscience and Technology, Southwest Petroleum University, Chengdu 610500, China; 202221000140@stu.swpu.edu.cn (X.W.); 198531010028@swpu.edu.cn (Y.J.); zlwang@swpu.edu.cn (Z.W.); 202199010147@swpu.edu.cn (Y.F.)

2 Collaborative Innovation Center of Shale Gas Resources and Environment, Chengdu 610500, China

* Correspondence: 202099010168@swpu.edu.cn

Abstract: Pore structure plays a critical role in evaluating shale "sweet spots". Compared to marine shale, lacustrine shale has more diverse lithofacies types and greater heterogeneity in pore structure due to frequently changing environmental conditions. Using methods such as mercury intrusion porosimetry (MIP), field emission scanning electron microscopy (FE-SEM), nuclear magnetic resonance (NMR), and X-ray diffraction (XRD), this study investigates the micropore structures and heterogeneity of different lithofacies in the Jurassic Dongyuemiao Member lacustrine shale. Image processing and multifractal theory were employed to identify the controlling factors of pore structure heterogeneity. The key findings are as follows. (1) Based on mineral content and laminae types, the lithofacies types of Dongyuemiao lacustrine shale are classified into four types: shell–laminae mixed shale (SLMS), silty–laminae clay shale (SLCS), clast–laminae clay shale (CLCS), and clay shale (CS). (2) Based on genesis, shale reservoirs' pore and permeability space are categorized into inorganic pores, organic pores, and micro-fractures. Inorganic pores consist of inter-particle pores and intra-particle pores. Pore size distribution curves for all four lithofacies exhibit two main peaks, with pore sizes concentrated in the ranges of 2–10 nm and 50–80 nm. Mesopores and macropores dominate, accounting for over 80% of the total pore volume. Mesopores are most developed in CLCS, representing 56.3%. (3) Quartz content is positively correlated with the multifractal dimension, while clay content shows a negative correlation. Higher quartz content, coupled with lower clay content, weakens pore structure heterogeneity. A negative correlation exists between total organic carbon (TOC) and the multifractal dimension, indicating that higher organic matter content enhances organic pore development and increases microscopic heterogeneity. (4) Porosity heterogeneity in SLMS is effectively characterized by D_0-D_{max}, while in the other three lithofacies, it is characterized by D_{min}-D_0. Permeability across all lithofacies correlates with D_0-D_{max}. In CS, SLMS, and SLCS, permeability is positively correlated with D_0-D_{max}, with higher values indicating greater permeability heterogeneity. In CLCS, permeability is negatively correlated with D_0-D_{max}, such that lower values reflect stronger heterogeneity.

Keywords: Sichuan Basin; lacustrine shale; pore structure; microscopic heterogeneity; multifractal method

1. Introduction

Shale oil and gas exploration originated in North America. By 2009, shale gas production accounted for over 12% of total natural gas production in North America [1]. In

https://doi.org/10.3390/fractalfract9040265

recent years, the development of shale oil, particularly using Permian Basin marine shale as a reservoir, has also achieved remarkable success in North America. Consequently, the United States has transitioned from an energy importer to an energy exporter, achieving energy independence. China, like North America, possesses substantial shale oil and gas resources, especially in Paleozoic marine shale and Mesozoic–Cenozoic lacustrine shale formations [2–4]. After nearly a decade of scientific research, the development of marine shale gas in southern China has reached a significant scale [5,6], notably improving the country's energy consumption structure. Recently, inspired by the success of marine shale gas development, the vast oil and gas potential in lacustrine shale has been increasingly recognized in the Ordos Basin, Bohai Bay Basin, and Sichuan Basin in China [7–9].

Shale, as a unique porous medium, comprises both organic and inorganic components, making it highly complex and irregular [10,11]. The heterogeneity of lacustrine shale is primarily reflected in two aspects: organic geochemical heterogeneity and inorganic mineral heterogeneity. The organic geochemical characteristics of different shales vary significantly, with total organic carbon (TOC) content ranging from 0.5% to 15% and vitrinite reflectance (Ro) ranging from 0.5% to 1.3%. The complexity of its mineral composition is evident in the varying dominant mineral types of different lacustrine shales. For example, the Lucaogou Formation in Jimusarsag is predominantly composed of felsic minerals, with clay mineral content often below 8%. In contrast, the Ziliujing Formation in the Sichuan Basin features comparable amounts of felsic minerals and clay minerals [12–14]. This compositional heterogeneity in lacustrine shale results in a more complex pore structure compared to marine shale [15,16]. The shale pore structure significantly influences the occurrence and permeability mechanisms of hydrocarbons [17,18]. Therefore, effectively characterizing pore structure heterogeneity is a critical factor in reservoir evaluation and enhancing hydrocarbon recovery [19,20].

Various qualitative and quantitative techniques have been used to characterize the pore structure of fine-grained sedimentary rocks [21,22]. Qualitative methods, such as field emission scanning electron microscopy (FE-SEM), transmission electron microscopy (TEM), and computed tomography (CT), are widely applied to study the morphology and connectivity of micro and nanoscale pore networks. Common quantitative methods for determining porosity and permeability include N_2/CO_2 physical adsorption (LTCA and LTNA), mercury intrusion capillary pressure (MIP), small-angle/ultra-small-angle neutron scattering (SANS), and nuclear magnetic resonance (NMR) [23–25]. The pore structure of coal, granite, and fine-grained sedimentary rocks can be analyzed qualitatively and quantitatively using the methods mentioned above.

Coal porosity and permeability are commonly measured through mercury intrusion porosimetry (MIP), liquid nitrogen adsorption, and nuclear magnetic resonance (NMR). Research shows that the porosity of coal reservoirs is generally below 2%, contributing a small fraction of total porosity, while bedrock porosity accounts for about 80%. Coal permeability ranges from 0.0002×10^{-15} to 0.6525×10^{-15} m^2, mostly falling within the medium-to-low range [26]. Han et al. characterized coal pore structures at different metamorphic stages using MIP and N_2 adsorption, integrating Frenkel–Halsey–Hill and Menger fractal models to define pore size distributions. Cai et al. analyzed coal porosity using low-temperature nitrogen adsorption/desorption isotherms, assessed coal compressibility across different ranks, and evaluated pore structures via MIP and N_2 adsorption at −196 °C; Zhao et al. combined scanning electron microscopy (SEM), low-temperature nitrogen adsorption (LTNA), high-pressure mercury injection (HPMI), and rate-controlled mercury injection (RCMI) techniques to throat size and pore distribution in tight sandstone reservoirs; and Longinos et al. used N_2 adsorption and NMR to study fracture behavior in granite cooled with liquid nitrogen [27–30]. Although each method has specific application scenarios and limitations, combining multiple techniques can effectively characterize the full-scale pore system of shale. Fractal theory, as an effective powerful tool for evaluating the complexity and heterogeneity of porous media, is increasingly applied to analyze shale's complex pore systems. By integrating fractal dimensions with macroscopic and microscopic pore structure characteristic parameters, the correlations between fractal dimensions and these parameters can be thoroughly examined. The Dongyuemiao shale of the Jurassic Ziliujing Formation in the Sichuan Basin has a high clay mineral content (~50%), well-developed shell interlayers, and advanced thermal evolution (~1.5% in the condensate gas generation stage). This terrestrial shale has undergone prolonged diagenesis and hydrocarbon generation, with production wells yielding both oil and gas at the wellhead [31–33]. The lacustrine shale of the Lianggaoshan Formation in the Sichuan Basin has a high TOC content (avg. 1.57%) and porosity (avg. 4.21%), along with significant oil and gas content. Well-developed fine sand layers make it more favorable for hydrocarbon accumulation than the Dongyuemiao section [34–36]. Using digital 3D rock models, some researchers have investigated the evolution of fractal dimensions' porosity [37]. Wang et al. (2015) evaluated the microporous structure of lacustrine shale using gas adsorption dating and fractal theory [38]. These studies demonstrate a strong correlation between fractal dimensions and factors such as mineral composition, specific surface area, and adsorption volume. However, a single fractal model is insufficient for understanding the heterogeneity of pore structures and fails to distinguish samples with identical fractal dimensions but differing pore size distributions [39]. The multifractal approach divides complex fractal structures into multiple regions with uniform singular intensities and generalized fractal dimensions. By analyzing these singular intensities and generalized fractal dimensions region by region, specific multifractal characteristics can be identified. This study focuses on the lacustrine shale of the Dongyuemiao Member. Previous research on shale reservoir pore structures has primarily relied on gas adsorption and scanning electron microscopy. As a complementary method, multifractal analysis holds significant potential for characterizing pore structure heterogeneity. This study employs X-ray diffraction, scanning electron microscopy, and nuclear magnetic resonance. Using image processing and multifractal theory, it analyzes the pore structure characteristics and heterogeneity of different lithofacies, providing theoretical support for lacustrine shale reservoir characterization and sweet spot prediction.

2. Geological Setting

The Sichuan Basin is a typical craton basin located in the western part of the Upper Yangtze Block (Figure 1a), covering an area of approximately 260,000 km^2. During the Early to Middle Jurassic period, most regions of the Sichuan Basin were dominated by shore-shallow lake, semi-deep lake, and deep lake facies, experiencing four lake transgressions during this period (Figure 1a). Organic-rich shale is primarily thought to have developed in semi-deep and deep lake facies under anoxic conditions [40,41]. From bottom to top, four sets of organic-rich shales were formed in the Zhenzhuchong Member, Dongyuemiao Member, Da'anzhai Member, and Lianggaoshan second Member, respectively. The Dongyuemiao Member is further divided into three sub-members: the first, second, and third sub-member. The first sub-member is further subdivided into four sections: the first, second, third, and fourth sections (Figure 1b). Except for the second sub-member, the other intervals are predominantly made up of shale. The organic matter in the shale is mainly Type II, with localized development of Type III. The vitrinite reflectance (Ro) is mostly greater than 1.2%, indicating that the shale has reached a high-maturity evolutionary stage and is predominantly characterized by gas generation [42].

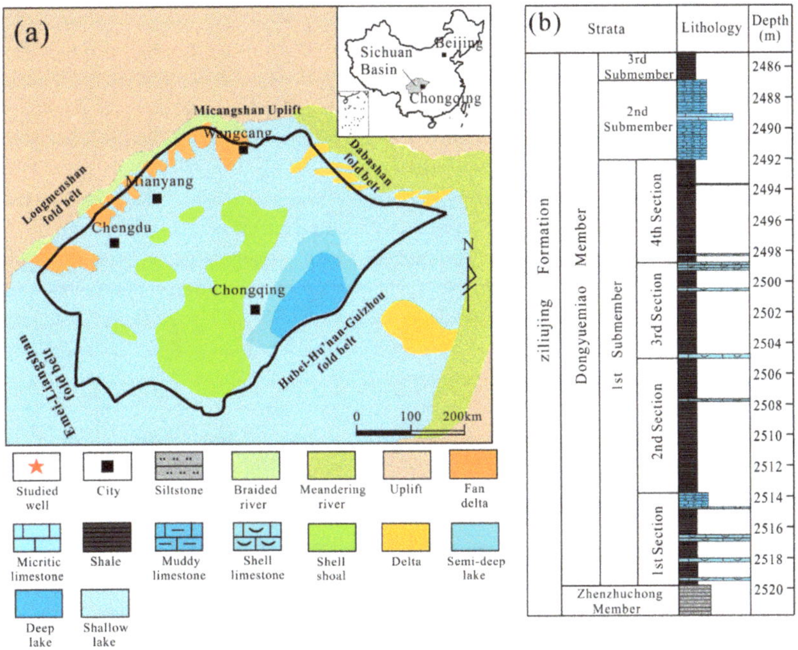

Figure 1. (**a**) Sedimentary environment of the Sichuan Basin during Dongyuemiao Member deposition (modified from [37,38]). (**b**) Stratigraphic column of the Jurassic Dongyuemiao Member.

3. Method and Process

3.1. TOC Content, Mineral Component, and Lithofacies Classification

Samples with varying heterogeneity from layers 1–4 of the Dongyuemiao section were selected for experiments. Total organic carbon (TOC) is a key indicator for assessing organic matter abundance in shale. In this study, TOC content was measured using the CS-230 carbon–sulfur analyzer from the American Lico Company (La Porte, IN, USA). Before analysis, shale samples were ground into a 200-mesh powder. Inorganic carbon was removed by treating the samples with 12.5% dilute hydrochloric acid, followed by rinsing with distilled water to eliminate any residual acid. The sample powder was dried at 60–80 °C then placed in a carbon–sulfur analyzer for complete combustion (>930 °C). TOC content was determined by measuring the CO_2 generated during combustion. Mineral composition was analyzed using the D8A25 diffractometer from Germany Bruce Company (Hamburg, Germany); after shale samples were ground into 200-mesh powder, the instrument parameters used were as follows: Cu target, voltage 40 kV, current 40 mA, step size 0.02°, scanning angle 3~45°, scanning speed 0.8 s/step, and emission, scattering, and reception slits of 0.6 mm, 3 mm, and 0.5 mm, respectively. Scanning electron microscopy (SEM) was used to observe microscopic pores and fractures in the shale. FEI Quanta 650 FE GFE-SEM from the American FEI Company (Hillsboro, OR, USA) was used for qualitatively analysis. Before SEM imaging, shale samples were cut into 2 cm × 2 cm × 1 cm rectangular blocks, polished with an IM 4000 argon ion polisher from Hitachi, Ltd., Tokyo, Japan, and coated to enhance conductivity, thus improving the resolution of micropore images.

Lithofacies represent a comprehensive reflection of sedimentary rocks and environments, encapsulating the petrological and sedimentary characteristics of rocks as well as the macroscopic heterogeneity features of shale, including lithology, mineral composition, and other attributes [43–45]. Extensive research has been conducted on shale lithofacies classification and characteristics. For example, Allix et al. (2010) classified shale lithofacies based on the content of clay minerals, carbonate rocks, and felsic minerals [46]. Wang et al. (2012) incorporated organic matter abundance as one of the criteria for lithofacies classification based on mineral composition [47]. Shi et al. (2020) further refined lithofacies classification by considering organic matter abundance, mineralogical facies, and sedimentary structures [48]. A prominent feature of lacustrine shale, particularly in the Dongyuemiao Member, is the presence of high-frequency, variable-thickness siliceous and calcareous laminae. In this study, lithofacies classification was conducted by grouping minerals into three categories: calcareous minerals (calcite and dolomite), siliceous minerals (quartz and feldspar), and clay minerals. Using a 50% mineral content threshold, shale was classified as mixed shale (all three mineral contents < 50%), clayey shale (with clay mineral content > 50%), and silty shale. By combining this mineralogical classification with laminae types, lithofacies were effectively categorized.

The main methods for characterizing the micropore structure of shale include low-temperature nitrogen adsorption, high-pressure mercury intrusion porosimetry, and nuclear magnetic resonance. These methods have their own advantages and disadvantages and are suited to different research needs.

3.2. Characterization Method of Microscopic Pore Structure

3.2.1. Mercury Intrusion Porosimetry (MIP)

The mercury intrusion porosimetry (MIP) method is based on the capillary model, which assumes that porous media consist of capillaries with varying diameters [49]. The analysis was conducted using the IV 9500 mercury intrusion porosimeter from America Micromeritics Company (Norcross, GA, USA). Mercury, as a non-wetting phase, does not wet the surface of rocks, whereas air or mercury vapor within the in rocks acts as the wetting phase. During the process, mercury is injected into the rock pores under pressure, displacing the wetting phase. When the injection pressure equals the capillary force, the corresponding capillary radius represents the pore throats. The volume of mercury entering the pores reflects the volume of rock pores connected by the throat. By continuously varying the injection pressure, the pore size distribution curve and capillary pressure curve can be obtained. The calculation formula is as follows:

$$P_c = \frac{2\sigma cos\theta}{r} \tag{1}$$

In the formula, P_c represents capillary pressure (unit: Mpa). σ is the interfacial tension between mercury and air, with σ = 480 dyn/cm; θ is the wetting angle between mercury and rock. For θ = 140°, $cos\theta$ = 0.765, and r represents the pore radius (unit: μm). The relationship between pore radius (r) and capillary pressure (P_c) is given by

$$r = \frac{0.735}{P_c} \tag{2}$$

MIP data in this study were obtained using the American Auto Pore IV 9500 fully automatic mercury intrusion porosimeter. Prior to testing, samples were dried to a constant weight at 105 °C. The experiment involved two main processes: pressurized mercury injection and depressurized mercury removal. The maximum experimental pressure reached 200 mPa.

3.2.2. Low-Temperature N_2 Adsorption (LTNA)

The low-temperature gas adsorption method is a commonly used technique for characterizing the specific surface area and pore size distribution of porous materials [50].

Nitrogen (N_2) is widely used as the detection gas for this purpose. The analysis was conducted using the ASAP 2460 specific surface area and porosity analyzer from America Micromeritics Company. Before the analysis, the sample was degassed at 110 °C for 10 h to remove moisture and volatile substances. After degassing, it was transferred to the analysis station, and adsorption experiments were performed using 99.99% pure nitrogen as the adsorbent at −195.85 °C. The relative pressure (P/P_0) was gradually increased from 0.01 to the saturated vapor pressure (P/P_0 = 0.998).

3.2.3. Low-Field Nuclear Magnetic Resonance (NMR)

Saturated water NMR technology relies on the nuclear magnetic resonance (NMR) relaxation behavior of spin hydrogen nuclei in rock fluids under a uniformly static magnetic field and radio frequency fields [51]. The analysis was conducted using the PQ001-050 nuclear magnetic resonance instrument from China Niumag Corporation (Suzhou, China). In this study, samples were saturated with distilled water for 48 h before measurement. The main frequency was set to 12 MHz. NMR relaxation times include longitudinal relaxation time (T_1) and transverse relaxation time (T_2). Under low-magnetic-field conditions, T_2 measurement is significantly faster than T_1 while providing equivalent pore size distribution analysis. Therefore, T_2 is traditionally used for pore size distribution analysis. The measured T_2 relaxation time comprises three components, expressed by the following formula:

$$\frac{1}{T_2} = \frac{1}{T_{2b}} + \frac{1}{T_{2s}} + \frac{1}{T_{2d}} \tag{3}$$

where T_{2s} represents surface relaxation time, T_{2d} represents diffusion relaxation time, and T_{2b} represents bulk relaxation time. Diffusion relaxation (T_{2d}) arises from the self-diffusion of fluid molecules, caused by proton spin diffusion through strong internal field gradients. However, in a uniform magnetic field, the magnetic field gradient does not exist, so the diffusion relaxation time (T_{2d}) is not considered in this experiment. The bulk relaxation time (T_{2b}) of the fluid depends on its properties, including chemical composition, viscosity, etc. In this experiment, the saturated fluid is water. Since the bulk relaxation time of water is relatively long (usually between 2000 and 3000 ms) and far exceeds the surface relaxation time of pores, $1/T_{2b}$ can also be ignored. Thus, the Formula (3) simplifies to

$$\frac{1}{T_2} = \frac{1}{T_{2s}} = \rho\frac{s}{v} \tag{4}$$

where ρ represents the surface transverse relaxation strength (μm/ms), which is influenced by the physical properties of the rock. S represents the pore surface area (cm^2). The ratio of pore surface area to pore volume (v) is closely related to pore size. Therefore, the Formula (4) can be simplified as follows:

$$\frac{1}{T_2} = \rho\frac{F_s}{r} \tag{5}$$

where F_s is the pore geometry factor. For spherical pores, F_s = 3, and r is the pore radius (nm). Based on this formula, it can be inferred that there is a linear relationship between T_2 the spectrum and pore size. Consequently, the Formula (5) can be further simplified as

$$r = CT_2 \tag{6}$$

In this formula, C represents a constant, and its value depends on the relaxation properties of the pore surface. The NMR experimental instrument used was a low-field nuclear magnetic resonance device manufactured by Suzhou Niu mag Analytical Instrument Company. For the NMR tests, the shale samples were first processed into cylindrical plunger samples. These samples were then dried at 110 °C for 24 h. After being evacuated for 12 h, the shale plunger samples were saturated with aqua pura under a pressure of 25 MPa for 48 h. After the saturation process, the plunger samples were removed and left to stand in the saturated fluid for 12 h before obtaining the NMR T_2 signals. The saturated fluid used was a $20{,}000 \times 10^{-6}$ potassium chloride solution at a saturation pressure of 25 MPa. The NMR test sequence employed was CPMG, with the following parameters: main frequency: 12 MHz, echo time (TE): 0.1 ms, number of echoes (NECH): 10,000, number of scans (NS): 64, waiting time (TW): 3000 ms.

3.3. Multifractal Theory

The box-counting method is a commonly used technique for analyzing and obtaining the multifractal spectra of porous media [52]. It assumes that a set of boxes or sublayers of varying lengths divide the total aperture layer length (L) into equal-scale linear segments (ε), defined as $\varepsilon = L \times 2^{-k}$ (where k is a positive integer) [53–56]. In non-uniform porous media, for a given scale ε, the probability of data distribution in the *ith* subinterval is defined as $P_i(\varepsilon)$.

$$P_i(\varepsilon) = \frac{m_i(\varepsilon)}{\Sigma_{i=1}^{N(\varepsilon)}} = \frac{m_i(\varepsilon)}{M(\varepsilon)} \tag{7}$$

$$\Sigma_{i=1}^{N(\varepsilon)} P_i(\varepsilon) = 1 \tag{8}$$

where ε is an arbitrary scale (i.e., the box size in box counting) using which the porous media is examined; $m_i(\varepsilon)$ represents the number of pixels or mass in any box, i, at size ε; $M(\varepsilon) = \sum_{i=1}^{N(\varepsilon)} m_i(\varepsilon)$ is the total mass or sum of pixels in all boxes for this ε; $N(\varepsilon)$ is the total number of non-empty boxes for each ε; and $P_i(\varepsilon)$ represents the probability of mass m_i at i relative to the total mass for a box.

When the interval becomes sufficiently small, the relationship between $P_i(\varepsilon)$ and ε can be expressed as

$$P_i(\varepsilon) \propto \varepsilon^{-a_i} \tag{9}$$

a_i represents the Lipschitz–Hölder index or singularity index, indicating the singularity intensity of the *ith* subinterval at scale ε. For multifractals, the target object can be divided into $N(\varepsilon)$ equally long subintervals for a given ε, where $N(\varepsilon)$ is expressed as

$$N(\varepsilon) \propto \varepsilon^{-f(a)} \tag{10}$$

where a measures the local scaling behavior of a measure in the system at scale ε, and $f(a)$ is the fractal spectrum, or singularity spectrum, that characterizes the distribution of singularity strengths a within the system. The shape of $f(a)$, typically a convex curve, provides insights into the system's heterogeneity; a narrow $f(a)$ indicates low variability in singularity strengths, whereas a wide $f(a)$ signifies high variability.

A complex fractal can be divided into subsets represented by different α values, where $f(a)$ describes the fractal characteristics of these subsets. The fractal spectrum, represented by the $a \sim f(a)$ relationship, serves as a fundamental descriptor of the local features of multifractals. The formula for defining the moment function is

$$M(\varepsilon) = \sum_{i=1}^{N(\varepsilon)} m_i(\varepsilon) \tag{11}$$

$$f(\alpha) = \frac{\log log \sum_{i=1}^{N(\varepsilon)} u_i(\varepsilon) \log log u_i(\varepsilon)}{\log log \varepsilon} \tag{12}$$

$$a(q) = \frac{\log log \sum_{i=1}^{N(\varepsilon)} u_i(\varepsilon) \log log P_i(\varepsilon)}{\log log \varepsilon} \tag{13}$$

Let $u_i(q,\varepsilon)$ denote the uniform partition function, which is defined as (14)

$$u_i(q,\varepsilon) = \frac{p_i(\varepsilon)^q}{\sum_{i=1}^{N(\varepsilon)} p_i(\varepsilon)^q} \tag{14}$$

where q represents the multifractal dimension of an object at different scales, ranging from -10 to 10 in unit steps. The probability distribution function $u(q,\varepsilon)$ is then given by

$$u(q,\varepsilon) = \Sigma_{i=1}^{N(\varepsilon)} p_i(\varepsilon)^q \sim \varepsilon^{\tau(q)} \tag{15}$$

where $\varepsilon^{\tau(q)}$ represents the mass scale function, and $\tau(q)$ is the scaling exponent or mass exponent. The relationship between $\tau(q)$ and q reflects the system's heterogeneity; it is linear for homogeneous fractals and nonlinear for multifractals. In Formula (15), $\varepsilon^{\tau(q)}$ is the mass scale function, expressed as

$$\tau(q) = -\lim_{\varepsilon \to 0} \frac{\Sigma_{i=1}^{N(\varepsilon)} p_i^q(\varepsilon)}{log\varepsilon} \tag{16}$$

Based on Formulas (15) and (16), the singularity strength $a(q)$ and the mass index $\tau(q)$ can be transformed into the following formulas using the Legendre transformation, yielding the following expression:

$$\alpha(q) = \frac{\tau(q)}{q} \tag{17}$$

$$f(\alpha) = q\alpha(q) - \tau(q) \tag{18}$$

Therefore, the generalized dimension (D_q) can be expressed by the following formula:

$$D_q = \frac{1}{q-1} \frac{\log log\left(\sum_{i=1}^{N(\varepsilon)} P_i(\varepsilon)^q\right)}{\log log\varepsilon} = \frac{\tau(q)}{q-1}, q \neq 1 \tag{19}$$

For $q > 0$ and $q = 1$, D_1 can be calculated using the following formula:

$$D_1 = \lim_{\varepsilon \to 0} \frac{\Sigma_{i=1}^{N(\varepsilon)} p_{i(\varepsilon)} log p_i(\varepsilon)}{log\varepsilon} \tag{20}$$

$\tau(q)$ can also be calculated using the following formula:

$$\tau(q) = (q-1)D_q \tag{21}$$

In the multifractal model of the T_2 spectrum, T_2 is considered the total interval (I) of the multifractal. Based on the chosen ε value, the total interval is divided into $N(\varepsilon) = L/\varepsilon = 2^k$ subintervals of equal length. To ensure each subinterval contains at least one value, k is set to 7 based on the T_2 length, resulting in 128 subintervals being divided. Each subinterval contains two values, yielding 256 points after T_2 interpolation.

4. Results

4.1. Lithofacies Type

The total organic carbon (TOC) content of lacustrine shale in the Dongyuemiao Member ranges from 0.57% to 4.03%, with a measurement error of δ = ±5% and an average of 1.71% (Table 1), The primary minerals in lacustrine shale are quartz (10.2–30.6%, avg. 23.3%), with a measurement error of δ = ±15%, and clay minerals (25.4–66.8%, avg. 53.1%), with a measurement error of δ = ±10%, which together constitute 76.4% of the total mineral content. In this study, lacustrine shale samples are classified into four lithofacies types (Figure 2a): shell–laminae mixed shale (SLMS) (Figure 2b), silty–laminae clay shale (SLCS) (Figure 2c), clast–laminae clay shale (CLCS) (Figure 2d), and clay shale (CS) (Figure 2e). Among these, CS has the highest TOC content, averaging 2.19%, while SLCS has the lowest TOC content, averaging 1.31%. Using representative samples from four lithofacies for XRD **patterns** (Figure 3), testing X-ray diffraction (XRD) results reveal variations in mineral composition across the four lithofacies; SLCS has the highest clay content, averaging 64.15% (averaging 64.15%), while SLMS has the lowest (averaging 41.5%). CS has the highest quartz content (averaging 27.2%), whereas SLMS has the lowest (averaging 21.7%).

Table 1. Mineral composition and organic matter content of different lithofacies of lacustrine shale.

Sample	Lithofacies	TOC (%) (δ = ±5%)	Clay (%) (δ = ±10%)	Quartz (%) (δ = ±15%)
WellC, 2651.71 m	SLMS	1.77	38.9	19.2
WellB, 2498.7 m	SLMS	1.25	50.9	23.6
WellA, 2951.72 m	SLMS	1.06	25.4	14.3
WellB, 2515.98 m	SLMS	1.69	50.9	29.6
WellA, 2937.66 m	CLCS	1.53	68.2	24.2
WellA, 2940.08 m	CLCS	1.85	65.3	25.2
WellB, 2947.4 m	CLCS	4.03	68.6	26.3
WellB, 2507.62 m	CLCS	1.94	56.8	28.3
WellC, 2655.22 m	CLCS	1.63	64.9	29.9
WellB, 2493.09 m	CS	1.53	67.5	25
WellB, 2494.22 m	CS	4.03	66.8	25.1
WellB, 2509.7 m	CS	1.89	63.3	28.4
WellB, 2513.74 m	CS	1.79	56.7	30.6
WellB, 2503.37 m	SLCS	1.2	63.9	26.7
WellA, 2944.06 m	SLCS	1.42	64.4	24.6

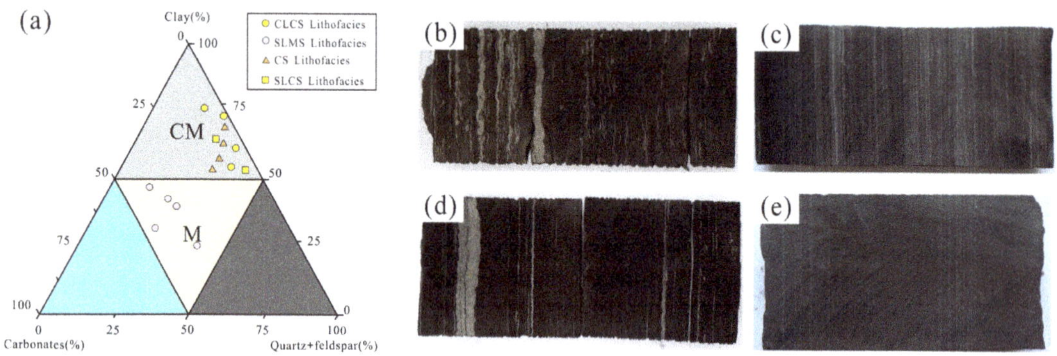

Figure 2. (**a**) Ternary lithofacies diagram of the Jurassic Dongyuemiao Member lacustrine shale. (**b**) Shell–laminae mixed shale (SLMS). (**c**) Silty–laminae clay shale (SLCS). (**d**) Clast–laminae clay shale (CLCS). (**e**) Clay shale (CS). M: mixed shale, and CM: clay shale.

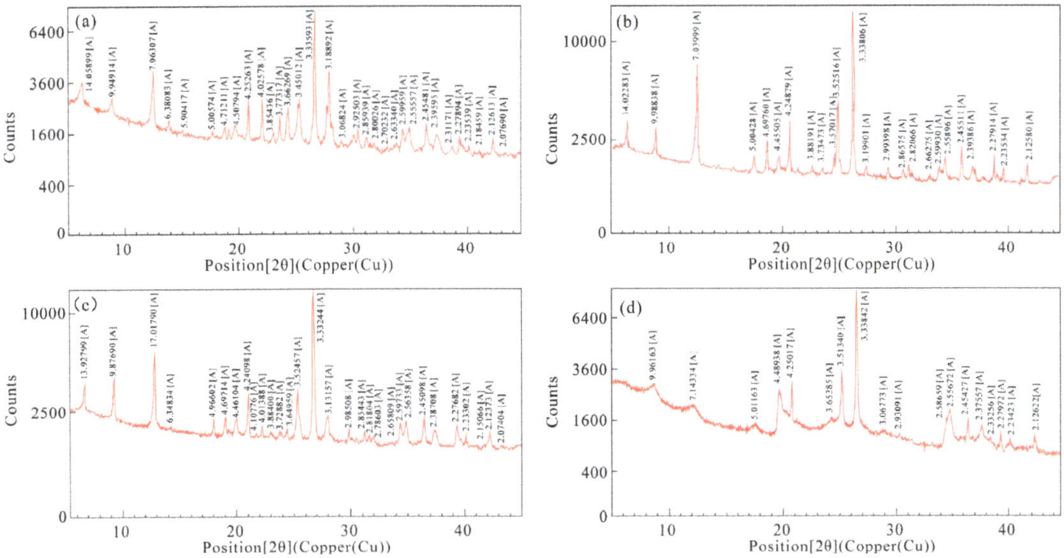

Figure 3. XRD patterns of representative samples of different lithofacies. (**a**) SLMS lithofacies (WellC, 2651.71 m). (**b**) CLCS lithofacies (WellA, 2940.08 m). (**c**) CS lithofacies. (WellB, 2494.22 m) (**d**) SLCS lithofacies (WellB, 2503.37 m).

4.2. Microscopic Pore Types and Morphology

FE-SEM images reveal numerous pores within or between mineral particles, including organic pores, inorganic pores, and micro-fractures (Figure 4). The organic matter in CLCS primarily consists of dense fragments of higher plants, with rare occurrences of solid bitumen. Inorganic pores are highly developed, predominantly comprising inter-particle (Inter-P) pores between clay platelets. These pores are large and exhibit good connectivity. A small number of intra-particle (Intra-P) pores within pyrite framboids are also observed. Micro-fractures are present between clay minerals. In SLMS, the organic matter is compact, and organic pore development is minimal. Organic pores are mostly isolated, narrow, and elongated. The development of inorganic pores is significant, including inter-P pores between clay platelets and dissolution pores within carbonate particles. Micro-fractures primarily consist of bedding fractures, with additional micro fractures observed inside shells.

The organic matter content inside CS is high, primarily composed of vitrinite. However, organic pore development is limited, with most organic pores being isolated and circular. Inorganic pores are developed, mainly consisting of Inter-P pores between clay platelets. Bedding fractures dominate the micro-fractures, with a few microcracks observed at the edges of some organic matter. In SLCS, the degree of organic pore development is relatively high. Organic pores are mostly or partially circular. Inorganic pores arehighly developed, mainly comprising inter-P pores within clay minerals with larger pore sizes. Micro-fractures are predominantly bedding fractures.

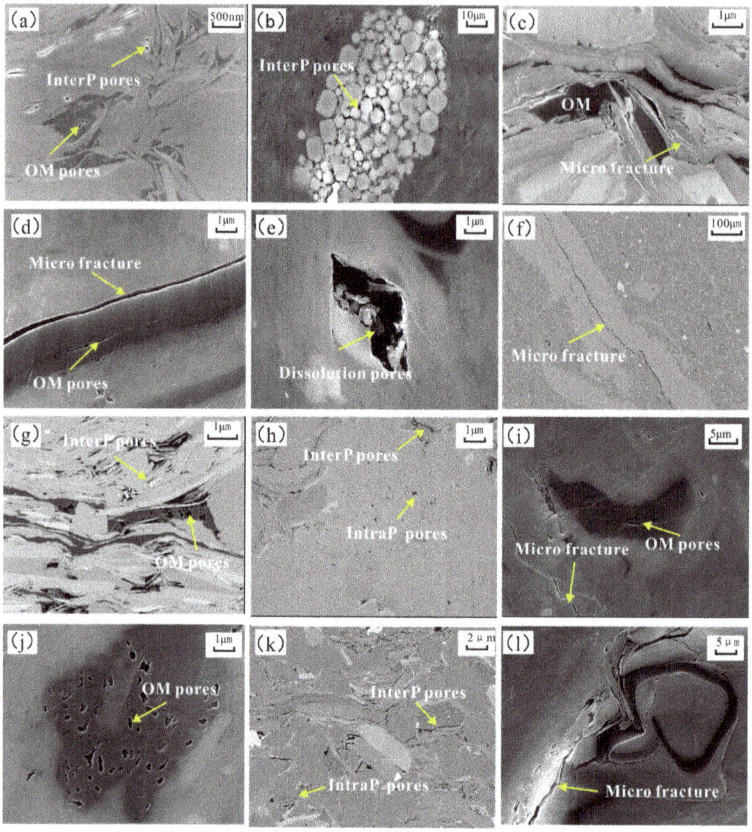

Figure 4. FE-SEM images of different lithofacies. (**a**) Inter-P pores and OM pores of sample Well A, 2937.66 m. (**b**) Inter-P pores of sample Well A, 2940.08 m. (**c**) Micro-fracture of sample Well A, 2940.08 m. (**d**) OM pores and micro-fracture of sample Well C, 2651.71 m. (**e**) Dissolution pores of sample Well B, 2498.7 m. (**f**) Micro-fractures of sample Well A, 2951.72 m. (**g**) Inter-P pores and OM pores of sample Well B, 2493.09 m. (**h**) Inter-P pores and Intra-P pores of sample Well B, 2494.22 m. (**i**) OM pores and micro-fractures of sample Well B, 2509.7 m. (**j**) OM pores of sample Well B, 2503.37 m. (**k**) Inter-P pores and Intra-P pores of sample Well A, 2944.06 m. (**l**) Micro-fracture of sample Well B, 2503.37 m.

4.3. Pore Size Distribution

4.3.1. MIP Characterization

The MIP primarily analyzes macropores [57]. The morphology of the MIP curve reflects the pore development and connectivity across different pore throat segments. While the MIP curves vary significantly between lithologies, their overall trend remains consistent (Figure 5). In terms of total mercury input, SLCS exhibits the highest value, indicating the most extensive pore development, followed by SLMS and CLCS, with CS having the lowest input. Initially, the steep curve slope and large mercury injection volume in CS suggest significant macropore development. As pressure increases, mercury progressively infiltrates smaller pores, stabilizing the curve without clear plateaus. When the injection pressure exceeds 10,000 MPa, the SLCS, SLMS, and CLCS curves show a sharp upward trend, while the CS curve remains relatively flat. The final mercury injection for CS is generally smaller than for other lithofacies. In the experiment, mercury saturation at the maximum pressure indicates the point at which mercury can no longer penetrate pores.

During pressure release, mercury begins to escape, with the remaining mercury saturation representing the amount trapped in the sample. Mercury removal efficiency is calculated using the following formula:

$$w = \left(1 - \frac{S_r}{S_{max}}\right) \times 100\%$$

where w represents the mercury removal efficiency (unit: %). S_{max} is the mercury saturation at the maximum mercury injection rate (unit: %), and S_r is the residual mercury saturation (unit: %). Based on the mercury intrusion–mercury removal curves, the mercury removal efficiency is determined, which provides insights into pore connectivity. CLCS has the highest mercury removal efficiency, at 25.51%, suggesting it has the best pore connectivity among the four lithofacies. SLMS follows, with a mercury removal efficiency of 22.78%. SLCS and CS have the lowest mercury removal efficiencies, at 21.47% and 21.48%, respectively, with no significant difference between them.

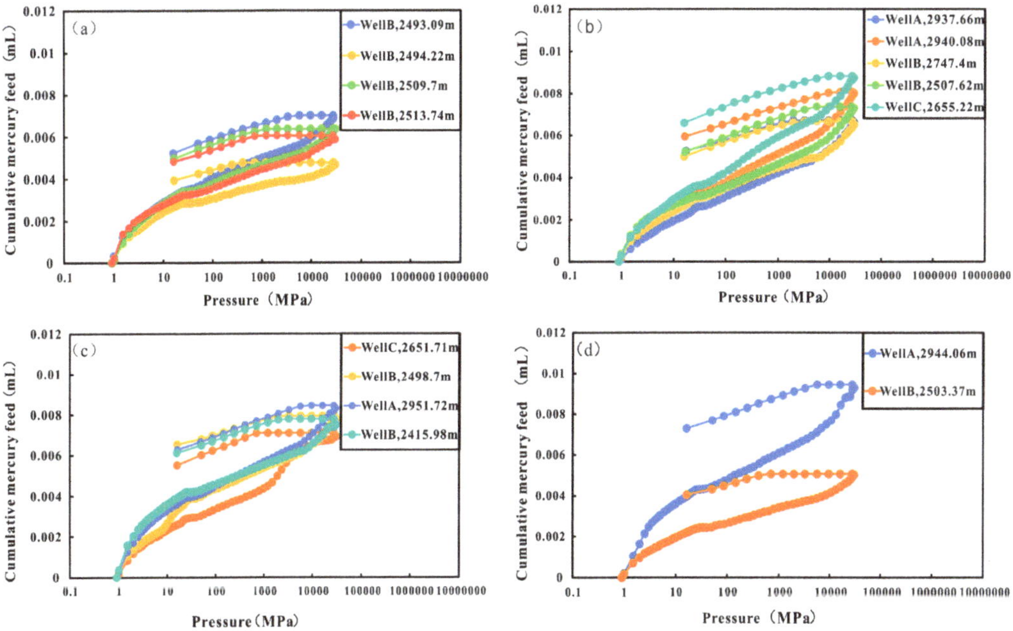

Figure 5. MIP curves of different lithofacies. (**a**) CS lithofacies. (**b**) CLCS lithofacies. (**c**) SLMS lithofacies. (**d**) SLCS lithofacies.

4.3.2. LTNA Characterization

LTNA is mainly used for analyzing micropores and mesopores. The LTNA curve shape reflects the partial pore size distribution within shale samples (Figure 6a). The curves for different lithofacies all exhibit an inverse "S" shape. The adsorption curve shows a slow upward trend in the low-pressure area, with a clear inflection point between relative pressures of 0.4–0.6, followed by a rapid increase in the high-pressure area. Due to variations in pore structure and morphology, the slope of the adsorption curve increases, and the saturation adsorption plateau is not well defined at high relative pressure (Figure 6b). This indicates that the desorption and adsorption curves do not overlap, resulting in a significant hysteresis loop (Figure 6c). According to the Kelvin equation, a hysteresis loop forms when capillary condensation occurs and the relative pressure (P/P_0) exceeds 0.5.

Based on the IUPAC classification method, the desorption curves for all four lithofacies display distinct inflection points within the relative pressure range of 0.4 to 0.6. These characteristics correspond to the H3 hysteresis loop (Figure 6d), suggesting that the pores in all four lithofacies are predominantly narrow slit-shaped and wedge-shaped pores.

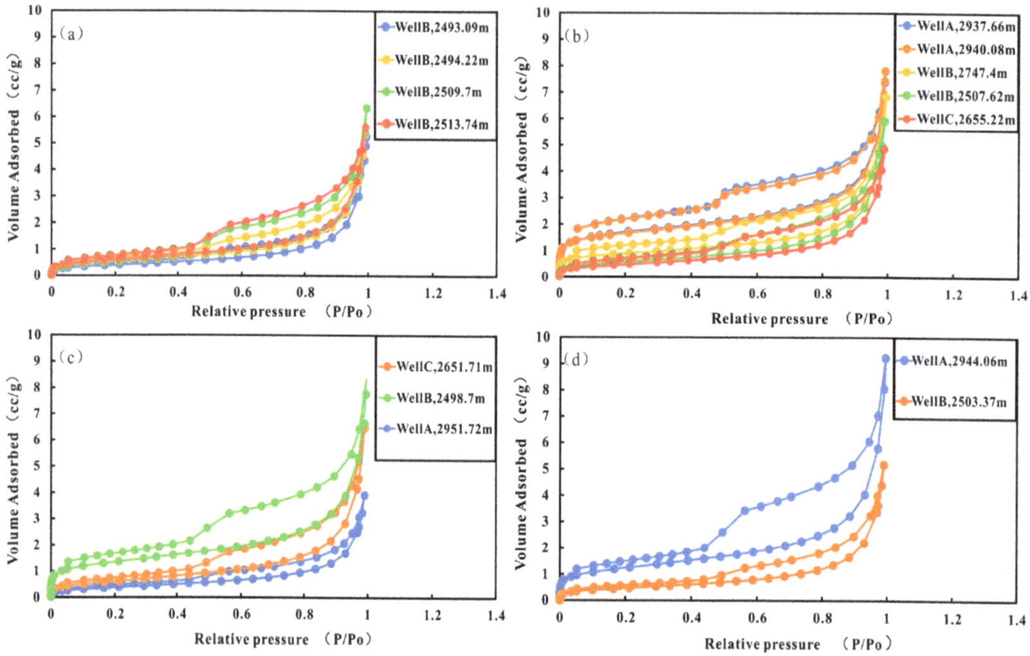

Figure 6. LTNA curves of different lithofacies. (**a**) CS. (**b**) CLCS. (**c**) SLMS. (**d**) SLCS.

4.3.3. NMR Characterization

In this study, four types of T_2 curves were identified, each directly reflecting different pore size distributions [58]. The T_2 spectrum curve of CS exhibits a three-peak shape (Figure 7a), The first peak, on the far left, represents the pore size distribution of micropores, corresponding to relaxation times between 0.03 and 0.1 ms. The second peak, with relaxation times between 0.3 and 5 ms, represents micropores and mesopores. The third peak, with relaxation times ranging from 15 to 60 ms, corresponds to the pore size distribution of macropores. Among the three peaks, the second peak is the most prominent, indicating a higher abundance of micropores and mesopores in CS. The T_2 curve of CLCS also displays a three-peak shape (Figure 7b). Compared to CS, CLCS shows better micropore development but less macropore development. The T_2 curve of SLMS exhibits a bimodal shape (Figure 7c). The first peak, with relaxation times ranging from 0.03 to 8 ms, reflects the pore sizes of micropores to mesopores. The second peak, with relaxation times between 50 and 150 ms, represents the pore size distribution of macropores. In comparing CS and CLCS, SLMS shows reduced development of micropores, mesopores, and macropores. The T_2 curve of SLCS also exhibits a bimodal shape (Figure 7d). However, compared to SLMS, the development of micropores, mesopores, and macropores is further diminished. Compared to the "three-peak" pore size distribution, the bimodal micropore structure shows significantly reduced complexity.

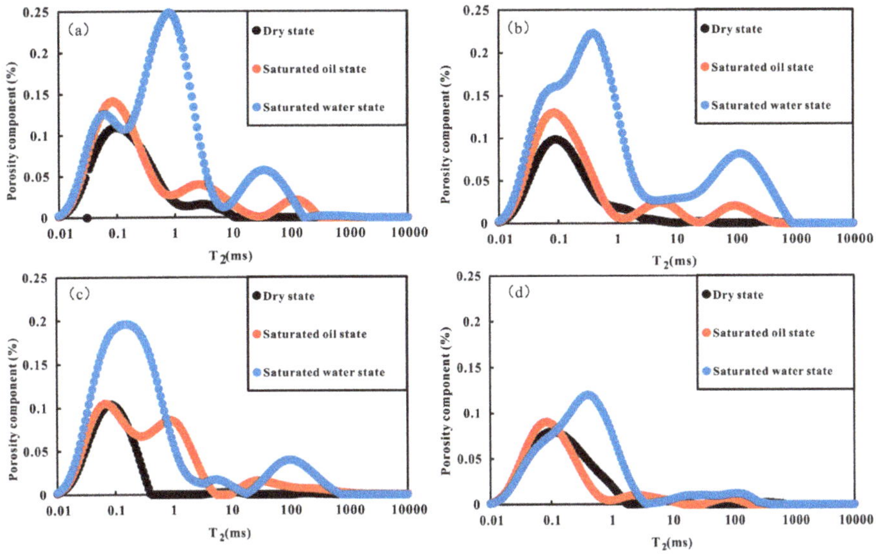

Figure 7. Distribution curves of NMR T_2 spectra for different lithofacies. (**a**) CS lithofacies. (**b**) CLCS lithofacies. (**c**) SLMS lithofacies. (**d**) SLCS lithofacies.

4.3.4. Full Pore Size Characterization of Shale

LTNA is advantageous for characterizing pore size distributions larger than 1 nm, while MIP is better suited to analyzing mesoporous and macroporous structures. According to previous research, pores smaller than 2 nm are referred to as micropores, of 2–50 nm as mesopores, and those larger than 50 nm as macropores [50]. Therefore, in this study, LTNA results were used to characterize micropores larger than 1 nm, and MIP data were employed for larger pore distributions that LTNA cannot detect. The combination of these two methods enabled comprehensive full pore size characterization of shale (Figure 8a–d). As shown in Table 2, the statistical results indicate that the pore volume of the four lithofacies is predominantly composed of mesopores and macropores, with similar contributions from both.

Table 2. Pore structure parameters of each lithofacies.

Sample	Lithofacies	Total MIP Intrusion Volume (mL/g) ($\delta = \pm 1\%$)	Nitrogen Adsorption Quantity (cm^3/g STP)	Micropore (mm^3/g)	Mesopore (mm^3/g)	Macropore (mm^3/g)	Total Pore Volume (mm^3/g)
WellC, 2651.71 m	SLMS	0.0024	9.15	0.86	5.27	5.82	11.95
WellB, 2498.7 m	SLMS	0.0069	6.47	0.06	1.67	4.6	6.33
WellA, 2951.72 m	SLMS	0.0077	8.35	0.15	2.63	6.22	9
WellB, 2515.98 m	SLMS	0.0051	4.3	0.22	6.16	5.93	12.31
WellA, 2937.66 m	CLCS	0.0087	9.25	1.29	4.12	6.49	11.9
WellA, 2940.08 m	CLCS	0.0066	7.83	1.3	4.21	4.65	10.16
WellB, 2947.4 m	CLCS	0.0046	5.36	0.52	4.11	5.66	10.29
WellB, 2507.62 m	CLCS	0.0065	5.92	0.21	5.31	4.83	10.35
WellC, 2655.22 m	CLCS	0.0073	4.86	0.16	4.49	5.14	9.79
WellB, 2493.09 m	CS	0.007	5.26	0.09	3.52	5.35	8.96
WellB, 2494.22 m	CS	0.011	5.9	0.42	5.25	8.27	13.94
WellB, 2509.7 m	CS	0.0056	5.38	0.19	4.82	3.87	8.88
WellB, 2513.74 m	CS	0.0063	6.35	0.13	4.52	4.98	9.63
WellB, 2503.37 m	SLCS	0.005	5.18	0.67	5.94	6.68	13.29
WellA, 2944.06 m	SLCS	0.0092	9.25	0.12	4.08	3.71	7.91

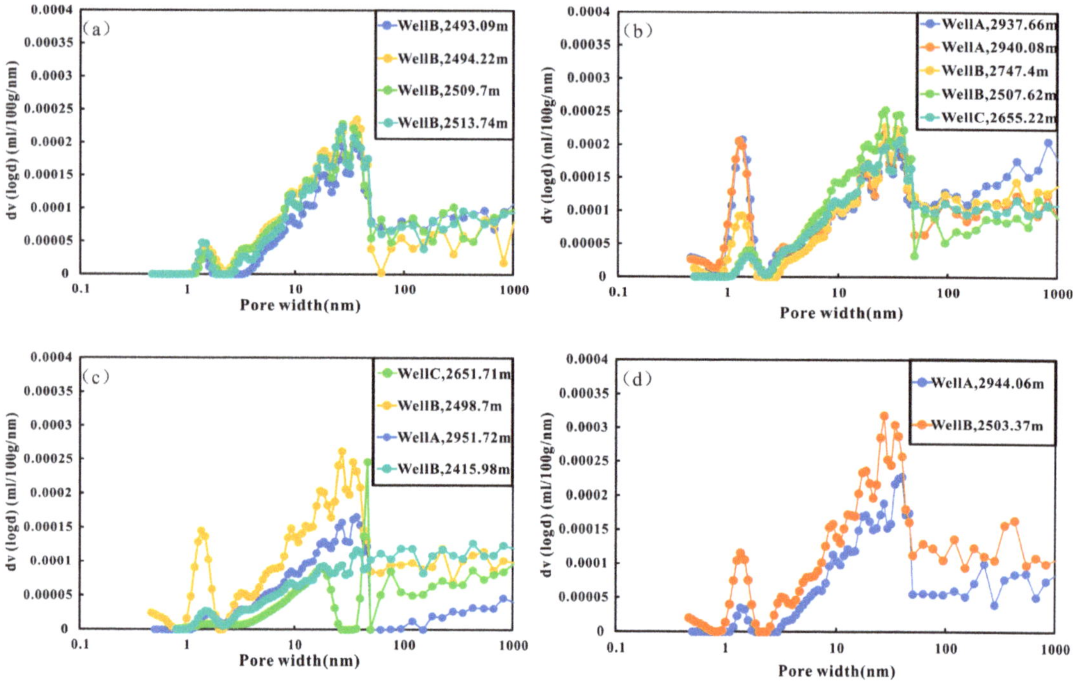

Figure 8. Pore size distribution curves of different lithofacies. (**a**) CS lithofacies. (**b**) CLCS lithofacies. (**c**) SLMS lithofacies. (**d**) SLCS lithofacies.

4.4. Fractal Dimension Based on NMR Results

Wang et al. analyzed the pore structure of shale nanopores using fractal theory and multifractal theory and calculated the fractal dimension and multifractal parameter capillary pressure under different fractal models [59]. The results showed that shale nanopores have multifractal characteristics, and multifractal parameters can reflect asymmetry in pore size, concentration, and pore size distribution. Shales with similar multifractal parameters have similar pore size distributions (Δa, Δf, a_0, a_1, a_2). The information dimension D_1 and correlation dimension D_2 are positively correlated with the size of shale nanopores. As the information dimension D_1 decreases, the contribution of pore size distribution becomes greater, and there is a strong negative correlation between the information dimension D_1 and the fractal dimension of the three-dimensional capillary model. Based on the T_2 spectral distribution and the multifractal model, the NMR multifractal characteristics under saturated water conditions were analyzed. The multifractal parameters for each lithofacies are presented in Table 3. The default setting for the multifractal moment order is a continuous interval between −10 and 10, assuming the statistical interval order is q = 1. Figure 9 shows the generalized fractal dimension spectra (D_q~q) for different lithofacies, displaying an overall inverse S-shaped curve. The rate at which D_q decreases with order (q) reflects the heterogeneity of the system. A faster decrease indicates stronger heterogeneity. Among the four lithofacies, CLCS exhibits the highest multifractal dimension difference (ΔD), signifying the strongest heterogeneity. As q increases, D_q demonstrates two distinct trends:

When $q > 0$, the multifractal parameters decrease significantly, highlighting the pore structure in high-probability regions;

When $q < 0$, the multifractal parameters decrease more gradually, reflecting the pore structure in low-probability regions.

Table 3. Multifractal characteristic parameters of each lithofacies.

Sample	Lithofacies	D_{min}	D_{max}	D_0	D_1	D_2	D_0-D_{max}	D_{min}-D_0	ΔD	a_{max}	a_{min}	$a0$	Δa
WellB, 2503.37 m	SLCS	1.78	0.82	1	0.91	0.86	0.18	0.78	0.96	2.1	0.78	1	1.3
WellA, 2944.06 m	SLCS	1.73	0.84	1	0.92	0.87	0.16	0.73	0.89	1.89	0.75	1	1.14
WellC, 2651.71 m	SLMS	1.36	0.8	1	0.89	0.84	0.2	0.36	0.56	1.64	0.89	1	0.75
WellB, 2498.7 m	SLMS	1.62	0.81	1	0.91	0.86	0.19	0.62	0.81	1.83	0.77	1	1.06
WellA, 2951.72 m	SLMS	2.05	0.8	1	0.9	0.85	0.2	1.05	1.25	1.88	0.78	1	1.1
WellB, 2515.98 m	SLMS	1.33	0.75	1	0.85	0.79	0.25	0.33	0.58	1.43	0.78	1	0.65
WellA, 2937.66 m	CLCS	1.87	0.85	1	0.91	0.88	0.15	0.87	1.02	2.15	0.77	1	1.38
WellA, 2940.08 m	CLCS	1.74	0.78	1	0.93	0.84	0.22	0.74	0.96	1.91	0.75	1	1.16
WellB, 2947.4 m	CLCS	1.8	0.83	1	0.91	0.85	0.17	0.8	0.97	1.8	0.77	1	1.03
WellB, 2507.62 m	CLCS	1.7	0.81	1	0.92	0.85	0.19	0.7	0.89	2.1	0.78	1	1.32
WellC, 2655.22 m	CLCS	1.87	0.86	1	0.95	0.89	0.14	0.87	1.01	2.06	0.78	1	1.28
WellB, 2493.09 m	CS	1.65	0.83	1	0.93	0.87	0.17	0.65	0.82	2.34	0.78	1	1.56
WellB, 2494.22 m	CS	2	0.93	1	0.96	0.95	0.07	1	1.07	2.36	0.75	1	1.41
WellB, 2509.7 m	CS	1.56	0.83	1	0.9	0.85	0.17	0.56	0.73	1.89	0.79	1	1.1
WellB, 2513.74 m	CS	1.63	0.82	1	0.87	0.84	0.18	0.63	0.81	1.49	0.76	1	0.73

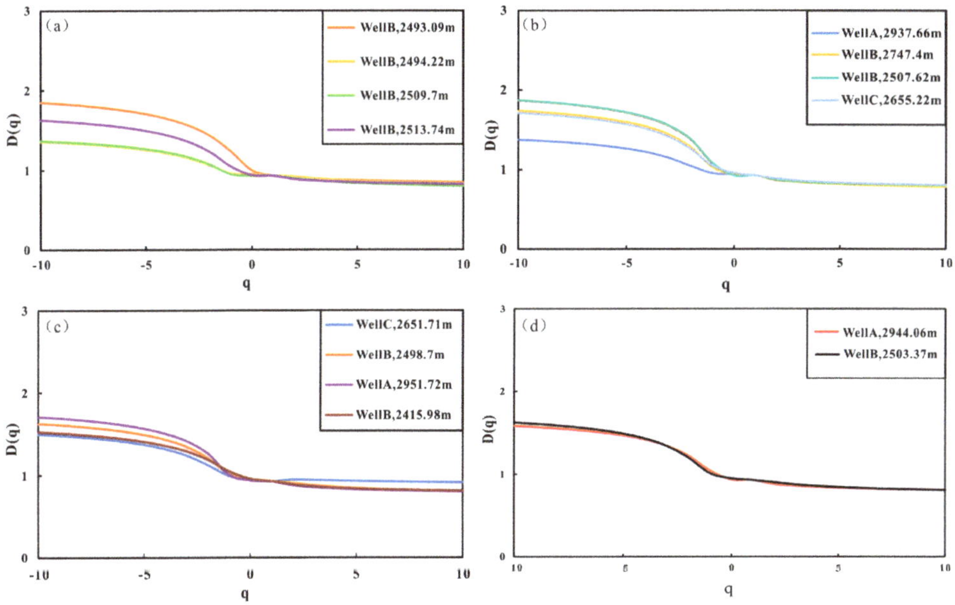

Figure 9. Multifractal spectra of different lithofacies. (**a**) CS lithofacies. (**b**) CLCS lithofacies. (**c**) SLMS lithofacies. (**d**) SLCS lithofacies.

Using q = 0 as the boundary, (D_{min}-D_0) represents the heterogeneity of macropores. (D_0-D_{min}) indicates the heterogeneity of micropores and mesopores.

For multifractal spectra, the Δa value serves as an indicator of heterogeneity; larger Δa values signify more complex pore distributions and greater heterogeneity. Figure 10 shows the fractal dimension spectra for different lithofacies, with the peak values (a_0, f(a_0)) serving as dividing points; (a_0-a_{min}) is more suitable for evaluating the pore structure in low-probability regions, such as macropores. Conversely, (a_{max}-a_0) better describes the pore structure in high-probability regions, such as micropores. The fractal spectrum distribution of f(a) vs. a in Figure 10 is convex (as expected), indicating the multifractal nature of porosity, and it is generally asymmetric. On the left side of (a_0, f(a_0)), the fractal spectrum distribution f(a) increases sharply with the increase in a. On the right side, f(a)

gradually decreases as a increases. Among the four rock phases, CLCS exhibits the largest decrease in the generalized fractal dimension (Δa), indicating a highly complex pore size distribution and significant heterogeneity within the pore space.

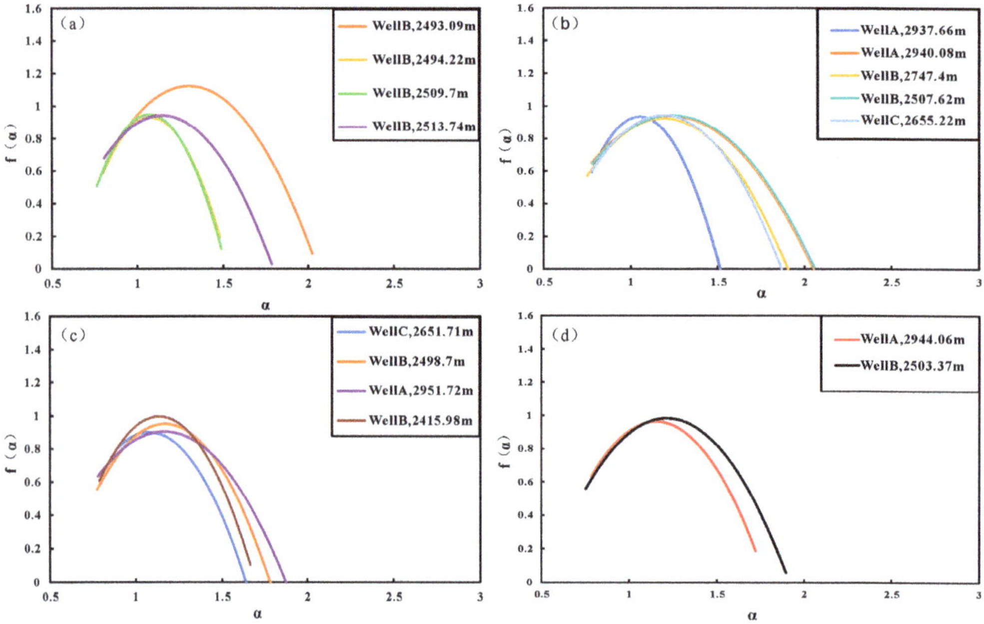

Figure 10. Generalized fractal dimension spectra of different lithofacies. (**a**) CS lithofacies. (**b**) CLCS lithofacies. (**c**) SLMS lithofacies. (**d**) SLCS lithofacies.

The f(a) distribution in the CS and CLCS lithofacies is right-skewed, suggesting that large-scale, high-probability features (e.g., macropores) contribute more significantly to the system's heterogeneity. In contrast, the SLMS and SLCS lithofacies exhibit a more symmetrical f(a), indicating no predominant micro- or macroporosity but rather a more uniform distribution between them (Figure 8). The width of the f(a) spectrum reflects the range of singularity strengths (a), which are associated with the local scaling behaviors of the pore distribution. A narrow f(a) implies uniform porosity with low heterogeneity, as observed in the CS-Well B (2509.7 m) and CLCS-Well A (2937.66 m) samples, while a wide f(a) signifies a highly heterogeneous system with diverse pore sizes and distributions, as seen in the CS-Well B (2493.09 m), CLCS-Well B (2507.62 m), and Well A (2940.08 m) samples.

5. Discussion

5.1. Controlling Factors of Heterogeneity in Pore Structure

Multifractal theory reveals an intrinsic relationship between the heterogeneity of pore structure and the varying effects of different components and organic matter content in rocks on pore architecture. Li et al. (2024) believe that the more brittle minerals and the less clay minerals in the mineral composition, the more complex the pore structure of shale and the stronger its heterogeneity [60]. Wang et al. (2022) [22] found that the heterogeneity parameter ΔD is negatively correlated with quartz content, indicating that quartz enrichment reduces heterogeneity. When simultaneously positively correlated with clay minerals, the enrichment of clay minerals enhances heterogeneity [22]. The results show that the heterogeneity parameters of CS are significantly negatively correlated with quartz content (R^2 = 0.93) (Figure 11a). This suggests that high quartz content promotes the preservation of pores, resulting in SLCS having a well-developed pore structure, lower heterogeneity, and consequently, a smaller fractal dimension. In contrast, the heterogeneity parameters of CLCS and SLMS exhibit a positive correlation with clay content (R^2 = 0.57 and 0.74) (Figure 11b,c). Clay minerals are susceptible to alteration during diagenesis, and compaction often destroys the pore spaces within and between clay minerals. These processes lead to a more complex pore structure, increased heterogeneity, and higher fractal dimensions. The heterogeneity parameters of SLCS are significantly positively correlated with TOC (R^2 = 0.54) (Figure 11d). Previous studies [60] suggest that higher TOC content leads to more developed internal micro- and mesopores, resulting in a more uniform pore size distribution and reduced heterogeneity. However, this study argues that increased organic matter abundance increases shale permeability to some extent. However, the pore size distribution of micropores and mesopores developed within organic matter is more complex, contributing to increased heterogeneity, as reflected by higher fractal dimensions. In the future, the principal component analysis (PAC) method can be used to further analyze the main controlling factors of heterogeneity in shale reservoir pore structures.

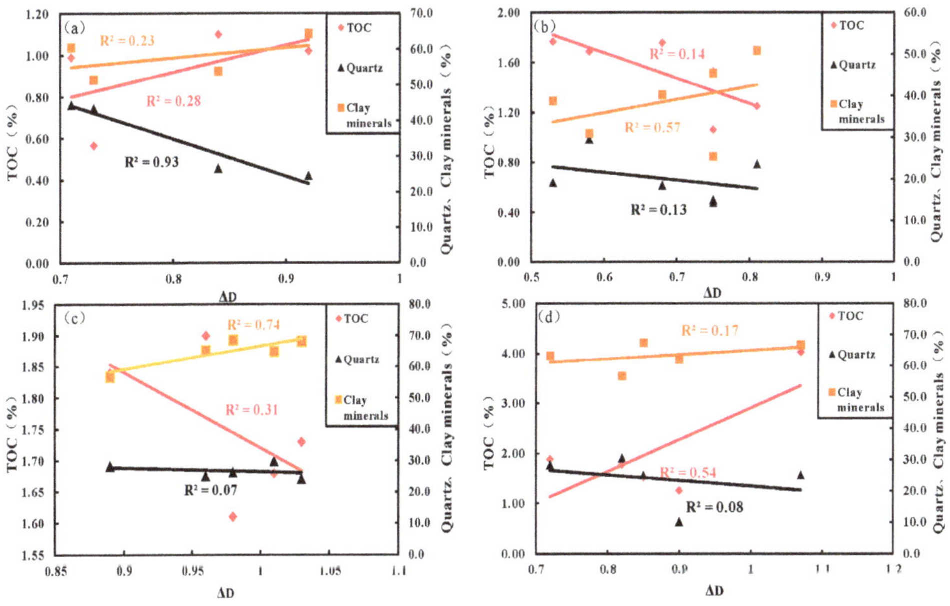

Figure 11. Correlation diagram between multifractal parameters of different lithofacies and mineral compositions (TOC): (**a**) CS. (**b**) CLCS. (**c**) SLMS. (**d**) SLCS.

5.2. *Multifractal Parameters Reflecting Heterogeneity in Porosity and Permeability*

Physical parameters comprehensively reflect the shale pore system, with pore gapping and permeability being the most critical factors influencing storage capacity and flow. These properties determine reservoir quality [61]. Establishing relationships between physical properties and multifractal parameters helps in understanding the heterogenous nature of shale reservoirs. Previous studies have found that the low-probability parameter (D_{min}-D_0) is positively correlated with porosity, while the high-probability parameter (D_0-D_{max}) is negatively correlated with permeability, thus characterizing the heterogeneity of porosity and permeability [22]. The correlations between the porosity of CS, CLCS, SLCS, and (D_{min}-D_0) are significant (R^2 = 0.92, 0.55, and 0.99) (Figure 12a,b,d). (D_{min}-D_0) is effective in characterizing the heterogeneity of porosity in these three lithofacies (Figure 12c). A larger (D_{min}-D_0) value indicates porosity heterogeneity. In contrast, (D_0-D_{max}) is more suitable for characterizing the heterogeneity of porosity in SLMS, where a high value represents strong heterogeneity and a low value indicates the opposite. The permeability of the four lithofacies is most correlated with (D_0-D_{max}) and ΔD (R^2 = 0.65, 0.60, 0.55 and 0.74). The permeability of CS, SLMS, and SLCS is positively correlated with both (D_0-D_{max}) (R^2 = 0.62, 0.34, and 0.88) (Figure 13a,c,d), while higher values indicate stronger heterogeneity in permeability. Conversely, the permeability of CLCS is negatively correlated with (D_0-D_{max}) (Figure 13b), with lower values indicating stronger heterogeneity in permeability.

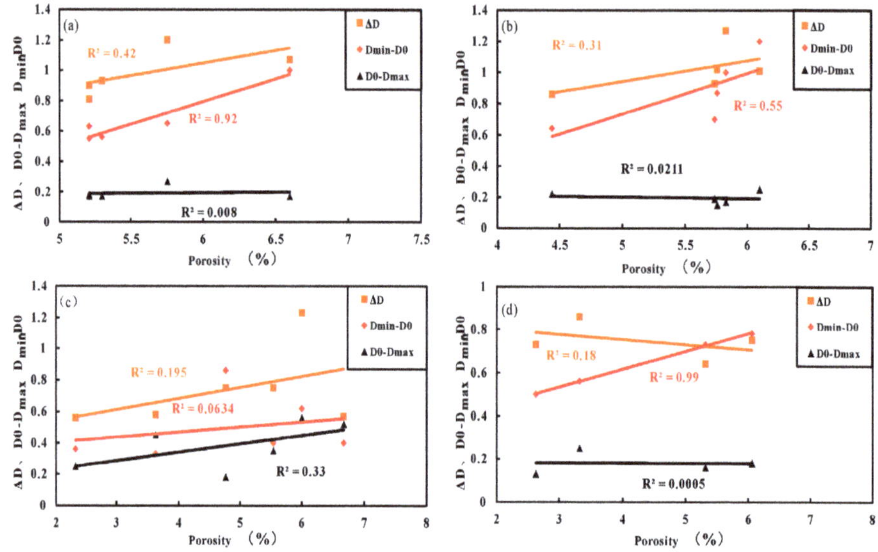

Figure 12. Correlation between multifractal parameters and porosity in different lithofacies. (**a**) CS lithofacies. (**b**) CLCS lithofacies. (**c**) SLMS lithofacies. (**d**) SLCS lithofacies.

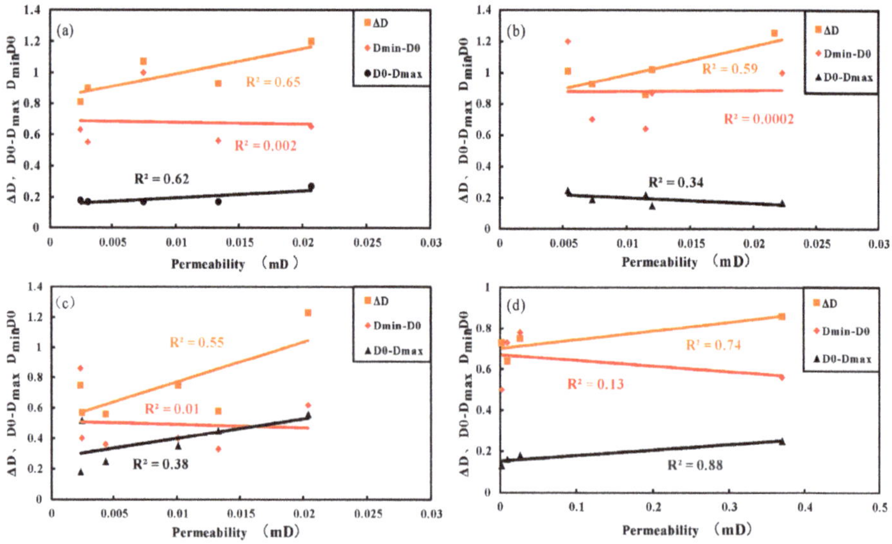

Figure 13. Correlation between multifractal parameters and permeability in different lithofacies. (**a**) CS lithofacies. (**b**) CLCS lithofacies. (**c**) SLMS lithofacies. (**d**) SLCS lithofacies.

6. Conclusions

Based on mineral content and laminae types, Dongyuemiao lacustrine shale lithofacies can be classified into four types: shell–laminae mixed shale (SLMS), silty–laminae clay shale (SLCS), clast–laminae clay shale (CLCS), and clay shale (CS).

The micro-reservoir and permeability spaces in lacustrine shale consist of inorganic pores, organic pores, and micro-fractures. Inorganic pores include inter-particle and intra-particle pores. Full pore size distribution analysis reveals two main peaks across all lithofacies, with pore sizes concentrated in the 2–10 nm and 50–80 nm ranges. Mesopores and macropores dominate total pore volume in lacustrine shale.

Heterogeneity in CS is primarily controlled by quartz content; higher quartz reduces heterogeneity. In SLMS and CLCS, heterogeneity is controlled by clay mineral content, with greater clay content leading to stronger heterogeneity. In SLCS, heterogeneity is mainly controlled by TOC content, where higher TOC increases heterogeneity. In SLMS, porosity heterogeneity is best characterized by (D_0-D_{max}), while in the other three lithofacies, it is defined by (D_{min}-D_0). Permeability in all lithofacies correlates most strongly with (D_0-D_{max}) and ΔD. In CS, SLMS, and SLCS, permeability is positively correlated with (D_0-D_{max}), meaning higher values indicate stronger heterogeneity in permeability. In CLCS, however, permeability is negatively correlated with (D_0-D_{max}), where lower values indicate stronger heterogeneity in permeability.

Author Contributions: Conceptualization, X.W. and Y.G.; methodology, X.W.; validation, Z.W.; formal analysis, Y.G.; investigation, Y.F. and Y.J.; data curation, Y.G.; writing—original draft preparation, X.W. and Y.G.; writing—review and editing, Y.J. and Y.F.; visualization, Z.W.; supervision, Y.J.; project administration, Z.W. All authors have read and agreed to the published version of the manuscript.

Funding: This research was supported by the National Natural Science Foundation of China (No. 42272171) and the National Natural Science Foundation of China (No. 42302166).

Institutional Review Board Statement: Not applicable.

Informed Consent Statement: Not applicable.

Data Availability Statement: Data are contained within the article.

Conflicts of Interest: The authors declare that they have no known competing financial interests or personal relationships that might have influenced the work presented in this article.

References

1. Zou, C.; Dong, D.; Wang, S.; Li, J.; Li, X.; Wang, Y.; Li, D.; Cheng, K. Geological characteristics, formation mechanism and resource potential of shale gas in China. *Pet. Explor. Dev.* **2010**, *37*, 641–653. [CrossRef]
2. Jiang, S.; Tang, X.; Cai, D.; Xue, G.; He, Z.; Long, S.; Peng, Y.; Gao, B.; Xu, Z.; Dahdah, N. Comparison of marine, transitional, and lacustrine shales: A case study from the Sichuan Basin in China. *J. Pet. Sci. Eng.* **2017**, *150*, 334–347. [CrossRef]
3. Xu, Q.; Liu, B.; Ma, Y.; Song, X.; Wang, Y.; Chen, Z. Geological and geochemical characterization of lacustrine shale: A case study of the Jurassic Da'anzhai member shale in the central Sichuan Basin, southwest China. *J. Nat. Gas Sci. Eng.* **2017**, *47*, 124–139. [CrossRef]
4. Yang, C.; Zhang, J.; Tang, X.; Ding, J.; Zhao, Q.; Dang, W.; Chen, H.; Su, Y.; Li, B.; Lu, D. Comparative study on micro-pore structure of marine, terrestrial, and transitional shales in key areas, China. *Int. J. Coal Geol.* **2017**, *171*, 76–92. [CrossRef]
5. Fu, Y.; Jiang, Y.; Dong, D.; Hu, Q.; Lei, Z.; Peng, H.; Gu, Y.; Ma, S.; Wang, Z.; Yin, X.; et al. Microscopic pore-fracture configuration and gas-filled mechanism of shale reservoirs in the western Chongqing area, Sichuan Basin, China. *Pet. Explor. Dev.* **2021**, *48*, 916–927. [CrossRef]
6. Fu, Y.; Jiang, Y.; Hu, Q.; Luo, T.; Li, Y.; Zhian, L.; Wang, Z.; Yin, X. Fracturing flowback fluids from shale gas wells in western chongqing: Geochemical analyses and relevance for exploration & development. *J. Nat. Gas Sci. Eng.* **2021**, *88*, 103821.
7. He, J.; Deng, H.; Ma, R.; Wang, R.; Wang, Y.; Li, A. Reservoir Characteristics of the Lower Jurassic Lacustrine Shale in the Eastern Sichuan Basin and Its Effect on Gas Properties: An Integrated Approach. *Energies* **2020**, *13*, 4495. [CrossRef]

8. Shu, Y.; Bao, H.; Zheng, Y.; Chen, M.; Lu, Y.; Liu, H.; Peng, W.; Zhou, L.; Ma, Y.; Wen, Y.; et al. Lithofacies Types, Assemblage Characteristics, and Sedimentary Evolution Model of Lacustrine Shale in Dongyuemiao Formation of Fuxing Area. *Front. Earth Sci.* **2021**, *9*, 772581. [CrossRef]
9. Li, Q.; Wu, S.; Xia, D.; You, X.; Zhang, H.; Lu, H. Major and trace element geochemistry of the lacustrine organic-rich shales from the Upper Triassic Chang 7 Member in the southwestern Ordos Basin, China: Implications for paleoenvironment and organic matter accumulation. *Mar. Pet. Geol.* **2020**, *111*, 852–867. [CrossRef]
10. Gu, Y.; Li, X.; Qi, L.; Li, S.; Jiang, Y.; Fu, Y.; Yang, X. Sedimentology and Geochemistry of the Lower Permian Shanxi Formation Shan 23 Submember Transitional Shale, Eastern Ordos Basin, North China. *Front. Earth Sci.* **2022**, *10*, 859845. [CrossRef]
11. Gu, Y.; Cai, G.; Hu, D.; Wei, Z.; Liu, R.; Han, J.; Fan, Z.; Hao, J.; Jiang, Y. Geochemical and Geological Characterization of Upper Permian Linghao Formation Shale in Nanpanjiang Basin, SW China. *Front. Earth Sci.* **2022**, *10*, 883146. [CrossRef]
12. Qiu, Z.; He, J. Depositional environment changes and organic matter accumulation of Pliensbachian-Toarcian lacustrine shales in the Sichuan basin, SW China. *J. Asian Earth Sci.* **2022**, *232*, 105035. [CrossRef]
13. Li, P.; Liu, Z.; Nie, H.; Liang, X.; Li, Q.; Wang, P. Heterogeneity Characteristics of Lacustrine Shale Oil Reservoir Under the Control of Lithofacies: A Case Study of the Dongyuemiao Member of Jurassic Ziliujing Formation, Sichuan Basin. *Front. Earth Sci.* **2021**, *9*, 736544. [CrossRef]
14. Qiu, Z.; Shi, Z.; Dong, D.; Lu, B.; Zhang, C.; Zhou, J.; Wang, H.; Xiong, B.; Pang, Z.; Guo, H. Geological characteristics of source rock and reservoir of tight oil and its accumulation mechanism: A case study of Permian Lucaogou Formation in Jimusar sag, Junggar Basin. *Pet. Explor. Dev.* **2016**, *43*, 928–939. [CrossRef]
15. Wei, G.; Wang, W.; Feng, L.; Tan, X.; Yu, C.; Zhang, H.; Zhang, Z.; Wang, S. Geological Characteristics and Exploration Prospect of Black Shale in the Dongyuemiao Member of Lower Jurassic, the Eastern Sichuan Basin, China. *Front. Earth Sci.* **2021**, *9*, 765568. [CrossRef]
16. Chen, D.; Zhang, J.; Wang, X.; Lan, B.; Li, Z.; Liu, T. Characteristics of Lacustrine Shale Reservoir and Its Effect on Methane Adsorption Capacity in Fuxin Basin. *Energy Fuels* **2018**, *32*, 11105–11117. [CrossRef]
17. Loucks, R.G.; Reed, R.M.; Ruppel, S.C.; Hammes, U. Spectrum of pore types and networks in mudrocks and a descriptive classification for matrix-related mudrock pores. *AAPG Bull.* **2012**, *96*, 1071–1098. [CrossRef]
18. Ross, D.J.K.; Marc Bustin, R. The importance of shale composition and pore structure upon gas storage potential of shale gas reservoirs. *Mar. Pet. Geol.* **2009**, *26*, 916–927. [CrossRef]
19. Fu, H.; Wang, X.; Zhang, L.; Gao, R.; Li, Z.; Xu, T.; Zhu, X.; Xu, W.; Li, Q. Investigation of the factors that control the development of pore structure in lacustrine shale: A case study of block X in the Ordos Basin, China. *J. Nat. Gas Sci. Eng.* **2015**, *26*, 1422–1432. [CrossRef]
20. Jiang, F.; Chen, D.; Chen, J.; Li, Q.; Liu, Y.; Shao, X.; Hu, T.; Dai, J. Fractal Analysis of Shale Pore Structure of Continental Gas Shale Reservoir in the Ordos Basin, NW China. *Energy Fuels* **2016**, *30*, 4676–4689. [CrossRef]
21. Jiang, Y.; Fu, Y.; Lei, Z.; Gu, Y.; Qi, L.; Cao, Z. Experimental NMR analysis oil and water imbibition during fracturing in Longmaxi Shale, SE Sichuan Basin. *J. Jpn. Pet. Inst.* **2019**, *62*, 1–10. [CrossRef]
22. Wang, Z.; Jiang, Y.; Fu, Y.; Lei, Z.; Xu, C.; Yuan, J.; Wen, R.; Wang, Z.; Gu, Y.; Yin, X. Characterization of pore structure and heterogeneity of shale reservoir from Wufeng Formation-Sublayers Long—11 in Western Chongqing Based on Nuclear Magnetic Resonance. *Earth Sci.* **2022**, *47*, 490–504.
23. Wang, Y.; Zhu, Y.; Wang, H.; Feng, G. Nanoscale pore morphology and distribution of lacustrine shale reservoirs: Examples from the Upper Triassic Yanchang Formation, Ordos Basin. *J. Energy Chem.* **2015**, *24*, 512–519. [CrossRef]
24. Chalmers, G.R.; Marc Bustin, R.; Power, M. Characterization of gas shale pore systems by porosimetry, pycnometry, surface area, and field emission scanning electron microscopy transmission electron microscopy image analyses: Examples from the Barnett, Woodford, Haynesville, Marcellus, and Doig units. *AAPG Bull.* **2012**, *96*, 1099–1119.
25. Fu, Y.; Jiang, Y.; Wang, Z.; Hu, Q.; Xie, J.; Ni, G.; Lei, Z.; Zhou, K.; Liu, X. Non-connected pores of the Longmaxi shale in southern Sichuan Basin of China. *Mar. Pet. Geol.* **2019**, *110*, 420–433. [CrossRef]
26. Wang, G.; Chen, X.; Cheng, W. Multi-scale characterization of coal pore and fractures and its influence on permeability— taking 14 large coal bases in China as examples. *J. Chongqing Univ.* **2024**, *47*, 4.
27. Han, W.; Zhou, G.; Gao, D.; Zhang, Z.; Wei, Z.; Wang, H.; Yang, H. Experimental analysis of the pore structure and fractal characteristics of different metamorphic coal based on mercury intrusion-nitrogen adsorption porosimetry. *Powder Technol.* **2020**, *362*, 386–398. [CrossRef]
28. Cai, Y.; Li, Q.; Liu, D.; Zhou, Y.; Lv, D. Insights into matrix compressibility of coals by mercury intrusion porosimetry and N_2 adsorption. *Int. J. Coal Geol.* **2018**, *200*, 199–212. [CrossRef]
29. Longinos, S.N.; Hazlett, R. Cryogenic fracturing using liquid nitrogen on granite at elevated temperatures: A case study for enhanced geothermal systems in Kazakhstan. *Sci. Rep.* **2024**, *14*, 160. [CrossRef]

30. Zhao, X.; Yang, Z.; Lin, W.; Xiong, S.; Luo, Y.; Wang, Z.; Chen, T.; Xia, D.; Wu, Z. Study on pore structures of tight sandstone reservoirs based on nitrogen adsorption, high-pressure mercury intrusion, and rate-controlled mercury intrusion. *J. Energy Resour. Technol.* **2019**, *141*, 112903. [CrossRef]
31. Shu, Z.; Shu, Y.; Chen, M. Lithofacies heterogeneity and reservoir pore development characteristics of continental shale: A case study of the Dongyuemiao shale of the Ziliujing Formation in the Sichuan Basin. *Bull. Geol. Sci. Technol.* **2024**, *43*, 1–15.
32. Liu, Z.; Hu, Z.; Liu, G. Source reservoir features and favorable enrichment interval evaluation methods of high mature continental shale: A case study of the Jurassic Dongyuemiao Member in the Fuxing area, eastern Sichuan Basin. *Nat. Gas Ind.* **2022**, *42*, 11–24.
33. Liu, H.; Li, X.; Wan, Y. Formation conditions and exploration and development potential of continental shale gas: A case of Dongyuemiao Member of the Jurassic in North Fuling area, eastern Sichuan Basin. *Mar. Orig. Pet. Geol.* **2020**, *25*, 148–154.
34. Hu, D.; Wei, Z.; Wei, X. Breakthrough in the exploration of continental shale oil/gas of Jurassic Lianggaoshan Formation in the Fuxing area of the Sichuan Basin and its inspiration. *Nat. Gas Ind.* **2025**, *45*, 1–13.
35. Wang, P.; Shen, B.; Liu, Z. Characteristics and main controlling factors of organic pore development in continental shales of the Lianggaoshan Formation in the Fuxing area, Sichuan Basin. *Pet. Geol. Exp.* **2024**, *46*, 499–509.
36. Shu, Z.; Zuou, L.; Li, X. Geological characteristics of gas condensate reservoirs and their exploration and development prospect in the Jurassic continental shale of the Dongyuemiao Member of Ziliujing Formation, Fuxing area, eastern Sichuan Basin. *Oil Gas Geol.* **2021**, *42*, 212–223.
37. Li, J.; Zheng, B. A New method for fractal characterization of microscopic pores and its application in shale reservoirs. *Nat. Gas Geosci.* **2015**, *35*, 52–59.
38. Wang, M.; Xue, H.; Tian, S.; Wilkins, R.; Wang, Z. Fractal characteristics of Upper Cretaceous lacustrine shale from the Songliao Basin, NE China. *Mar. Pet. Geol.* **2015**, *67*, 144–153. [CrossRef]
39. Liu, K.; Ostadhassan, M.; Kong, L. Fractal and multifractal characteristics of pore throats in the Bakken Shale. *Transp. Porous Media* **2019**, *126*, 579–598. [CrossRef]
40. Zhao, P.; Wang, Z.; Sun, Z.; Cai, L.; Wang, L. Investigation on the pore structure and multifractal characteristics of tight oil reservoirs using NMR Measurements: Permian Lucaogou Formation in Jimusaer Sag, Junggar Basin. *Mar. Pet. Geol.* **2017**, *86*, 1067–1081. [CrossRef]
41. Zheng, S.; Yao, Y.; Liu, D.; Cai, Y.; Liu, Y.; Li, X. Nuclear magnetic resonance T_2 cutoffs of coals: A novel method by multifractal analysis theory. *Fuel* **2019**, *241*, 715–724. [CrossRef]
42. Peng, J.; Hu, Z.; Feng, D.; Wang, Q. Sedimentology and sequence stratigraphy of lacustrine deep-water fine-grained sedimentary rocks: The Lower Jurassic Dongyuemiao Formation in the Sichuan Basin, Western China. *Mar. Pet. Geol.* **2022**, *146*, 105933. [CrossRef]
43. Zhou, Y.; Jiang, C.; Hu, D.; Wei, Z.; Wei, X.; Wang, D.; Hao, J.; Jiang, Y.; Gu, Y. Controlling Factors of Organic-Rich Lacustrine Shale in the Jurassic Dongyuemiao Member of Sichuan Basin, SW China. *Geofluids* **2023**, *152*, 3380389. [CrossRef]
44. Liang, C.; Jiang, Z.; Cao, Y.; Wu, J.; Wang, Y.; Hao, F. Sedimentary characteristics and origin of lacustrine organic-rich shales in the salinized Eocene Dongying Depression. *Geol. Soc. Am.* **2018**, *130*, 154–174. [CrossRef]
45. Chen, S.; Zhang, S.; Wang, Y.; Tan, M. Lithofacies types and reservoirs of Paleogene fine-grained sedimentary rocks Dongying Sag, Bohai Bay Basin, China. *Pet. Explor. Dev.* **2016**, *43*, 218–229. [CrossRef]
46. Allix, P.; Burnhama, F.; Fowler, T.; Herron, M.; Symington, B. Coaxing oil from shale. *Oilfield Rev.* **2010**, *22*, 4–15.
47. Wang, C.; Cheng, C.; Carr, T. Marcellus shale lithofacies prediction by multiclass neural network classification in the Appalachian Basin. *Math. Geosci.* **2012**, *44*, 975–1004. [CrossRef]
48. Shi, J.; Jin, Z.; Liu, Q.; Huang, Z. Lithofacies classification and origin of the Eocene lacustrine fine-grained sedimentary rocks in the Jiyang Depression, Bohai Bay Basin, Eastern China. *J. Asian Earth Sci.* **2020**, *194*, 104002. [CrossRef]
49. Wang, S.; Javadpour, F.; Feng, Q. Confinement correction to mercury intrusion capillary pressure of shale nanopores. *Sci. Rep.* **2016**, *6*, 20160. [CrossRef]
50. De, B. The Structure and Properties of Porous Materials. In Proceedings of the Tenth Symposium of the Colston Research Society, Bristol, UK, 24–27 March 1958; Butterworths: London, UK, 1958; Volume 389, pp. 68–94.
51. Hu, Y.; Guo, Y.; Zhang, J. A method to determine nuclear magnetic resonance cutoff value of tight sandstone reservoir based on multifractal analysis. *Energy Sci. Eng.* **2020**, *8*, 1135–1148. [CrossRef]
52. Ge, X.; Fan, Y.; Li, J. Pore structure characterization and classification using multifractal theory: An application in Santanghu Basin of Western China. *J. Pet. Sci. Eng.* **2015**, *127*, 297–304. [CrossRef]
53. Chhabra, A.; Jensen, R. Direct determination of the f(α) singularity spectrum. *Phys. Rev. Lett.* **1989**, *62*, 1327–1330. [CrossRef]
54. Posadas, A.N.D.; Giménez, D.; Bittelli, M.; Vaz, C.M.P.; Flury, M. Multifractal Characterization of Particle-Size Distributions. *Soil Sci. Soc. Am. J.* **2001**, *65*, 1361. [CrossRef]
55. Lopes, R.; Betrouni, N. Fractal and multifractal analysis: A review. *Med. Image Anal.* **2009**, *13*, 634–649. [CrossRef] [PubMed]
56. Yuan, Y.; Rezaee, R. Fractal analysis of the pore structure for clay bound water and potential gas storage in shales based on NMR and N_2 gas adsorption. *J. Pet. Sci. Eng.* **2019**, *177*, 756–765. [CrossRef]

57. Wang, W.; Song, Y.; Huang, J. Study on fractal characteristics of pore throat structure of tight sandstone by high pressure mercury injection wxperiment. *Bullet Geol. Sci. Technol.* **2021**, *40*, 22–30.
58. He, Y.; Mao, Z.; Xiao, L. An improved method of using NMR T_2 distribution to ecaluate pore size distribution. *Chin. J. Geophys.* **2005**, *45*, 373–378.
59. Wang, F.; Zai, Y. Fractal and multifractal characteristics of shale nanopores. *Results Physics.* **2021**, *25*, 104277. [CrossRef]
60. Li, Y.; Liu, D.; Feng, X.; Zhao, F.; Chen, Y.; Wang, Y.; Liu, R.; Du, W.; Fan, Q.; Song, Y.; et al. Heterogeneity characteristics and its controlling factors of marine shale reservoirs from the Wufeng-Longmaxi in the Northern Guizhou area. *Geol. China* **2024**, *51*, 780–798.
61. Ramandi, H.; Mostaghimi, P.; Armstrong, R. Porosity and Permeability Characterization of Coal: A Micro-Computed Tomography Study. *Int. J. Coal Geol.* **2016**, *154–155*, 57–68. [CrossRef]

fractal and fractional

MDPI

Article

Pore Fractal and Structure Analysis of Pore-Filling Chlorite in Continental Shales: A Case Study from the Qingshankou Formation in the Gulong Sag

Yuan Kang [1,2], Rukai Zhu [1,3,*], Kouqi Liu [2], Jingya Zhang [1,3] and Chang Liu [1,3]

1 Research Institute of Petroleum Exploration and Development (RIPED), China National Petroleum Corporation (CNPC), Beijing 100083, China; kangyuan@stu.pku.edu.cn (Y.K.); zhangjingya@petrochina.com.cn (J.Z.); liuchang2020911@petrochina.com.cn (C.L.)
2 Institute of Energy, School of Earth and Space Sciences, Peking University, Beijing 100871, China; kouqi.liu@pku.edu.cn
3 State Key Laboratory of Continental Shale Oil, Daqing 163712, China
* Correspondence: zrk@petrochina.com.cn

Abstract: The successful exploration and development of shale oil in the clay-rich Gulong shale have sparked increased research into the influence of clay minerals on shale reservoirs. However, compared to chlorite in sandstones, limited studies have focused on the occurrence of chlorite in continental shales and its effects on shale reservoir properties. This study offers a comprehensive analysis of chlorite in Gulong shale samples from three wells at different diagenetic stages. Four primary chlorite occurrences are identified in the Gulong shale: Type I, which is chlorite filling dissolved pores in carbonate; Type II, which is isolated chlorite; Type III, which is chlorite filling organic matter; and Type IV, which is chlorite filling authigenic microquartz. Types I and III chlorites exhibit higher porosity, offering more storage space for shale reservoirs. Chlorites of Types I, III, and IV, filled with other substances, display higher fractal dimensions, indicating more complex pore structures. These complex pores are favorable for oil adsorption but hinder oil seepage. The processes of organic matter expulsion and dissolution, which intensify with increasing diagenesis, promote the development of Types I and III chlorites, thereby positively influencing the shale reservoir porosity of Gulong shale. This study underscores the influence of chlorite occurrences on shale reservoir properties, providing valuable insights for the future exploration and development of shale oil and gas.

Keywords: Gulong shale; chlorite occurrence; pore structure; fractal characteristics

1. Introduction

In recent years, the Cretaceous Qingshankou Formation of the Gulong Sag in the Songliao Basin, known as the Gulong shale, has made breakthroughs in the exploration and development of shale oil [1–4]. The Gulong shale has a high clay mineral content (>40%) and a high organic matter content (>2%) [2]. As major minerals, clay minerals play a key role in shale reservoirs and are important components of micro- and nanoscale pore networks [5,6]. Chen, et al. (2016) [7] noted that clay minerals contribute to a total porosity of up to 16% in shale reservoirs. Moreover, certain clay minerals, such as montmorillonite, have large surface areas and strong adsorption capacities, which are crucial for preserving organic matter during sedimentation and diagenesis [8–11].

As a key clay mineral, chlorite is significantly influenced by diagenetic processes, making its study crucial for understanding shale diagenesis [12,13]. Furthermore, the tendency of chlorite precipitation within pore spaces highlights its significant impact on porosity, thus becoming an important element in studying shale reservoir evolution [14, 15]. In shales, chlorite also exists in diverse occurrences. He, et al. (2022) [16] and Sun, et al. (2024) [17] reported the formation of chlorite in the Gulong shale during early diagenesis, forming films along detrital grain surfaces in shales with high silt content. In the Nandan region, black shale chlorite appears as small particles (<1 μm) embedded within the matrix [18]. Different occurrences of chlorite may have varying effects on shale reservoir properties as it is in sandstone.

Currently, the influence of chlorite on shale pore spaces remains contentious. In clay minerals from the Longmaxi Formation marine shale, Li, et al. (2015) [19] observed a negative correlation between chlorite content and the specific surface area of micropores, suggesting a destructive effect of chlorite on shale micropores. Conversely, research by Chang, et al. (2022) [20] indicates a weak correlation between chlorite content and pore structure parameters or fractal dimensions. The effect of chlorite on shale reservoirs appears insignificant, possibly due to low chlorite content [20]. In transitional (marine–continental) shale, higher proportions of illite and chlorite in the Lower Yangtze Region's Permian shale correspond to reduced contributions to the pore system by clay minerals [21]. In the Linxing area's transitional shale, chlorite promotes the development of inorganic pores and complicates pore structure [22]. In continental shale, Hou, et al. (2022) [23] observed that clay minerals significantly contribute to pore development in the Gulong shale, with chlorite playing a particularly crucial role and exhibiting a positive correlation. Wang, et al. (2023) [2] discovered a nonlinear relationship between pore volume and chlorite content in Gulong shale. While certain forms of chlorite protect porosity, excessive growth can compromise pore integrity. Several scholars have proposed upper limits of chlorite content conducive to pore preservation, with suggested thresholds ranging from 5% to 10% [24–26]. Research on the Cretaceous Qingshankou Formation shale from the continental setting reveals that clay minerals reduce the pore fractal dimensions [27]. However, in the marine Bakken Formation shale, Longmaxi Formation shale, the Niutitang Formation shale, and the transitional Longtan Formation shale, the fractal dimension of the pores increases [20,28–30]. Chang, et al. (2022) [20] suggested that this difference may be related to the types of clay minerals present. For instance, fractal dimension exhibits a strong positive correlation with illite, while showing a negative correlation with kaolinite. These findings underscore the ongoing debate over clay's influence on shale pore spaces across marine, transitional, and continental settings, necessitating further research. Therefore, direct pore and fractal analysis of individual clay minerals is of significant importance.

FE-SEM can provide high-resolution (up to 100 nm and higher) 2D images [31]. Two-dimensional pore and fractal analysis provides an intuitive way to directly observe the morphological features and complexity of chlorite porosity, which is of significant importance for understanding the impact of clay minerals on reservoir pore structure. By analyzing the fractal dimension of the pores, this method allows for a quantitative description of the pore morphology, size distribution, and complexity. This, in turn, provides a scientific basis for assessing the pore structural characteristics and permeability of the shale reservoir. Due to its high brightness and distinct boundary features under FE-SEM, chlorite can be effectively analyzed for its fractal dimension and pore structure using FE-SEM. Currently, the analysis of the pore structure of clay minerals in the Gulong shale mainly focuses on the overall rock samples, with limited studies on the specific pore structure of individual clay minerals. Given that chlorite is easily identifiable under electron microscopy and has the characteristic of filling pores, it is necessary to conduct a detailed pore structure analysis of chlorite.

To investigate the effect of chlorite on shale reservoirs, this study examined samples from different maturity stages of Gulong shale, utilizing scanning electron microscopy (SEM), X-ray diffraction analysis (XRD), and two-dimensional (2D) porosity and fractal dimension analysis. It summarizes the various occurrences of chlorite in Gulong shale and studies the pore structures and evolutionary characteristics associated with different occurrences of chlorite. Detailed analysis of chlorite occurrences in shale and their effect on pore structures offers valuable insights into the clay evolution mechanisms and reservoir effects of continental shale.

2. Geological Setting

The Songliao Basin, situated in northeast China, comprises six primary tectonic units (Figure 1a,b) [32]. Gulong Sag, located in the western part of the central depression (Figure 1b), represents a long-standing deep-water sag. The sedimentary sequence within Gulong Sag consists of rift, depression, and inversion structural layers from bottom to top (Figure 1b) [1]. However, the deepest Gulong Sag has experienced no significant uplift [1,2]. The Qingshankou Formation is stratified into three members: Qing-1 Member (K_2qn^1), Qing-2 Member (K_2qn^2), and Qing-3 Member (K_2qn^3) (Figure 1c). Among these, the K_2qn^1 serves as a primary source rock and shale oil reservoir in Gulong Sag, characterized by substantial thickness and high organic carbon content. It averages 70–80 m in thickness, with total organic carbon ranging from 1% to 5%, averaging 2.9%. Vitrinite reflectance (Ro) varies between 0.72% and 1.5%, averaging 0.91% [2]. Diagenesis processes predominantly include mechanical compaction, pressure dissolution, cementation, and dissolution, occurring primarily in the middle diagenesis phase A2-B [7].

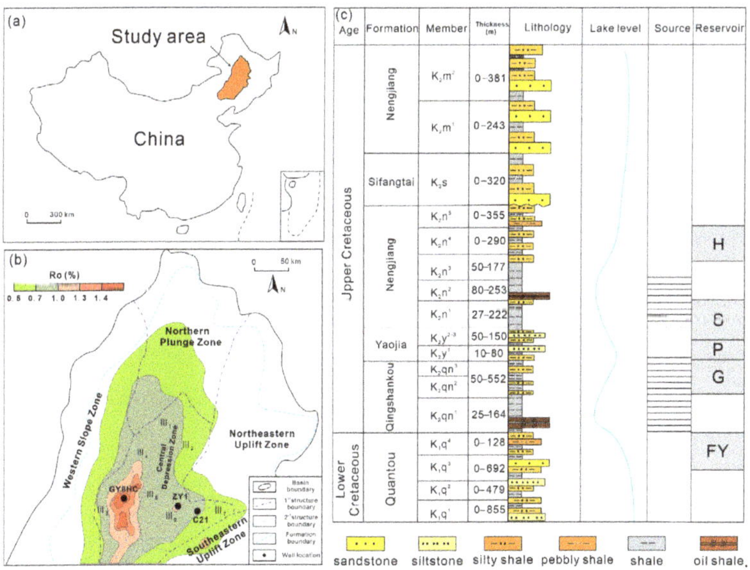

Figure 1. Geological background map of the research area. (**a**) Schematic location of the study area. (**b**) Structure units in northern Songliao Basin, superimposed by Ro contours of Gulong shale and sample locations. III_1, Heiyupao Sag; III_2, Mingshui Terrace; III_3, Longhupao-Daan Terrace; III_4, Qijia-Gulong Sag; III_5, Daqing Anticline; III_6, Sanzhao Sag; III_7, Chaoyangou Terrace. Modified from Wang, et al. (2023) [2]. (**c**) Cretaceous stratigraphic columns of the Songliao Basin, modified from Wang, et al. (2023) [2]. Stripe lines and yellow boxes in the column show source rocks and oil layers, respectively. H: Heidimiao oil layer; S: Saertu oil layer; P: Putaohua oil layer; G: Gaotaizi oil layer; FY: Fuyu oil layer.

The study primarily focuses on the K_2qn^1 and K_2qn^2 shales, renowned as the Gulong shale in China, which is a prominent area for terrestrial shale oil exploration [1,32]. The Gulong shale exhibits a broad thermal maturity spectrum ranging from immature to highly mature conditions (Figure 1b). Organic matter in the shale predominantly consists of lacustrine Type I kerogen, with organic macerals primarily composed of lamalginite and alginite-derived amorphous organic matter. Gulong shale predominantly comprises shale (>95% of total thickness), predominantly fine-grained (<0.0039 mm) clastic minerals such as quartz, clay minerals, feldspar, and minor calcite. Clay mineral content is generally high, while carbonate minerals are sparse and felsic minerals are prominent. Intermingled are thin layers of siltstone, dolomite, and shell limestone, typically 1–5 cm thick per layer. For instance, the Q1 oil layer measures 14.6 m in thickness, comprising 51 layers of gray-black shale (13.36 m), eight layers of siltstone (0.52 m), four layers of dolomite (0.49 m), and one layer of shell limestone (0.23 m). Organic carbon content in the Q1 layer ranges from 0.8% to 4.5%, averaging 2.6%. Mineral content analysis indicates that the Gulong shale primarily consists of felsic, clayey, and mixed shale (Figure 1c), with clay shale exhibiting comparatively higher total organic carbon (TOC) content.

3. Samples and Methods

3.1. Samples

The core samples were obtained from the Qingshankou Formation shale of GY 8 wells with high maturity (Ro = 1.3–1.5%), ZY1 well with medium maturity (Ro = 0.9–1.0%), and C21 well with low maturity (Ro = 0.6–0.7%) [2]. Comprehensive whole-rock, clay analysis, and SEM-EDS analysis were carried out on all shale samples.

3.2. Methods

(1) X-ray diffraction (XRD)

XRD analysis of shale samples was conducted on both whole-rock (bulk) and clay fractions (<2 μm) using a TTR-type X-ray diffractometer manufactured by Rigaku (Japan). Prior to analysis, samples were pulverized to a 200-mesh powder. Whole-rock XRD patterns were acquired over a range of 5 to 65° with a step size of 0.02° and a scanning speed of 2°/min, utilizing Cu-Kα radiation (λ = 1.5418 Å). Experimental conditions included a 40 kV applied voltage and a 40 mA current, with a scattering slit of 1° and a receiving slit of 0.3 mm. Before analyzing clay minerals via XRD, carbonates and organic matter were removed using 0.3 M acetic acid and 10% hydrogen peroxide (H_2O_2), respectively. Clay fractions were isolated through suspension precipitation and subsequently transferred to silicon wafers to prepare oriented samples. The identification and quantification of clay minerals involved multiple treatments, including air drying (25 °C) (N sheet), ethylene glycol saturation (E sheet), and heating (550 °C) (T sheet). XRD data were collected over a 2.5–30° 2θ range. Mineral phases were identified and quantitatively assessed in weight percent (wt.%) using the Pearson VII fitting function on the Jade 6.0 software platform. Quantitative phase analysis was accomplished via Rietveld refinement, employing customized clay mineral structure models. The mineral compositions and relative mineral percentages of the shale samples were determined according to the Chinese oil and gas industry standard SY/T 5163-2018 [33].

(2) Field-Emission Scanning Electron Microscopy (FE-SEM)

All shale samples were cut into relatively flat 10 × 10 × 5 mm squares and manually polished with 800 mesh, 1000 mesh, and 2000 mesh sandpaper, respectively. A Leica argon ion polishing instrument was used to polish the circumferences of the samples to ensure smooth surfaces before carbon plating. After polishing, the sample surface was treated with carbon coating. The morphology, occurrence, dissolution, and pore characteristics of minerals and OM were imaged at a temperature and humidity of 22 °C and 40%, respectively. FE-SEM observation was performed using an FEI Helios Nano-Lab 650 (America) equipped with an energy-dispersive spectrometer (EDS). FE-SEM imaging was operated with a chamber pressure of 8–10 mbar, a working distance of 3–15 mm, and a voltage of 5–15 kV. EDS was operated with a working distance of 8–10 mm, a voltage of 10–20 kV, and a detection depth of 1–5 μm. EDS spectra were then used to determine mineral types.

(3) Pore structure Measurement and Fractal Analysis

The 2D pore structure and fractal dimension analysis were conducted using ImageJ 1.54d software (see Figure 2). The 2D porosity refers to the porosity of a chlorite aggregate, which consists of visible needle-like chlorite particles clustered together. First, we extracted the outline of the chlorite aggregate and then used the difference in grayscale values between pores and chlorite to extract the pores of the chlorite aggregate (Figure 2). The ratio of the pore area of the chlorite aggregate to the total area of the chlorite aggregate represents the chlorite porosity in this study (see formula (1)). This calculation provides a quantitative measure of the porosity specific to the chlorite aggregate under study.

$$Porosity = \frac{Pore\ Area}{Chlorite\ Assemble\ Area} \tag{1}$$

Figure 2. Analysis flowchart for 2D porosity and pore fractal dimension analysis.

In addition, the pore size and pore aspect ratio were statistically analyzed, with the aspect ratio defined as the ratio of the long axis to the short axis of the pores.

Fractal dimension analysis by Image J was implemented in the box-counting method, which involves several steps after binarizing the image (Figure 2). The basic principle of two-dimensional fractal dimension is based on the formula (2) [34]. With the Image J software, this is mainly realized through the following steps: First, the image is covered with boxes of various sizes (ε), and then each box is checked to count how many contain at least one foreground pixel (pixel value = 1). The number of such boxes ($N(\varepsilon)$) is calculated for each box size ε. Next, a double logarithmic graph of ε versus $N(\varepsilon)$ is plotted. Finally, a linear regression analysis is performed on the log-log plot to determine the fractal dimension D, where the slope of the regression line (typically D) provides the fractal dimension.

$$D = \lim_{\varepsilon \to \infty} \frac{\ln N\varepsilon(F)}{-\ln \varepsilon} \tag{2}$$

D is the fractal dimension of F; F is the object being measured, ε is the size of the box; and N is the number of similar elements calculated for each box size ε.

4. Results

4.1. Mineralogy

As can be seen in Figure 3, the collected shale samples consist mainly of quartz, feldspar, and clay minerals, of which the clay minerals are mainly I/S mixed layer, illite, and chlorite. According to Hou, et al. (2022) [23] and Wang, et al. (2023) [2], clay minerals in the Gulong shale are dominated by illite (11–94%, 63% on average), followed by illite/smectite mixed layer (I/S) (3–68%, 22% on average), chlorite (1–59%, 12% on average) and kaolinite (1–26%, 7.4% on average). Notably, the content of chlorite increases with increasing depth, reflecting thermal maturity change [2,23]. As one of the major clay mineral types in the Gulong shale, coupled with its tendency to fill pore growths, it is important to study the effect of chlorite on shale reservoirs.

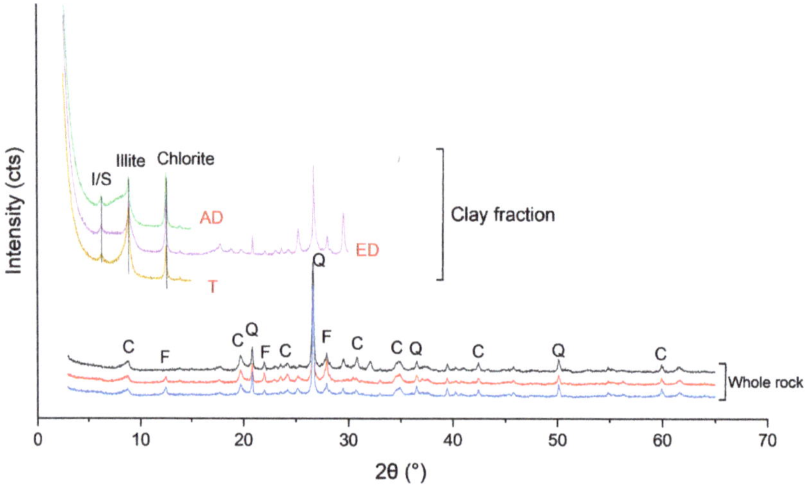

Figure 3. Representative XRD patterns of the investigated samples (whole-rock and <2 μm clay fraction). Q—quartz, F—feldspar, C—clay, AD—air-dried powder, ED—ethylene glycolate powder.

4.2. Chlorite Occurrence

Under SEM, four types of chlorite occurrences were identified (Figure 4): Type I is the chlorite filling the dissolved pore of carbonate, Type II is the isolated chlorite, Type III is the chlorite filling organic matter, and Type IV is the chlorite filling authigenic quartz. These four chlorite types exhibit distinct crystallographic states and pore characteristics. Type I chlorite shows coexistence with carbonate minerals, linked to carbonate dissolution pores (Figure 4a_1–a_3,f). This type of chlorite displays irregularly shaped pores, including dissolution pores, as well as encapsulated or triangular pores between chlorite crystals. Type II chlorite appears as needle-like or hair-like chlorite filling intergranular pores supported by particles such as feldspar and quartz (Figure 4b_1–b_3,e). The interior of these chlorite intergranular pores appears relatively clean, predominantly structured with triangular or slit-shaped pores where chlorite crystals interlock. Type III chlorite features acicular chlorite crystals filling intergranular pores enriched with flowing organic matter (Figure 4c_1–c_3). Type IV chlorite reveals microcrystalline quartz wrapped by chlorite, with triangular intergranular pores between the wrapped quartz and chlorite (Figure 4d_1–d_3,g). The wrapped microcrystalline quartz typically exhibits well-defined authigenic morphology, with grain sizes generally in the range of a few microns.

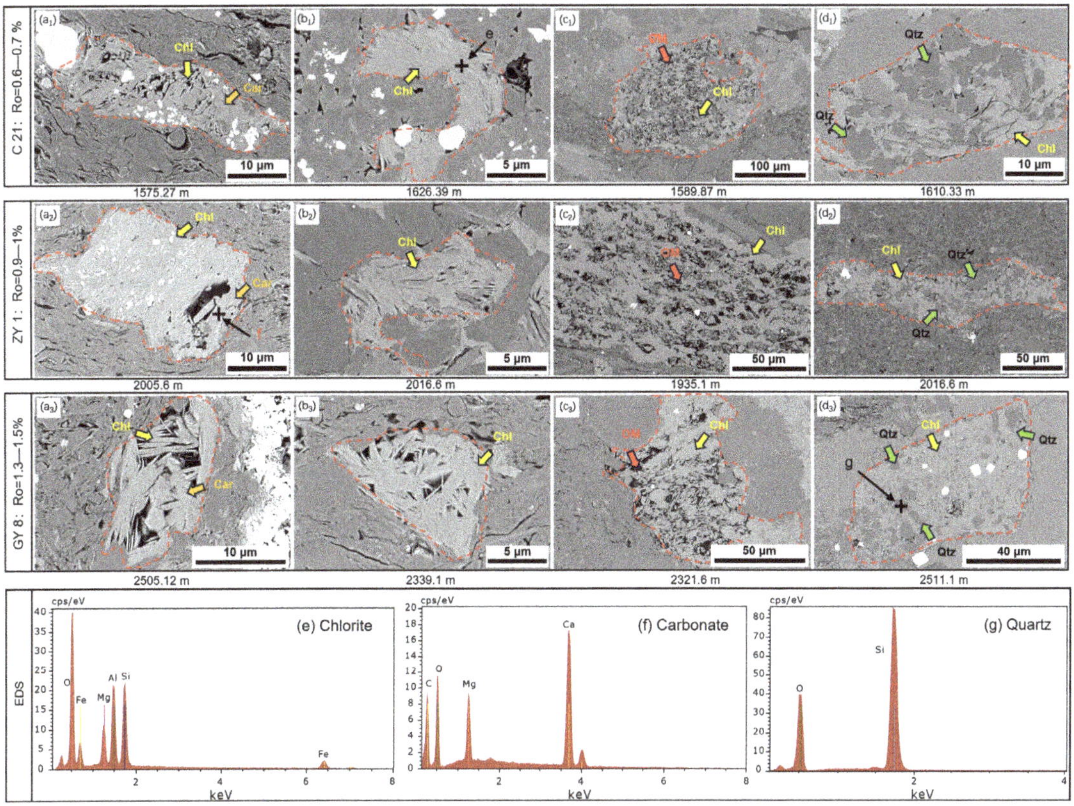

Figure 4. Occurrence of chlorite of three different maturity shale samples. (**a$_1$–a$_3$**) The chlorite filling the dissolved pore of carbonate. (**b$_1$–b$_3$**) The isolated chlorite. (**c$_1$–c$_3$**) The chlorite filling organic matter. (**d$_1$–d$_3$**) Chl—chlorite, (**e**) EDS pattern of chlorite, (**f**) EDS pattern of carbonate, (**g**) EDS pattern of quartz, Car—carbonate, Qtz—quartz, OM—organic matter.

The distribution of intergranular pores in the four types of chlorites also varies. According to the International Union of Pure and Applied Chemistry (IUPAC), the classification of pore size is as follows: micropores (pore size < 2 nm), mesopores (pore size ranging from 2 to 50 nm), macropores (pore size > 50 nm) [35]. As shown in Figure 5, Type I chlorite predominantly consists of macropores, with 16.1% of mesopores developed, mainly distributed between 50 and 500 nm. Type II chlorite is also dominated by macropores, with 19.3% of mesopores developed, and the pore size mainly ranges from 100 to 500 nm. Type III chlorite is composed entirely of macropores, while Type IV chlorite predominantly features macropores and mesopores, with the main pore size ranging from 2 to 100 nm.

4.3. Distribution of Pore Structure and Fractal Dimension of Four Types of Chlorite

In total, we counted the occurrence characteristics of chlorite in 118 samples from three wells and analyzed the pore structure and fractal dimension of chlorite in four occurrences (Figure 6).

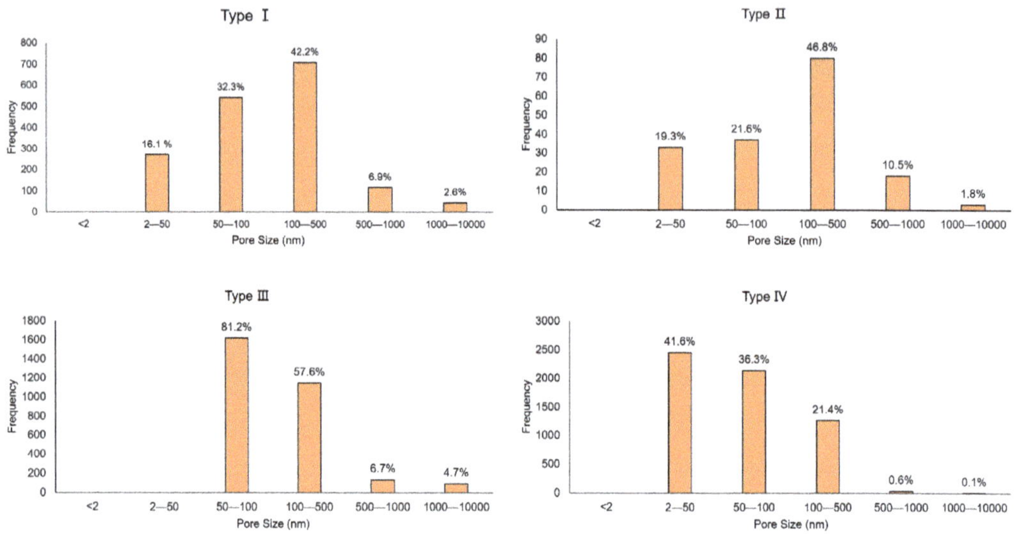

Figure 5. The pore size distribution of four types of chlorite.

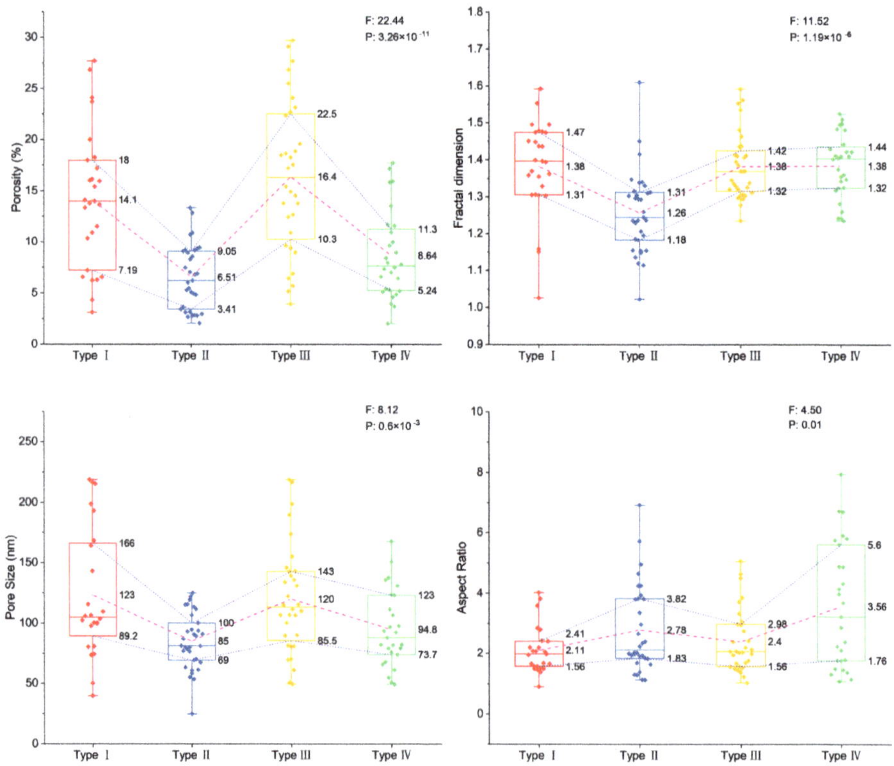

Figure 6. Pore structure and fractal dimension of four types of chlorite. The dashed line represents the connection between the average and the data range from 25–75% of pore structure parameters. P and F are the result of One-Way Analysis of Variance (ANOVA). If the p values are less than 0.05, it can be concluded that there is a significant difference between the group means.

The pore structures of the four chlorite types exhibit distinct characteristics (Figure 6). Porosity analysis reveals that Type I and Type III chlorites possess higher porosity, with average values of 14.1% and 16.4%, respectively. Type II and Type IV chlorites have lower porosity, with average values of 6.51% and 6.64%, respectively. Type IV chlorite shows relatively higher porosity compared to Type II, possibly due to support from authigenic quartz.

Two-dimensional fractal analysis results indicate that Type II chlorite has the lowest fractal dimension, averaging 1.26 (Figure 6). Type I, Type II, and Type III chlorites exhibit relatively higher fractal dimensions, with average values of 1.38 each. Among these, Type I chlorite shows the highest overall fractal dimension, possibly reflecting the complexity of the dissolution processes involved.

The pore size distribution shows that Type I and Type III chlorites have the largest average pore sizes, with Type I chlorite having an average pore size of 123 nm and Type III chlorite having an average pore size of 120 nm (Figure 6). Type II and Type IV chlorites have smaller pore sizes, with average pore sizes of 85 nm and 94.8 nm, respectively. In terms of the aspect ratio of the pores, Type I and Type III chlorites have smaller length-to-width ratios, averaging 2.11 and 2.4, respectively, indicating more regular pore shapes (Figure 6). In contrast, Type II and Type IV chlorites have larger length-to-width ratios, averaging 2.78 and 3.56, respectively, suggesting that their pores are more elongated.

5. Discussion

5.1. Genesis of Pore-Filling Chlorite

Type I chlorite exhibits associated occurrences within dissolution pores filled with carbonate minerals (Figure 4a_1–a_3). Wang, et al. (2023) [2] found that chlorite and carbonate show significant associations in the Gulong shale. During carbonate dissolution, Fe^{2+} and Mg^{2+} ions are released, facilitating the co-occurrence of chlorite and carbonate [36–39]. As dissolution intensifies, carbonate gradually dissolves completely, allowing chlorite to fill the pores progressively [2].

Type II chlorite grows between mineral grains such as feldspar and quartz (Figure 4b_1–b_3). Crystallization of chlorite on the surfaces of feldspar or quartz particles occurs when the ion concentration and activation energy conditions are met in the external fluid [12,15,39,40]. Pores within this chlorite type remain unfilled by other substances, possibly due to rapid fluid exchange rates in the larger intergranular pores [41–43].

Type III chlorite's interstitial pores are filled with organic matter (Figure 4c_1–c_3). During thermal maturation, organic matter expelled during hydrocarbon generation is retained within pre-existing authigenic chlorite's triangular or slit-like pores. According to He, et al. (2023) [44], hydrocarbon charging peaks in the Gulong Sag during moderate maturity stages, favoring the early-formed chlorite interstitial pores to be filled with organic matter. Alteration in the chemical composition of fluid after charging may halt further growth of this type of chlorite [45].

Type IV chlorite's interstitial pores develop microcrystalline quartz (Figure 4d_1–d_3). During the transformation of minerals like smectite into chlorite, quartz precipitates (see formula (3)) [46]. When fluid transport of Si^{4+} is hindered, microcrystalline quartz often forms in association with authigenic crystals [46,47]. Limited by crystalline space and time, this quartz crystal typically exhibits small grain sizes but distinct hexagonal self-growth patterns [46,48].

$$2.4Ca_{0.1}Na_{0.2}Fe_{1.1}MgAlSi_{3.6}O_{10}(OH)_2 \text{ (Smectite)} + 0.88H_2O + 1.44H^+ \Rightarrow Fe_{2.6}Mg_{2.2}Al_{2.4}Si_{2.8}O_{10}(OH)_8 \text{ (Chlorite)} + 0.24Ca^{2+} + 0.48Na^+ + 0.04Fe^{2+} + 0.20Mg^{2+} + 5.84SiO_2 \text{ (Quartz)}$$

The pore structures and characteristics of these four chlorite types are influenced by the different diagenetic processes they undergo. Thus, they may have distinct contributions and evolutionary characteristics to shale reservoirs [49,50].

5.2. Evolution of Pore Structure and Fractal Dimension for Pore-Filling Chlorite

The fractal dimension can characterize the complexity or irregularity of pore distribution [28]. A higher fractal dimension typically indicates a more intricate and irregular pore network, which may influence the shale's ability to retain or transmit fluids—an essential factor in hydrocarbon reservoir studies [51–53]. In contrast, a lower fractal dimension suggests a simpler, smoother pore structure, often associated with a less interconnected pore network and potentially lower permeability [20].

In analyzing the porosity characteristics of four chlorite types, it is evident that Type I and Type III chlorites exhibit higher porosity (Figure 6). This implies that dissolution processes and the thermal evolution of organic matter may promote the formation of interstitial pores within chlorite crystals. The cementation between chlorite and quartz inhibits the development of interstitial pores. Compared to Type II chlorite, the other three types display larger fractal dimensions and more complex porosity (Figure 6). The filling of other substances leads to increased complexity in the internal pore structure of chlorite. Additionally, as shown in Figure 6, Type I and Type III chlorites exhibit more regular pore shapes, while Type II and Type IV chlorites have more elongated pores. The elongated pores in Type II and Type IV chlorites are less conducive to the percolation of oil and gas fluids.

The porosity of Type I chlorite initially increases and then gradually decreases with maturity (Figure 7), which may be due to the early thermal evolution of organic matter. The substantial organic acid release and subsequent dissolution of carbonate minerals increases the porosity. As organic matter matures, organic acid release diminishes, fluid pH shifts from acidic to mildly alkaline, and residual ions crystallize rapidly within pores [2,16,54], leading to chlorite filling and gradual porosity reduction. Liu, et al. (2019) [55] reported the existence of a change in fluid pH from acidic to moderately alkaline in the Gulong shale. Fractal characteristics of Type I chlorite show minimal variation with maturity (Figure 7), indicating dissolution processes have negligible impact on its pore structure.

Porosity and fractal characteristics of Type II chlorite show minimal change with thermal evolution (Figures 7 and 8). A P value greater than 0.05 suggests no significant difference between the three box charts (Figures 7 and 8). This may result from the gradual occupation of mineral surfaces available for chlorite crystallization. Once chlorite fills available pore space, subsequent crystallization sites become scarce, resulting in minimal changes in porosity and pore structure during thermal evolution [15].

As can be seen in Figures 7 and 8, the porosity and fractal dimension of Type III chlorite increase gradually with maturity. During organic matter thermal evolution, kerogen begins hydrocarbon expulsion, progressively filling interstitial chlorite pores and expanding organic matter-occupied spaces, thereby enlarging chlorite pores continuously [56,57]. Organic matter occupies pore spaces, shielding interstitial chlorite pores from compaction and cementation effects of other minerals. Increased organic matter injection leads to more complex pore shapes, making Type III chlorite pores more intricate as maturity progresses (Figure 8).

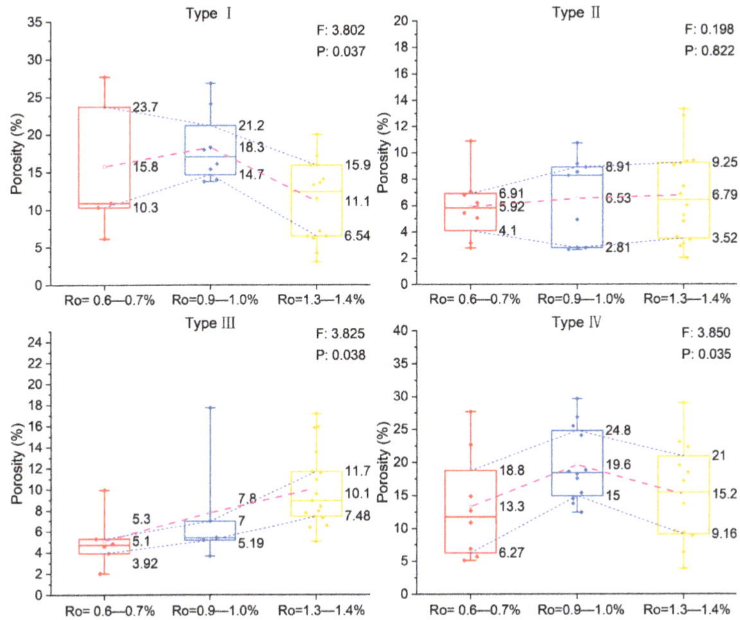

Figure 7. The evolution of porosity character of four types of chlorite. The dashed line represents the connection between the average and the data range from 25–75% of porosity. If the p values for the four types of porosity and Fractal dimension are less than 0.05, it can be concluded that there is a significant difference between the group means.

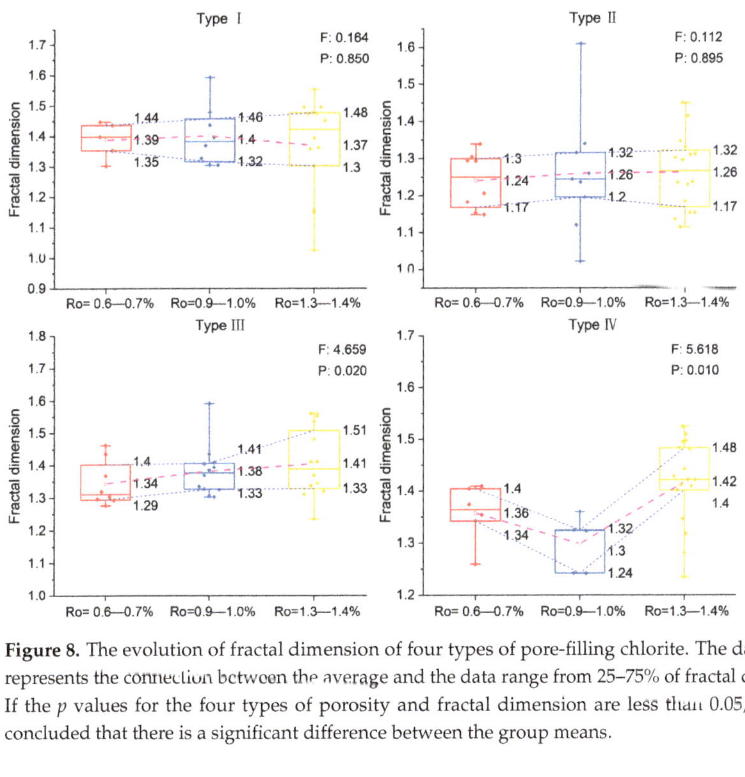

Figure 8. The evolution of fractal dimension of four types of pore-filling chlorite. The dashed line represents the connection between the average and the data range from 25–75% of fractal dimension. If the p values for the four types of porosity and fractal dimension are less than 0.05, it can be concluded that there is a significant difference between the group means.

Type IV chlorite exhibits an initial increase followed by a decrease in porosity (Figure 7). This may result from the authigenic microcrystalline quartz providing a framework support for chlorite [58]. Initially, chlorite forms numerous interstitial pores with authigenic quartz particles, facilitating early pore space development [58]. As chlorite continues to grow, these pore spaces gradually fill and are further compacted, resulting in a later-stage reduction in Type IV chlorite porosity. The fractal dimension of Type IV chlorite initially decreases and then increases, indicating a transition from simple to complex pore structures (Figure 8). The rigid support of regular, self-generated quartz contributes to simplified chlorite pores initially, but as chlorite grows and compaction takes effect, these pores become increasingly complex at higher maturity stages.

5.3. Pore Structure Versus Fractal Dimension of Pore-Filling Chlorite

As shown in Figure 9, the relationship between porosity and fractal dimension varies across different types of chlorites, each exhibiting distinct characteristics in their fitting curves. Type I chlorite shows a phenomenon where the fractal dimension first increases with porosity and then decreases after reaching a certain point (Figure 9a). This is due to the gradual dissolution of carbonate minerals: as porosity increases, the pore complexity decreases, and at high porosity, the carbonate is typically not replaced by chlorite, leading to a lower fractal dimension. Thus, high porosity in Type I is mainly attributed to dissolution processes. Type II chlorite, in contrast, demonstrates a positive correlation between porosity and fractal dimension, with both increasing together (Figure 9b). This behavior is primarily driven by the intercrystalline pores of chlorite, where higher porosity indicates a more complex pore structure, resulting in a higher fractal dimension. Type III chlorite exhibits no significant correlation between porosity and fractal dimension, showing an irregular curve that reflects a more complex and heterogeneous pore structure (Figure 9c). Finally, Type IV chlorite also shows a positive correlation between porosity and fractal dimension, similar to Type II (Figure 9d). The increase in porosity leads to more complex intercrystalline pores, which in turn increases the fractal dimension, indicating a more intricate pore structure.

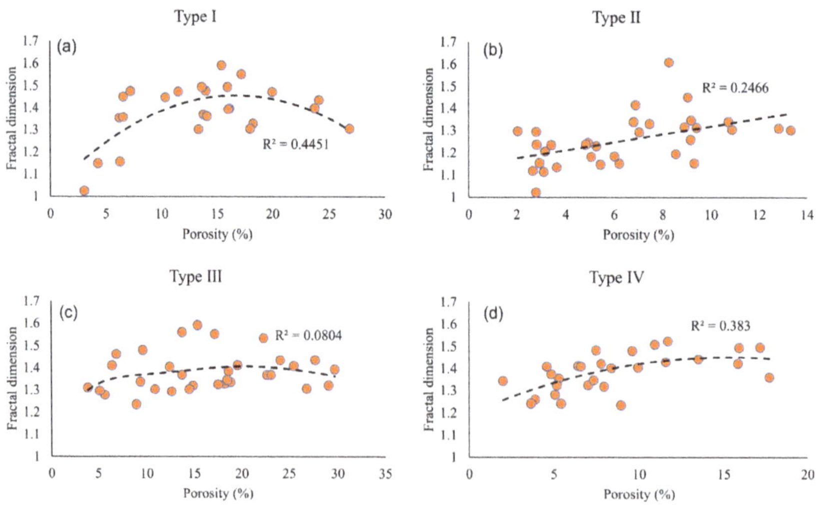

Figure 9. Porosity vs. fractal dimension of four types of pore-filling chlorite. (**a**) Porosity vs. fractal dimension of Type I chlorite, (**b**) Porosity vs. fractal dimension of Type II chlorite, (**c**) Porosity vs. fractal dimension of Type III chlorite, (**d**) Porosity vs. fractal dimension of Type IV chlorite.

In summary, the four types of chlorites differ in the relation between porosity and fractal dimension. Type I shows a decreasing fractal dimension at high porosity due to carbonate dissolution. Type II and Type IV display a positive correlation indicating more complex pore structures, while Type III shows a complex, irregular relationship without a clear trend.

Currently, studies based on BET average pore size obtained through nitrogen adsorption reveal a negative correlation between pore size and fractal dimension, with a general understanding that smaller pores lead to more complex pore structures (Figure 10b) [28,59,60]. However, in the chlorite samples from the study area, the fractal dimension and pore size show a positive correlation (Figure 10a). This suggests that larger chlorite pores possess higher pore complexity. The injection of organic matter and the dissolution of carbonate minerals lead to the expansion of intergranular pores. At the same time, this results in a more dispersed pore size distribution, making the pore structure more complex. As seen in Figure 6, chlorite with no fillings generally corresponds to lower fractal dimensions and smaller average pore sizes, whereas chlorite with fillings tends to have higher fractal dimensions and more dispersed pore size distributions. Therefore, for chlorite particles, an increase in the average pore size can lead to a more complex pore structure, primarily due to the larger average pore size indicating higher pore dispersion.

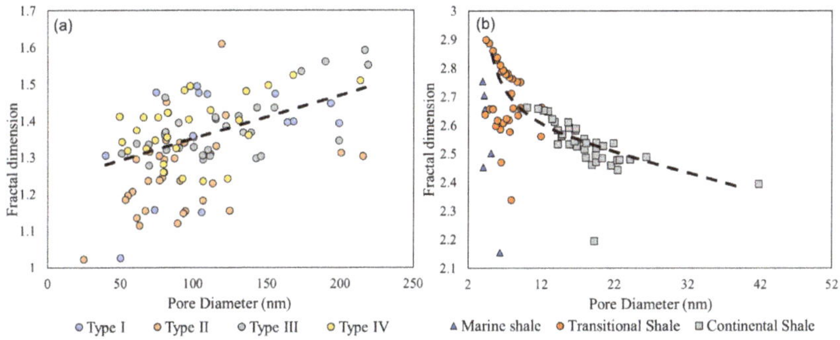

Figure 10. The pore size vs. fractal dimension of four types of pore-filling chlorite. (**a**) Pore diameter vs. fractal dimension of this study, (**b**) Pore diameter vs. fractal dimension of marine, transitional and continental shale, Data are from Jiang, et al. (2016) [61], Li, et al. (2019) [28], and Wang, et al. (2021) [59].

5.4. Evolution Model of Pore-Filling Chlorite

Based on the findings of these studies, the four chlorite types undergo distinct evolutions during diagenesis (Figure 1). Enhanced dissolution of carbonate promotes chlorite growth. Research by Liu, et al. (2019) [55] indicates that fluids in the Gulong Sag transitioned from early acidic to later mildly alkaline conditions, conducive to chlorite growth. Additionally, early dissolution released Fe^{2+} and Mg^{2+} that facilitated rapid chlorite growth in later stages, gradually occupying the crystal outlines of carbonate (Figure 11). Type II chlorite, lacking sufficient nucleation sites, shows minimal changes during diagenesis (Figure 11). Conversely, Type III chlorite, influenced by increasing organic matter expulsion, fills more interstitial spaces, resulting in stagnant growth, increased porosity, and complexification of interstitial pores (Figure 11). Type IV chlorite evolves through diagenesis with concurrent growth of authigenic quartz, leading to accelerated growth of chlorite in later stages and a rapid reduction in porosity at high maturity levels (Figure 11). Compression by authigenic quartz complicates the interstitial pore structure of chlorite at advanced maturity stages.

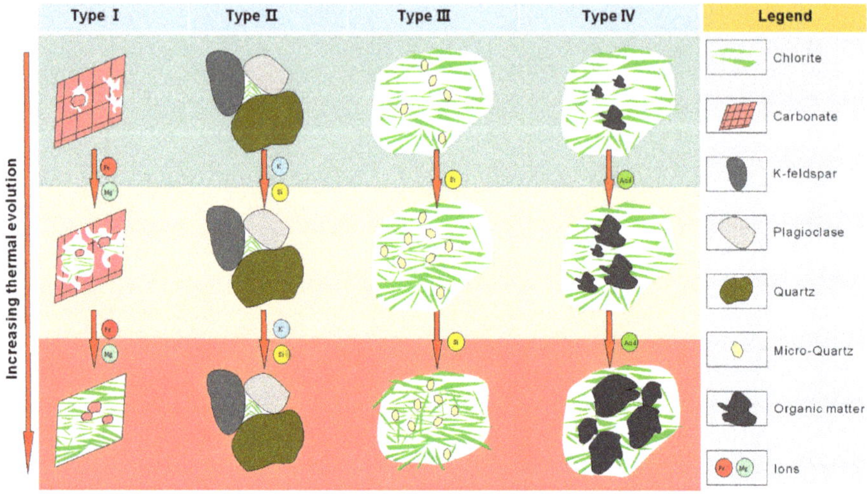

Figure 11. Model of the evolution of the four types of chlorites.

Researchers have found that fractal dimension is related to organic carbon content and microporous development [28,51,62,63]. Fractal dimension reflects the development degree of shale micropores. The more developed the micro-pore, the smaller the average pore size, the larger the specific surface area, and the larger the fractal dimension [62]. Yang, et al. (2014) [62] reported that the higher the organic carbon content, the larger the fractal dimension. The fractal dimension has different effects on the storage and transportation of hydrocarbons. The larger the fractal dimension, the pore structure tends to be more complex, which is favorable to the adsorption and storage of hydrocarbons, but not conducive to fluid seepage [28,62]. The Gulong shale is extensively developed with a significant amount of chlorite, which typically occurs in a pore-filling occurrence (Figures 2 and 4). Type I and Type III chlorite have higher porosity, and these two types of yielding chlorite have higher contribution to the pore space. In addition, the presence of carbonate, cementation of authigenic quartz, and injection of organic matter can lead to chlorite pore complexity (Figure 6), which may adsorb hydrocarbons into the pore space and thus be detrimental to hydrocarbon seepage [47,64].

5.5. Chlorite in Gulong Shale Reservoir Characteristics

In the study of marine, continental, and transitional shale reservoirs, it has been found that the relationship between shale pore fractal dimensions and clay minerals is complex. In the study by Wang, et al., (2023) [2], it is observed that chlorite in Gulong shale exhibits a transition from initially negatively affecting porosity to later exerting a positive influence when its content exceeds 5%. This shift is likely associated with changes in chlorite occurrence. With increasing diagenesis, enhanced dissolution and hydrocarbon generation processes lead to an increasing presence of Type I and Type III chlorites due to carbonate dissolution and organic matter maturation, thereby preserving intercrystalline pores in chlorite. This phenomenon is supported by findings from Hou, et al., (2022) [23] regarding a positive correlation between chlorite and organic matter in Gulong shale and from Kang, et al., (2024) [4] regarding a negative correlation between chlorite and carbonate. Consequently, the transition in chlorite pore effects in Gulong shale may be attributed to an initial dominance of Type II and Type IV chlorites, followed by a gradual shift toward Type I and Type III chlorites, resulting in an initially negative and subsequently positive correlation with porosity. Furthermore, Type I and Type III chlorites exhibit higher 2D fractal dimensions. The increasing presence of these chlorites complicates shale pore structures, potentially hindering gas permeation while enhancing gas adsorption in Gulong shale reservoirs.

In summary, shales at or after the peak of hydrocarbon production may develop more Type I and Type III chlorite, and chlorite may contribute more positively to shale reservoirs at this stage. Therefore, during shale oil and gas exploration and development in shale reservoirs where a large amount of chlorite is developed, chlorite occurrence can be an important factor for consideration. The practical implications of this approach for hydrocarbon exploration and reservoir evaluation are significant. Understanding the role of chlorite, especially Type I and Type III, in shale reservoirs can guide more accurate reservoir assessments. Chlorite's impact on pore structure, along with fractal analysis, helps identify areas with higher storage capacity and better pore connectivity, which directly influence fluid flow and hydrocarbon extraction efficiency.

6. Limitations and Future Research

This paper conducts a comparative analysis of the two-dimensional porosity and pore fractality of chlorite in shale samples with varying occurrences, aiming to deepen our understanding of the contribution of chlorite to shale reservoirs. By focusing on a large number of samples, this study provides valuable insights into the relationship between chlorite's microstructure and its role in shaping the overall pore characteristics of shale. However, the image analysis method utilized in this study is limited by inherent issues related to image scale. Specifically, it is challenging to obtain large-scale images that can adequately represent the entire shale sample. As a result, the impact of factors such as depth on the pore characteristics of chlorite remains unresolved in the current analysis. To address this limitation, future work should focus on conducting large-scale analyses that incorporate a more comprehensive view of chlorite's pore characteristics throughout the shale sample. In recent years, numerous scholars have delved into the scale problem associated with shale image analysis. Techniques such as box-counting and Z-contrast criterion have been proposed to facilitate the study of representative elementary areas in shale [65–67]. These approaches help overcome the challenge of capturing the full diversity of pore structures within the rock. Moreover, the 2D fractal employed in this study provides a clear and intuitive representation of the planar morphology of the pores. However, this approach is inherently limited in that it does not offer a complete understanding of the 3D pore structure, which is crucial for fully characterizing shale reservoir.

In light of these limitations, FIB/SEM and X-ray CT scanning emerge as a promising technique to address the issue of observing the pore structure in three dimensions [68]. FIB/SEM can produce such 3D images at high resolution (1 nm) [69], but it is labor-intensive, costly, and suffers from reconstruction uncertainty [70]. Three-dimensional imaging would allow for a more accurate representation of the complex spatial arrangement of pores and better capture the heterogeneous nature of the pore network. While X-ray CT scanning holds great potential, challenges remain in terms of mineral identification, resolution, and scale determination [68,71]. Therefore, further development of these methods, particularly in the context of accurately identifying minerals and establishing reliable scaling protocols, is essential to make three-dimensional pore analysis a viable and effective tool for future research in shale petrophysics [66].

By overcoming these technical limitations and incorporating three-dimensional analysis into the study of chlorite and shale porosity, we can gain a more comprehensive understanding of the factors influencing pore structure and ultimately improve our ability to model and predict the behavior of shale reservoirs.

7. Conclusions

This article analyzes different occurrences of chlorite in Gulong shale and their evolution using XRD, SEM, 2D porosity, and fractal analysis. It also investigates the impact of chlorite on shale reservoirs. The main conclusions drawn are as follows:

(1) Through SEM analysis, four types of chlorite occurrences were identified in Gulong shale: Type I fills dissolved carbonate pores, Type II exists without filling, Type III fills organic matter, and Type IV fills authigenic micro quartz.

(2) Porosity analysis reveals that Type I and Type III chlorites exhibit higher porosity, with average values of 14.1% and 16.4%, respectively. In contrast, Type II and Type IV chlorites show lower porosity, averaging 6.51% and 6.64%, respectively.

(3) Results from two-dimensional fractal analysis indicate that Type II chlorite has the lowest fractal dimension, averaging 1.26. Type I, Type II, and Type III chlorites exhibit relatively higher fractal dimensions for pore structures, each averaging 1.38. The presence of other substances contributes to pore complexity.

(4) As thermal maturation progresses, Type I and Type II chlorites show little change, while Type III chlorites gradually increase in porosity due to hydrocarbon expulsion from organic matter, leading to enhanced pore fractality and complexity. Type IV chlorites initially decrease in porosity, followed by an increase, with pore fractality increasing initially and then decreasing. During diagenetic evolution, quartz and chlorite continue to grow, and fluid pH shifts from acidic to neutral or alkaline, promoting rapid growth of late-stage chlorites. Consequently, the porosity of this chlorite type decreases rapidly in high maturity stages. Additionally, compression from authigenic quartz results in increased complexity of intercrystalline chlorite pores at high maturity stages.

(5) The transition in pore effects of chlorites in Gulong shale may be attributed to an initial dominance of Type II and Type IV chlorites, gradually shifting toward the predominance of Type I and Type III chlorites. This shift leads to chlorites initially showing a negative correlation with porosity and later transitioning to a positive correlation.

Author Contributions: Y.K., R.Z. and K.L., conceptualization; Y.K., J.Z. and C.L., investigation; Y.K., writing—original draft preparation; Y.K. and K.L., writing—review and editing; R.Z. and K.L., supervision; R.Z. and K.L., project administration; R.Z., funding acquisition. All authors have read and agreed to the published version of the manuscript.

Funding: This research was funded by the National Natural Science Foundation of China (Project 42072187) and Heilongjiang Province open competition projects: "Research on the diagenetic dynamic evolution process and its coupling relationship with pores and fractures" (101022hx0001b22016).

Data Availability Statement: The data presented in this study are available on request from the corresponding author.

Acknowledgments: We thank Daqing Oilfield Academician Workstation and the National Key Laboratory of Continental Shale Oil (NKL-CSO) for providing the cored rock samples and primary data. We thank the associate editor and the reviewers for their useful feedback that improved this paper.

Conflicts of Interest: Authors Yuan Kang, Rukai Zhu, Jingya Zhang and Chang Liu were employed by Research Institute of Petroleum Exploration and Development (RIPED), China National Petroleum Corporation (CNPC). The remaining authors declare that the research was conducted in the absence of any commercial or financial relationships that could be construed as a potential conflict of interest.

References

1. Sun, L.; Cui, B.; Zhu, R.; Wang, R.; Feng, Z.; Li, B.; Zhang, J.; Gao, B.; Wang, Q.; Zeng, H.; et al. Shale oil enrichment evaluation and production law in Gulong Sag, Songliao Basin, NE China. *Pet. Explor. Dev.* **2023**, *50*, 505–519. [CrossRef]
2. Wang, F.; Feng, Z.; Wang, X.; Zeng, H. Effect of organic matter, thermal maturity and clay minerals on pore formation and evolution in the Gulong Shale, Songliao Basin, China. *Geoenergy Sci. Eng.* **2023**, *223*, 211507. [CrossRef]
3. Zhao, W.; Bian, C.; Li, Y.; Zhang, J.; He, K.; Liu, W.; Zhang, B.; Lei, Z.; Liu, C.; Zhang, J.; et al. Enrichment factors of movable hydrocarbons in lacustrine shale oil and exploration potential of shale oil in Gulong Sag, Songliao Basin, NE China. *Pet. Explor. Dev.* **2023**, *50*, 520–533. [CrossRef]
4. Kang, Y.; Liu, K.; Zhu, R.; Yin, G.; Zhang, J.; Zhang, S. The evolution of clay mineral and its indication of hydrocarbons under overpressure: An example from the shale of the Qingshankou formation in the Gulong Sag. *Pet. Sci.* **2024**, *21*, 3867–3883. [CrossRef]
5. Curtis, M.E.; Cardott, B.J.; Sondergeld, C.H.; Rai, C.S. Development of organic porosity in the Woodford Shale with increasing thermal maturity. *Int. J. Coal Geol.* **2012**, *103*, 26–31. [CrossRef]
6. Nelson, P.H. Pore-throat sizes in sandstones, tight sandstones, and shales. *Am. Assoc. Pet. Geol. Bull.* **2009**, *93*, 329–340. [CrossRef]
7. Chen, S.; Han, Y.; Fu, C.; Zhang, H.; Zhu, Y.; Zuo, Z. Micro and nano-size pores of clay minerals in shale reservoirs: Implication for the accumulation of shale gas. *Sediment. Geol.* **2016**, *342*, 180–190. [CrossRef]
8. Kennedy, M.J.; Löhr, S.C.; Fraser, S.A.; Baruch, E.T. Direct evidence for organic carbon preservation as clay-organic nanocomposites in a Devonian black shale; from deposition to diagenesis. *Earth Planet. Sci. Lett.* **2014**, *388*, 59–70. [CrossRef]
9. Pichevin, L.; Bertrand, P.; Boussafir, M.; Disnar, J.-R. Organic matter accumulation and preservation controls in a deep sea modern environment: An example from Namibian slope sediments. *Org. Geochem.* **2004**, *35*, 543–559. [CrossRef]
10. Zhao, T.; Xu, S.; Hao, F. Differential adsorption of clay minerals: Implications for organic matter enrichment. *Earth-Sci. Rev.* **2023**, *246*, 104598. [CrossRef]
11. Ji, L.; Zhang, T.; Milliken, K.L.; Qu, J.; Zhang, X. Experimental investigation of main controls to methane adsorption in clay-rich rocks. *Appl. Geochem.* **2012**, *27*, 2533–2545. [CrossRef]
12. Beaufort, D.; Rigault, C.; Billon, S.; Billault, V.; Inoue, A.; Inoue, S.; Patrier, P. Chlorite and chloritization processes through mixed-layer mineral series in low-temperature geological systems—A review. *Clay Miner.* **2015**, *50*, 497–523. [CrossRef]
13. Patrier, P.; Beaufort, D.; Azzam, F.; Blaise, T.; Portier, E.; Brigaud, B.; Clerc, S. New insights on diagenetic chlorite and its source material in turbiditic sandstones of contrasted reservoir quality in the Lower Cretaceous Agat formation (Duva oil and gas field, northern Norwegian North Sea). *Mar. Pet. Geol.* **2023**, *152*, 106221. [CrossRef]
14. Ma, P.J.; Lin, C.Y.; Zhang, S.Q.; Dong, C.M.; Xu, Y.F. Formation of chlorite rims and the impact of pore-lining chlorite on reservoir quality: A case study from Shiqianfeng sandstones in upper Permian of Dongpu Depression, Bohai Bay Basin, eastern China. *Aust. J. Earth Sci.* **2017**, *64*, 825–839. [CrossRef]
15. Worden, R.H.; Griffiths, J.; Wooldridge, L.J.; Utley, J.E.P.; Lawan, A.Y.; Muhammed, D.D.; Simon, N.; Armitage, P.J. Chlorite in sandstones. *Earth-Sci. Rev.* **2020**, *204*, 103105. [CrossRef]
16. He, W.; Liu, B.; Sun, M.; Wang, L.; Zhang, J.; Yasin, Q.; Tian, S.; Gao, S.; Ukaomah, C.F. Pore types, genesis, and evolution model of lacustrine oil-prone shale: A case study of the Cretaceous Qingshankou Formation, Songliao Basin, NE China. *Sci. Rep.* **2022**, *12*, 17210. [CrossRef] [PubMed]
17. Sun, L.D.; Wang, F.L.; Bai, X.F.; Feng, Z.H.; Shao, H.M.; Zeng, H.S.; Gao, B.; Wang, Y.C. Discovery of nano organo-clay complex pore-fractures in shale and its scientific significance: A case study of Cretaceous Qingshankou Formation shale, Songliao Basin, NE China. *Pet. Explor. Dev.* **2024**, *51*, 813–825. [CrossRef]
18. Zhao, H.; Ling, K.; Qin, S.; Lei, M.; Wen, H. Modes of occurrence of lithium in black shale in the Nandan area, Guangxi, SW China: Implications for clay-type resources. *Ore Geol. Rev.* **2023**, *157*, 105409. [CrossRef]
19. Li, H.; Liu, D.; Peng, P.; Wang, Q. Tectonic Impact on Reservoir Character of Chongqing and Its Neighbor Area. *Nat. Gas Geosci.* **2015**, *26*, 1705–1711.
20. Chang, J.Q.; Fan, X.D.; Jiang, Z.X.; Wang, X.M.; Chen, L.; Li, J.T.; Zhu, L.; Wan, C.X.; Chen, Z.X. Differential impact of clay minerals and organic matter on pore structure and its fractal characteristics of marine and continental shales in China. *Appl. Clay Sci.* **2022**, *216*, 13. [CrossRef]
21. Cao, T.; Song, Z.; Luo, H.; Zhou, Y.; Wang, S. Pore system characteristics of the Permian transitional shale reservoir in the Lower Yangtze Region, China. *J. Nat. Gas Geosci.* **2016**, *1*, 383–395. [CrossRef]
22. Li, G.; Qin, Y.; Li, G.; Wu, M.; Liu, H. Influence of reservoir properties on gas occurrence and fractal features of transitional shale from the Linxing area, Ordos Basin, China. *Arab. J. Geosci.* **2022**, *15*, 250. [CrossRef]
23. Hou, L.; Wu, S.; Jing, Z.; Jiang, X.; Yu, Z.; Hua, G.; Su, L.; Yu, C.; Liao, F.; Tian, H. Effects of types and content of clay minerals on reservoir effectiveness for lacustrine organic matter rich shale. *Fuel* **2022**, *327*, 125043. [CrossRef]
24. Li, Y.; Li, S.; Mou, W.; Yan, C. Influences of clay minerals on physical properties of Chang 6 tight sandstone reservoir in Jiyuan area, Ordos Basin. *Nat. Gas Geosci.* **2017**, *28*, 1043–1053.

25. Huang, S.J.; Xie, L.W.; Zhang, M.; Wu, W.-H.; Shen, L.C.; Liu, J. Formation mechanism of authigenic chlorite and relation to preservation of porosity in nonmarine Triassic reservoir sandstones, Ordos Basin and Sichuan Basin, China. *J. Chengdu Univ. Technol. (Sci. Technol. Ed.)* **2004**, *31*, 273–281.
26. Lu, Z.; Qing, C. Geological factors affecting reservoir permeability of the Upper Shaximiao Formation in the Xinchang gas field, western Sichuan. *Sediment. Geol. Tethyan Geol.* **2001**, *21*, 57–63.
27. Wang, M.; Xue, H.; Tian, S.; Wilkins, R.W.T.; Wang, Z. Fractal characteristics of Upper Cretaceous lacustrine shale from the Songliao Basin, NE China. *Mar. Pet. Geol.* **2015**, *67*, 144–153. [CrossRef]
28. Li, Y.; Wang, Z.; Pan, Z.; Niu, X.; Yu, Y.; Meng, S. Pore structure and its fractal dimensions of transitional shale: A cross-section from east margin of the Ordos Basin, China. *Fuel* **2019**, *241*, 417–431. [CrossRef]
29. Ma, X.; Guo, S.; Shi, D.; Zhou, Z.; Liu, G. Investigation of pore structure and fractal characteristics of marine-continental transitional shales from Longtan Formation using MICP, gas adsorption, and NMR (Guizhou, China). *Mar. Pet. Geol.* **2019**, *107*, 555–571. [CrossRef]
30. Wang, X.; Hou, J.; Li, S.; Dou, L.; Song, S.; Kang, Q.; Wang, D. Insight into the nanoscale pore structure of organic-rich shales in the Bakken Formation, USA. *J. Pet. Sci. Eng.* **2020**, *191*, 107182. [CrossRef]
31. Zhu, H.; Huang, C.; Ju, Y.; Bu, H.; Li, X.; Yang, M.; Chu, Q.; Feng, H.; Qiao, P.; Qi, Y.; et al. Multi-scale multi-dimensional characterization of clay-hosted pore networks of shale using FIBSEM, TEM, and X-ray micro-tomography: Implications for methane storage and migration. *Appl. Clay Sci.* **2021**, *213*, 106239. [CrossRef]
32. Sun, L.; Liu, H.; He, W.; Li, G.; Zhang, S.; Zhu, R.; Jin, X.; Meng, S.; Jiang, H. An analysis of major scientific problems and research paths of Gulong shale oil in Daqing Oilfield, NE China. *Pet. Explor. Dev.* **2021**, *48*, 527–540. [CrossRef]
33. *SY/T 5163-2018*; X-ray Diffraction Analysis Method of Clay Minerals and Common Non-Clay Minerals in Sedimentary Rocks. National Energy Administration: Beijing, China, 2018.
34. Peng, R.; Xie, H.; Ju, Y. Computation Method of Fractal Dimension for 2-D Digital Image. *J. China Univ. Min. Technol.* **2004**, *33*, 19–24.
35. White, R.; Budarin, V.; Luque, R.; Clark, J.; Macquarrie, D. Tuneable porous carbonaceous materials from renewable resources. *Chem. Soc. Rev.* **2009**, *38*, 3401–3418. [CrossRef]
36. Stricker, S.; Jones, S.J. Enhanced Porosity Preservation by Pore Fluid Overpressure and Chlorite Grain Coatings in the Triassic Skagerrak, Central Graben, North Sea, UK. In *Conference on Reservoir Quality of Clastic and Carbonate Rocks-Analysis, Modelling and Prediction*; Geological Soc Publishing House: London, UK, 2014. [CrossRef]
37. Zhu, H.; Zhong, D.; Yao, J.; Sun, H.; Niu, X.; Liang, X.; You, Y.; Li, X. Alkaline diagenesis and its effects on reservoir porosity: A case study of Upper Triassic Chang 7 Member tight sandstone in Ordos Basin, NW China. *Pet. Explor. Dev.* **2015**, *42*, 56–65. [CrossRef]
38. Webb, A.D.; Dickens, G.R.; Oliver, N.H.S. Carbonate alteration of the Upper Mount McRae Shale beneath the martite-microplaty hematite ore deposit at Mount Whaleback, Western Australia. *Miner. Depos.* **2004**, *39*, 632–645. [CrossRef]
39. Griffiths, J.; Worden, R.H.; Wooldridge, L.J.; Utley, J.E.P.; Duller, R.A. Compositional variation in modern estuarine sands: Predicting major controls on sandstone reservoir quality. *AAPG Bull.* **2019**, *103*, 797–833. [CrossRef]
40. Wunderlin, C.A.; Collo, G.; Ezpeleta, M.; Reinoso Carbonell, V.V.; Nóbile, J.C.; Ciccioli, P.L.; Faudone, S. Authigenic and detrital clay minerals as indicators of paleoenvironmental and postdepositional evolution in a Cretaceous–Cenozoic succession from Argentine Central Andes. *Sediment. Geol.* **2022**, *437*, 106179. [CrossRef]
41. Zhu, R.; Li, M.; Yang, J.; Zhang, S.; Cai, Y.; Cao, Y.; Kang, Y. Advances and trends of fine-grained sedimentology. *Oil Gas Geol.* **2022**, *43*, 251–264.
42. Wilson, M.J.; Shaldybin, M.V.; Wilson, L. Clay mineralogy and unconventional hydrocarbon shale reservoirs in the USA. I. Occurrence and interpretation of mixed-layer R3 ordered illite/smectite. *Earth-Sci. Rev.* **2016**, *158*, 31–50. [CrossRef]
43. Goral, J.; Panja, P.; Deo, M.; Andrew, M.; Linden, S.; Schwarz, J.-O.; Wiegmann, A. Confinement Effect on Porosity and Permeability of Shales. *Sci. Rep.* **2020**, *10*, 49. [CrossRef] [PubMed]
44. He, W.; Zhu, R.; Cui, B.; Zhang, S.; Meng, Q.; Bai, B.; Feng, Z.; Lei, Z.; Wu, S.; He, K.; et al. The geoscience frontier of Gulong shale oil: Revealing the role of continental shale from oil generation to production. *Engineering* **2023**, *28*, 79–92. [CrossRef]
45. Ye, C.; Yang, Y.; Fang, X.; Hong, H.; Wang, C.; Yang, R.; Zhang, W. Chlorite chemical composition change in response to the Eocene-Oligocene climate transition on the northeastern Tibetan Plateau. *Palaeogeogr. Palaeoclimatol. Palaeoecol.* **2018**, *512*, 23–32. [CrossRef]
46. Anjos, S.M.C.; De Ros, L.F.; Silva, C.M.A. Chlorite Authigenesis and Porosity Preservation in the Upper Cretaceous Marine Sandstones of the Santos Basin, Offshore Eastern Brazil. In *Clay Mineral Cements in Sandstones*; Wiley: Hoboken, NJ, USA, 1999; pp. 289–316. [CrossRef]

47. Zhao, C.; Jiang, Y.; Yang, H.; Wang, L. The genesis of authigenic minerals and the porosity evolution of various lithologies and their implications for identifying high-quality reservoirs in the fourth member of Xujiahe Formation (Northeast-Sichuan Basin, China). *J. Pet. Sci. Eng.* **2022**, *212*, 110261. [CrossRef]
48. Li, K.; Xi, K.; Cao, Y.; Niu, X.; Wu, S.; Feng, S.; You, Y. Chlorite authigenesis and its impact on reservoir quality in tight sandstone reservoirs of the Triassic Yanchang formation, southwestern Ordos basin, China. *J. Pet. Sci. Eng.* **2021**, *205*, 108843. [CrossRef]
49. Wang, Y.; Wang, Z.; Zhang, Z.; Yao, S.; Zhang, H.; Zheng, G.; Luo, F.; Feng, L.; Liu, K.; Jiang, L. Recent techniques on analyses and characterizations of shale gas and oil reservoir. *Energy Rev.* **2024**, *3*, 100067. [CrossRef]
50. Li, X.; Zhu, H.; Zhang, K.; Li, Z.; Yu, Y.; Feng, X.; Wang, Z. Pore characteristics and pore structure deformation evolution of ductile deformed shales in the Wufeng-Longmaxi Formation, southern China. *Mar. Pet. Geol.* **2021**, *127*, 104992. [CrossRef]
51. Yang, F.; Ning, Z.; Wang, Q.; Kong, D.; Peng, K.; Xiao, L. Fractal Characteristics of Nanopore in Shales. *Nat. Gas Geosci.* **2014**, *25*, 618–623.
52. Yang, R.; He, S.; Yi, J.; Hu, Q. Nano-scale pore structure and fractal dimension of organic-rich Wufeng-Longmaxi shale from Jiaoshiba area, Sichuan Basin: Investigations using FE-SEM, gas adsorption and helium pycnometry. *Mar. Pet. Geol.* **2016**, *70*, 27–45. [CrossRef]
53. Fu, H.; Yan, D.; Yao, C.; Su, X.; Wang, X.; Wang, H.; Li, Y. Pore Structure and Multi-Scale Fractal Characteristics of Adsorbed Pores in Marine Shale: A Case Study of the Lower Silurian Longmaxi Shale in the Sichuan Basin, China. *J. Earth Sci.* **2022**, *33*, 1278–1290. [CrossRef]
54. Liu, B.; Wang, Y.; Tian, S.; Guo, Y.; Wang, L.; Yasin, Q.; Yang, J. Impact of thermal maturity on the diagenesis and porosity of lacustrine oil-prone shales: Insights from natural shale samples with thermal maturation in the oil generation window. *Int. J. Coal Geol.* **2022**, *261*, 104079. [CrossRef]
55. Liu, B.; Wang, H.; Fu, X.; Bai, Y.; Bai, L.; Jia, M.; He, B. Lithofacies and depositional setting of a highly prospective lacustrine shale oil succession from the Upper Cretaceous Qingshankou Formation in the Gulong sag, northern Songliao Basin, northeast China. *AAPG Bull.* **2019**, *103*, 405–432. [CrossRef]
56. Huang, D. Advances in hydrocarbon generation theory (I): Generation and evolution model for immature oils and hydrocarbon. *J. Pet. Sci. Eng.* **1999**, *22*, 121–130. [CrossRef]
57. Vandenbroucke, M.; Largeau, C. Kerogen origin, evolution and structure. *Org. Geochem.* **2007**, *38*, 719–833. [CrossRef]
58. Zhao, J.; Jin, Z.; Jin, Z.; Wen, X.; Geng, Y. Origin of authigenic quartz in organic-rich shales of the Wufeng and Longmaxi Formations in the Sichuan Basin, South China: Implications for pore evolution. *J. Nat. Gas Sci. Eng.* **2017**, *38*, 21–38. [CrossRef]
59. Wang, S.W.; Li, X.J.; Xue, H.T.; Shen, Z.H.; Chen, L.Y. Fractal characteristics of shale pore structure and its influence on seepage flow. *R. Soc. Open Sci.* **2021**, *8*. [CrossRef]
60. Jiang, Z.; Zhang, W.; Liang, C.; Wang, Y.; Liu, H.; Chen, X. Basic characteristics and evaluation of shale oil reservoirs. *Pet. Res.* **2016**, *1*, 149–163. [CrossRef]
61. Jiang, F.J.; Chen, D.; Chen, J.; Li, Q.W.; Liu, Y.; Shao, X.H.; Hu, T.; Dai, J.X. Fractal Analysis of Shale Pore Structure of Continental Gas Shale Reservoir in the Ordos Basin, NW China. *Energy Fuels* **2016**, *30*, 4676–4689. [CrossRef]
62. Yang, F.; Ning, Z.; Liu, H. Fractal characteristics of shales from a shale gas reservoir in the Sichuan Basin, China. *Fuel* **2014**, *115*, 378–384. [CrossRef]
63. Yang, Y.; Wang, B.; Yuan, Q.; Huang, D.; Peng, H. Characterization, factors, and fractal dimension of pore structure of fly ash-based geopolymers. *J. Mater. Res. Technol.* **2023**, *26*, 3395–3407. [CrossRef]
64. Liu, Z.; Han, J.; Yang, H.; Lv, J.; Dong, S. A new model for coal gas seepage based on fracture-pore fractal structure characteristics. *Int. J. Rock Mech. Min. Sci.* **2024**, *173*, 105626. [CrossRef]
65. Medina, F.J.; Jausoro, I.; Floridia Addato, M.A.; Rodriguez, M.J.; Tomassini, F.G.; Caneiro, A. On the evaluation of Representative Elementary Area for porosity in shale rocks by Field Emission Scanning Electron Microscopy. *Energy* **2022**, *253*, 124141. [CrossRef]
66. Arif, M.; Mahmoud, M.; Zhang, Y.; Iglauer, S. X-ray tomography imaging of shale microstructures: A review in the context of multiscale correlative imaging. *Int. J. Coal Geol.* **2021**, *233*, 103641. [CrossRef]
67. Cosenza, P.; Prêt, D.; Fauchille, A.-L.; Hedan, S. Representative elementary area of shale at the mesoscopic scale. *Int. J. Coal Geol.* **2019**, *216*, 103316. [CrossRef]
68. Saif, T.; Lin, Q.; Bijeljic, B.; Blunt, M.J. Microstructural imaging and characterization of oil shale before and after pyrolysis. *Fuel* **2017**, *197*, 562–574. [CrossRef]
69. Goral, J.; Andrew, M.; Olson, T.; Deo, M. Correlative core- to pore-scale imaging of shales. *Mar. Pet. Geol.* **2020**, *111*, 886–904. [CrossRef]

70. Kelly, S.; El-Sobky, H.; Torres-Verdín, C.; Balhoff, M.T. Assessing the utility of FIB-SEM images for shale digital rock physics. *Adv. Water Resour.* **2016**, *95*, 302–316. [CrossRef]
71. Saif, T.; Lin, Q.; Butcher, A.R.; Bijeljic, B.; Blunt, M.J. Multi-scale multi-dimensional microstructure imaging of oil shale pyrolysis using X-ray micro-tomography, automated ultra-high resolution SEM, MAPS Mineralogy and FIB-SEM. *Appl. Energy* **2017**, *202*, 628–647. [CrossRef]

fractal and fractional

MDPI

Article

Evolution of Pore Structure and Fractal Characteristics in Transitional Shale Reservoirs: Case Study of Shanxi Formation, Eastern Ordos Basin

Yifan Gu [1,2,*], Xu Wu [1,2], Yuqiang Jiang [1,2], Quanzhong Guan [3], Dazhong Dong [4] and Hongzhan Zhuang [1,2]

1 School of Geoscience and Technology, Southwest Petroleum University, Chengdu 610500, China; 202221000140@stu.swpu.edu.cn (X.W.); 198531010028@swpu.edu.cn (Y.J.); 202411000047@stu.swpu.edu.cn (H.Z.)

2 Collaborative Innovation Center of Shale Gas Resources and Environment, Chengdu 610500, China

3 College of Energy (College of Modern Shale Gas Industry), Chengdu University of Technology, Chengdu 610059, China; guanqz@cdut.edu.cn

4 PetroChina Research Institute of Petroleum Exploration and Development, Beijing 100083, China; ddz@petrochina.com.cn

* Correspondence: 202099010168@swpu.edu.cn

Abstract: The fractal dimension quantitatively describes the complexity of the shale pore structure and serves as a powerful tool for characterizing the evolution of shale reservoirs. Thermal simulation experiments were conducted on the low-maturity transitional shale from the Shanxi Formation in the Ordos Basin. The initial samples consisted mainly of quartz (39.9%) and clay minerals (49.9%) with moderate-to-good hydrocarbon generation potential. Samples from different thermal maturation stages were analyzed through geochemical, mineralogical, and pore structure experiments to reveal the evolution of mineral compositions and pore structure parameters. The fractal dimensions of the pore structure were calculated using both the FHH and capillary bundle models. Correlation coefficients and principal component analysis (PCA) were employed to explore the factors influencing the fractal dimension and its evolutionary patterns during reservoir development. The results indicate that (1) with increasing thermal maturity, the quartz content gradually increases while the contents of clay minerals, carbonate minerals, pyrite, and feldspar decrease. (2) The evolution of porosity follows five stages: a slow decrease (0.78 < Ro < 1.0%), a rapid increase (1.0% < Ro < 2.0%), a relatively stable phase (2.0% < Ro < 2.7%), a rapid rise (2.7% < Ro < 3.2%), and a slow decline (Ro > 3.2%). The evolution of the pore volume (PV) and specific surface area (SSA) indicates that the proportion of pores in the 5–20 nm and 20–60 nm ranges gradually increases while the proportion of pores smaller than 5 nm decreases. (3) The fractal dimension of shale pores (D_1, average value 2.61) derived from the FHH model is higher than D_2 (average value 2.56). This suggests that the roughness of pore surfaces is greater than the complexity of the internal pore structure at various maturities. The D_M distribution range calculated from the capillary bundle model was broad (between 2.47 and 2.94), with an average value of 2.84, higher than D_1 and D_2. This likely indicates that larger pores have more complex structures. (4) D_1 shows a strong correlation with porosity, PV, and SSA and can be used to reflect pore development. D_2 correlates well with geochemical parameters (TOC, HI, etc.) and minerals prone to diagenetic alteration (carbonates, feldspar, and pyrite), making it useful for characterizing the changes in components consumed during pore structure evolution. (5) Based on the thermal maturation process of organic matter, mineral composition, diagenesis, and pore structure evolution, an evolutionary model of the fractal dimension for transitional shale was established.

Keywords: Ordos Basin; transitional shale; pore structure evolution; heterogeneity; multifractal dimension

1. Introduction

Fractal theory was introduced by Mandelbrot in 1967 [1,2]. Unlike traditional Euclidean geometry, this theory is commonly used to analyze materials or patterns that exhibit self-similarity [1,2]. In recent years, fractal theory has been widely applied in the study of unconventional oil and gas reservoirs (such as shale, coalbeds, and tight sandstones), enabling the quantitative assessment of micro-reservoir heterogeneity [3–6]. Previous studies had utilized methods such as SEM imaging, nitrogen adsorption (LTNA), mercury intrusion (MICP), small-angle scattering (SANS), and nuclear magnetic resonance (NMR) to conduct fractal analysis [7–9]. Additionally, many researchers have attempted to apply models such as the Frenkel–Halsey–Hill (FHH), Neimark, Brunauer–Emmett–Teller (BET), and Menger sponge models to study fractal properties of the reservoir space in different types of shale [10–14]. Currently, the most popular approach involves combining multiple experimental techniques to obtain the full pore size distribution of the pore structure [6,15,16]. CO_2 adsorption experiments and LTNA data are commonly used for the fractal characterization of micropores and mesopores [6,17,18]. MICP experiments and NMR data are used to characterize the fractal features of larger pores and microfractures, as well as of other large-scale spaces [8,9,19,20]. Furthermore, conducting mathematical analysis using the fractal dimensions of different pore types, along with pore structures and shale compositions, helps identify the main controlling factors of fractal dimensions. This approach also supports the development of models to evaluate pore heterogeneity [21–23].

The transitional shale in the Ordos Basin holds significant resource potential and is considered a key contributor to the future growth of shale gas production in China due to its suitability for large-scale exploration and development [24,25]. Compared to marine shale, transitional shale typically contains a higher clay content and exhibits greater variation in thickness, stronger reservoir heterogeneity, and more pronounced differences in gas content, making the development strategy more complex [26–28]. As hydrocarbon generation progresses, the pore structure of shale and its controlling factors undergo significant changes. These variations pose major challenges for predicting sweet spots and guiding efficient reservoir development [29,30]. The fractal dimension provides a quantitative measure of the heterogeneity and complexity of shale pore structures. It also serves as a bridge linking microscopic pore characteristics (such as the pore size distribution and connectivity) with reservoir properties, lithology, and geochemical parameters [4,5,19,20]. A higher fractal dimension indicates greater pore heterogeneity. This often corresponds to larger internal surface areas and higher surface energy, which can hinder hydrocarbon flow and reduce recovery efficiency [16]. Therefore, studying the pore structure and fractal characteristics of transitional shale is essential for accurately evaluating reservoir quality, predicting productive zones, and enhancing hydrocarbon recovery.

To investigate the pore structure characteristics, fractal features, and controlling factors of shale at different stages of evolution, this study selected low-maturity transitional shale from the Permian Shanxi Formation in the Eastern Ordos Basin for thermal simulation experiments. Shale samples of varying maturities were subjected to fractal analysis, enabling a comprehensive quantitative evaluation of heterogeneity throughout the micro-evolution process. Initially, experiments such as mineral composition analysis, geochemical parameter assessment, scanning electron microscopy (SEM), and nitrogen adsorption were used to analyze the pore structure characteristics of the transitional shale. Next, the fractal dimension of the shale was calculated using the FHH model based on LTNA and the mercury intrusion capillary bundle model. Finally, through correlation analysis and principal component analysis, the relationship between the pore structure and fractal dimension of the transitional shale was revealed, and the factors influencing the fractal dimension were discussed. The variations in fractal dimension across different evolutionary stages and their underlying mechanisms were explained.

2. Geological Setting

The Ordos Basin is a typical cratonic basin in northern China (Figure 1a), consisting of six tectonic units [24]. The Ordovician-to-Early-Carboniferous depositional processes are absent in the Ordos Basin due to the effects of the Caledonian orogeny. During the Benxi-Formation-to-Taiyuan-Formation deposition, the Ordos Basin underwent continuous subsidence and experienced multiple marine transgressions. Under an epicontinental sea setting, thick coal-bearing transitional marine-continental strata were formed. Transitional strata are the primary targets for coalbed methane and shale gas exploration in the Ordos Basin [27]. During the deposition of the Shanxi Formation, seawater gradually receded from both the eastern and western sides of the Ordos Basin, transitioning from a marine to a terrestrial environment (Figure 1b). The Shanxi Formation is in conformable contact with both the underlying Taiyuan Formation and the overlying Lower Shihezi Formation. The upper lithology of the Taiyuan Formation consists of bioclastic limestone or microcrystalline limestone, which is indicative of an open platform environment in a shallow water continental shelf setting. The lower part of the Xiashihezi Formation primarily consists of medium sandstone, characteristic of a terrestrial braided river sedimentary system. This study focused on the transitional shale of the Permian Shanxi Formation [24].

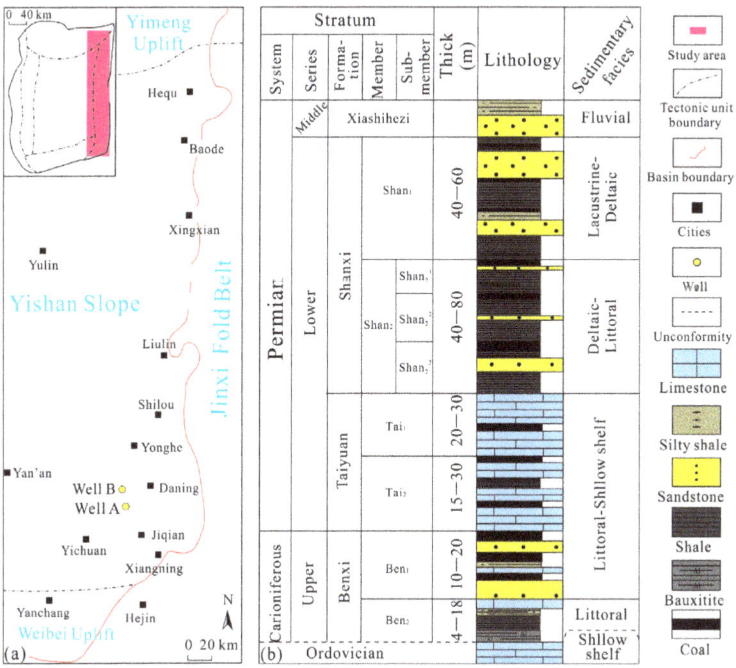

Figure 1. (**a**) The study area in the southeastern Ordos Basin. (**b**) Generalized stratigraphy of Formation in the eastern Ordos Basin (modified from Kuang et al., 2020 [24]).

3. Materials, Experiments and Theories

3.1. Sample

To thoroughly investigate the impact of thermal evolution on pore structure and its fractal characteristics, a low-maturity transitional shale from a field outcrop in the Eastern Ordos Basin was selected as the sample for this simulation experiment (Figure 1a). The low-maturity transitional shale belongs to the Shanxi Formation and was formed in a lagoonal environment. The initial sample has a vitrinite reflectance (Ro) of 0.78% and a total organic carbon (TOC) content of 2.87%.

In this study, a closed system was used for the thermal simulation experiments. The experimental equipment was the DK-II type strata–porosity–thermal press hydrocarbon simulation instrument, independently developed by the Wuxi Research Institute of Sinopec. Detailed experimental procedures and specific equipment parameters can be found in previous study reports [31]. This study involved seven sets of parallel experiments, with the specific experimental parameters provided in Table 1.

Table 1. Simulation parameter information of samples.

Sample	Temperature (°C)	Static Rock Pressure (MPa)	Formation Pressure (MPa)	Holding Time (h)
S1	325	50	21	48
S2	350	68	25	48
S3	400	75	30	48
S4	450	85	34	48
S5	500	90	36	48
S6	550	95	38	48
S7	575	100	40	48

3.2. Experiments

After the thermal simulation experiments, seven solid products (S1–S7) were obtained. The seven solid products, along with the initial sample (So), underwent a series of tests, including TOC analysis, vitrinite reflectance (Ro), rock pyrolysis, gas adsorption (LTNA), and mercury intrusion porosimetry (MICP). TOC content was measured using the CS744-MHPC carbon–sulfur analyzer. Ro was determined using a vitrinite reflectance meter and a Zeiss ScopeA1 polarizing microscope. These experiments were conducted at the China Sichuan Keyuan Testing Center of Engineering Technology Co., Ltd. (Chengdu, China). The Rock-Eval pyrolysis instrument from Aitch company (Paris, France) was used to measure the free hydrocarbons (S_1), kerogen cracking hydrocarbons (S_2), and the temperature of maximum pyrolysis product yield (Tmax) of the initial samples. The calculation method for the Hydrogen Index (HI) followed the approach proposed by Espitalié [32]. Mineral composition analysis was conducted at the China Petroleum Exploration and Development Research Institute using a Rigaku X-ray diffractometer from Tokyo, Japan. The NMR T_2 spectra were obtained using the NMRC12-010V low-temperature nanopore analyzer, manufactured by Numa in Suzhou, China. The plunger samples in dry, deionized water-saturated, dodecane-saturated, and Mn^{2+}-saturated states were subjected to signal measurement. Detailed operational procedures and experimental theories can be found in Wang et al. (2022) [33]. Nitrogen adsorption (LTNA) experiments were carried out using the Autosorb-IQ3 fully automated surface area and pore distribution analyzer, produced by Contech (Goddard, KS, USA). Under a vacuum at 110 °C, the samples were degassed for 12 h before undergoing LTNA experiments. The density functional theory (DFT) model was used to calculate the micropore and meso/macropore surface areas, volumes, and pore size distributions. Mercury intrusion capillary pressure (MICP) tests were conducted using the AutoPore IV 9520 fully automated mercury intrusion instrument, produced by Contech (USA). This experiment was carried out at China University of Geosciences. Argon ion polishing and FE-SEM observations were performed at Southwest Petroleum University. Each shale sample was first cut into small cubes measuring 1 cm $\times$ 1 cm $\times$ 1 cm. Next, the argon ion polishing system was used to flatten and smooth the surfaces of the samples. After processing, the samples were examined using the FEI Quanta 650 FEG field emission scanning electron microscope (using SEM).

3.3. Fractal Dimension Calculation

3.3.1. Fractal Dimension Calculation Based on LTNA Data

Based on LTNA data, the modified FHH (Frenkel–Halsey–Hill) theory by Pfeifer [12] effectively characterizes the fractal features of pores in shale at different scales. There are two methods for calculating the fractal dimension using the FHH model. The first method is based on the van der Waals force mechanism. The second method is based on the capillary condensation mechanism [34]. The fractal dimension calculation method based on capillary condensation is more suitable for studying the heterogeneity of porous media. The specific calculation formula is as follows:

$$lnV = C + (D-3)lnln\frac{p_0}{p} \quad (1)$$

In this formula, V represents the volume of gas adsorbed at an absolute pressure P (unit: mL/g). p_0 represents the saturation vapor pressure (unit: MPa). D represents the fractal dimension. C is a constant. During the calculation, a scatter plot is created with ln (lnp_0/p) on the x-axis and lnV on the y-axis. The fitted slope is then converted into the desired value of D. Based on the FHH model, the calculated fractal dimension ranges from 2 to 3. A higher fractal dimension indicates a more complex and irregular pore system.

The double logarithmic curve typically exhibits a distinct inflection point. Using this inflection point as a boundary, the fractal dimensions for different adsorption stages can be calculated. Most researchers choose the point where the relative pressure (p/p_0) is between 0.45 and 0.5 as the inflection point [6,17]. This paper adopts the same inflection point, dividing D into two ranges: D_1 ($p/p_0 < 0.45$) and D_2 ($p/p_0 > 0.45$).

3.3.2. Fractal Dimension Calculation Based on MICP Data

The fractal dimension derived from the capillary bundle model shows a stronger correlation with shale porosity and pore structure parameters. Therefore, the capillary bundle model is often used in single fractal studies of MICP pore size distributions [15,20]. The specific calculation formula is as follows:

$$\log(1 - S_{Hg}) = (D-3) \times logP_c - (D-3) \times logP_{min} \quad (2)$$

In the formula, S_{Hg} represents the proportion of mercury volume entering the pores (unit: %), which is equivalent to the pore volume fraction. p_0 represents the capillary pressure (unit: MPa), which is equivalent to pore diameter. Therefore, the above formula is often converted into the following equation [35]:

$$logS = (3-D) \times logr + (D-3) \times logr_{max} \quad (3)$$

In this formula, r represents the pore radius (unit: nm). r_{max} represents the maximum pore radius (unit: nm). s represents the cumulative pore volume fraction corresponding to the pore radius (r). D represents the fractal dimension of the porous medium. Therefore, the fractal dimension (D) can be calculated from the slope of the lg (dV/dp) vs. lgp curve.

To differentiate the fractal dimension calculated from LTNA data, the fractal dimension based on MICP data will be labeled as DM. Additionally, considering the lower accuracy of MICP in measuring small pores and taking into account the pore size range of the study subject [8,16,20], this paper focuses on fitting the fractal dimension of pores within the 50 nm to 1000 nm range.

4. Results

4.1. Mineralogy and Geochemical Characteristics of Shale

The geochemical parameters of the initial sample and the post-thermal simulation samples are shown in Table 2. The TOC values ranged from 2.07 to 2.87, with an average of 2.44 (Figure 2a). The cross-plot of TOC and (S_1 + S_2) confirms that the initial sample (S_0) had moderate-to-good hydrocarbon generation potential [36].

Table 2 presents the mineral composition of the initial sample and the post-simulation samples. The initial sample was primarily composed of quartz (39.9%) and clay minerals (49.9%). The carbonate mineral content was low (6.3%), mainly consisting of siderite (4.7%) and dolomite (1.6%). The clay minerals in the initial sample were mainly kaolinite, with no smectite present. As thermal evolution increased, the quartz content gradually rose. The contents of clay minerals, carbonate minerals, pyrite, and feldspar all gradually decreased (Figure 2b). The illite content gradually increased while the kaolinite content decreased until it reached zero. The contents of I/S and chlorite both slightly decreased after reaching their peak values (Figure 2b).

4.2. Porosity and Pore Structure Characteristics

4.2.1. Porosity Characterization

Although the solid samples after simulation were crushed, the NMR signal intensity of the product of unit mass and TOC could be used to characterize porosity [37]. The NMR signal intensity after saturation with water is shown in Figure 3a. The NMR signal intensity showed a progressive change at different thermal evolution stages. As thermal evolution increased, porosity changes occurred in five stages: slow decrease (0.78 < Ro < 1.0%), rapid increase (1.0% < Ro < 2.0%), relative stability (2.0% < Ro < 2.7%), rapid increase (2.7% < Ro < 3.2%), and slow decrease (Ro > 3.2%). Using SEM and Image-J software (v1.8.0), the total pixel area of organic pores, organic-matter-related microcracks, inorganic pores, and microcracks within a 200 um × 200 um range was extracted. The proportions of different types of spaces were characterized (Figure 3b). The results indicate that the proportion of organic pores reaches its maximum early in the over-mature stage and then gradually decreases.

4.2.2. LTNA and MICP Curves

The mercury intrusion–extrusion curves for samples S0, S1, and S2 show no distinct inflection points, and the total mercury intrusion was low (Figure 4a). This suggests that these samples were relatively dense, primarily consisted of small pores, and had poor connectivity. For other samples, once the mercury intrusion pressure exceeded 10 MPa, the rate and volume of mercury intrusion significantly increased. This indicates that these samples had good connectivity and larger average pore diameters. The total mercury intrusion, inflection points, and rising rates differed significantly across samples, indicating variations in the pore size distribution and total pore volume [30,38].

Table 2. Geochemical and mineral composition parameter variations in the simulated samples.

Sample	Temperature (°C)	Ro (%)	TOC (%)	$S_1 + S_2$ (mg/g)	HI (mg/g TOC)	Quartz (%)	Feldspar (%)	Clay (%)	Dolomite (%)	Siderite (%)	Pyrite (%)	Kaolinite (%)	I/S (%)	Illite (%)	Chlorite (%)
S0	Initial	0.78	2.87	2.82	122	39.9	1.4	49.9	2.1	4.7	2.0	85	3	9	3
S1	350	0.85	2.75	2.64	98	40.6	1.2	49.4	2.4	4.2	2.2	77	3	16	4
S2	375	1.09	2.6	2.22	66	42.3	1.2	47.9	2.2	4.6	1.8	73	4	19	4
S3	400	1.52	2.43	1.17	48	46.5	0.8	45.6	1.5	4.2	1.4	63	8	24	5
S4	450	2.13	2.34	0.28	14	48.5	0.7	44.5	1.3	3.4	1.6	35	18	41	6
S5	500	2.76	2.31	0.12	6	49.5	0.6	44.2	1.1	3.2	1.4	23	22	44	11
S6	550	3.27	2.07	0.09	4	50.4	0.5	43.3	1.0	3.6	1.2	3	16	71	10
S7	575	3.86	2.12	0.04	1	52.0	0.6	42.4	1.1	3.0	0.9	0	13	76	11

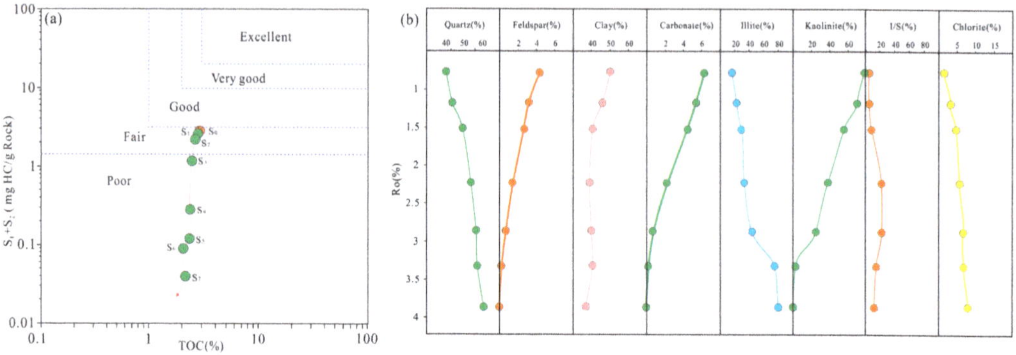

Figure 2. (**a**) TOC versus (S1 + S2) cross-plot showing that the quality of samples (the evaluation criterion is from El Nady et al., 2015 [36]). (**b**) Mineral composition and content variation data of the simulated samples.

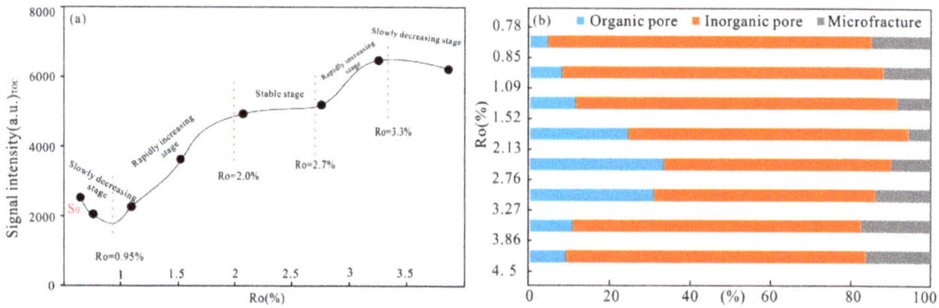

Figure 3. (**a**) Porosity evolution characteristics and (**b**) proportions of different types of pores of simulated samples.

The LTNA curves of the post-simulation samples exhibit inverted "S" shapes. However, the adsorption curve shapes differed significantly between samples of different maturity, suggesting that both diagenesis and hydrocarbon generation processes jointly control changes in the pore structure during evolution. The LTNA isotherm hysteresis loop shapes of the post-simulation samples are classified into two types according to IUPAC: H3 and H4 (Figure 4b). The H4 type corresponds to the S0 and S1 samples. A narrow hysteresis loop indicates that narrow fissure pores dominate in the shale, favoring the adsorption and accumulation of shale gas but hindering flow. The H3 type corresponds to the S3, S4, and S5 samples, with wider hysteresis loops than in S0 and S1. This indicates that slit pores dominated by layered structures prevail, resulting in poor pore connectivity, although the average pore diameter is relatively large [30,38,39].

4.2.3. Pore Size Distribution and Pore Structure Parameters

The pore volume (PV) and specific surface area (SSA) distributions of shale samples were derived from the integration of the MICP and LTNA curves. MICP was used to characterize the pore size distribution in the range of 50–1000 nm. LTNA was used to characterize the pore size distribution in the range of 1–50 nm. To minimize the effect of TOC on pore structure, the PV and SSA values of different samples were normalized based on their TOC values. The PV distribution ranged from 0.0027 to 0.0263 cm^3/g (with an average of 0.0115 cm^3/g), and the SSA distribution ranged from 1.097 to 5.309 m^2/g (with an average of 2.718 m^2/g). Figure 4c,d show that as the thermal evolution increased, both the PV of 10–100 nm pores and the SSA of 10–80 nm pores significantly increased.

To characterize changes in the PV and SSA across different pores, the pore size distribution was divided into six intervals: <5 nm, 5–20 nm, 20–60 nm, 60–120 nm, 120–500 nm, and 500–1000 nm. The PV and SSA distributions for different pore sizes are shown in Table 3. The PV and SSA of the initial samples were primarily contributed by pores smaller than 5 nm. As the simulation temperature increased, the PV and SSA of 5–20 nm and 20–60 nm pores gradually increased (Figure 5a,b).

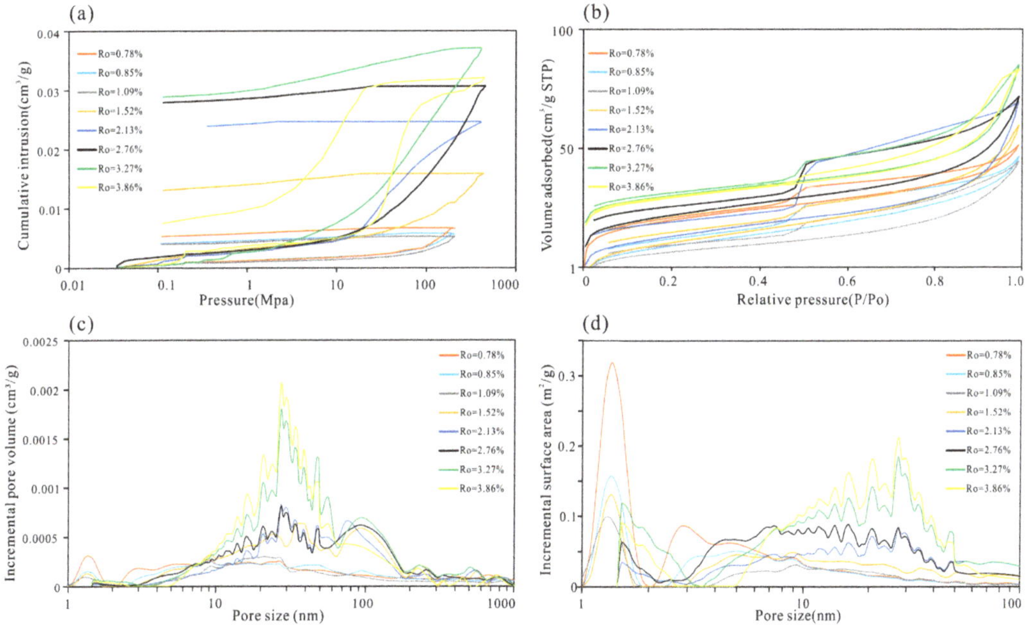

Figure 4. (**a**) Mercury curve, (**b**) adsorption–desorption curve, (**c**) pore volume, and (**d**) specific surface area distribution of simulated samples.

4.3. SEM Image Analysis

4.3.1. Organic Pores and Organic-Matter-Related Microfractures

The morphology and development of organic pores in transition shale vary significantly at different stages of thermal evolution, being influenced by the organic matter content, type, and thermal maturity [40,41]. At the low-maturity stage (Figure 6a), organic matter is distributed in striped and lumped forms within the shale. Organic pores on the surface of the organic matter are not observed. In the early maturity stage (Figure 6b), kerogen degradation produces oil, which causes the release of a small amount of retained oil from the kerogen surface, forming large elliptical organic pores. However, these organic pores are isolated from one another [42]. Additionally, after the degradation of the kerogen surface components, volume shrinkage occurs. Narrow microfractures form at the edges and within the kerogen (Figure 6b) [41,43]. In the late-maturity stage (Figure 6c), retained hydrocarbons on the kerogen surface are released, forming a large number of small organic pores [44]. In the early high-maturity stage (Figure 6d), retained hydrocarbons and kerogen, filled in inorganic pores, continue to undergo cracking. On one hand, the inorganic pores that were filled earlier are released [44,45], while organic pores ranging from 50 to 200 nm form within the solid bitumen [41,42]. In the late high-maturity stage (Figure 6e), the sizes of the organic pores continue to increase, reaching 100–400 nm. These organic pores form a honeycomb-like internal structure. Microfractures smaller than 80 nm in width form at the edges of the organic matter [46]. In the over-mature stage (Figure 6f), retained hydrocarbons are gradually consumed [42]. Organic pores cease to form after a vitrinite reflectance (Ro) of 2.7%. Organic pores formed in the early stages are protected and supported by the adsorption of retained oil and wet gas. As a result, organic pores formed in the early stages are not significantly affected by strong compaction. In contrast, after Ro reaches 3.2% (Figure 6g), wet gas and retained oil have been fully converted into methane. Intense compaction causes circular organic pores and organic-matter-related microfractures to be squeezed and reduced in size [45,46].

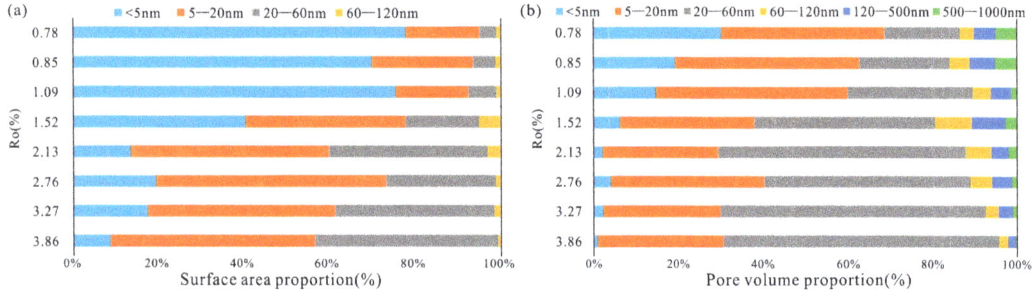

Figure 5. (**a**) Specific surface area and (**b**) pore volume proportion of simulated samples.

4.3.2. Inorganic Pores and Microfractures

From the low-maturity to the early maturity stages, compaction significantly impacts pore development, leading to the close packing of grains (or particles). The number of residual primary intergranular pores rapidly decreases [41] (Figure 7a). Compaction is mainly reflected in the deformation of inorganic minerals and the formation of microfractures within them (Figure 7b,c,h). Additionally, intense compaction causes the deformation of organic matter, organic pores, and organic-matter-related microfractures (Figure 6h). In the mature-to-high-maturity stages, after kerogen is cracked, it releases a large quantity of organic acids, which leads to the dissolution of minerals such as feldspar, carbonate minerals, and pyrite. A large number of intraparticle and intergranular dissolution pores are formed (Figure 7d,e). In the over-mature stage, the pH of the diagenetic environment increases significantly, and early-formed quartz is also dissolved, forming narrow-shaped pores. However, these pores are limited in size and number [47] (Figure 7f). Replacement processes are primarily seen in the mutual transformation of clay minerals, accompanied by the formation of many secondary pores. The transformation of clay minerals occurs at various thermal evolution stages. Under relatively acidic conditions during the low-to-high-maturity stages, feldspar transforms into kaolinite and montmorillonite (I/S) and kaolinite transform into illite (Figure 7a). During the transformation, siliceous cement is precipitated [48]. In the over-mature stage, under relatively alkaline and high-temperature conditions, kaolinite is largely transformed into illite, I/S, and chlorite (Figure 7f,g). Additionally, during the transformation of clay minerals, a widespread contraction of mineral volume occurs, and microfractures form at the mineral edges (Figure 7a,g) [38,48].

Table 3. Pore structure parameters of simulated samples.

Sample ID	Ro	Pore Volume (cm^3/g)							Specific Surface Area (m^2/g)				
		<5 nm	5–20 nm	20–60 nm	60–120 nm	120–500 nm	500–1000 nm	Total	<5 nm	5–20 nm	20–60 nm	60–120 nm	Total
S0	0.78	0.00111	0.00140	0.00065	0.00012	0.00019	0.00018	0.0037	1.415	0.311	0.072	0.017	1.814
S1	0.85	0.00061	0.00138	0.00057	0.00015	0.00019	0.00017	0.0032	0.908	0.303	0.067	0.016	1.294
S2	1.09	0.00040	0.00123	0.00081	0.00012	0.00013	0.00004	0.0027	1.040	0.231	0.087	0.014	1.373
S3	1.52	0.00040	0.00207	0.00277	0.00056	0.00052	0.00018	0.0065	0.445	0.411	0.186	0.055	1.097
S4	2.13	0.00024	0.00309	0.00661	0.00069	0.00046	0.00024	0.0113	0.311	1.073	0.844	0.069	2.298
S5	2.76	0.00054	0.00492	0.00657	0.00070	0.00064	0.00017	0.0135	0.653	1.811	0.845	0.037	3.346
S6	3.27	0.00054	0.00689	0.01558	0.00073	0.00084	0.00028	0.0249	0.935	2.336	1.960	0.078	5.309
S7	3.86	0.00020	0.00787	0.01711	0.00059	0.00045	0.00010	0.0263	0.446	2.522	2.215	0.031	5.214

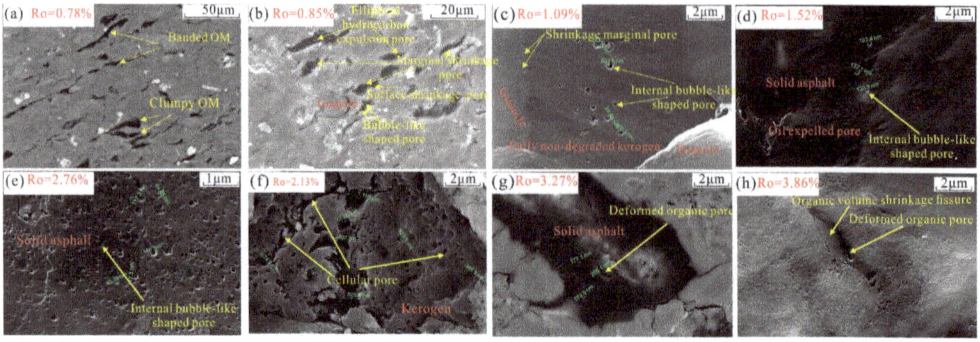

Figure 6. SEM characteristics of organic pores of simulated samples.

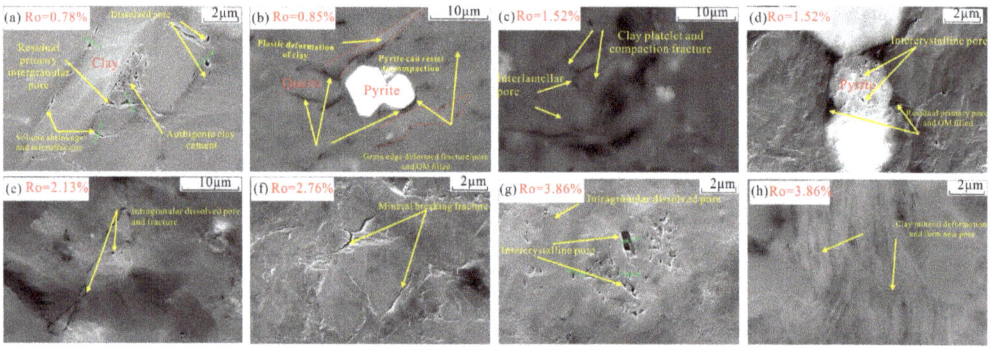

Figure 7. SEM characteristics of inorganic pores of simulated samples.

4.4. Multifractal Characteristics

4.4.1. Fractal Dimension Based on LTNA

The FHH fractal analysis showed that the adsorption volume of the shale samples had a logarithmic relationship with relative pressure. Two distinct adsorption behaviors were observed within the pressure ranges of P/Po = 0.01–0.45 and 0.45–0.98, indicating the dual fractal nature of the shale pore system, the slope of curves (a)–(d) in Figure 8 changes when passing through the same position, which proves this point. D_1 represents the fractal dimension in the low-pressure region (P/Po < 0.45), which corresponds to monolayer and multilayer adsorption and filling in micropores [16]. It is used to describe the irregularity, roughness, and complexity of the pore surface [17,20]. D2 denotes the fractal dimension in the high-pressure range (P/Po > 0.45). As pressure increases, capillary condensation occurs within the pores. Thus, D2 reflects the complexity and heterogeneity of the pore volume and structure.

The results show that D1 values for all shale samples ranged from 2.56 to 2.70, with an average of 2.61 (Table 4). D2 values ranged from 2.45 to 2.72, with an average of 2.56. The correlation coefficients for both D1 and D2 models exceeded 0.97, indicating a strong fractal behavior in the pore structures across all samples. The higher average value of D1 compared to D2 suggests that, across different maturity levels, the pore surface was more irregular and complex than the internal pore structure.

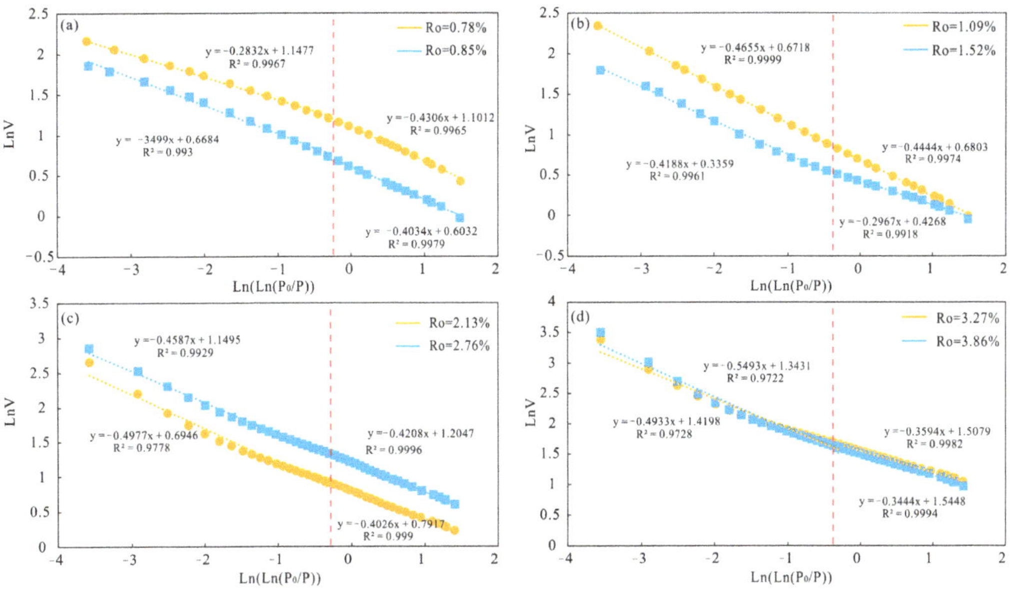

Figure 8. Fractal characteristics of pore structures of shale samples based on nitrogen adsorption (LTNA).

Table 4. Fractal parameters of simulated sample. DM calculation method was used with reference to Li et al., 2017 [20].

Sample ID	Ro	0.01 < P/P0 < 0.45		0.45 < P/P0 < 0.98		HPMI (50 nm< r <1000 nm)				
		D1	R^2	D2	R^2	DM1	R^2	DM2	R^2	DM
S0	0.78	2.57	0.997	2.72	0.998	2.92	0.990	2.96	0.995	2.94
S1	0.85	2.60	0.998	2.65	0.993	2.90	0.993	2.94	0.997	2.92
S2	1.09	2.56	0.997	2.53	1.00	2.91	0.996	2.96	0.996	2.94
S3	1.52	2.70	0.992	2.58	0.996	2.87	0.998	2.95	0.989	2.91
S4	2.13	2.60	1.00	2.50	0.978	2.61	0.987	2.96	0.989	2.72
S5	2.76	2.58	1.00	2.54	0.993	2.25	0.995	2.96	0.894	2.47
S6	3.27	2.64	0.998	2.51	0.973	2.77	0.999	2.94	0.989	2.87
S7	3.86	2.66	1.00	2.45	0.972	2.86	1.00	2.97	0.979	2.92

4.4.2. Fractal Dimension Based on MICP

MICP is primarily used to characterize macropores (pore diameter > 50 nm). The fitting results from the MICP model indicate that the pore diameter of shale samples follows a logarithmic relationship with cumulative pore volume. Two distinct adsorption behaviors are observed when the pore diameter (r) ranges from 40 to 80 nm and 80 to 1000 nm. This phenomenon demonstrates that the shale samples possess dual fractal characteristics in their pore structures, similarly the slope of curves (a)–(d) in Figure 9 changes when passing through the same position, which proves this point. DM1 represents the fractal dimension in the small pore diameter range (r < 80 nm) while DM2 represents the fractal dimension in the large pore diameter range (r > 80 nm). The method for calculating DM is based on the approach proposed by Li et al. (2017) [20]. In this study, the DM values ranged widely, from 2.47 to 2.94, with an average of 2.84. The average DM value was higher than both D1 and D2, indicating that the larger pores had a more complex pore structure. The fractal dimension of sample S5 was the smallest while S0 and S2 exhibited higher fractal dimensions.

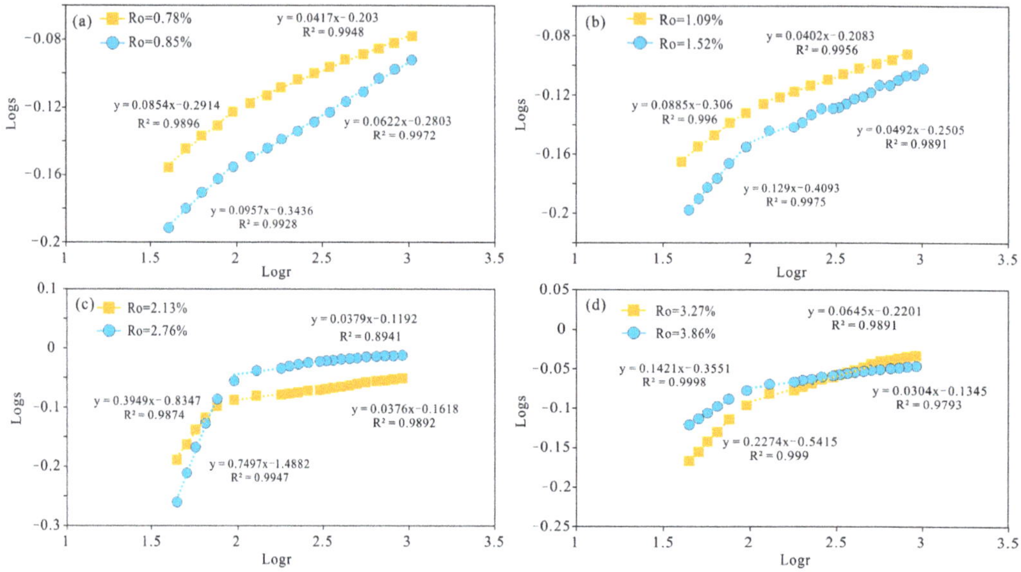

Figure 9. Fractal characteristics of shale pores based on high-pressure mercury injection data.

5. Discussion

5.1. Relationship Between Fractal Dimension and Porosity and Pore Structure

Porosity and pore structure differences are key indicators of reservoir heterogeneity. There are variations in the relationships between the SSA, PV, and fractal dimensions across shale samples of different maturities (Figure 10a). This article has explored the influence of fractal dimensions D_1 and D_2 as representative parameters of the pore surface and pore internal complexity, porosity, or thermal evolution degree on the complexity of shale internal pores. A weak negative correlation was observed between D_1 and D_2 for all samples, indicating an inverse relationship between the complexity and irregularity of pore surfaces and pore spaces across different thermal evolution stages. Except for pores smaller than 5 nm, porosity showed a positive correlation with the SSA and PV. Porosity was positively correlated with D_1 and negatively correlated with D_2, suggesting that higher porosity was associated with greater roughness and complexity in the pore surface. This was accompanied by a decrease in the complexity of the internal pore space and an increase in pore development. Additionally, a positive correlation was found between the PV and SA for all pores, indicating consistent trends in the variation of PV and SSA. However, the PV and SSA values for pores smaller than 5 nm were negatively correlated with those for pores larger than 5 nm. This indicates that the development of pores smaller than 5 nm decreases with increasing maturity (Figure 5).

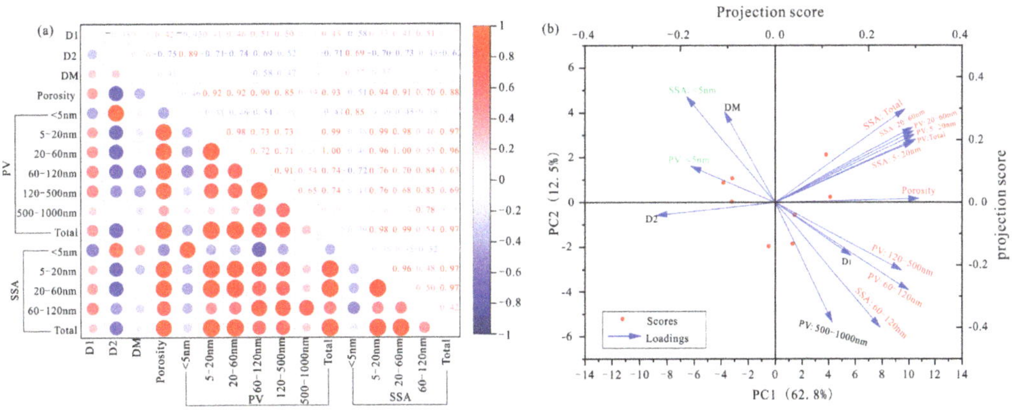

Figure 10. (**a**) Correlation coefficient matrix of fractal dimension and pore structure parameters. (**b**) Principal component analysis (PCA) of fractal dimension and pore structure variables of simulated samples; arrows represent variables.

There was a positive correlation between the pore volume (PV) and specific surface area (SSA) of >5 nm pores and total pores with D_1, particularly for pores in the 20–500 nm range. However, D_2 showed a positive correlation only with the PV and SSA of <5 nm pores. The PV and SSA of <5 nm pores were negatively correlated with D_1 while they showed a positive correlation with D_2. In the mature-to-highly-mature stage, the proportion of <5 nm pores was high (Figure 5). The retained hydrocarbons released from the kerogen during pyrolysis acted to fill the pores. Additionally, compaction led to a reduction in pore size and a decrease in the number of pores, particularly for <5 nm pores (Figures 6 and 7). These factors resulted in a decrease in surface roughness and complexity for <5 nm pores. The number of <5 nm pores was also reduced. However, the spatial distribution of <5 nm pores was more complex. As thermal evolution progressed, the number of >5 nm pores increased, especially for pores in the 20–60 nm range. This also resulted in an increase in surface roughness and complexity for >5 nm pores. The spatial distribution complexity of >5 nm pores decreased. The strong correlation between the 20–60 nm pores and D_1 and D_2 further supports this view (Figure 10a). Notably, DM showed a weak positive correlation only with D_1, D_2, and the SSA of >5 nm pores (Figure 10a).

Principal component analysis (PCA) was used to explore the relationship between porosity, pore structure, and fractal parameters (D_1, D_2, D_M) with the methodology outlined in [49]. The loadings were determined based on the projection lengths of the arrows on the principal components (PCs). This indicated the relationship between the original variables and the principal components, describing the extent of each variable's influence on the components. The relationship between the original variables was determined through the angle between the arrows or the projection lengths of the arrows on the principal components [49,50]. PC1 explained 62.8% of the variance while PC2 explained only 12.5%, indicating that PC1 represented the most significant features of the sample (Figure 10b). The porosity, along with the PV and SSA values of >5 nm pores (variables in red), had long and closely related projection lengths on the PC1 axis, suggesting that these variables represented the majority of the features in PC1. Therefore, PC1 could be defined as the degree of pore development. D_1 had a positive projection on PC1, but its projection length was small. This suggests that D_1 (roughness and irregularity of pore surfaces) had a general positive correlation with pore development. In contrast, D_2 and D_M had negative projections on PC1, with D_2 having a larger projection length. This indicates a stronger negative correlation between D_2 (complexity of pore space) and pore development. However, the projection length of D_M was smaller, indicating its limited ability to characterize pore development.

5.2. Controlling Factors of Fractal Dimension

During the thermal evolution of shale, both organic matter evolution and the diagenetic transformation of inorganic minerals alter the pore structure [38,40], thereby affecting the fractal dimension. Both D_2 and D_M showed positive correlations with geochemical parameters and the mineral content (Figure 11a). However, D_2 had a higher correlation coefficient, suggesting it was more suitable for explaining the impact of organic–inorganic diagenesis on pore structure. In contrast, D_1 showed a weak negative correlation with geochemical parameters and mineral content. There was a highly significant correlation between geochemical parameters and the mineral content, which could be attributed to their complementary changes during pore structure evolution (Figure 11a).

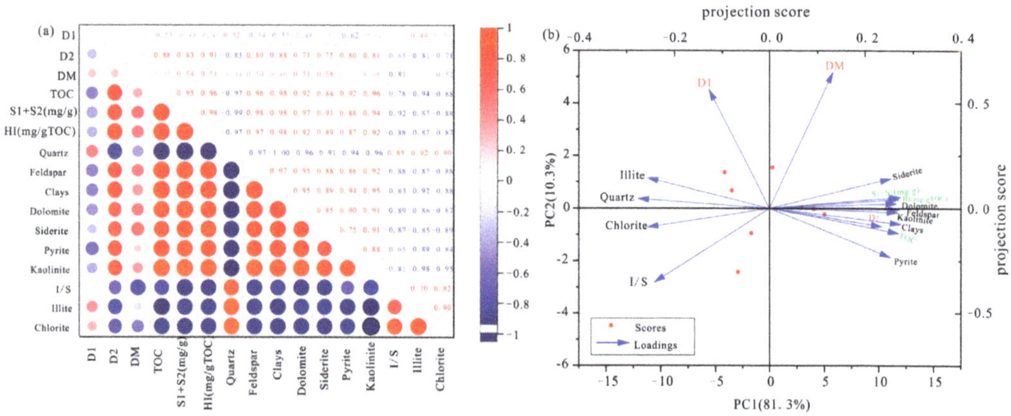

Figure 11. (**a**) Correlation coefficient matrix of fractal dimension, geochemical parameters, and mineral compositions. (**b**) PCA of fractal dimension, geochemical parameters, and mineral compositions variables of simulated samples; arrows represent variables.

The PCA method was also used to explore the relationship between geochemical parameters, mineral composition, and fractal dimensions. The variance explained by PC1 was 81.3% while PC2 accounted for only 10.3% (Figure 11b), indicating that PC1 represented the most significant features of the samples. Figure 11b shows that both mineral composition (black text) and geochemical parameters (green text) had long projections along the PC1 axis, indicating that these variables accounted for most of the characteristics of PC1. Additionally, variables on the right half of the PC1 axis corresponded to the "depleted" components, which gradually decreased in content during pore structure evolution. The variables on the left half of the PC1 axis represented the "generated" components (Figure 2). D_2 had a large projection on the right half of the PC1 axis and a very small angle with the "depleted" component variables. Therefore, the positive direction of PC1 could be defined as the consumption of unstable components during pore structure evolution or as an increase in the complexity and irregularity of the pore structure. Similarly, the negative direction of PC1 could represent the generated components (including illite, quartz, etc.) during pore structure evolution, indicating a decrease in the complexity and irregularity of the pore structure. However, D_1 and D_M had much larger projections on the PC2 axis (Figure 11b), suggesting that they were less effective in indicating the impacts of mineral composition and geochemical parameters on fractal dimensions.

5.3. Evolution Model of Fractal Dimension

The above results indicate that the degree of pore development can be represented by D_1, followed by D_2. During pore structure evolution, the "depleted" components (such as the TOC, carbonates, feldspars, etc.) are represented by D_2. By combining organic matter evolution, mineral composition, and diagenetic processes [30,39], the fractal dimension evolution of the pore structure in transitional shale is revealed (Figure 12).

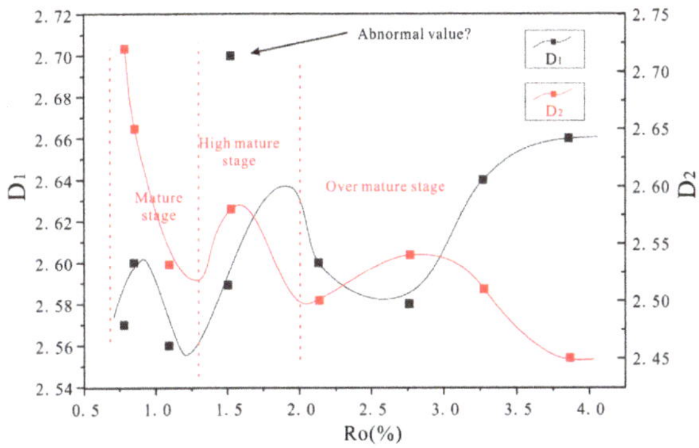

Figure 12. Cross-plot of fractal dimension D_1 and D_2 with maturity. The division of evolutionary stages is shown with references to He et al., 2022 [51].

(1) Immature-to-low-mature stage (Ro < 0.7%): Compaction causes inorganic minerals and organic matter to become tightly packed. The pore system is dominated by residual primary intergranular pores. Pore development gradually declines. As a result, D_1 exhibits low values, indicating smooth pore surfaces with low complexity and high self-similarity. In the early stage of pore structure evolution, the "depleted" components in shale have not yet undergone mineral transformation. D_2 shows relatively high values, reflecting the irregular development of various pore types and greater internal pore complexity.

(2) Mature stage (0.7% < Ro < 1.3%): In the early part of the mature stage, the pyrolysis of kerogen releases hydrocarbons, causing a shrinkage in the kerogen volume. At the same time, complex organic pores and microfractures form on the surface and edges of the kerogen. The complexity of the pore surface increases, and D_1 shows a slight rise. As hydrocarbon generation intensifies, the residual inorganic and organic pores become filled with adsorbed and retained oil, leading to a significant decrease in pore quantity during this stage. As the pore volume (PV) and specific surface area (SSA) of most pores decrease, D_1 also declines overall. In the later part of the mature stage, the cracking and expulsion of retained oil releases some inorganic pores, while solid bitumen forms, enriched with nanoscale organic pores. The number of pores significantly increases, likely due to the transformation of 1–5 nm pores into pores greater than 5 nm in size. The proportion of organic pores continues to rise. A significant number of different pore types are formed, resulting in a marked increase in D_1. Organic acids released during organic matter cracking promote dissolution and the transformation of clay minerals. As organic matter, feldspar, carbonate minerals, and clay minerals are progressively consumed, the complexity of the pore structure gradually decreases, and D_2 continues to decline.

(3) Highly mature stage (1.3% < Ro < 2.0%): Kerogen and retained oil are gradually cracked into light oil and wet gas. A significant number of organic pores form on the surface of organic matter, leading to a marked increase in D_1. Inorganic pores filled with retained oil are released, which may result in a more complex internal pore

structure, causing a notable increase in D_2. As the oil generation capacity of kerogen gradually ceases, organic acids are no longer released, and the diagenetic environment shifts to an alkaline one. Although dissolution is weakened, the transformation of kaolinite significantly intensifies. Inorganic pores related to clay minerals (especially those in the 5–60 nm range) increase significantly, and pore development continues to advance. As a result, D_1 continues to rise while D_2 gradually declines.

(4) Over-mature stage (Ro > 2.0%): Alkyl side chains of kerogen are cracked to form light hydrocarbons (C_1–C_4). The residual oil and wet gas are cracked into methane. The cracking of oil generates large amounts of methane, which helps maintain pore pressure. As a result, the number and size of organic pores reach their maximum values at Ro = 2.7% and then stabilize. In the early part of the over-mature stage, the scale of inorganic pores slightly decreases due to compaction. Due to the weak thermal evolution of organic matter and limited inorganic diagenesis, the complexity of pore surfaces gradually decreases, with D_1 reaching its minimum value at this stage. In the later part of the over-mature stage, as kaolinite fully transforms into illite, the proportion of inorganic pores and microfractures increases. As a result, D_1 significantly increases while D2 decreases.

6. Conclusions

In this study, transitional shale samples at various stages of thermal maturity were analyzed to investigate their geochemical characteristics, mineral composition, and pore structure. The evolution of the pore structure and its fractal dimensions was determined using the FHH (Frenkel–Halsey–Hill) and capillary bundle models. Correlation analysis and principal component analysis were employed to identify the controlling factors and evolutionary patterns of pore structure fractal dimensions.

(1) The initial samples used for thermal simulation were primarily composed of quartz (39.9%) and clay minerals (49.9%), with moderate-to-good hydrocarbon generation potential. As thermal maturity increased, the quartz content gradually rose while the amounts of clay minerals, carbonate minerals, pyrite, and feldspar decreased. The pore evolution process consists of five stages: a slow decrease phase (0.78 < Ro < 1.0%), a rapid increase phase (1.0% < Ro < 2.0%), a relatively stable phase (2.0% < Ro < 2.7%), a rapid increase phase (2.7% < Ro < 3.2%), and a slow decrease phase (Ro > 3.2%).

(2) The evolution of the pore volume (PV) and specific surface area (SSA) indicated a gradual increase in the proportion of pores in the 5–20 nm and 20–60 nm ranges while the proportion of pores smaller than 5 nm decreased. The D_1 value obtained using the FHH model (average = 2.61) was higher than the D_2 value (average = 2.56). This suggests that, at various stages of maturity, the roughness of pore surfaces is greater than the complexity of the internal pore structure. The D_M distribution calculated using the capillary bundle model had a broad range (2.47 to 2.94), with an average value (2.84) higher than both D_1 and D_2, possibly indicating that larger pores exhibit more complex pore structures.

(3) D_1 can be used to indicate the extent of pore development. D_2 can be used to characterize the changes in consumed components during the pore development process. Based on the thermal evolution of organic matter, mineral composition, diagenesis, and pore structure evolution, an evolutionary model of the fractal dimensions for transitional shales has been established.

Author Contributions: Conceptualization, X.W. and Y.G.; methodology, Y.J.; validation, Q.G.; formal analysis, Y.G.; investigation, H.Z.; data curation, Q.G.; writing—original draft preparation, X.W. and Y.G.; writing—review and editing, Y.J. and D.D.; visualization, H.Z.; supervision, D.D. and Y.J.;

project administration, D.D. and Y.J. All authors have read and agreed to the published version of the manuscript.

Funding: This research was funded by the National Natural Science Foundation of China, grant number 42202192, and supported by Sichuan Science and Technology Program, grant number 2024NSFSC0811.

Institutional Review Board Statement: Not applicable.

Informed Consent Statement: Not applicable.

Data Availability Statement: Data are contained within the article.

Conflicts of Interest: Author Dazhong Dong was employed by PetroChina Research Institute of Petroleum Exploration and Development. The remaining authors declare that the research was conducted in the absence of any commercial or financial relationships that could be construed as a potential conflict of interest.

References

1. Mandelbrot, B. How long is the coast of Britain? Statistical self-similarity and fractional dimension. *Science* **1967**, *156*, 636–638. [CrossRef] [PubMed]
2. Mandelbrot, B. *Fractals: Form, Chance and Dimension*; Freeman: San Francisco, CA, USA, 1977.
3. Li, J.; Zheng, B. A new method for fractal characterization of microscopic pores and its application in shale reservoirs. *Nat. Gas Ind.* **2016**, *35*, 52–59, (In Chinese with English abstract).
4. Peng, N.; He, S.; Hu, Q.; Zhang, B.; He, X.; Zhai, G.; He, C.; Yang, R. Organic nanopore structure and fractal characteristics of Wufeng and lower member of Longmaxi shales in southeastern Sichuan, China. *Mar. Pet. Geol.* **2019**, *103*, 456–472. [CrossRef]
5. Huang, W.; Lu, S.; Hersi, O.; Wang, M.; Deng, S.; Lu, R. Reservoir spaces in tight sandstones: Classification, fractal characters, and heterogeneity. *J. Nat. Gas Sci. Eng.* **2017**, *46*, 80–92. [CrossRef]
6. Guan, Q.; Dong, D.; Sun, S.S.; Hu, L.; Qi, L.; Li, C.; Chen, C.; Chen, X. Fractal characteristics of organic-rich shale pore structure and its geological implications: A case study of the Lower Silurian Longmaxi Formation in the Weiyuan block, Sichuan Basin. *Nat. Gas Ind.* **2024**, *44*, 108–118, (In Chinese with English abstract).
7. Guo, X.; Huang, Z.; Zhao, L.; Han, W.; Ding, C.; Sun, X.; Yan, R.; Zhang, T.; Yang, X.; Wang, R. Pore structure and multi-fractal analysis of tight sandstone using MIP, NMR and NMRC methods: A case study from the Kuqa depression, China. *J. Pet. Sci. Eng.* **2019**, *178*, 544–558. [CrossRef]
8. Ma, X.; Guo, S.; Shi, D.; Zhou, Z.; Liu, G. Investigation of pore structure and fractal characteristics of marine continental transitional shales from Longtan Formation using MICP, gas adsorption, and NMR (Guizhou, China). *Mar. Pet. Geol.* **2019**, *107*, 555–571. [CrossRef]
9. Sun, W.; Zuo, Y.; Wu, Z.; Liu, H.; Xi, S.; Shui, Y.; Wang, J.; Liu, R.; Lin, J. Fractal analysis of pores and the pore structure of the Lower Cambrian Niutitang shale in northern Guizhou province: Investigations using NMR, SEM and image analyses. *Mar. Pet. Geol.* **2019**, *99*, 416–428. [CrossRef]
10. Avnir, D.; Jaroniec, M. An isotherm equation for adsorption on fractal surfaces of heterogeneous porous materials. *Langmuir* **1989**, *5*, 1431–1433. [CrossRef]
11. Neimark, A.V.; Unger, K.K. Method of discrimination of surface fractality. *J. Colloid Interface Sci.* **1993**, *158*, 412–419. [CrossRef]
12. Pfeifer, P.; Wu, Y.; Cole, M.W.; Krim, J. Multilayer adsorption on a fractally rough surface. *Phys. Rev. Lett.* **1989**, *62*, 1997–2000. [CrossRef] [PubMed]
13. Washburn, E.W. The dynamics of capillary flow. *Phys. Rev.* **1921**, *2*, 273–283. [CrossRef]
14. Zhang, B.; Li, S. Determination of the surface fractal dimension for porous media by mercury porosimetry. *Ind. Eng. Chem. Res.* **1995**, *34*, 1383–1386. [CrossRef]
15. He, T.; Zhou, Y.; Chen, Z.; Zhang, Z.; Xie, H.; Shang, Y.; Cui, G. Fractal Characterization of the Pore-Throat Structure in Tight Sandstone Based on Low-Temperature Nitrogen Gas Adsorption and High-Pressure Mercury Injection. *Fractal Fract.* **2024**, *8*, 356. [CrossRef]
16. Zhang, C.; Guan, P.; Zhang, J.; Liang, X.; Ding, X.; You, Y. A Review of the Progress on Fractal Theory to Characterize the Pore Structure of Unconventional Oil and Gas Reservoirs. *Acta Sci. Nat. Univ. Pekin.* **2023**, *59*, 897–908, (In Chinese with English abstract).
17. Qian, C.; Li, X.; Shen, W.; Zhang, Q.; Guo, W.; Hu, Y.; Yue, C.; Jia, Y. Study on the Pore Structure and Fractal Characteristics of Different Lithofacies of Wufeng–Longmaxi Formation Shale in Southern Sichuan Basin, China. *ACS Omega* **2022**, *7*, 8724–8738. [CrossRef]

18. Liu, R.; Wei, Z.; Jia, A.; He, S.; Hou, Y.; He, Q.; Wang, T.; Zeng, Y.; Yang, R. Fractal Characteristics of Pore Structure in Deep Over pressured Organic Rich Shale in Wufeng-Longmaxi Formation in Southeast Sichuan and Its Geological Significance. *Earth sci.* **2023**, *48*, 1496–1516, (In Chinese with English abstract).
19. Zheng, S.; Yao, Y.; Liu, D.; Cai, Y.; Liu, Y. Characterizations of full-scale pore size distribution, porosity and permeability of coals: A novel methodology by nuclear magnetic resonance and fractal analysis theory. *Int. J. Coal Geol.* **2018**, *196*, 148–158. [CrossRef]
20. Li, P.; Zheng, M.; Bi, H.; Wu, S.; Wang, X. Pore throat structure and fractal characteristics of tight oil sandstone: A case study in the Ordos Basin, China. *J. Pet. Sci. Eng.* **2017**, *149*, 665–674. [CrossRef]
21. Chang, J.; Fan, X.; Jiang, Z.; Wang, X.; Chen, L.; Li, J.; Zhu, L.; Wan, C.; Chen, Z. Differential impact of clay minerals and organic matter on pore structure and its fractal characteristics of marine and continental shales in China. *Appl. Clay Sci.* **2022**, *216*, 106334. [CrossRef]
22. Li, X.; Jiang, Z.; Jiang, S.; Li, Z.; Song, Y.; Jiang, H.; Qiu, H.; Cao, X.; Miao, Y. Various controlling factors of matrix-related pores from differing depositional shales of the Yangtze Block in south China: Insight from organic matter isolation and fractal analysis. *Mar. Pet. Geol.* **2020**, *111*, 720–734. [CrossRef]
23. Li, Y.; Liu, D.; Feng, X.; Zhao, F.; Chen, Y.; Wang, Y.; Lin, R.; Du, W.; Fan, Q.; Song, Y.; et al. Heterogeneity characteristics and its controlling factors of marine shale reservoirs from the Wufeng–Longmaxi Formation in the Northern Guizhou area. *Geol. China* **2024**, *51*, 780–798, (In Chinese with English abstract).
24. Kuang, L.C.; Dong, D.Z.; He, W.Y.; Wen, S.M.; Sun, S.S.; Li, S.X.; Qiu, Z.; Liao, X.W.; Li, Y.; Wu, J.; et al. Geological characteristics and development potential of transitional shale gas in the east margin of the Ordos Basin, NW China. *Pet. Explor. Dev.* **2020**, *47*, 471–482. [CrossRef]
25. Dong, D.Z.; Wang, Y.M.; Li, X.J.; Zou, C.N.; Guan, Q.Z.; Zhang, C.C.; Huang, J.L.; Wang, S.F.; Wang, H.Y.; Liu, H.L.; et al. Breakthrough and prospect of shale gas exploration and development in China. *Nat. Gas Ind. B* **2016**, *3*, 43825. [CrossRef]
26. Wu, J.; Wang, H.; Shi, Z.; Wang, Q.; Zhao, Q.; Li, S.; Liu, D.; Sun, S.; Qiu, Z. Favorable lithofacies types and genesis of marine-continental transitional black shale: A case study of Permian Shanxi Formation in the eastern margin of Ordos Basin, NW China. *Pet. Explor. Dev.* **2021**, *48*, 1137–1149. [CrossRef]
27. Zhang, Q.; Zhang, L.F.; Wang, Y.M.; Xiao, Y.F.; Liu, D.; Liu, W.; Li, S.X.; Li, X.T. Characteristics and controlling factors of transitional shale gas reservoirs: An example from Permian Shanxi Formation, Daning-Jixian block, Ordos Basin, China. *J. Nat. Gas Geosci.* **2022**, *7*, 147–157. [CrossRef]
28. Li, X.; Jiang, Z.X.; Wang, S.; Wu, F.; Miao, Y.N.; Wang, X.M.; Wang, H.X.; Liu, X.B. Differences of marine and transitional shales in the case of dominant pore types and exploration strategies, in China. *J. Nat. Gas Sci. Eng.* **2022**, *103*, 104628. [CrossRef]
29. Guo, S.B.; Mao, W.J. Division of diagenesis and pore evolution of a Permian Shanxi shale in the Ordos Basin, China. *J. Pet. Sci. Eng.* **2019**, *182*, 106351. [CrossRef]
30. Lu, C.G.; Xiao, X.M.; Gai, H.F.; Feng, Y.; Li, G.; Meng, G.M.; Gao, P. Nanopore structure characteristics and evolution of type III kerogen in marine-continental transitional shales from the Qinshui basin, northern China. *Geoenergy Sci. Eng.* **2023**, *221*, 211413. [CrossRef]
31. Ma, Z.L.; Zheng, L.J.; Li, Z.M. The Thermocompression Simulation Experiment of Source Rock Hydrocarbon Generation and Expulsion in Formation Porosity. *Acta Sedimentol. Sin.* **2012**, *30*, 955–963, (In Chinese with English abstract).
32. Espitalié, J.; Laporte, J.L.; Madec, M.; Marquis, F.; Leplat, P.; Paulet, J.; Boutefeu, A. Rapid method for source rocks characterization and for determination of petroleum potential and degree of evolution. *Rev. Inst. Fr. Pet.* **1977**, *32*, 23–42. [CrossRef]
33. Wang, Z.; Jiang, Y.; Fu, Y.; Lei, Z.; Xu, C.; Yuan, J.; Wen, R.; Wang, Z.; Gu, Y.; Yin, X. Characterization of Pore Structure and Heterogeneity of Shale Reservoir from Wufeng Formation-Sublayers Long-11 in Western Chongqing Based on Nuclear Magnetic Resonance. *Earth Sci.* **2022**, *47*, 490–504, (In Chinese with English abstract).
34. Jiang, F.; Chen, D.; Chen, J.; Li, Q.; Liu, Y.; Shao, X.; Hu, T.; Dai, X. Fractal analysis of shale pore structure of continental gas shale reservoir in the Ordos Basin, NW China. *Energy Fuels* **2016**, *30*, 4676–4689. [CrossRef]
35. He, C.; Zhao, Y.; Yu, L.; Lu, L.; Liu, X.; Pan, A.; Li, C. Pore structure and fractal characteristics of deep shale gas reservoirs in the Permian Dalong Formation, northeastern Sichuan Basin. *Pet. Geol. Exp.* **2024**, *46*, 263–277, (In Chinese with English abstract).
36. El Nady, M.M.; Ramadan, F.S.; Hammad, M.M.; Lotfy, N.M. Evaluation of organic matters, hydrocarbon potential and thermal maturity of source rocks based on geochemical and statistical methods: Case study of source rocks in Ras Gharib oilfield, central Gulf of Suez, Egypt. *Egypt. J. Pet.* **2015**, *24*, 203–211. [CrossRef]
37. Zhou, B. The Applications of NMR Relaxometry, NMR Cryoporometry, and FFC NMR to Nanoporous Structures and Dynamics in Shale at Low Magnetic Fields. *Energy Fuels* **2018**, *32*, 8897–8904. [CrossRef]
38. Xu, L.W.; Yang, K.J.; Wei, H.; Liu, L.F.; Li, X.; Chen, L.; Xu, T.; Wang, X.M. Diagenetic evolution sequence and pore evolution model of Mesoproterozoic Xiamaling organic-rich shale in Zhangjiakou, Hebei, based on pyrolysis simulation experiments. *Mar. Pet. Geol.* **2021**, *132*, 105233. [CrossRef]
39. Wang, G.P.; Jin, Z.J.; Zhang, Q.; Zhu, R.K.; Tang, X.; Liu, K.Q.; Dong, L. Effects of clay minerals and organic matter on pore evolution of the early mature lacustrine shale in the Ordos Basin, China. *J. Asian Earth Sci.* **2023**, *246*, 105516. [CrossRef]

40. Milliken, K.L.; Rudnicki, M.; Awwiller, D.N.; Zhang, T. Organic matter-hosted pore system, Marcellus Formation (Devonian), Pennsylvania. *AAPG Bull.* **2013**, *97*, 177–200. [CrossRef]
41. Loucks, R.G.; Reed, R.M. Scanning-Electron-Microscope Petrographic Evidence for Distinguishing Organic-Matter Pores Associated with Depositional Organic Matter versus Migrated Organic Matter in Mudrocks. Presented at the Convention of the Gulf Coast Association of Geological Societies American Association of Petroleum Geologists Sectional Meeting, Lafayette, Louisiana, 5–7 October 2014.
42. Loucks, R.G.; Reed, R.M.; Ruppel, S.C.; Jarvie, D.M. Morphology, genesis, and distribution of nanometer-scale pores in siliceous mudstones of the Mississippian Barnett shale. *J. Sediment. Res.* **2009**, *79*, 848–861. [CrossRef]
43. Bernard, S.; Wirth, R.; Schreiber, A.; Schulz, H.M.; Horsfield, B. Formation of nanoporous pyrobitumen residues during maturation of the Barnett Shale (Fort Worth Basin). *Int. J. Coal Geo.* **2014**, *127*, 114–115. [CrossRef]
44. Gao, Z.Y.; Xuan, Q.X.; Hu, Q.H.; Jiang, Z.X.; Liu, X.X. Pore structure evolution characteristics of continental shale in China as indicated from thermal simulation experiments. *AAPG Bull.* **2021**, *105*, 2159–2180. [CrossRef]
45. Zhao, Y.M.; Gao, P.; Zhou, Q.; Xiao, X.M.; Xing, Y.J.; Liu, W. A Review of the Heterogeneity of Organic-Matter-Hosted Pores in Shale Reservoirs. *Energies* **2022**, *15*, 8805. [CrossRef]
46. Pommer, M.; Milliken, K. Pore types and pore-size distributions across thermal maturity, Eagle Ford Formation, southern Texas(Review). *AAPG Bull.* **2015**, *99*, 1713–1744. [CrossRef]
47. Xu, L.W.; Yang, K.J.; Zhang, L.F.; Liu, L.F.; Jiang, Z.X.; Li, X. Organic-induced nanoscale pore structure and adsorption capacity variability during artificial thermal maturation: Pyrolysis study of the Mesoproterozoic Xiamaling marine shale from Zhangjiakou, Hebei, China. *J. Pet. Sci. Eng.* **2021**, *202*, 108502. [CrossRef]
48. Peltonen, C.; Marcussen, Ø.; Bjørlykke, K.; Jahren, J. Clay mineral diagenesis and quartz cementation in mudstones: The effects of smectite to illite reaction on rock properties. *Mar. Petrol. Geol.* **2009**, *26*, 887–898. [CrossRef]
49. Xue, J.; Lee, C.; Wakeham, S.G.; Armstrong, R. Using principal components analysis (PCA) with cluster analysis to study the organic geochemistry of sinking particles in the ocean. *Org. Geochem.* **2011**, *42*, 356–367. [CrossRef]
50. Gotelli, N.J.; Ellison, A.M. *A Primer of Ecological Statistics*; Sinauer Associates: Sunderland, MA, USA, 2004.
51. He, Q.B.; Chen, S.J.; Li, S.X.; Guo, B.Z.; Lu, J.G.; Li, Y.; Li, X.G.; Zhao, L.P.; Ma, Z.W. Organic geochemical characteristics and hydrocarbon generation mechanism of marine-continental transitional organic-rich shale: A case study from the Shanxi formation in the eastern margin of the Ordos Basin. *J. Pet. Sci. Eng.* **2022**, *219*, 111116. [CrossRef]

fractal and fractional

MDPI

Article

Multifractal Analysis of Tight Sandstone Using Micro-CT Methods: A Case from the Lower Cretaceous Quantou Formation, Southern Songliao Basin, NE China

Lei Li [1], Zhongcheng Li [2], Haotian Han [2], Chao Liu [1,*], Yilin Li [3], Wanchun Zhao [1,4], Jianyi Wang [2] and Zhidong Bao [3,*]

[1] State Key Laboratory of Continental Shale Oil, Northeast Petroleum University, Daqing 163318, China; lei634616@hotmail.com (L.L.); zhaowanchun@nepu.edu.cn (W.Z.)
[2] Research Institute of Exploration and Development, PetroChina Jilin Oilfield Company, Jilin 138000, China; lizc-jl@petrochina.com.cn (Z.L.); han-ht@petrochina.com.cn (H.H.); wangjianyi5211@163.com (J.W.)
[3] College of Geosciences, China University of Petroleum-Beijing, Beijing 102249, China; liyilinamo@163.com
[4] Key Laboratory of Oil & Gas Reservoir and Underground Gas Storage Integrity Evaluation of Heilongjiang Province, Daqing 163318, China
* Correspondence: lcyxdz@163.com (C.L.); baozhd@cup.edu.cn (Z.B.)

Abstract: The relationships between the pore structure and a single fractal or specific region have been widely reported. However, the intrinsic relationship between multifractal parameters and physical properties have remained uncertain. In this study, micro-computed tomography scanning technology and high-pressure mercury injection technologies were applied to determine the pore structures of tight sandstone at different scales. Subsequently, the multifractal theory was applied to quantitatively evaluate the multiscale pore structure heterogeneity. An evident linear relationship exists between $logX(q,\varepsilon)$ and $log(\varepsilon)$, indicating the pore structure of tight sandstones exhibits significant multifractal characteristics. Multifractal parameters, including $\Delta(\alpha)$, ΔD, $D_{min} - D_0$, and $D_0 - D_{max}$, can serve as sensitive indicators to assess the multiscale pore structure heterogeneity. In particular, the relative development degree of large-scale pores (>10 μm) can be reflected by $D_{min} - D_0$, which has strong heterogeneity. The heterogeneity of the multiscale structure is closely linked to the mineral components of tight sandstone reservoirs, and the heterogeneity of small-scale pores (1–10 μm) is stronger by clay mineral enrichment. Furthermore, the part of the pore structure corresponding to the combination of pore size range of 10 to 20 μm and throat size range of 20 to 40 μm in a low probability measure area may dominate the permeability of tight sandstone. The findings enhance the understanding of pore structure heterogeneity and broaden the application of multifractal theory.

Keywords: pore structure; multifractal analyses; micro-CT; tight sandstones; Songliao Basin

1. Introduction

A scientific framework for quantitatively characterizing the heterogeneous distribution of pore sizes has been provided by fractal geometry theory [1]. Since then, numerous studies have shown that fractal theory effectively explains the irregular, unstable, and complex nature of pore structures in reservoir rocks [2–5]. Single-fractal theory is inadequate for fully characterizing the heterogeneity of pore systems across all scales [6,7]. Therefore, multifractal theory is required to capture the irregularities and multiscale distribution of pore sizes.

The storage and percolation capacities of tight sandstone reservoirs have always been a popular topic in petrophysical studies, which are of great significance to hydrocarbon exploration and development. As indicated by existing studies, the storage and percolation capacities are closely related to the pore structure characteristics [8,9]. Therefore, pore structure characterizations are required to obtain an understanding of the storage and percolation capacities of sandstone reservoirs. The pore geometrical shape, type, and size distribution determine the petrophysical properties as well as the percolation and accumulation of hydrocarbon in sandstone reservoirs. To date, over 30 parameters have been proposed to be able to affect pore structure, including commonly used ones such as porosity and permeability [6,10,11]. These parameters primarily evaluate the pore structure via a macroscopic perspective, given the essential role of micro-pore structure evaluation in oil and gas exploration and migration [8,12]. Therefore, advanced analytical techniques have been introduced to accurately assess multiscale pore structure [12,13]. Three techniques have been commonly applied to assess micro-pore structure: (1) imaging methods (including scanning electron microscopy [SEM] and X-ray computed tomography [X-ray CT]), (2) fluid invasion methods (including high-pressure mercury injection [HPMI], low-pressure gas [N_2, CO_2] physisorption, and nuclear magnetic resonance [NMR]), and (3) radiation detection methods [13,14]. However, these methods are difficult in detecting isolated pores in tight sandstones and lack micro-pore structure visualization, which can potentially compromise the results [11].

X-ray CT technology stands out as one of the few non-destructive methods capable of providing three-dimensional (3D) high-resolution insights into the internal structure of tight sandstones [15–17]. The high-flux and high-brightness synchrotron radiation of micro-CT enables 4D dynamic observations of pore structure in tight sandstones and rapid acquisition, which make micro-CT the appropriate choice to spatially characterize the petrophysical parameters of tight sandstones [17,18]. The application of micro-CT is restricted due to its resolution limitation, leading to an underestimation of pore sizes less than 1 μm. Additionally, it is constrained by noise and high expense [19]. As indicated above, the extremely wide pore size distribution of tight sandstone reservoirs cannot be solely determined by any single method. The comprehensive pore structure characterizations of tight sandstone reservoirs, including the pore size distribution, pore geometry, and connectivity, relies on the integration of multiple experimental methods.

In this study, micro-computed tomography scanning technology and high-pressure mercury injection (HPMI) technologies were applied to determine the pore structures of tight sandstone at different scales. Subsequently, the multifractal theory was applied to quantitatively evaluate the multiscale pore structure heterogeneity. This study aims to (1) examine the multifractal characteristics across multiscale pore structure and (2) explore the relationship between multifractal parameters and physical property. The findings provide a better understanding of pore structure heterogeneous and broaden the application of multifractal theory.

2. Geological Setting

The Songliao Basin is the largest Mesozoic–Cenozoic continental oil and gas basin in Northeast China (Figure 1a), extending in a north–northeastern orientation and covering an area of approximately 26 × 10^4 km^2 [20,21]. It is a large composite sedimentary basin characterized by tectonic complexity. The Songliao Basin can be subdivided into seven first class tectonic zones (Figure 1b). The study area, as one of the most oil rich areas, belongs to the Central Depression Zone (Figure 1c). The Central Depression Zone basement is comprised of late Paleozoic to early Mesozoic metamorphic rocks and contemporaneous igneous rocks. The basin hosts primarily Cretaceous strata, reaching a maximum thickness of 7000 m [22] (Figure 1d).

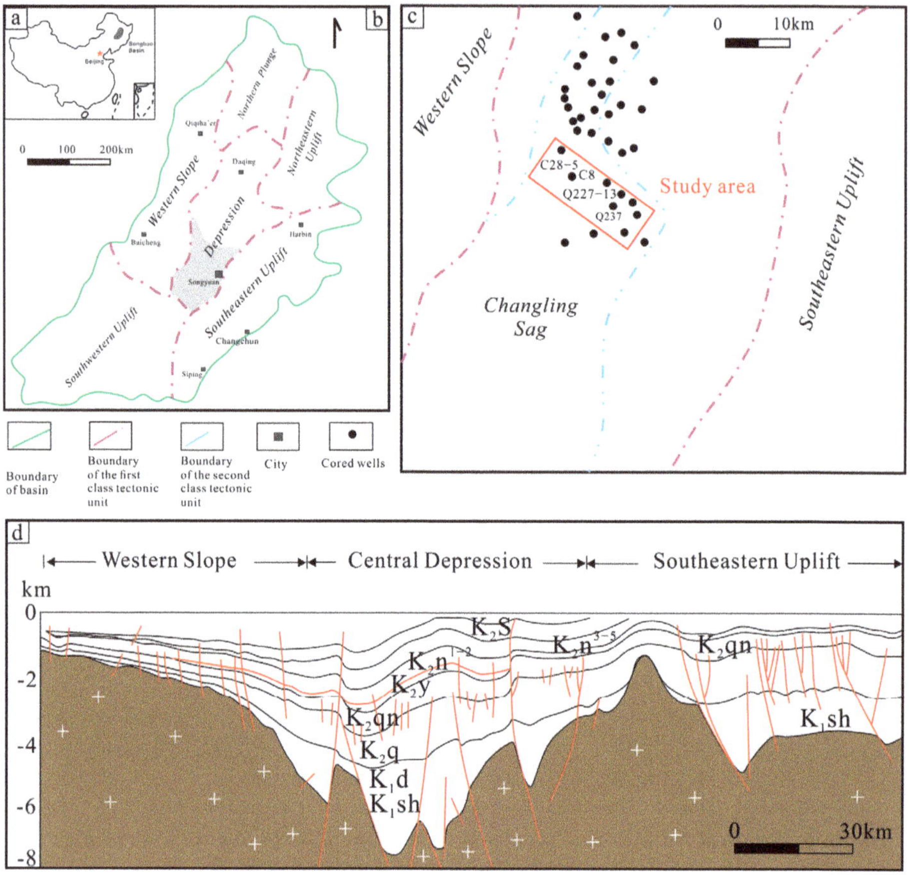

Figure 1. (**a**) Location of the Songliao Basin [20]. (**b**) The first class tectonic unit of the Songliao Basin [23]. (**c**) The second class tectonic units of the study area and well locations. (**d**) Structural cross-section of the Songliao Basin [22,23].

The sandstones in the fourth member of the Quantou Formation (K_1q^4), commonly known as the Fuyu Sandstone, constitute one of the primary oil-bearing layers in the southern Songliao Basin. The upper section of K_1q^4 consists of interbedded gray-black mudstone, gray-brown siltstone, and argillaceous siltstone of varying thicknesses, while the lower section is predominantly composed of purplish-red mudstone interbedded with grayish-white siltstone (Figure 2a). Deposited during the sag phase, these strata range in thickness from 70 m to 100 m, centered around a unified subsidence area. The reservoirs within the Quantou Formation are primarily of fine sandstone and siltstone, developed in shallow-water deltaic distributary channel deposits (Figure 2b) [24].

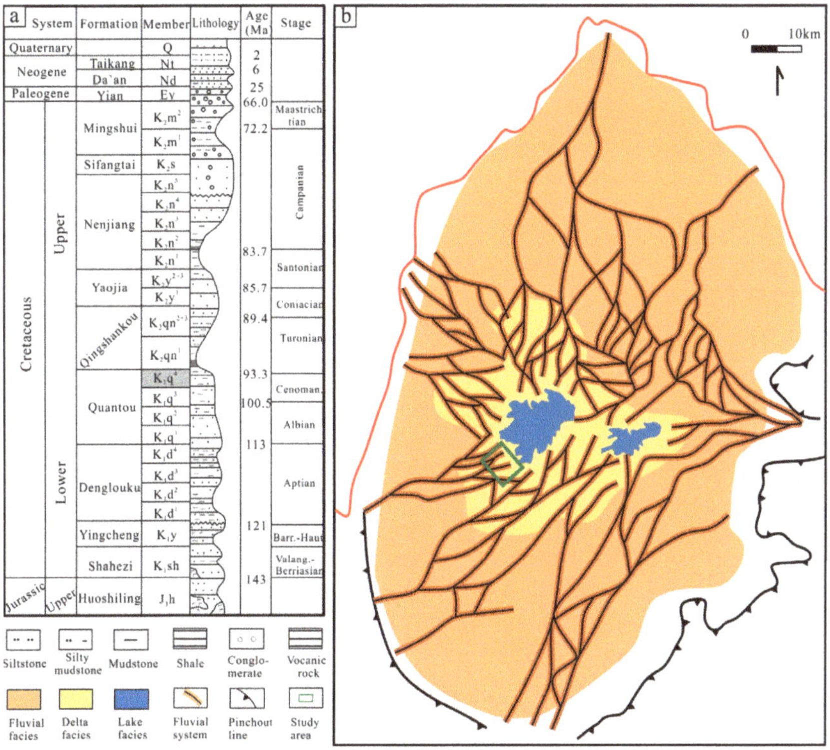

Figure 2. (**a**) Lithostratigraphy and sequence stratigraphy [25]. (**b**) Depositional settings of K_1q^4 in the Songliao Basin (modified from [26]).

3. The Experimental Work and Multifractal Theory

3.1. Sample Preparation and Experiments

The tight sandstone samples were collected from the Yuzi Well and Gudian areas of the Changling Sag, southern Songliao Basin (Figure 1c). In general, the tight sandstone reservoirs of the K_1q^4 are highly tight, with low porosity (ranging from 2 to 14%, average 8.54%) and permeability (ranging from 0.01 to 5×10^{-3} μm^2, average 0.493×10^{-3} μm^2) [24]. The basic information of representative tight sandstone samples is summarized in Table 1. For this experiment, X-ray diffraction (XRD), thin section, SEM, HPMI, and micro-CT methods were used to obtain the mineral composition, the pore-throat type, as well as the micro-pore structure, pore size distributions, and pore-throat network of tight sandstone samples, respectively.

Table 1. Basic information of tight sandstone samples in the K_1q^4. Data are consistent with measurements reported in reference [24].

No.	Well	Depth/m	Lithology	Porosity/%	Permeability/$\times 10^{-3}$ μm^2
S1	C8	2390.01–2390.21	siltstone	4.34	0.19
S2	C8	2391.53–2391.73	siltstone	7.82	0.12
S3	C28-5	2221.67–2221.77	siltstone	7.29	0.25
S4	Q227-13	1678.69–1678.78	fine sandstone	9.84	0.69
S5	Q237	1665.12–1665.20	fine sandstone	12.65	0.71
S6	Q237	1684.47–1684.57	fine sandstone	6.91	0.29

3.1.1. XRD

The samples were first ground using an agate mortar and subsequently sieved to obtain powder with a particle size finer than 200 mesh. Approximately 2 to 3 g of each sample were used for the analysis. XRD analysis was performed using a Bruker D8 Advance AA25 diffractometer. Cu Kα radiation ($\lambda = 1.5406$ Å) was employed, with the X-ray tube operated at 40 kV and 40 mA. Data were collected over a 2θ range of 5° to 70°, with a step size of 0.02° and a counting time of 0.5 s per step. Phase identification was carried out using the PDF-4+ database provided by the International Centre for Diffraction Data [27]. Phase identification was performed using the PDF-4+ database, and quantitative mineral contents were calculated using the Rietveld refinement method implemented in TOPAS v5.0 [28], which is widely accepted for clay-bearing and heterogeneous sedimentary rocks [29].

3.1.2. Thin Sections and SEM

Thin sections enable a direct observation of pore-throat type, size, and number in tight sandstones [30,31]. In order to visually observe the pore-throat type, the samples of tight sandstone were impregnated with blue resin under high pressure [20]. SEM analysis was performed on a Nova Nano SEM 450 field emission scanning electron microscope. Prior to imaging, the samples were sputter-coated with a thin layer of gold to enhance electrical conductivity and prevent charging under the electron beam. Observations were conducted under high vacuum conditions. For each representative sample, 3 to 5 regions were randomly selected, and a total of 15 high-resolution images were acquired to document microstructural and pore characteristics.

3.1.3. HPMI

Before testing, the tight sandstone samples were dried at 105 °C until a constant weight was reached. Each cylindrical core plug was then sealed in a closed dilatometer and evacuated to remove any residual air or moisture. HPMI experiments were carried out using the CMS-300 mercury intrusion instrument manufactured by Temco (Houston, TX, USA) and the AutoPore IV 9505 instrument manufactured by Micromeritics Instrument Corporation (Norcross, GA, USA). A stepwise pressure increase method was adopted during the mercury injection process to ensure that pressure equilibrium was achieved at each increment. This approach helps to minimize potential hysteresis effects and enhances the accuracy of the pore size distribution. The maximum injection pressure applied was 200 MPa, allowing access to a wide range of pore throat sizes. The mercury injection rate was controlled at 0.1 mL/min, and each pressure step was held for approximately 30 s to allow for pore saturation before proceeding to the next level. For each of the six samples (S1 to S6), one representative measurement was performed on an individual core plug [32].

3.1.4. Micro-CT

For this experiment, tight sandstone samples were processed to a diameter of 2.5 cm and a length of 3.0 cm. Micro-CT scanning was conducted using the GE Phoenix Nanotom S digital nano X-ray system, manufactured by General Electric, Wunstorf, Germany. The scanning was performed at a voltage of 110 kV, a current of 100 μA, and a resolution of 1.29 μm. Image processing is performed by subjecting micro-CT data to a series of algorithms to accurately characterize tight sandstone samples [14]. In addition, image processing uses background equalization, noise reduction, and pore boundary sharpening to improve image resolution (Figure 3). The reconstruction of pore volume and the segmentation of pore space were accomplished through 3D visualization using Volume Graphics Studio Max (version 3.4) and Avizo 8.0 imaging software.

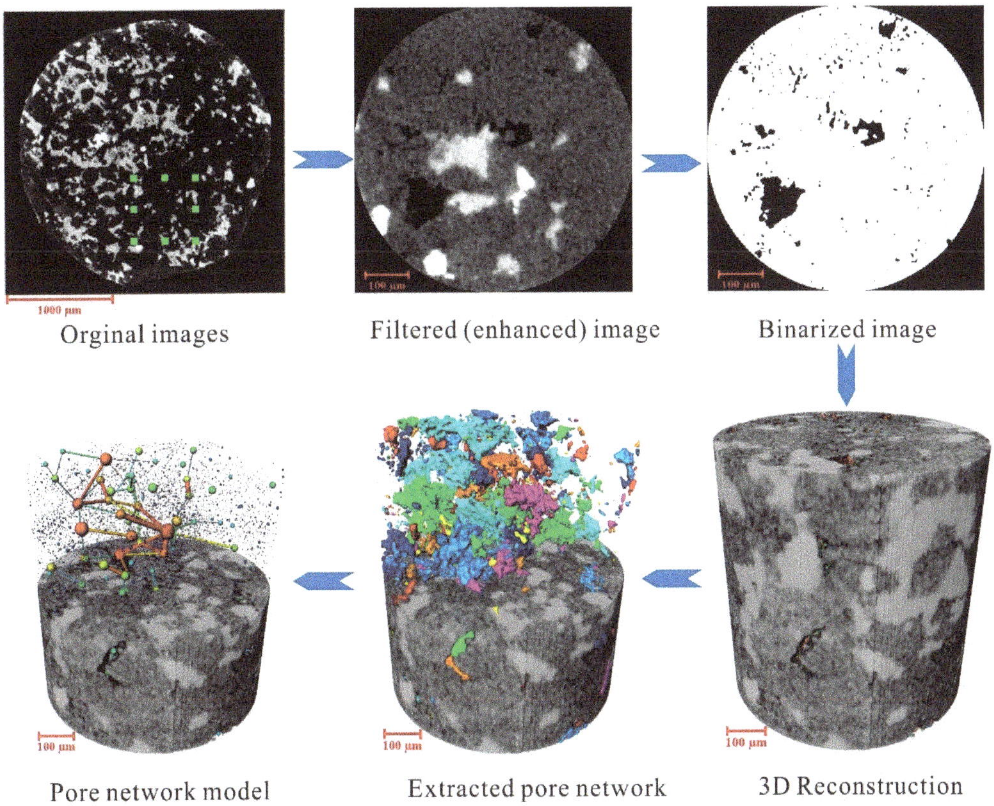

Figure 3. Workflow of micro-CT image processing and 3D visualization model of pore structure.

The pore network model is a simplified representation that visualizes the intricate pore space within porous media [33]. In this study, the tight sandstone pore network model was extracted using the central axis method (Figure 4). The fundamental concept of the pore network model involves dividing the complex pore space of porous media into multiple units, where the narrow and elongated pore spaces are defined as throats, and the connections between these throats are designated as equivalent pores, commonly represented by "spheres" and "sticks" [14,33]. The "spheres" are viewed as pores. At the same time, "sticks" are considered throats by quantitatively measuring and statistically analyzing the "spheres", which involves conducting statistical analysis of the pore size distribution characteristics in the samples. While this approach provides a manageable and intuitive framework for visualizing pore connectivity, it inevitably simplifies the geometrical and physicochemical complexity of the actual pore structure. Such simplification may limit the accuracy of fluid flow predictions in highly heterogeneous or anisotropic systems. Nevertheless, for the purpose of comparative analysis and statistical characterization, this level of abstraction is generally considered sufficient and widely accepted in reservoir-scale studies [34,35].

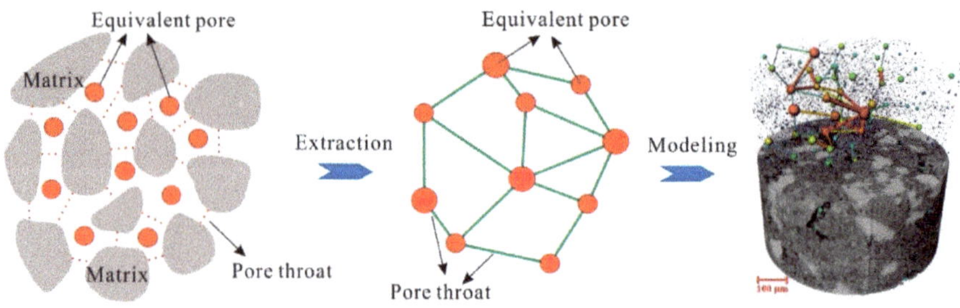

Figure 4. Schematic of 3D visualization of pore network model from micro-CT [11,33].

3.2. Multifractal Theory

This study employed the box-counting method, using cubic boxes with side length ε pixels to cover the reconstructed 3D image, forming a regular and orderly grid. The formation of the grid adopted a binary scale method [1,36,37]. The ε varies with along 2^k ($k = 0, 1, 2, 3, \ldots$), and the count of boxes for multiscale pore size distributions at a specified minor scale is denoted as $N(\varepsilon)$, thus:

$$N(\varepsilon) \propto^{-D}$$

where D is the fractal dimension; it can be considered that the studied object has fractal characteristics.

The proportion of pore in each box presented as $p_i(\varepsilon)$.

$$p_i(\varepsilon) = \frac{M_i\ (\varepsilon)}{M}$$

where $M_i\ (\varepsilon)$ is the pore volume in the i_{th} box. M is the total pore volume.

The weighted sum of the q-order moments of the proportion of pore in each box is obtained to yield the distribution. The formula $X(q, \varepsilon)$ is as follows:

$$X(q, \varepsilon) = \sum_{i=1}^{N(\varepsilon)} p_i^q$$

There is a relationship between the partition function $X(q, \varepsilon)$ and ε:

$$N(q, \varepsilon)\ \sim \varepsilon^{-\tau(q)}$$

The quality index $\tau(q)$ can be obtained by the ratio of logX(q, ε) and log(ε), the quality index $\tau(q)$ is related to the generalized dimensions $D(q)$: $\tau(q) = (1 - q)D(q)$, $D(q)$ can be calculated as:

$$D(q) = \lim_{\varepsilon \to 0} \frac{1}{1-q} \frac{\sum_{i=1}^{N(\varepsilon)} p_i^q}{log\varepsilon}$$

where q is the probability density weight index, which is an integer ranging from $-\infty$ to $+\infty$. In the case of $q = 1$, D_1 can be calculated as [38]:

$$D_1 = \lim_{\varepsilon \to 0} \frac{\log \sum_{i=1}^{N(\varepsilon)} p^i \log(p^i)}{log\varepsilon}$$

In multifractals, $P_i(\varepsilon)$ and ε exist:

$$P_i(\varepsilon) = \varepsilon^{\alpha_i}$$

where α_i is the singularity index, which represents the pore size distributions in the i_{th} box.

The count of boxes for multiscale pore size distribution with a specified minor scale is counted as $N_\varepsilon(\alpha)$, thus:

$$N_\varepsilon(\alpha) \propto \varepsilon^{-f(\alpha)}$$

where $f(\alpha)$ represents the singularity spectrum, reflecting the fractal dimension of subintervals with identical singularity strength [16].

The $f(\alpha)$ function can be obtained through Legendre transformation [39]:

$$\alpha(q) = \frac{d\tau(q)}{dq}$$

$$f(\alpha) = q\alpha(q) - \tau(q)$$

Generally, the multifractal parameters is often used to calculate the quality index $\tau(q)$, singularity spectra $f(\alpha)$, generalized dimensions $D(q)$ [16,38].

4. Results

4.1. Mineralogical Compositions

The XRD data reveals that quartz, feldspar, carbonate, and clay minerals predominantly comprise tight sandstone samples in the K_1q^4 (Table 2). Quartz is the predominant original component, ranging from 26.2% to 36.2% (mean: 32.27%). Feldspar content varies between 6.7% and 32.8% (mean: 18.97%). Carbonate content ranges from 9.5% to 17.7% (mean: 12.65%), encompassing calcite, ankerite, and dolomite.

Table 2. XRD of tight sandstones in the K_1q^4.

No.	Quartz (%)	Feldspar (%)	Calcite (%)	Ankerite (%)	Dolomite (%)	Siderite (%)	Pyrite (%)	Clay (%)
S1	26.2	10.5	5.20	3.20	0.00	1.40	2.40	51.10
S2	30.6	12.1	4.30	0.00	0.00	2.90	2.30	47.80
S3	35.8	6.7	2.40	3.90	0.00	1.40	2.40	47.40
S4	36.2	21.5	3.20	0.00	6.20	1.10	0.60	31.20
S5	30.4	32.8	6.70	0.00	9.10	1.60	0.30	19.10
S6	34.4	30.2	1.20	0.00	13.20	0.60	0.30	20.10

4.2. Pore Morphology

Thin section and SEM images reveal three types of pores: primary intergranular pore, dissolution pore, and intercrystalline pore (Figure 5). The primary intergranular pore were left from the deposition period (Figure 5a). Quantitative analysis of SEM images reveals that intergranular pores account for approximately 10–20% of total visible pore types, with typical sizes ranging from 5 to 15 μm. The dissolution pore formed during the middle diagenetic stage, including feldspar dissolution (Figure 5b) pore and intergranular dissolution pore (Figure 5c). Feldspar dissolution pores, formed through intense rock–water reactions, are primarily located within feldspar grains. The dissolution pore is widely developed and has good pore connectivity (Figure 5b). Dissolution pores are more abundant, comprising 30–40%, with sizes ranging from 2 to 20 μm, depending on the extent of feldspar alteration. Intercrystalline pores in clay minerals are the dominant type of pore (Figure 5e). Intercrystalline pores, primarily associated with clay minerals, make up 40–60% of total pore observations but are significantly smaller, typically less than 1 μm in diameter (Figure 5f). Additionally, various throat types, including shrunken-pore throat (Figure 5d), bent-sheet-shaped throat (Figure 5c), micro-throat (Figure 5c), and sheet-shaped throat (Figure 5d), are observed in thin sections.

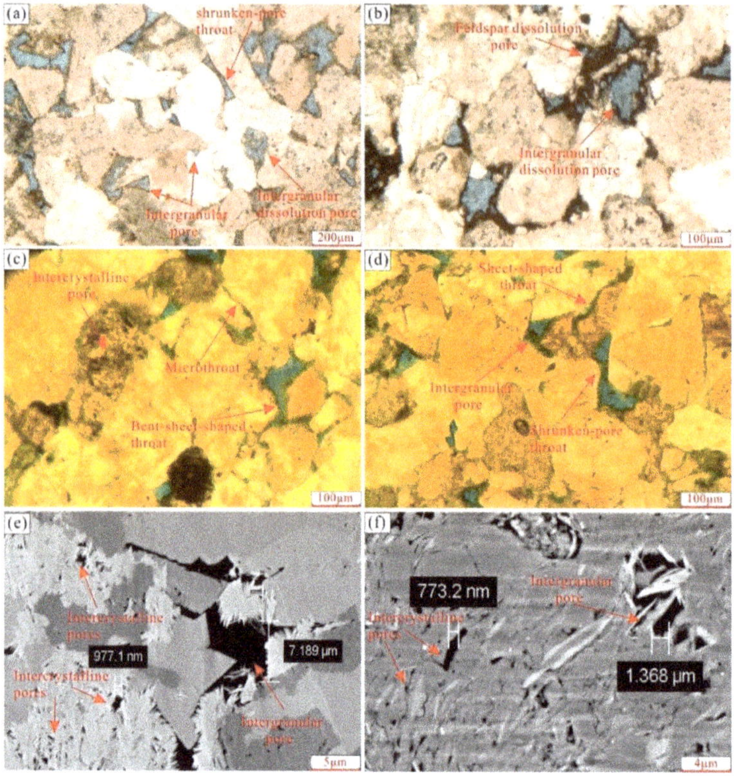

Figure 5. Typical pore types and throat types in plane-polarized light images of tight sandstone samples in the K_1q^4. (**a**) Intergranular pore and shrunken-pore throat, Well H87-7, 2306.59 m. (**b**) Feldspar dissolution pores, Well H87-7, 2308.48 m. (**c**) Bent-sheet shaped throat, Well H87-7, 2308.48 m. (**d**) Sheet-shaped throat, Well R53, 2204.61 m. (**e**) Intercrystalline pores in clay minerals, Well C45, 2110.26 m. (**f**) Intergranular distribution of illite, Well Q262, 2320.95 m.

4.3. Mercury Intrusion Curves

The mercury intrusion curves in HPMI experiments can reflect pore-throat connectivity and distribution characteristics. The physical properties of the tight sandstone reservoirs in the K_1q^4 vary greatly, with displacement pressure distributed between 0.250 and 4.076 MPa. The maximum mercury saturation mainly ranges from 33.139% to 96.355%, with a large variation in curve parameters and overall low mercury withdrawal efficiency (Figure 6), indicating strong heterogeneity in the tight sandstone reservoirs. For samples 5 and 6, the platform period is obvious, with coarse skewness, indicating good pore-throat connectivity and sorting. Samples 3, 4, and 2 show no platform period, indicating poor pore-throat connectivity and sorting. Sample 1 exhibits secondary mercury intrusion when the pressure increases to a certain value, suggesting a complex pore structure where mercury may enter large pores through small throats (similar to an ink-bottle effect).

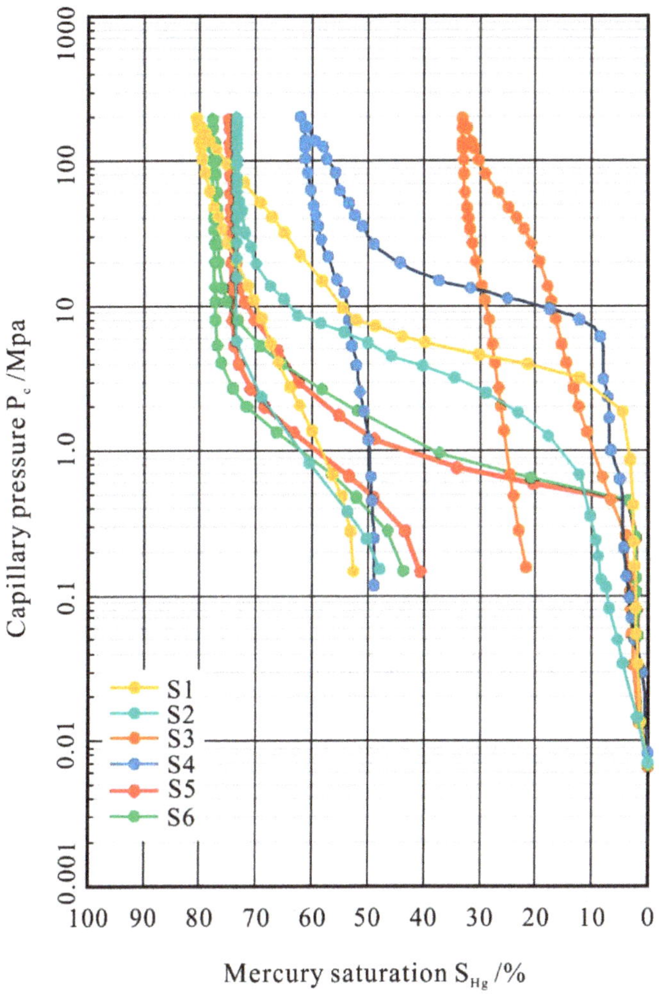

Figure 6. Characteristics of capillary pressure curve of HPMI.

4.4. Three-Dimensional Micro-CT Models

Since micro-CT scanning captures the spatial location and grayscale data of each critical unit within the sample, the 3D spatial distribution of the pore-throat is achievable post-precise identification [40]. Within the Volume Graphics Studio Max system, binary segmentation was conducted on a 3D tight sandstone model measuring 2.5 cm in diameter and 3.0 cm in height, utilizing the pore-throat grayscale threshold (0 to 19,000) derived from the digital terrain model to distinguish the pore-throat (Figure 7a). Following that, the measurement analysis function is employed to gather data on the pore-throat, which are then color-coded based on the size of the pore-throat unit bodies. Interconnected pores are considered a singular unit throughout the coloring process (Figure 7b,c).

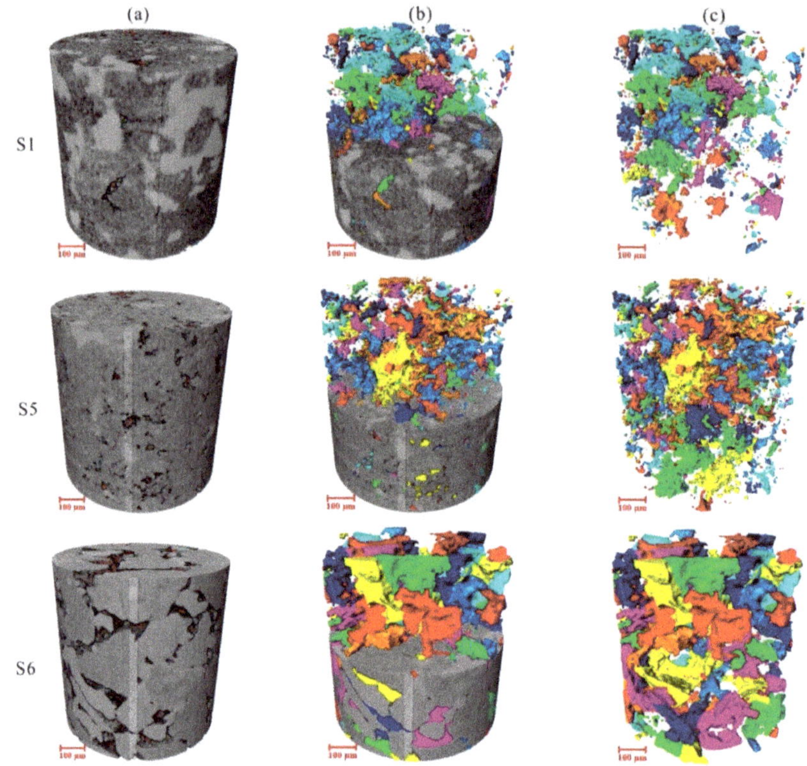

Figure 7. Micro-pore structure of tight sandstones from the micro-CT analysis in the K_1q^4. (**a**) The 3D pore network reconstruction. (**b**) The 3D local visualization of pore connectivity. (**c**) The 3D visualization of pore connectivity.

The central axis method is used to extract the pore network model of the 3D pore structure of tight sandstone based on micro-CT scanning (Figure 4). As shown in Figure 8, the upper part exhibits a sheet-like structure formed by the superposition of cylindrical rods (Figure 8a). In contrast, the lower part shows an isolated relationship between pores with a few throat connections (Figure 8b). The small number of throats and short extensions hinder the interconnection between pores in the sample. The throats are most developed in sample 6, with many cylindrical rods representing the throat densely superimposed on the model, forming a spider web-like structure. The throat development position aligns well with the hole position, indicating the sample has strong seepage capacity.

The pore structure of tight sandstones exhibits the following characteristics (Figure 8): (1) Pore shape: pore shapes are often isolated, but some appear banded. (2) Pore size distributions: the pore displays a multiscale of pore size distributions, characterized by uneven layering and localized development in the vertical direction. (3) Pore-throat connectivity: small-scale pores (1–10 μm) appear isolated, whereas large-scale pores (>10 μm) with tubular connections may communicate with micropores.

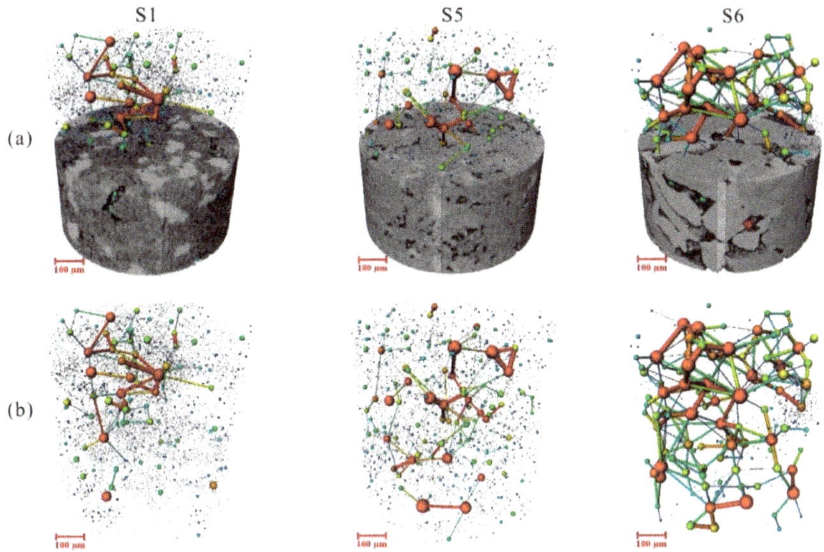

Figure 8. 3D visualization of pore network model from the micro-CT scanning of tight sandstone in the K_1q^4. (**a**) The 3D local visualization of pore network model. (**b**) The 3D visualization of pore network model.

5. Discussion

Shallow water deltaic deposits in the Quantou Formation of the Songliao Basin have been identified as significant hydrocarbon reservoirs. However, exploration and development of these low-permeability, tight sandstone reservoirs are complicated by factors such as narrow river channels, thin sand bodies, and strong reservoir heterogeneity [24]. Consequently, a comprehensive evaluation of reservoir pore structures is critical to overcoming the challenges and enhancing the effectiveness of exploration and development efforts [21,24,41]. Advanced techniques such as SEM, low-pressure gas adsorption (N_2, CO_2), MICP, and NMR are employed to estimate the fractal dimension of pore structures [42,43]. However, due to the limitations of constant-rate mercury injection pressures, the single fractal model better represents the distribution of small pore throats, while the fractal dimensions calculated from the HPMI method mainly capture small pore throat distributions [5]. Therefore, the heterogeneity for a full-scale pore system cannot be simply characterized by a single fractal dimension. Additionally, the intrinsic relationships between multifractal parameters, pore structure, and physical properties have remained uncertain.

5.1. Pore System Characteristics of Tight Sandstone Reservoirs

5.1.1. Pore Size Distribution

The distribution characteristics of pore radius obtained from HPMI, particularly for pore-throats smaller than 1 μm (Figure 9). Samples 5 and 6 exhibit the highest distribution frequency at 0.4 μm, with approximately 25% and 20%, respectively. The pore size distributions of samples 3 and 4 are more dispersed, with sample 3 exhibiting multiple frequency peaks within the range of 0.06 μm to 4 μm, while sample 4 shows a bimodal distribution at 0.16 μm and 1 μm. The distribution frequency of sample 1 is relatively low, primarily concentrated at 0.4 μm, whereas the pore size distributions of sample 2 are broader, with a peak around 0.6 μm. These characteristics indicate significant differences in the pore structures among the samples, which have important implications for fluid flow and reservoir performance.

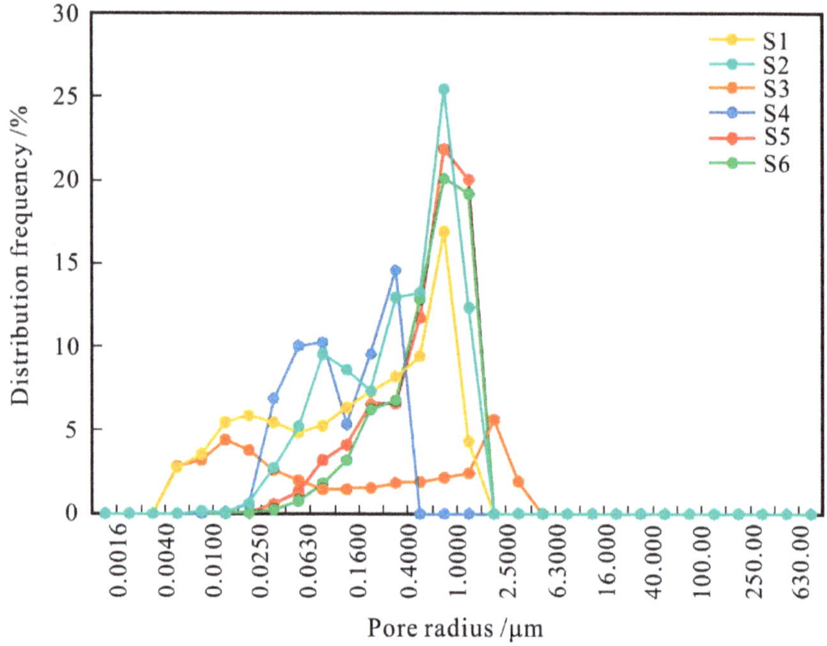

Figure 9. Distribution characteristics of pore radius obtained from HPMI.

The pore size distributions of various pore types with different scales were obtained by micro-CT, indicating strong pore structure heterogeneity of the tight sandstone samples (Figure 10). For pore sizes of 1–10 μm, the pore size distributions exhibited a unimodal distribution with a peak in the pore size of 1–2 μm and then rapidly declined as the pore size increased. This indicates that the pores in tight sandstone are primarily concentrated in the small size range, with a high degree of consistency in pore structures within this range across all samples. For pore sizes greater than 10 μm, the distribution frequency of all samples significantly decreased and approached zero, indicating the scarcity of large pores in these samples. Accordingly, we define three pore size categories: nano-scale (<1 μm), small-scale (1–10 μm), and large-scale (>10 μm). This classification is informed by analytical resolution thresholds, statistical distribution characteristics, and the functional implications of pore connectivity. Specifically, nano-scale pores are mainly detected by HPMI and SEM due to micro-CT resolution limits, small-scale pores represent the dominant high-frequency domain across samples, and large-scale pores, despite their low abundance, play a key role in enhancing permeability owing to their better connectivity with throats [11]. The pore size distributions characteristics indicate that tight sandstone samples are predominantly composed of small pores, with relatively few large pores. This pore structure has significant implications for reservoir performance and fluid flow characteristics. The high-frequency distribution of small pores may restrict fluid flow, while the relative scarcity of large pores suggests low permeability in these rocks.

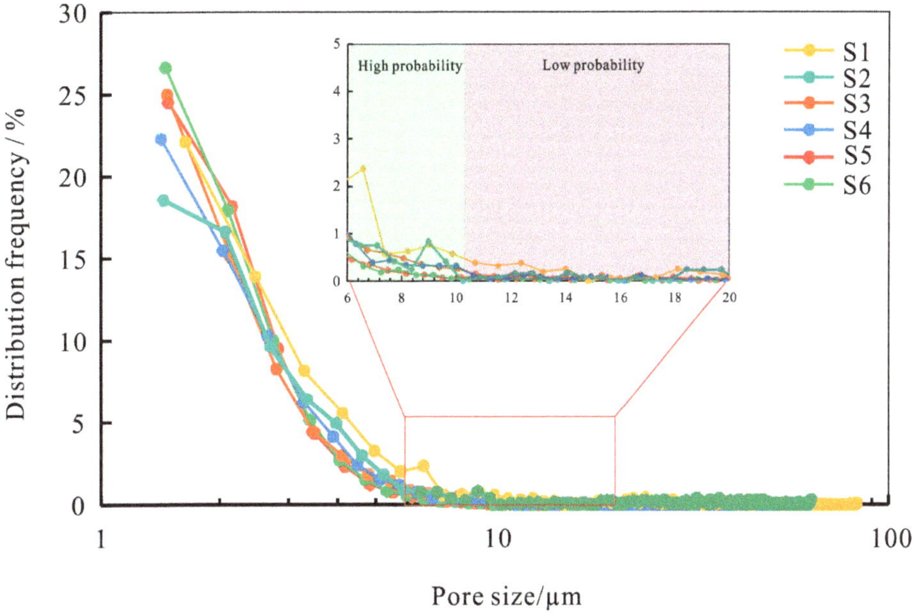

Figure 10. Distribution characteristics of pore radius obtained from micro-CT.

5.1.2. Pore Connectivity

Physical properties of connected pores and throats (fractures), including diameter, count proportion, and throat length, collectively govern the permeability of porous media [44]. Table 3 presents several parameters that characterize the scale of pore development and the distribution features, including the voxel count of pores, area, volume, and the proportion of connected pores. We can see that the pore size in tight sandstone spans a large scale, and that the distribution of those pore sizes is uneven, with a relatively complex distribution of pores (Table 3).

Figure 11 shows connected pore voxel count, volumes, and areas of tight sandstone, indicating that the smaller the connected pore size, the greater the number of pores, with the voxel count exponentially increasing as the size decreases. It is worth noting that the connected pore voxels count of large-scale pores (>10 μm) is mostly less than 10. Although the connected pore voxels count of small-scale pores (1–10 μm) is predominant, the connected pore volume and area are relatively small. Conversely, while the number of pores with connected pores of large-scale pores (>10 μm) is small, the volume and area of these pores are large, and both the volume and area of these connected pores increase with an increase. For instance, samples 5 and 6 feature only 9 and 13 voxel counts in the large-scale pores (>10 μm). Despite this, their corresponding volumes are 205,038.81 μm^3 and 287,707.84 μm^3, and their areas are 84,136.05 μm^2 and 93,388.77 μm^2, respectively.

Table 3. The total pores, closed pores, and connected pores subsection distribution were computed by micro-CT of tight sandstones in the K_1q^4.

No.	Pore Size (μm)	Total Pores			Closed Pores			Connected Pores			Connected Pores Proportion	
		Voxels Count	Volumes /μm^3	Areas /μm^2	Voxels Count	Volumes /μm^3	Areas /μm^2	Voxels Count	Volumes /μm^3	Areas /μm^2	Volumes	Areas
	1–2	8335	117,782.36	205,770.12	8193	116,116.53	202,759.29	142	1665.83	3010.83	1.41%	1.46%
	2–3	4662	289,852.33	369,030.20	4535	281,931.05	358,951.09	127	7921.28	10,079.11	2.73%	2.73%
	3–4	1731	289,523.89	295,763.68	1651	276,190.81	282,052.21	80	13,333.08	13,711.46	4.61%	4.64%
	4–5	621	227,435.22	201,123.33	585	213,758.66	188,666.16	36	13,676.55	12,457.17	6.01%	6.19%
	5–6	256	173,501.81	136,826.30	237	160,439.21	126,410.63	19	13,062.60	10,415.67	7.53%	7.61%
	6–7	128	143,411.68	103,184.85	116	129,808.11	93,018.01	12	13,603.57	10,166.85	9.49%	9.85%
	7–8	76	131,031.73	82,624.81	70	120,922.97	76,200.67	6	10,108.76	6424.14	7.71%	7.78%
	8–9	44	113,469.66	67,596.48	38	98,022.09	58,972.12	6	15,447.57	8624.35	13.61%	12.76%
	9–10	22	78,972.38	39,991.28	18	64,495.11	32,562.76	4	14,477.27	7428.52	18.33%	18.58%
S5	10–11	21	102,918.69	55,810.80	15	74,569.52	39,265.61	6	28,349.17	16,545.19	27.55%	29.65%
	11–12	13	82,716.21	43,245.58	9	57,368.11	29,691.39	4	25,348.10	13,554.19	30.64%	31.34%
	12–13	10	79,822.47	37,244.55	8	64,188.14	29,090.07	2	15,634.33	8154.48	19.59%	21.89%
	13–14	13	134,382.71	58,804.19	12	124,591.66	55,514.33	1	9791.05	3289.85	7.29%	5.59%
	14–15	7	85,880.42	29,111.20	5	60,229.64	19,677.14	2	25,650.78	9434.06	29.87%	32.41%
	15–16	5	79,281.50	29,440.52	4	64,731.25	22,970.98	1	14,550.26	6469.53	18.35%	21.97%
	16–17	5	79,281.50	32,192.50	4	73,305.12	25,505.38	1	5976.38	6687.11	7.54%	20.77%
	17–18	5	110,734.78	48,231.80	4	86,852.87	40,792.45	1	23,881.91	7439.35	21.57%	15.42%
	18–19	3	77,233.56	31,460.01	1	25,977.08	13,003.42	2	51,256.48	18,456.59	66.37%	58.67%
	19–20	3	96,251.07	30,294.76	2	63,301.55	19,643.88	1	32,949.52	10,650.89	34.23%	35.16%
	1–2	548	7715.20	13,412.42	501	7073.34	12,290.67	47	641.86	1121.76	8.32%	8.36%
	2–3	350	22,557.41	28,044.51	328	21,275.83	26,429.65	22	1281.57	1614.86	5.68%	5.76%
	3–4	161	28,366.35	26,974.31	142	24,648.28	23,549.24	19	3718.07	3425.07	13.11%	12.70%
	4–5	103	39,196.39	30,939.30	93	35,370.99	28,101.40	10	3825.40	2837.90	9.76%	9.17%
	5–6	61	42,482.98	29,813.39	54	37,728.06	26,450.01	7	4754.92	3363.38	11.19%	11.28%
	6–7	39	45,776.00	28,830.38	35	41,293.71	25,896.72	4	4482.29	2933.66	9.79%	10.18%
	7–8	21	34,198.90	19,555.06	19	31,135.58	18,046.79	2	3063.33	1508.27	8.96%	7.71%
	8–9	13	33,909.10	16,717.22	10	26,870.11	13,151.42	3	7038.99	3565.79	20.76%	21.33%
	9–10	14	49,959.89	24,831.22	11	40,140.94	19,198.60	3	9818.96	5632.62	19.65%	22.68%
S6	10–11	7	32,807.85	16,136.72	5	22,892.29	10,930.56	2	9915.56	5206.16	30.22%	32.26%
	11–12	10	62,120.89	27,582.33	6	37,652.92	16,740.58	4	24,467.96	10,841.75	39.39%	39.31%
	12–13	6	49,818.21	22,236.36	4	33,864.02	14,596.28	2	15,954.19	7640.08	32.02%	34.36%
	13–14	4	42,510.88	15,965.30	3	33,026.81	12,520.39	1	9484.07	3444.91	22.31%	21.58%
	14–15	3	38,249.70	12,594.01	2	25,036.83	8389.07	1	13,212.87	4204.94	34.54%	33.39%
	15–16	1	79,281.50	6463.29	0	0.00	0.00	1	15,632.19	6463.29	100.00%	100.00%
	16–17	1	15,632.19	4761.29	0	0.00	0.00	1	15,632.19	4761.29	100.00%	100.00%
	17–18	1	21,649.36	5884.28	0	0.00	0.00	1	21,649.36	5884.28	100.00%	100.00%
	18–19	4	106,563.79	26,574.31	1	27,666.53	6533.34	3	78,897.26	20,040.97	74.04%	75.41%
	19–20	4	123,655.73	39,967.66	1	30,877.98	9860.40	3	92,777.75	30,107.26	75.03%	75.33%

Furthermore, we compared samples 5 and 6 from the same well; both showcase similar mineralogy and physical properties (Table 1). We analyzed the throat size, area distributions, and channel length (Figure 12). The voxel numbers of throats decrease as the throat size increases. The throat areas primarily exhibit a throat size distribution ranging from 0 to 20 μm, and the average throat length appears to be randomly distributed. In sample 6, the number of throats within the 20 to 40 μm size range surpasses that in sample 5, leading to a pore-throat representation exhibiting increased connectivity, a larger cross-sectional diameter, and a higher permeability than sample 5 (Table 1). The pore-throat in sample 6 is globally connected, whereas the reservoirs in Sample 5 are locally connected. These observations imply that the percolation capability of sample 6 overshadows that of sample 5. Therefore, the throat size range of 20 to 40 μm may significantly escalate the overall connectivity and seepage capacity of the pores within the range of sizes.

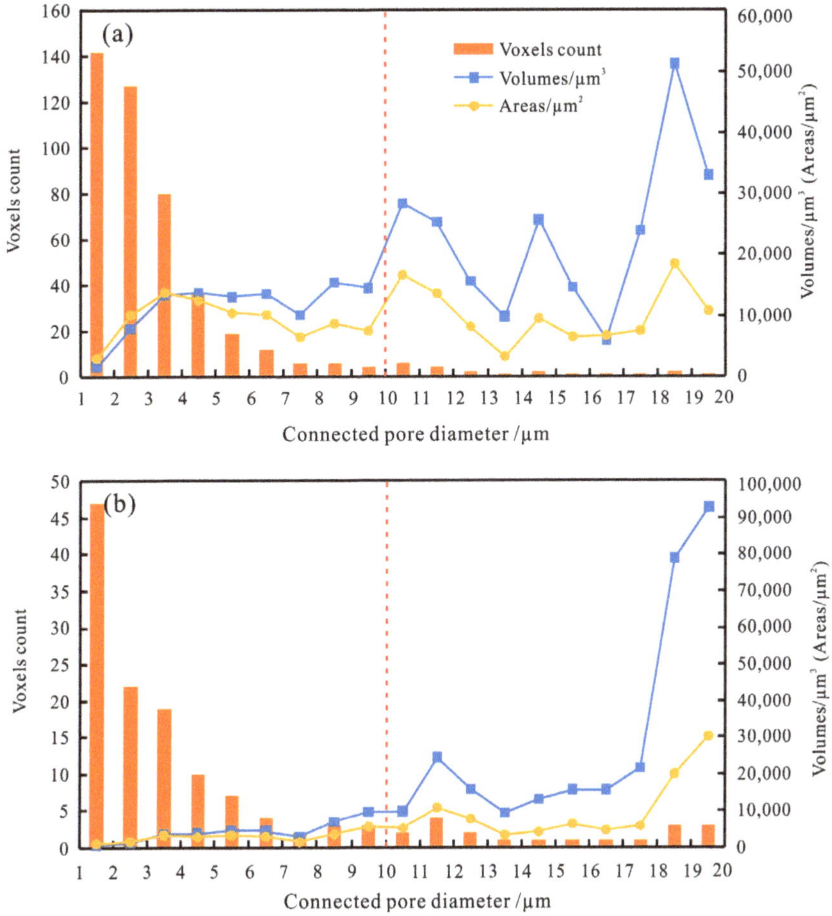

Figure 11. The connected pore voxels count, volumes, areas of tight sandstones in the K_1q^4. (**a**) Sample 5; (**b**) Sample 6.

5.2. Multifractal Analyses of Tight Sandstone Reservoirs

5.2.1. Multiscale Pore Structure Determination

The multifractal theory is an effective method for characterizing the complex degrees and heterogeneities of pore structure. Figure 13 shows a relationship between $logX(q,\varepsilon)$ and $log(\varepsilon)$ with q ranging from -10 to 10 for the tight sandstone sample derived from micro-CT. An evident linear relationship exists between $logX(q,\varepsilon)$ and $log(\varepsilon)$, indicating the pore structure of tight sandstones exhibits significant multifractal characteristics [38]. A nearly monotonic decrease in $D(q)$ is observed within the interval of $-10 \leq q \leq 10$. Furthermore, a stronger correlation when q is close to 0, reflecting that the pore structure of the tight sandstone has a strongly heterogeneous distribution.

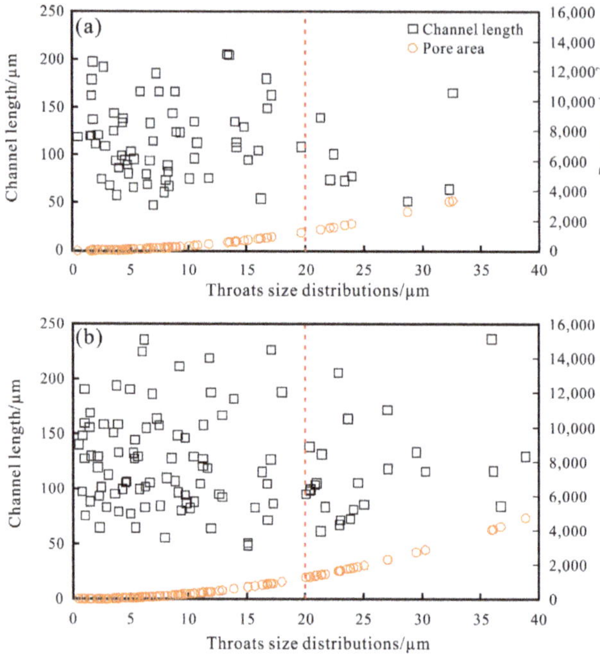

Figure 12. Throat size distributions, area distributions, and channel length of tight sandstones in the K_1q^4. (**a**) Sample 5; (**b**) Sample 6.

As the current pore classification schemes are not suitable for describing the pore types of tight sandstone, and consequently resulting in misestimates of the storage and percolation capacities, it is essential that a specific classification scheme is developed for tight sandstone reservoirs. We classify the samples based on pore size distributions, pore connectivity, and fractal dimension into three categories: the nano-scale pores (<1 μm), the small-scale pores (1–10 μm), and the large-scale pores (>10 μm).

Based on the pore size distributions obtained from HPMI (Figure 10), pore size distributions smaller than 1 μm (i.e., nanopores) are mostly intercrystalline pores formed between clay aggregates, which can be defined as intercrystalline nanopores size distributions less than 1 μm (Figure 5e). The pore size distribution analysis shows a unimodal distribution for pores in the range of 1.08–38.72 μm, with a peak in the 1–10 μm range, and little variation in diameters greater than 10 μm (Figure 11). The correlation between pore volume and pore area was assessed for small-scale pores (1–10 μm) and large-scale pores (>10 μm). Small-scale pores (1–10 μm) exhibited a positive correlation between pore surface area and volume, characterized by high correlation coefficients (R^2 = 0.9814) (Figure 14a). Conversely, for large-scale pores (>10 μm), a positive correlation between pore surface area and volume was observed, but with a lower correlation coefficient (R^2 = 0.5735) (Figure 14b). The complexity of small-scale pores (1–20 μm) demonstrated relatively uniform characteristics across samples compared to large-scale pores (>10 μm).

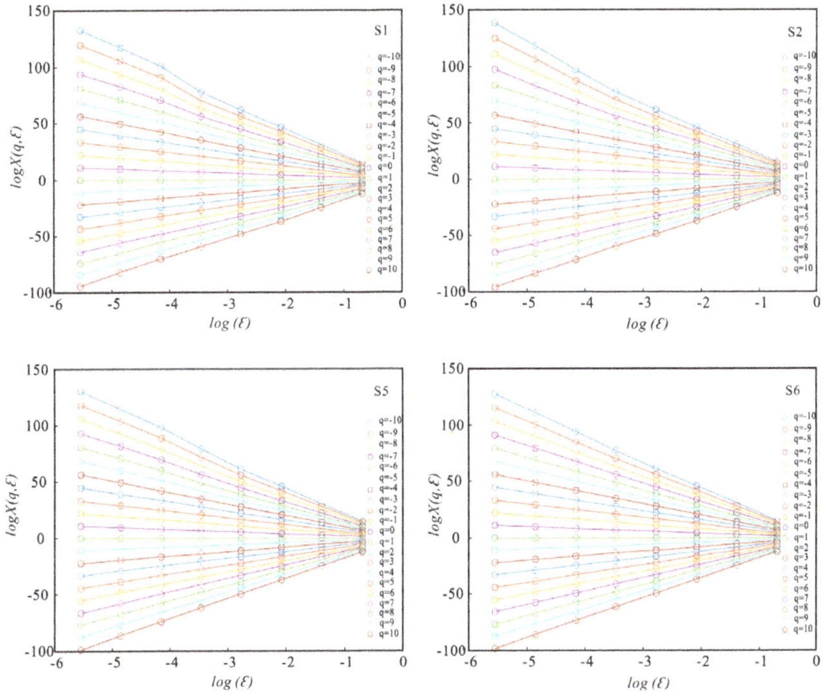

Figure 13. Relationships between $\log X(q, \varepsilon)$ and $\log(\varepsilon)$ from multifractal analyses of tight sandstones in the K_1q^4.

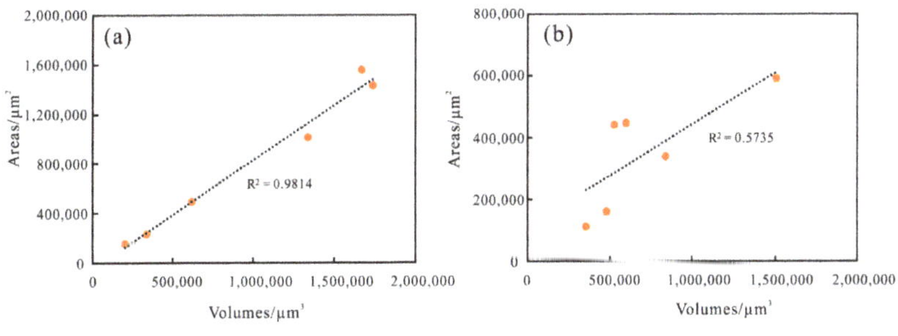

Figure 14. Correlation analysis between volumes and areas of tight sandstones in the K_1q^4. (**a**) Small-scale pores (1–10 μm), (**b**) Large-scale pores (>10 μm).

The multifractal analysis divides the pore size distributions into high probability measure areas and low probability measure areas. The small-scale pores (1–10 μm) have a high distribution frequency, while the large-scale pores (>10 μm) have a low distribution frequency. Consequently, the small-scale pores (1–10 μm) are identified as the high probability measure area, while large-scale pores (>10 μm) correspond to the low probability measure area.

5.2.2. Multifractal Analyses for Multiscale Pore Structure

The generalized dimensions of the sample are shown in Figure 15. It can be seen that the $D(q)$ spectra exhibit typical anti-S curves (Figure 15a). The generalized dimensions ΔD reflects the degree of variation of the local characteristics of the pore. The greater the value of ΔD, the stronger the heterogeneity of pore structure [38]. When $q = -10$, the generalized dimensions take the maximum value D_{max}; when $q = 10$, the generalized dimensions take the minimum value D_{min}. Using $q = 0$ as the boundary, the multifractal parameter $D_{min} - D_0$ represents the width of the left branch of the generalized dimensions, corresponding to the heterogeneity of lower probability measure areas. The multifractal parameter $D_0 - D_{max}$ represents the width of the right branch, corresponding to the heterogeneity of high probability measure areas.

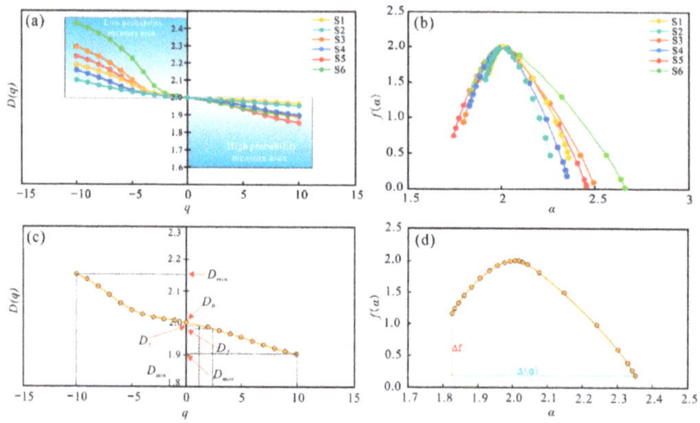

Figure 15. Generalized dimensions (**a**,**c**) and singularity spectra (**b**,**d**) of tight sandstones in the K_1q^4.

The $f(\alpha)$ of tight sandstone exhibits left hook shapes, which indicates that a large probability subset occupies a dominant position [45] (Figure 15b). The multifractal parameter $\Delta(\alpha)$ represents the width of the multifractal spectrum, offering insights into the heterogeneity development of multiscale pore structure. The multifractal parameter Δf reflects the shape characteristics of the multifractal spectrum and characterizes the degree of symmetry in pore size distributions [16]. Figure 15c,d illustrate the graphical meanings of multifractal parameters.

Table 4 summarizes the characteristic multifractal parameters of the pore structure in tight sandstone. Note that the multifractal parameters $D_0 - D_{max}$ and $D_{min} - D_0$ ranged from 0.0323 to 0.1454, and 0.1402 to 0.3275, respectively, suggesting that low probability measure areas exhibit stronger heterogeneity of tight sandstone reservoirs.

Table 4. Characteristic multifractal parameters of tight sandstones in the K_1q^4.

No.	D_1	ΔD	$D_0 - D_1$	$\Delta(\alpha)$	Δf	$D_0 - D_{max}$	$D_{min} - D_0$
S1	1.9875	0.3820	0.0125	0.7127	0.7080	0.0323	0.2366
S2	1.9887	0.5341	0.0113	0.8242	1.3173	0.1066	0.3275
S3	1.9977	0.4003	0.0023	0.6980	0.8391	0.1078	0.2925
S4	1.9913	0.2543	0.0087	0.5243	0.9802	0.0668	0.1556
S5	1.9994	0.2197	0.0006	0.4276	1.1887	0.1454	0.1874
S6	1.9988	0.1471	0.0012	0.3505	1.0669	0.0988	0.1402

5.3. The Research Significance of Characterizing Pore Structure Based on Multi-Fractal Dimension

5.3.1. The Relationship Between Multifractal Characteristics and Rock Physical Properties

Physical property parameters serve as a comprehensive reflection of the pore system, where porosity and permeability emerge as crucial indicators for assessing tight sandstone reservoirs' storage capacity and seepage potential [46]. Figure 16a shows the correlation of multifractal parameters and porosity; there was a positive relationship between porosity and $D_0 - D_{max}$ (R^2 = 0.5775), significantly higher than that of multifractal parameters $D_{min} - D_0$ (R^2 = 0.0881). The multifractal parameter $D_0 - D_{max}$ represents the width of the right branch, corresponding to the heterogeneity of high probability measure areas. The positive correlation between porosity and $D_0 - D_{max}$ within a specific range for small-scale pores (1–10 μm) indicates that $D_0 - D_{max}$ is a sensitive indicator for the porosity of tight sandstone. Furthermore, small-scale pores (1–10 μm) situated in the high probability measure areas substantially influence porosity. We also found that the voxel count, volumes, and areas of small-scale pores (1–10 μm) were distributed intensely at the high probability measure area. Therefore, small-scale pores (1–10 μm) possess a more robust ability to characterize porosity heterogeneity.

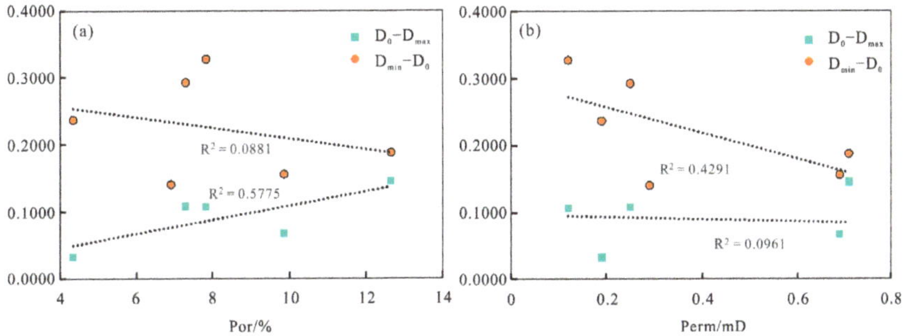

Figure 16. Correlation analysis of multifractal parameters with porosity (**a**) and permeability (**b**).

The multifractal parameters demonstrate a correlation with permeability. In contrast to porosity, the correlation between multifractal parameters $D_{min} - D_0$ and permeability is significantly higher than that of multifractal parameters $D_0 - D_{max}$, and has a negative relationship between porosity with high correlation coefficient (R^2 = 0.4291) (Figure 16b). This phenomenon shows that $D_{min} - D_0$ is a sensitive indicator for permeability. The multifractal parameter $D_{min} - D_0$ represents the width of the left branch of the generalized dimensions, corresponding to the heterogeneity of lower probability measure areas. Therefore, the low probability measure area substantially influences the permeability of tight sandstone. Additionally, we observed a pore size range of 10 to 20 μm and a throat size range of 20 to 40 μm in the low probability measure area. Consequently, the part of the micro-pore corresponding to the combination of pore size range of 10 to 20 μm and throat size range of 20 to 40 μm in a low probability measure area may dominate the permeability of tight sandstone.

5.3.2. The Relationship Between Multifractal Parameters and Mineral Composition

Previous studies have demonstrated a significant correlation between the pore structure of tight sandstone reservoirs and mineral composition [38]. We investigated the impact of mineral composition on pore heterogeneity by examining the correlation between multifractal parameters and mineral composition (Figure 17). The feldspar mineral in the long stone is negatively correlated with $\Delta(\alpha)$, ΔD, and $D_{min} - D_0$, suggesting that the higher the feldspar content, the lower heterogeneity of pore structure. Furthermore, a strong negative value of $D_{min} - D_0$ suggests that the segment of large-scale pores (>10 μm) in the low probability measure area primarily influences the heterogeneity of pore size distributions. The diagenetic stage of K_1q^4 is the middle, where secondary dissolution pores are relatively common [24]. Dissolution alters the micro-pore structure, increasing the number of large pores. As a result, feldspar promotes the formation of relatively larger-scale macropores, contributing to the homogeneity of larger pores.

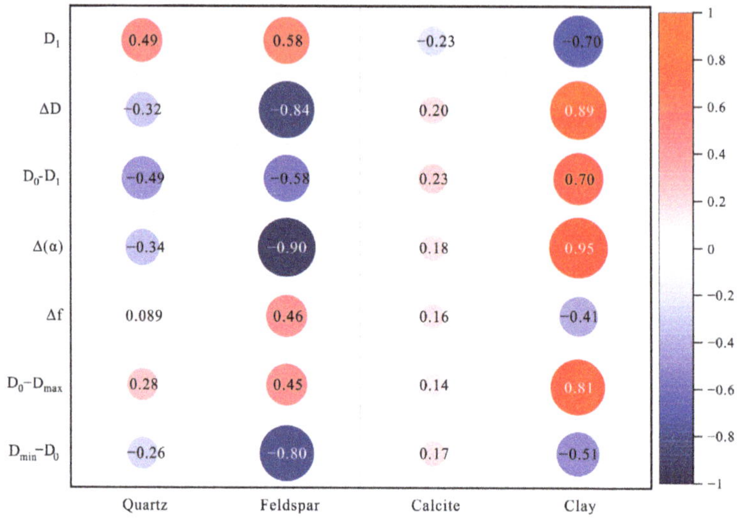

Figure 17. Pearson correlation matrix between tight sandstones compositions and fractal parameters.

Clay minerals positively correlate with $\Delta(\alpha)$ andΔD, suggesting that clay minerals govern the complexity and dispersion of pore size distributions. The correlation between clay minerals and $D_0 - D_{max}$ is stronger than that of $D_{min} - D_0$. Pore size distribution reveals that small-scale pores (1–10 μm) represent high probability measure areas, while the large-scale pores (>10 μm) correspond to lower probability measure areas. Thus, clay minerals significantly control small-scale pores (1–10 μm) structure heterogeneity. This may be because clay minerals fill the pore space during diagenesis, resulting in the disappearance of pores, especially macropores (Figure 5e). The subsequent transformation and volume shrinkage of clay minerals in the later diagenetic process rapidly created numerous interface pores and diagenetic contraction joints, thereby promoting the formation of pores, particularly nanopores (Figure 5f).

It should be noted that diagenetic processes play a critical role in the evolution of pore structures. However, the present study primarily focuses on establishing a statistical correlation between multifractal heterogeneity and mineralogical composition. Although this work centers on the physical complexity of pore networks as characterized by multifractal analysis, future studies should incorporate geochemical parameters, such as pore fluid chemistry and organic matter content, to further clarify their influence on pore structure development and fluid flow behavior.

6. Conclusions

Multiscale pore structure characteristics of tight sandstones were investigated by integrating HPMI and micro-CT 3D pore network modeling, with multifractal theory applied to quantitatively assess pore structure heterogeneity. The results revealed a multiscale pore size distribution, wherein small pores (1–10 μm) tend to be isolated, while larger pores (>10 μm) with tubular connections may facilitate communication with micropores. The pore structures exhibited clear multifractal behavior, and the extracted multifractal parameters proved to be sensitive indicators for characterizing heterogeneity across different scales. In addition, the pore structure was closely related to compositional variation; tight sandstones enriched in feldspar displayed improved pore connectivity, whereas clay mineral enrichment was associated with increased micropore heterogeneity. Notably, pore structure heterogeneity exerted a significant influence on permeability, with pores sized

10–20 μm and throats 20–40 μm in the low probability measure zone potentially playing a dominant role in controlling reservoir permeability.

Author Contributions: Conceptualization, L.L. and Z.B.; methodology, Z.L. and L.L.; software, C.L. and L.L.; validation, Y.L.; formal analysis, H.H., J.W., and W.Z.; writing—original draft preparation, L.L. and Z.B.; writing—review and editing, Z.B. and C.L. All authors have read and agreed to the published version of the manuscript.

Funding: This research was funded by the funding project of Heilongjiang Provincial Natural Science Foundation Project (LH2022D014), and the National Natural Science Foundation of China (52074088, 52474036, 52174022).

Data Availability Statement: Data are available from the authors upon request.

Conflicts of Interest: Authors Zhongcheng Li, Haotian Han and Jianyi Wang were employed by PetroChina Jilin Oilfield Company. The remaining authors declare that the research was conducted in the absence of any commercial or financial relationships that could be construed as a potential conflict of interest.

References

1. Mandelbrot, B.B. *The Fractal Geometry of Nature*; W. H. Freeman: New York, NY, USA, 1982.
2. Broseta, D.; Barré, L.; Vizika, O.; Shahidzadeh, N.; Guilbaud, J.; Lyonnard, S. Capillary condensation in a fractal porous medium. *Phys. Rev. Lett.* **2001**, *86*, 5313. [CrossRef] [PubMed]
3. Li, Y.; Wang, Z.; Pan, Z.; Niu, X.; Yu, Y.; Meng, S. Pore structure and its fractal dimensions of transitional shale: A cross-section from east margin of the Ordos Basin, China. *Fuel* **2019**, *241*, 417–431. [CrossRef]
4. Xin, Y.; Wang, G.; Liu, B.; Ai, Y.; Cai, D.; Yang, S.; Liu, H.; Xie, Y.; Chen, K. Pore structure evaluation in ultra-deep tight sandstones using NMR measurements and fractal analysis. *J. Pet. Sci. Eng.* **2022**, *211*, 110180. [CrossRef]
5. Wu, Y.; Liu, C.; Ouyang, S.; Luo, B.; Zhao, D.; Sun, W.; Awan, R.S.; Lu, Z.; Li, G.; Zang, Q. Investigation of pore-throat structure and fractal characteristics of tight sandstones using HPMI, CRMI, and NMR methods: A case study of the lower Shihezi Formation in the Sulige area, Ordos Basin. *J. Pet. Sci. Eng.* **2022**, *210*, 110053. [CrossRef]
6. Anovitz, L.M.; Cole, D.R.; Jackson, A.J.; Rother, G.; Littrell, K.C.; Allard, L.F.; Pollington, A.D.; Wesolowski, D.J. Effect of quartz overgrowth precipitation on the multiscale porosity of sandstone: A (U) SANS and imaging analysis. *Geochim. Cosmochim. Acta* **2015**, *158*, 199–222. [CrossRef]
7. Wang, J.; Zhang, J.; Xiao, X.; Chen, Y.N.; Han, D. Pore Structure and Fractal Characteristics of Inter-Layer Sandstone in Marine–Continental Transitional Shale: A Case Study of the Upper Permian Longtan Formation in Southern Sichuan Basin, South China. *Fractal Fract.* **2024**, *9*, 11. [CrossRef]
8. Shanley, K.W.; Cluff, R.M. The evolution of pore-scale fluid-saturation in low-permeability sandstone reservoirs. *AAPG Bull.* **2015**, *99*, 1957–1990. [CrossRef]
9. Wang, W.; Lin, C.; Zhang, X. Fractal Characteristics of Pore Throat and Throat of Tight Sandstone Sweet Spot: A Case Study in the East China Sea Basin. *Fractal Fract.* **2024**, *8*, 684. [CrossRef]
10. Hu, Q.; Zhang, Y.; Meng, X.; Li, Z.; Xie, Z.; Li, M. Characterization of micro-nano pore networks in shale oil reservoirs of Paleogene Shahejie Formation in Dongying Sag of Bohai Bay Basin, East China. *Pet. Explor. Dev.* **2017**, *44*, 720–730. [CrossRef]
11. Liu, P.; Nie, B.; Zhao, Z.; Li, J.; Yang, H.; Qin, C. Permeability of micro-scale structure in coal: Insights from μ-CT image and pore network modelling. *Gas Sci. Eng.* **2023**, *111*, 204931. [CrossRef]
12. Yang, R.; He, S.; Yi, J.; Hu, Q. Nano-scale pore structure and fractal dimension of organic-rich Wufeng-Longmaxi shale from Jiaoshiba area, Sichuan Basin: Investigations using FE-SEM, gas adsorption and helium pycnometry. *Mar. Pet. Geol.* **2016**, *70*, 27–45. [CrossRef]
13. Sun, M.; Zhao, J.; Pan, Z.; Hu, Q.; Yu, B.; Tan, Y.; Sun, L.; Bai, L.; Wu, C.; Blach, T.P. Pore characterization of shales: A review of small angle scattering technique. *J. Nat. Gas Sci. Eng.* **2020**, *78*, 103294. [CrossRef]
14. Liu, P.; Nie, B.; Zhao, Z.; Zhao, Y.; Li, Q. Characterization of ultrasonic induced damage on multi-scale pore/fracture in coal using gas sorption and μ-CT 3D reconstruction. *Fuel* **2023**, *332*, 126178. [CrossRef]
15. Zhou, S.; Liu, D.; Cai, Y.; Yao, Y.; Li, Z. 3D characterization and quantitative evaluation of pore-fracture networks of two Chinese coals using FIB-SEM tomography. *Int. J. Coal Geol.* **2017**, *174*, 41–54. [CrossRef]
16. Soto-Gomez, D.; Perez-Rodriguez, P.; Juiz, L.V.; Paradelo, M.; Lopez-Periago, J.E. 3D multifractal characterization of computed tomography images of soils under different tillage management: Linking multifractal parameters to physical properties. *Geoderma* **2020**, *363*, 114129. [CrossRef]

17. Liu, Q.; Sun, M.; Sun, X.; Liu, B.; Ostadhassan, M.; Huang, W.; Chen, X.; Pan, Z. Pore network characterization of shale reservoirs through state-of-the-art X-ray computed tomography: A review. *Gas Sci. Eng.* **2023**, *113*, 204967. [CrossRef]
18. Zhu, H.; Huang, C.; Ju, Y.; Bu, H.; Li, X.; Yang, M.; Chu, Q.; Feng, H.; Qiao, P.; Qi, Y. Multi-scale multi-dimensional characterization of clay-hosted pore networks of shale using FIBSEM, TEM, and X-ray micro-tomography: Implications for methane storage and migration. *Appl. Clay Sci.* **2021**, *213*, 106239. [CrossRef]
19. Alyafei, N.; Mckay, T.J.; Solling, T.I. Characterization of petrophysical properties using pore-network and lattice-Boltzmann modelling: Choice of method and image sub-volume size. *J. Pet. Sci. Eng.* **2016**, *145*, 256–265. [CrossRef]
20. Zhang, L.; Bao, Z.; Dou, L.; Xu, Q. Diagenetic alterations related to sedimentary architecture of deltaic distributary channels in red beds of the Cretaceous Yaojia Formation, Songliao Basin. *J. Pet. Sci. Eng.* **2021**, *203*, 108564. [CrossRef]
21. Peng, H.; Jiao, Y.; Fu, X.; Wu, L.; Guo, X.; Wang, Q.; Liu, C. Provenance and uranium source tracing for uranium-bearing series in the south of Songliao Basin: Evidence from zircon UPb chronology and lithogeochemistry. *J. Geochem. Explor.* **2025**, *272*, 107703. [CrossRef]
22. Feng, Z.; Jia, C.; Xie, X.; Zhang, S.; Feng, Z.; Cross, T.A. Tectonostratigraphic units and stratigraphic sequences of the nonmarine Songliao basin, northeast China. *Basin Res.* **2010**, *22*, 79–95.
23. Li, L.; Bao, Z.; Li, L.; Li, Z.; Ban, S.; Li, Z.; Wang, T.; Li, Y.; Zheng, N.; Zhao, C. The source and preservation of lacustrine shale organic matter: Insights from the Qingshankou Formation in the Changling Sag, Southern Songliao Basin, China. *Sediment. Geol.* **2024**, *466*, 106649. [CrossRef]
24. Xi, K.; Cao, Y.; Jahren, J.; Zhu, R.; Bjørlykke, K.; Zhang, X.; Cai, L.; Hellevang, H. Quartz cement and its origin in tight sandstone reservoirs of the Cretaceous Quantou formation in the southern Songliao basin, China. *Mar. Pet. Geol.* **2015**, *66*, 748–763. [CrossRef]
25. Zhang, Q.; Wu, X.; Radwan, A.E.; Wang, B.; Wang, K.; Tian, H.; Yin, S. Diagenesis of continental tight sandstone and its control on reservoir quality: A case study of the Quan 3 member of the cretaceous Quantou Formation, Fuxin uplift, Songliao Basin. *Mar. Pet. Geol.* **2022**, *145*, 105883. [CrossRef]
26. Feng, Z.; Zhang, S.; Feng, Z. Discovery of "Enveloping surface of oil and gas overpressure migration" in the Songliao Basin and its bearings on hydrocarbon migration and accumulation mechanisms. *Sci. China Earth Sci.* **2012**, *55*, 2005–2017. [CrossRef]
27. Liu, B.; Wang, H.; Fu, X.; Bai, Y.; Bai, L.; Jia, M.; He, B. Lithofacies and depositional setting of a highly prospective lacustrine shale oil succession from the Upper Cretaceous Qingshankou Formation in the Gulong sag, northern Songliao Basin, northeast China. *AAPG Bull.* **2019**, *103*, 405–432. [CrossRef]
28. Bish, D.L.; Howard, S.A. Quantitative phase analysis using the Rietveld method. *Appl. Crystallogr.* **1988**, *21*, 86–91. [CrossRef]
29. Środoń, J.; Drits, V.A.; McCarty, D.K.; Hsieh, J.C.; Eberl, D.D. Quantitative X-ray diffraction analysis of clay-bearing rocks from random preparations. *Clays Clay Miner.* **2001**, *49*, 514–528. [CrossRef]
30. Loucks, R.G.; Reed, R.M.; Ruppel, S.C.; Jarvie, D.M. Morphology, genesis, and distribution of nanometer-scale pores in siliceous mudstones of the Mississippian Barnett Shale. *J. Sediment. Res.* **2009**, *79*, 848–861. [CrossRef]
31. Milliken, K.L.; Rudnicki, M.; Awwiller, D.N.; Zhang, T. Organic matter–hosted pore system, Marcellus formation (Devonian), Pennsylvania. *AAPG Bull.* **2013**, *97*, 177–200. [CrossRef]
32. Zhang, F.; Jiang, Z.; Sun, W.; Li, Y.; Zhang, X.; Zhu, L.; Wen, M. A multiscale comprehensive study on pore structure of tight sandstone reservoir realized by nuclear magnetic resonance, high pressure mercury injection and constant-rate mercury injection penetration test. *Mar. Pet. Geol.* **2019**, *109*, 208–222. [CrossRef]
33. Klaver, J.; Desbois, G.; Littke, R.; Urai, J.L. BIB-SEM characterization of pore space morphology and distribution in postmature to overmature samples from the Haynesville and Bossier Shales. *Mar. Pet. Geol.* **2015**, *59*, 451–466. [CrossRef]
34. Ramkumar, M.; Nagarajan, R.; Santosh, M. Advances in sediment geochemistry and chemostratigraphy for reservoir characterization. *Energy Geosci.* **2021**, *2*, 308–326. [CrossRef]
35. Soleimani, M.; Jodeiri Shokri, B. 3D static reservoir modeling by geostatistical techniques used for reservoir characterization and data integration. *Environ. Earth Sci.* **2015**, *74*, 1403–1414. [CrossRef]
36. Peyton, R.L.; Gantzer, C.J.; Anderson, S.H.; Haeffner, B.A.; Pfeifer, P. Fractal dimension to describe soil macropore structure using X ray computed tomography. *Water Resour. Res.* **1994**, *30*, 691–700. [CrossRef]
37. Zeleke, T.B.; Si, B.C. Characterizing scale-dependent spatial relationships between soil properties using multifractal techniques. *Geoderma* **2006**, *134*, 440–452. [CrossRef]
38. Li, R.; Hou, X.; Chen, L.; Fang, H.; Zheng, C. Multifractal investigation on multi-scale pore structure heterogeneity of high rank coal reservoirs. *Nat. Resour. Res.* **2022**, *31*, 1665–1685. [CrossRef]
39. Halsey, T.C.; Jensen, M.H.; Kadanoff, L.P.; Procaccia, I.; Shraiman, B.I. Fractal measures and their singularities: The characterization of strange sets. *Phys. Rev. A* **1986**, *33*, 1141. [CrossRef]
40. Taud, H.; Martinez-Angeles, R.; Parrot, J.F.; Hernandez-Escobedo, L. Porosity estimation method by X-ray computed tomography. *J. Pet. Sci. Eng.* **2005**, *47*, 209–217. [CrossRef]

41. Cao, Z.; Lin, C.; Dong, C.; Ren, L.; Han, S.; Dai, J.; Xu, X.; Qin, M.; Zhu, P. Impact of sequence stratigraphy, depositional facies, diagenesis and CO_2 charge on reservoir quality of the lower cretaceous Quantou Formation, Southern Songliao Basin, China. *Mar. Pet. Geol.* **2018**, *93*, 497–519. [CrossRef]
42. Zhou, N.; Wang, M.; Lu, S.; Dodd, T.J.; Liu, W.; Guan, Y. Evolution of fractal pore structure in sedimentary rocks. *Earth Space Sci.* **2022**, *9*, e2021E–e2167E. [CrossRef]
43. Zhang, P.; Misch, D.; Meng, Q.; Sachsenhofer, R.F.; Liu, Z.; Jia, J.; Gao, F.; Bechtel, A.; Hu, F. Lateral changes of organic matter preservation in the lacustrine Qingshankou Formation (Cretaceous Songliao Basin, NE China): Evidence for basin segmentation. *Int. J. Coal Geol.* **2022**, *254*, 103984. [CrossRef]
44. Giffin, S.; Littke, R.; Klaver, J.; Urai, J.L. Application of BIB–SEM technology to characterize macropore morphology in coal. *Int. J. Coal Geol.* **2013**, *114*, 85–95. [CrossRef]
45. Wang, F.; Yang, K.; Zai, Y. Multifractal characteristics of shale and tight sandstone pore structures with nitrogen adsorption and nuclear magnetic resonance. *Pet. Sci.* **2020**, *17*, 1209–1220. [CrossRef]
46. Ramandi, H.L.; Mostaghimi, P.; Armstrong, R.T.; Saadatfar, M.; Pinczewski, W.V. Porosity and permeability characterization of coal: A micro-computed tomography study. *Int. J. Coal Geol.* **2016**, *154*, 57–68. [CrossRef]

fractal and fractional

Article

Pore Structure Quantitative Characterization of Tight Sandstones Based on Deep Learning and Fractal Analysis

Xinglei Song [1,2], Congjun Feng [1,2,*], Teng Li [3,4,5], Qin Zhang [6], Jiaqi Zhou [1,2] and Mengsi Sun [7]

1 State Key Laboratory of Continental Evolution and Early Life, Northwest University, Xi'an 710069, China
2 Department of Geology, Northwest University, Xi'an 710069, China
3 School of Petroleum Engineering, Xi'an Shiyou University, Xi'an 710065, China
4 Engineering Research Center of Development and Management for Low to Ultra-Low Permeability Oil & Gas Reservoirs in West China, Ministry of Education, Xi'an 710065, China
5 Xi'an Key Laboratory of Tight Oil (Shale Oil) Development, Xi'an 710065, China
6 PetroChina Research Institute of Petroleum Exploration & Development, Beijing 100083, China
7 School of Petroleum Engineering and Environmental Engineering, Yan'an University, Yan'an 716000, China
* Correspondence: fengcj@nwu.edu.cn; Tel.: +86-18729221068

Abstract: Sandstone reservoirs exhibit strong heterogeneity and complex microscopic pore structures, presenting challenges for quantitative characterization. This study investigates the Chang 8 tight sandstone reservoir in the Jiyuan, Ordos Basin through analyses of its physical properties, high-pressure mercury injection (HPMI), casting thin sections (CTS), and scanning electron microscopy (SEM). Deep learning techniques were employed to extract the geometric parameters of the pores from the SEM images. Fractal geometry was applied for the combined quantitative characterization of pore parameters and fractal dimensions of the tight sandstone. This study also analyzed the correlations between the fractal dimensions, sample properties, pore structure, geometric parameters, and mineral content. The results indicate that the HPMI-derived fractal dimension (D_{MIP}) reflects pore connectivity and permeability. D_{MIP} gradually increases from Type I to Type III reservoirs, indicating deteriorating pore connectivity and increasing reservoir heterogeneity. The average fractal dimensions of the small and large pore-throats are 2.16 and 2.52, respectively, indicating greater complexity in the large pore-throat structures. The SEM-derived fractal dimension (D_{SEM}) reflects the diversity of pore shapes and the complexity of the micro-scale geometries. As the reservoir quality decreases, the pore structure becomes more complex, and the pore morphology exhibits increased irregularity. D_{MIP} and D_{SEM} values range from 2.21 to 2.49 and 1.01 to 1.28, respectively, providing a comprehensive quantitative characterization of multiple pore structure characteristics. The fractal dimension shows negative correlations with permeability, porosity, median radius, maximum mercury intrusion saturation, mercury withdrawal efficiency, and sorting factor, while showing a positive correlation with median and displacement pressures. Among these factors, the correlations with the maximum mercury intrusion saturation and sorting factor are the strongest ($R^2 > 0.8$). Additionally, the fractal dimension is negatively correlated with pore circularity and major axis length, but positively correlated with pore perimeter, aspect ratio, and solidity. A higher proportion of circular pores and fewer irregular or long-strip pores correspond to lower fractal dimensions. Furthermore, mineral composition influences the fractal dimension, showing negative correlations with feldspar, quartz, and chlorite concentrations, and a positive correlation with carbonate content. This study provides new perspectives for the quantitative characterization of pore structures in tight sandstone reservoirs, enhances the understanding of low-permeability formation reservoir performance, and establishes a theoretical foundation for reservoir evaluation and exploration development in the study area.

 https://doi.org/10.3390/fractalfract9060372

Keywords: tight sandstone; pore structure; deep learning; fractal dimension; HPMI

1. Introduction

Continuous advancements in oil and gas exploration have positioned unconventional energy sources, particularly tight sandstone reservoirs, as a central focus of the industry, highlighting their critical importance for future exploration and development [1–3]. Unlike traditional reservoirs, these oil and gas reservoirs exhibit suboptimal features, complex pore structures, limited connectivity, and pronounced heterogeneity, all of which considerably affect the reservoir's storage capacity and flow characteristics [4–6]. Therefore, a thorough analysis of the micro-pore-throat structure of tight sandstone reservoirs is crucial for precise reservoir characterization and enhanced oil and gas development efficiency.

Fractal theory, introduced by Mandelbrot (1983), provides a statistical framework for characterizing the irregularity and self-similarity of intricate porous media [7]. Studies have shown that porous rocks exhibit self-similar pore structures that are consistent across various scales [8]. Therefore, the fractal dimension (D), an important indicator of pore system complexity, provides a reliable basis for assessing reservoir quality and conducting petrophysical analyses [9]. A higher fractal dimension typically indicates a more complex pore architecture. In recent years, fractal theory has been extensively utilized in petroleum exploration and production, particularly in the analysis of the micro-pore-throat structures of reservoirs [10,11]. In pore characterization, obtaining the fractal dimension using nuclear magnetic resonance T_2 spectra [12,13], nitrogen adsorption models [14,15], scanning electron microscopy (SEM) image analysis [16], and high-pressure mercury injection (HPMI) [17,18] has become a common approach. SEM image analysis allows for the direct examination of pore shape and structure, whereas HPMI offers advantages such as simplicity, cost-effectiveness, high precision, and extensive pore-throat identification capabilities [19,20]. Combining both methods facilitates a more comprehensive study of the fractal features and structural characteristics of pores. Researchers have quantitatively analyzed the fractal features and pore structures of sandstones [21,22], shales [23,24], coal rocks [25,26], and carbonates [27] by combining the HPMI and SEM. However, current SEM image analysis largely relies on manual operations, which are time-consuming, labor-intensive, and prone to subjectivity based on the researcher's experience, making quantitative research weak and uncertain. Although general image processing software, such as Image-Pro and JMicro Vision, can assist in the analysis, they are not suitable for bulk processing or automated analysis of sandstone sample images [28,29]. With advancements in computer technology, deep learning-based image recognition techniques have significantly improved pore extraction and morphological analyses. However, most existing studies are limited to simple pore extraction, classification, or partial mineral identification, with limited attention to the quantitative characterization of pore structures and geometric parameters [30–33]. In addition, pore recognition in tight sandstone still faces several challenges: (1) tight sandstone pores have poor permeability, and their complex and diverse pore structures hinder effective image recognition; and (2) different study objects have varying applicability, requiring the selection of an appropriate learning model based on sample characteristics. For example, the fully convolutional network model is suitable for coal rock fracture identification [34], the Mask R-CNN model is suitable for rock pore identification [35], and the U-Net model is commonly employed for micro-pore extraction and identification [36]. In summary, previous studies have focused on traditional methods to extract pores from SEM images, and this is inefficient and not highly accurate. Although some studies have applied deep learning techniques, their effectiveness is limited, and different models are required for different regions. Moreover, the relationships between pore structure parameters, geometric parameters, mineral composition, fractal dimension, and reservoir quality in tight sandstone reservoirs require further investigation.

Based on the aforementioned challenges, this study examined the Chang 8 tight sandstone reservoir in the Jiyuan, Ordos Basin. Using SEM image-derived pore parameters and HPMI experimental data from 16 core samples, pore structures were extracted with an improved U-Net model to calculate and compare the two types of fractal dimensions. This study quantitatively characterizes the pore structure of tight sandstone and analyzes its heterogeneity in both planar and spatial domains. It further explores the correlations among the fractal dimension, reservoir quality, pore structure parameters, geometric parameters, and mineral composition. The findings reveal the geological significance of the fractal dimension in characterizing pore structures in continental tight sandstone reservoirs. This study provides precise, quantitative descriptions of pore characteristics in tight sandstone reservoirs, while presenting novel approaches and methodologies for reservoir evaluation and exploration.

2. Geological Setting

The Ordos Basin, China's second-largest sedimentary basin, spans five provinces—Shaanxi, Gansu, Shanxi, Ningxia, and Inner Mongolia—covering an area of 2.5×10^5 km^2 and containing an estimated 6.2×10^{12} m^3 of resources, offering significant exploration potential [37–39]. It comprises six major structural units: the Yimeng Uplift, Wei Bei Uplift, Tianhuan Depression, Western Margin Thrust Belt, Northern Shaanxi Slope, and Jinxi Folding Belt [40]. The study area, Jiyuan, is situated in the central-western Ordos Basin, covering 1305 km^2. It spans from Dunwa in the west to Haojianshan in the east, and from Shangwan in the north to Jiyuan in the south (Figure 1a). The confirmed petroleum reserves in the region amount to 8×10^8 t, with an annual crude oil output of 7×10^6 t, establishing it as the largest oil field in terms of both reserves and production within the Mesozoic strata of the Ordos Basin. The Chang 8 reservoir is the primary production target in this region. The proven favorable oil-bearing area spans 1500 km^2, with a wide distribution and an average single-well oil layer thickness of 15–20 m, providing favorable conditions for exploration and development [41–43].

Alternating layers of gray sandstones and black mudstones define the Chang 8 reservoir, which is located in the lower to middle portion of the Upper Triassic Yanchang Formation. The depositional environment consists of underwater distributary channels and interdistributary bays, representing a typical shallow-water delta system (Figure 1b) [44,45]. Casting thin section (CTS) analysis indicate that the Chang 8 reservoir consists predominantly of fine-grained feldspathic lithic sandstone and lithic feldspathic sandstone, with a smaller proportion of feldspathic sandstone (Figure 2). The clastic constituents mainly include quartz, feldspar, and lithic fragments. The quartz content is from 20.52% to 57.32%, averaging 32.18%; the feldspar content is from 19.95% to 57.58%, averaging 35.56%; and the lithic fragments are from 7.85% to 28.45%, averaging 21.75%. Overall, the Chang 8 reservoir in the study area exhibits favorable reservoir conditions and significant exploration potential. Its rock types and mineral composition significantly influence the pore structure and reservoir characteristics, providing geological evidence for further quantitative characterization of pore features.

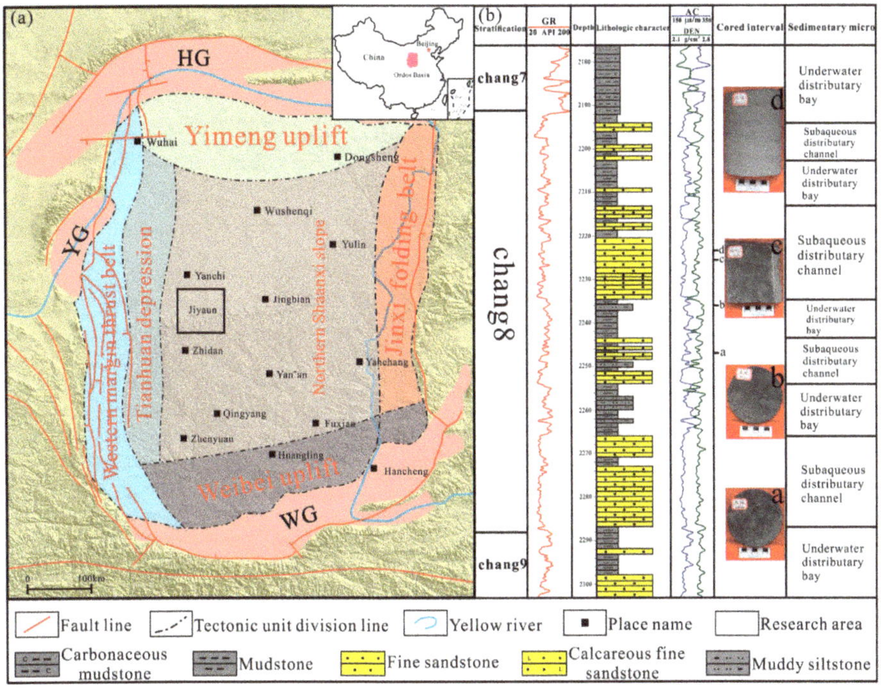

Figure 1. (**a**) Location of the research region (Modified from Yang [46]). (**b**) Column diagram of the Chang 8.

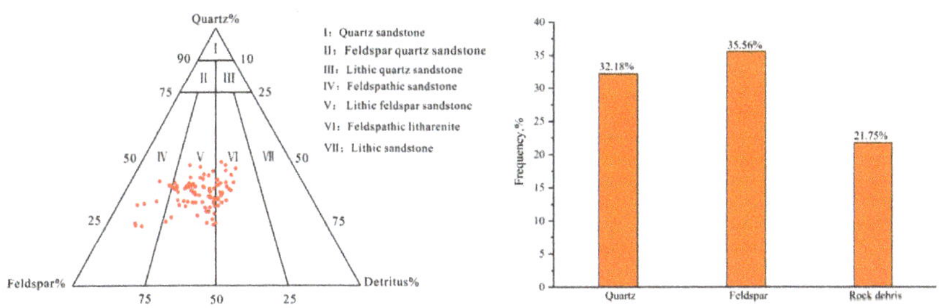

(**a**) (**b**)

Figure 2. Lithologic triangle diagram and clastic composition diagram of the study area. (**a**) Sandstone triangle classification of Chang 8. (**b**) Relative content histogram of detrital component of Chang 8.

3. Materials, Experiments, and Methods

3.1. Materials

This research examined 16 core samples collected from the Chang 8 reservoir in Jiyuan, Ordos Basin, China. All specimens were composed of fine-grained sandstones. The samples were subjected to oil washing, gas permeability testing, and porosity calculations using the weighing method to determine the reservoir's physical parameters (Table 1). The samples were obtained from depths ranging from 2206.42 to 2265.07 m. Porosity values ranged from 2.4% to 12.4%, and the grade difference was 10, with an average of 7.86%, whereas permeability ranged from 0.008×10^{-3} μm^2 to 0.24×10^{-3} μm^2, and the grade difference was 0.232, with an average of 0.11×10^{-3} μm^2. A total of 200 high-resolution images were obtained through CTS and SEM analyses. HPMI experiments were conducted on all 16 samples to determine the microscopic pore-throat characteristics.

Table 1. Information on Experimental Samples.

Sample	Depth (m)	Diameter (cm)	Length (cm)	Porosity (%)	Permeability (10^{-3} μm^2)	Lithology
X1	2218.63	2.50	5.20	9.29	0.24	Fine lithic feldspar sandstone
X2	2230.21	2.48	5.20	11.54	0.23	Fine lithic feldspar sandstone
X3	2208.42	2.50	5.10	9.08	0.17	Fine feldspar lithic sandstone
X4	2236.86	2.49	5.00	7.16	0.11	Fine lithic feldspar sandstone
X5	2255.65	2.50	5.20	12.40	0.21	Fine lithic feldspar sandstone
X6	2231.73	2.50	5.20	10.43	0.20	Fine feldspar lithic sandstone
X7	2252.71	2.50	5.20	5.65	0.08	Fine lithic feldspar sandstone
X8	2240.06	2.48	5.10	6.68	0.06	Fine lithic feldspar sandstone
X9	2213.66	2.50	5.00	7.50	0.09	Fine lithic feldspar sandstone
X10	2227.31	2.50	5.20	9.55	0.07	Fine feldspar lithic sandstone
X11	2206.42	2.50	5.20	9.99	0.06	Fine lithic feldspar sandstone
X12	2259.82	2.50	5.20	7.40	0.05	Fine lithic feldspar sandstone
X13	2260.54	2.48	5.00	6.52	0.06	Fine feldspar lithic sandstone
X14	2253.39	2.50	5.20	2.41	0.01	Fine feldspar lithic sandstone
X15	2261.07	2.50	5.20	7.04	0.02	Fine lithic feldspar sandstone
X16	2265.07	2.50	5.10	3.09	0.02	Fine feldspar lithic sandstone

3.2. *Experiments*

3.2.1. HPMI

HPMI is a direct and efficient technique for assessing microporous structures and is widely employed in the evaluation of tight sandstone reservoirs [47]. In this study, cylindrical rock samples with a diameter of 2.5 cm were selected, ensuring that the surfaces were intact and crack-free. The experiment was conducted using the Auto Pore V9600 fully automatic mercury injection apparatus, which covers a measuring range of 5 nm to 1200 μm. Mercury was injected continuously, with a volume accuracy of less than 0.1 μL for both injection and withdrawal. The experimental procedures followed the national standard GB/T29171-2012. The smallest detectable pore-throat radius was 0.0036 μm, the highest capillary pressure was 206.82 MPa, and the maximum mercury inlet pressure was 96.45 MPa.

3.2.2. CTS and SEM

CTS, prepared using staining agents, enables the direct examination of pore architecture and mineral compositions in rock samples [48]. In this investigation, samples were sliced into 2 cm × 2 cm sections and stained under vacuum conditions. A Leica DM4500PFCHG-013 polarizing microscope was used to examine 16 thin sections of tight sandstone, following the national standard SY/T5368-2000. SEM is widely used to analyze sample morphology at a high resolution and to characterize pore structures and mineral compositions. It can be used to observe the microscopic pore-throat development characteristics of the reservoir [49]. The JSM-7500F field emission scanning electron microscope was used in this study. The secondary electron image resolution of the scanning electron microscope was 1 nm, and the magnification was 20–300,000×. All experiments were conducted at the State Key Laboratory of Continental Evolution and Early Life, Northwest University.

3.3. Methods

3.3.1. Fractal Dimension Based on HPMI

Fractal theory has evolved into a mature framework, enabling the quantitative characterization of pore types, pore structures, and reservoir physical properties [50]. Typically, the fractal dimensions calculated using the HPMI method range between 2 and 3. A larger fractal dimension indicates greater complexity in the pore structure and stronger reservoir heterogeneity [51]. In this study, the fractal dimensions of the pore-throats were calculated using HPMI data from 16 samples.

Fractal geometry theory states that the number of objects $N(r)$ at a specific scale r adheres to a power-law relationship:

$$N(r) \propto r^{-D} \tag{1}$$

It was assumed that the reservoir pores exhibited fractal characteristics. The pore-throat radius is the linear scale r of the measurement, the number of pores is converted based on the pore-throat radius r, and the mercury inlet volume V_{Hg} is N (r). For HPMI, the number of pores N (r) can be obtained using the capillary model, as follows:

$$N(r) = \frac{V_{Hg}}{\pi r^2 l} \tag{2}$$

Combining Equations (1) and (2), we obtain:

$$\frac{V_{Hg}}{\pi r^2 l} \propto r^{-D} \tag{3}$$

$$V_{Hg} \propto r^{2-D} \tag{4}$$

where l is the length of capillary, μm; V_{Hg} is the volume of mercury, μm^3; r is the radius of pore-throat, μm.

According to the Young–Laplace equation:

$$P_c = \frac{2\sigma \cos\theta}{r} \tag{5}$$

where P_{c} is the mercury injection pressure, MPa; σ is the interfacial tension, N/m; and θ is the contact angle, °.

Substituting Equation (5) into Equation (4) yields

$$V_{Hg} \propto P_c^{D-2} \tag{6}$$

Mercury saturation S_{Hg} in the sample is defined as

$$S_{Hg} = \frac{V_{Hg}}{V_p} \tag{7}$$

Assuming that l is constant, from Equations (6) and (7), the relationship between S_{Hg} and P_{c} is obtained as

$$S_{Hg} = \alpha P_c^{D-2} \tag{8}$$

where S_{Hg} is the mercury saturation, %; V_{p} is the total pore volume of the sample, μm^3; and α is a constant, indicating that the mercury saturation in the rock sample follows a power-law relationship with the capillary pressure

Taking the logarithm of both sides of Equation (8):

$$\ln S_{Hg} = (D-2)\ln P_c + C \tag{9}$$

This suggests that if the pore structure of the reservoir rock samples has fractal features, $\ln S_{Hg}$ and $\ln P_c$ demonstrate a linear correlation in a double-logarithmic coordinate system. Here, C represents the intercept, and D represents the fractal dimension, which may be obtained from the slope of the line.

3.3.2. Enhanced U-Net Model

U-Net, proposed by Ronneberger in 2015, is a deep learning model based on convolutional neural networks [52]. It achieves efficient image segmentation through skip connections and a fully convolutional network decoding stage. In this study, DropBlock and Batch Normalization layers were integrated into the convolutional blocks of the original U-Net model to construct an improved reconstruction block, enhancing both recognition accuracy and computational speed (Figure 3b). DropBlock is a regularization technique that, unlike traditional Dropout, randomly masks blocks of neurons in the network, reducing redundant connections and improving the model's generalization ability. Batch Normalization is an essential technique for regularization and accelerating training. Normalizing the input distribution to the intermediate layers improves both the training speed and stability of deep networks. Additionally, the skip connections in the original U-Net model were replaced with a feature pyramid network, merging features from different resolutions to enhance feature extraction, thus forming an improved U-Net architecture (Figure 3a).

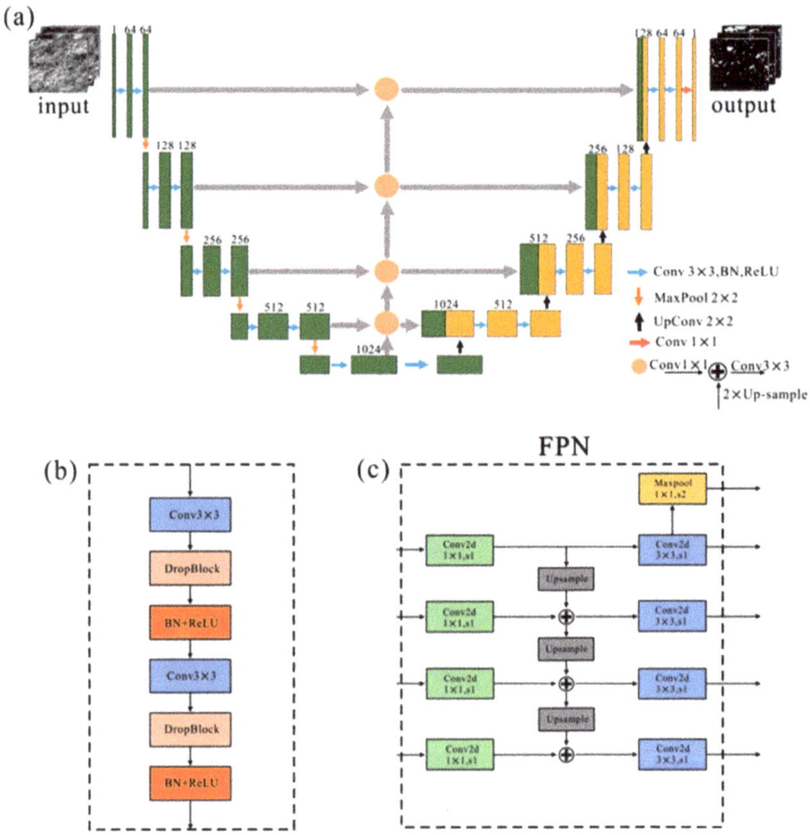

Figure 3. Pore recognition model. (**a**) Diagram of the enhanced U-Net model. (**b**) Schematic of the reconstruction convolutional block. (**c**) Feature pyramid network.

The primary operational steps were as follows: (1) Dataset preprocessing: The original SEM images were cropped to 512 × 512 pixels using Adobe Photoshop™ for computer recognition. The dataset was then partitioned into training and test sets in a 7:3 ratio. ImageJ 1.54d software was used to identify the training set and convert it to grayscale. In deep learning, larger and more diverse training samples improve the accuracy. Therefore, the Python Imgaug data augmentation library was used to apply elastic transformations, random cropping, scaling, and rotation to increase the diversity of the training dataset and improve model robustness. During the training, image augmentation techniques, such as rotate (limit = 30, p = 0.5), elastic transform (alpha = 50, sigma = 10, p = 0.2), and horizontalflip (p = 0.5), were applied to enhance the diversity of the training dataset and improve the model's robustness. (2) Backbone feature extraction: A 512 × 512 image was input with two 3 × 3 convolutions and corresponding 2 × 2 max pooling at each step. The initial number of feature channels was 64, which doubled after each downsampling step, resulting in five effective feature layers. (3) The feature values at five depths were input into a feature pyramid network (Figure 3c). The network builds a feature fusion path with bottom-up, top-down, and lateral connections, along with convolutional fusion, to combine feature maps at different scales. Upsampling was performed using 2 × 2 convolutions, reducing the number of feature channels by half at each step. (4) Prediction: The final effective feature layers are used to classify and identify each feature point.

The model was trained and evaluated on a deep learning workstation running Windows 10, equipped with a 13th Gen Intel(R) Core(TM) i9-13980HX @2.20 GHz processor, 128 GB of RAM, and NVIDIA GeForce RTX 3060 GPU. The deep learning model was built in Python 3.12, using the PyTorch 1.7 framework. During training, the Adam optimizer was employed for the loss function. The learning rate was set to 0.0001 to ensure model stability during the learning process and minimize oscillation near the optimal solution. The batch size was configured at 8, and the epoch was set to 100, ensuring that the model had sufficient iterations to learn the training data to gradually converge to a better state. This model is particularly suitable for image identification applications in fine-grained sedimentary rocks with complex microstructures. Compared to other models, it performs better and can automatically extract pore geometric parameters from SEM images in batches [53–57]. The model's IOU reaches 0.91. The AP is comparatively high, reaching 0.945, indicating strong performance in pore recognition. The model achieves a precision of 0.951, an MAE of 0.108, and a Dice score of 0.872.

3.3.3. Fractal Dimension Based on SEM

Tight sandstone, a porous medium in sedimentary rocks, exhibits pronounced fractal characteristics in its pores. SEM images revealed the pore features of the tight sandstone, with the extracted pore boundaries being closed. Because the pores captured by SEM are 2D images, the fractal dimension derived from the SEM images cannot exceed 2. According to Voss, the fractal dimension of the pore morphology can be determined by utilizing the relationship between the pore area and the perimeter [58]. The relationship between the pore area and the perimeter for pores exhibiting fractal properties is expressed as

$$\lg P = \frac{D}{2}\lg A + C \tag{10}$$

where P is the pore perimeter (μm) extracted from the SEM images, A is the pore area (μm^2), D is the fractal dimension, and C is a constant.

In this study, deep learning technology was used to extract pore parameters from 16 samples, yielding 7258 valid pores. The geometric parameters of these pores were analyzed, and each sample's fractal dimension was calculated using Equation (10).

4. Results

4.1. Pore Structure Characterization

4.1.1. Pore Structure Characteristics Based on HPMI

Based on the analysis of the physical property parameters and HPMI data from 16 samples in the study area, the samples were categorized into three types (Table 2). Type I samples exhibited favorable physical qualities, with average permeability and porosity values of 0.21×10^{-3} μm^2 and 9.98%, respectively. The capillary pressure curve displayed an elongated plateau, skewed toward the bottom left. The mean displacement pressure and median pressure were moderate at 0.76 MPa and 6.15 MPa, respectively. The mean sorting factor was 2.20, signifying satisfactory sorting quality and a relatively uniform throat size distribution. The throat type was sheet-like (Figure 4f). The pores were mainly intergranular and feldspar dissolution pores, with some affected by compaction and dissolution, forming irregular and long-strip pores. Type II samples exhibited moderate physical properties, with an average porosity of 7.87% and a permeability of 0.07×10^{-3} μm^2. The capillary pressure curve displayed a slow increase followed by a clear horizontal plateau. The average displacement pressure and median pressure exceeded those of the Type I samples, recorded at 1.20 MPa and 8.85 MPa, respectively. The mean sorting factor was 1.91, indicating moderate sorting. The pores were primarily residual intergranular and feldspar dissolution pores, predominantly long strips in shape. A small number of irregular micropores were observed, along with constricted throats (Figure 4b,e). Type III samples had the least advantageous physical qualities, averaging permeability of 0.03×10^{-3} μm^2 and porosity of 5.29%, both of which were markedly inferior to those of Type I and II samples. The horizontal platform of the capillary pressure curve was not obvious, and the curve was steep. The average displacement and median pressures were the largest, at 2.55 MPa and 52.43 MPa, respectively. The average sorting factor was the lowest at 1.41, indicating poor sorting. The pores mainly consisted of intercrystalline pores and micropores, with intercrystalline pores forming between clay minerals, such as chlorite. These pores were characterized by a complex and variable structure with poor connectivity. Many small throats with poor storage capacities and high heterogeneities were observed (Figure 4c).

Table 2. Classification criteria for typical samples of Chang 8 in the study area.

Group	Sample Number	Φ (%)	K (10^{-3} μm^2)	Displacement Pressure	Median Radius	Median Pressure	Coefficient of Variation	Sorting Factor	Skewness	Maximum Mercury Saturation (%)	Mercury Withdrawal Efficiency (%)
I	X1	9.29	0.24	0.95	0.11	8.35	16.36	2.15	0.17	92.49	41.72
	X2	11.54	0.23	0.61	0.12	8.11	14.26	2.14	0.33	89.89	32.48
	X3	9.08	0.17	0.76	0.14	5.77	16.95	2.09	0.44	86.83	39.29
	X4	7.16	0.11	0.69	0.14	5.39	16.67	2.18	0.46	87.39	38.44
	X5	12.40	0.21	0.80	0.11	4.36	17.84	2.41	1.68	91.51	40.71
	X6	10.43	0.20	0.73	0.15	4.91	16.81	2.23	1.58	89.98	37.57
II	X7	5.65	0.08	1.19	0.09	7.55	15.45	2.03	0.43	88.32	32.81
	X8	6.68	0.06	1.54	0.05	8.76	13.99	1.85	0.50	85.08	36.76
	X9	7.50	0.09	0.96	0.09	6.67	14.25	1.88	0.60	84.45	32.50
	X10	9.55	0.07	0.78	0.10	9.80	14.69	1.89	0.65	85.33	38.90
	X11	9.99	0.06	1.51	0.10	11.48	15.85	1.91	0.45	82.62	39.21
III	X12	7.40	0.05	2.26	0.03	27.06	8.42	0.96	-0.02	75.80	26.21
	X13	6.52	0.06	4.40	0.02	52.41	8.54	1.37	-0.05	77.52	23.22
	X14	2.41	0.01	0.95	0.14	90.79	4.94	1.28	0.48	73.45	24.99
	X15	7.04	0.02	2.29	0.02	35.55	9.26	1.60	0.05	79.61	29.71
	X16	3.09	0.02	2.86	0.02	56.36	8.96	1.83	0.01	75.90	28.57

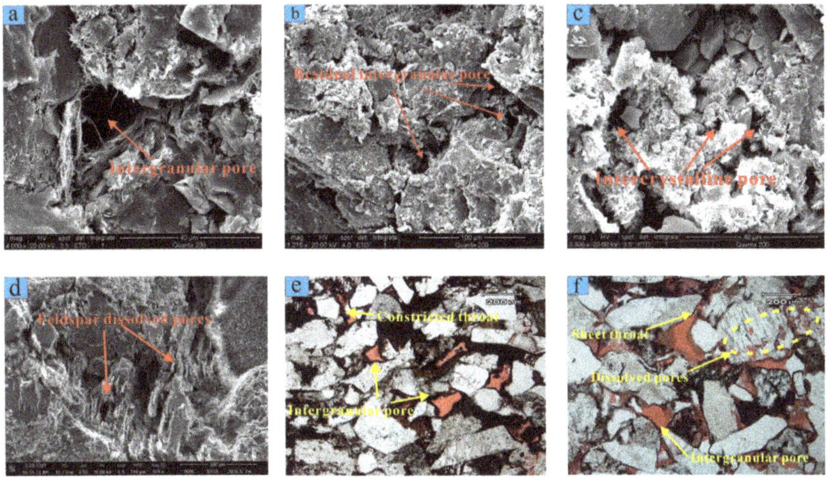

Figure 4. Pore and throat types of Chang 8 reservoir. (**a**) Type I: Intergranular pore, locally developed quartz, X3, 2208.42 m, SEM image, 4000×; (**b**) Type II: Residual intergranular pore, X8, 2240.06 m, SEM image, 1215×; (**c**) Type III: Intercrystalline pore, developed quartz, chlorite fill pores, X14, 2253.39 m, SEM image, 3806×; (**d**) Type I: Feldspar dissolution pore, X1, 2218.63 m, SEM image, 1215×; (**e**) Type II: Constricted throat, intergranular pore, X10, 2227.31m, CTS image, plane-polarized light, 4 × 10; (**f**) Type I: Sheet throat, dissolution and intergranular pore, developed quartz, X2, 2230.21 m, CTS image, plane-polarized light, 4 × 10.

4.1.2. Pore Structure Characteristics Based on SEM

SEM images provide pore-shape-related parameters that are essential for quantitatively assessing pore structure types, distribution characteristics, and heterogeneity. In this study, 150 SEM images were acquired from 16 samples. Using an enhanced U-Net deep learning model, the geometric parameters of each pore were automatically extracted from the SEM images, including the perimeter, circularity, major axis, aspect ratio, and solidity (Table 3). The mean values of these parameters were then calculated for each sample. The analysis revealed that the pore geometric parameters of the tight sandstone reservoir samples exhibited significant variation and uneven distribution, indicating substantial reservoir heterogeneity and complex pore structures. There were six Type I samples (X1–X6). The pore perimeter varied between 57.08 μm and 129.56 μm, averaging 97.05 μm. Circularity ranged from 0.47 to 0.51, with an average of 0.51. The major axis ranged between 122.65 μm and 187.25 μm, averaging 160.68 μm. The aspect ratio varied from 1.33 to 2.16, with a mean of 1.71. Solidity varied from 0.21 to 0.45, with a mean of 0.34. There were five Type II samples (X7–X11). The pore perimeter ranged from 112.08 μm to 434.96 μm, with an average of 188.16 μm. The average circularity was 0.40, with a range of 0.39 to 0.47. The major axis ranged from 81.9 μm to 142.65 μm, with a mean of 104.13 μm. The average aspect ratio varied from 1.52 to 2.41, with an average of 2.01. The average solidity was 0.53, with a range of 0.42 to 0.63. There were five Type III samples (X12–X16). The pore perimeter ranged between 289.68 um and 497.92 um, with an average of 411.97 um. Circularity varied from 0.32 to 0.35, with an average of 0.34. The major axis ranged between 62.1 um and 105.6 um, with a mean of 78.3 um. The aspect ratio varied from 2.21 to 2.78, with an average of 2.45. Solidity ranged between 0.67 and 0.73, averaging 0.70. The data indicated relatively high average values for the perimeter, major axis, aspect ratio, and solidity across all samples, whereas the average circularity remained comparatively low. This indicates that tight sandstone pores are primarily irregular, exhibiting a scarcity of circular pores, which aligns with the low porosity and permeability characteristic of tight sandstone. Longitudinal analysis revealed that Type I samples exhibited the highest average circularity and major axis, followed by Type II and Type III. This suggests that reservoir quality is positively correlated with both circularity and the major axis. The larger these parameters, the better the pore permeability. The greater the pore circularity, the closer the pore shape is to that of a circle. Circular pores typically exhibit a uniform distribution, enabling better connectivity and fluid flow. The major axis refers to the long axis of the outer ellipse of the pore, which directly affects the pore size and connectivity. When the length of the major axis increased, the pore size also increased, and the connectivity was enhanced. Conversely, Type I samples displayed the lowest average pore perimeter, aspect ratio, and solidity, with Type II showing intermediate values and Type III exhibiting the highest values. This indicates that the reservoir quality is inversely correlated with the pore perimeter, aspect ratio, and solidity. The smaller these parameters, the higher the pore permeability. A larger pore perimeter indicates a more complex pore structure, increasing the fluid flow path tortuosity and reducing permeability. The aspect ratio, which represents the proportion of the long-to-short axis length of the pore's outer ellipse, affects fluid seepage; higher values indicate narrower pores and longer seepage channels, which impede fluid flow. Solidity indicates the degree of pore inward curvature; higher values correspond to more concave pore shapes, reduced circularity, and restricted inter-pore fluid flow.

Table 3. Results of Pore Geometry Parameters Extracted by Deep Learning.

Group	Sample Number	Φ (%)	K (10^{-3} μm^2)	Perimeter (μm)	Circularity	Major Axis (μm)	Aspect Ratio	Solidity
I	X1	9.29	0.24	66.64 6.37~659.15	0.56 0.31~0.97	187.25 2.84~394.5	1.57 1.01~5.78	0.34 0.06~0.45
	X2	11.54	0.226	113.44 2.69~856.46	0.53 0.22~0.96	156.85 1.334~358.3	1.33 1.09~4.95	0.26 0.05~0.37
	X3	9.08	0.174	57.08 4.36~417.35	0.49 0.14~0.94	131.30 1.34~256.78	1.60 1.06~6.25	0.43 0.08~0.49
	X4	7.16	0.11	89.68 9.03~874.24	0.47 0.079~0.93	122.65 2.07~173.16	2.16 1~3.05	0.45 0.04~0.57
	X5	12.4	0.212	129.56 6.27~958.28	0.51 0.079~0.95	184.10 2.84~294.5	1.52 1.01~3.78	0.21 0.05~0.43
	X6	10.43	0.199	125.88 8.34~946.83	0.51 0.07~0.93	181.95 1.8~276.9	2.05 1.05~4.92	0.36 0.10~0.38
II	X7	5.65	0.075	137.32 9.38~1176.45	0.40 0.097~0.86	91.90 1.28~109.56	2.415 1.05~2.10	0.55 0.36~0.88
	X8	6.68	0.06	130.8 10.69~1208.68	0.36 0.052~0.88	81.90 1.21~112.31	2.29 1.09~4.13	0.63 0.26~0.84
	X9	7.5	0.085	112.08 14.83~1157.15	0.47 0.026~0.91	102 0.98~168.51	1.52 1.09~3.6	0.57 0.33~0.74
	X10	9.55	0.068	125.64 12.34~1298.58	0.39 0.092~0.89	142.65 0.93~143.5	1.89 1.09~2.65	0.42 0.36~0.69
	X11	9.99	0.057	434.96 9.34~1498	0.37 0.067~0.83	102.2 1.02~150.6	1.92 1.01~3.49	0.48 0.37~0.73
III	X12	7.4	0.053	497.92 8.04~1343.24	0.34 0.058~0.87	105.6 1.05~129.61	2.37 1.03~5.54	0.73 0.36~0.95
	X13	6.52	0.059	399.76 5.34~1576.17	0.35 0.064~0.82	68.90 0.89.48~89.9	2.78 1.05~4.17	0.67 0.27~0.96
	X14	2.41	0.008	398.24 6.31~1035.56	0.32 0.047~0.82	80.35 0.95~96.96	2.55 1.01~5.86	0.68 0.29~0.95
	X15	7.04	0.017	474.24 7.56~1498.47	0.33 0.06~0.79	77.55 0.45~92.82	2.35 1.09~4.6	0.71 0.32~0.93
	X16	3.09	0.022	289.68 5.83~1312.49	0.34 0.098~0.72	62.10 0.34~88.73	2.21 1.14~5.23	0.70 0.27~0.82

Complex pore networks, characterized by multiple types and cross-scale variations, determine the conditions for tight oil occurrence. The configuration of pore structures in tight sandstone reservoirs is a critical criterion for evaluating tight oil accumulation and serves as a fundamental factor influencing reservoir physical properties. Based on the pore parameters and shapes identified by the enhanced U-Net model, the pore shapes are classified into three categories: circular, random irregular, and long-strip pores (Figures 5 and 6). Circular pores display regular shapes, high circularity (0.65–0.95), aspect ratios close to 1 (1.15–1.3), and low solidity (0.1–0.35). These pores are primarily found in primary intergranular pores, soluble minerals, and isolated pores in the cement. Their development is controlled by weak compaction, uniform dissolution, and the sealing effect of the cement. In contrast, the random irregular pores display diverse shapes, often resembling triangles, polygons, honeycomb patterns, and grids. They are characterized by low circularity (0.3–0.6), relatively large aspect ratios (1.35–1.70), and high solidity (0.4–0.6). These pores are found in both brittle and clay minerals and are mainly influenced by the original sedimentary structures, dissolution, and compaction. Long-strip pores are characterized by being narrow and slender, with the lowest circularity (0.1–0.3), highest aspect ratio (greater than 2.4), and greatest solidity (0.7–1.0). This type of pore predominantly occurs in the intergranular spaces of brittle minerals, such as kaolinite, quartz, and feldspar. It also appears in highly compacted intergranular pores formed by the directional arrangement of microfractures and flaky minerals, such as mica and chlorite.

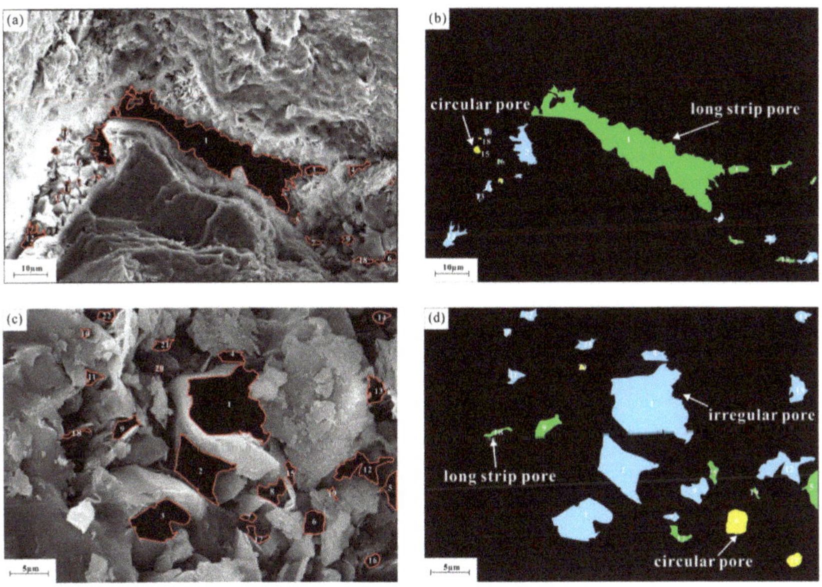

Figure 5. *Cont.*

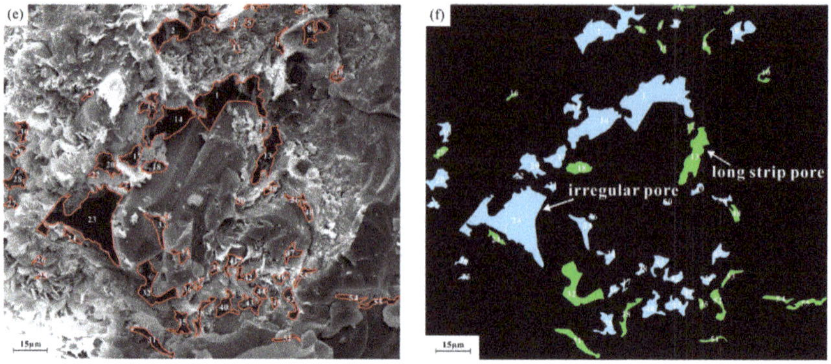

Figure 5. Sample pore extraction effect diagram. (**a**) X5 pore recognition; (**b**) X5 pore extraction; (**c**) X10 pore recognition; (**d**) X10 pore extraction; (**e**) X15 pore recognition; (**f**) X15 pore extraction.

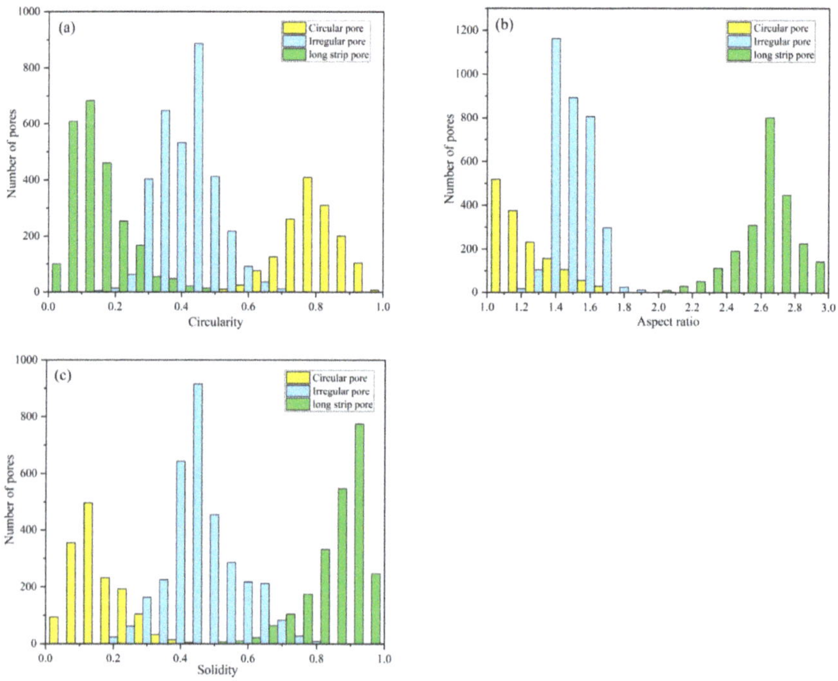

Figure 6. Types and characteristics of pore shapes. (**a**) Circularity distribution statistics of different pore shape. (**b**) Aspect ratio distribution statistics of different pore shape. (**c**) Solidity distribution statistics of different pore shape.

4.2. Fractal Characterization

4.2.1. Fractal Dimensions Obtained from HPMI Data

Fractal analysis was conducted using HPMI data from 16 sandstone samples based on the $\ln S_{Hg}$-$\ln P_c$ relationship plot (Figure 7), where the slope (k = *D*-2) indicates the fractal dimension. The results indicate that all samples show distinct segmentation in their pore fractal characteristics. The slope of the first section is the largest, corresponding to a fractal dimension between 4.0 and 9.0. The slopes of the second and third sections are small, with fractal dimensions ranging from 2.0 to 3.0. Pores are generally considered to exhibit fractal characteristics when the fractal dimension is within the range of 2.0 to 3.0. When it exceeds 3.0, the pores do not exhibit these characteristics. Therefore, the first segment is considered invalid and cannot serve as a metric for assessing the pore structure of the sample. The pore structure of tight sandstones exhibits distinct binary properties [59]. When the pore-throat radius is smaller than the peak pore-throat radius, the pore structure resembles a capillary type, where the pore radius is close to the throat radius, resulting in a small slope for the $\ln P_c$ and $\ln S_{Hg}$ fitting line. The mercury intrusion volume increases slowly with the pore-throat radius and is determined by the throat radius. At this stage, the fractal dimension of the pore-throat is between 2 and 3, and the pore-throat exhibits characteristics between two-dimensional and three-dimensional extension, where the capillary length (l) increases with the capillary radius (r). However, the growth rate of l is slower than that of r. The pore structure exhibits some degree of self-similarity, indicating fractal characteristics. When the pore-throat radius exceeds the peak pore-throat radius, the pore structure resembles a bead-chain model, where the pore radius is significantly larger than the throat radius, and the slope of the $\ln P_c$ and $\ln S_{Hg}$ fitting line is larger. As the pore-throat radius increases, the mercury intrusion volume increases rapidly, and the contribution of the pore to the pore-throat volume exceeds that of the throat. At this point, the fractal dimension of the pore-throat is greater than 3, and the pore-throat exhibits three-dimensional non-geometric extension, lacking fractal characteristics. The distribution of the throat radius in each sample is uneven, with varying peak values that are closely related to permeability and porosity. Specifically, Type I samples exhibit the largest peak throat radius at approximately 0.39 μm, corresponding to a capillary pressure of 1.85 MPa. Type II samples have a smaller peak throat radius than Type I samples, approximately 0.22 μm, corresponding to a capillary pressure of 3.57 MPa. Type III samples have the smallest peak throat radius at approximately 0.06 μm, corresponding to a maximum capillary pressure of 11.67 MPa. The second and third sections of the fractal plots for the Type I and II samples show obvious segmentation. The overall fractal dimension of the macropores and micropores ranges between 2.11 and 2.7, demonstrating a dual-fractal characteristic (Figure 4). The inflection point occurs at $P_c = 6.82$ MPa, distinguishing small pores ($r < 0.1$ μm) from large pores ($r > 0.1$ μm). The porosity ratios of the small and large pores were computed individually, and the overall fractal dimension (D_{MIP}) of the whole pore space was derived from the weighted average of the porosity of each pore space (Equation (11)). For Type I samples (X1–X6), the fractal dimension (D_1) for small pores varies between 2.11 and 2.19, with a mean of 2.15. The correlation factor (R^2) varies from 0.937 to 0.977, with a mean of 0.964. The fractal dimension (D_2) for large pores varies between 2.34 and 2.47, with a mean of 2.41. The R^2 values range from 0.963 to 0.986, with an average of 0.979. D_{MIP} values vary from 2.21 to 2.28, with an average of 2.24. For Type II samples (X7–X11), D_1 for small pores ranges from 2.12 to 2.24, averaging 2.18. The R^2 values range from 0.953 to 0.992, with an average of 0.973. D_2 for large pores extends from 2.58 to 2.70, with an average of 2.65. The corresponding R^2 values range from 0.987 to 0.994, with an average of 0.990. D_{MIP} extends from 2.31 to 2.37, with a mean of 2.34. For the Type III samples (X12–X16), the second and third segments of the fractal curve do not show distinct segmentation, and no obvious inflection point is observed. The fitted curve is an approximate straight line, with R^2 values ranging between 0.989 and 0.996, and a maximum average of 0.993. D_{MIP} values range from 2.39 to 2.50, averaging 2.44, with most values exceeding 2.40. This is attributed to the poor permeability and porosity of Type III reservoirs, which are dominated by small pores with small pore-throat radius. These reservoirs exhibit relatively simple pore structures and poor overall connectivity, thereby exhibiting a single fractal characteristic. D_{MIP} increases sequentially from Type I to III reservoirs, indicating that declining reservoir quality corresponds to increased pore structure complexity, greater heterogeneity, and reduced pore connectivity. Furthermore, the fact that D_2 exceeds D_1 suggests that large pore-throat structures are more complex, aligning with previous research findings [60–63].

$$D_{MIP} = D_1 \times \frac{\varphi_1}{\varphi_1 + \varphi_2} + D_2 \times \frac{\varphi_2}{\varphi_1 + \varphi_2} \tag{11}$$

where D_1 and D_2 are the fractal dimensions corresponding to the small and large pore sections, respectively; φ_1 and φ_2 are the porosity proportions of the small and large pores, respectively.

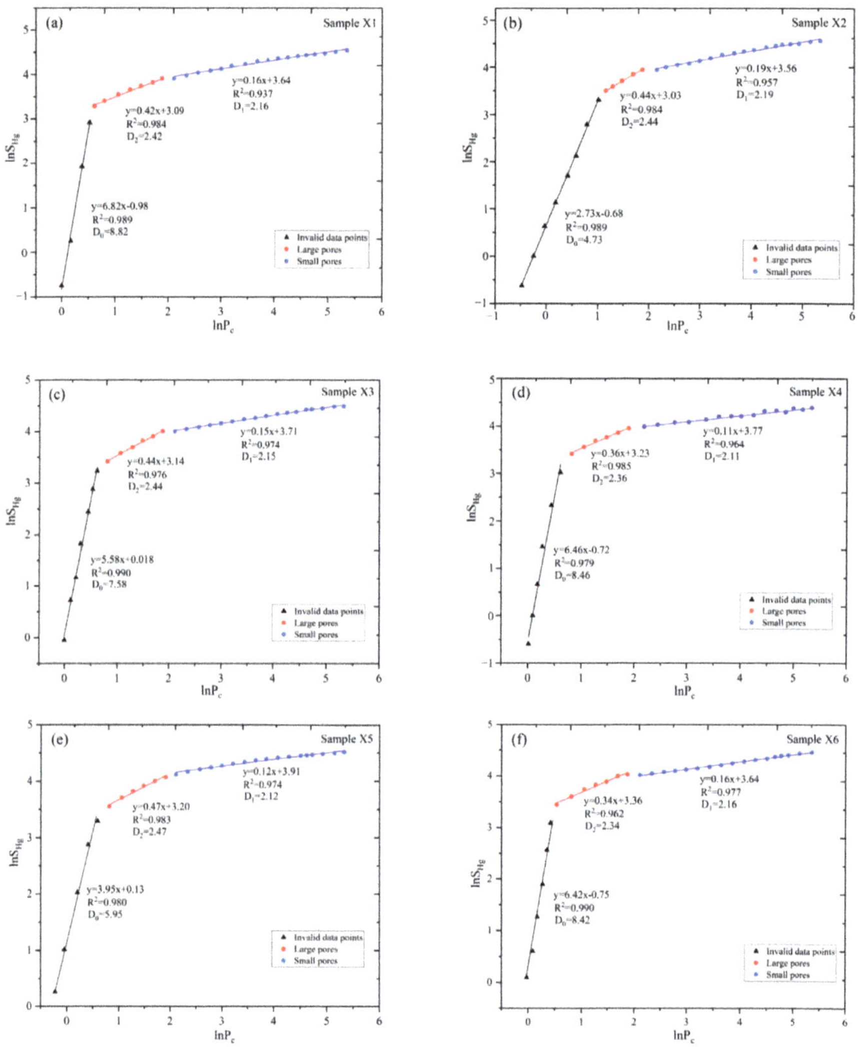

Figure 7. *Cont.*

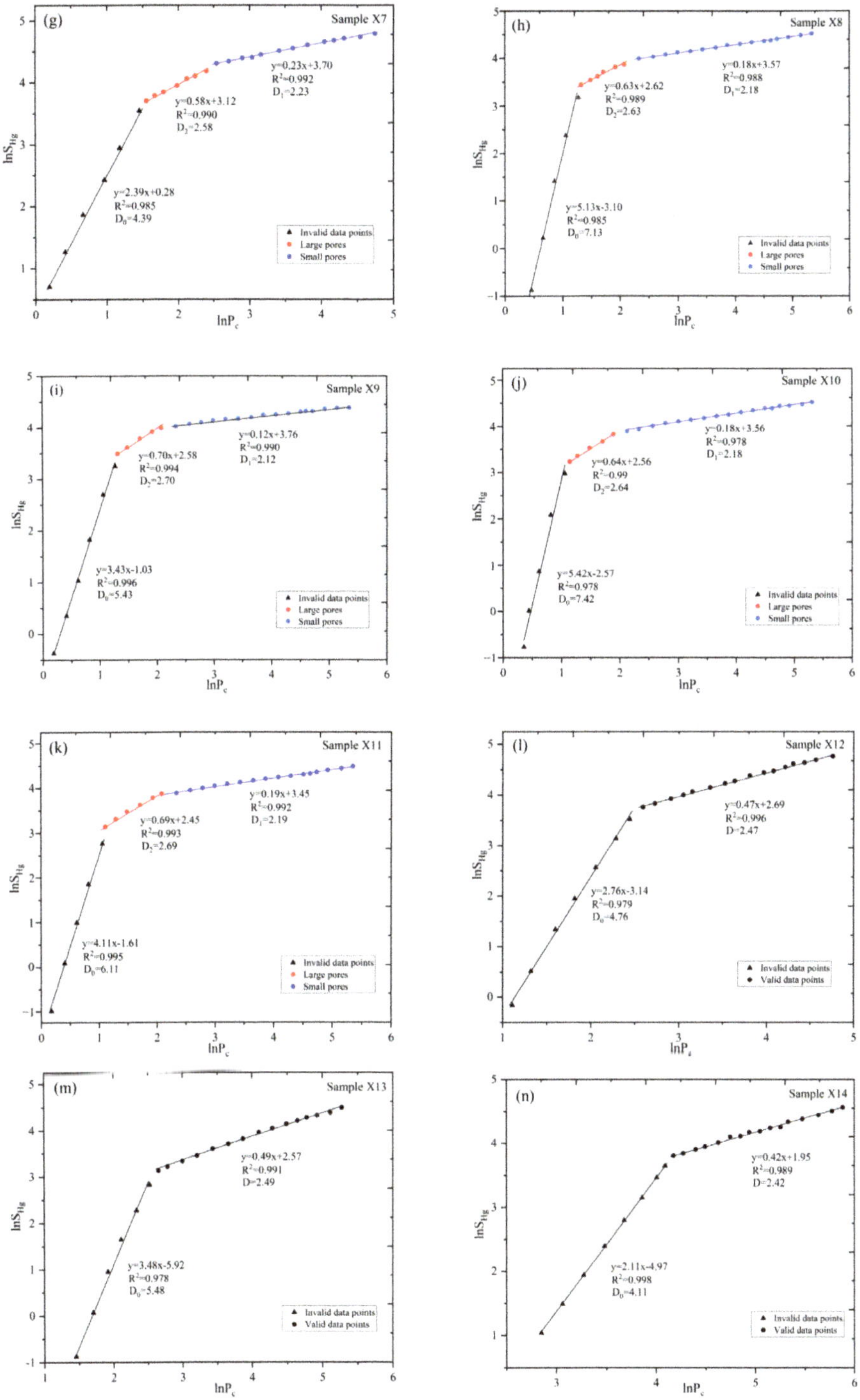

Figure 7. *Cont.*

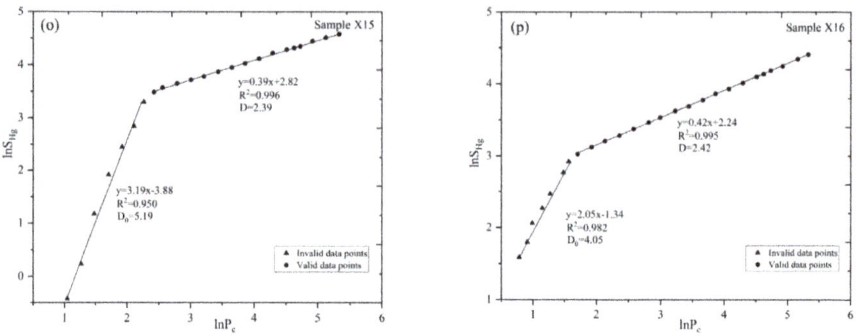

Figure 7. Relationship diagram between lg (P_c) and lg (S_{Hg}), based on HPMI. (**a**) Fractal dimensions of Sample X1; (**b**) Fractal dimensions of Sample X2; (**c**) Fractal dimensions of Sample X3; (**d**) Fractal dimensions of Sample X4; (**e**) Fractal dimensions of Sample X5; (**f**) Fractal dimensions of Sample X6; (**g**) Fractal dimensions of Sample X7; (**h**) Fractal dimensions of Sample X8; (**i**) Fractal dimensions of Sample X9; (**j**) Fractal dimensions of Sample X10; (**k**) Fractal dimensions of Sample X11; (**l**) Fractal dimensions of Sample X12; (**m**) Fractal dimensions of Sample X13; (**n**) Fractal dimensions of Sample X14; (**o**) Fractal dimensions of Sample X15; (**p**) Fractal dimensions of Sample X16.

4.2.2. Fractal Dimensions Obtained from SEM

According to fractal theory, the fractal dimension calculated from the SEM images should range between 1 and 2. The calculated results were within this range, indicating that they possessed fractal geometric significance. Using deep learning methods to extract pore structures from the SEM images, a double-logarithmic coordinate plot of the pore area (*A*) and perimeter (*P*) was created for the 16 sandstone samples (Figure 8). The two parameters showed a strong linear relationship ($R^2 > 0.9$), with the overall fractal dimension (D_{SEM}) ranging from 1.01 to 1.28 and an average of 1.14. This demonstrates that the pore morphology of the experimental samples exhibited clear fractal characteristics. For Type I samples (X1–X6), D_{SEM} ranged from 1.01 to 1.08, averaging 1.04. For Type II samples (X7–X11), it ranged from 1.15 to 1.17, averaging 1.15. For Type III samples (X12–X16), D_{SEM} varied from 1.19 to 1.28, with a mean of 1.24. The findings demonstrate that Type I reservoir samples possess the lowest D_{SEM} values, followed by Type II, with Type III having the largest values. This suggests that deteriorating reservoir quality correlates with increasing pore structure complexity and irregular pore shape.

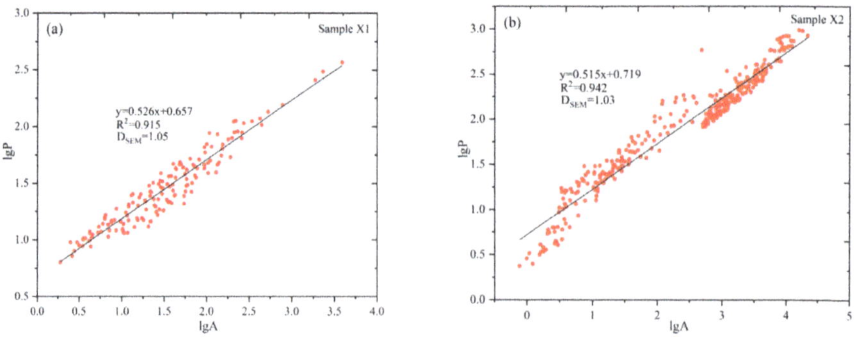

Figure 8. *Cont.*

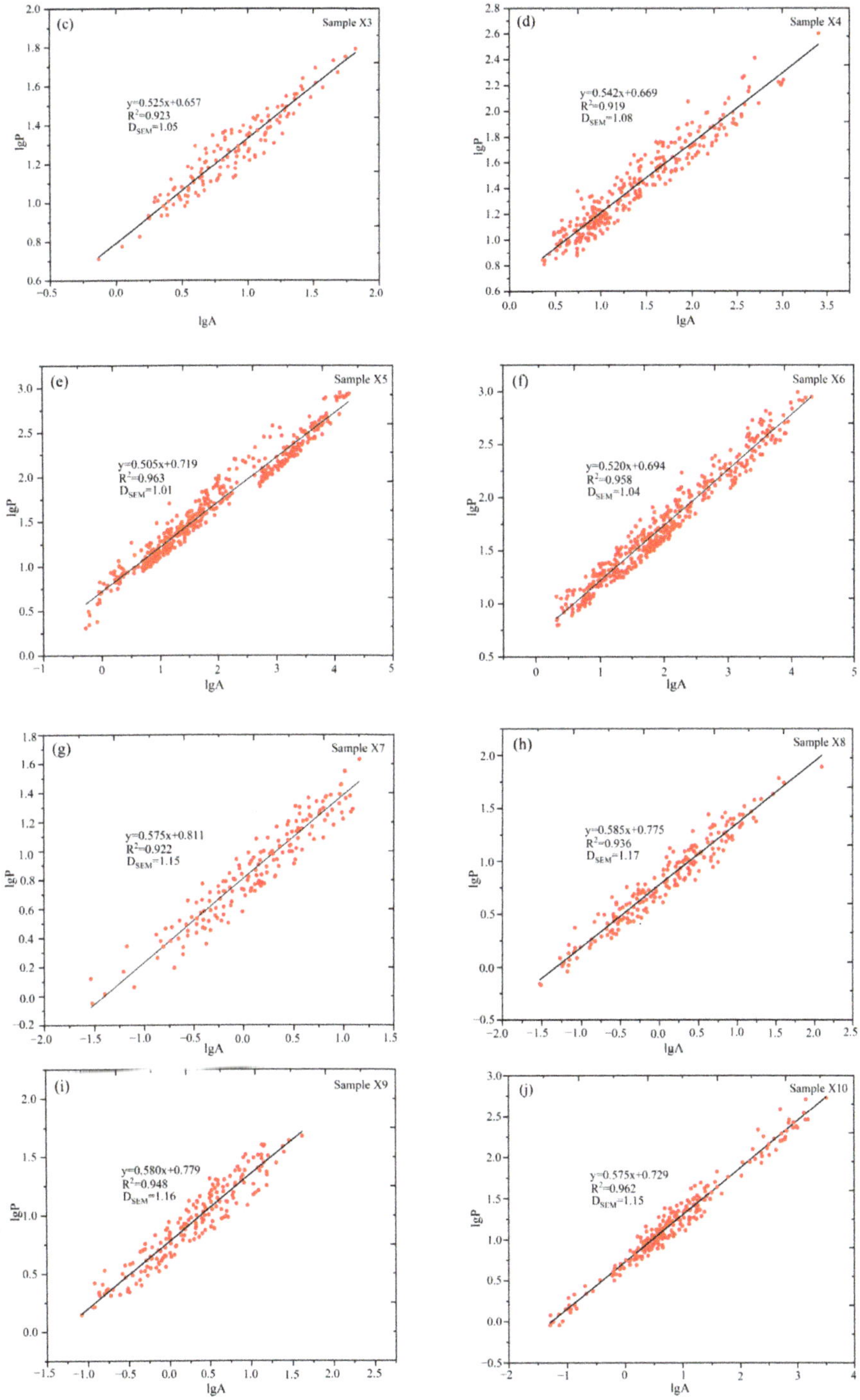

Figure 8. *Cont.*

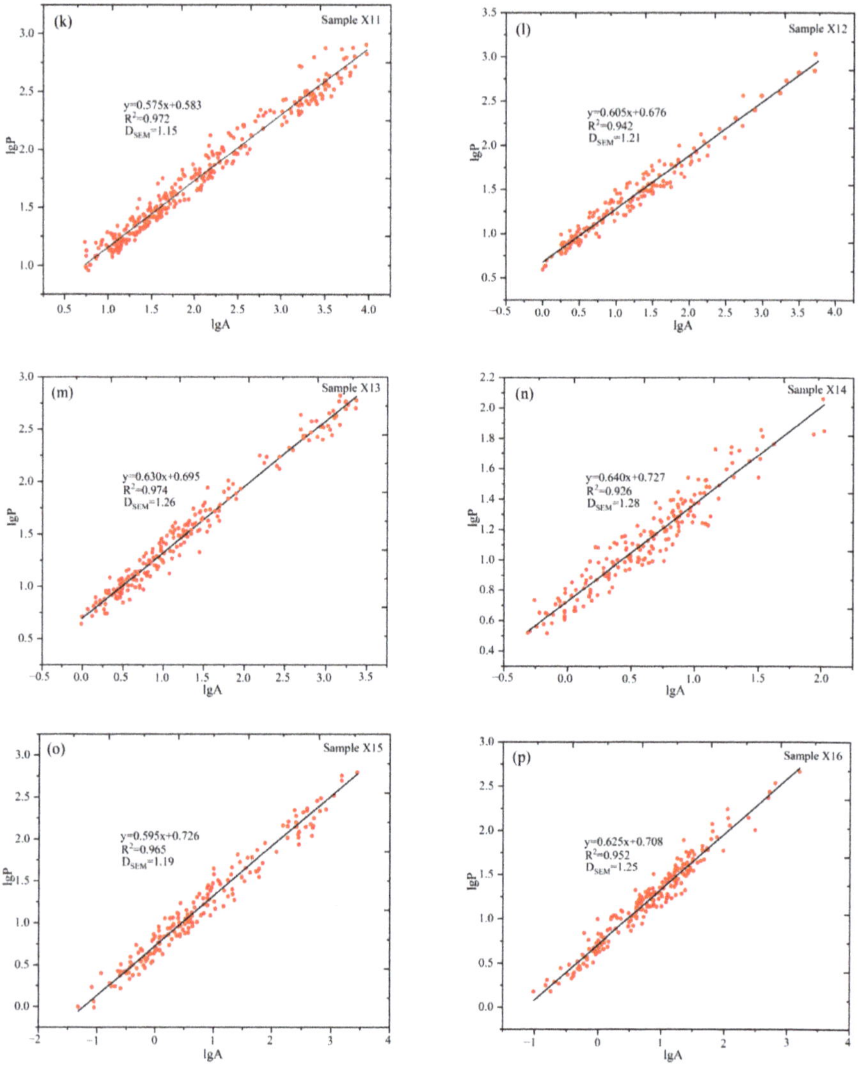

Figure 8. Fractal dimensions calculated by SEM images using deep learning.(**a**) Fractal dimensions of images from Sample X1; (**b**) Fractal dimensions of images from Sample X2; (**c**) Fractal dimensions of images from Sample X3; (**d**) Fractal dimensions of images from Sample X4; (**e**) Fractal dimensions of images from Sample X5; (**f**) Fractal dimensions of images from Sample X6; (**g**) Fractal dimensions of images from Sample X7; (**h**) Fractal dimensions of images from Sample X8; (**i**) Fractal dimensions of images from Sample X9; (**j**) Fractal dimensions of images from Sample X10; (**k**) Fractal dimensions of images from Sample X11; (**l**) Fractal dimensions of images from Sample X12; (**m**) Fractal dimensions of images from Sample X13; (**n**) Fractal dimensions of images from Sample X14; (**o**) Fractal dimensions of images from Sample X15; (**p**) Fractal dimensions of images from Sample X16.

5. Discussion

5.1. Comparison of Fractal Dimensions from HPMI and SEM Data

As discussed in Section 4.2.1, the effectiveness of the fractal dimensions derived from the HPMI data strongly correlates with the pore-throat radius, owing to capillary forces during the experiment. When the pore-throat radius is smaller than the peak radius, the pore structure approximates the capillary model, with the pore radius approaching the throat radius. The mercury intrusion volume increases slowly as the pore-throat radius increases, with the intrusion volume dominated by the throat radius [64]. At this stage, the D_{MIP} values range from 2.21 and 2.49, indicating a certain degree of self-similarity among pores and obvious fractal characteristics. When the pore-throat radius exceeds the peak radius, the D_0 values range from 4.05 to 8.82, which is substantially greater than 3, implying that the pore structure lacks fractal characteristics. During the HPMI process, the high pressure can cause deformation or compression of the pores. As a result, the actual pore structure becomes simpler than that obtained through mercury intrusion, leading to an underestimation of the fractal dimension. Moreover, due to the limitations of HPMI imposed by the maximum intrusion pressure, smaller pore-throats cannot be detected, and some tiny pore-throats are overlooked. This may lead to an underestimation of the complexity of the pore structure and the intricate connectivity between pores, ultimately affecting the accuracy of the fractal dimension.

In contrast, SEM images can provide a direct and comprehensive view of pore morphology and geometric parameters, with their measurement range not limited by the pore-throat radius or pressure factors. The D_{SEM} values range from 1.01 to 1.28, effectively reflecting the complexity and diversity of the pore morphology. However, the SEM method has its limitations: it cannot quantify pore-throat size distributions and often overlooks pore connectivity features. The observed differences between D_{MIP} and D_{SEM} are primarily due to differences in their underlying theoretical principles and mathematical models. The HPMI-derived fractal model is based on variations in the capillary pressure curve, primarily reflecting the connectivity and permeability of the pores. The SEM-derived fractal model is based on changes in pore morphology and size (such as pore area and perimeter), primarily reflecting the pore morphology and micro-geometric features.

In summary, although both the HPMI and SEM methods have advantages and disadvantages, and their respective fractal dimensions fall within different ranges, they illuminate the complexity and heterogeneity of pore structures at varying scales. Therefore, the combined use of HPMI and SEM data enables a more comprehensive quantitative characterization of the multiple features of the pore structure, providing strong support for advanced studies on the pore properties of tight reservoirs.

5.2. Relationship Between D and Pore Structure Parameters

Fractal dimension is closely correlated with reservoir heterogeneity and pore structure complexity, significantly influencing the reservoir's physical properties. Typically, a larger fractal dimension indicates greater reservoir heterogeneity, poorer pore connectivity, more complex pore-throat structures, and diminished physical properties. The results show that both D_{SEM} and D_{MIP} exhibit strong negative correlations with porosity and permeability (Figure 9a,b). As the fractal dimension increases, both porosity and permeability decrease, indicating that complex pore structures with high fractal dimensions reduce rock permeability. Conversely, samples with large porosities, owing to the presence of large primary or dissolution pores, exhibit better pore connectivity and simpler pore structures. The median radius, indicative of the pore radius at 50% saturation of the non-wetting phase, may be estimated as the mean pore radius of the sample. It exhibits an inverse association with the fractal dimension (Figure 9c). This is because as the pore-throat radius increases, the reservoir pore-throat distribution becomes more concentrated, resulting in a larger storage space and more regular pore surfaces. Consequently, the reservoir's pore structure becomes less complex, heterogeneity decreases, and physical properties improve. Both the maximum mercury saturation and mercury withdrawal efficiency, which reflect inter-pore connectivity, show significant negative correlations with the fractal dimension (Figure 9d,e). This indicates that complex pore structures affect the mercury intrusion and withdrawal processes, leading to a decrease in fluid permeability and deterioration in the physical properties of the reservoir. A higher fractal dimension indicates a more complex, finer, and irregularly structured pore system, resulting in poorer connectivity between pores. This prevents the mercury that enters the rock sample from being completely expelled, thereby reducing the mercury withdrawal efficiency. The sorting factor, which reflects the consistency of the pore-throat size distribution, shows a strong negative correlation with the fractal dimension (Figure 9f). It shows that the better the sorting of tight sandstone, the simpler the pore structure and the lower the fractal dimension. The median pressure, which refers to the capillary pressure corresponding to 50% saturation of the non-wetting phase, shows a positive correlation with the fractal dimension. A higher median pressure implies a denser rock sample with more complex pore structures, thus a higher fractal dimension (Figure 9g). Similarly, the displacement pressure, which refers to the capillary pressure corresponding to the maximum connected pore in the pore system and serves as a key indicator of the performance of sandstone reservoirs, shows a strong positive correlation with the fractal dimension. A higher fractal dimension indicates a more complex pore structure and increased displacement pressure (Figure 9h).

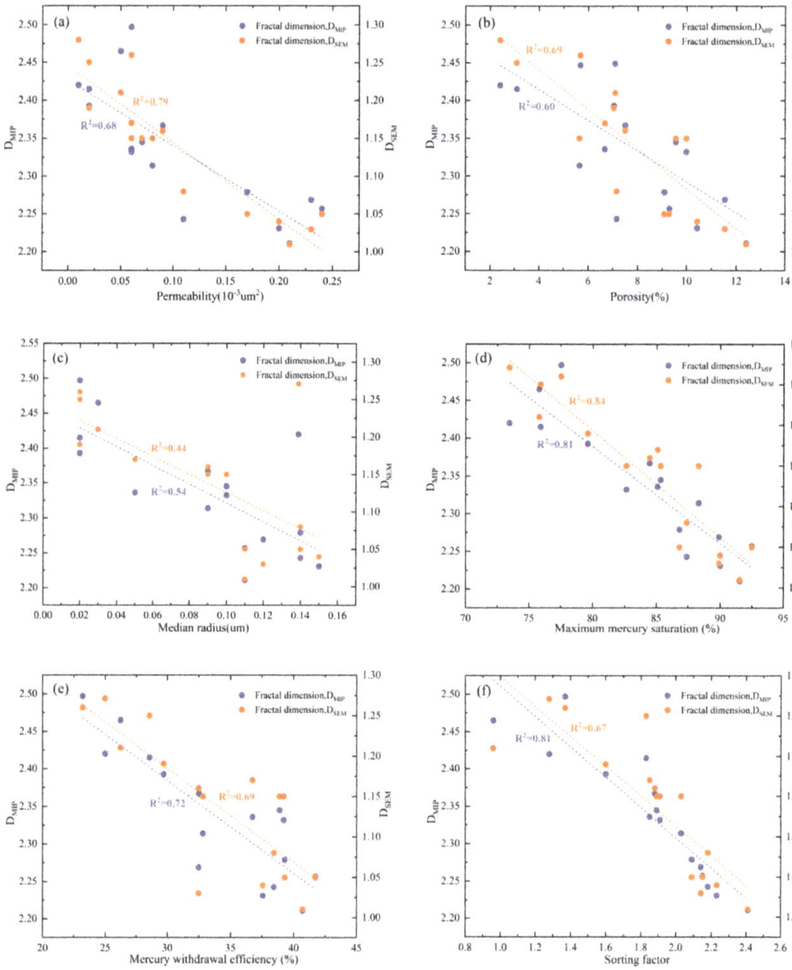

Figure 9. *Cont.*

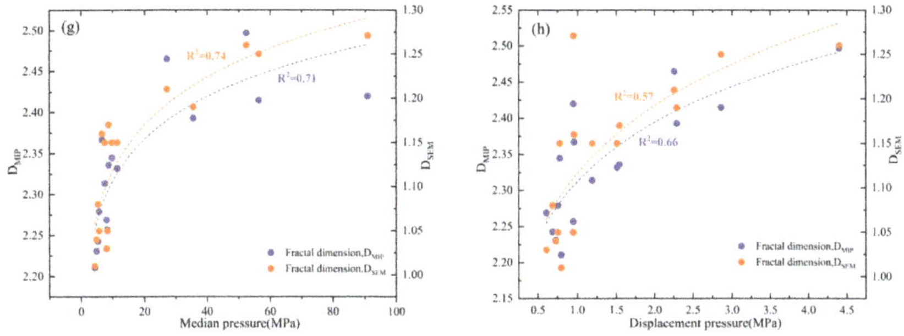

Figure 9. The relationship between fractal dimension and pore structure parameters. (**a**) Permeability; (**b**) porosity; (**c**) median radius; (**d**) maximum mercury saturation; (**e**) mercury withdrawal efficiency; (**f**) sorting factor; (**g**) median pressure; (**h**) displacement pressure.

In conclusion, the fractal dimension demonstrates a significant correlation with the pore structure parameters of the reservoir. It effectively reflects the complexity and heterogeneity of the pore space. Lower fractal dimensions indicate smoother pore surfaces, more uniform pore distribution, higher connectivity, and reduced reservoir heterogeneity, resulting in greater storage capacity and permeability. This study reveals that maximum mercury saturation and the sorting factor exhibit the strongest correlation with the fractal dimension ($R^2 > 0.8$). These two factors, representing pore interconnectivity and uniformity of the pore-throat size distribution, exert the most significant influence on the fractal dimension. Additionally, the complexity of the reservoir pore architecture is influenced by multiple variables, including median pressure, displacement pressure, median radius, and mercury withdrawal efficiency.

5.3. Relationship Between D and Pore Geometry Parameters

Figure 10 shows the relationship between the fractal dimension and pore geometric characteristics. A strong negative correlation was observed between the fractal dimension and both the circularity and length of the major axis. Increased circularity signifies that the pore shape more closely approximates a circle, and circular pores generally exhibit a more uniform distribution, leading to enhanced connection and fluid flow. The length of the major axis, which denotes the principal axis of the circumscribed circle of the pore, affects pore connectivity and fluid flow paths. An augmentation in main axis length corresponds with increased pore size and improved connectivity among pores. Therefore, when the circularity and major axis length increase, the reservoir pore structure becomes more simplistic, resulting in a reduced pore fractal dimension. In contrast, solidity has a positive association with the fractal dimension. Solidity refers to the degree of concavity and convexity of the pores. The stronger the solidity, the more pronounced the concave pore shape, resulting in a lower circularity and an increased complexity of the pore configuration. The pore perimeter also exhibits a positive correlation with the fractal dimension. A longer perimeter typically indicates more complex pore edges with additional curved channels, resulting in a more intricate pore structure and reduced pore connectivity. Lower reservoir permeability is associated with higher levels of heterogeneity. The aspect ratio had a positive connection with the fractal dimension. Increased aspect ratios result in narrower pores, reduced pore size uniformity, and elevated fractal dimensions, consequently decreasing reservoir permeability. Overall, the pores in tight sandstones tend to have longer perimeters, lower circularities, and greater solidities, which correspond to larger fractal dimensions. This demonstrates the high heterogeneity and complex pore structure of these reservoirs.

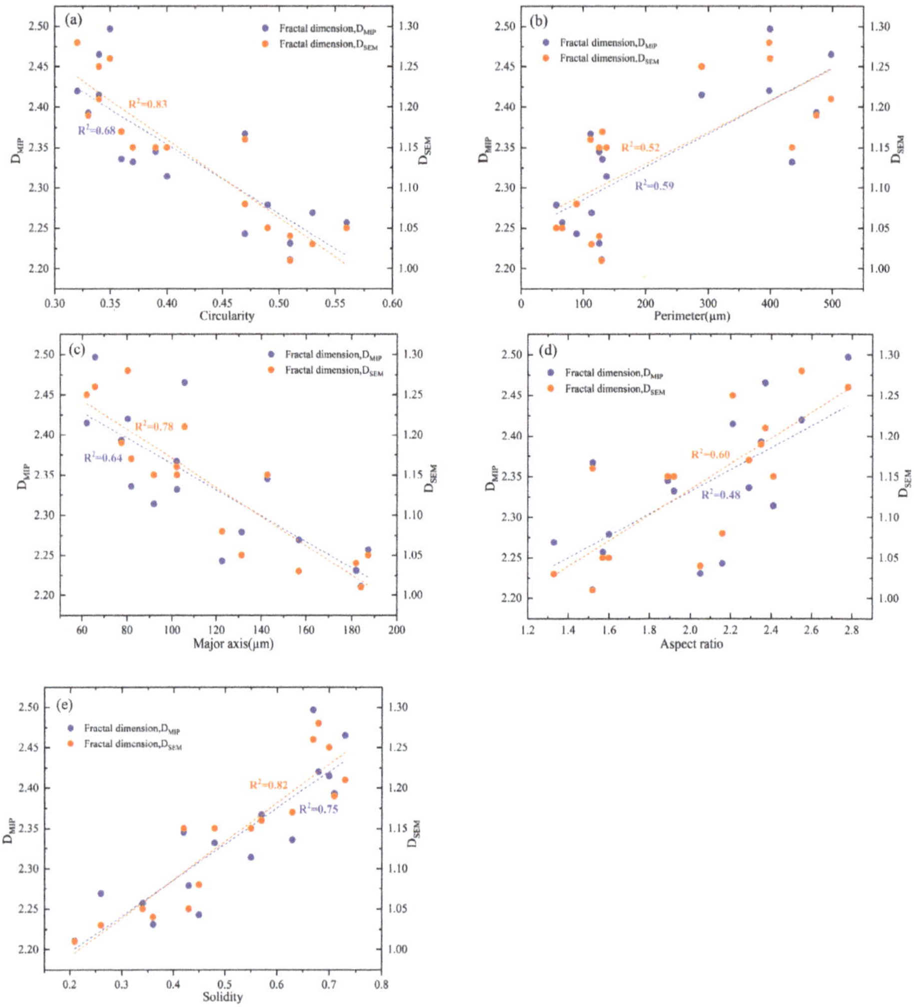

Figure 10. Relationship between fractal dimension and pore geometric parameters. (**a**) Circularity; (**b**) perimeter; (**c**) major axis; (**d**) aspect ratio; (**e**) solidity.

5.4. Relationship Between D and Pore Morphology

Based on the classification of pore shape features described in Section 4.1.2, the proportions of different pore morphologies, along with D_{MIP} and D_{SEM}, were statistically analyzed for each sample (Figure 11). This study found that pore shapes and fractal dimensions vary across reservoir types. Overall, circular pores constituted a relatively small proportion (21.05%) of the three reservoir rock sample types, whereas irregular pores (45.74%) and long-strip pores (33.21%) constituted a larger proportion. The overall fractal dimensions were relatively high (average D_{MIP} = 2.34; average D_{SEM} = 1.14). These findings indicate that the tight sandstone reservoirs in the study area exhibited complex pore structures, predominantly characterized by irregular pores. Poor pore connectivity and significant heterogeneity were the main factors contributing to poor reservoir properties. A vertical comparison among the reservoir types showed that Type I reservoirs (X1–X6) had the lowest fractal dimensions (average D_{MIP} = 2.24; average D_{SEM} = 1.04). Circular pores accounted for the largest proportion of these samples (average 26%), whereas irregular pores (average 47.83%) and long-strip pores (average 26.17%) had smaller proportions. Type II reservoir samples (X7–X11) exhibited higher fractal dimensions than Type I (average D_{MIP} = 2.34; average D_{SEM} =1.15). In these samples, the proportion of circular pores was smaller than that in Type I (average 20.6%), whereas the proportions of irregular pores (average 50.4%) and long-strip pores (average 29%) were slightly higher than those in Type I reservoirs. Type III reservoir samples (X12–X16) had the highest fractal dimensions (average D_{MIP} = 2.44; average D_{SEM} = 1.24). The proportion of circular pores was the lowest in these samples, an average of 11.2%, which was much smaller than that in Types I and II, whereas the proportions of irregular pores (average 52.6%) and long-strip pores (average 36.2%) continued to increase. In summary, the fractal dimension gradually increases as circular pores decrease and irregular and long-strip pores increase. This results in more complex pore structures, reduced connectivity, increased heterogeneity, and poor reservoir properties. Therefore, the reservoir quality can be quantitatively assessed by analyzing the distribution of pore shape characteristics.

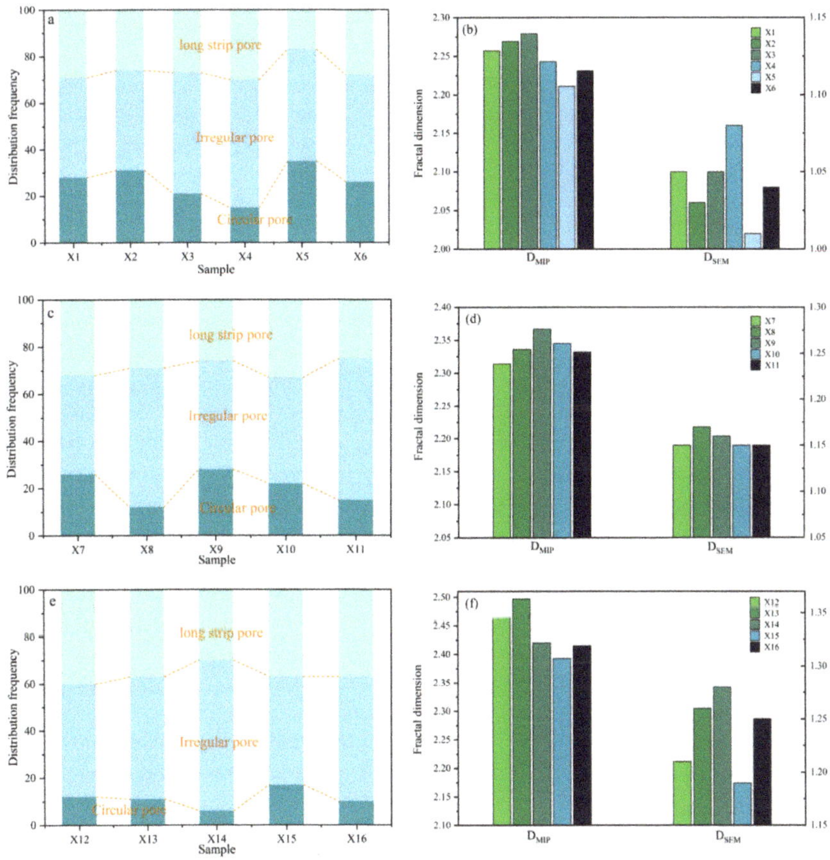

Figure 11. Comparison of pore morphology characteristics of different reservoir type samples. (**a**) The proportion of pores with different shapes in Type I. (**b**) The distribution of fractal dimension in Type I; (**c**) The proportion of pores with different shapes in Type II; (**d**) The distribution of fractal dimension in Type II; (**e**) The proportion of pores with different shapes in Type III; (**f**) The distribution of fractal dimension in Type III.

A horizontal comparison revealed that the pore shape proportions influenced the fractal dimension. In Type I reservoirs, sample X5 had the lowest fractal dimension (D_{MIP} = 2.21, D_{SEM} = 1.01), with the highest proportion of circular pores (35%) and the lowest proportion of long strip pores (17%). In Type III reservoirs, sample X15 exhibited the lowest fractal dimension (D_{MIP} = 2.39, D_{SEM} = 1.19), with circular pores accounting for the largest proportion (17%) and irregular pores accounting for a relatively smaller proportion (46%). This suggests that the presence of more circular pores—indicating better circularity and simpler pore structures—correlates with lower fractal dimensions. Conversely, a higher proportion of irregular and long-strip pores— indicating complex structures and poor connectivity—correlates with higher fractal dimensions. Additionally, the pore shape appears to have a greater influence on D_{SEM} than on D_{MIP}. For instance, Type II sample X8 had the highest D_{SEM} (1.17), with the lowest proportion of circular pores (12%) and higher proportions of irregular (59%) and long-strip pores (29%). Similarly, the Type III sample X14 had the highest D_{SEM} (1.28), with only 6% circular pores and high proportions of irregular (64%) and elongated pores (30%). The relationship between circularity and each fractal dimension (Figure 10a) showed that circularity correlated more strongly with D_{SEM} (R^2_{SEM} = 0.83) than with D_{MIP} (R^2_{MIP} = 0.68). This is because the D_{SEM} is calculated directly from the geometric parameters of the pores, such as the pore area and perimeter, which are directly related to the pore shape. In contrast, the D_{MIP} is derived using a mathematical model based on mercury pressure–volume relationships and segmented fitting of the intrusion curve, which cannot directly quantify pore boundary complexity.

5.5. Relationship Between D and Mineral Composition

Differences in the mineral composition directly influence the sedimentation and diagenetic processes of a reservoir, which is reflected in the heterogeneity of its pore structure. The fractal dimension signifies the intricacy and variability of the pore architecture. The examination of the link between mineral composition and fractal dimension (Figure 12) demonstrated a significant negative relationship between quartz concentration and fractal dimension ($R^2 > 0.6$): a higher quartz content corresponds to a lower fractal dimension. As a key framework mineral in tight sandstones, a higher quartz content enhances the sandstone's resistance to compaction, thereby preserving intergranular pores and improving pore-throat connectivity. This leads to reduced heterogeneity and complexity. Feldspar content exhibited a weak negative correlation with the fractal dimension ($R^2 < 0.5$), indicating a limited influence on reservoir pore structure complexity. Feldspar is easily dissolved, creating dissolution pores. As the feldspar content increased, the dissolution pores also increased, enhancing pore connectivity and permeability, thereby reducing heterogeneity. However, dissolution pores facilitate the formation of clay minerals, which subsequently fill the pores and throats, thereby reducing pore space and increasing the structural complexity and heterogeneity. A higher carbonatite content promotes carbonate cementation, which reduces pore size, blocks the throat, and increases heterogeneity, resulting in a higher fractal dimension. In contrast, the chlorite content had a weak negative correlation with the fractal dimension ($R^2 < 0.4$). A higher chlorite content corresponds to a lower fractal dimension. Chlorite formation occupies the pore space, reducing the storage capacity and permeability. However, chlorite forms a cementing film on mineral particles, which inhibits compaction and helps preserve the pore structure, thereby reducing heterogeneity. Additionally, increased chlorite content creates uniform tiny pores that further lower the fractal dimension.

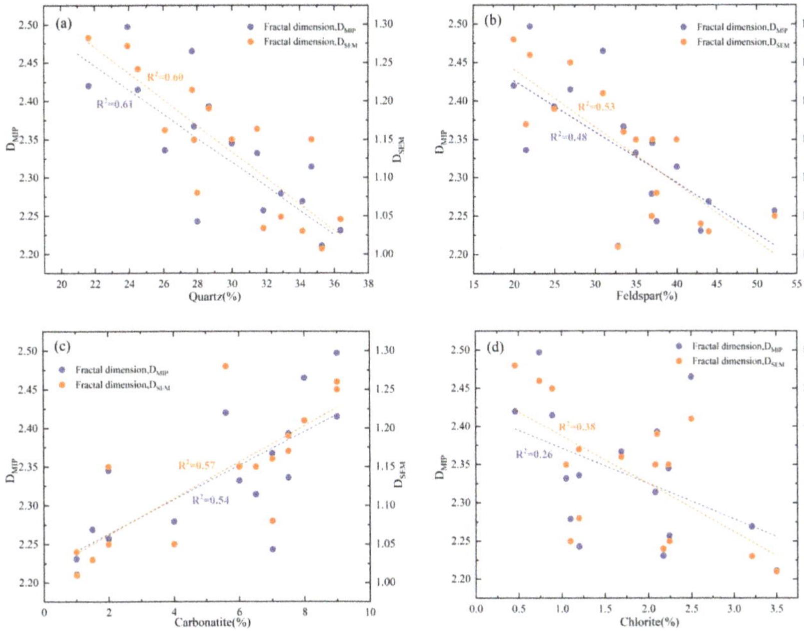

Figure 12. The relationship between fractal dimension and mineral composition. (**a**) Quartz; (**b**) feldspar; (**c**) carbonatite; (**d**) chlorite.

In summary, various mineral components substantially influence the complexity of the reservoir pore structure. The combined effects of these mineral components reflect the reservoir's physical properties. In addition, changes in the mineral composition indicate reservoir connectivity, providing a foundation for heterogeneity assessments. Therefore, analyzing mineral content is essential for predicting pore structure, connectivity, and heterogeneity characteristics.

5.6. Geological Significance of D

The D_{MIP} revealed that the pore structure of tight sandstone reservoirs exhibits distinct dual-fractal properties, which expose its complexity and variability in multiscale space. D_{MIP} characterizes the connectivity of the three-dimensional pore throat structure and the tortuosity of flow paths, thereby quantitatively characterizing reservoir heterogeneity. Integrating SEM imaging with mercury injection experiments enables the qualitative identification of pore morphology and quantitative extraction of structural parameters. Particularly, with the introduction of deep learning techniques, a large amount of pore information can be efficiently extracted from SEM images. Pore-type boundaries can be automatically recognized and classified. D_{SEM} quantifies the geometric complexity and morphological diversity of pore boundaries. D_{MIP} primarily characterizes the spatial connectivity and permeability of pores, whereas D_{SEM} captures the heterogeneity and complexity of pore morphology on a 2D plane. Combining both methods enables the characterization of the reservoir pore structure across multiple scales and enhances the understanding of fluid flow mechanisms and reservoir performance. This integrated approach provides technical support for the classification, evaluation, and development of tight sandstone reservoirs, as well as for rock physics modeling.

Furthermore, significant correlations were observed between the fractal dimension and pore structure parameters, geometric characteristics, and mineral composition. By thoroughly exploring these relationships, key factors influencing the pore structure can be identified and used to predict the heterogeneity of the reservoir and its evolution trend. The pore characteristics of tight sandstone are influenced by geological processes, including the sedimentary environment, diagenesis, and mineral composition. Sedimentation primarily affects the macroscopic characteristics of the pore structure. Dense sandstones typically consist of sand grains, with the pores between the grains forming a large-scale pore-throat system. The particle size and arrangement of the initial sediments determine these large pore-throats. Different sedimentary environments result in pore-throat systems of varying scales, with grain size distribution, particle shape, and deposition mode all influencing pore distribution characteristics. Diagenesis, through processes such as compaction, cementation, dissolution, and mineral precipitation, forms smaller-scale pore-throat systems. The Chang 8 reservoir is primarily composed of a grey sandstone and dark mudstone interbedded sequence, with the depositional environment being mainly underwater distributary channels and interdistributary bays, representing a typical shallow-water deltaic sedimentary system. The lithology is dominated by fine-grained feldspathic sandstone and feldspar-rich lithic sandstone, with complex detrital components, low compositional maturity, and low to moderate structural maturity. The burial depth is greater than 2200 m, and compaction reduces the original pore space of the reservoir. Cementation destroys some primary porosity and damages particular secondary dissolution pores, leading to a finer pore-throat structure. As a result, the reservoir pore structure is complex, and the fractal dimension is high. Therefore, in the future, performing fractal dimension calculations and spatial distribution analysis on wellbore profiles and regional planes will effectively reveal the mechanisms by which sedimentary-diagenetic processes influence pore structure evolution and reservoir quality.

Fractal-derived pore classifications can effectively guide reservoir management strategies. In Type I and Type II reservoirs, due to the higher porosity and the coexistence of large and small pore-throat systems, development should separately address the characteristics of both pore types. In regions with higher porosity, the focus should be on developing large pore-throats using conventional high-pressure fracturing techniques for extraction. In areas with lower porosity or where micropores dominate, techniques such as acidizing and micro-fracture formation should be used to enhance the extraction of micropores. A phased development strategy can also be employed: first, large pore-throat systems can be accessed through high-pressure, large-volume fracturing, followed by micro-fracturing or acid fracturing to enhance micropore connectivity, achieving more comprehensive reservoir development. Type III reservoirs, with poor permeability, are mainly composed of tiny pores. The pore-throat radius are small, the pore structure is simple, and the connectivity is poor, displaying clear monofractal characteristics. The development strategy for such reservoirs should focus on using small-volume, high-pressure fracturing techniques combined with acid fracturing technology for extraction. Small-volume, high-pressure fracturing can effectively prevent damage to the pore structure. In contrast, acid fracturing can dissolve carbonate minerals in the reservoir, improving micropore connectivity and the reservoir's mining effect.

6. Conclusions

The complex pore structure of tight sandstone reservoirs poses significant challenges for their quantitative characterization. This study employed experimental tests, deep learning technology, and fractal geometry theory to quantitatively analyze the pore structure characteristics of Chang 8 tight sandstone in the Jiyuan area of the Ordos Basin and to further explore the factors affecting the fractal dimension. The main conclusions are as follows:

(1) HPMI reveals the complex and highly heterogeneous nature of the pore structure in tight sandstone reservoirs, demonstrating distinct fractal segmentation. Fractal dimension correlates with reservoir quality. The D_{MIP} values for the Type I, II, and III reservoirs increased sequentially, indicating that as reservoir quality decreased, pore complexity and heterogeneity increased, and connectivity decreased. Additionally, the average fractal dimensions of the small and large pore-throats were 2.16 (D_1) and 2.52 (D_2), respectively, indicating greater complexity in large pore-throat structures.

(2) The deep learning-based SEM pore extraction technique effectively quantified the pore structure of tight sandstone. Analysis of the fractal dimension and geometric parameters reveals the pore morphology and its microscopic complexity. The study found that the pore structure became more complex as the reservoir quality decreased, and the pore morphology became more irregular. Specifically, the D_{SEM} increased as the with the deterioration of reservoir quality. Type I reservoirs had the smallest fractal dimension, followed by Type II, with Type III showing the highest values. Additionally, reservoir quality exhibits strong correlations with pore geometric features; negative correlations with pore perimeter, aspect ratio, and solidity; and positive correlations with circularity and major axis length.

(3) The fractal dimensions D_{MIP} and D_{SEM} of the studied reservoirs ranged from 2.21 to 2.49 and from 1.01 to 1.28, respectively. This difference may be due to the distinct theoretical principles and computational models underlying the two methods. D_{MIP} reflects changes in pore connectivity and permeability, whereas D_{SEM} focuses on pore morphology and microscopic geometric characteristics. Collectively, these two methods reveal the pore structure's complexity and heterogeneity at different scales and characterize its multiple features.

(4) A significant correlation exists between the fractal dimension of tight sandstone reservoirs and their structural characteristics, geometric shape parameters, and mineral composition. The fractal dimension demonstrates an inverse relationship with permeability, porosity, median radius, maximum mercury saturation, mercury withdrawal efficiency, and sorting factor, while showing a positive link with median pressure and displacement pressure. Among these, the maximum mercury saturation and sorting factor demonstrate a strong correlation with the fractal dimension ($R^2 > 0.8$). Additionally, the fractal dimension is negatively correlated with pore circularity and major axis length and positively correlated with perimeter, aspect ratio, and solidity. Pores with higher circularity and simpler structures correspond to lower fractal dimensions, whereas irregular and long-strip pores with complex structures and poor connectivity correspond to higher fractal dimensions. In terms of mineral composition, the fractal dimension has a negative association with the concentrations of feldspar, quartz, and chlorite, and a positive correlation with carbonate content.

(5) The pore structure of dense sandstone reservoirs exhibits distinct dual-fractal characteristics, revealing its complexity and heterogeneity across multiple scales. Differences in the sedimentary environment and diagenesis jointly control the reservoir's pore structure characteristics. Therefore, the reservoir management strategy must fully account for differences in pore types, especially in Type I and II reservoirs, where the dual-fractal characteristics of the pore structure significantly impact development design.

Author Contributions: X.S.: Investigation, Formal analysis, Conceptualization, Data Curation, Writing—original draft; C.F.: Writing—review and editing, Supervision, Funding acquisition, Methodology; T.L.: Writing—review and editing, Supervision, Funding acquisition; Q.Z.: Supervision, Data curation; J.Z.: Investigation, Resources, Data curation; M.S.: Supervision, Project administration, Investigation. All authors have read and agreed to the published version of the manuscript.

Funding: This work is funded by the Natural Science Basic Research Plan in Shaanxi Province of China (Grant No. 2017JM4013; Grant No. 2020JQ-798; 2024JC-YBMS-256) and the Open Fund (PLN2023-21) of the National Key Laboratory of Oil and Gas Reservoir Geology and Exploitation (Southwest Petroleum University).

Data Availability Statement: The data that support the findings of this study are available from the corresponding author upon reasonable request. All the data can be found in the paper.

Conflicts of Interest: The authors declare that there are no conflicts of interest regarding the publication of this paper.

Nomenclature

Acronym	Description	Unit
P	Pore perimeter	μm
V_{Hg}	Volume of mercury	μm^3
l	Length of capillary	μm
r	Radius of pore-throat	μm
θ	Contact angle	°
σ	Interfacial tension	N/m
S_{Hg}	Mercury saturation	%
V_p	Total pore volume of the sample	μm^3
P_c	Mercury injection pressure	MPa
A	Pore area	μm^2
φ_1	Porosity proportions of the small pores	%
φ_2	Porosity proportions of the large pores	%
D_{MIP}	Fractal dimension of MICP	
D_{SEM}	Fractal dimension of SEM	
D_1	Fractal dimension of small pores	
D_2	Fractal dimension for large pores	
$N(r)$	Number of pores	

References

1. Zhou, N.; Lu, S.; Wang, M.; Huang, W.; Xiao, D.; Jiao, C.; Wang, J.; Tian, W.; Zhou, L.; Chen, F.; et al. Limits and grading evaluation criteria of tight oil reservoirs in typical continental basins of China. *Pet. Explor. Dev.* **2021**, *48*, 1089–1100. [CrossRef]
2. Zhao, W.; Hu, S.; Hou, L.; Yang, T.; Li, X.; Guo, B.; Yang, Z. Types and resource potential of continental shale oil in China and its boundary with tight oil. *Pet. Explor. Dev. Online* **2020**, *47*, 1–11. [CrossRef]
3. Pang, X.; Jia, C.; Wang, W.; Chen, Z.; Li, M.; Jiang, F.; Hu, T.; Wang, K.; Wang, Y. Buoyance-Driven Hydrocarbon Accumulation Depth and Its Implication for Unconventional Resource Prediction. *Geosci. Front.* **2021**, *12*, 101133. [CrossRef]
4. Xie, J.; Zhang, J.; Xiang, W.; Fang, Y.; Xue, Y.; Cao, J.; Tian, R. Effect of Microscopic Pore Structures on Ultrasonic Velocity in Tight Sandstone with Different Fluid Saturation. *Pet. Sci.* **2022**, *19*, 2683–2694. [CrossRef]
5. Wen, Y.; Ni, G.; Zhang, X.; Zheng, Y.; Wang, G.; Wang, Z.; Huang, Q. Fine characterization of pore structure of acidified anthracite based on liquid intrusion method and Micro-CT. *Energy* **2023**, *263*, 125639. [CrossRef]
6. Gao, H.; Wang, C.; Cheng, Z.; Li, T.; Dou, L.; Zhao, K.; Xue, J.; Luo, K. Effect of pressure pulse stimulation on imbibition displacement within a tight sandstone reservoir with local variations in porosity. *Geoenergy Sci. Eng.* **2023**, *226*, 211811. [CrossRef]
7. Pfeifer, P.; Avnir, D. Chemistry in noninteger dimensions between two and three. I. Fractal theory of heterogeneous surfaces. *J. Chem. Phys.* **1983**, *79*, 3558–3565. [CrossRef]
8. Zhang, Z.; Qin, Y.; Yi, T.; You, Z.; Yang, Z. Pore structure characteristics of coal and their geological controlling factors in Eastern Yunnan and Western Guizhou, China. *ACS Omega* **2020**, *5*, 19565–19578. [CrossRef]
9. Guo, Y.; Zhou, L.; Guo, F.; Chen, X.; Wu, J.; Zhang, Y. Pore structure and fractal characteristic analysis of gasification-coke prepared at different high-temperature residence times. *ACS Omega* **2020**, *5*, 22226–22237. [CrossRef]
10. Huang, W.; Lu, S.; Hersi, O.; Wang, M.; Deng, S.; Lu, R. Reservoir Spaces in Tight Sandstones: Classification, Fractal Characters, and Heterogeneity. *Nat. Gas. Sci. Eng.* **2017**, *46*, 80–92. [CrossRef]
11. Zhao, Y.; Wang, C.; Ning, L.; Zhao, H.; Bi, J. Pore and Fracture Development in Coal under Stress Conditions Based on Nuclear Magnetic Resonance and Fractal Theory. *Fuel* **2022**, *309*, 122112. [CrossRef]

12. Wu, Y.; Liu, C.; Ouyang, S.; Luo, B.; Zhao, D.; Sun, W.; Awan, R.; Lu, Z.; Li, G.; Zang, Q. Investigation of Pore-Throat Structure and Fractal Characteristics of Tight Sandstones Using HPMI, CRMI, and NMR Methods: A Case Study of the Lower Shihezi Formation in the Sulige Area, Ordos Basin. *Pet. Sci. Eng.* **2022**, *210*, 110053. [CrossRef]
13. Wang, G.; Han, D.; Qin, X.; Liu, Z.; Liu, J. A comprehensive method for studying pore structure and seepage characteristics of coal mass based on 3D CT reconstruction and NMR. *Fuel* **2020**, *281*, 118735. [CrossRef]
14. Xu, S.; Yang, Z.; Wu, S.; Wang, L.; Wei, W.; Yang, F.; Cai, J. Fractal Analysis of Pore Structure Differences Between Shale and Sandstone Based on the Nitrogen Adsorption Method. *Nat. Resour. Res.* **2022**, *31*, 1759–1773. [CrossRef]
15. Zhang, J.; Li, X.; Wei, Q.; Sun, K.; Zhang, G.; Wang, F. Characterization of Full-Sized Pore Structure and Fractal Characteristics of Marine–Continental Transitional Longtan Formation Shale of Sichuan Basin, South China. *Energy Fuels* **2017**, *31*, 10490–10504. [CrossRef]
16. Karin, P.; Chammana, P.; Oungpakornkaew, P.; Rungsritanapaisan, P.; Amornprapa, W.; Charoenphonphanich, C.; Sriprapha, K. Impact of Soot Nanoparticle Size and Quantity on Four-Ball Steel Wear Characteristics Using EDS, XRD and Electron Microscopy Image Analysis. *Mater. Res. Technol.* **2022**, *16*, 1781–1791. [CrossRef]
17. Lu, Y.; Liu, J.; Tang, J.; Ao, X.; Li, H.; Zhou, J.; Sun, X. Pore Changes of Slickwater-Containing Shale under Supercritical CO_2 Treatment. *Fuel* **2022**, *312*, 122775. [CrossRef]
18. Yang, Y.; Wang, B.; Yuan, Q.; Huang, D.; Peng, H. Characterization, Factors, and Fractal Dimension of Pore Structure of Fly Ash-Based Geopolymers. *Mater. Res. Technol.* **2023**, *26*, 3395–3407. [CrossRef]
19. Gao, H.; Luo, K.Q.; Wang, C.; Li, T.; Cheng, Z.L.; Dou, L.B.; Zhao, K.; Zhang, N.; Liu, Y.L. Impact of dissolution and precipitation on pore structure in CO_2 sequestration within tight sandstone reservoirs. *Pet. Sci.* **2025**, *22*, 868–883. [CrossRef]
20. Gao, H.; Cao, J.; Wang, C.; He, M.; Dou, L.; Huang, X.; Li, T. Comprehensive characterization of pore and throat system for tight sandstone reservoirs and associated permeability determination method using SEM, ratecontrolled mercury and high pressure mercury. *J. Pet. Sci. Eng.* **2018**, *174*, 514–524. [CrossRef]
21. Qiao, J.; Zeng, J.; Jiang, S.; Zhang, Y.; Feng, S.; Feng, X.; Hu, H. Insights into the pore structure and implications for fluid flow capacity of tight gas sandstone: A case study in the upper paleozoic of the Ordos Basin. *Mar. Pet. Geol.* **2020**, *118*, 104439. [CrossRef]
22. Dou, W.; Liu, L.; Jia, L.; Xu, Z.; Wang, M.; Du, C. Pore structure, fractal characteristics and permeability prediction of tight sandstones: A case study from Yanchang Formation, Ordos Basin, China. *Mar. Pet. Geol.* **2021**, *123*, 104737. [CrossRef]
23. Wang, P.; Jiang, Z.; Ji, W.; Zhang, C.; Yuan, Y.; Chen, L.; Yin, L. Heterogeneity of intergranular, intraparticle and organic pores in Longmaxi shale in Sichuan Basin, South China: Evidence from SEM digital images and fractal and multifractal geometries. *Mar. Pet. Geol.* **2016**, *72*, 122–138. [CrossRef]
24. Ojha, S.P.; Misra, S.; Tinni, A.; Sondergeld, C.; Rai, C. Relative permeability estimates for Wolfcamp and Eagle Ford shale samples from oil, gas and condensate windows using adsorption-desorption measurements. *Fuel* **2017**, *208*, 52–64. [CrossRef]
25. Yu, S.; Bo, J.; Fengli, L.; Jiegang, L. Structure and Fractal Characteristic of Micro- and Meso-Pores in Low, Middle-Rank Tectonic Deformed Coals by CO_2 and N_2 Adsorption. *Microporous Mesoporous Mater.* **2017**, *253*, 191–202. [CrossRef]
26. Mendhe, V.A.; Bannerjee, M.; Varma, A.K.; Kamble, A.D.; Mishra, S.; Singh, B.D. Fractal and pore dispositions of coal seams with significance to coalbed methane plays of East Bokaro, Jharkhand, India. *Nat. Gas. Sci. Eng.* **2017**, *38*, 412–433. [CrossRef]
27. Zhang, K.; Pang, X.; Zhao, Z.; Shao, X.; Zhang, X.; Li, W.; Wang, K. Pore structure and fractal analysis of Lower Carboniferous carbonate reservoirs in the Marsel area, Chu-Sarysu basin. *Mar. Pet. Geol.* **2018**, *93*, 451–467. [CrossRef]
28. Zhu, F.; Hu, W.; Cao, J.; Sun, F.; Liu, Y.; Sun, Z. Micro/Nanoscale Pore Structure and Fractal Characteristics of Tight Gas Sandstone: A Case Study from the Yuanba Area, Northeast Sichuan Basin, China. *Mar. Pet. Geol.* **2018**, *98*, 116–132. [CrossRef]
29. Xie, S.; Cheng, Q.; Ling, Q.; Li, B.; Bao, Z.; Fan, P. Fractal and multifractal analysis of carbonate pore-scale digital images of petroleum reservoirs. *Mar. Pet. Geol.* **2010**, *27*, 476–485. [CrossRef]
30. Zhang, H.; Wu, W. Shale content prediction of well logs based on CNN-BiGRU-VAE neural network. *J. Earth Syst. Sci.* **2023**, *132*, 139. [CrossRef]
31. Du, Y.; Chen, J.; Zhang, T. Reconstruction of Three-Dimensional Porous Media Using Deep Transfer Learning. *Geofluids* **2020**, *2020*, 6641642. [CrossRef]
32. Argilaga, A. Fractal Informed Generative Adversarial Networks (FI-GAN): Application to the generation of X-ray CT images of a self-similar partially saturated sand. *Comput. Geotech.* **2023**, *158*, 105384. [CrossRef]
33. Chen, Z.; Liu, X.; Yang, J.; Little, E.; Zhou, Y. Deep learning-based method for SEM image segmentation in mineral characterization, an example from Duvernay Shale samples in Western Canada Sedimentary Basin. *Comput. Geosci.* **2020**, *138*, 104450. [CrossRef]
34. Ge, Y.; Xu, X.; Yang, S.; Zhou, Q.; Shen, F. A review of data augmentation methods for sequence data. *J. Front. Comput. Sci. Technol.* **2021**, *15*, 827.
35. Xue, D.; Tang, Q.; Wang, A.; Yi, H.; Zhang, C.; Geng, C.; Zhou, H. Microscopic phase state FCN intelligent recognition and fractal reconstruction of coal rocks. *Rock. Soil. Mech.* **2020**, *39*, 1203–1221. [CrossRef]

36. Wang, D.; Wang, L.; Qin, Y.; Wei, L.; Cao, T.; Li, W.; Li, L.; Chen, X.; Xia, Y. Fracture recognition and 3D reconstruction of coal-rock composite based on VRA-UNet network. *Coal Sci. Technol.* **2025**, *53*, 95–107. [CrossRef]
37. Song, X.; Gao, H.; Feng, C.; Yi, P.; Wang, C.; Li, T. Analysis of the influence of micro-pore structure on oil occurrence using nano-CT scanning and nuclear magnetic resonance technology: An example from Chang 8 tight sandstone reservoir, Jiyuan, Ordos Basin. *Processes* **2023**, *11*, 1127. [CrossRef]
38. Song, X.; Feng, C.; Li, T.; Zhang, Q.; Pan, X.; Sun, M.; Ge, Y. Quantitative classification evaluation model for tight sandstone reservoirs based on machine learning. *Sci. Rep.* **2024**, *14*, 20712. [CrossRef]
39. Qu, Y.; Sun, W.; Tao, R.; Luo, B.; Chen, L.; Ren, D. Pore-throat structure and fractal characteristics of tight sandstones in Yanchang Formation, Ordos Basin. *Mar. Pet. Geol.* **2020**, *120*, 104573. [CrossRef]
40. Wang, J.Q.; Deng, J.X.; Xu, Z.H.; Xia, H.; Yan, L.L. Rock physics properties and influencing factors of subsalt carbonate reservoir of Ordovician Majiagou Formation in central Ordos Basin. *Pet. Sci.* **2024**, *21*, 3965–3980. [CrossRef]
41. Li, C.; Liu, G.; Cao, Z.; Sun, M.; You, Y.; Liu, N. Oil charging pore throat threshold and accumulation effectiveness of tight sandstone reservoir using the physical simulation experiments combined with NMR. *J. Pet. Sci. Eng.* **2022**, *208*, 109338. [CrossRef]
42. Tong, Q.; He, D.; Xia, Z.; Huang, J.; Li, Y.; Xu, F.; Guo, S. Research on sand body architecture at the intersection of a bidirectional sedimentary system in the Jiyuan area of Ordos Basin. *Sci. Rep.* **2023**, *13*, 12261. [CrossRef] [PubMed]
43. Huang, X.; Li, T.; Wang, X.; Gao, H.; Ni, J.; Zhao, J.; Wang, C. Distribution characteristics and its influence factors of movable fluid in tight sandstone reservoir: A case study from Chang8 oil layer of Yanchang Formation in Jiyuan oilfield, Ordos Basin. *Acta Pet. Sin.* **2019**, *40*, 557.
44. Liu, H.; Zhao, Y.; Luo, Y.; Xiao, G.; Meng, Y.; Zhou, S.; Shao, L. Origin of the reservoir quality difference between Chang 8 and Chang 9 Member sandstones in the Honghe Oil Field of the Southern Ordos Basin, China. *J. Pet. Sci. Eng.* **2020**, *185*, 106668. [CrossRef]
45. Zhang, K.; Liu, R.; Liu, Z. Sedimentary sequence evolution and organic matter accumulation characteristics of the Chang 8-Chang 7 members in the Upper Triassic Yanchang Formation, southwest Ordos Basin, central China. *J. Pet. Sci. Eng.* **2021**, *196*, 107751. [CrossRef]
46. Yang, P.; Ren, Z.; Fu, J.; Bao, H.; Xiao, H.; Shi, Z.; Wang, K.; Zhang, Y.; Liu, W.; Li, W. A Tectono-Thermal Perspective on the Petroleum Generation, Accumulation and Preservation in the Southern Ordos Basin, North China. *Pet. Sci.* **2024**, *21*, 1459–1473. [CrossRef]
47. Fu, S.; Fang, Q.; Li, A.; Li, Z.; Han, J.; Dang, X.; Han, W. Accurate characterization of full pore size distribution of tight sandstones by low-temperature nitrogen gas adsorption and high-pressure mercury intrusion combination method. *Energy Sci. Eng.* **2021**, *9*, 80–100. [CrossRef]
48. Li, C.; Chen, G.; Li, X.; Zhou, Q.; Sun, Z. The occurrence of tight oil in the Chang 8 lacustrine sandstone of the Huaqing area, Ordos Basin, China: Insights into the content of adsorbed oil and its controlling factors. *J. Nat. Gas. Geosci.* **2022**, *7*, 27–37. [CrossRef]
49. Liu, Y.; Dong, X.; Chen, Z.; Hou, Y.; Luo, Q.; Chen, Y. A novel experimental investigation on the occurrence state of fluids in microscale pores of tight reservoirs. *J. Pet. Sci. Eng.* **2021**, *196*, 107656. [CrossRef]
50. Yang, M.; Pan, Y.; Feng, H.; Yan, Q.; Lu, Y.; Wang, W.; Qi, Y.; Zhu, H. Fractal Characteristics of Pore Structure of Longmaxi Shales with Different Burial Depths in Southern Sichuan and Its Geological Significance. *Fractal Fract.* **2024**, *9*, 2. [CrossRef]
51. Wang, J.; Zhang, J.; Xiao, X.; Chen, Y.N.; Han, D. Pore Structure and Fractal Characteristics of Inter-Layer Sandstone in Marine-Continental Transitional Shale: A Case Study of the Upper Permian Longtan Formation in Southern Sichuan Basin, South China. *Fractal Fract.* **2024**, *9*, 11. [CrossRef]
52. Ronneberger, O.; Fischer, P.; Brox, T. U-net: Convolutional networks for biomedical image segmentation. In Proceedings of the Medical Image Computing and Computer-Assisted Intervention–MICCAI 2015: 18th International Conference, Proceedings, Part III 18, Munich, Germany, 5–9 October 2015; Springer: Berlin/Heidelberg, Germany, 2015; pp. 234–241. [CrossRef]
53. Xi, Z.; Tang, S.; Zhang, S.; Qi, Y.; Wang, X. Evaluation of mechanical properties of porous media materials based on deep learning: Insights from pore structure. *Fuel* **2024**, *371*, 131923. [CrossRef]
54. Huo, F.; Li, H.; Dong, H.; Ren, W. An improved Swin transformer for sandstone micro-image classification. *Geoenergy Sci. Eng.* **2025**, *247*, 213680. [CrossRef]
55. Gao, Y.; Yu, Z.; Chen, W.; Yin, Q.; Wu, J.; Wang, W. Recognition of rock materials after high-temperature deterioration based on SEM images via deep learning. *J. Mater. Res. Technol.* **2023**, *25*, 273–284. [CrossRef]
56. Jin, H.; Cao, L.; Kan, X.; Sun, W.; Yao, W.; Wang, X. Coal petrography extraction approach based on multiscale mixed-attention-based residual U-net. *Meas. Sci. Technol.* **2022**, *33*, 075402. [CrossRef]
57. Liu, Y.; Zhang, Q.; Zhang, N.; Lv, J.; Gong, M.; Cao, J. Enhancement of thin-section image using super-resolution method with application to the mineral segmentation and classification in tight sandstone reservoir. *J. Pet. Sci. Eng.* **2022**, *216*, 110774. [CrossRef]

58. Voss, R.F.; Laibowitz, R.B.; Alessandrini, E.I. Fractal geometry of percolation in thin gold films. In *Scaling Phenomena in Disordered Systems*; Springer Science & Business Media: Berlin/Heidelberg, Germany, 1991; pp. 279–288.
59. Zhu, H.; Zhang, T.; Zhong, D.; Li, Y.; Zhang, J.; Chen, X. Characteristics of the binary pore structure in tight sandstone reservoirs. *Pet. Explor. Dev.* **2019**, *46*, 1297–1306. [CrossRef]
60. He, T.; Zhou, Y.; Chen, Z.; Zhang, Z.; Xie, H.; Shan, Y.; Cui, G. Fractal characterization of the pore-throat structure in tight sandstone based on low-temperature nitrogen gas adsorption and high pressure mercury injection. *Fractal Fract.* **2024**, *8*, 356. [CrossRef]
61. Zhang, H.; Li, X.; Liu, J.; Wang, Y.; Guo, L.; Wu, Z.; Tian, Y. A Fractal Characteristics Analysis of the Pore Throat Structure in Low-Permeability Sandstone Reservoirs: A Case Study of the Yanchang Formation, Southeast Ordos Basin. *Fractal Fract.* **2025**, *9*, 224. [CrossRef]
62. Zhang, H.; Guo, L.; Wu, Z.; Ma, J. Pore-throat structure, fractal characteristics and permeability prediction of tight sandstone: The Yanchang Formation, Southeast Ordos Basin. *Sci. Rep.* **2024**, *14*, 27913. [CrossRef]
63. Lu, Y.; Liu, K. Pore structure characterization of eocene Low-Permeability sandstones via fractal analysis and machine learning: An example from the dongying depression, bohai bay basin, China. *ACS Omega* **2021**, *6*, 11693–11710. [CrossRef]
64. Li, P.; Zheng, M.; Bi, H.; Wu, S.; Wang, X. Pore throat structure and fractal characteristics of tight oil sandstone: A case study in the Ordos Basin, China. *J. Pet. Sci. Eng.* **2017**, *149*, 665–674. [CrossRef]

fractal and fractional

Article

Structure and Fractal Characteristics of Organic Matter Pores in Wufeng–Lower Longmaxi Formations in Southern Sichuan Basin, China

Quanzhong Guan [1,*], Dazhong Dong [2], Bin Deng [1], Cheng Chen [1], Chongda Li [1], Kun Jiao [1], Yuehao Ye [1], Haoran Liang [1] and Huiwen Yue [1]

1 College of Energy, Chengdu University of Technology, Chengdu 610059, China
2 PetroChina Research Institute of Petroleum Exploration and Development, Beijing 100083, China
* Correspondence: muchang503@126.com

Abstract: Organic matter pores constitute a significant storage space in shale gas reservoirs, contributing to approximately 50% of the total porosity. This study employed a comprehensive approach, utilizing scanning electron microscopy, low-pressure N_2 adsorption, thermal analysis, image statistics, and fractal theory, to quantitatively characterize the structure and complexity of organic matter pores in the Wufeng–lower Longmaxi Formations (WLLFs). The WLLFs exhibit a high organic matter content, averaging 3.20%. Organic matter pores are typically well-developed, predominantly observed within organic matter clusters, organic matter–clay mineral complexes, and the internal organic matter of pyrite framboid. The morphology of these pores is generally elliptical and spindle-shaped, with the primary pore diameter displaying a bimodal distribution at 10~40 nm and 100~160 nm, potentially influenced by the observational limit of scanning electron microscopy. Shales from greater burial depths within the same gas well contain more organic matter pores; however, the development of organic matter pores in deep gas wells is roughly equivalent to that in medium and shallow gas wells. Fractal dimension values can be utilized to characterize the complexity of organic matter pores, with organic matter macropores ($D_{>50}$) being more complex than organic matter mesopores ($D_{2\text{–}50}$), which in turn are more complex than organic matter micropores ($D_{<2}$). The development of macropores and mesopores is a key factor in the heterogeneity of organic matter pores. The complexity of organic matter pores in the same well increases gradually with the burial depth of the shale, and the complexity of organic matter pores in deep gas wells is roughly equivalent to that in medium and shallow gas wells. The structure and fractal characteristics of organic matter pores in shale are primarily controlled by components, diagenesis, tectonism, etc. The lower Longmaxi shale exhibit a high biogenic quartz content and robust hydrocarbon generation from organic matter. This composition effectively shields organic matter pores from multi-directional extrusion, leading to the formation of macropores and mesopores without specific orientation. High-quality shale sections (one and two sublayers) have relatively high fractal dimension $D_{2\text{–}50}$ and $D_{>50}$ values of organic matter pores and gas content. Consequently, the quality parameters of shale and fractal dimension characteristics can be comprehensively evaluated to identify high-quality shale sections.

Keywords: organic matter pores; fractal characteristics; Wufeng–Longmaxi shales; Sichuan Basin

1. Introduction

The "North American Shale Revolution" and advancements in horizontal well and hydraulic fracturing technologies have brought the vast potential of shale oil and gas resources to the fore [1,2]. Given China's significant reliance on foreign oil and gas, there are substantial security concerns. However, the robust development of shale gas can effectively mitigate issues related to resource scarcity, environmental degradation, and economic growth [3]. The basins in China contain three types of organic-rich shales: marine facies, marine–continental transition facies, and continental facies [4]. At present, the industrial production of marine facies shale gas has been realized, with a projected total output of 25 billion cubic meters in 2023. The primary producing strata are the black shales of the WLLFs in the southern region of the Sichuan Basin.

The primary distinction between shale and conventional reservoirs lies in the presence of micro–nano organic matter pores [5–7]. Since their discovery, organic matter pores have attracted significant attention. Research on shale organic matter pores predominantly centers on their origin, preservation, and characterization [8–15]. The formation of organic matter pores is influenced by the content, type, and thermal evolution level of the organic matter. Organic matter with a moderate content (total organic carbon (TOC) < 5.5%), favorable type (Types I and II_1), and a higher degree of thermal evolution (Ro) typically exhibit well-developed pores [10,11,14,16–18]. The effective preservation of these pores post-formation is also vital, primarily governed by diagenetic processes and mineral composition [19–21]. Autogenic minerals establish rigid frameworks during diagenesis, which, coupled with overpressure formed by volume expansion after organic matter pyrolysis, prevent the collapse or even erasure of organic matter pores [22–24]. Organic matter pore characterization typically involves scanning electron microscopy, gas adsorption, and rock physics model analysis methods [25–28]. However, each testing method has its limitations, making it challenging to employ a unified parameter to characterize the complexity of the organic matter pore structure [8,26,29].

In recent years, the application of fractal theory on organic matter pores in shales has gradually emerged [30,31]. The fractal dimension, as an effective indicator describing the complexity and irregularity of pore structure, can reveal its intrinsic relationship with the storage and migration performance of shale gas. Research indicates that the fractal characteristics of pore surfaces are closely related to the adsorption capacity of shale gas, while the fractal structure of pore networks has a significant impact on the seepage process [19,24,26,32]. However, the strong heterogeneity of shale leads to significant differences in the pore structure of shale organic matter in different regions. In addition, the diversity of fractal calculation methods and the applicability of models under complex geological conditions still need further optimization.

Previous studies have shown that the organic matter pores in the "sweet spot" of the WLLFs account for about 50%, which are the main storage spaces for free gas and adsorbed gas, and have been used as one of the preferred indicators for target layers [18,33]. The complexity of these organic matter pores, however, remains to be quantitatively characterized, as well as the factors that control it. This study employs a comprehensive approach, utilizing scanning electron microscopy, low-pressure N_2 adsorption, thermal analysis, image statistical processing, and fractal theory on the WLLFs. The objectives are to (1) digitally characterize the development characteristics of organic matter pores, such as pore size and direction; (2) quantitatively analyze the complexity of these pores; and (3) identify the controlling factors of the heterogeneity of organic matter pores and their geological significance.

2. Geologic Setting and Well Description

The Sichuan Basin, situated on the northwestern periphery of the Yangtze Plate and encompassed by several mountain ranges, is a substantial superimposed basin [2,34]. Since the Precambrian era, the Sichuan Basin has experienced numerous tectonic activities, resulting in approximately 10 km of strata deposition and the formation of several organic-rich shale sets, including the Doushantuo Formation (Ediacaran), Qiongzhusi Formation (Cambrian), Wufeng Formation–Longmaxi Formation (Ordovician to Silurian), Wujiaping Formation (Permian), and Ziliujing Formation (Jurassic) [35,36]. At present, the Wufeng Formation–Longmaxi gas shales in southern Sichuan are being produced on a large scale, leading to the formation of significant shale gas fields such as Changning-Weiyuan, Fuling, and Luzhou.

The Sichuan Basin is a secondary craton basin within the Yangtze Plate, which was located in the equatorial sea area on the northern edge of the Gondwana continent during the late Ordovician period. The convergent extrusion from the southeast Cathaysian plate caused uplift along the southern margin of the Yangtze land, and it connected with the ancient Kangdian land in the southwest to form the ancient Yunnan-Guizhou-Guangxi land; at the same time, the passive continental margin in the north subducted towards the North China plate, and the scope of the ancient uplift in central Sichuan on the western edge of the plate also continuously expanded [37]. This process has led to the formation of a semi-enclosed stagnant sea basin, sandwiched by ancient uplifts in central Sichuan and central Guizhou-Xuefeng, a phenomenon referred to as the "three uplifts sandwiched by one depression" [1,22]. From the Late Ordovician to the Early Silurian period, large-scale transgression deposited a series of organic-rich shale strata—the Wufeng Formation and Longmaxi Formation—on the deep-water shelf (Figure 1). These shales are consistently distributed throughout the Sichuan Basin, characterized by their substantial sedimentary thickness and high organic matter content, each with a sedimentary center located in the eastern and southern Sichuan Basin [2].

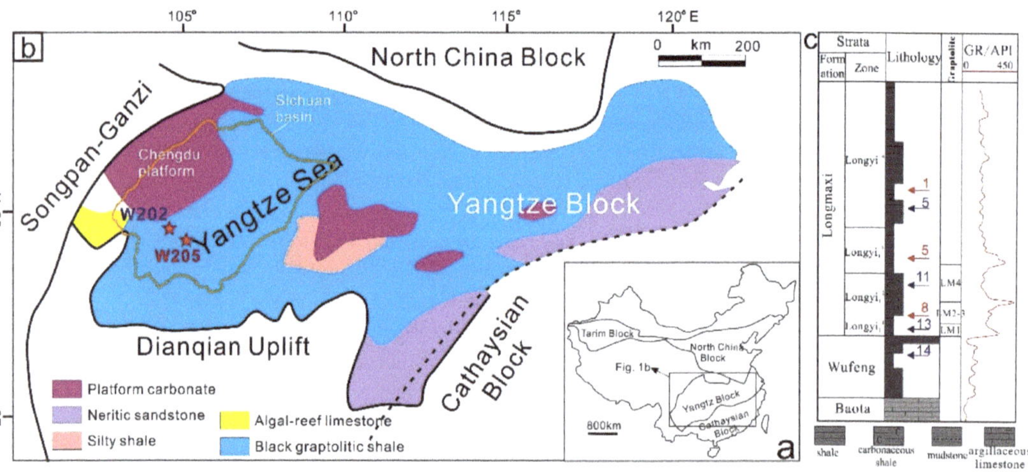

Figure 1. Location of the southern Sichuan research area (**b**) in the Sichuan Basin (**a**) and study wells stratum histogram (**c**) (modified after Ref. [34]). Samples marked with blue arrows and red arrows are from the W202 well and W205 well, respectively.

The Weiyuan shale gas field, situated in the southern region of the Sichuan Basin (Figure 1b), is primarily characterized by the Weiyuan anticline, where the top strata have undergone erosion [34,38]. Wells W202 and W205 are both located on the southeastern wing of this anticline. The depths of the organic-rich shale sections from the Wufeng Formation to the Longmaxi Formation in wells W202 and W205 range from 2560 m to 2580 m and 3687.5 m to 3710 m, respectively (Figure 1c). The bottom boundary of these formations is marked by the nodular limestone of the Baota Formation. During the Wufeng period, underwater uplifts resulted in the absence of the Wufeng black shale and Guanyinqiao layer in well W205. However, both wells have successfully developed their primary shale sections. The lithology of the Longmaxi Formation changes from bottom to top as carbonaceous shale, calcareous shale, and clay shale, which are laterally continuous in the southern Sichuan area.

3. Samples and Analytical Methods

3.1. Samples

In the Weiyuan shale gas field, 18 shale core samples were procured from the WLLFs in wells W202 (depth < 3500 m) and W205 (depth > 3500 m). Given the necessity for multiple supporting experiments on the samples, and considering their lithology, stratigraphic position, and sampling depth data, efforts were made to reduce the influence of heterogeneity within individual samples. Consequently, a subset of 7 samples was chosen for a comprehensive analysis (Figure 1c).

3.2. TOC Content and Mineralogy

All samples were subjected to crushing and processing in accordance with relevant industry standards prior to TOC content and mineral composition analyses. The TOC contents were ascertained using an LECO carbon–sulfur analyzer, boasting a precision of 1%. The mineralogical compositions were identified utilizing a Brucker D8 Advance X-ray diffractometer operating at 40 kV and 30 mA, employing Cu radiation. For comprehensive details on the experimental procedures, please refer to reference [39].

3.3. Field Emission-Scanning Electron Microscopy (FE-SEM) for Organic Matter Pore Analysis

The organic porosity in shale oil/gas reservoirs predominantly exists at the micro–nano scale. To facilitate a more precise identification and observation of these pores, samples are initially subjected to physical and argon ion polishing treatments prior to analysis via a scanning electron microscope. It is important to note that all samples are deliberately not gold or carbon plated to avoid any potential interference with the accurate observation of organic porosity. For more information on the imaging process, please refer to reference [40].

To effectively extract information from organic matter pores in shale, JMicroVision 1.2.7 software is employed for processing. The relatively large grayscale values of both the organic matter and its pores enable a clear distinction from inorganic minerals and pores. Consequently, JMicroVision can accurately determine parameters such as the area, perimeter, length, width, equivalent circle diameter, azimuth angle, and the elongation of the organic matter pores (as shown in Figure 2).

3.4. Low-Pressure N_2 Gas Adsorption for OM Extraction

The procedures for organic matter extraction and nitrogen adsorption experiments can be found in Guan et al., 2019 [10]. The BET-BJH model was employed to determine pore volumes, pore size distributions, and specific surface areas during low-pressure N_2 adsorption.

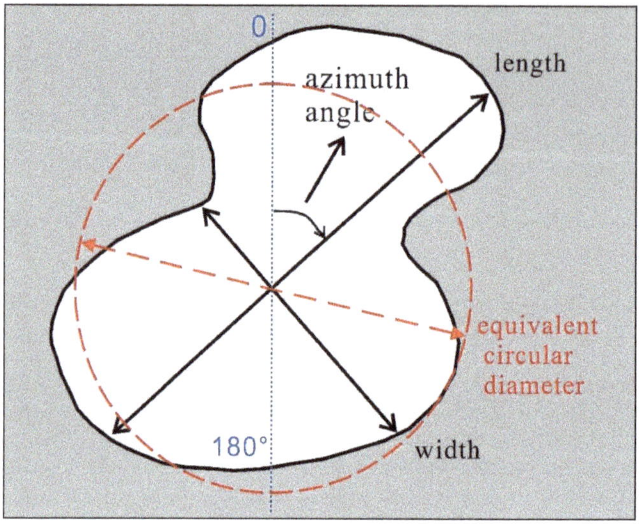

Figure 2. Definitions and measurement methods of parameters.

3.5. Fractal Theory

Organic matter pores in shale, ranging from a few nanometers to several micrometers, exhibit complex and highly heterogeneous morphologies due to the combined effects of diagenesis and tectonics. These characteristics make them challenging to characterize using traditional Euclidean geometry. Consequently, fractal theory has been employed to analyze such intricate and heterogeneous objects, including organic matter pores [19,26]. Fractals capture the self-similarity of an object's local parts to its overall structure, which can be quantitatively represented by the fractal dimension [41]. This dimension provides a measure of the complexity and self-similarity of fractal objects, offering insights into how they occupy space. Common methods for calculating the fractal dimension include the Frenkel–Halsey–Hill (FHH) model, the micropore fractal model, and image analysis.

In this research, the FHH model is employed to fractalize the low-pressure N_2 adsorption curves of organic matter extraction, subsequently calculating the fractal dimension D. The mathematical formula for the FHH model is presented below:

$$\ln(V/V_0) = k + (D - 3)\ln[\ln(P_0/P)]$$

where V is the volume of gas adsorption at equilibrium pressure P; V_0 is the monolayer coverage volume; k is a constant; P_0 is the saturation pressure; P is the equilibrium pressure; and D is the fractal dimension.

4. Results

4.1. TOC Content and Mineral Composition

In the study area, 52 core samples from five wells were analyzed for TOC content and XRD. However, due to the aforementioned reasons, only seven representative samples were selected for this study. The TOC content of the WLLFs in the Weiyuan Shale Gas Field varies between 1.05% and 8.10%, with an average of 3.20%. In both wells, it was observed that the TOC content increases progressively with depth, as presented in Table 1.

The WLLF components are predominantly quartz and clay minerals, with average concentrations of 23.29% and 38.0% respectively, accounting for more than 60% of all components. The content of quartz in shale exhibits a complementary relationship with that of clay minerals, particularly within the "sweet intervals", where the quartz content exceeds 35% (Table 1). The remaining minerals include feldspar, calcite, dolomite, and pyrite. Some shale layers have a higher concentration of carbonate rock minerals, such as sample 2-14, potentially due to their location within a calcium-rich deep-water shelf sedimentary system.

Table 1. Total organic carbon contents and mineral compositions of the shale samples from Wufeng and Longmaxi Formations in the two wells.

Sample No.	TOC %	Clay %	Quartz %	Feldspar %	Calcite %	Dolomite %	Pyrite %	Anhydrite %	Siderite %
2-5	2.54	22	21	9	28	17	1	2	0
2-11	2.48	14	68	6	7	3	2	0	0
2-13	6.06	34	41	8	3	7	2	2	3
2-14	2.29	7	25	2	35	21	10	0	0
5-1	1.05	31	39	17	5	2	2	3	1
5-5	3.00	28	39	9	13	7	2	2	0
5-8	2.84	27	33	9	21	6	2	2	0

4.2. FE-SEM Observation and Analyses

Using SEM to observe a large number of shale samples from the WLLFs in the study area, it was found that organic matter pores are generally well-developed (Figure 3). In the shale "sweet interval", organic matter pores account for more than 50% of the total pores and are the main storage space for shale gas. The shapes of organic matter pores in the WLLFs vary, mainly being elliptical and spindle-shaped (Figure 3), with some showing directionality (Figure 3e), possibly due to tectonic compression. Under SEM observation, the occurrence of organic matter pores is diverse, mainly developed in independent organic matter clusters (Figure 3a,e,f), organic matter–clay mineral complexes (Figure 3d), and the internal organic matter of pyrite framboid (Figure 3b,c). The diameter of organic matter pores varies greatly, with the smallest reaching about 1nm, mainly affected by the resolution of the instrument [8,29]. The organic matter pores in shale have strong heterogeneity; even within the same organic matter, there are significant differences in the size, shape, and even organic matter pore development (Figure 3).

To mitigate the impact of heterogeneity, a large number of SEM observations were conducted on the organic matter within each sample to analyze the development characteristics of organic matter pores. Based on factors such as the degree and clarity of organic matter pore development in SEM images, 85 photographs were selected for analysis, with statistical results shown in Table 2. The area distribution of shale organic matter pores under the SEM microscope in the study area ranged from 49.38 nm^2 to 45.50 μm^2. The existence of ultra-large organic matter pores may be due to the fusion of a large number of "organic matter primary pores" during the hydrocarbon evolution process of organic matter (Figure 3a,d,f). The perimeter of organic matter pores ranged from 18.30 nm to 62,729.13 nm, with the complexity of pore morphology determining the size of the perimeter. The long and short axis sizes of organic matter pores were used to calculate the elongation rate of the pores, quantitatively characterizing the degree of change in pore morphology. Herein, the diameter of organic matter pores is represented by the equivalent circle diameter, distributed between 7.93 nm and 7608.62 nm, but mainly concentrated between 10~40 nm and 100~160 nm, as shown in Figure 4. Overall, various parameters of shale organic matter pores show a trend of gradually increasing with the increase in sample burial depth.

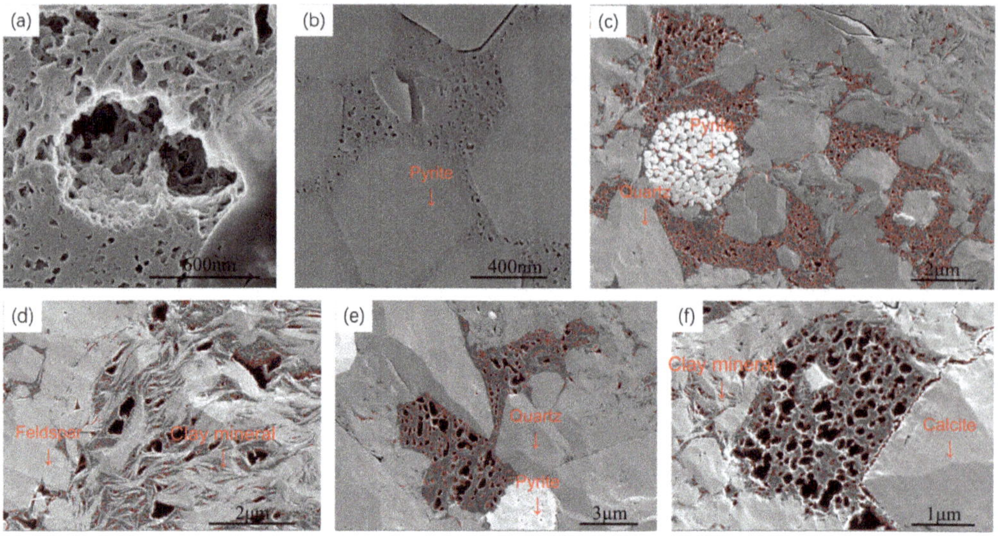

Figure 3. HIM-SEM images of organic matter pores. (**a**) Organic matter pores with different pore sizes. Large organic matter pores have formed by the fusion of small ones. The large organic matter pores are complex, containing many small organic matter pores that are interconnected, 2-11. (**b**) Organic matter present in the intercrystallite pore of pyrite framboid, with well-developed small organic matter pores, 2-5. (**c**) Organic matter distributed in contiguous patches. Internal pores have developed but with strong heterogeneity. Organic matter pores in the upper left corner are larger, 2-13. (**d**) Organic matter–clay composite distributed in flakes. Ultra-large organic matter pore development, 5-5. (**e**) Individual organic matter aggregates, with intricately developed pores with directionality compressed by rigid minerals, 2-14. (**f**) Single organic matter clusters, with highly developed pores, ellipsoidal and non-directional, 5-8.

4.3. Pore Size Distribution, Pore Volume, and Special Surface Area (SSA)

Generally speaking, low-pressure N_2 adsorption experiments are relatively difficult for characterizing pores smaller than 2 nm. However, in conjunction with the SEM observations used, due to the inherent resolution of SEM and the identification level of JMicroVision software [28], this study adopts low-pressure N_2 adsorption experiments to analyze organic matter pores. The summary curves of low-pressure N_2 adsorption experiments on shale organic matter extracts are shown in Figure 5. The adsorption curve of each sample has three stages: a rapid rise under $P/P_0 < 0.1$, a slow rise from 0.1 to 0.9, and a sharp rise above 0.9. According to the IUPAC classification, all samples' adsorption curves exhibit distinct type II and type IV characteristics, indicating that the structure of the extracted organic matter is primarily composed of micropores and mesopores, with a small amount of microfractures or macropores [42]. Moreover, due to capillary condensation, the N_2 adsorption–desorption curves of each sample form a "hysteresis loop". These "hysteresis loops" are essentially close to types H3 and H4 in IUPAC classification, suggesting that the main pore types in the organic matter extracts are plate-slit pores, ink bottle pores, and tubular pores, with the main pore diameters being in the micropore–mesopore (type H4) or macropore (type H3) range.

Table 2. Quantitative organic matter pore parameters from FE-SEM images for shale samples of two wells.

Sample No.	Pores	Area/nm^2	Perimeter/nm	Length/nm	Width/nm	Equivalent Circular Diameter/nm	Elongation
2-5	1798	277.00–153,739.61 (2580.92)	33.62–5595.64 (184.25)	5.26–1185.76 (65.74)	4.08–488.47 (35.16)	14.56–560.52 (45.01)	0.09–1.00 (0.59)
2-11	1709	650.36–2,087,905.77 (33,272.35)	66.43–14,565.63 (744.00)	8.06–3393.23 (253.60)	4.49–1695.35 (132.44)	28.78–1630.46 (167.10)	0.05–1.00 (0.58)
2-13	1042	226.76–238,962.07 (5347.66)	39.22–4761.47 (242.49)	7.41–1143.86 (86.03)	4.76–467.86 (46.90)	16.99–551.59 (59.66)	0.12–1.00 (0.57)
2-14	247	8910.67–45,500,000.00 (228,316.75)	245.88–62,729.13 (1083.17)	29.85–30,418.80 (449.83)	26.93–1599.36 (159.12)	106.51–7608.62 (239.24)	0.02–1.00 (0.53)
5-1	436	142.40–32,566.75 (1886.11)	31.43–2724.89 (215.87)	10.67–634.58 (63.18)	4.98–256.18 (34.26)	13.47–203.63 (38.71)	0.20–1.00 (0.59)
5-5	405	49.38–27,577.32 (1376.73)	18.30–2520.00 (174.39)	6.67–397.78 (52.41)	3.65–237.62 (29.48)	7.93–187.83 (33.19)	0.20–0.96 (0.63)
5-8	1307	326.53–18,971.43 (1518.06)	47.07–1386.75 (169.13)	5.71–424.35 (57.52)	5.71–165.77 (32.37)	20.39–155.42 (39.99)	0.16–1.00 (0.59)

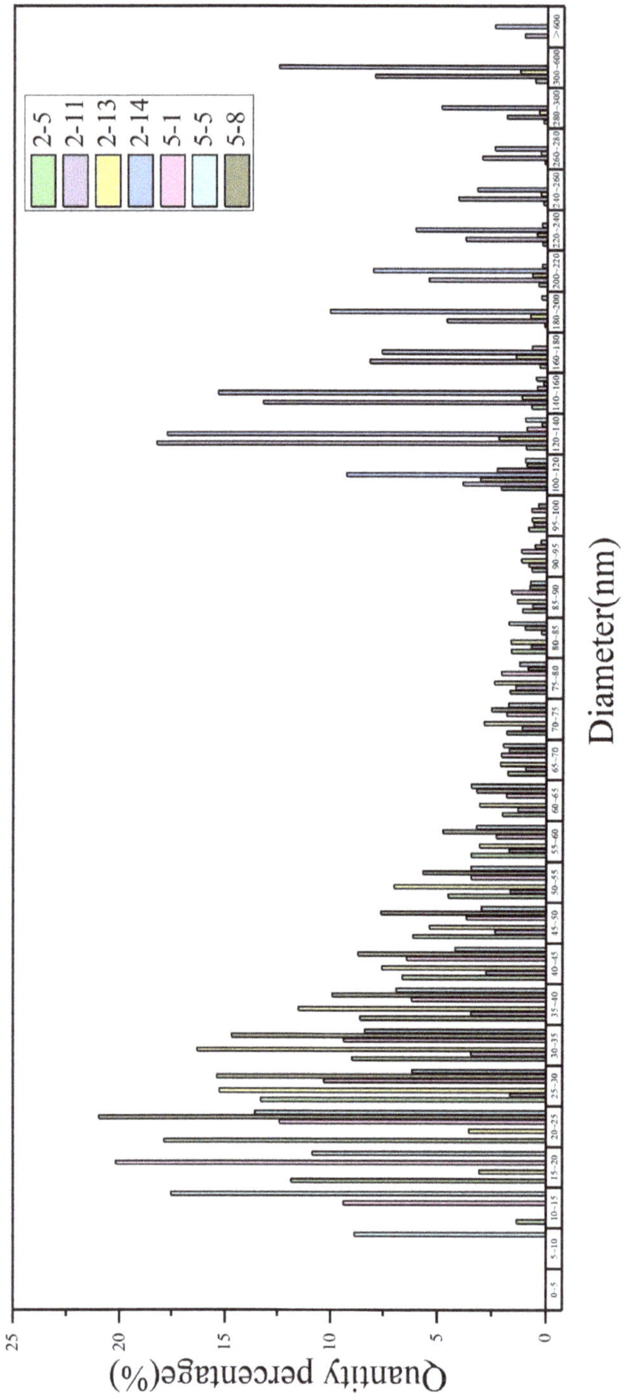

Figure 4. Organic matter pore size distribution from different layers and wells in the study area.

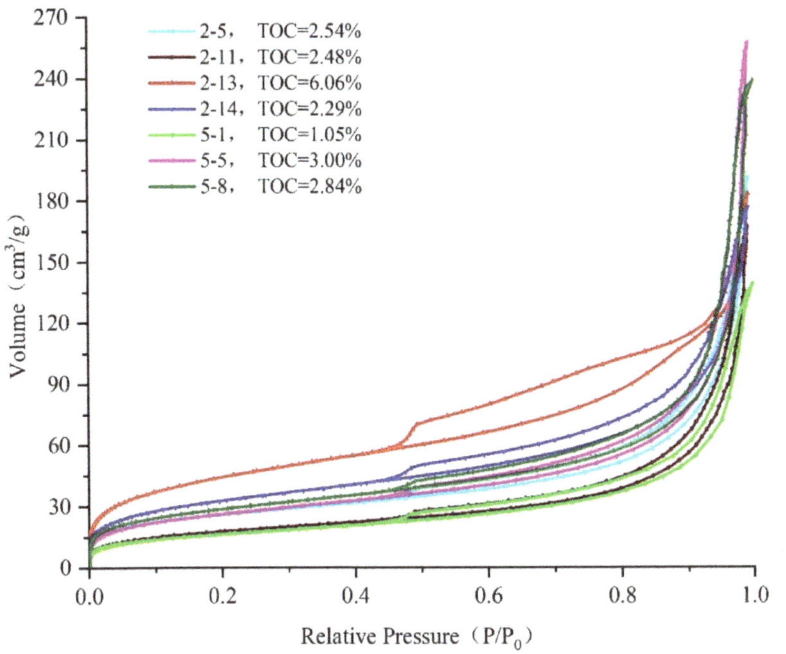

Figure 5. Low pressure N_2 adsorption–desorption isotherms collected for shale organic matter extracts in two wells.

The experimental results show that the SSA and pore volume of shale organic matter extracts are 60.1268~163.0092 m^2/g and 0.2151~0.3972 mL/g, respectively, with an average of 101.28 m^2/g and 0.2986 mL/g, all of which are significantly higher than the overall SSA and pore volume of the shale (Table 3). The average diameter distribution of organic matter pores is between 10.64 and 22.18 nm, which is consistent with the main organic matter pore distribution observed under the SEM microscope (Figure 4). However, the experimental analysis shows that the development of organic matter pores mainly concentrates on 2~4 nm, indicating that a certain amount of macropores develop in shale organic matter pores, and also reveals that both SEM and JMicroVision recognition and analysis of organic matter pores are constrained by their own conditions [8,29].

Table 3. Quantitative pore parameters of the shale samples in two wells from low-pressure N_2 gas adsorption analyses.

Sample No.	SSA (m^2/g)	Pore Volume (mL/g)	Mean Radius (nm)	Mode Radius (nm)
2-5	96.4838	0.295448	9.14	2.89
2-11	65.783	0.257775	10.24	2.97
2-13	163.0092	0.282815	5.32	2.88
2-14	121.4183	0.272646	6.72	3.12
5-1	60.1268	0.215191	8.55	2.9
5-5	96.734	0.397203	11.09	2.95
5-8	105.4049	0.368997	9.82	2.85

4.4. Fractal Analysis

The organic matter pores, at the micro–nano level, can be quantitatively characterized through various analytical tests. However, these tests often struggle to fully highlight the complexity and heterogeneity of organic matter pores. A more precise method for characterizing this complexity is fractal dimension analysis. Figure 6 illustrates the key process of calculating the fractal dimension using the FHH model, based on low-pressure N_2 adsorption curves. To ensure consistency with the pore size distribution of the low-pressure N_2 adsorption curve, we investigate the characteristics and significance of the fractal dimension under varying pore size distribution conditions. This study employs the Kelvin equation to calculate the radius of pores under relative pressure, thereby obtaining the fractal dimension characteristics of <2 nm, 2~50 nm, and >50 nm.

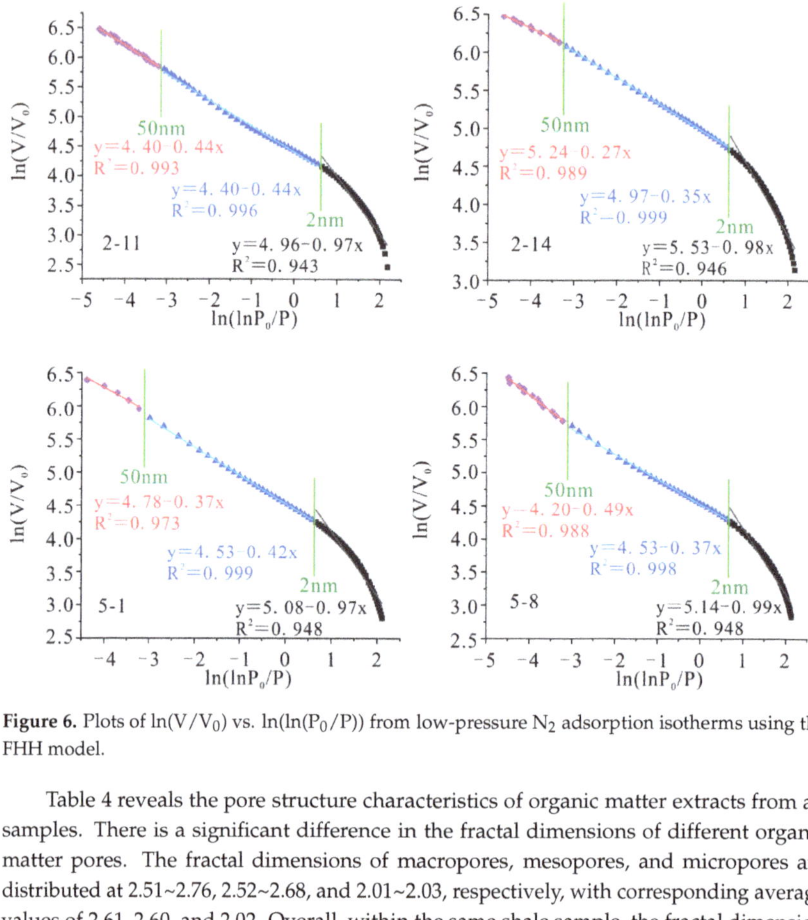

Figure 6. Plots of $\ln(V/V_0)$ vs. $\ln(\ln(P_0/P))$ from low-pressure N_2 adsorption isotherms using the FHH model.

Table 4 reveals the pore structure characteristics of organic matter extracts from all samples. There is a significant difference in the fractal dimensions of different organic matter pores. The fractal dimensions of macropores, mesopores, and micropores are distributed at 2.51~2.76, 2.52~2.68, and 2.01~2.03, respectively, with corresponding average values of 2.61, 2.60, and 2.02. Overall, within the same shale sample, the fractal dimension of organic matter pores is $D_{>50} > D_{2-50} > D_{<2}$, possibly due to the more complex structural features of macropores in organic matter pores, followed by mesopores, while micropores are closer to elliptical. Longitudinally, well W202 also shows an increasing trend in fractal dimensions with depth, possibly due to factors such as TOC content and tectonic activities. However, the change in the fractal dimensions of deep shale in well W205 is not significant, possibly due to sampling distribution issues. Comparing the two wells indicates that the complexity of the pore structure of deep shale organic matter pores is roughly equivalent to that of mid-shallow shale pores.

Table 4. Fractal dimensions obtained from the FHH model of Wufeng–Longmaxi shales from two wells.

Sample No.	$Slope_{>50}$	$R^2_{>50}$	$D_{>50}$	$Slope_{2-50}$	R^2_{2-50}	D_{2-50}	$Slope_{<2}$	$R^2_{<2}$	$D_{<2}$
2-5	−0.39	0.991	2.61	−0.38	0.996	2.52	−0.99	0.949	2.01
2-11	−0.45	0.993	2.55	−0.44	0.996	2.56	−0.97	0.943	2.03
2-13	−0.24	0.962	2.76	−0.32	0.992	2.68	−0.98	0.961	2.02
2-14	−0.27	0.989	2.73	−0.35	0.999	2.65	−0.98	0.946	2.02
5-1	−0.37	0.973	2.63	−0.42	0.999	2.58	−0.97	0.948	2.03
5-5	−0.53	0.992	2.47	−0.41	0.996	2.59	−0.99	0.936	2.01
5-8	−0.49	0.988	2.51	−0.37	0.998	2.63	−0.99	0.948	2.01

5. Discussion

5.1. Effect of Shale Compositions on Organic Matter Pore Structure

The organic matter and mineral compositions of shale significantly influence the pore structure characteristics of organic matter pores [10,22,43,44]. These pores, which primarily form during the kerogen generation and hydrocarbon expulsion process, as well as the secondary cracking of crude oil, constitute the main storage space in marine shale reservoirs [13,18]. They also include a small quantity of primary biological pores [45]. Organic matter is the material basis for the formation of organic matter pores. This study reveals that TOC content has a certain positive correlation with D_{2-50} and $D_{>50}$, and a negative correlation with $D_{<2}$, indicating that the organic matter pores are mainly organic mesopores and macropores in the WLLFs, which is consistent with observations under SEM (Figures 3 and 4), as shown in Figure 7. This may be due to the fusion of generated organic matter micropores into mesopores or macropores. However, there is a good positive correlation between quartz and $D_{<2}$, which should be related to biogenic quartz at the bottom of the Longmaxi Formation. Previous studies have revealed that the biogenic quartz content in the organic-rich shale of the Longmaxi Formation exceeds 50%, with small particle size, and particles filled with organic matter [22,43]. These types of organic matter clumps are relatively small, with limited internal development space, mainly honeycomb-like organic matter pores. The clay mineral content is roughly equivalent to that of quartz, but they are complementary. The figure shows that clay minerals have basically no correlation with $D_{>50}$, D_{2-50}, and $D_{<2}$, possibly because the organic matter–clay mineral complex promotes the development of organic matter pores but is difficult to preserve [19,46].

The shale in the study area contains relatively few other minerals, but all of them, except for feldspar, have varying degrees of impact on the organic matter pore structure (Figure 7). Organic acids emitted during the hydrocarbon generation and expulsion can dissolve carbonate minerals, resulting in relatively large dissolution pores. In the later stage, some of these pores are filled by migrating oil/bitumen, leading to organic matter pore development with secondary cracking [43,44,47]. Therefore, calcite and dolomite show a negative correlation with $D_{<2}$, and even dolomite shows a positive correlation with $D_{>50}$. Most of the pyrite framboid is filled by organic matter, and the inter-crystalline pores of pyrite framboid are generally larger. The internal organic matter pores have a high degree of development and large diameters, showing a positive correlation with both $D_{>50}$ and D_{2-50} (Figures 3 and 7).

5.2. Effect of Diagenesis on Organic Matter Pore Structure

The diagenesis of shale is a highly intricate process, involving physical, chemical, and biological interactions [17,22,43]. The most notable of these are compaction, dissolution, and hydrocarbon generation and explosion from organic matter, with the dissolution having been previously detailed in 5.1. This section aims to investigate the influence of compaction and hydrocarbon generation and explosion on the pore structure of the organic matter itself.

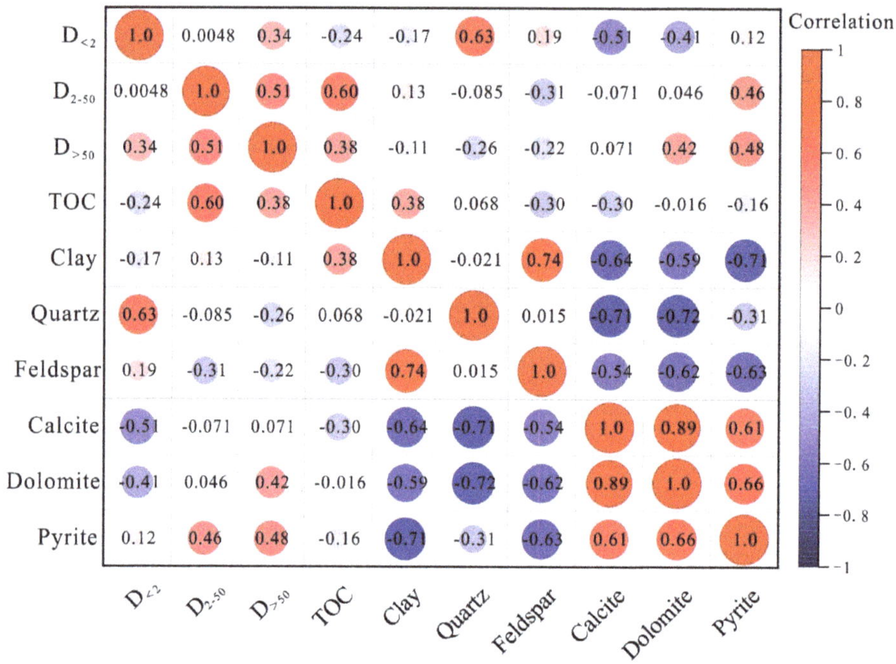

Figure 7. Correlation diagram between shale components and fractal dimension values of different organic matter pores.

Research suggests that organic matter inherently possesses low hardness and high plasticity, facilitating its flattening during the compaction process [23,48]. This results in the formation of elongated pores within the organic matter. However, observations from wells W202 and W205 show that the elongation rates of organic matter pores in both shallow and deep shale layers are concentrated between 0.4 and 0.7, unaffected by depth (Figure 8). On the contrary, the development of organic matter pores is better within some deeply buried shales (Figures 3 and 4). This may be due to a large amount of rigid minerals forming an effective framework that inhibits compaction [10,22,43]; on the other hand, during the hydrocarbon generation and expulsion process within the shale's organic matter, there is a sharp decrease in mass (Figure 9), but a rapid increase in volume (oil/gas density < water density < kerogen density), causing general overpressure within the pores, effectively counteracting some or even all of the compaction effects [23,49].

The hydrocarbon generation from organic matter is the primary process of pore formation in organic matter, except for some internal pores within primitive organisms [45]. Organic matter is a high molecular weight polymer, and changes occur within its molecules during the hydrocarbon generation process, resulting in changes to its own elastic modulus. Through the application of TG-DSC analysis, it was determined that phase transitions could occur during the heating reaction process of organic matter. These transitions could potentially lead to glass transitions, thereby enhancing resilience performance. Coupled with the previously mentioned overpressure, this comprehensive function can effectively protect the pores in organic matter.

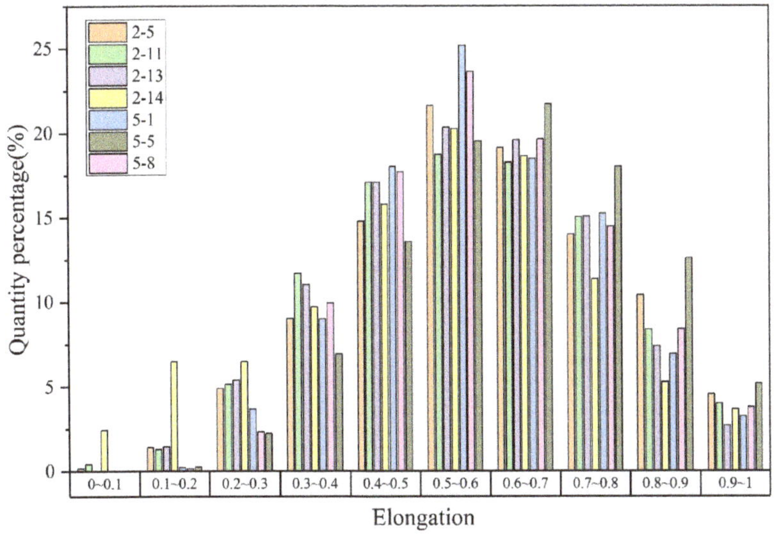

Figure 8. Organic matter pore elongation value distribution from different layers and wells in the study area.

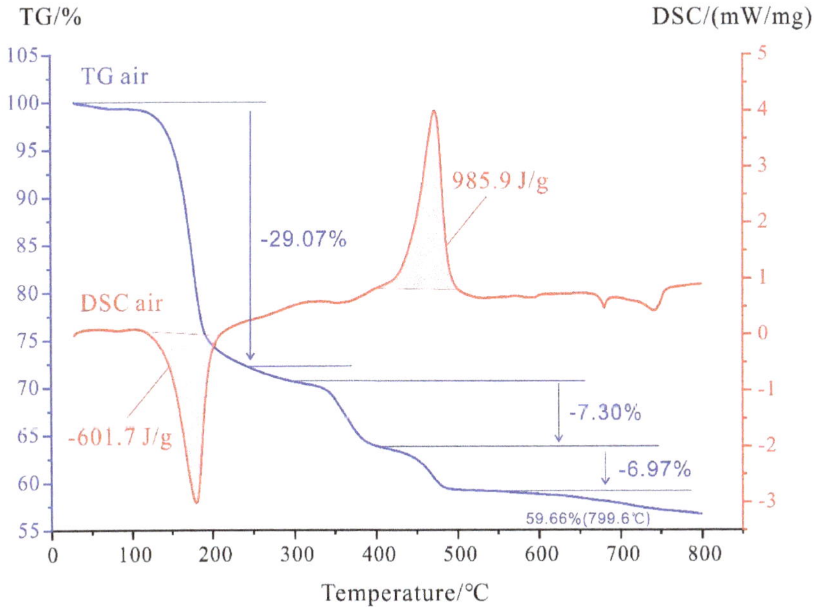

Figure 9. TG and DSC curves for sample 2-5 under air atmosphere. TG—blue solid line and DSC—red solid line.

5.3. Effect of Tectonism on Organic Matter Pore Structure

The Sichuan Basin has experienced numerous tectonic events, with the WLLFs having been subjected to late Caledonian, Hercynian, Indosinian, and Yanshan-Himalayan orogenic movements [36,50]. These have had a certain impact on organic matter pore structure. Core observation of the Wufeng Formation–Longmaxi Formation shale revealed evidence of multiple tectonic activities in both the Wufeng Formation and the bottom of the Longmaxi Formation (Figure 10). The Wufeng Formation shale core exhibited high-angle fractures, some of which were intersected by another set of fractures, all filled with multi-stage vein bodies, suggesting multi-stage tectonic-fluid action. Low-angle and high-angle fractures were both present at the bottom of the Longmaxi Formation shale, likely formed due to slippage at the shale's base, and step phenomena were observed on the core cross-section (Figure 10a,b). However, the upper part of the Longmaxi Formation was less affected by tectonic activity.

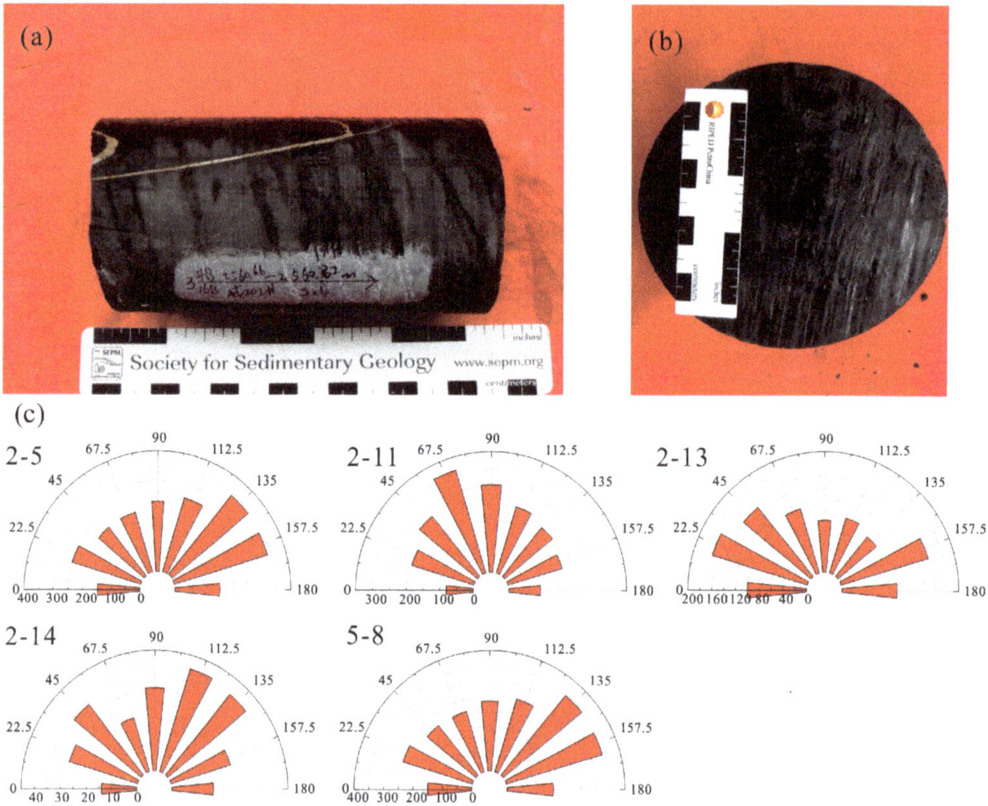

Figure 10. Representative core photographs and rose diagrams. (**a**) High-angle fractures, filled by minerals, from the lower Longmaxi Formation, 2-5. (**b**) Low-angle translational fractures, steps, and mirror-smooth features, with obvious scratches, 2-13. (**c**) Rose diagrams of organic matter pores in length directions; the long axis of organic matter pores is distributed in all directions, but there is a slight dominant direction.

Tectonic compression can cause changes in the morphology of organic matter pores, with a certain directionality in their long axes. Through the analysis of SEM images, it is found that tectonic activity has a weak effect on the organic matter pores (Figure 10c). The organic matter pore size in the lower Longmaxi Formation does not have obvious directionality, and it is larger than that in the upper part (Figure 10, Table 3). However, all organic matter pores have a certain elongation rate, indicating that they are indeed affected by tectonism (Figure 8, Table 2). On the other hand, some pressure is offset by the shale itself (such as pore pressure increasing by the hydrocarbon generation of organic matter, autogenic mineral-forming rigid frameworks, etc.), allowing for an effective preservation of organic matter pores (Figure 3).

5.4. Relationship Between Fractal Characteristics of Organic Matter Pores and Shale Gas Occurrence

Natural gas primarily exists in shale reservoirs in two forms: free or adsorbed. Parameters of pore structure such as the specific surface area and pore volume can effectively reflect the storage capacity of shale reservoirs. Fractal dimension is an important indicator characterizing the non-homogeneity of porous media surfaces and pore structures. A higher fractal dimension indicates a more complex pore structure, which is often not conducive to gas permeation and diffusion but is favorable for gas storage. The larger the fractal dimension value, the rougher the surface of the shale pores, increasing adsorption sites and hindering gas diffusion. As the fractal dimension value increases, the complexity of the internal pore space in shale increases, the connectivity between pores becomes more tortuous, and the permeation capability weakens. Therefore, without other factors, shale gas reservoirs with high fractal dimension values have relatively weak self-permeation capabilities, enhancing the enrichment effect of shale gas.

Compared with the shallow shale reservoirs in the Wufeng Formation–Longmaxi Formation of the study area, the deep shale reservoirs have roughly equivalent parameters, and the distribution range of fractal dimension values of organic matter pores is almost the same. Combining gas content data, it is found that the high-quality shale sections (one and two sublayers) in the Weiyuan area have relatively high fractal dimension D_{2-50} and $D_{>50}$ values of organic matter pores and gas content, indicating that shale with high silica content has a complex three-dimensional "pore-crack" connectivity system (Figures 1 and 11). This is related to the fact that the marine shale in the study area is rich in biogenic silica and has its own overpressure. The high TOC content facilitates the formation of a rich network of organic matter pores, while the overpressure effectively preserves the development of internal pores. In addition, simulation experiments also show that the 3~4 nm pores in the deep shale reservoirs are mainly free gas; thus, a relatively large pore volume contains more free gas, which is beneficial for large-scale mining in the later stage. Moreover, siliceous shale is mainly composed of brittle minerals, which can easily form a complex pore-crack system through fracturing modification in the later stage.

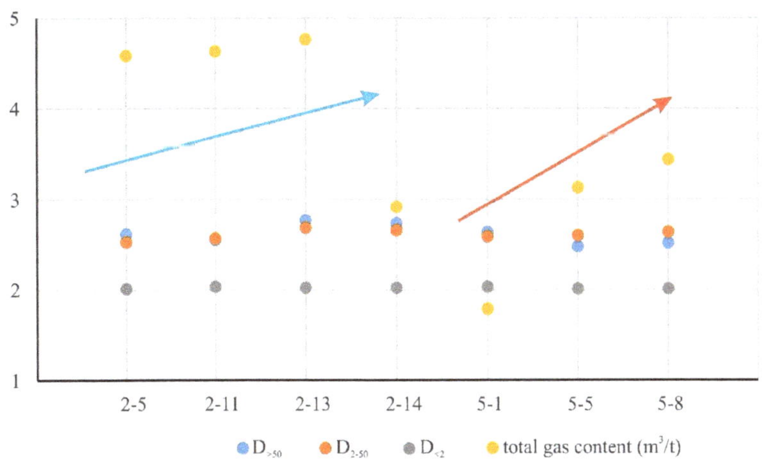

Figure 11. The relationship between the fractal dimension of shale organic matter pores and gas content. 2-14 is special, from the non-producing interval.

6. Conclusions

Based on the above comprehensive research, the following conclusions are drawn:

(1) The organic matter pores of the WLLFs in the Weiyuan Shale Gas Field exhibit strong heterogeneity. Organic matter pores are generally found in organic matter clusters, organic matter–clay composites, and the internal organic matter of pyrite framboid. Morphologically, they are mainly elliptical and spindle-shaped, with significant differences in pore arrangement direction, and a small amount show orientation. The structure of organic matter pores in shale at different depths varies greatly.

(2) The organic matter pore structures in the WLLFs are relatively complex, with fractal dimensions $D_{>50} > D_{2\text{–}50} > D_{<2}$, mainly influenced by mesopores and macropores. The organic matter pores are primarily mesopores and macropores, with the main pore diameter distribution in the range of 10~40 nm and 100~160 nm, unaffected by the depth of shale burial. However, the fractal dimension values of organic matter pores show an increasing trend with burial depth, and the macropore diameter of organic matter pores at the bottom of the Longmaxi Formation shale relatively increases.

(3) The complexity of organic matter pore structure is influenced by various factors such as the composition of authigenic minerals, the physicochemical properties of organic matter, diagenesis, and tectonism. TOC contents, quartz, and clay minerals have a significant impact on the structure of organic matter pores. Compaction and tectonic activities can destroy organic matter pores, mainly reflected in the directional characteristics of organic matter pore development. However, the hydrocarbon generation and explosion of organic matter and its own phase transition, as well as the framework of rigid minerals, play a good protective role in preserving organic matter pores.

(4) The organic-rich siliceous shale deposited during the 1st and 2nd sublayers has a high TOC content, large fractal dimension values of organic matter pores, and complex organic matter pore structures, forming a three-dimensional connected pore system. Therefore, it has both numerous gas adsorption sites to store adsorbed gas and a large pore space to store free gas, resulting in high gas content. This is the "sweet spot" for shale gas enrichment. Therefore, the quality parameters of shale and fractal dimension characteristics can be comprehensively evaluated to identify high-quality shale sections.

Author Contributions: Conceptualization, Q.G., D.D., and B.D.; methodology, Q.G., Y.Y., K.J., H.L., and H.Y.; software, Q.G., C.L., and C.C.; validation, Q.G.; formal analysis, Q.G., C.L., and C.C.; investigation, Q.G., C.L., and C.C.; resources, Q.G.; data curation, Q.G., C.L., and C.C.; writing—original draft preparation, Q.G.; writing—review and editing, Q.G.; visualization, Q.G.; supervision, Q.G. and D.D.; project administration, Q.G. All authors have read and agreed to the published version of the manuscript.

Funding: This research was funded by the National Natural Science Foundation of China, grant number 42202192, and supported by the Sichuan Science and Technology Program, grant number 2024NSFSC0811.

Data Availability Statement: All data can be found in the paper.

Conflicts of Interest: The authors declare no conflicts of interest.

References

1. Zou, C.; Dong, D.; Wang, Y.; Li, X.; Huang, J.; Wang, S.; Guan, Q.; Zhang, C.; Wang, H.; Liu, H.; et al. Shale Gas in China: Characteristics, Challenges and Prospects (II). *Shiyou Kantan Yu Kaifa/Pet. Explor. Dev.* **2016**, *43*, 166–178. [CrossRef]
2. Dong, D.; Liang, F.; Guan, Q.; Jiang, Y.; Zhou, S.; Yu, R.; Gu, Y.; Zhang, S.; Qi, L.; Liu, Y. Development Model and Identification of Evaluation Technology for Wufeng Formation–Longmaxi Formation Quality Shale Gas Reservoirs in the Sichuan Basin. *Nat. Gas Ind. B* **2023**, *10*, 165–182. [CrossRef]
3. WANG jian; GUO Qiulin; ZHAO Chenlei; WANG Yuman; YU Jingdu; LIU Zhuangxiaoxue; CHEN Ningsheng Potentials and Prospects of Shale Oil-Gas Resources in Major Basins of China. *Acta Pet. Sin.* **2023**, *44*, 2033–2044.
4. Zhang, J.; Li, Z.; Wang, D.; Xu, L.; Li, Z.; Niu, J.; Chen, L.; Sun, Y.; Li, Q.; Yang, Z.; et al. Shale Gas Accumulation Patterns in China. *Nat. Gas Ind. B* **2023**, *10*, 14–31. [CrossRef]

5. Loucks, R.G.; Reed, R.M.; Ruppel, S.C.; Hammes, U. Spectrum of Pore Types and Networks in Mudrocks and a Descriptive Classification for Matrix-Related Mudrock Pores. *Am. Assoc. Pet. Geol. Bull.* **2012**, *96*, 1071–1098. [CrossRef]
6. Curtis, J.B. Fractured Shale-Gas Systems. *Am. Assoc. Pet. Geol. Bull.* **2002**, *86*, 1921–1938. [CrossRef]
7. Sun, L.; Fu, D.; Wu, Y.; Wang, Z. The Current Study and Development of Organic Matter Pores in Organic-Rich Shale. *Arab. J. Geosci.* **2022**, *15*, 62. [CrossRef]
8. Bai, B.; Wang, G.; Shang, F.; Zhou, H.; Liu, Y.; Shi, Z. Effects of SEM Image Resolution on Organic Pore Structure Estimation of Shale Reservoirs. *Gas Sci. Eng.* **2024**, *121*, 205170. [CrossRef]
9. Chen, Z.; Song, Y.; Jiang, Z.; Liu, S.; Li, Z.; Shi, D.; Yang, W.; Yang, Y.; Song, J.; Gao, F.; et al. Identification of Organic Matter Components and Organic Pore Characteristics of Marine Shale: A Case Study of Wufeng-Longmaxi Shale in Southern Sichuan Basin, China. *Mar. Pet. Geol.* **2019**, *109*, 56–69. [CrossRef]
10. Guan, Q.; Lü, X.; Dong, D.; Cai, X. Origin and Significance of Organic-Matter Pores in Upper Ordovician Wufeng-Lower Silurian Longmaxi Mudstones, Sichuan Basin. *J. Pet. Sci. Eng.* **2019**, *176*, 554–561. [CrossRef]
11. Katz, B.J.; Arango, I. Organic Porosity: A Geochemist' s View of the Current State of Understanding. *Org. Geochem.* **2018**, *123*, 1–16. [CrossRef]
12. Wang, G. Deformation of Organic Matter and Its Effect on Pores in Mud Rocks. *Am. Assoc. Pet. Geol. Bull.* **2020**, *104*, 21–36. [CrossRef]
13. Teng, J.; Liu, B.; Mastalerz, M.; Schieber, J. Origin of Organic Matter and Organic Pores in the Overmature Ordovician-Silurian Wufeng-Longmaxi Shale of the Sichuan Basin, China. *Int. J. Coal Geol.* **2022**, *253*, 103970. [CrossRef]
14. Liu, B.; Mastalerz, M.; Schieber, J. SEM Petrography of Dispersed Organic Matter in Black Shales: A Review. *Earth-Sci. Rev.* **2022**, *224*, 103874. [CrossRef]
15. Nie, H.; Jin, Z.; Zhang, J. Characteristics of Three Organic Matter Pore Types in the Wufeng-Longmaxi Shale of the Sichuan Basin, Southwest China. *Sci. Rep.* **2018**, *8*, 7014. [CrossRef]
16. Hu, G.; Pang, Q.; Jiao, K.; Hu, C.; Liao, Z. Development of Organic Pores in the Longmaxi Formation Overmature Shales: Combined Effects of Thermal Maturity and Organic Matter Composition. *Mar. Pet. Geol.* **2020**, *116*, 104314. [CrossRef]
17. Pang, Q.; Hu, G.; Hu, C.; Zhang, C. Variation of Organic Pore Structure With Maceral Types in the Longmaxi Shale, Sichuan Basin. *Front. Earth Sci.* **2021**, *9*, 715278. [CrossRef]
18. Borjigin, T.; Lu, L.; Yu, L.; Zhang, W.; Pan, A.; Shen, B.; Wang, Y.; Yang, Y.; Gao, Z. Formation, Preservation and Connectivity Control of Organic Pores in Shale. *Shiyou Kantan Yu Kaifa/Pet. Explor. Dev.* **2021**, *48*, 687–699. [CrossRef]
19. Chang, J.; Fan, X.; Jiang, Z.; Wang, X.; Chen, L.; Li, J.; Zhu, L.; Wan, C.; Chen, Z. Differential Impact of Clay Minerals and Organic Matter on Pore Structure and Its Fractal Characteristics of Marine and Continental Shales in China. *Appl. Clay Sci.* **2022**, *216*, 106334. [CrossRef]
20. Ma, Z.; Tang, X.; Deng, E.; Liu, Y.; Wang, Y.; Zhang, J.; Liang, G.; Zhang, J.; Luo, H. Lithofacies and Its Controls on the Organic Matter-Hosted Pores in the Cambrian Gas-Rich Niutitang Shale in Upper Yangtze Plate, China. *J. Pet. Sci. Eng.* **2022**, *218*, 111052. [CrossRef]
21. Milliken, K.L.; Ergene, S.M.; Ozkan, A. Quartz Types, Authigenic and Detrital, in the Upper Cretaceous Eagle Ford Formation, South Texas, USA. *Sediment. Geol.* **2016**, *339*, 273–288. [CrossRef]
22. Guan, Q.; Dong, D.; Zhang, H.; Sun, S.; Zhang, S.; Guo, W. Types of Biogenic Quartz and Its Coupling Storage Mechanism in Organic-Rich Shales: A Case Study of the Upper Ordovician Wufeng Formation to Lower Silurian Longmaxi Formation in the Sichuan Basin, SW China. *Shiyou Kantan Yu Kaifa/Pet. Explor. Dev.* **2021**, *48*, 700–709. [CrossRef]
23. Zhang, K.; Song, Y.; Jia, C.; Jiang, Z.; Han, F.; Wang, P.; Yuan, X.; Yang, Y.; Zeng, Y.; Li, Y.; et al. Formation Mechanism of the Sealing Capacity of the Roof and Floor Strata of Marine Organic-Rich Shale and Shale Itself, and Its Influence on the Characteristics of Shale Gas and Organic Matter Pore Development. *Mar. Pet. Geol.* **2022**, *140*, 105647. [CrossRef]
24. Yu, Y.; Wang, Z.; Zhang, K.; Feng, X.; Cheng, M. Effects of Shear on Development Characteristics of Organic Matter Pores in Shale: A Case Study of Shale in the Niutitang Formation of the Well XAD1. *J. Pet. Sci. Eng.* **2022**, *211*, 110166. [CrossRef]
25. Guan, Q.; Dong, D.; Zhang, H.; Zhang, S.; Lyu, X.; Wang, Y. Characterizing the Characteristics of Natural Fractures in Shale Based on the Modified Petrophysical Model. *Nat. Gas Ind. B* **2021**, *8*, 325–333. [CrossRef]
26. Zhu, B.; Meng, J.; Pan, R.; Song, C.; Zhu, Z.; Jin, J.; Zhang, C.; Liu, L.; Xu, L.; Zhou, X. Multi-Scale Characterization of Organic Matter Pore Space in Deep Marine Shale Combined with Mathematical Morphology. *Geoenergy Sci. Eng.* **2023**, *223*, 211539. [CrossRef]
27. Zhang, W.; Hu, W.; Borjigin, T.; Zhu, F. Pore Characteristics of Different Organic Matter in Black Shale: A Case Study of the Wufeng-Longmaxi Formation in the Southeast Sichuan Basin, China. *Mar. Pet. Geol.* **2020**, *111*, 33–43. [CrossRef]
28. Pommer, M.; Milliken, K. Pore Types and Pore-Size Distributions across Thermal Maturity, Eagle Ford Formation, Southern Texas. *Am. Assoc. Pet. Geol. Bull.* **2015**, *99*, 1713–1744. [CrossRef]
29. Wu, J.; Luo, C.; Zhong, K.; Li, Y.; Li, G.; Du, Z.; Yang, J. Innovative Characterization of Organic Nanopores in Marine Shale by the Integration of HIM and SEM. *Energy* **2023**, *282*, 128390. [CrossRef]

30. Guan, Q.; Dong, D.; Sun, S.; Hu, L.; Qi, L.; Li, C.; Chen, C.; Chen, X. Fractal Characteristics of Organic-Rich Shale Pore Structure and Its Geological Implications: A Case Study of the Lower Silurian Longmaxi Formation in the Weiyuan Block, Sichuan Basin. *Nat. Gas Ind.* **2024**, *44*, 108–118. [CrossRef]
31. Peng, N.; He, S.; Hu, Q.; Zhang, B.; He, X.; Zhai, G.; He, C.; Yang, R. Organic Nanopore Structure and Fractal Characteristics of Wufeng and Lower Member of Longmaxi Shales in Southeastern Sichuan, China. *Mar. Pet. Geol.* **2019**, *103*, 456–472. [CrossRef]
32. Li, X.; Wang, P.; Wang, S.; Jin, C.; Zhang, C.; Zhao, R.; Jiang, Z. Influence of the Duration of Tectonic Evolution on Organic Matter Pore Structure and Gas Enrichment in Marine Shale: A Case Study of the Lower Silurian Longmaxi Shale in Southern China. *Front. Earth Sci.* **2022**, *9*, 787697. [CrossRef]
33. Ma, Y.; Cai, X.; Zhao, P. China's Shale Gas Exploration and Development: Understanding and Practice. *Shiyou Kantan Yu Kaifa/Pet. Explor. Dev.* **2018**, *45*, 561–574. [CrossRef]
34. Wang, S.; Dong, D.; Wang, Y.; Li, X.; Huang, J.; Guan, Q. Sedimentary Geochemical Proxies for Paleoenvironment Interpretation of Organic-Rich Shale: A Case Study of the Lower Silurian Longmaxi Formation, Southern Sichuan Basin, China. *J. Nat. Gas Sci. Eng.* **2016**, *28*, 691–699. [CrossRef]
35. Dong, D.; Gao, S.; Huang, J.; Guan, Q.; Wang, S.; Wang, Y. Discussion on the Exploration & Development Prospect of Shale Gas in the Sichuan Basin. *Nat. Gas Ind. B* **2015**, *2*, 9–23. [CrossRef]
36. Wei, Z.; Chu, R.; Xie, J.; Bao, F.; Zeng, S.; Sheng, M.; Zeng, Q. Crustal Structure in the Weiyuan Shale Gas Field, China, and Its Tectonic Implications. *Tectonophysics* **2022**, *837*, 229449. [CrossRef]
37. Guo, X.; Hu, D.; Li, Y.; Wei, Z.; Wei, X.; Liu, Z. Geological Factors Controlling Shale Gas Enrichment and High Production in Fuling Shale Gas Field. *Pet. Explor. Dev.* **2017**, *44*, 513–523. [CrossRef]
38. Feng, Z.; Dong, D.; Tian, J.; Qiu, Z.; Wu, W.; Zhang, C. Geochemical Characteristics of Longmaxi Formation Shale Gas in the Weiyuan Area, Sichuan Basin, China. *J. Pet. Sci. Eng.* **2018**, *167*, 538–548. [CrossRef]
39. Guan, Q.; Dong, D.; Wang, S.; Huang, J.; Wang, Y.; Lu, H.; Zhang, C. Preliminary Study on Shale Gas Microreservoir Characteristics of the Lower Silurian Longmaxi Formation in the Southern Sichuan Basin, China. *J. Nat. Gas Sci. Eng.* **2016**, *31*, 382–395. [CrossRef]
40. Wu, J.; Yuan, Y.; Niu, S.; Wei, X.; Yang, J. Multiscale Characterization of Pore Structure and Connectivity of Wufeng-Longmaxi Shale in Sichuan Basin, China. *Mar. Pet. Geol.* **2020**, *120*, 104514. [CrossRef]
41. Mandelbrot, B. How Long Is the Coast of Britain? Statistical Self-Similarity and Fractional Dimension. *Science* **1967**, *156*, 636–638. [CrossRef] [PubMed]
42. Sing, K.S.W. Reporting Physisorption Data for Gas/Solid Systems with Special Reference to the Determination of Surface Area and Porosity (Recommendations 1984). *Pure Appl. Chem.* **1985**, *57*, 603–619. [CrossRef]
43. Milliken, K.L.; Olson, T. Silica Diagenesis, Porosity Evolution, and Mechanical Behavior In Siliceous Mudstones, Mowry Shale (Cretaceous), Rocky Mountains, USA. *J. Sediment. Res.* **2017**, *87*, 366–387. [CrossRef]
44. Ko, L.T.; Ruppel, S.C.; Loucks, R.G.; Hackley, P.C.; Zhang, T.; Shao, D. Pore-Types and Pore-Network Evolution in Upper Devonian-Lower Mississippian Woodford and Mississippian Barnett Mudstones: Insights from Laboratory Thermal Maturation and Organic Petrology. *Int. J. Coal Geol.* **2018**, *190*, 3–28. [CrossRef]
45. WU, J.; Zhou, W.; Sun, S.; Zhou, S.; Shi, Z. Graptolite-Derived Organic Matter and Pore Characteristics in the Wufeng-Longmaxi Black Shale of the Sichuan Basin and Its Periphery. *Acta Geol. Sin.* **2019**, *93*, 982–995. [CrossRef]
46. İnan, S.; Al Badairy, H.; İnan, T.; Al Zahrani, A. Formation and Occurrence of Organic Matter-Hosted Porosity in Shales. *Int. J. Coal Geol.* **2018**, *199*, 39–51. [CrossRef]
47. Ko, L.T.; Loucks, R.G.; Zhang, T.; Ruppel, S.C.; Shao, D. Pore and Pore Network Evolution of Upper Cretaceous Boquillas (Eagle Ford-Equivalent) Mudrocks: Results from Gold Tube Pyrolysis Experiments. *Am. Assoc. Pet. Geol. Bull.* **2016**, *100*, 1693–1722. [CrossRef]
48. Eliyahu, M.; Emmanuel, S.; Day-Stirrat, R.; Macaulay, C.I. Mechanical Properties of Organic Matter in Shales Mapped at the Nanometer Scale. *Mar. Pet. Geol.* **2014**, *59*, 294–304. [CrossRef]
49. Guo, T. Key Geological Issues and Main Controls on Accumulation and Enrichment of Chinese Shale Gas. *Pet. Explor. Dev.* **2016**, *43*, 349–359. [CrossRef]
50. Zhu, H.; Ju, Y.; Qi, Y.; Huang, C.; Zhang, L. Impact of Tectonism on Pore Type and Pore Structure Evolution in Organic-Rich Shale: Implications for Gas Storage and Migration Pathways in Naturally Deformed Rocks. *Fuel* **2018**, *228*, 272–289. [CrossRef]

fractal and fractional

MDPI

Article

Heterogeneity of Deep Tight Sandstone Reservoirs Using Fractal and Multifractal Analysis Based on Well Logs and Its Correlation with Gas Production

Peiqiang Zhao [1,2,*], Qiran Lv [3], Yi Xin [4] and Ning Wu [5]

1 State Key Laboratory of Petroleum Resources and Engineering, China University of Petroleum (Beijing), Beijing 102249, China
2 Beijing Key Laboratory of Earth Prospecting and Information Technology, China University of Petroleum (Beijing), Beijing 102249, China
3 Daqing Branch, China Petroleum Group Logging Co., Ltd., Daqing 163000, China; 15663745381@wo.cn
4 R&D Center for Ultra-Deep Complex Reservoir Exploration and Development, China National Petroleum Corporation, Korla 841000, China; xinyi-tlm@petrochina.com.cn
5 Sixth Oil Production Pant, PetroChina Changqing Oilfield Company, Xi'an 710021, China; wuning_cq@petrochina.com.cn
* Correspondence: pqzhao@cup.edu.cn

Abstract: Deep tight sandstone reservoirs are characterized by low porosity and permeability, complex pore structure, and strong heterogeneity. Conducting research on the heterogeneity characteristics of reservoirs could lay a foundation for evaluating their effectiveness and accurately identifying advantageous reservoirs, which is of great significance for searching for "sweet spot" oil and gas reservoirs in tight reservoirs. In this study, the deep tight sandstone reservoir in the Dibei area, northern Kuqa depression, Tarim Basin, China, is taken as the research object. Firstly, statistical methods are used to calculate the coefficient of variation (CV) and coefficient of heterogeneity (TK) of core permeability, and the heterogeneity within the reservoir is evaluated by analyzing the variations in the reservoir permeability. Then, based on fractal theory, the fractal and multifractal parameters of the GR and acoustic logs are calculated using the box dimension, correlation dimension, and the wavelet leader methods. The results show that the heterogeneity revealed by the box dimension, correlation dimension, and multifractal singular spectrum calculated based on well logs is consistent and in good agreement with the parameters calculated based on core permeability. The heterogeneity of gas layers is comparatively weaker, while that of dry layers is stronger. In addition, the fractal parameters of GR and the acoustic logs of three wells with the oil test in the study area were analyzed, and the relationship between reservoir heterogeneity and production was determined: As reservoir heterogeneity decreases, production increases. Therefore, reservoir heterogeneity can be used as an indicator of production; specifically, reservoirs with weak heterogeneity have high production.

Keywords: fractals and multifractals; heterogeneity; deep tight sandstone; Tarim Basin

1. Introduction

In recent years, an increasing number of researchers have begun to focus on deep and ultra-deep oil and gas resources [1–3]. In China's oil exploration, formations in the eastern plains and coastal areas with depths between 3500 and 4500 m are classified as "deep formations", while those exceeding 4500 m are referred to as "ultra-deep formations". In western China, formations with depths between 4500 and 6000 m are defined as "deep formations", and those exceeding 6000 m are called "ultra-deep formations" [4,5]. Ultra-deep oil and gas exploration in China is primarily concentrated in the Tarim, Junggar, and Sichuan Basins. In the Kuqa Sag of the Tarim Basin, well Bozi-9 has achieved high production rates of 41.82×10^4 m^3 of gas per day and 115.15 m^3 of oil per day at a depth of 7700 m in the Cretaceous Bashkirchik Formation sandstone. The Triassic Karamay Formation of well Zheng-10 in the central Junggar Basin has been tested for 7530 m^3 of gas per day and 78.18 m^3 of oil per day at a depth of 6800 m. In those regions, deep and ultra-deep resources show great potential and play crucial roles in supplementing conventional oil and gas reserves [6,7].

Deep tight sandstone reservoirs typically exhibit complex pore structures, poor connectivity, and strong heterogeneity [8]. These reservoirs have undergone complex sedimentation, diagenesis, and post-depositional tectonic modifications during long-term geological evolution, resulting in heterogeneous spatial distributions and internal property variations [9]. The ambiguous characterization of heterogeneity in deep tight sandstone reservoirs hinders successful natural gas exploration and development.

Studies on reservoir heterogeneity originate from geological core analysis, where parameters such as variation coefficients and range values are used to assess heterogeneity of permeability. Other studies have employed statistical methods such as variance analysis, standard deviation calculations, Lorenz coefficients, and Dikstra–Persson coefficients (VDP) for core data or well logs to quantify heterogeneity [10,11]. Lake and Jensen (1991) compared dynamic and static calculation methods to analyze the differences between Lorenz coefficients and VDP coefficients, describing their correlation with heterogeneity [12]. Fitch et al. (2013) evaluated the Lorenz coefficient as a metric for quantifying the heterogeneity of core data [13]. They analyzed how population variability impacts heterogeneity measurements and applied this approach to wireline-derived porosity and permeability data to compare reservoir unit heterogeneity. Hosseinzadeh and Tavakoli (2024) employed the Lorenz coefficient to calculate heterogeneity and documented primary sedimentary and secondary diagenetic alterations through thin section analysis [14]. Farooq et al. (2019) investigated the definition and quantification of the term heterogeneity through a suite of statistical measures, including Lorenz and VDP coefficients in Tal Formation carbonate rocks [15]. These metrics enable the interpretation of variations in well log data, allowing for a comparison of heterogeneity within single and multi-layered reservoir units. Tavoosi et al. (2021) employed multiple parameters, including CV, VDP, and the Lorenz coefficient, for a quantitative assessment of different scales of heterogeneity of carbonate sequences [16]. Davies et al. (2023) proposed a new methodology for quantifying heterogeneity on any continuous data and quantitatively analyzed the relationships between heterogeneity and the properties of carbonate formations in the North Sea [17]. The methods mentioned above have been successfully used in different study areas.

Furthermore, fractal dimensions have emerged as critical features for characterizing heterogeneity. Mostly, fractal geometry analysis based on core data in a laboratory is a promising method to interpret the pore structures in geoscience. The pore space of rocks has been investigated based on its fractal geometry, a method proposed by Mandelbrot [18]. According to fractal theory, the pore volumes of rocks have self-similar features [19]. The fractal dimension (D) is a vital parameter to characterize the regularity of self-similar structures and to indicate the complexity and heterogeneity of pore geometry [20–22]. Billi and Storti (2004) reported on the fractal dimensions of grain size distribution in carbonate clastic rocks that originated from the core of the regional strike–slip fault zone in the foreland of the Southern Apennine Mountains in Italy, reporting values ranging from 2.09 to 2.93 [23]. Based on SEM images, Xie et al. (2010) studied the fractal characteristics of Jurassic marine carbonate reservoir samples in the western Hubei and eastern Sichuan regions of China, determining the fractal dimensions of high gas-bearing reservoirs [24]. Xia et al. (2025) revealed the influence of pore–throat fractal characteristics on macroscopic reservoir permeability by analyzing MICP-derived capillary pressure curves [25]. Liu et al. (2004) used homogenization theory to establish a two-dimensional numerical model of rock and introduced the Weibull statistical method to simplify the modeling of heterogeneity [26]. Liu et al. (2016) made advances to a fractal capillary pressure model by introducing a "matchstick-like" tubular pore model specifically for evaluating the coal rock structures [27]. Zhou et al. (2023) systematically demonstrated the correlation between experimental fractal features and reservoir quality and established a fractal parameter-based methodology for reservoir quality prediction [28]. Liu et al. (2023) and Peng et al. (2020) achieved quantitative evaluations of reservoir heterogeneity intensity through fractal dimension extraction and inflection point analyses of the pore distribution in capillary pressure curves [29,30].

Studies have also indicated that calculating the fractal dimension of well logs provides an effective method for determining reservoir heterogeneity. In a single fractal analysis of well logs, box counting and correlation dimensions are the most commonly used. Liu et al. applied these methods to acoustic logs for fracture identification [31], and Wen integrated multiple fractal computation approaches (including acoustic log analysis) and established correlations between fractal dimensions and heterogeneity through the inversion of logs [32]. Mou studied the fractal characteristics of well logs in the Liaohe oil field using the box counting method, revealed the corresponding relationship between volcanic rock lithology and fractal dimension, and verified the feasibility of predicting volcanic rock lithology based on the fractal dimension of well logs [33].

Multifractal theory enhances heterogeneity characterization by introducing concepts such as generalized fractal dimensions and singularity spectra, which quantify both local and global structural variations. The box counting method is commonly used to calculate the multifractal parameters based on physical and log curves. Zhao et al. (2017) and Yan et al. (2017) investigated pore structures through a multifractal analysis of NMR data and by controlling factors of heterogeneity in pore structures [34,35]. Zhao et al. (2019) analyzed NMR and low-temperature N_2 adsorption experimental data and calculated the multifractal dimensions to evaluate the heterogeneity of pore structure [36]. Lv et al. (2025) also used the box dimension method to analyze the array sound wave pattern and achieved fracture identification [37]. However, the box counting method is not typically suitable for analyzing one-dimensional log curves, while wavelet-based fractal approaches, including wavelet and detrended fluctuation analysis (DFA), are employed more often for computing multifractal characteristics of one-dimensional log curves.

Lozada-Zumaeta et al. (2012) used the R/S (rescaled range) fractal model and interpolation method to estimate the spatial distribution of rock physics properties, such as effective porosity, permeability, and shale volume [38]. Partovi and Sadeghnejad (2017) proposed an automatic well-to-well comparison method based on fractal parameters [39]. Wavelet transform has been used to analyze the gamma, density, and acoustic logs, and automatic boundary detection has been achieved on an Iranian oil field. López et al. (2007) implemented a wavelet-driven fractal analysis combined with waveform classifiers to determine the fractal parameters across diverse log types [40]. Subhakar et al. (2016) conducted comparative fractal/multifractal studies using detrended fluctuation analysis (DFA) and wavelet transforms, employing multifractal logging features to identify geological boundaries while benchmarking the results against those of wavelet methods [41]. Amoura et al. (2022) applied the wavelet-derived fractal dimensions to velocity logs, interpreting lithology and heterogeneity through multifractal spectrum morphology and positional characteristics [42]. Rahul Prajapati (2024) used the Hurst index based on the wavelet leader method in the multifractal to quantify the variation in rock properties in shale exploration intervals [43]. Those studies indicate that the wavelet leader method could be used to characterize heterogeneity.

In the above studies, many scholars have used fractal parameters to characterize the heterogeneity of reservoirs, but the application of heterogeneity in oil and gas exploration and development is relatively rare. In this study, to comprehensively evaluate the heterogeneity of deep tight sandstone reservoirs, different methods, including statistical methods based on core permeability and fractal and multifractal analyses on well logs, were used. This study begins with a concise overview of the gas-bearing formations in the target Dibei region in the Tarim Basin, China, and then provides a detailed exposition of two methods: (1) heterogeneity characterization through a core analysis and (2) computational methodologies employing fractal dimension quantification. The calculated fractal dimensions based on well logs are integrated with core analytical data to systematically analyze reservoir heterogeneity in the target region. Furthermore, the qualitative correlation between fractal parameters and production was newly analyzed and found.

2. Geological Setting

This study took place in the Kuqa Formation located at the northern margin of the Tarim Basin, West China. The Kuqa Depression, a foreland basin developed on the Paleozoic folded basement, exhibits an overall north-high–south-low topography influenced by southward thrusting of the South Tianshan orogenic belt. The Kuqa area forms a transitional region between the South Tianshan structural zone and the Tarim Plate (Figure 1). The gas layers primarily studied here are the Jurassic Ahe Formation tight gas layers, with almost all formations extending to depths exceeding 4500 m. The overlying Jurassic Yangxia Formation consists of dark mudstone and a small amount of coal-bearing strata, which can be stably distributed and can form a good cap rock. The underlying Triassic Taliqike Formation also included dark mudstone and exists as the bottom plate of the gas reservoir.

The Yiqikelike structural belt of the Kuqa Depression exhibits favorable petroleum geological conditions. The Dibei gas reservoir demonstrates extensive contiguous distribution, where strong reservoir heterogeneity facilitates hydrocarbon accumulation in localized reservoirs independent of structural traps. Both the underlying Tariqike Formation and overlying Yangxia Formation contain thick, widespread coal seams with high organic matter abundance, serving dual roles as source rocks and effective seals. The tight sandstone reservoirs have undergone prolonged diagenetic evolution and tectonic modifications, exhibiting intricate pore structures and pronounced heterogeneity due to polygenetic controls (sedimentary, diagenetic, and tectonic) [44]. Stratigraphically, it comprises three units from top to bottom: sandy conglomerates interbedded with mudstone (J_1a^1), upper sandy conglomerates (J_1a^2), and lower sandy conglomerates (J_1a^3). Depositionally, the Ahe Formation represents braided river delta plain subfacies with low compositional maturity, dominated by (pebbly) coarse-to-medium sandstones. The thickness of the sand ranges from 250 m to 300 m. Vertically, it manifests as the superposition of multiple river channel sediments, with the thickness of a single river channel being about 1–2 m, while the cumulative thickness of multiple river channels can reach about 20 m. Laterally, frequent channel migrations result in continuous planar distribution [45,46].

Spatially, the source rocks and the Jurassic Ahe reservoirs form a "sandwich-style" interlayer (Figure 2), constituting an optimal source–reservoir–cap rock assemblage. This tightly arranged oil and gas injection system improves the efficiency of oil and gas drainage, which is beneficial for oil and gas trapping in tight sandstone reservoirs. This is a key factor in the formation of natural gas reservoirs in the northern part of Kuqa [47,48].

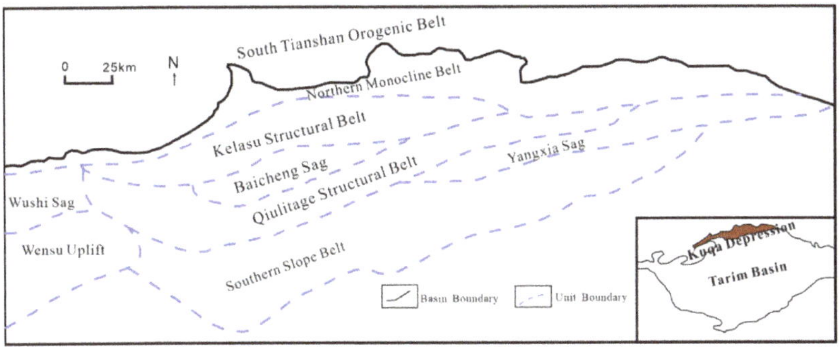

Figure 1. Structural position of the Dibei gas reservoir (modified from Li et al., 2024 [49]).

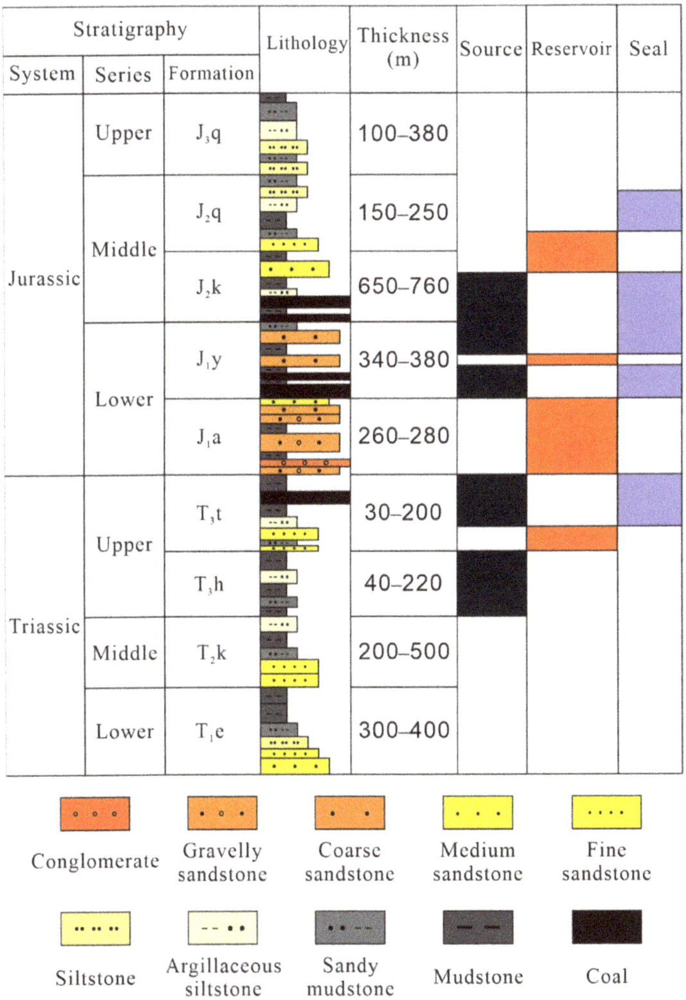

Figure 2. Stratigraphic column of hydrocarbon storage and cover combination in the northern Kuqa structural belt (modified from Li et al., 2024 [49]).

The distribution of core porosity and permeability is shown in Figure 3. Porosity ranges from 4% to 8%, with an average value of 5.94%. The permeability is mainly distributed from 0.1 mD to 5 mD, with an average of 0.82 mD, constituting a tight reservoir.

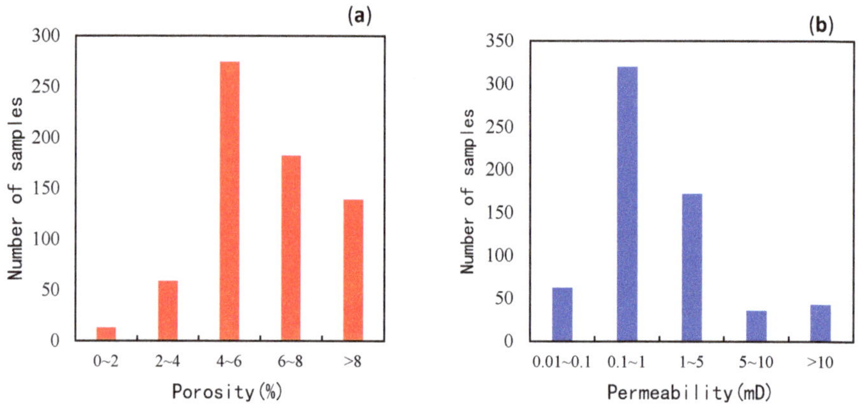

Figure 3. Distribution of porosity and permeability in the Dibei area. (**a**) Porosity distribution histogram; (**b**) Permeability distribution histogram.

Four wells were used in this study, as their wellbore conditions were relatively good and the quality of logging data was high. In addition, two of the wells have rich core data. The information on the four well logs is shown in Table 1.

Table 1. The base data for four wells.

Wells	YN4	YN2C	DB101	DB102
Depth intervals (m)	285	32	297	284
Average core permeability (mD)	0.704	0.44	/	0.04
Oil test (m^3/day)	/	67,320 (gas)	5851 (gas)	16,328 (gas)

3. Calculation of Heterogeneity

3.1. Heterogeneity Based on Core Permeability

The intralayer heterogeneity of reservoir layers is affected by particle size and the mud interlayer, which further affects the physical properties of the reservoir. Core permeability characterizes heterogeneity by calculating the coefficient of heterogeneity (*TK*) and coefficient of variation (*CV*), which collectively delineate the development characteristics of reservoir heterogeneity [50]. The coefficient of variation is defined as the ratio of standard deviation (*SD*) to mean permeability (*MN*), while the coefficient of heterogeneity represents the ratio of maximum permeability (*MAX*) to mean permeability (*MN*). The mathematical expressions are as follows:

$$CV = \left(\frac{SD}{MN}\right) \tag{1}$$

$$TK = \left(\frac{MAX}{MN}\right) \tag{2}$$

where *CV* is the coefficient of variation, *SD* is the standard deviation, *MN* is the average value, *TK* is the coefficient of heterogeneity, and *MAX* is the maximum value of the core permeability.

The calculation results were further compared with fractal dimension calculations.

3.2. Methods for Calculating Fractal Dimensions

3.2.1. Box Dimension Method

Mandelbrot established fractal theory and applied nonlinear concepts such as self-similarity and fractal dimension to characterize these complex phenomena [18]. The fractal dimension method effectively quantifies the complexity and self-similarity of signals. When applied to well log curves, the magnitude of the fractal dimension can indicate the intensity of heterogeneity. In practical calculations, the fractal dimension is typically defined as the limit value when the measurement scale δ approaches zero. To formalize this, suppose a planar curve F is enclosed within a bounded interval. By progressively reducing the unit measurement length δ and recording the corresponding coverage count $M_\delta(F)$, the dimension of F is determined by the power-law relationship of $M_\delta(F)$, as follows:

$$M_\delta(F) \sim C\delta^{-S} \tag{3}$$

In general, taking the logarithm of both sides of Equation (3) and finding the difference, the box dimension is taken as the value when δ tends toward zero. The formula is as follows:

$$dim_B S = \lim_{\delta \to 0} \frac{\log M_\delta(F)}{-\log \delta} \tag{4}$$

In the calculation of box dimension, curve F is placed within squares with side lengths δ, and then the number of boxes $N_\delta(F)$ with different side lengths δ that intersect at F is counted. As the value of δ decreases gradually, the value of $N_\delta(F)$ increases gradually. The logarithm of $N_\delta(F)$ with respect to F is the box dimension S. In practical applications, the box dimension is typically determined by the slope of the graph formed by the intersection of the functions $\log N_\delta$ (F) and $-\log\delta$.

Figure 4 shows an example using the GR log data from well YN4 at 4456–4467 m to calculate the box dimension. The number of boxes is calculated for each scale, and the logarithms of the scales and box counts are plotted on the x- and y-axes, respectively, to create a contour plot. The slope of the fitted line in the figure represents the box dimension of the GR log.

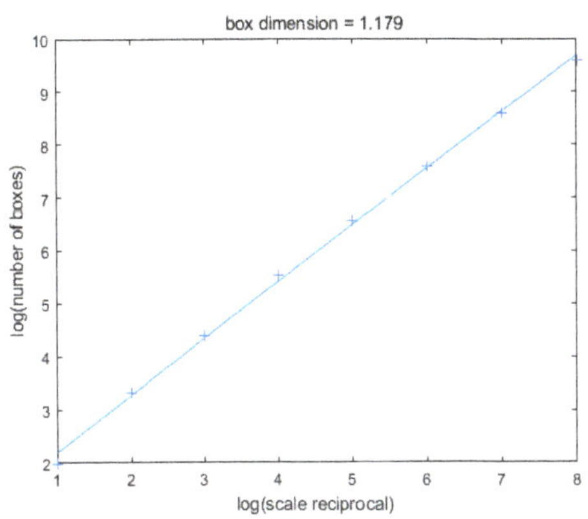

Figure 4. The cross-plot of box dimension calculations.

3.2.2. Correlation Dimension Method

The method of calculating the correlation dimension involves the use of a computational approach: delay embedding phase space reconstruction. This study specifically employs the G-P algorithm proposed by Grassberger and Procaccia [51,52]. The correlation dimension also characterizes nonlinear systems in the phase space and describes the self-similarity of signals.

Firstly, the time delay and embedding dimension of the data need to be determined. The delay time τ determines the interval between vector points, while the embedding dimension m determines the length of the vectors. The autocorrelation method is used to calculate the autocorrelation function $C(\tau)$ of the time series. When $C(\tau)$ is less than $\frac{C(0)}{e}$ (e is the natural logarithm), the corresponding τ is considered the optimal delay time. The formula is as follows:

$$C(\tau) = \frac{1}{N}\sum_{i=1}^{n-\tau} x(i)\cdot x(i+\tau) \tag{5}$$

where n represents the number of data points and τ denotes the delay time.

The embedding dimension is determined using the false nearest neighbor (FNN) method. The data are reconstructed, and the results of the vector space reconstruction are as follows:

$$y_t = \left\{x_t, x_{t+\tau}, \cdots, x_{t+(m-1)\tau}\right\} \tag{6}$$

For each point y_t, its nearest neighbor point y_k is found in the m-dimensional vector space. The formula for the Euclidean distance ratio between the m-dimensional and $(m + 1)$ dimensional vector spaces is as follows:

$$R_d = \frac{\sqrt{\sum_i^{m+1}\left(y_t^{(i)} - y_k^{(i)}\right)^2}}{\sqrt{\sum_i^{m}\left(y_t^{(i)} - y_k^{(i)}\right)^2}} \tag{7}$$

where R_d represents the distance variation ratio and m denotes the embedding dimension. If R_d exceeds the set threshold (usually set to 10), the point will be flagged as a false nearest neighbor.

The optimal embedding dimension m is determined by gradually increasing the dimension until the number of false nearest neighbors is minimized. Using the delay time and embedding dimension calculated with the above method, the data are transformed into an m-dimensional vector space. The constructed vector space is as follows:

$$y_i = \left\{x_t, x_{t+\tau}, \cdots, x_{t+(m-1)\tau}\right\}, (i = 1, 2, \cdots, N - (m-1)\tau) \tag{8}$$

The Euclidean distance between any two vectors in the aforementioned vector space is then calculated. Assuming that any two vectors in the embedded space are denoted as y_i and y_j (i, j = 1, 2, ..., K), the formula for calculating their Euclidean distance R_{ij} is as follows:

$$R_{ij} = |y_i - y_j| = \sqrt{\sum_{k=1}^{N}\left(x_{k+(i-1)\tau} - x_{k+(j-1)\tau}\right)^2} \tag{9}$$

For an arbitrarily given scale ξ, let $N(\xi)$ represent the number of distances satisfying $R_{ij} < \xi$ (where $i \neq j$), and let the number of vectors be $K = N - (m - 1)\tau$. The total number of distances is $K (K - 1)$. The correlation integral is defined as the ratio of the number of distances meeting the condition $R_{ij} < \xi$ to the total number of distances.

$$C(\xi) = \frac{N(\xi)}{K(K-1)} \tag{10}$$

The method of calculating $C(\xi)$ is as follows:

$$C(\xi) = \frac{1}{K(K-1)}\sum_{i=1}^{K}\sum_{j=1}^{K}\theta(\xi - R_{ij})(i \neq j) \tag{11}$$

$$\theta(\xi - R_{ij}) = \begin{cases} 0, \text{when } \xi - R_{ij} < 0 \\ 1, \text{when } \xi - R_{ij} \geq 0 \end{cases} \quad (12)$$

The formula for calculating the correlation dimension is as follows:

$$D = \lim_{\xi \to 0} \frac{\log C(\xi)}{\log \xi} \quad (13)$$

Similarly to the box dimension, the slope is generally used to replace the correlation dimension in practical applications. The fractal dimension expresses the self-similarity of data complexity, which can be well used to judge the changes of well logging curves caused by the non-homogeneity of a pore structure.

Figure 5 uses the acoustic log data from well YN4 at 4467–4472 m as an example, creating a plot with the correlation integral and neighborhood radius as the x- and y-axes, respectively. In the plot, the boundaries of the neighborhood radius remain within an approximately linear region. The slope of the fitted line in the plot represents the correlation dimension of the acoustic log.

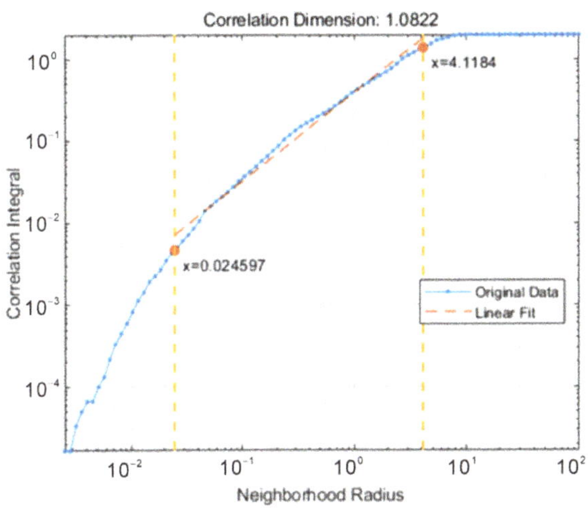

Figure 5. The cross-plot of correlation dimension calculation.

3.2.3. Multifractal Calculation Based on Wave Leader Method

A multifractal analysis based on the box counting of NMR log data could characterize the properties at each depth. In this study, to describe the heterogeneity of a section of the reservoir, the wave leader method based on one-dimensional logs is utilized. The acoustic and GR logs in the one-dimensional logs are selected as the objects of calculation, with the fractal dimensions of these two logs representing intralayer heterogeneity and microscopic heterogeneity, respectively.

Multifractal computation under the wavelet transform framework primarily involves a multiscale analysis of signals to reveal their complex, non-stationary statistical properties. For a signal $X(t_0)$ that is locally bounded, if there exist a constant $C > 0$ and a polynomial P satisfying $\deg(P) < \alpha$, then the Hölder exponent at point t_0 is defined as follows [53]:

$$h(t_0) = sup\{a : X \in C^a(t_0)\} \quad (14)$$

However, directly calculating multifractal properties from the definition of the Hölder exponent is challenging. To address this challenge, the wavelet leader multifractal method employs the wavelet transform as its foundation. This approach shifts the analysis to signal increments and replaces wavelet structural coefficients with local maxima of wavelet transforms. The algorithm offers distinct advantages, including simplified computational procedures and enhanced robustness against noise [54].

Let $\Psi_0(t)$ be a mother wavelet function with compact support. For all $k = 0, 1, \ldots, N_\Psi - 1$ satisfies the conditions $\int t^k \Psi_0(t)dt \equiv 0$ and $\int t^{N_\Psi} \Psi_0(t)dt \neq 0$, with a vanishing moment $N_\Psi \geq 1$. The function set $\{\Psi_{j,k}(t)\}$, generated through the translation and dilation of $\{\Psi_{j,k}(t) = 2^{-\frac{j}{2}} \Psi_0(2^{-j}t - k), j \in Z, k \in Z\}$, forms an orthonormal basis in $L^2(R)$. The discrete wavelet transform (DWT) of a sequence X is defined as follows:

$$d_x(j,k) = \int \Psi_{j,k} X(t) 2^{-j} \Psi_0\left(2^{-j}t - k\right) dt \tag{15}$$

Let $\lambda = \lambda_{j,k} = \left[k2^j, k(k+1)2^j\right]$ be defined as a dyadic interval (of scale 2^j), and 3λ be expressed as the union of λ and its two adjacent binary intervals on the same scale j, as specified in Equation (16). Under this constraint, the wavelet leader number expression is as follows:

$$3\lambda_{j,k} = \lambda_{j,k-1} \cup \lambda_{j,k} \cup \lambda_{j,k+1} \tag{16}$$

$$L_X(j,k) \equiv L_\lambda = \sup_{[k2^j,(k+1)2^j] \subset 3\lambda_{j,k}} |d_X(j,k)| \tag{17}$$

L_X (j, k) be the maximum wavelet coefficient in the range of $3\lambda_{j,k}$ at all scales, which is used as the wavelet leader number. The scale function (ζ_L) of the structural function (S_L) is calculated through the wavelet leader number, and the calculation formula is as follows:

$$S_L(q,j) = \frac{1}{n} \sum_{k-1}^{n_j} |L_X(j,k)|^q \tag{18}$$

$$\zeta_L(q) = \liminf_{j \to +\infty} \frac{\log(S_L(q,j))}{\log(2^j)} \tag{19}$$

Finally, the multifractal spectrum ($f(\alpha) - \alpha$) can be derived by applying the Legendre transform:

$$\alpha = \frac{d\zeta_L(q)}{dq} \tag{20}$$

$$f(\alpha) = \inf_{q \neq 0}(1 + \alpha - \zeta_L(q)) \tag{21}$$

4. Results and Analysis

4.1. Heterogeneity Analysis Based on Core Permeability

Macroscopic heterogeneities are categorized into interlaminar, intralayer, and lateral heterogeneity, typically resulting from differences in secondary environments, material sources, or geological processes. Interlayer reservoir heterogeneity refers to the macroscopic heterogeneity caused by differences between secondary layers or rock formations. Interlaminar heterogeneity refers to the macroscopic heterogeneity in the horizontal direction within the same sand body segment, describing the geometry, scale, continuity, and variations in porosity and permeability of the reservoir sand body. Intralayer heterogeneity refers to the differences in the vertical direction within the same sandstone layer, including intralayer grain size rhythms and stratification structures, which can also be influenced by thin clay interlayers.

Previous geological studies have shown that the main reservoirs in the DIBEI region consist of braided river delta plain sediments, characterized by poor lateral connectivity within an effective pore space and vertical differences between layers. This study focuses on intralayer heterogeneity, which primarily reflects the changes in the fluid flow capability within individual reservoir units in the vertical direction and is mainly influenced by rock pore structure and discontinuous thin shale layers.

To evaluate intralayer heterogeneity, the core permeability of wells YN2C and YN4 was analyzed using the CV and TK. Figures 6 and 7 present the well logs of wells YN2C and YN4, respectively. In Figure 6, the depth is shown in the first track; the second track presents the natural gamma ray (GR); the third track presents the acoustic (DT), compensated neutron porosity (CNL) and bulk density (DEN) logs; the fourth track presents the resistivity logs, including deep, medium induction resistivity and microsphere-focusing resistivity (Rxo); and the fifth track presents the permeability value obtained from the core experiment. The last track presents the interpretation intervals, including the gas, dry, and water–gas layers. In Figure 7, the fourth track presents the three resistivity curves obtained via array resistivity logging at different measurement depths, and the other curves are the same as those in Figure 6. The two figures have five interpretation intervals based on the well logs, including two dry, one water–gas, and two gas layers. As seen in the figures, the water–gas and dry layers exhibit significantly greater variability in core permeability compared with the gas layers. The CV and TK values of the core permeability of the five intervals are calculated as shown in Table 2. The fluid types of intervals were interpreted based on the well logs.

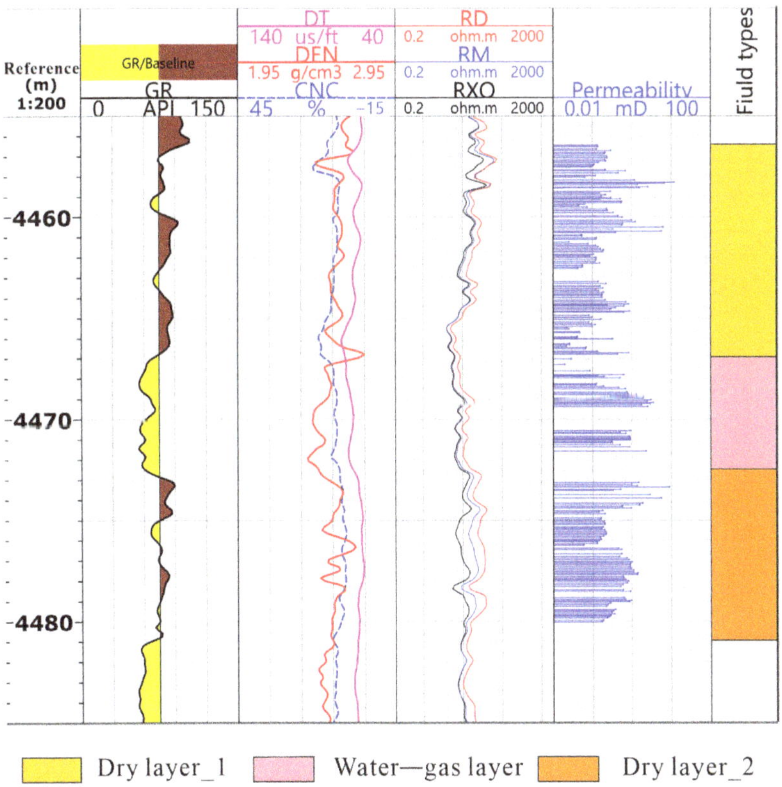

Figure 6. Conventional logs and core permeability in the water gas and dry layers.

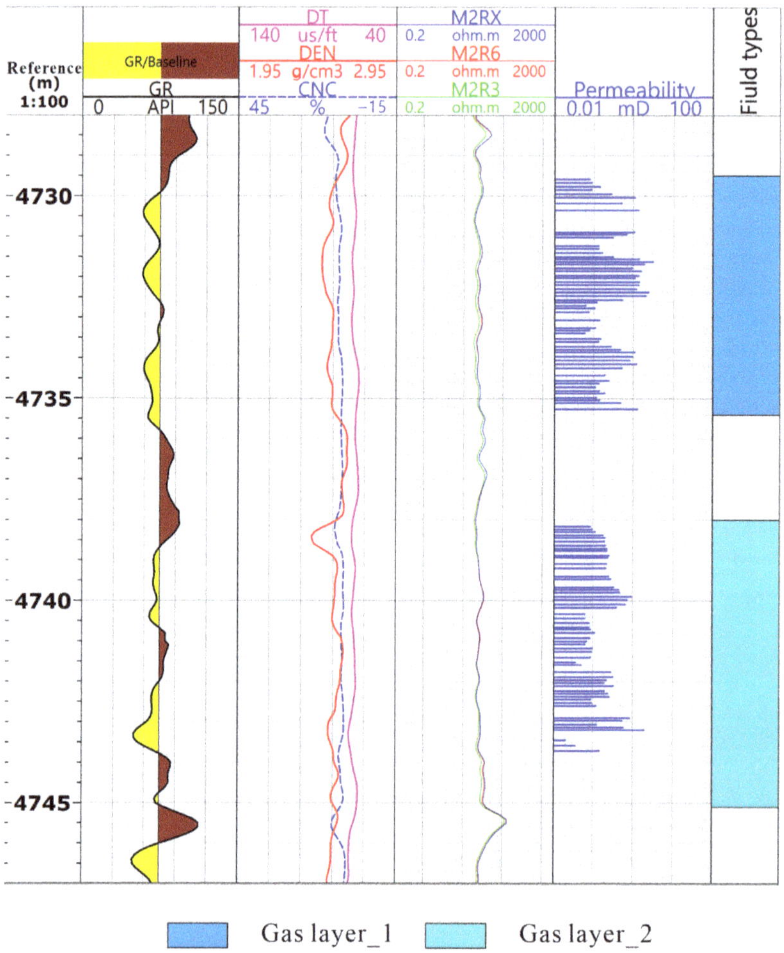

Figure 7. Conventional logs and core permeability in the gas layers.

Table 2. Parameters of the heterogeneity of core permeability. The fluid types of the intervals were interpreted based on the well logs.

	Dry Layer _1	Water–Gas Layer	Dry Layer _2	Gas Layer _1	Gas Layer _2
Wells	YN4	YN4	YN4	YN2C	YN2C
Thickness (m)	10.5	5.5	8.5	5.9	7.1
Number of data	101	44	88	57	70
MN (mD)	0.57	1.38	0.61	0.65	0.26
MAX (mD)	11.9	9.07	5.61	3.09	1.81
SD	1.51	1.65	0.80	0.68	0.29
TK	20.84	6.56	8.93	4.71	6.84
CV	2.65	1.19	1.32	1.04	1.11
Heterogeneity	Strong	Moderate	Strong	Weak	Weak

The results show that the gas layers exhibit the lowest CV and TK values, while the two dry layers have the highest CV and TK values. The CV and TK values of the water–gas layer are in the middle. These values indicate that the dry layers exhibit the strongest heterogeneity and the gas layers exhibit the lowest heterogeneity.

4.2. Fractal Analysis of Well Logs

In the multifractal spectrum, if the $f(\alpha)-\alpha$ spectrum exhibits a unimodal bell-shaped curve and the $\tau(q)-q$ relationship deviates from linearity, the signal is confirmed to possess multifractal characteristics. The spectrum width, denoted as $\Delta\alpha$, quantifies the heterogeneity of the multifractal structure by measuring the difference between the maximum and minimum singularity exponents. The spectrum width is calculated as follows:

$$\Delta\alpha = \alpha_{max} - \alpha_{min} \tag{22}$$

A larger $\Delta\alpha$ indicates greater variability in the geological unit's scaling properties, reflecting a stronger heterogeneity. The natural gamma ray (GR) log reflects rhythmic variations in the sedimentary grain sizes, which are indicative of vertical grain-size distribution patterns, a key manifestation of reservoir heterogeneity. Strong fluctuations in the GR log correlate with great intralayer heterogeneity, making it a robust tool for evaluating intralayer heterogeneity. The multifractal singularity spectrum associated with three types of layers is illustrated in Figure 8. Figure 8a shows the scaling exponent plot, where the three curves are nonlinear convex functions, confirming the multifractal characteristics of the GR log in the three types of layers. If $q > 0$, large fluctuations play a dominant role, and $\alpha(10)$ is approximately the singularity index corresponding to the maximum fluctuation. If $q < 0$, small fluctuations play a dominant role, and $\alpha(-10)$ is approximately the singularity index corresponding to the minimum fluctuation. In Figure 8b, the dry layer exhibits a larger interval between $\alpha(-10)$ and $\alpha(0)$, indicating higher variability. Furthermore, the $\alpha(-10)$ value reflects an expanded dynamic range of the log, with more complex small fluctuations in the log. Figure 8c displays the singularity spectrum ($f(\alpha) - \alpha$), where a broader spectrum width ($\Delta\alpha$) signifies greater complexity in lithological variations within the reservoir, directly corresponding to stronger heterogeneity. The multifractal parameters are calculated in Table 3. In addition, the fractal and correlation dimensions of GR logs are calculated. As can be seen from Table 3, a strong correlation exists in the three parameters (fractal dimension, correlation dimension, and multifractal spectrum width). The greater the spectrum width, the greater the values of the fractal and correlation dimensions. By comparing the fractal dimensions with the CV and TK values of core permeability, the CV and TK values in those intervals were found to have similar trends to those of the fractal dimension and multifractal spectra. Therefore, the heterogeneity of the gas layer is the weakest, and the heterogeneity of the dry layer is the strongest.

Table 3. Fractal parameters based on the GR logs.

	Dry Layer _1	**Water–Gas Layer**	**Dry Layer _2**	**Gas Layer _1**	**Gas Layer _2**
Wells	YN4	YN4	YN4	YN2C	YN2C
Thickness (m)	10.5	5.5	8.5	5.9	7.1
Box dimension	1.179	1.138	1.208	1.082	1.081
Correlation dimension	1.401	1.365	1.368	0.917	1.205
$\Delta\alpha$	1.480	1.235	1.314	0.958	1.290
α (−10)	2.495	2.162	2.202	1.990	2.094
α (0)	1.385	1.283	1.355	1.617	1.498
α (10)	1.015	0.927	0.888	1.032	0.805
Heterogeneity	Strong	Moderate	Strong	Weak	Weak

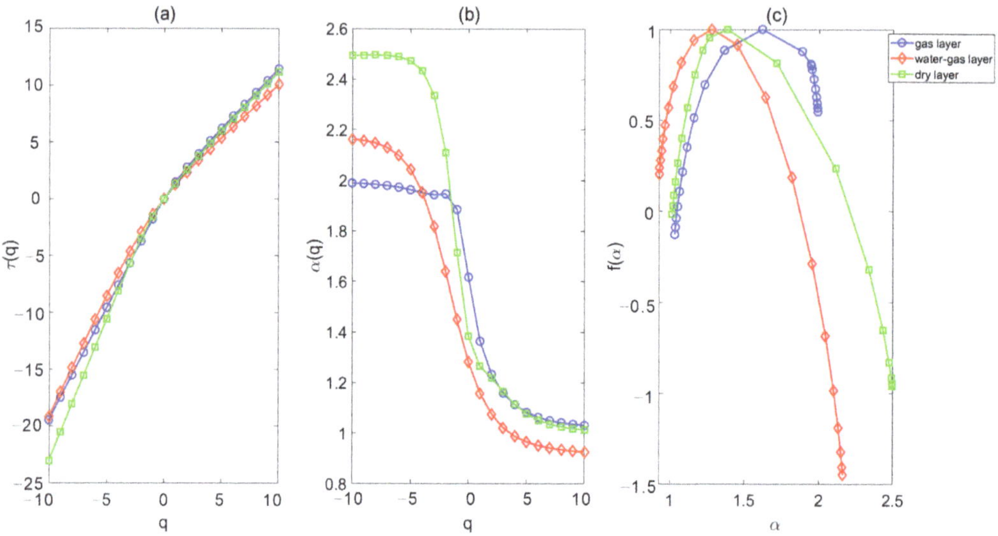

Figure 8. Multifractal characteristics of the GR logs of three types of layers in YN4 and YN2C: (**a**) variation in the scaling coefficient with q order; (**b**) variation in the Hurst index with q order; and (**c**) multifractal singular spectrum of three types of reservoirs. The values of Δα are 0.958 in the gas layer and 1.48 in the dry layer, while α(0) is 1.385 in the gas layer and 1.498 in the dry layer, respectively.

Microscopic heterogeneity is mainly determined by factors such as the pore structure and the interstitial material of the rocks and is mainly related to the size and uniformity of the pore–throats of the rocks, as well as the configuration and connectivity of the pores and throats. In well logging, the acoustic log could indicate the effective porosity, where the response is more pronounced in connected pores. Therefore, acoustic logs were used to characterize the microscopic heterogeneity.

As shown in Figure 9a, the acoustic logs of the three reservoir intervals all exhibit multifractal characteristics. In Figure 9b, the values of parameters for the dry layer are the same as in Figure 8b. In Figure 9c, the width of the singular spectrum of the acoustic log for the dry layer is the largest. The box dimensions, correlation dimensions, and multifractal singular spectrum widths for the three intervals are calculated separately and presented in Table 4. The three parameters obtained from acoustic logs are consistent with the results obtained from the GR log, that is, the microscopic heterogeneity of the dry layer is the largest, that of the water–gas layer is in the middle, and that of the gas layer is the smallest.

Table 4. Fractal parameters based on the acoustic logs.

	Dry Layer _1	Water–Gas Layer	Dry Layer _2	Gas Layer _1	Gas Layer _2
Wells	YN4	YN4	YN4	YN2C	YN2C
Thickness (m)	10.5	5.5	8.5	5.9	7.1
Box dimension	1.158	1.081	1.173	1.108	1.140
Correlation dimension	1.175	1.044	0.975	0.978	1.057
Δα	1.627	1.396	1.728	0.863	0.628
α (−10)	1.956	2.246	1.791	1.825	1.461
α (0)	0.947	1.171	1.037	1.394	1.102
α (10)	0.330	0.850	0.063	0.962	0.833
Heterogeneity	Strong	Moderate	Strong	Weak	Weak

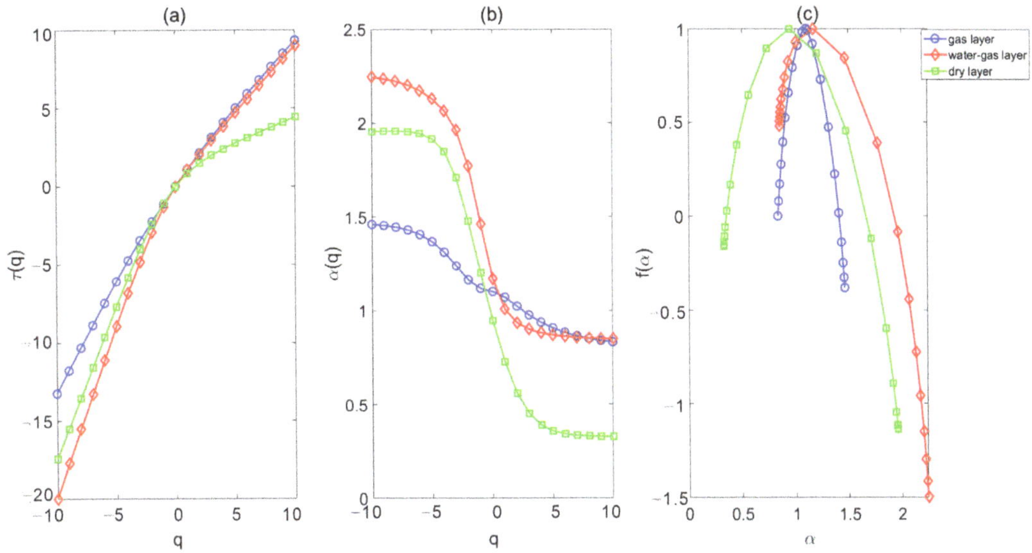

Figure 9. Multifractal characteristics of the acoustic log curves: (**a**) variation in the scale coefficient with q order; (**b**) variation in the Hurst index with q order; and (**c**) multifractal singular spectrum of three types of reservoirs. The values of $\Delta\alpha$ are 0.863 in the gas layer and 1.627 in the dry layer, while $\alpha(0)$ is 1.394 in the gas layer and 0.947 in the dry layer, respectively.

4.3. Correlation Between Heterogeneity and Production

Three wells (YN2C, DB102, and DB101) with varying production capacities were analyzed. Figures 10–12 show the well logs, calculated permeabilities, and oil test intervals. To evaluate the relationship between heterogeneity and production, the heterogeneity of the main reservoir sections of wells YN2C, DB102, and DB101 was analyzed using the box counting dimension, correlation dimension, and wavelet leader methods. In these figures, the first track presents the depth; the second track shows the lithology provided by mud logging; the third track presents the lithology analysis data; the fourth track presents the natural gamma ray (GR), caliper (CAL), and bit size (BIT); the fifth track presents the acoustic (DT), compensated neutron porosity (CNL), and density (DEN) logs; the sixth track presents three resistivity curves obtained via array resistivity logging at different measurement depths; the seventh track presents the permeability values calculated from the well logs; the eighth track shows the interpreted intervals, including for the gas and poor gas layers; the ninth track shows the evaluation results of heterogeneity; and the last track is the results of well testing.

Figures 13 and 14 present the calculated multifractal curves of GR and the acoustic logs, respectively. As can be seen from the figures, the calculated results of the two logs share similar characteristics. The scale exponent of Figures 13a and 14a of the three types of reservoirs in both curves is a nonlinear convex function, indicating that the reservoirs have multifractal features. In Figures 13b and 14b, the $\alpha(-10)$ of well DB101 is relatively high, at 1.932 and 2.11 (the first value is from Figure 10, and the second value is from Figure 11), suggesting a high degree of curve fluctuation. Figures 13c and 14c shows the multifractal singularity spectrum. Well DB101 has the widest spectrum, with values of 1.763 and 1.695, while well YN2C has the narrowest, with values of 1.044 and 1.248. In the figures, the values of $\alpha(0) - \alpha(10)$ of the three wells are similar, between 0.4 and 0.7, indicating that the fluctuations in the three wells are similar on a larger scale. The values of $\alpha(-10) - \alpha(0)$ of well DB101 are large, at 1.24 and 1.25, which is related to the small fluctuations in the logging signal, suggesting a high complexity in small fluctuations for the well with low gas production.

The box dimension, correlation dimension, and multifractal spectrum width of the three wells are shown in Tables 5 and 6. The results show that the fractal dimension of wells with high gas production is small, while that of wells with low gas production is large. In Table 5, compared with DB101, DB102 has a higher correlation dimension and box dimension, indicating that DB102 has higher geometric complexity but lower data correlation, whereas DB101 has lower geometric complexity but higher data correlation. In Table 6, the correlation dimension of YN2C is higher than that of DB102, suggesting that the relationship between the acoustic log curves of YN2C and DB102 is the same as the relationship between the GR logs of the aforementioned DB102 and DB101. The data show that YN2C has the highest production and lowest heterogeneity, suggesting that lower heterogeneity may be associated with higher production. Table 7 summarizes the correlation between the multifractal spectrum width and production based on the oil test results.

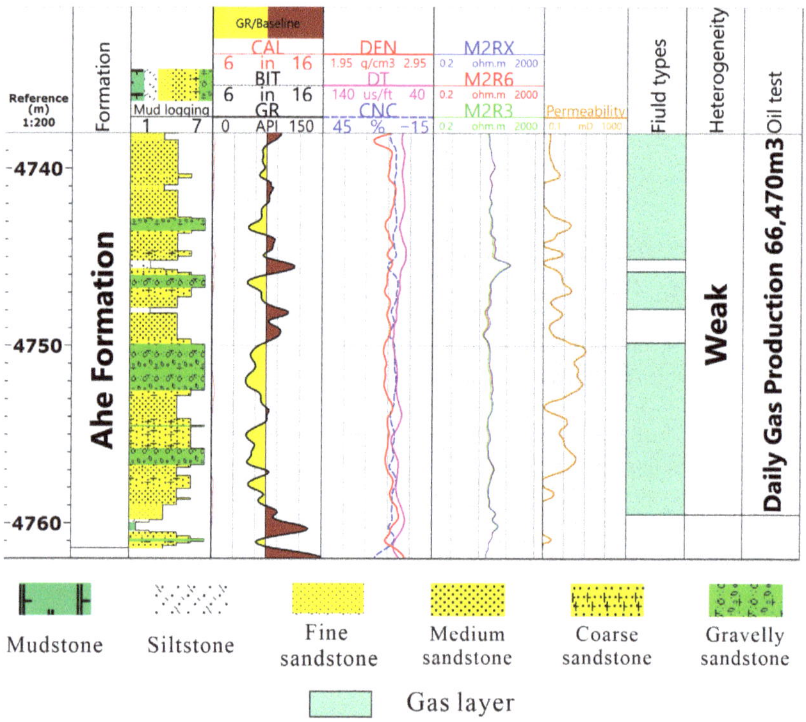

Figure 10. Analysis of the heterogeneity and results of oil testing in well YN2C.

Table 5. Fractal parameters of GR logs in three wells.

	YN2C	DB102	DB101
Box dimension	1.228	1.232	1.334
Correlation dimension	1.581	1.686	1.659
$\Delta\alpha$	1.044	1.305	1.762
$\alpha\,(-10)$	1.691	1.562	1.932
$\alpha\,(0)$	1.169	0.908	0.694
$\alpha\,(10)$	0.648	0.257	0.169

Table 6. Fractal parameters of acoustic logs in three wells.

	YN2C	DB102	DB101
Box dimension	1.130	1.187	1.342
Correlation dimension	1.137	1.079	1.799
$\Delta\alpha$	1.248	1.450	1.695
α (−10)	2.165	2.449	2.110
α (0)	1.517	1.667	0.855
α (10)	0.916	1.000	0.414

Table 7. Correlation between the multifractal spectrum width and production based on the oil test results.

	YN2C	DB102	DB101
$\Delta\alpha$ based on GR logs	1.044	1.305	1.763
$\Delta\alpha$ based on acoustic logs	1.248	1.450	1.695
Oil test (m^3/day)	67,320 (gas)	16,328 (gas)	5851 (gas)

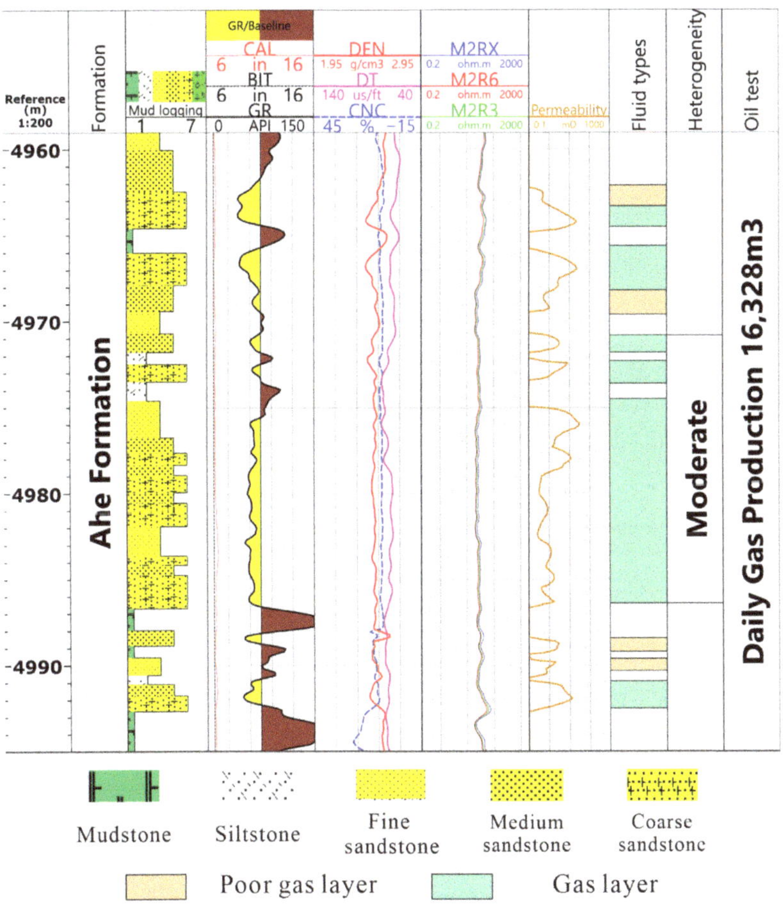

Figure 11. Analysis of the heterogeneity and results of oil testing in well DB102.

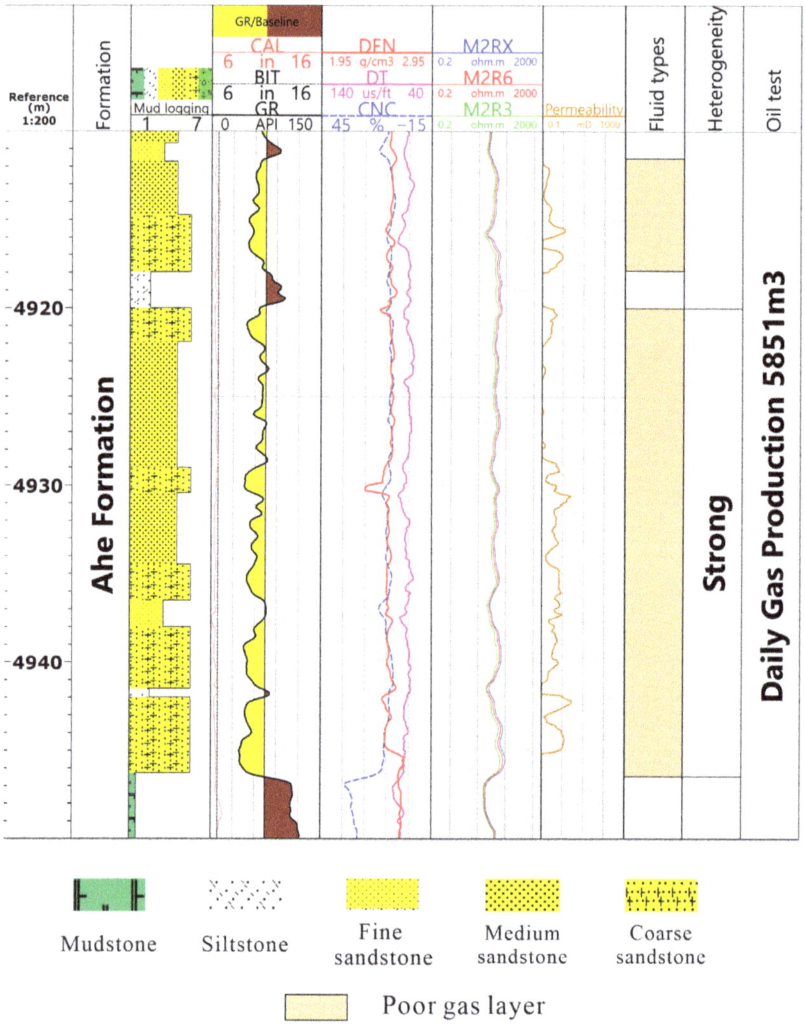

Figure 12. Analysis of the heterogeneity and results of oil testing in well DB101.

The analysis of the three wells showed that, in terms of macroscopic heterogeneity, the values of box dimension, correlation dimension, and multifractal spectrum width indicate that the stronger the heterogeneity, the lower the production. In terms of microscopic heterogeneity, the results from single-fractal calculations show that low productivity and dry layers exhibit similar levels of heterogeneity compared with high gas production zones, but the multifractal analysis reveals that the stronger the heterogeneity, the lower the production.

The fractal and multifractal dimensions of GR and acoustic logs demonstrate that gas reservoirs exhibit low macro- and microscopic heterogeneity compared with dry layers. The geological investigations indicate that the gas reservoirs in well YN2C were deposited in a distributary channel of a braided river delta plain environment, predominantly composed of stable distributary channel microfacies. These channels experienced consistent hydrodynamic conditions during deposition, leading to better sediment sorting and cleaner sandstone layers. In contrast, dry zones and mudstones are associated with floodplain deposits, which contain more interlayers and consequently stronger heterogeneity.

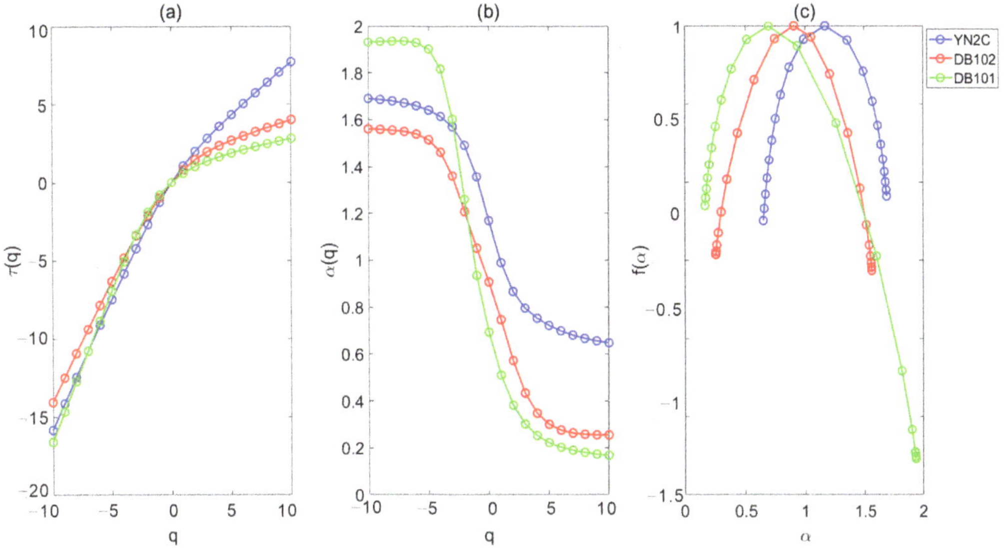

Figure 13. Results of the GR log calculations in three wells: (**a**) variation in the scaling coefficient with q order; (**b**) variation in the Hurst index with q order; and (**c**) multifractal singular spectrum of three types of gas layers. The values of $\Delta\alpha$ are 1.044 in well YN2C and 1.763 in well DB101, while $\alpha(0)$ is 1.169 and 0.694, respectively.

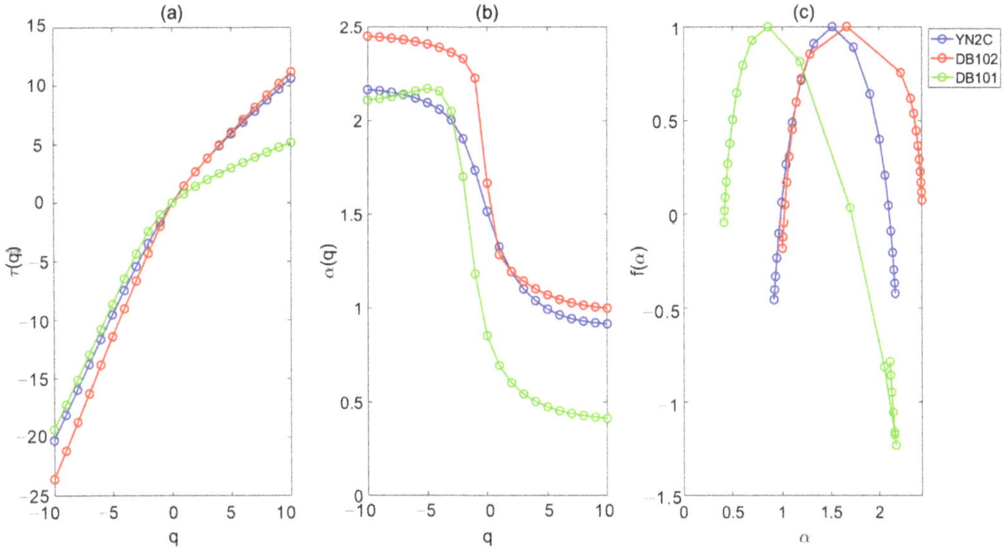

Figure 14. Results of the acoustic log calculations in three wells: (**a**) variation in the scale coefficient with q order; (**b**) variation in the Hurst index with q order; and (**c**) multifractal singular spectrum of three types of gas layers. The values of $\Delta\alpha$ are 1.248 in well YN2C and 1.695 in well DB101, while $\alpha(0)$ is 1.517 and 0.855, respectively.

At the microscopic scale, the stable depositional environment of gas reservoirs resulted in quartz grains being less compacted during diagenesis compared with argillaceous particles in dry layers reservoirs. This preserved more regular pore structures, as evidenced by small and uniform variations in the effective porosity on acoustic logs. Consequently, gas reservoirs display lower heterogeneity and smaller fractal dimensions in the acoustic log analysis.

5. Conclusions

Based on core permeability, GR, and acoustic logs, the heterogeneity of a deep tight sandstone reservoir was analyzed using statistical methods and fractal and multifractal theory. The following conclusions are obtained:

(1) In the analysis of core permeability, the heterogeneity of gas layers is small, and that of dry layers is the largest.

(2) The box dimension, correlation dimension, and multifractal parameter of gas layers, dry layers, and water–gas layers were calculated using well logs. The fractal dimension of the GR log reflects the intralayer heterogeneity within the layer, while the fractal dimension of the acoustic log indicates microscopic heterogeneity. This analysis also examines the two types of heterogeneity in different reservoirs. The two heterogeneity and gas content results were consistent. The calculation results also have good consistency with the results based on core permeability. Layers with high gas content exhibit lower fractal dimensions and weaker heterogeneity, while the dry layers exhibit a larger fractal dimension and stronger heterogeneity.

(3) The fractal dimensions and multifractal parameters of the wells with gas production were calculated using GR and acoustic logs, and the heterogeneity of different wells was analyzed. As a result, it was determined that the weaker the heterogeneity, the higher the production. Therefore, the reservoir heterogeneity could be used as an indicator for production estimation.

Author Contributions: Conceptualization, P.Z.; validation, Q.L., Y.X. and N.W.; formal analysis, N.W.; investigation, P.Z. and Q.L.; data curation, Y.X.; writing—original draft, P.Z.; writing—review and editing, P.Z.; funding acquisition, P.Z. All authors have read and agreed to the published version of the manuscript.

Funding: This paper is supported by the Beijing Municipal Natural Science Foundation (1242025); National Natural Science Foundation of China (U24B6001); and CNPC Innovation Found (2024DQ02-0151).

Data Availability Statement: The data presented in this study are available on request from the corresponding author.

Acknowledgments: The authors would like to thank the respected reviewers for their kind comments and the editorial office for their advice.

Conflicts of Interest: Qiran Lv, Yi Xin and Ning Wu are employed by China Petroleum Group Logging Co., Ltd., China National Petroleum Corporation and Sixth Oil Production Pant, PetroChina Changqing Oilfield Company, respectively. The remaining authors declare that the research was conducted in the absence of any commercial or financial relationships that could be construed as a potential conflict of interest.

References

1. Hong, D.; Cao, J.; Wu, T.; Dang, S.; Hu, W.; Yao, S. Authigenic Clay Minerals and Calcite Dissolution Influence Reservoir Quality in Tight Sandstones: Insights from the Central Junggar Basin, NW China. *Energy Geosci.* **2020**, *1*, 8–19. [CrossRef]
2. Wang, Q.; Yang, H.; Yang, W. New progress and future exploration targets in petroleum geological research of ultra-deep clastic rocks in Kuqa Depression, Tarim Basin, NW China. *Pet. Explor. Dev.* **2025**, *52*, 79–94. [CrossRef]

3. Wang, G.-X.; Wu, W.; Li, Q.; Liu, W.-Q.; Zhou, Y.-S.; Liang, S.-Q.; Sui, Y.-P. Sequence stratigraphy and sedimentary evolution from Late Cretaceous to Quaternary in the Romney 3D seismic area, deep-water Taranaki Basin (New Zealand). *Pet. Sci.* **2025**, *22*, 1854–1875. [CrossRef]
4. Tao, K.; Cao, J.; Chen, X.; Nueraili, Z.; Hu, W.; Shi, C. Deep Hydrocarbons in the Northwestern Junggar Basin (NW China): Geochemistry, Origin, and Implications for the Oil vs. Gas Generation Potential of Post-Mature Saline Lacustrine Source Rocks. *Mar. Pet. Geol.* **2019**, *109*, 623–640. [CrossRef]
5. Mao, Z.; Zeng, L.; Liu, G.; Gao, Z.; Tian, H.; Liao, Q.; Zhang, Y. Fractures and Their Effectiveness in the Deep Jurassic Tight Sandstone Reservoir in the Southern Margin of the Junggar Basin. *Pet. Geol.* **2020**, *41*, 1212–1221. [CrossRef]
6. Lai, J.; Li, D.; Bai, T.; Zhao, F.; Ai, Y.; Liu, H.; Cai, D.; Wang, G.; Chen, K.; Xie, Y. Reservoir Quality Evaluation and Prediction in Ultra-Deep Tight Sandstones in the Kuqa Depression, China. *J. Struct. Geol.* **2023**, *170*, 104850. [CrossRef]
7. Wang, J.Y.; Wang, B.; Hu, Z.Q.; Shang, F.; Liu, D.; Li, Z.; Qiu, Q.; Song, Z.; Hu, Z. Genetic Mechanism of High-Quality Reservoirs in Deep-Ultra-Deep Clastic Rocks: A Case Study of the Permian-Triassic in the Central Junggar Basin, NW China. *Pet. Nat. Gas Geol.* **2025**, *46*, 151–166.
8. Xiao, D.; Lu, S.; Lu, Z.; Huang, W.; Gu, M. Combining Nuclear Magnetic Resonance and Rate-Controlled Porosimetry to Probe the Pore-Throat Structure of Tight Sandstones. *Pet. Explor. Dev.* **2016**, *43*, 1049–1059. [CrossRef]
9. Schmitt, M.; Fernandes, C.P.; Cunha Neto, J.A.B.; Wolf, G.F. Characterization of Pore Systems in Seal Rocks Using Nitrogen Gas Adsorption Combined with Mercury Injection Capillary Pressure Techniques. *Mar. Pet. Geol.* **2013**, *39*, 138–149. [CrossRef]
10. Fitch, P.J.; Lovell, M.A.; Davies, S.J.; Pritchard, T.; Harvey, P.K. An Integrated and Quantitative Approach to Petrophysical Heterogeneity. *Mar. Pet. Geol.* **2015**, *63*, 82–96. [CrossRef]
11. Zheng, C.; Xie, J.; Wang, J.K.; Zhao, X.; Duan, Y.; Sun, Y. Application of Lorenz Coefficient in Reservoir Heterogeneity Evaluation. *J. Shandong Univ. Sci. Technol.* **2018**, *37*, 103–110. [CrossRef]
12. Lake, L.; Jensen, J. A Review of Heterogeneity Measures Used in Reservoir Characterization. *Situ* **1991**, *15*, 409–439.
13. Fitch, P.; Davies, S.; Lovell, M.; Pritchard, T. Reservoir Quality and Reservoir Heterogeneity: Petrophysical Application of the Lorenz Coefficient. *Petrophysics* **2013**, *54*, 465–474.
14. Hosseinzadeh, M.; Tavakoli, V. Analyzing the Impact of Geological Features on Reservoir Heterogeneity Using Heterogeneity Logs: A Case Study of Permian Reservoirs in the Persian Gulf. *Geoenergy Sci. Eng.* **2024**, *237*, 212810. [CrossRef]
15. Farooq, U.; Ahmed, J.; Ali, S.; Pakistan, M.; Siddiqi, F.; Kazmi, S.A.A.; Mushir, K. Heterogeneity in the Petrophysical Properties of Carbonate Reservoirs in Tal Block. In Proceedings of the SPWLA 60th Annual Logging Symposium, The Woodlands, TX, USA, 15–19 June 2019. [CrossRef]
16. Tavoosi, I.P.; Mehrabi, H.; Rahimpour-Bonab, H.; Ranjbar-Karami, R. Quantitative Analysis of Geological Attributes for Reservoir Heterogeneity Assessment in Carbonate Sequences; A Case from Permian–Triassic Reservoirs of the Persian Gulf. *J. Pet. Sci. Eng.* **2021**, *200*, 108356. [CrossRef]
17. Davies, A.; Cowliff, L.; Simmons, M.D. A Method for Fine-Scale Vertical Heterogeneity Quantification from Well Data and Its Application to Siliciclastic Reservoirs of the UKCS. *Mar. Pet. Geol.* **2023**, *149*, 106077. [CrossRef]
18. Mandelbrot, B.B. *The Fractal Geometry of Nature*; W.H. Freeman: San Francisco, CA, USA, 1982.
19. Ghanbarian, B.; Hunt, A.G. *Fractals: Concepts and Applications in Geosciences*; CRC Press: Boca Raton, FL, USA, 2017.
20. Jafari, A.; Babadagli, T. Relationship between Percolation–Fractal Properties and Permeability of 2-D Fracture Networks. *Int. J. Rock Mech. Min. Sci. Géoméch. Abstr.* **2013**, *60*, 353–362. [CrossRef]
21. Pia, G.; Sanna, U. An Intermingled Fractal Units Model to Evaluate Pore Size Distribution Influence on Thermal Conductivity Values in Porous Materials. *Appl. Therm. Eng.* **2014**, *65*, 330–336. [CrossRef]
22. Karimpouli, S.; Tahmasebi, P. 3D Multi-Fractal Analysis of Porous Media Using 3D Digital Images: Considerations for Heterogeneity Evaluation. *Geophys. Prospect.* **2019**, *67*, 1082–1093. [CrossRef]
23. Billi, A.; Storti, F. Fractal Distribution of Particle Size in Carbonate Cataclastic Rocks from the Core of a Regional Strike-Slip Fault Zone. *Tectonophysics* **2004**, *384*, 115–128. [CrossRef]
24. Xie, S.; Cheng, Q.; Ling, Q.; Li, B.; Bao, Z.; Fan, P. Fractal and Multifractal Analysis of Carbonate Pore-Scale Digital Images of Petroleum Reservoirs. *Mar. Pet. Geol.* **2010**, *27*, 476–485. [CrossRef]
25. Xia, W.; Xi, K.; Xin, H.; Ma, W.; Zhao, H.; Feng, S.; Dan, W. The Influence of Pore Throat Heterogeneity and Fractal Characteristics on Reservoir Quality: A Case Study of Chang 8 Member Tight Sandstones, Ordos Basin. *Unconv. Resour.* **2025**, *5*, 100123. [CrossRef]
26. Liu, H.; Roquete, M.; Kou, S.; Lindqvist, P.-A. Characterization of Rock Heterogeneity and Numerical Verification. *Eng. Geol.* **2004**, *72*, 89–119. [CrossRef]
27. Liu, P.; Yuan, Z.; Li, K. An Improved Capillary Pressure Model Using Fractal Geometry for Coal Rock. *J. Pet. Sci. Eng.* **2016**, *145*, 473–481. [CrossRef]
28. Zhou, Y.; Xu, J.; Lan, Y.; Zi, H.; Cui, Y.; Chen, Q.; You, L.; Fan, X.; Wang, G. New Insights into Pore Fractal Dimension from Mercury Injection Capillary Pressure in Tight Sandstone. *Geoenergy Sci. Eng.* **2023**, *228*, 212059. [CrossRef]

29. Liu, Y.; Pang, X.Q.; Ding, C.; Chen, D.; Li, M. Pore Structure and Fractal Characteristics of Tight Sandstone Reservoirs in Yan 10 Member of Wuqi Area. *Sci. Technol. Eng.* **2023**, *23*, 12474–12483.
30. Peng, J.; Han, H.; Xia, Q.; Li, B. Fractal Characteristic of Microscopic Pore Structure of Tight Sandstone Reservoirs in Kalpintag Formation in Shuntuoguole Area, Tarim Basin. *Pet. Res.* **2020**, *5*, 1–17. [CrossRef]
31. Liu, Y.; Liu, Y.; Zhang, Q.; Li, C.; Feng, Y.; Wang, Y.; Xue, Y.; Ma, H. Petrophysical Static Rock Typing for Carbonate Reservoirs Based on Mercury Injection Capillary Pressure Curves Using Principal Component Analysis. *J. Pet. Sci. Eng.* **2019**, *181*, 106175. [CrossRef]
32. Wen, H.M. Research on Fractal Well Logging Interpretation Theory and Methods. Ph.D. Thesis, Chengdu University of Technology, Chengdu, China, 2003.
33. Mou, D.; Wang, Z.W. Fractal Dimension of Well Logging Curves Associated with the Texture of Volcanic Rocks. In Proceedings of the Industrial and Control Engineering. *Ind. Control. Eng.* **2014**, 278–281. [CrossRef]
34. Zhao, P.; Wang, Z.; Sun, Z.; Cai, J.; Wang, L. Investigation on the Pore Structure and Multifractal Characteristics of Tight Oil Reservoirs Using NMR Measurements: Permian Lucaogou Formation in Jimusaer Sag, Junggar Basin. *Mar. Pet. Geol.* **2017**, *86*, 1067–1081. [CrossRef]
35. Yan, J.-P.; He, X.; Geng, B.; Hu, Q.-H.; Feng, C.-Z.; Kou, X.-P.; Li, X.-W. Nuclear magnetic resonance T2 spectrum multifractal characteristics and pore structure evaluation. *Appl. Geophys.* **2017**, *14*, 205–215. [CrossRef]
36. Zhao, P.; Wang, X.; Cai, J.; Luo, M.; Zhang, J.; Liu, Y.; Rabiei, M.; Li, C. Multifractal Analysis of Pore Structure of Middle Bakken Formation Using Low Temperature N_2 Adsorption and NMR Measurements. *J. Pet. Sci. Eng.* **2019**, *176*, 312–320. [CrossRef]
37. Lv, H.; Guo, J.; Gu, B.; Zhu, Z.; Zhang, Z. A Fracture Identification Method Using Fractal Dimension and Singular Spectrum Entropy in Array Acoustic Logging: Insights from the Asmari Formation in the M Oilfield, Iraq. *J. Appl. Geophys.* **2025**, *233*, 105631. [CrossRef]
38. Lozada-Zumaeta, M.; Arizabalo, R.; Ronquillo-Jarillo, G.; Coconi-Morales, E.; Rivera-Recillas, D.; Castrejón-Vácio, F. Distribution of Petrophysical Properties for Sandy-Clayey Reservoirs by Fractal Interpolation. *Nonlinear Process. Geophys.* **2012**, *19*, 239–250. [CrossRef]
39. Partovi, S.M.A.; Sadeghnejad, S. Fractal Parameters and Well-Logs Investigation Using Automated Well-to-Well Correlation. *Comput. Geosci.* **2017**, *103*, 59–69. [CrossRef]
40. López, M.; Aldana, M. Facies Recognition Using Wavelet Based Fractal Analysis and Waveform Classifier at the Oritupano-A Field, Venezuela. *Nonlinear Process. Geophys.* **2007**, *14*, 325–335. [CrossRef]
41. Subhakar, D.; Chandrasekhar, E. Reservoir Characterization Using Multifractal Detrended Fluctuation Analysis of Geophysical Well-Log Data. *Stat. Mech. Its Appl.* **2016**, *445*, 57–65. [CrossRef]
42. Amoura, S.; Gaci, S.; Barbosa, S.; Farfour, M.; Bounif, M. Investigation of Lithological Heterogeneities from Velocity Logs Using EMD-Hölder Technique Combined with Multifractal Analysis and Unsupervised Statistical Methods. *J. Pet. Sci. Eng.* **2022**, *208*, 109588. [CrossRef]
43. Prajapati, R.; Kumar, R.; Singh, U.K. Assessment of Reservoir Heterogeneities and Hydrocarbon Potential Zones Using Wavelet-Based Fractal and Multifractal Analysis of Geophysical Logs of Cambay Basin, India. *Mar. Pet. Geol.* **2024**, *160*, 106633. [CrossRef]
44. Wang, Q.H.; Zhang, R.H.; Yang, X.Z.; Chen, X.; Xiao, D.; Wang, B.; Wu, C.; Yu, H. Major Breakthrough and Geological Significance of Tight Sandstone Gas Ex-ploration in Jurassic Ahe Formation of Dibei Area, Eastern Kuqa Depression. *Acta Pet. Sinica.* **2022**, *43*, 1049–1064. [CrossRef]
45. Zhao, M.J.; Zhang, S.C. Genetic Types and Accumulation Conditions of Natural Gas in Tarim Basin. *China Pet. Explor.* **2001**, *6*, 27–31.
46. Wei, G.; Wang, K.; Zhang, R.; Wang, B.; Yu, C. Structural Diagenesis and High-Quality Reservoir Prediction of Tight Sandstones: A Case Study of the Jurassic Ahe Formation of the Dibei Gas Reservoir, Kuqa Depression, Tarim Basin, NW China. *J. Asian Earth Sci.* **2022**, *239*, 105399. [CrossRef]
47. Yang, Y.-B.; Xiao, W.; Zheng, L.-L.; Lei, Q.-H.; Qin, C.-Z.; He, Y.-A.; Liu, S.-S.; Li, M.; Li, Y.-M.; Zhao, J.-Z.; et al. Pore Throat Structure Heterogeneity and Its Effect on Gas-Phase Seepage Capacity in Tight Sandstone Reservoirs: A Case Study from the Triassic Yanchang Formation, Ordos Basin. *Pet. Sci.* **2023**, *20*, 2892–2907. [CrossRef]
48. Liang, D.G.; Jia, C.Z. Achievements and Prospect Prediction of Natural Gas Exploration in Tarim Basin. *Nat. Gas Ind.* **1999**, *19*, 3–12.
49. Li, D.; Wang, G.; Bie, K.; Lai, J.; Lei, D.-W.; Wang, S.; Qiu, H.-H.; Guo, H.-B.; Zhao, F.; Zhao, X.; et al. Formation Mechanism and Reservoir Quality Evaluation in Tight Sandstones under a Compressional Tectonic Setting: The Jurassic Ahe Formation in Kuqa Depression, Tarim Basin, China. *Pet. Sci.* **2024**, *22*, 998–1020. [CrossRef]
50. Hu, H.; Wang, G.C.; Wang, Y.Z.; Huang, R. Yan'an Oilfield Fuchang Development Unit Chang 2 Heterogeneity Study. *J. Hebei GEO Univ.* **2024**, *47*, 34–42. [CrossRef]
51. Grassberger, P.; Procaccia, I. Measuring the Strangeness of Strange Attractors. *Nonlinear Phenom.* **1983**, *9*, 189–208. [CrossRef]

52. Liao, Z.; Wang, L.; He, Z. Feature Extraction and Classification of Mining Microseismic Signals Based on EEMD and Correlation Dimension. *Gold Sci. Technol.* **2020**, *28*, 585–594.
53. Tan, Z.; Huang, J.; Ye, C. Identification of major turning points in the Chinese stock market: A Bayesian approach based on improved wavelet leader method and approximation techniques. *China Manag. Sci.* **2018**, *26*, 12–24. [CrossRef]
54. Zhang, L.; Li, R.J.; Liu, X.L. Analysis of Stock Market Effectiveness and Risk Detection Based on Wavelet Leader Multifractal Method. *Chin. J. Manag. Sci.* **2014**, *22*, 17–26.

fractal and fractional

MDPI

Article

Pore-Throat Structure, Fractal Characteristics, and Main Controlling Factors in Extremely Low-Permeability Sandstone Reservoirs: The Case of Chang 3 Section in Huachi Area, Ordos Basin

Huanmeng Zhang [1], Chenyang Wang [2], Jinkuo Sui [2], Yujuan Lv [2], Ling Guo [1,*] and Zhiyu Wu [1,*]

1 Department of Geology, Northwest University, Xi'an 710069, China; 202033002@stumail.nwu.edu.cn
2 The Second Oil Production Plant, PetroChina Changqing Oilfield Company, Qingcheng 745100, China
* Correspondence: lguogeology@163.com (L.G.); 20224004@nwu.edu.cn (Z.W.)

Abstract: The pore-throat structure of the extremely low-permeability sandstone reservoir in the Huachi area of the Ordos Basin is complex and highly heterogeneous. Currently, there are issues such as unclear understanding of the micro-pore-throat structural characteristics, primary controlling factors of reservoir quality, and classification boundaries of the reservoir in the study area, which seriously restricts the exploration and development effectiveness of the reservoir in this region. It is necessary to use a combination of various analytical techniques to comprehensively characterize the pore-throat structure and establish reservoir classification evaluation standards in order to better understand the reservoir. This study employs a suite of analytical and testing techniques, including cast thin sections (CTS), scanning electron microscopy (SEM), cathodoluminescence (CL), X-ray diffraction (XRD), as well as high-pressure mercury injection (HPMI) and constant-rate mercury injection (CRMI), and applies fractal theory for analysis. The research findings indicate that the extremely low-permeability sandstone reservoir of the Chang 3 section primarily consists of arkose and a minor amount of lithic arkose. The types of pore-throat are diverse, with intergranular pores, feldspar dissolution pores, and clay interstitial pores and microcracks being the most prevalent. The throat types are predominantly sheet-type, followed by pore shrinkage-type and tubular throats. The pore-throat network of low-permeability sandstone is primarily composed of nanopores (pore-throat radius $r < 0.01$ μm), micropores ($0.01 < r < 0.1$ μm), mesopores ($0.1 < r < 1.0$ μm), and macropores ($r > 1.0$ μm). The complexity of the reservoir pore-throat structure was quantitatively characterized by fractal theory. Nanopores do not exhibit ideal fractal characteristics. By splicing high-pressure mercury injection and constant-rate mercury injection at a pore-throat radius of 0.12 μm, a more detailed characterization of the full pore-throat size distribution can be achieved. The average fractal dimensions for micropores (Dh2), mesopores (Dc3), and macropores (Dc4) are 2.43, 2.75, and 2.95, respectively. This indicates that the larger the pore-throat size, the rougher the surface, and the more complex the structure. The degree of development and surface roughness of large pores significantly influence the heterogeneity and permeability of the reservoir in the study area. Dh2, Dc3, and Dc4 are primarily controlled by a combination of pore-throat structural parameters, sedimentary processes, and diagenetic processes. Underwater diversion channels and dissolution are key factors in the formation of effective storage space. Based on sedimentary processes, reservoir space types, pore-throat structural parameters, and the characteristics of mercury injection curves, the study area is divided into three categories. This classification provides a theoretical basis for predicting sweet spots in oil and gas exploration within the study area.

Keywords: extremely low-permeability sandstone reservoirs; pore size distribution; fractal theory; reservoir heterogeneity; Ordos Basin

1. Introduction

Extremely low-permeability sandstone reservoirs generally experience destructive diagenetic processes such as compaction and cementation, which lead to a reduction in pore and throat radius, and a decrease in porosity and permeability. The pore-throat structure becomes more complex, with irregular shapes, poorer connectivity, and increased heterogeneity of the sandstone. These factors collectively increase the difficulty of oil and gas exploration and development [1,2]. Under the condition of similar porosity, there is a significant difference in permeability, which is an important factor affecting the exploration and development of this type of reservoir [3–5]. The pore-throat is the primary storage space and migration pathway for oil and gas, and they are the key factor determining the development of "sweet spots" in extremely low-permeability sandstone reservoirs. Therefore, an accurate understanding of the pore-throat structure of such reservoirs is of great significance for clarifying reservoir quality, oil saturation, and predicting oil well production. Currently, mercury injection and fractal theory are commonly used to quantitatively characterize the complexity of reservoir pore-throat structures [6–8]. Previous studies have predominantly employed either high-pressure mercury injection (HPMI) or constant rate mercury injection (CRMI) individually to investigate the fractal characteristics of reservoir pore-throat structures. However, HPMI has a shielding effect on larger pores and is more suitable for characterizing smaller pores. The high-pressure mercury injection curve is a composite curve of pores and throats, making it difficult to distinguish between them [9–11]. The maximum pressure of CRMI is 6.2 MPa, which is unable to characterize pores smaller than 0.12 μm, but it can separate pores and throats, allowing for the determination of their respective quantities and pore sizes [12–14]. Therefore, a single technique can only reflect the structural characteristics of part of the pore-throat system. This study combines HPMI with CRMI, and applies fractal theory. It also integrates direct observation methods of pore-throat structures, such as cast thin sections and scanning electron microscopy, to conduct a full-size pore-throat structure evaluation of the extremely low-permeability sandstone reservoirs in the Huachi area [15].

The Triassic Yanchang Formation sandstones in the Ordos Basin serve as a typical example for the study of extremely low-permeability sandstone reservoirs. Many scholars have successively conducted research on the characteristics of extremely low-permeability sandstone reservoirs, diagenetic processes, mechanisms of tightness, and factors influencing the formation and distribution of high-quality reservoirs [16–19]. It has been recognized that the sedimentary environment and petrological characteristics control the original pore features, while diagenesis modifies these original pores. Both of them jointly control the development of sweet spots [20,21]. The pore-throat structure includes several aspects such as the geometry, size, connectivity, and distribution characteristics of pores and throats. Sedimentary environment and diagenetic processes are the primary factors that determine the grain composition, size, sorting, cements, and maturity in clastic rocks, thereby influencing the heterogeneity of the reservoir. High-energy depositional environments typically form thick-bedded sandstones (>1 m) characterized by coarse-grained, well-sorted sediments, favorable reservoir properties, and relatively weak heterogeneity. In contrast, low-energy depositional environments often develop sand–mud interbedded lithofacies with finer-grained particles. These vertically intercalated mudstone layers act as flow barriers, significantly reducing permeability and resulting in strong heterogeneity. Additionally, sandstone composition directly influences reservoir fluid flow behavior and heterogeneity. Quartz grains exhibit strong compaction resistance, which helps preserve pore spaces. In contrast, lithic sandstones are more susceptible to compaction and diagenetic alterations, leading to modified pore-throat structures and enhanced heterogeneity [22,23]. Strong compaction is the most significant factor in destroying the pore-throat structure of the reservoir, significantly increasing the degree of heterogeneity. The impact of different cements and dissolution processes on reservoir heterogeneity is more complex [24]. The Yanchang Formation is the main oil-producing horizon of tight sandstone oil in the Ordos Basin. Among them, the Chang 3 oil layer group in the Huachi area is an important productive layer, characterized by a large oil-bearing area, thick oil layers, and tight reservoirs, and it belongs to a typical low-permeability tight oil reservoir [25]. Due to the overall low development degree of the Chang 3 section strata in the Yanchang Formation of the Ordos Basin, there has been very little research on the Chang 3 section by predecessors. This has led to an unclear understanding of the heterogeneity and micro-pore-throat structural characteristics of the Chang 3 reservoir in the Huachi area, which seriously restricts the exploration and development of oil and gas in this region.

This paper takes the extremely low-permeability sandstone of the Chang 3 section in the southern part of the Huachi area in the Ordos Basin as the research object. By employing analytical techniques such as CTS, SEM, XRD, HPMI, and CRMI, this study quantitatively characterizes the pore-throat size distribution, fractal characteristics, and their controlling factors of the reservoirs in the study area [26–30]. Integrating sedimentary characteristics and diagenetic processes, the Chang 3 reservoirs in the study area are classified. The research findings will provide effective methods and a theoretical basis for the evaluation of extremely low-permeability sandstone reservoirs and the prediction of favorable zones.

2. Geological Setting

The Ordos Basin is a multi-cycle sedimentary basin characterized by stable subsidence and migrating depressions. It is part of the North China Basin. In the late Mesozoic era, it gradually separated from the North China Basin and evolved into a large inland basin. The Triassic system forms an asymmetric north–south oriented rectangular basin with a steep and narrow western flank and a broad and gentle eastern flank. The basin margins are marked by well-developed faults and folds, while the internal structure is relatively simple, with gently dipping strata (generally with a dip angle of less than 1°) (Figure 1a). Based on the current tectonic morphological characteristics of the basin, it can be divided into six first-order tectonic units, namely: the Yimeng Uplift, the Weibei Uplift, the Tianhuan Syncline, the Yishan Slope, the Jinxi Torsion Fold, and the Western Margin Thrust Belt. Overall, the basin's tectonics have been long-term stable, characterized by overall uplift, continuous subsidence, broad and gentle slopes, and low uplifts. The formation and evolution of the Ordos Basin, as well as the sedimentary characteristics of the Yanchang Formation, are mainly influenced by the surrounding Yinshan and Qinling orogenic belts, as well as the ancient uplifts of Lvliang, Longxi, Haiyuan, and Qianlisshan. The sediments within the basin primarily originate from three directions: northeast, southwest, and northwest, with significant differences in the types of parent rock assemblages. The lake basin's sedimentary evolution has gone through the entire process of formation, development, climax, decline, and extinction. Starting in the early Late Triassic, the basin began to subside and accept sediments, forming the Yanchang Formation, a lake-delta facies oil-bearing strata sequence over 1000 m thick. Based on sedimentary cycles and oil-bearing properties, the Yanchang Formation can be divided from bottom to top into 10 oil-bearing groups [31–33]. Among them, during the Chang 3 sedimentation period, the lake water rapidly receded and became shallower, with a large accumulation and progradation of clastic materials at the basin margins, resulting in a sedimentary body thickness of 90–110 m (Figure 1c) (modified by Lv, 2023) [34]. The lake basin is characterized by this oscillatory uplift and subsidence evolution, which has led to the development of multiple reservoir-cap rock assemblages in the vertical direction within the Yanchang Formation, facilitating the formation of large deltaic lithologic oil reservoirs [35,36].

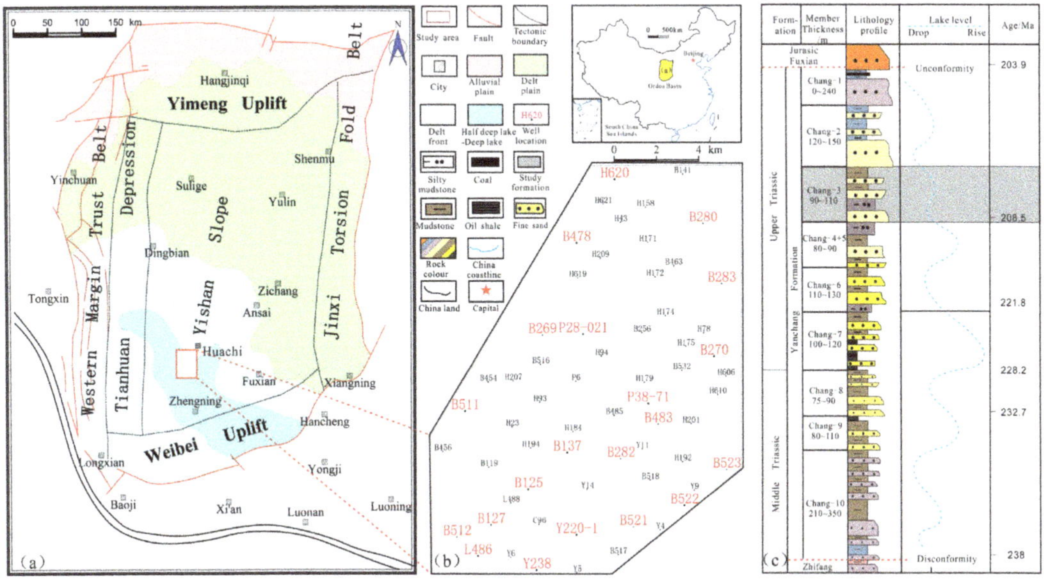

Figure 1. Tectonic position of the study area (**a**), sampling points (**b**), stratigraphic and lithological details of the Triassic Yanchang Formation in the Huachi area (**c**).

The Huachi area is located in the middle to southwestern part of the Yishan Slope in the Ordos Basin. During the Yanchang period, it was situated in the central part of the lake basin and represents the region with the largest sedimentary area during the Chang 3 period in the Ordos Basin. The surface topography of the area is characterized by typical loess plateau landforms with undulating terrain. The sandstone types of the Chang 3 reservoir in the study area are mainly lithic feldspar sandstone and feldspar lithic sandstone, with a high content of feldspar in the clastic components. The reservoirs are characterized by complex lithic and matrix compositions, low compositional maturity of rock components, moderate structural maturity, and a high degree of diagenesis.

3. Materials and Methods

3.1. Sampling and Processing

A total of over 50 representative core samples were evenly collected from the extremely low-permeability sandstones of the Triassic Yanchang Formation in the southern part of the Huachi area. The sampling depth ranged from 1493 to 1840 m (Table 1). The samples covered all sandstone types in the target formation. These samples were processed into thin sections and cylindrical specimens with a diameter of 2.5 cm and a length of approximately 7 cm. Since rock samples may be contaminated or contain impurities during sampling, they must be cleaned and dried to ensure the accuracy of the results. Using gentle cleaning methods, vacuum drying, and controlled drying temperatures can effectively prevent the sample pre-treatment process from affecting the rock's pore-throat structure. Prior to the experiments, all samples were cleaned with dichloromethane (Tianjin Fuyu Fine Chemical

Co., Ltd., Tianjin, China) to achieve the volatilization and dissolution of organic compounds. Subsequently, the samples were placed in a Memmert UN 110 constant temperature drying oven (Schwabach, Bavaria, Germany) for vacuum drying for 24 h at 120 °C. The effect of the pre-treatment process on the samples was verified by measuring differences in sample quality and observing changes in the pore-throat structure before and after treatment using CT scans.

Table 1. Sample location and lithology.

Sample	Well Name	Depth/m	Lithology
B125	B125	1742.5	Fine-grained arkose
B127-1	B127	1584.9	Medium-fine-grained arkose
B137	B137	1598.4	Medium-fine-grained arkose
B269-1	B269	1699.7	Fine-grained arkose
B269-2	B269	1699.8	Fine-grained arkose
B269-3	B269	1700	Fine-grained arkose
B269-4	B269	1700.2	Fine-grained arkose
B270	B270	1801.12	Very fine, fine-grained arkose
B270-1	B270	1741	Very fine, fine-grained arkose
B270-2	B270	1741.1	Fine-grained arkose
B270-4	B270	1800.7	Fine-grained arkose
B270-6	B270	1801.8	Fine-grained arkose
B280	B280	1714.02	Fine-grained arkose
B282	B282	1802.4	Fine-grained lithic arkose
B283-1	B283	1704.9	Fine-grained arkose
B283-3	B283	1705.6	Fine-grained arkose
B283-4	B283	1705.9	Fine-grained arkose
B283-6	B283	1718.6	Fine-grained arkose
B478	B478	1837	Fine-grained arkose
B478-3	B478	1837.7	Fine-grained arkose
B483-1	B483	1812.5	Fine-grained arkose
B483-2	B483	1812.9	Fine-grained arkose
B483-3	B483	1814	Medium fine-grained arkose
B483-4	B483	1816.1	Fine-grained arkose
B511	B511	1832.1	Fine-grained arkose
B511-2	B511	1833	Very fine. fine-grained arkose
B511-3	B511	1833.1	Very fine, fine-grained arkose
B512-1	B512	1761.2	Very fine, fine-grained arkose
B512-2	B512	1762.7	Medium-grained arkose
B521-1	B521	1600.8	Fine-grained feldspathic siltstone
B521-2	B521	1601.5	Fine-grained feldspathic siltstone
B522	B522	1652.3	Very fine, fine-grained arkose
B523-1	B523	1820.6	Very fine, fine-grained arkose
B523-2	B523	1821	Fine-medium grained arkose
B523-3	B523	1822.3	Argillaceous very fine, fine-grained lithic arkose
B523-5	B523	1822.7	Fine-grained arkose
H172-1	H172	1797.5	Fine-grained arkose
H172-2	H172	1801.5	Medium fine-grained arkose
H172-3	H172	1803.5	Medium-grained arkose
H184	H184	1815.6	Fine-grained arkose
H620	H620	1813.5	Very fine, fine-grained arkose
H620-1	H620	1820.8	Fine-grained arkose
H620-2	H620	1821.6	Medium fine-grained arkose
H620-3	H620	1822.6	Medium fine-grained arkose
L486	L486	1622.8	Argillaceous very fine, fine-grained lithic arkose
P28-021	P28-021	1750.4	Fine-grained arkose
P38-71	P38-71	1679.8	Fine-grained arkose
Y220-1	Y220-1	1705.9	Fine-grained lithic arkose
Y238	Y238	1757.13	Medium fine-grained feldspathic litharenite
Y53-1	Y53	1493	Fine-grained feldspathic litharenite

Thin sections were prepared for cast thin section (CTS) analysis (45 samples), X-ray diffraction (XRD) (35 samples), scanning electron microscopy (SEM) (27 samples), and cathodoluminescence (CL) (27 samples). Cylindrical samples were used for physical property measurements (50 samples) and heavy mineral analysis (12 samples). Given the extremely high heterogeneity of the reservoir in the study area, the goal was to facilitate the accurate characterization of the full-size pore-throat structure of the same rock sample by combining the results of high-pressure mercury injection (HPMI) and constant-rate mercury injection (CRMI) experiments. Rock core samples were evenly selected from 12 wells within the study area for mercury intrusion analysis (sample names are designated by well numbers (Figure 1b). After being cut in half, these samples were respectively used for the two types of mercury injection experiments.

3.2. Experimental Measurements

CTS analysis can determine the mineral composition, matrix material, pore types, and cement types. It can also achieve quantitative characterization of pore size, throat size, and grain size [37]. XRD can directly analyze the mineral composition and relative content of rocks, thereby determining the lithology of the rocks [38]. SEM is widely used to identify and authenticate different types of rocks and minerals, and to observe and analyze the microstructure of samples [39]. CL can detect minerals that are not easily observed by other optical microscopes [40]. Mercury injection experiments can determine the corresponding pore size and pore volume by measuring the amount of mercury entering the pores under different external pressures, thereby calculating the pore size distribution, which can effectively characterize the size, shape, and structural features of pores [41].

The CTSs of 45 samples were examined using an Olympus-cx21 polarizing microscope (Olympus, Tokyo, Japan) according to the Chinese Oil and Gas Industry Standard (SY/T) 5368-2000 [42]. According to standard SY/T 5163-2018 [43], XRD testing of whole rock was accomplished using the basic intensity contrast method with an X-ray diffractometer model AL-Y3500 (Aolong, Dandong, China). According to standard SY/T 5162-2014 [44], SEM observations were performed using a Tescan S-3200N field emission electron microscope. According to standard SY/T 5916-2013 [45], CL testing of rock flakes was accomplished using a CL8200 MK5 cathodoluminescence instrument (Cambridge Image Technology Ltd., Hatfield, UK). According to standard SY/T 5336-2006 [46], the helium porosity and pulse permeability were measured by using CMS-300 overburden pore penetrometer (Core Lab, Houston, TX, USA). This type of testing is particularly suitable for low-permeability samples that do not obey Darcy's law.

According to the standard SY/T 5346-2005 [47], HPMI experiments was conducted using the Autopore IV 9505 mercury injection instrument (Micromeritics, Atlanta, GA, USA). The experiment includes both the mercury injection under pressure and the mercury withdrawal under depressurization, with the maximum pressure reaching 200 MPa. The smallest identifiable pore-throat radius is approximately 3.6 nm. CRMI experiment was conducted using an ASPE730 constant-rate mercury injection instrument (Raykol, Xiamen, China), with a maximum injection pressure of 6.21 MPa and the smallest identifiable pore-throat radius being about 0.12 μm. Considering the high salinity of formation water in the study area, residual salt particles after the evaporation of moisture in natural cores may clog pore-throats. In accordance with the requirements of the standard SY/T 5336-2006 (Practices for core analysis), all rock samples were subjected to salt-washing treatment with methanol prior to mercury injection testing. The samples were then dried at 105 °C until a constant weight was achieved. Subsequently, the samples were placed in a sealed dilatometer, evacuated, and then tested.

CTS, XRD, SEM, CL, and pore-permeability tests were conducted at the State Key Laboratory of Continental Dynamics, Northwest University, P. R. China. HPMI and CRMI tests were completed at Beijing Qingchen Huanyu Petroleum Geological Technology Co. Ltd. (Beijing, China).

3.3. Fractal Theory

Fractal theory is widely used to quantitatively characterize the pore-throat structural features of rocks that exhibit self-similarity [48,49]. The ideal fractal dimension typically ranges from 2 to 3. A higher fractal dimension indicates a more complex pore-throat structure, while a lower fractal dimension indicates a more homogeneous structure [50]. This study, based on fractal theory, utilizes the mercury injection/withdrawal curves obtained from high-pressure mercury injection (HPMI) and constant rate mercury injection (CRMI) analyses of extremely low-permeability sandstone samples. It calculates the fractal dimension for each sample by integrating the characteristics of the pore size distribution. Fractal theory indicates that the number of pore-throats N(>r) with a radius greater than r can be expressed as follows [51,52]:

$$N(>r) = \int_{r}^{r_{max}} P(r)dr = ar^{-D} \tag{1}$$

where "r_{max}" represents the maximum of the pore-throat radius, "P(r)" denotes the distribution density function of pore-throat radius, "a" is a proportionality constant, and "D" represents the fractal dimension.

By taking the derivative with respect to "r", the distribution density function of pore-throat radius "P(r)" is derived.

$$P(r) = \frac{dN(>r)}{dr} = a'r^{-D-1} \tag{2}$$

where "a'" is a proportionality constant, which is equivalent to -D×a.

By converting Equation (2) to Equation (3), the total volume "V(<r)" of pores with a radius less than "r" can be represented:

$$V(<r) = \int_{r_s}^{r} P(r)ar^3dr = \frac{-a^2D}{3-D}\left(r^{3-D} - r_{min}^{3-D}\right) \tag{3}$$

The total pore volume (V) can be listed as follows:

$$V = \frac{-a^2D}{3-D}\left(r_{max}^{3-D} - r_{min}^{3-D}\right) \tag{4}$$

By converting Equations (3) and (4) into Equation (5), one can derive the cumulative volume fraction V(c) of pore-throats with radius less than "r(s)".

$$V(c) = \frac{V(<r)}{V} = \frac{r^{3-D} - r_{min}^{3-D}}{r_{max}^{3-D} - r_{min}^{3-D}} \tag{5}$$

Because "rmin" is far less than "rmax" Equation (5) can be simplified as follows:

$$V(c) = \left(\frac{r}{r_{max}}\right)^{3-D} \tag{6}$$

Taking a logarithm on both sides of Equation (6), the formula is changed as follows:

$$\log V(c) = \log(1 - S_{Hg}) = (3-D)\log r - (3-D)\log r_{max} \tag{7}$$

In the mercury injection experiment, the wetting phase is air, and the non-wetting phase is mercury, so logV(c) can be expressed as $\lg(1 - S_{Hg})$, where S_{Hg} is mercury saturation. If the pore size distribution conforms to fractal theory, a linear relationship exists between $\log(1 - S_{Hg})$ and logr. The Hurst exponent (H) can then be determined as the slope of this straight line. Then the fractal dimension (D) is equal to 3-H.

4. Results

4.1. Mineral Composition and Petrophysical Characteristics

Core observations and grain size analyses indicate that the Chang 3 reservoir in the study area mainly consists of grey and light grey fine-grained sandstones and siltstones. The primary grain sizes range from 0.05 mm to 0.45 mm, with an average of 0.24 mm. The grain size frequency curve is unimodal with a sharp peak, indicating a relatively simple composition of the sediments in the study area. The steep shape of the probability cumulative curve suggests a narrow grain size distribution range and moderate to good sorting. The probability cumulative curve mainly forms two subpopulations of jumps and suspensions, indicating poor rounding, predominantly sub-angular (Figure 2a). Based on the thin-section identification and the sandstone classification ternary diagram, the sandstone types of the Chang 3 reservoir in the study area are mainly arkose and a small amount of lithic arkose (Figure 2b). Feldspar is commonly found in potash feldspar. The total content of clastic materials ranges from 80.5% to 91.5%, with an average of 86.1%. The average proportions of feldspar, quartz, and lithic fragments are 51.0%, 24.5%, and 10.7%, respectively (Figure 2c).

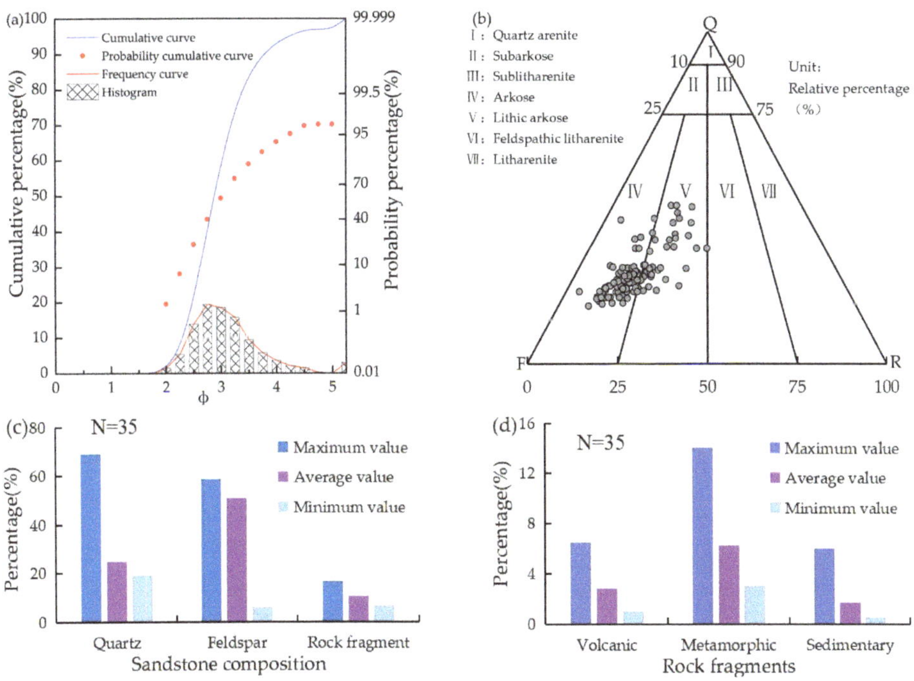

Figure 2. Reservoir characteristics of the Chang 3 member of the Yanchang Formation study area, Ordos Basin. N—number of samples. (**a**) Probability map of grain size. (**b**) Triangulation of the extended Chang 3 reservoir. (**c**) Histogram of sandstone composition. (**d**) Histogram of rock chip content.

The clastic components are predominantly feldspar, followed by quartz and lithic fragments (Figure 3a,b). The types of rock fragments are primarily metamorphic rock (average 6.23%), followed by volcanic rock (average 2.79%) and sedimentary rock (average 1.69%) (Figure 2d). Metamorphic rock fragments consist of quartzite, schist, phyllite, slate, and metamorphic sandstone. Igneous rock fragments mainly include effusive rocks and aphanitic rocks. Sedimentary rock fragments are mainly dolomite. According to the quantitative analysis by X-ray diffraction, the highest content of clay minerals is chlorite, with an average relative content of 57.7%. The second most abundant is the illite-montmorillonite interlayer, with an average relative content of 26.9%. Illite has the smallest proportion, with an average relative content of 15.4%. No illite zone was observed in the sandstone reservoir of the study area. Cementation is strongly developed, mainly consisting of clay mineral and carbonate cementation, with a small amount of siliceous cementation. SEM images reveal intergranular chlorite cementation (Figure 3c) and mixed illite-montmorillonite and micritic calcite cementation (Figure 3d). The main types of cementation include pore-lining, overgrowth-pore, and pore-lining-pore cementation.

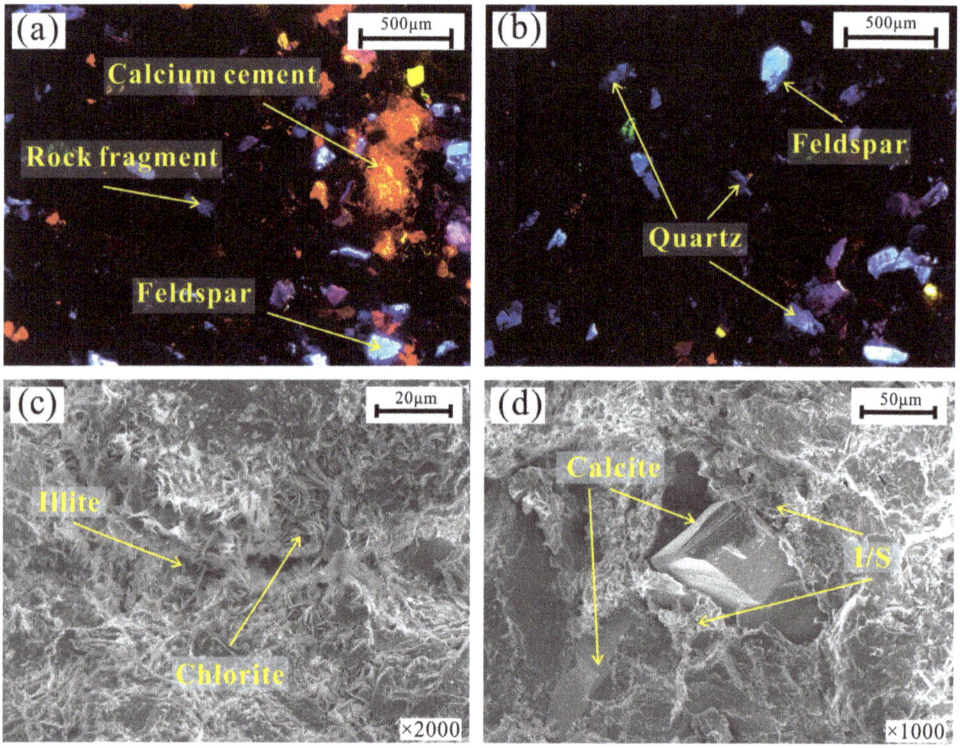

Figure 3. Mineral composition in the Huachi area based on CL and SEM. (**a**) CL thin section image showing feldspar and rock Fragment. (**b**) CL thin section image showing quartz and feldspar. (**c**) SEM image showing illite and chlorite. (**d**) SEM image showing chlorite and illite/smectite formation.

Physical property analysis indicates that the porosity of the Chang 3 section in the Huachi area mainly ranges from 10% to 18%, with an average of 12.01% (Figure 4a). Permeability is primarily distributed between 0.2 and 2 mD, with a secondary range of 1.5 to 5 mD, and an average value of 1.54 mD (Figure 4b). Due to the high content of chlorite in the reservoir, it can block pores and throats. This leads to poor reservoir connectivity and reduced permeability. The reservoir in the study area exhibits extremely strong heterogeneity and generally poor physical properties. However, due to the development of dissolution pores and micro-fractures, some local areas have high porosity and high permeability characteristics. During the Chang 3 period, the sedimentary environment of the study area was a delta front subfacies, characterized by the development of underwater distributary channels, underwater levees, and interdistributary bays. The average porosity and permeability of the underwater distributary channels are 13.6% and 1.45 mD, respectively. For the underwater levees, the average porosity and permeability are 10.4% and 0.35 mD, respectively. The interdistributary bays have average porosity and permeability values of 9.9% and 0.31 mD, respectively. Sedimentary processes are key factors influencing the distribution of sand bodies and controlling the porosity and permeability of the reservoir. Within different sedimentary facies, porosity and permeability show a clear exponential positive correlation (Figure 4c), indicating that the reservoir's permeability is significantly influenced by pore-throat size. The weaker correlation between porosity and permeability in underwater distributary channels suggests that the pore-throat structure in this type of reservoir is more complex and heterogeneous. Overall, the sandstone reservoir of the Chang 3 section in the Huachi area is a typical low-porosity, extremely low-permeability sandstone reservoir.

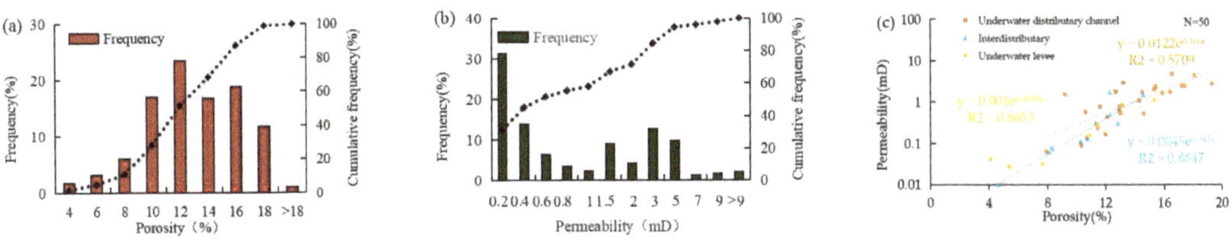

Figure 4. Petrophysical characteristics of the study area. (**a**) Distribution graph of porosity. (**b**) Distribution graph of permeability. (**c**) Porosity and permeability relationship graph (showing an index correlation).

4.2. Characteristics of the Pore-Throat Network

Based on the comprehensive analysis of CTS and SEM images from 27 cored wells in the study area, the extremely low-permeability sandstone reservoir of the Chang 3 section mainly develops three types of pores: intergranular pores, intragranular dissolution pores, clay interstitial pores, and microcracks (Figure 5).

The predominant pore type in the Chang 3 reservoir of the study area is residual intergranular pores, which exist between the framework grains. Their pore sizes are primarily concentrated in the range of 50 to 70 μm. These pores vary in size and have relatively regular shapes, often appearing as nearly triangular, quadrangular, and irregular forms (Figure 5a). Dissolution pores mainly include feldspar and lithic dissolution pores (Figure 5b). Feldspar dissolution occurs along unstable joints and twin seams, with some dissolution pores being filled with micritic calcite (Figure 5d). These dissolution pores often exhibit irregular reticulate, honeycomb, or elongated shapes, and in some cases, complete dissolution leads to the formation of moldic pores. Chlorite cementation can be observed on grain surfaces and between grains (Figure 5e). The pore sizes of these dissolution pores generally range from 10 to 30 μm. Lithic dissolution pores are relatively common in the Chang 3 section of the study area. They are mainly formed by the selective dissolution of soluble components within lithic grains, resulting in honeycomb-like dissolution pores, with lithic grains often being partially dissolved. The pore sizes generally range from 5 to 20 μm, with large dissolution pores visible locally, and their distribution is uneven (Figure 5f). Intergranular pores are well-developed in the reservoir, mainly consisting of kaolinite and halite intergranular pores (Figure 5g). The pore diameters typically range from 1 to 5 μm, with good sorting. These pores make a certain contribution to the reservoir's permeability.

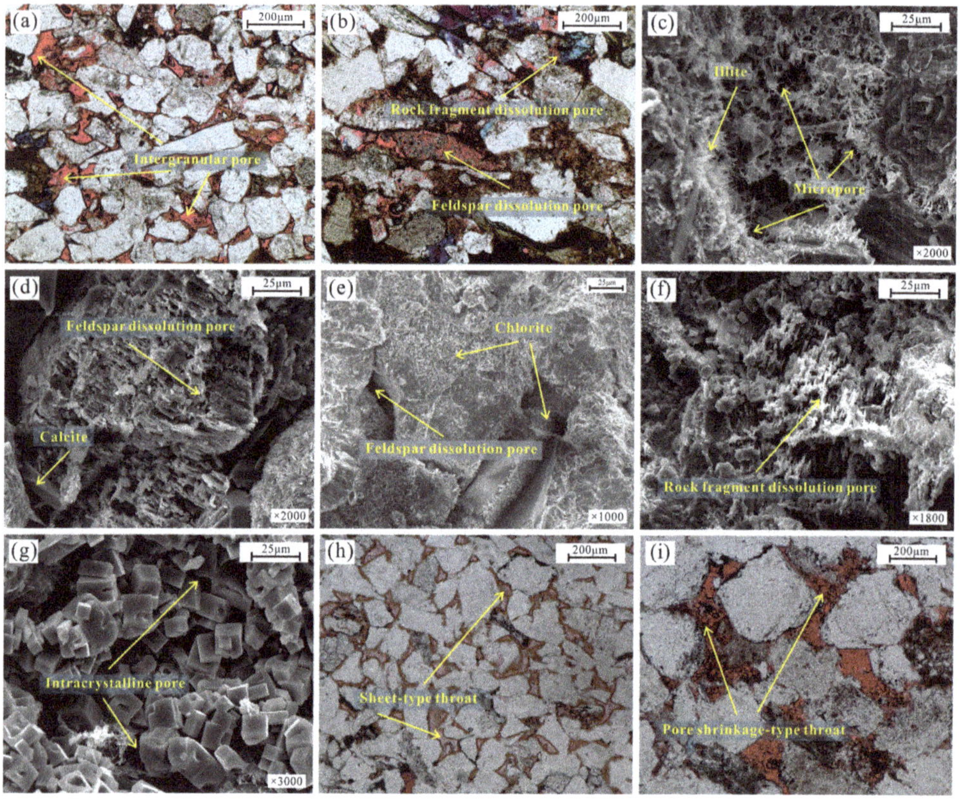

Figure 5. Pore-throat structure in the Huachi area based on CL and SEM. (**a**) CTS image showing intergranular pores developed a sheet-shaped throat. (**b**) CTS image showing feldspar and rock fragment dissolution pores developed. (**c**) SEM image showing intergranular filamentous illite clay and micropores. (**d**) SEM image showing feldspar dissolution pores filled with clay-crystal calcite. (**e**) SEM image showing feldspar irregular dissolution and chlorite colluvium. (**f**) SEM image showing rock fragment leachate dissolution pores. (**g**) SEM image showing intracrystalline pores. (**h**) CTS image showing sheet-type throat. (**i**) CTS image showing pore shrinkage-type throat.

Based on the size and shape of throats, the predominant throat type identified in the sandstone reservoir of the study area is sheet-type throats (Figure 5h), followed by pore shrinkage-type throats and bundled throats (Figure 5i).

4.3. Characteristics of the Pore-Throat Distribution

In the study area, 12 samples were selected for HPMI and CRMI tests. The reservoir properties and pore-throat structural parameters in the study area exhibit a wide range of variations, and the mercury withdrawal efficiency is relatively low. These characteristics indicate that the reservoir in the study area has strong heterogeneity. The rock samples were classified into three types (Figure 6) based on an integrated analysis of high-pressure mercury injection curve trends (Figure 7a), pore-throat radius distribution patterns (Figure 7b), pore-throat structural parameters, and rock physical properties (Table 2).

Table 2. Parameters from the CRMI experiment.

Sample	Porosity (%)	Permeability (mD)	Sorting Coefficient	Skewness	Structural Coefficient (φ)	HMIP Entry Pressure (MPa)	HMIP Maximum Mercury Saturation (%)	CMIP Entry Pressure (MPa)	CMIP Maximum Mercury Saturation (%)
B522	15.297	1.080	1.62	0.30	1100.96	0.30	98.56	0.30	50.95
B270	18.100	4.360	1.08	0.58	1252.07	0.18	97.92	0.18	63.95
B478	16.400	1.700	2.43	0.23	383.55	0.34	96.60	0.34	51.67
P38-71	17.200	2.390	0.75	0.41	534.49	0.21	96.85	0.21	58.46
H620	15.886	1.594	1.52	0.18	357.57	0.30	96.48	0.30	66.85
B511	13.000	0.532	1.82	0.10	372.36	0.28	96.68	0.68	53.11
P28-021	13.105	0.828	1.26	0.16	161.94	0.30	98.74	0.30	62.59
Y220-1	12.300	0.500	2.46	0.36	17.45	1.36	93.15	0.36	57.63
B137	7.905	0.062	2.04	0.28	11.85	1.26	94.95	1.26	53.37
Y238	8.109	0.060	1.94	0.15	16.52	1.12	95.93	1.12	50.95
B125	4.608	0.010	2.04	−0.07	0.15	3.51	91.44	3.51	16.37
L486	4.096	0.040	2.23	−0.33	7.93	2.99	90.17	1.99	32.55

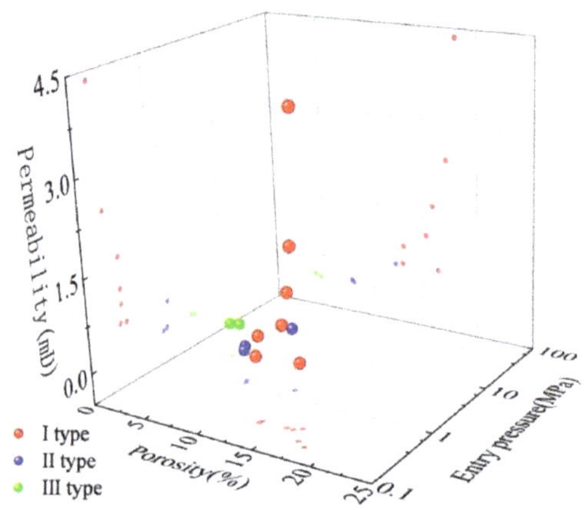

Figure 6. Sample classification results based on high-pressure mercury injection test parameters.

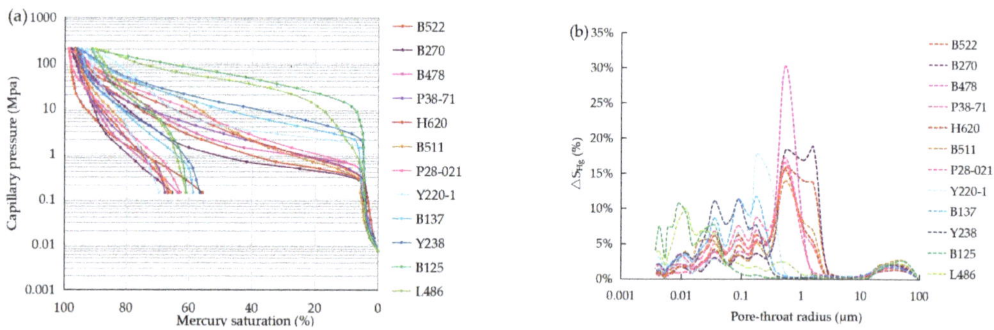

Figure 7. Capillary pressure curves, pore-throat size distribution curves for sandstone samples in the study area. (**a**) Typical high-pressure mercury injection experiment curves. (**b**) The characteristics of the pore radius of the high-pressure mercury injection experiment.

Type I samples are characterized by capillary pressure curves that are low, flat, and elongated. The average maximum mercury injection saturation is 97.4%. The entry pressure is less than 1 MPa, with an average value of 0.28 MPa. The predominant storage space consists of intergranular pores, which account for 69% of the total porosity. Feldspar dissolution pores are the second most abundant, making up 27% of the total porosity. Throat types are predominantly pore-shrinkage type. Porosity is greater than 12%, with an average value of 15.6%. Permeability is greater than 0.5 mD, with an average value of 1.8 mD. The average skewness is 0.28, indicating a positively skewed distribution of pore-throat sizes. This suggests that the reservoir has good storage and permeability capabilities. The sorting coefficient mostly ranges from 1 to 2.5, with an average value of 1.5, indicating good sorting. The structural coefficient varies significantly and is always greater than 100. This indicates that the pore-throat tortuosity is high and the pore-throat structure is complex in this type of reservoir. The pore-throat radius distribution curve of type I samples exhibits a weak bimodal shape, with the main peak at 0.54 μm and the secondary peak at 1.57 μm.

The angle between the mercury injection and withdrawal curves of type II samples is small. The average maximum mercury injection saturation is 94.7%. The entry pressure is high, ranging from 1 to 2 MPa, with an average value of 1.25 MPa. The pore space of this type of reservoir is dominated by open and semi-closed pores. The pore sizes are medium, with fine-microfine throats being predominant. Porosity ranges from 7% to 12%, with an average value of 9.44%. Permeability ranges from 0.05 to 0.5 mD, with an average value of 0.21 mD. The average skewness is 0.27, indicating a positively skewed distribution of pore-throat sizes. The sorting coefficient mostly ranges from 1 to 2.5, with an average value of 2.15, indicating good sorting. The structural coefficient ranges from 10 to 100, indicating that this type of reservoir has moderate pore-throat tortuosity and good connectivity. The pore-throat radius distribution of type II samples is relatively broad and exhibits a multi-modal shape, mainly distributed between 0.04 and 0.18 μm.

The capillary pressure curves of type III samples are characterized by being high, steep, and short, with an average maximum mercury injection saturation of 90.8%. The entry pressure is greater than 2 MPa, with an average value of 3.25 MPa. The sorting is medium-good, and the pore-throat sizes exhibit a negatively skewed distribution. This indicates poor storage and permeability capabilities of the reservoir. The structural coefficient is less than 10, indicating that this type of reservoir has low pore-throat tortuosity and typically develops isolated micro- and nanopores and microfine throats. The pore-throat connectivity is poor. Porosity is less than 7%, with an average value of 4.35%. Permeability is less than 0.05 mD, with an average value of 0.03 mD. The pore-throat radius distribution of type III samples shows a distinct bimodal pattern, with the primary peak at 0.009 μm and the secondary peak at 0.036 μm.

CRMI experiments can effectively distinguish between pores and throats [53]. Figure 8 shows the CRMI curves of typical samples with three types of pore-throat structures. In the early stages of mercury injection, the amount of mercury entering the throat increased rapidly, while the amount of mercury entering the pore was minimal. The shape of the throat capillary pressure curve coincided with that of the total capillary pressure curve. Subsequently, the pore mercury injection increased, and the total capillary pressure curve deviated from the throat capillary pressure curve. When the pore mercury feed no longer increases, the pore mercury feed curve steepens and completely deviates from the total capillary pressure curve. This stage is the pore-throat transition zone. As the mercury injection pressure continues to increase further, the mercury saturation in throats increases continuously, while the mercury saturation in pores no longer changes. The trend of the capillary pressure curve for throats becomes consistent with that of the total capillary pressure curve. This indicates that the pores in the rock sample have been completely filled with mercury, and any subsequent increase in mercury is entirely from the throats. This stage is referred to as the throat zone [16]. The mercury injection curves of all three types feature both the pore-throat transition zone and the throat zone. The mercury saturation in throats is significantly higher than that in pores, indicating that there are many isolated pores surrounded by extremely narrow throats [15].

For type I samples, the average entry pressure, maximum mercury injection saturation, and maximum connected pore-throat radius are 0.33 MPa, 58.23%, and 2.59 μm, respectively. In these samples, there is a large difference between the mercury saturation in pores and throats (Figure 8a). For type II samples, the average entry pressure, maximum mercury injection saturation, and maximum connected pore-throat radius are 0.91 MPa, 53.98%, and 1.08 μm, respectively (Figure 8b). For type III samples, these values are 2.75 MPa, 24.46%, and 0.48 μm, respectively. The difference in mercury saturation between pores and throats in these samples is smaller than in the other two types (Figure 8c). This indicates that, given similar pore radii and porosity, the larger the throats, the higher the mercury saturation. Throats have a significant impact on extremely low-permeability sandstone reservoirs.

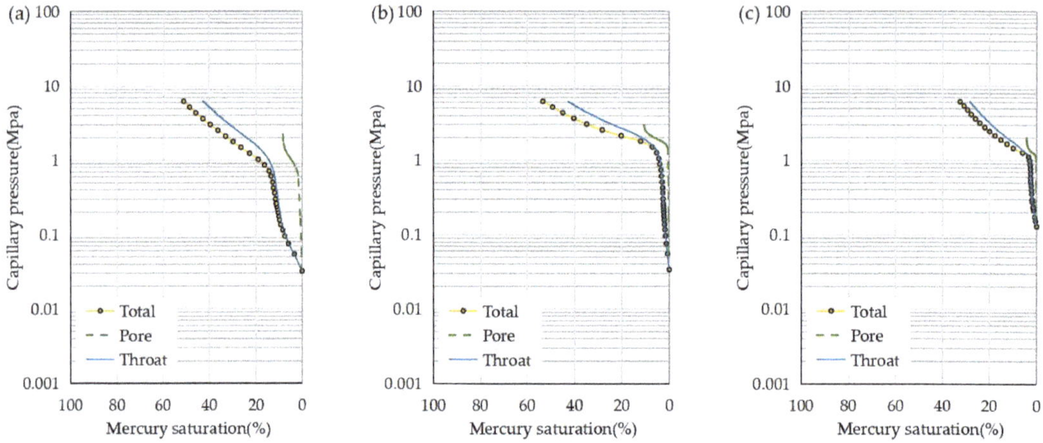

Figure 8. Typical constant-rate mercury injection curves for three types of samples in the study area. (**a**) Constant-rate mercury injection curves for type I samples (B522). (**b**) Constant-rate mercury injection curves for type II samples (B137). (**c**) Constant-rate mercury injection curves for type III samples (L486).

The pore radius distribution range of the three types of samples is basically consistent, all falling between 100 and 250 μm. The throat radius distribution of different samples shows significant differences, with no obvious pattern in the peak values, and the average throat radius ranges from 0.38 to 8.0 μm. The strong variability in throat distribution leads to abnormally large pore-throat ratios, with the average pore-throat radius ratio ranging from 17.25 to 489.01 (Figure 9). This further indicates that the pore-throat distribution in the extremely low-permeability sandstone reservoirs of the study area is complex and highly heterogeneous.

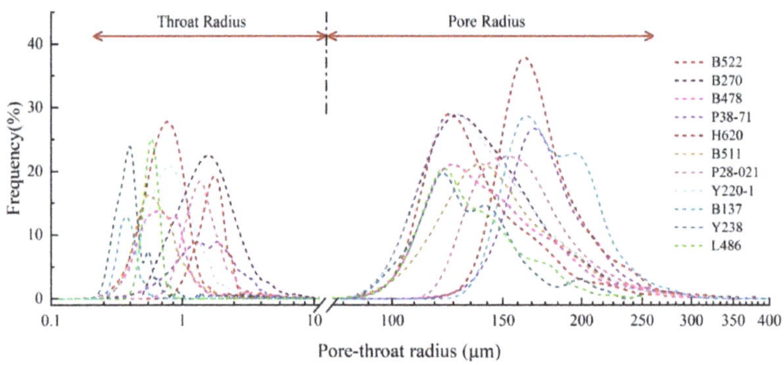

Figure 9. The characteristics of the pore radius of the constant-rate mercury injection experiment.

4.4. Fractal Features

4.4.1. Fractal Characteristics of HPMI and CRMI

In this study, the pore-throat types were classified into four categories using the Xoaoth classification standard: nanopores ($r < 0.01$ μm), micropores ($0.01 < r < 0.1$ μm), mesopores ($0.1 < r < 1$ μm), and macropores ($r > 1$ μm) [54]. The high-pressure mercury pressing and constant velocity mercury pressing data of 12 samples were processed to plot scatter plots of $\lg(1 - S_{Hg})(>r)$ and $\lg(r)$. Superimposed pore-throat radius distribution curves were fitted to the segmented pore-throat fractal curves for different sizes. Fractal curves of three typical samples are shown, respectively (Figure 10). The fractal dimensions of different sizes of the pore-throats were calculated by substituting the fitted slopes into Equation (7). The fractal dimensions based on high-pressure mercury compression tests were defined as Dh1 (micropores), Dh2 (nanopores), Dh3 (mesopores), and Dh4 (macropores). The fractal dimension based on the constant velocity mercury pressure test is defined as Dc3 (mesopores) and Dc4 (macropores).

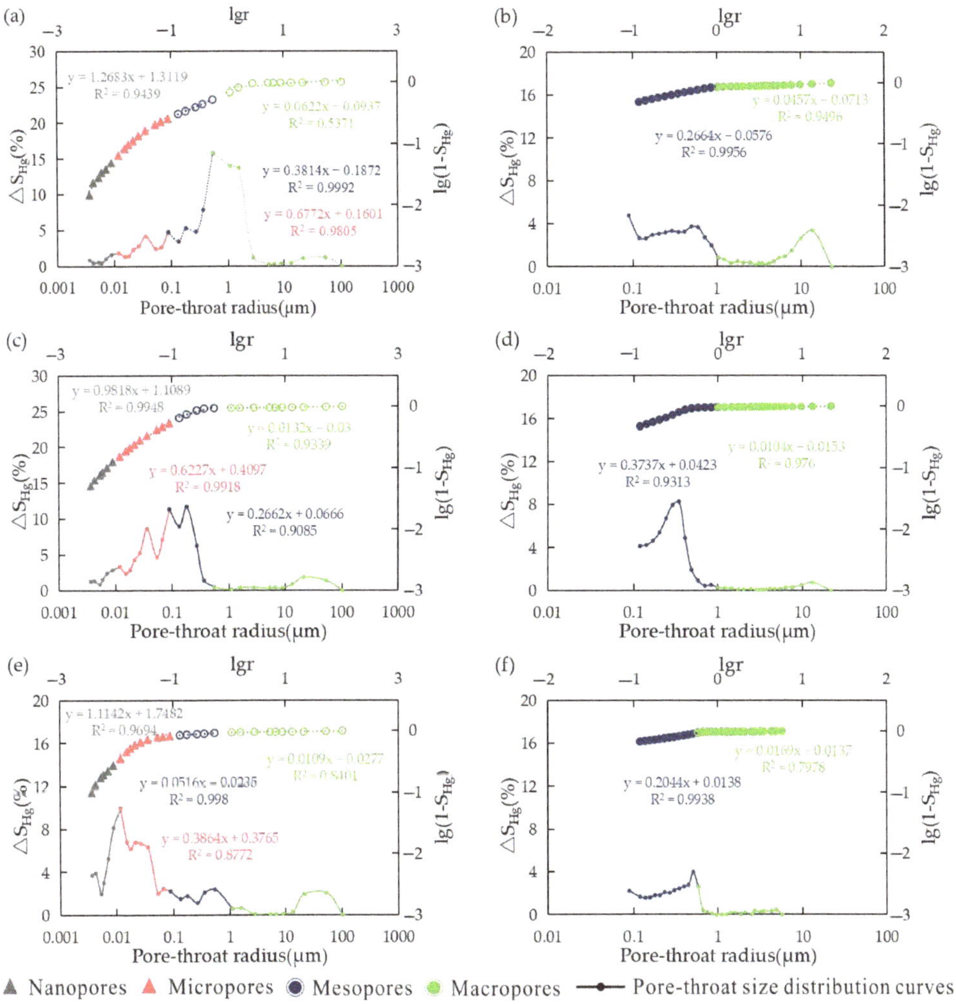

Figure 10. Fractal characteristic and pore-throat radius distribution curves of three types of rock samples. (**a**) B522 sample HPMI fractal fitting curve. (**b**) B522 sample CRMI fractal fitting curve. (**c**) B137 sample HPMI fractal fitting curve. (**d**) B137 sample CRMI fractal fitting curve. (**e**) L486 sample HPMI fractal fitting curve. (**f**) L486 sample CRMI fractal fitting curve.

By comparing the fractal dimension fitting curves of HPMI and CRMI, it is evident that HPMI characterizes mesopores and macropores with fewer data points and lower accuracy relative to CRMI. This indicates that HPMI has a shielding effect on larger pores and is more suitable for characterizing smaller pores (Figure 10a,c,e). The maximum pressure for CRMI is 6.2 MPa, with a constant injection rate, which can eliminate the "hempskin effect" and compensate for the shortcomings caused by high pressure, such as mineral compaction and wetting hysteresis. It is more suitable for characterizing the distribution of larger pore-throats. However, mercury cannot enter the storage space with a radius less than 0.12 μm, and therefore cannot characterize the pore-throat structure at the micronano scale (Figure 10b,d,f) [55]. For instance, the correlation of the fractal dimension fitting curve for macropores obtained by HPMI for type I sample B522 (R^2 = 0.54) (Figure 10a) is significantly lower than that obtained by CRMI (R^2 = 0.95) (Figure 10b).

Therefore, combining HPMI for characterizing nanopores and micropores with CRMI for characterizing mesopores and macropores can provide a detailed characterization of the full-scale pore-throat structural features of extremely low-permeability sandstone reservoirs.

4.4.2. Combined HPMI and CRMI for Full-Size Pore-Throat Fractal Characterisation

The pore-throat radius distribution curves with a radius less than 0.12 μm obtained by HPMI were spliced with those with a radius greater than 0.12 μm obtained by CRMI. That is, the pore-throat radius distribution and fractal characteristics of nanopores and micropores are characterized by HPMI curves, while those of mesopores and macropores are characterized by CRMI curves. The full-size pore-throat radius distribution curves of 12 samples were plotted. In order to quantitatively characterise the complexity of the pore-throat structure and the microhomogeneity of the extremely low-permeability sandstone reservoir, scatter plots of lg(r) and $\lg(1 - S_{Hg})$ were drawn based on the full-size pore-throat radius distribution curves (Figure 11). As can be seen from the figure, the pore-throat radius splicing curves of the 12 samples are complete, with smooth splicing points, which can accurately characterize the pore-throat structural features of the reservoir. The fractal dimensions of pore-throats of various sizes were calculated by segmental curve fitting (Table 3).

Table 3. Fractal dimension calculation results of the Chang 2 sandstone samples in the Haojiaping area.

Sample	Dh_1	R^2	Dh_2	R^2	Dc_3	R^2	Dc_4	R^2
B522	1.732	0.944	2.323	0.981	2.726	0.999	2.954	0.950
B270	1.732	0.944	2.323	0.981	2.777	0.998	2.845	0.844
B478	1.990	0.975	2.504	0.985	2.725	0.995	2.959	0.981
P38-71	2.260	0.973	2.285	0.999	2.771	0.999	2.905	0.707
H620	2.195	0.982	2.268	0.980	2.636	0.997	2.927	0.738
B511	1.990	0.974	2.331	0.922	2.693	0.992	2.960	0.987
P28-021	0.967	0.900	2.380	0.988	2.679	0.998	2.928	0.830
Y220-1	1.957	0.923	2.722	0.995	2.711	0.999	2.935	0.935
B137	2.018	0.995	2.377	0.992	2.626	0.931	2.990	0.972
Y238	1.731	0.951	2.307	0.995	2.825	0.994	2.984	0.965
B125	1.323	0.969	2.688	0.927	2.971	0.823	2.972	0.966
L486	1.886	0.969	2.614	0.877	2.805	0.987	2.984	0.954

Fractal dimensions can quantitatively reflect the complexity and heterogeneity of pore-throat structures. The fractal dimensions of porous rocks typically range between 2 and 3. When the fractal dimension is closer to 2, it indicates that the reservoir pore surfaces are smoother and more uniformly distributed. Conversely, when the fractal dimension is closer to 3, it suggests that the reservoir pore-throat structure is more complex and the surfaces are rougher. The fractal dimension of nanopores (Dh1) is mostly less than 2, with an average value of 1.82. This suggests that the nanopore surfaces are smooth and approximate two-dimensional planes, exhibiting relatively simple structures. The pore-throat system demonstrates high structural homogeneity, being primarily composed of isolated pores. This indicates that nanopores do not exhibit ideal fractal characteristics.

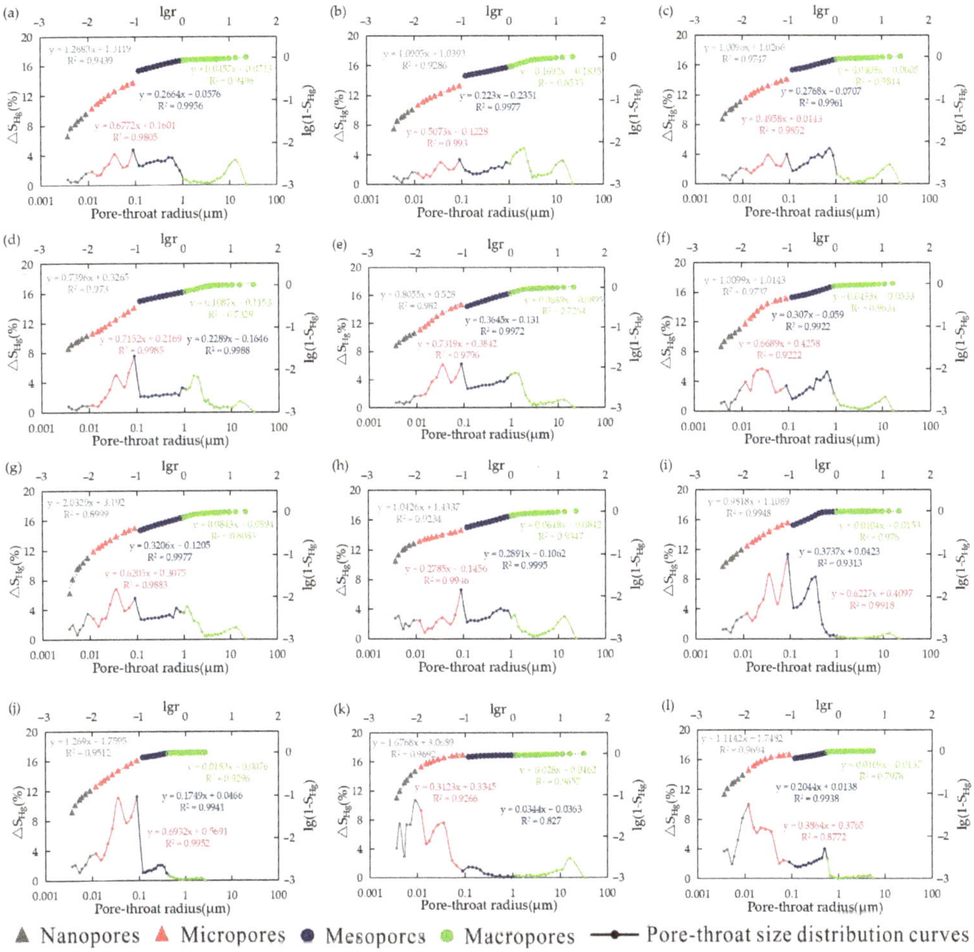

Figure 11. Combination of HPMI and CRMI to characterise the full pore size fractal dimension fitting curve for samples B522 (**a**), B270 (**b**), B478 (**c**), P38-71 (**d**), H620 (**e**), B511 (**f**), P28-021 (**g**), Y220-1 (**h**), B137 (**i**), Y238 (**j**), B125 (**k**), and L486 (**l**).

The pore-throat fractal dimensions of the 12 samples are shown in Table 2 (Table 2). The fractal dimension of micropores (Dh2) ranges from 2.268 to 2.722, with an average value of 2.43. The fractal dimension of mesopores (Dc3) ranges from 2.626 to 2.971, with an average value of 2.75. The fractal dimension of macropores (Dc4) ranges from 2.845 to 2.990, with an average value of 2.95. Dc4 is the highest, followed by Dc3, while Dh2 is the lowest. This indicates that the larger the pore-throat size, the rougher the surface, and the more complex the structure. It can be attributed to the large-scale pore-throats having a larger reservoir space due to the influence of solvation. The pore-throat spaces are deformed, resulting in stronger heterogeneity and higher fractal dimensions.

5. Discussion

5.1. Relationships Between Fractal Dimension and Pore-Throat Structure Parameters

In order to study the extent to which reservoir pore-throat structural parameters are affected by fractal dimension, scatter plots were used to represent the variation of pore-throat structural parameters under fractal dimension for different sizes of pore-throat. The correlation between fractal dimension and pore structure parameters was analysed (Figure 12).

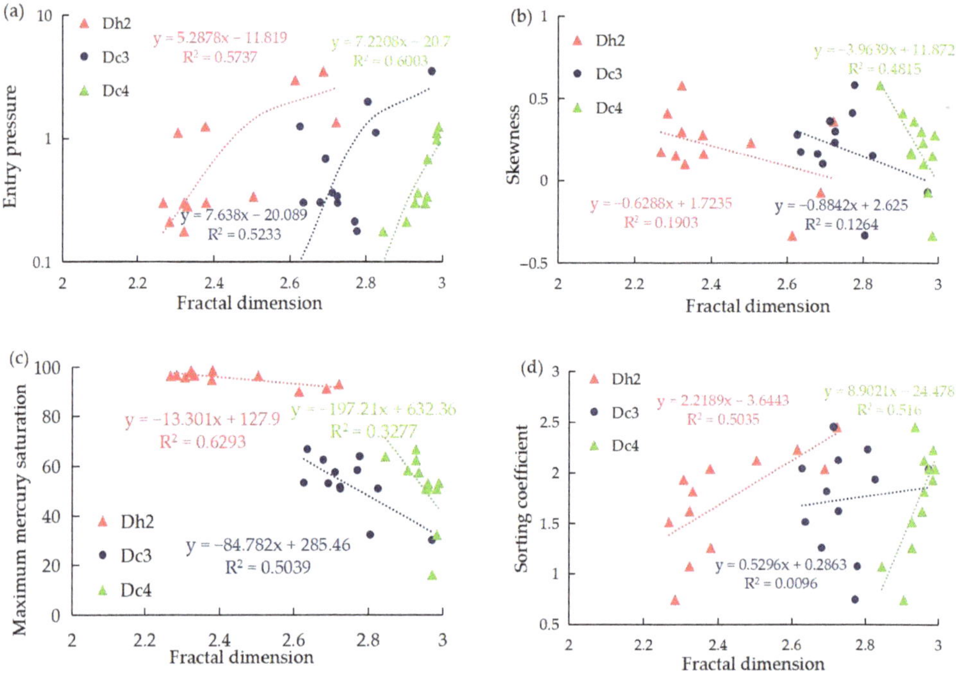

Figure 12. Correlation of fractal dimension with pore-throat structure parameters. (**a**) Correlation between fractal dimension and entry pressure. (**b**) Correlation between fractal dimension and skewness. (**c**) Correlation between fractal dimension and maximum mercury saturation. (**d**) Correlation between fractal dimension and sorting coefficient.

Entry pressure refers to the capillary pressure of the largest connected pore in the pore system. The magnitude of entry pressure reflects the concentration and size of the reservoir pore-throat. It is one of the main parameters for distinguishing the quality of rock storage performance [56]. Entry pressure shows a significant positive correlation with Dc3 and Dc4, with correlation coefficients of 0.5 and 0.6, respectively. It has a weak positive correlation with Dh2 (Figure 12a). The development of mesopores and macropores is a key factor influencing the pore-throat structure of the reservoir. Changes in the fractal dimension indicate alterations in the complexity of the reservoir's pore-throat structure. The higher the fractal dimension of mesopores and macropores, the more irregular the pore-throat structure of the reservoir, the greater the tortuosity, and the stronger the heterogeneity. The smaller the maximum connected throat radius of the reservoir, the lower the permeability, which leads to an increase in displacement pressure.

Skewness is a measure of asymmetry in the distribution of pore-throat sizes, with Skp ranging from −1 to 1. Skp = 0 indicates a symmetrical pore-throat distribution curve; Skp

< 0 is a fine skewness; Skp > 0 is a coarse skewness. Skewness showed a clear negative correlation with the Dc4 with a correlation coefficient of 0.5, and a weak negative correlation with the Dc3 and Dh2 (Figure 12b). It indicates that the symmetry of the distribution of micropores and mesopores has a certain influence on the fractal characteristics of the reservoir pore-throat structure. The asymmetry of the size distribution of macropores has a greater influence on the physical properties of the reservoir. The larger the fractal dimension, the more it reduces the effective reservoir space and permeation capacity of the reservoir. The pore-throat size distribution tends to be smaller, and the skewness decreases.

The maximum mercury injection saturation refers to the cumulative mercury saturation at the highest pressure during the experiment, which can reflect the storage capacity and heterogeneity of the reservoir. The maximum mercury injection saturation shows a negative correlation with micropores, mesopores, and macropores. Among them, the correlation with micropores and mesopores is more significant (Figure 12c). This indicates that the more complex the pore-throat structure, the weaker the storage capacity and homogeneity of the reservoir. The heterogeneity of micropores and mesopores is stronger, and, therefore, the connectivity of smaller pores and throats has a greater impact on the storage capacity of the reservoir.

The sorting coefficient indicates the uniformity of reservoir particle sizes and reflects the concentration of pore-throat distribution. The smaller the sorting coefficient, the more uniform the pore-throat distribution. The sorting coefficient shows a significant positive correlation with micropores and macropores. There is no obvious correlation with mesopores (Figure 12d). This indicates that the better the sorting, the more regular the pore-throat structure and the stronger the homogeneity. Micropores and macropores play a decisive role in the uniformity of extremely low-permeability sandstone reservoirs. The larger the fractal dimension, the stronger the heterogeneity of the reservoir, the more complex the pore-throat structure, and the poorer the sorting.

In summary, the fractal dimension of macropores shows a significant correlation with displacement pressure, skewness, sorting coefficient, and maximum mercury injection saturation. The fractal dimension of mesopores is only significantly correlated with displacement pressure and maximum mercury injection saturation. The fractal dimension of micropores only has a clear correlation with maximum mercury injection saturation and sorting coefficient. The analysis results indicate that the heterogeneity and permeability of the reservoir in the study area are more significantly influenced by the development degree and surface roughness of macropores. The connectivity of micropores and mesopores has a certain impact on the storage capacity of the reservoir.

5.2. The Main Controls on the Fractal Dimension

5.2.1. Sedimentation

Sedimentary processes control the distribution of reservoir sand bodies and the characteristics of sand-mud combinations. Different sedimentary facies exhibit distinct features in lithology, grain structure, and composition, which directly affect the microscopic pore-throat structure of the reservoir. During the Chang 3 period of the Yanchang Formation in the study area, the sedimentary environment was a delta front subfacies, characterized mainly by the development of underwater distributary channels, underwater levees, and interdistributary bays. By studying the distribution characteristics of sedimentary facies during the Chang 3 period in the study area, it can be found that type I samples are mainly distributed in the underwater distributary channel microfacies. Type II samples are primarily found in the underwater levee microfacies. Type III samples are predominantly developed in the interdistributary bay microfacies (Figure 13). Therefore, sedimentary facies have a certain controlling effect on the pore-throat structure of low-permeability sandstones.

This study quantitatively describes the relationship between sedimentary facies and fractal dimensions from the perspective of heavy mineral maturity. Heavy mineral maturity can be used to analyze the transportation distance of sediments, commonly represented by the proportion of zircon, tourmaline, and rutile in heavy minerals (ZTR = (zircon content + tourmaline content + rutile content)/heavy mineral content × 100). As the transportation distance increases, the content of stable heavy minerals relatively rises. The higher the maturity, the larger the ZTR value [57]. The average ZTR value of the reservoir in the study area is 54%, indicating a relatively low compositional maturity. The scatter plot of fractal dimensions versus mineral maturity shows a significant positive correlation between the fractal dimension of large pores and throats (D4) and ZTR (Figure 14). Micropores and mesopores show no obvious correlation with ZTR. Zircon, rutile, and other stable heavy minerals have strong resistance to weathering, leading to complex compositions, poor sorting, and complex pore-throat structures in sandstone reservoirs. The complexity of large pore-throat structures is more significantly influenced by the content of heavy minerals.

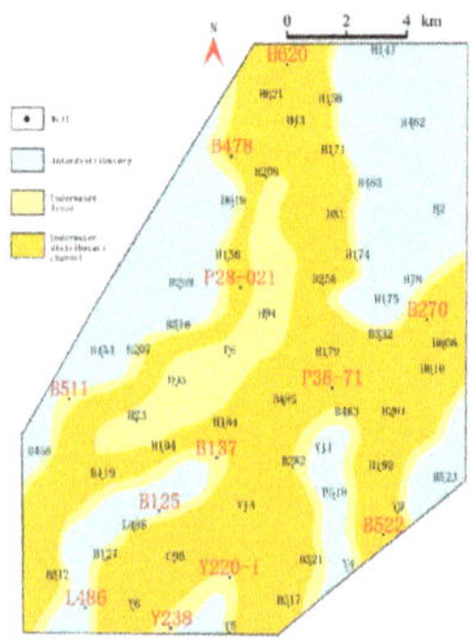

Figure 13. Characteristics of sedimentary microphase spreading in the long 3 section of the study area.

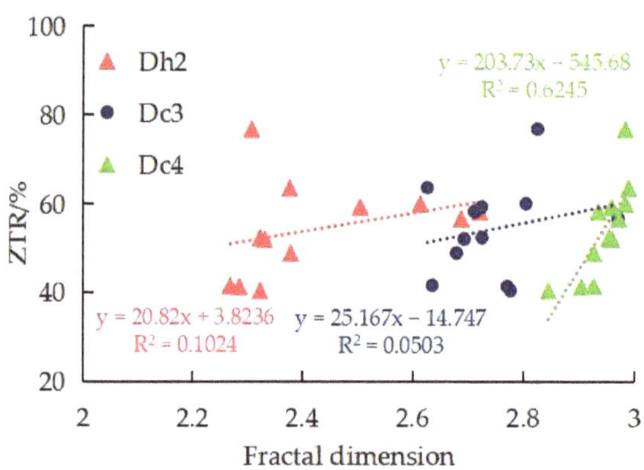

Figure 14. Correlation of fractal dimension with component maturity (ZTR).

5.2.2. Diagenesis

Based on thin-section and scanning electron microscope observations, the extremely low-permeability sandstone reservoirs in the study area are primarily influenced by compaction, cementation, and dissolution processes (Figures 3 and 5). Among these, compaction and cementation are destructive diagenetic processes and are the main controlling factors for the reduction of reservoir storage space.

(1) Compaction

The study area has a low content of rigid particles such as quartz and barite. The reservoir is susceptible to plastic deformation, has relatively weak resistance to compaction, and exhibits distinct point-to-line contacts of particles (Figure 5h,i). These factors result in a complex pore-throat structure with poor connectivity, thereby increasing the heterogeneity and fractal dimension of the reservoir [58].

(2) Cementation

The predominant cementation type is clay mineral filling cementation (Figure 3c,d). Different types and contents of clay minerals have different effects on pore-throat structure. In the study area, the development of chlorite coatings can effectively inhibit the secondary enlargement of quartz and compaction, thereby protecting the pore-throat structure. However, when the chlorite content is high, it can block pores, leading to reduced porosity and permeability, as well as increased reservoir heterogeneity and fractal dimension [59].

(3) Dissolution

Dissolution, on the other hand, is a constructive diagenetic process and is the primary means of forming secondary porosity. Solubility phenomena are widely observed in feldspar and rock fragments, as evidenced by thin sections and scanning electron microscopy. This improves reservoir connectivity, enhances pore space, and reduces fractal dimension. The study area reservoirs contain a relatively high amount of unstable feldspar minerals, and the dissolution of feldspar can generate a variety of different types, sizes, and complex structures of pores and throats (Figure 5d,e). If dissolution is intense, the pores formed are larger, resulting in a greater disparity in pore and throat radius within the reservoir, increased heterogeneity, and a larger fractal dimension.

Smaller pores and throats in the reservoir, due to their smaller radius, are less affected by diagenetic processes, and their pore-throat structures remain almost undeformed (Figure 5g). Therefore, smaller pore-throats have similar shapes, weaker heterogeneity, higher self-similarity, and correspondingly smaller fractal dimensions. Larger storage spaces in the reservoir, such as residual intergranular pores and feldspar dissolution pores, are strongly affected by diagenetic processes. Consequently, larger pores and throats undergo significant structural deformation, exhibit stronger heterogeneity, and have larger fractal dimensions.

5.3. *Reservoir Type Classification*

Based on sedimentary characteristics and reservoir space types, combined with pore-throat structural parameters and the characteristics of mercury injection curves, the extremely low-permeability sandstone reservoirs in the study area are divided into three categories (Table 4). Type I reservoirs are mainly found in the underwater distributary channel microfacies, with pore-throat structures dominated by large intergranular pores, dissolution pores, and pore-shrinkage-type throats. The maximum mercury injection saturation is greater than 96%. Type II reservoirs are primarily developed in the underwater levee microfacies, with pore-throat structures dominated by small to medium intergranular pores, a small number of dissolution pores, narrow sheet-type, and neck-shaped throats. The maximum mercury injection saturation ranges from 92% to 96%. Type III reservoirs are mainly developed in the interdistributary microfacies, with pore-throat structures dominated by intragranular pores, intergranular pores, and bundled throats. The maximum mercury injection saturation is less than 92%. Among them, type I and type II reservoirs are considered sweet spots for oil and gas exploration. By combining the full-size pore-throat fractal dimension characteristics to study the heterogeneity of pore-throat structures and to conduct quantitative evaluations, a more precise basis for reservoir classification is provided.

Table 4. Fractal characteristics of the pore-throat structure.

Parameters	Type I	Type II	Type III
Sedimentation	Underwater distributary channel	Underwater levee	Interdistributary
Pore types	Large intergranular pores, dissolution pores	Small-medium intergranular pores, dissolution pores	Intercrystalline pores, intraparticle pores
Throat types	Pore shrinkage-type, wide sheet-type throats	Narrow sheet-type, neck-shaped throats	Tubular throats
Permeability (mD)	>0.5	0.05~0.5	<0.05
Porosity (%)	>12	7~12	<7
Entry pressure (MPa)	<1	1~2	>2
Maximum mercury saturation (%)	>96	92~96	<92
Full-size pore-throat distribution characteristics and fractal dimension fitting curves	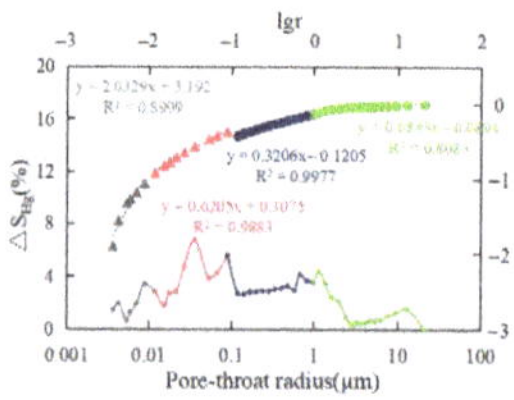	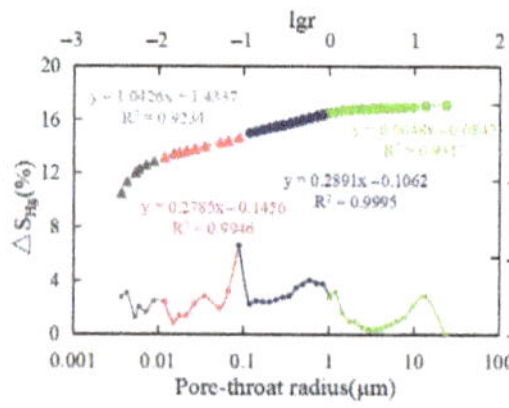	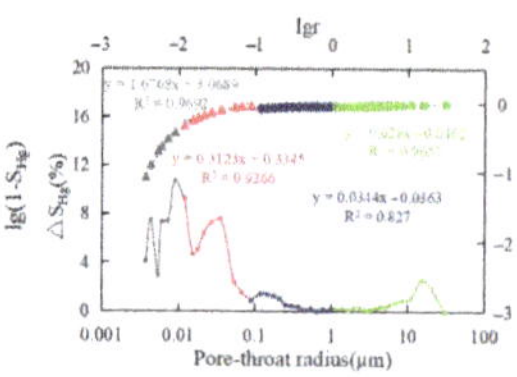
Typical HPMI curve	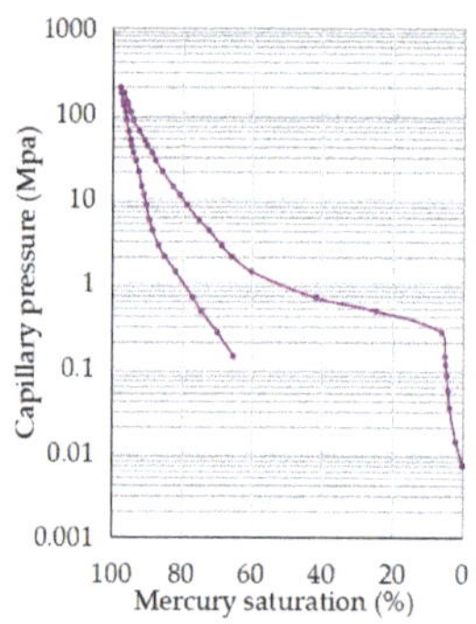	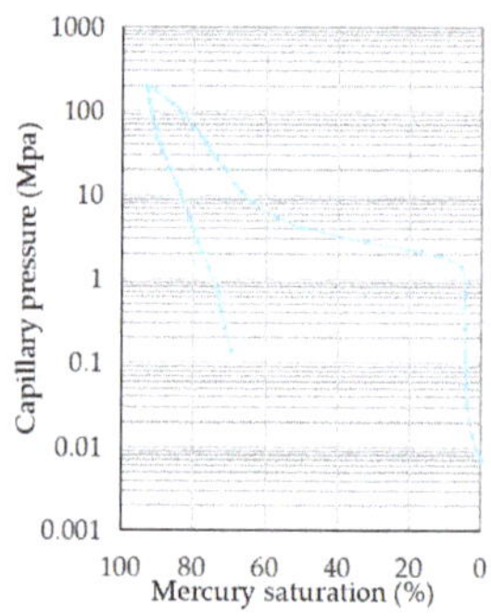	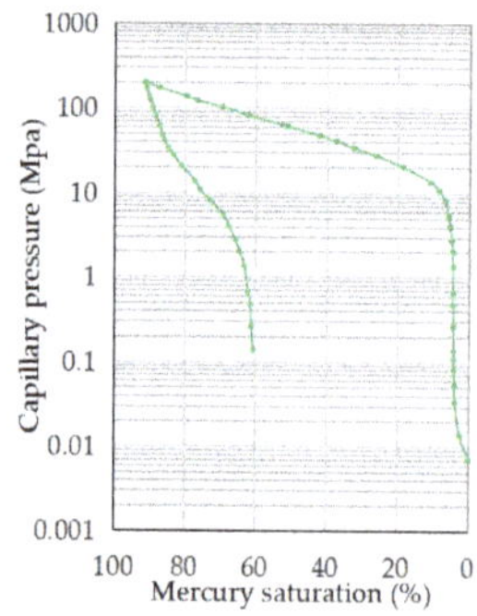
Typical CRMI curve	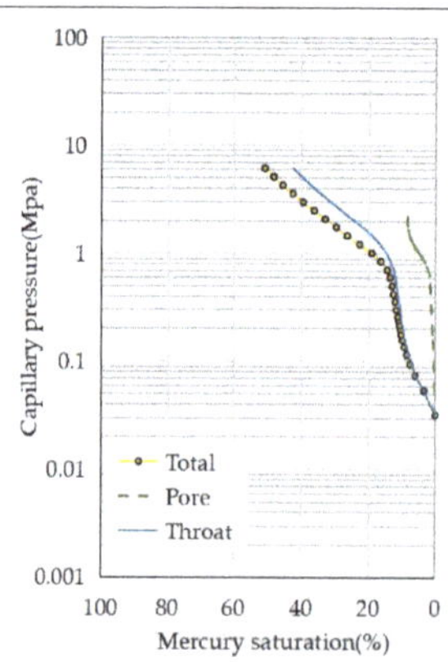	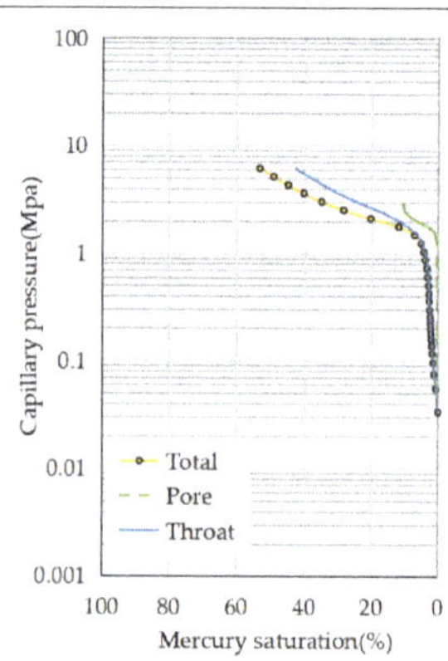	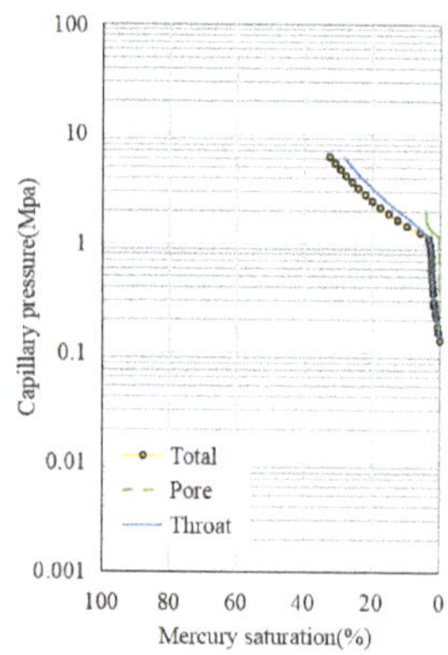

6. Conclusions

(1) The lithology of the Chang 3 reservoir in the study area mainly consists of arkose and a small amount of lithic arkose. It is characterized by simple sediment composition, medium to good sorting, poor rounding, intense diagenesis, and strong heterogeneity. The porosity of the reservoir ranges from 10% to 18%, with an average value of 12%. Permeability ranges from 0.2 to 2 mD, with an average value of 1.54 mD, making it a typical low-porosity, extremely low-permeability reservoir. Common pore types in the study area include intergranular pores, intragranular dissolution pores, and micropores. The main fillings are chlorite and illite. Throat types are predominantly tabular, followed by constricted and bundled throats.

(2) By splicing the pore-throat sizes obtained from high-pressure mercury injection (HPMI) and constant rate mercury injection (CRMI), the full-size pore radius distribution characteristics were derived. This method addresses the technical shortcomings of CRMI, which cannot characterise pores smaller than 0.12 μm in radius, and HPMI, which cannot finely characterise larger pore-throats. Based on fractal theory, the segmented fractal dimensions were calculated for each sample. The average fractal dimensions for micropores (Dh2), mesopores (Dc3), and macropores (Dc4) are 2.43, 2.75, and 2.95, respectively. This indicates that the larger the pore-throat size, the rougher the surface, and the more complex the structure.

(3) Dc4 shows significant correlation with displacement pressure, skewness, sorting coefficient, and maximum mercury injection saturation. Dh2 and Dc3 only have a clear correlation with maximum mercury injection saturation. This indicates that the heterogeneity and permeability of the reservoir are more significantly influenced by the development degree and surface roughness of macropores. The connectivity of micropores and mesopores has a certain impact on the storage capacity of the reservoir. Therefore, modifying larger pore-throats helps to increase the storage space and flow capacity of the reservoir.

(4) The sedimentary environment and diagenetic processes directly control the pore-throat structure of the reservoir, thereby influencing its fractal dimension. The fractal dimension of large pores and throats (Dc4) shows a positive correlation with compositional maturity (ZTR). As the transportation distance increases, the content of heavy minerals such as zircon, tourmaline, and rutile rises, further enhancing the complexity of the large pore-throat structure. The study area is subjected to compaction, cementation, and dissolution. Smaller pores and throats are less affected by diagenetic processes, with their structures remaining almost undeformed and having smaller fractal dimensions. Large pores and throats, on the other hand, are strongly influenced by diagenetic processes, resulting in significant deformation of their structures and larger fractal dimensions.

(5) The extremely low-permeability sandstone reservoirs in the study area are classified based on sedimentary characteristics, reservoir space types, pore-throat structural parameters, and the characteristics of mercury injection curves. Type I, type II, and type III reservoirs show distinct differences in physical properties, pore-throat size distribution, and mercury injection curve morphology. A classification basis for the Chang 3 reservoir in the Huachi area has been established, providing a theoretical basis for predicting sweet spots in oil and gas exploration in the study area. In addition, it will be necessary to further clarify and validate the basis for reservoir classification by combining fractal theory with full-pore-size analysis of a large number of samples.

Author Contributions: Conceptualization, H.Z. and L.G.; methodology, H.Z. and L.G.; software, H.Z., Y.L., and J.S.; validation, H.Z., C.W., and L.G.; formal analysis, H.Z. and L.G.; investigation, C.W.

and J.S.; resources, J.S., Y.L., and Z.W.; data curation, H.Z., Z.W., and L.G.; writing—original draft preparation, H.Z.; writing—review and editing, L.G.; visualization, Z.W.; supervision, Z.W.; project administration, Z.W.; funding acquisition, Z.W. All authors have read and agreed to the published version of the manuscript.

Funding: This research was funded by the National Natural Science Foundation of China (grant number 41302076).

Data Availability Statement: All data generated or analyzed during this study are included in this published article.

Acknowledgments: This work was supported by the State Key Laboratory of Continental Dynamics and The Second Oil Production Plant of PetroChina Changqing Oilfield Company. Finally, we would like to express our thanks to the reviewers of this paper.

Conflicts of Interest: Authors Chenyang Wang, Jinkuo Sui, and Yujuan Lv are employed by The Second Oil Production Plant of PetroChina Changqing Oilfield Company. The remaining authors declare that the research was conducted in the absence of any commercial or financial relationships that could be construed as potential conflicts of interest.

References

1. Ji, B.; Fang, J. An overview of efficient development practices at low permeability sandstone reservoirs in China. *Ener. Geo.* **2023**, *4*, 100179. [CrossRef]
2. Ouyang, S.; Sun, W.; Wu, Y.; Meng, Z.; Li, G.; Zhao, D. Seepage characteristics and infliuencing factors of low permeability-ultra low permeability reservoir: A case of Chang 6 reservoir in Houshi-Xinghe area of Ansai oilfield in Ordos Basin. *Geol. Sci. Technol. Inf.* **2019**, *38*, 199–207.
3. Zeng, L.; Liu, H. Influence of fractures on the development of low-permeability sandstone reservoirs: A case study from the Taizhao district, Daqing Oilfield, China. *J. Pet. Sci. Eng.* **2010**, *72*, 120–127. [CrossRef]
4. Chen, J.; Yao, J.; Mao, Z.; Li, Q.; Luo, A.; Deng, X.; Shao, X. Sedimentary and diagenetic controls on reservoir quality of low-porosity and low-permeability sandstone reservoirs in Chang101, upper triassic Yanchang Formation in the Shanbei area, Ordos Basin, China. *Mar. Pet. Geol.* **2019**, *105*, 204–221. [CrossRef]
5. Zhang, S.; Fang, Z. Permeability damage micro-mechanisms and stimulation of low-permeability sandstone reservoirs: A case study from Jiyang Depression, Bohai Bay Basin, China. *Pet. Explor. Dev.* **2020**, *47*, 374–382. [CrossRef]
6. Schmitt, M.; Fernandes, C.; Neto, J.; Wolf, F.; Santos, V. Characterization of pore systems in seal rocks using Nitrogen Gas Adsorption combined with Mercury Injection Capillary Pressure techniques. *Mar. Pet. Geol.* **2013**, *39*, 138–149. [CrossRef]
7. Wang, L.; Zhao, N.; Sima, L.; Meng, F.; Guo, Y. Pore Structure Characterization of the Tight Reservoir: Systematic Integration of Mercury Injection and Nuclear Magnetic Resonance. *Energ. Fuel.* **2018**, *32*, 7471–7484. [CrossRef]
8. Rezaee, R.; Saeedi, A.; Clennell, B. Tight gas sands permeability estimation from mercury injection capillary pressure and nuclear magnetic resonance data. *J. Pet. Sci. Eng.* **2012**, *88–89*, 92–99. [CrossRef]
9. Zhang, M.; Duan, C.; Li, G.; Fu, X.; Zhong, Q.; Liu, H.; Dong, Z. Determinations of the multifractal characteristics of the pore structures of low-, middle-, and high-rank coal using high-pressure mercury injection. *J. Pet. Sci. Eng.* **2021**, *203*, 108656. [CrossRef]
10. Kong, X.; Xiao, D.; Jiang, S.; Lu, S.; Sun, B.; Wang, J. Application of the combination of high-pressure mercury injection and nuclear magnetic resonance to the classification and evaluation of tight sandstone reservoirs: A case study of the Linxing Block in the Ordos Basin. *Nat. Gas Ind. B* **2020**, *7*, 433–442. [CrossRef]
11. Wei, Y.; Li, J.; Du, Y.; Liu, S.; Li, W.; Yang, J.; Feng, W.; Song, Z.; Zhang, Y. Classification Evaluation of Gas Shales Based on High-Pressure Mercury Injection: A Case Study on Wufeng and Longmaxi Formations in Southeast Sichuan, China. *Energ. Fuel.* **2021**, *35*, 9382–9395. [CrossRef]
12. Zhang, F.; Jiang, Z.; Sun, W.; Li, Y.; Zhang, X.; Zhu, L.; Wen, M. A multiscale comprehensive study on pore structure of tight sandstone reservoir realized by nuclear magnetic resonance, high pressure mercury injection and constant-rate mercury injection penetration test. *Mar. Pet. Geol.* **2019**, *109*, 208–222. [CrossRef]
13. Wang, J.; Zhang, J.; Shen, W.; Liu, H. Comparison of the Pore Structure of Ultralow-Permeability Reservoirs Between the East and West Subsags of the Lishui Sag Using Constant-Rate Mercury Injection. *J. Ocean. U. China* **2019**, *20*, 315–328. [CrossRef]
14. Lin, Y.; Zhang, J.; Liu, X.; Zhou, H. Pore structure features of reservoirs at late high water-cut stage, Lamadian Oilfield, Daqing, China. *Pet. Explor. Dev.* **2008**, *35*, 215–219. [CrossRef]

15. Chen, F.; Zhou, Y.; Ynag, D.; Wan, H. Study of pore structure of tight sandstone reservoir based on fractal theory: A case study from Chang 7tight sandstone of Yanchang formation in Qingcheng area, Ordos Basin. *J. China. U. Min. Techno.* **2022**, *51*, 941–955.
16. Wu, C.; Zhan, Z.; Li, W.; Wang, N.; Wu, L. The combination methods of mercury intrusion to characterize pore-throat characteristics in tight oil reservoir: A case study of Yanchangzu Chang 7 reservoir in Dingbian area of Shanbei. *Chin. J. Geol.* **2023**, *58*, 710–722.
17. Peng, P.; Ning, Z.; Qi, L.; He, B.; Zhao, H.; Niu, T. Research method of pore structure in tight reservoir. *Res. Eval. Dev.* **2014**, *4*, 26–32.
18. Huang, H.; Chen, L.; Sun, W.; Xiong, F.; Ji, W.; Jia, J.; Tang, X.; Zhang, S.; Gao, J.; Luo, B. Pore-throat structureand fractal characteristicsof shihezi formation tight gassandstone in the ordos basin, China. *Fractals* **2018**, *26*, 1840005. [CrossRef]
19. Zhang, H.; Guo, L.; Wu, Z.; Ma, J. Pore-throat structure, fractal characteristics and permeability prediction of tight sandstone: The Yanchang Formation, Southeast Ordos Basin. *Sci. Rep.* **2024**, *14*, 27913. [CrossRef]
20. Li, W.; Mu, L.; Zhao, L.; Li, J.; Wang, S.; Fan, Z.; Shao, D.; Li, C.; Shan, F.; Zhao, W.; et al. Pore-throat structure characteristics and its impact on the porosity and permeability relationship of Carboniferous carbonate reservoirs in eastern edge of Pre-Caspian Basin. *Pet. Explor. Dev.* **2020**, *47*, 1027–1041. [CrossRef]
21. Wang, W.; Lin, C.; Zhang, X. Fractal dimension analysis of pore throat structure in tight sandstone reservoirs of Huagang formation: Jiaxing area of east china sea basin. *Fractal Fract.* **2024**, *8*, 374. [CrossRef]
22. Jamil, M.; Siddiqui, N.A.; Rahman, A.H.B.A.; Ibrahim, N.A.; Ismail, M.S.B.; Ahmed, N.; Usman, M.; Gul, Z.; Imran, Q.S. Facies Heterogeneity and Lobe Facies Multiscale Analysis of Deep-Marine Sand-Shale Complexity in the West Crocker Formation of Sabah Basin, NW Borneo. *Appl. Sci.* **2021**, *11*, 5513. [CrossRef]
23. Zhang, X. The control of sandbody architecture on reservoir heterogeneities from the Shaximiao Formation of western Sichuan depression. Ph.D. Dissertation, Chengdu University of Technology, Chengdu, China, 2021.
24. Tian, L.; Song, F.; Zhang, S.; Yang, Y. Heterogeneity characteristics of deep tight sandstone reservoir and its controlling factors: A case study of Jurassic Qigu Formation in Yongjin Area, Junggar Basin. *J. Northeast. Pet. Univ.* **2024**, *48*, 29–40.
25. Mo, J. Particularly low permeability reservoir characteristics comprehensive evaluation of Chang3 reservoir in Huachi area, Ordos basin. Master's Dissertation, Northwest University, Xi'an, China, 2010; pp. 1–97.
26. Li, X.; Wei, W.; Wang, L.; Cai, J. Fractal dimension of digital 3D rock models with different pore structures. *Energies* **2022**, *15*, 7461. [CrossRef]
27. Jouini, M.; Alabere, A.; Alsuwaidi, M.; Morad, S.; Bouchaala, F.; Jallad, O. Experimental and digital investigations of heterogeneity in lower cretaceous carbonate reservoir using fractal and multifractal concepts. *Sci. Rep.* **2023**, *13*, 20306. [CrossRef]
28. Liu, M.; Xie, R.; Li, C.; Li, X.; Jin, G.; Guo, J. Determining the segmentation point for calculating the fractal dimension from mercury injection capillary pressure curves in tight sandstone. *J. Geophys. Eng.* **2018**, *15*, 1350–1362. [CrossRef]
29. Wu, Y.; Tahmasebi, P.; Lin, C.; Zahid, M.; Dong, C.; Golab, A.; Ren, L. A comprehensive study on geometric, topological and fractal characterizations of pore systems in low-permeability reservoirs based on SEM, MICP, NMR, and X-ray CT experiments. *Mar. Pet. Geol.* **2019**, *103*, 12–28. [CrossRef]
30. Rahner, M.; Halisch, M.; Fernandes, C.; Weller, A.; Santos, V. Fractal dimensions of pore spaces in unconventional reservoir rocks using X-ray nano- and micro-computed tomography. *J. Nat. Gas Sci. Eng.* **2018**, *55*, 298–311. [CrossRef]
31. Gao, Y. Research on Brittleness Evaluation and Main Controlling Factors of Tight Oil Reservoirs—A Case Study from Triassic Yanchang Formation Chang 7 Oil Formation in Dingbian Dongrengou Oilfield, Ordos Basin. Master's Dissertation, Northwest University, Xi'an, China, 2022; pp. 1–89.
32. Li, A.; Qiang, Q.; Xu, L.; Zhang, W.; Yin, S.; Xie, H. Chemical characteristics of formation water of Chang 2 oil formation of Yanchang Formation in Zhaike area, Ordos Basin and its indication to dense oil reservoir. *Nat. Gas Geosci.* **2022**, *33*, 1637–1647.
33. Wang, L.; Li, W.; Liu, Q.; Wang, D.; Zhang, M.; Bai, B. Lithofacies characteristics and sedimentary environment of Chang 7 black shale in the Yanchang Formation, Ordos Basin. *J. Palaeogr.* **2023**, *25*, 598–613.
34. Lv, Q.; Xin, H.; Wang, L.; Luo, S.; Dan, W.; Feng, S. Sedimentary types, characteristics and model of lacustrine fine-grained gravity flow in the Member 7 of Trassic Yanchang Formation in Ningxian area, Ordos Basin. *J. Palaeogeogr.* **2023**, *25*, 823–840.
35. Zhang, H. Study on the sedimentary facies of the member 4+5 and 3 of Triassic Yanchang Formation in Huachi area. Master's Dissertation, Northwest University, Xi'an, China, 2010; pp. 1–94.
36. Long, S.; Hou, Y.; Yang, C.; Guo, Y.; Zhang, J.; Zeng, Y.; Gao, N.; Li, S. Sequence stratigraphy and evolution of Triassic Chang 7 to Chang 3 members in Qingcheng area, southwestern Ordos Basin. *Lith. Res.* **2024**, *36*, 145–156.
37. Dong, S.; Zeng, L.; Xu, C.; Dowd, P.; Gao, Z.; Mao, Z.; Wang, A. A novel method for extracting information on pores from cast thin-section images. *Comput. Geosci.* **2019**, *130*, 69–83. [CrossRef]
38. Lindquist, W.; Venkatarangan, A.; Dunsmuir, J.; Wong, T. Pore and throat size distributions measured from synchrotron X-ray tomographic images of Fontainebleau sandstones. *J. Geophys. Res. Solid. Earth.* **2000**, *105*, 21509–21527. [CrossRef]

39. Yan, Q.; Zhang, Y.; Fu, H.; Jian, M.; Wang, J.; Sui, S.; Fu, H.; Hao, R.; Guo, M. High pressure mercury injection and scanning electron microscopy applied to characterize micro-and nano-scale pore throats in tight sandstone reservoirs: A case study of the fourth member of Shahejie Formation in Yi176 block, Zhanhua Sag, Bohai Bay Basin. *Pet. Geo. Exper.* **1999**, *104*, 292–301.
40. Coenen, T.; Haegel, N. Cathodoluminescence for the 21st century: Learning more from light. *Appl. Phys. Rev.* **2017**, *4*, 031103. [CrossRef]
41. Shao, X.; Pang, X.; Jiang, F.; Li, L.; Huyan, Y.; Zheng, D. Reservoir Characterization of Tight Sandstones Using Nuclear Magnetic Resonance and Incremental Pressure Mercury Injection Experiments: Implication for Tight Sand Gas Reservoir Quality. *Energ. Fuel.* **2017**, *31*, 10420–10431. [CrossRef]
42. *SY/T 5368-2000*; Thin Section Examination of Rock. Standards Press of China: Beijing, China, 2000.
43. *SY/T 5163-2018*; Thin Section Examination of Rock. Standards Press of China: Chengdu, China, 2018.
44. *SY/T 5162-2014*; Analytical Method of Rock Sample by Scanning Electron Microscope. Standards Press of China: Dongying, China, 2014.
45. *SY/T 5916-2013*; Identification Method of Rocks and Minerals by Cathode Luminescence. Standards Press of China: Beijing, China, 2013.
46. *SY/T 5336-2006*; Recommend Practices for Core Analysis. Standards Press of China: Beijing, China, 2006.
47. *SY/T 5346-2005*; Rock Capillary Pressure Measurement. Standards Press of China: Beijing, China, 2005.
48. Guo, R.; Xie, Q.; Qu, X.; Chu, M.; Li, S.; Ma, D.; Ma, X. Fractal characteristics of pore-throat structure and permeability estimation of tight sandstone reservoirs: A case study of Chang 7 of the Upper Triassic Yanchang Formation in Longdong area, Ordos Basin, China. *J. Pet. Sci. Eng.* **2020**, *184*, 106555. [CrossRef]
49. Zhong, X.; Zhu, Y.; Jiao, T.; Qi, Z.; Luo, J.; Xie, Y.; Liu, L. Microscopic pore throat structures and water flooding in heterogeneous low-permeability sandstone reservoirs: A case study of the Jurassic Yan'an Formation in the Huanjiang area, Ordos Basin, Northern China. *J. Asian Earth Sci.* **2021**, *219*, 104903. [CrossRef]
50. Wu, Y.; Liu, C.; Ouyang, S.; Luo, B.; Zhao, D.; Sun, W.; Awan, R.; Lu, Z.; Li, G.; Zang, Q. Investigation of pore-throat structure and fractal characteristics of tight sandstones using HPMI, CRMI, and NMR methods: A case study of the lower Shihezi Formation in the Sulige area, Ordos Basin. *J. Pet. Sci. Eng.* **2022**, *210*, 110053. [CrossRef]
51. Qu, Y.; Sun, W.; Tao, R.; Luo, B.; Chen, L.; Ren, D. Pore-throat structure and fractal characteristics of tight sandstones in Yanchang Formation, Ordos Basin. *Mar. Pet. Geol.* **2020**, *120*, 104573. [CrossRef]
52. Han, Z.; Zhou, Y.; Wei, F.; Zhang, Q.; Fan, P.; Deng, C.; Gong, J.; Cheng, X.; Ma, E.; Wu, H. Pore-Throat Structure and Fractal Characteristics of Tight Sandstones: A Case Study from the Chang 6_3 Sublayer, Southeast Ordos Basin. *Geofluids.* **2023**, *2023*, 3832452. [CrossRef]
53. Zhang, W.; Shi, Z.; Tian, Y.; Xie, D.; Li, W. The combination of high-pressure mercury injection and rate-controlled mercury injection to characterize the pore-throat structure in tight sandstone reservoirs. *FBOGF* **2021**, *28*, 14–20. [CrossRef]
54. Xoaoth, B. Song, S., Wang, Y., Eds. and Translators; *Coal and Gas Outburst*; China Industry Press: Beijing, China, 1966.
55. Zou, R.; Wang, L.; Luo, R.; Zhang, L.; Deng, H.; Meng, Z.; Zou, R.; Zhang, Y. Full—Range pore throat radius distribution and fractal characteristics of different types of reservoirs in deep carbonate gas reservoirs: A case study of Deng—4Member in Gaoshiti—Moxi Area, Sichuan Basin. *J. China. U. Min. Techno.* **2024**, *53*, 1218–1236.
56. Chen, F.; Zhang, X.; Liu, R.; Zhang, S.; Zhai, Z. Pore structure characteristics of deep Upper Paleozoic coal measure tight sandstone reservoirs in northeast Qinshui Basin and its reservoir geological significance. *J. Coal Sci. Eng.* **2024**, 1–16. Available online: https://link.cnki.net/urlid/11.2402.td.20241111.1527.007 (accessed on 21 April 2025).
57. Yang, G.; Wu, K.; Zhang, B.; Yin, Z.; Wang, J.; Wang, M.; He, X.; Xing, H.; Deng, L.; Zhang, K.; et al. Application of U-Pb isotopic dating and heavy mineral assemblage to source-sink system analysis of the Paleogene in western margin of Qaidam Basin. *Mar. Ori. Pet. Geo.* **2023**, *30*, 177–185.
58. Zhang, H.; Li, X.; Liu, J.; Wang, Y.; Guo, L.; Wu, Z.; Tian, Y. A Fractal Characteristics Analysis of the Pore Throat Structure in Low-Permeability Sandstone Reservoirs: A Case Study of the Yanchang Formation, Southeast Ordos Basin. *Fractal Fract.* **2025**, *9*, 224. [CrossRef]
59. Cai, L.; Yang, T.; Tian, J.; Yi, J.; Ren, Q. Advances in Studies of Development and Growth Mechanisms of Clay Minerals in Tight Sandstone Reservoirs. *J. Pet. Sci. Eng.* **2023**, *41*, 1859–1889.

fractal and fractional

MDPI

Article

The Pore Structure and Fractal Characteristics of Upper Paleozoic Coal-Bearing Shale Reservoirs in the Yangquan Block, Qinshui Basin

Jinqing Zhang [1,2,3], Xianqing Li [2,3,*], Xueqing Zhang [2,3,4], Xiaoyan Zou [2,3,5], Yunfeng Yang [6] and Shujuan Kang [1,*]

1 Institute of Geology and Geophysics, Chinese Academy of Sciences, Beijing 100029, China; zhangjinqing@mail.iggcas.ac.cn
2 State Key Laboratory for Fine Exploration and Intelligent Development of Coal Resources, China University of Mining and Technology (Beijing), Beijing 100083, China
3 College of Geoscience and Surveying Engineering, China University of Mining and Technology (Beijing), Beijing 100083, China
4 School of Emergency Management and Safety Engineering, North China University of Science and Technology, Tangshan 063210, China
5 College of Urban, Rural Planning and Architectural Engineering, Shangluo University, Shangluo 726000, China
6 Suzhou Zhongke Dixing Innovation Technology Institute, Suzhou 215163, China
* Correspondence: lixq@cumtb.edu.cn (X.L.); ksj@mail.iggcas.ac.cn (S.K.)

Abstract: The investigation of the pore structure and fractal characteristics of coal-bearing shale is critical for unraveling reservoir heterogeneity, storage-seepage capacity, and gas occurrence mechanisms. In this study, 12 representative Upper Paleozoic coal-bearing shale samples from the Yangquan Block of the Qinshui Basin were systematically analyzed through field emission scanning electron microscopy (FE-SEM), high-pressure mercury intrusion, and gas adsorption experiments to characterize pore structures and calculate multi-scale fractal dimensions (D_1–D_5). Key findings reveal that reservoir pores are predominantly composed of macropores generated by brittle fracturing and interlayer pores within clay minerals, with residual organic pores exhibiting low proportions. Macropores dominate the total pore volume, while mesopores primarily contribute to the specific surface area. Fractal dimension D_1 shows a significant positive correlation with clay mineral content, highlighting the role of diagenetic modification in enhancing the complexity of interlayer pores. D_2 is strongly correlated with the quartz content, indicating that brittle fracturing serves as a key driver of macropore network complexity. Fractal dimensions D_3–D_5 further unveil the synergistic control of tectonic activity and dissolution on the spatial distribution of pore-fracture systems. Notably, during the overmature stage, the collapse of organic pores suppresses mesopore complexity, whereas inorganic diagenetic processes (e.g., quartz cementation and tectonic fracturing) significantly amplify the heterogeneity of macropores and fractures. These findings provide multi-scale fractal theoretical insights for evaluating coal-bearing shale gas reservoirs and offer actionable recommendations for optimizing the exploration and development of Upper Paleozoic coal-bearing shale gas resources in the Yangquan Block of the Qinshui Basin.

Keywords: pore characteristics; fractal dimension; Qinshui Basin; Yangquan Block; Upper Paleozoic; coal-bearing shale

1. Introduction

Shale gas, as a critical unconventional natural gas resource, plays a pivotal role in mitigating energy supply–demand imbalances and optimizing energy structures through its large-scale development [1,2]. Globally, countries such as the United States, China, and Argentina have achieved commercial shale gas production [3,4]. In 2024, U.S. shale gas output reached 1.02~1.08 × 10^{12} m^3, while China ranked second with 2.60~2.80 × 10^{10} m^3 [5–7]. However, over 90% of China's production is concentrated in the Longmaxi Formation of the Sichuan Basin, highlighting an urgent need to explore new stratigraphic intervals to ensure sustainable resource development and diversification [8–10].

In recent years, China's shale gas exploration has expanded into continental, transitional facies and highly mature strata [11,12]. Breakthroughs in non-marine shales such as the Lianggaoshan Formation in the Sichuan Basin demonstrate significant potential for new stratigraphic targets [13–15]. The Qinshui Basin, a critical coal-bearing basin in North China [16,17], hosts multiple Upper Paleozoic marine–continental transitional coal-bearing shale intervals (e.g., Shanxi and Taiyuan Formations), characterized by substantial thickness (>30 m per layer), a high total organic carbon content (TOC) (average TOC = 3.82%), and moderate thermal maturity, positioning them as promising targets for shale gas resource expansion [18–20].

The complexity and heterogeneity of pore structures in shale reservoirs directly govern gas storage and seepage capacity [21,22], while fractal dimensions serve as critical quantitative parameters to characterize pore geometry, offering essential guidance for reservoir evaluation and development optimization [23–25]. Fractal geometry, pioneered by French mathematician Mandelbrot in the 1970s, transcends the limitations of traditional Euclidean geometry in describing regular morphologies and provides a mathematical framework for analyzing natural complex and irregular structures, such as mountain contours and coastline configurations [26]. The fractal dimension, a core parameter in fractal theory, overcomes the integer constraints of conventional topological dimensions by employing non-integer values to quantify the self-similarity and spatial filling capacity of porous systems. For instance, a higher fractal dimension signifies greater pore structure complexity and stronger heterogeneity. The application of fractal theory to pore structure studies is particularly imperative for coal-bearing shale reservoirs. These reservoirs exhibit pronounced multi-scale pore systems and high heterogeneity, which challenge traditional Euclidean geometry in quantifying intricate topological relationships. Fractal dimensions, by measuring pore surface roughness and structural irregularity, reveal the spatial distribution patterns of pore networks, thereby establishing a theoretical foundation for evaluating seepage capacity, adsorption–desorption behavior, and reservoir performance. This approach bridges the gap between complex natural pore systems and quantitative characterization, enabling robust assessments of coal-bearing shale gas reservoirs under varying geological and diagenetic conditions [25,26].

Current methodologies, including Micro-nano Computed Tomography (Micro-nano CT), scanning electron microscopy (SEM), nuclear magnetic resonance (NMR), and fluid intrusion techniques [27–29], have been widely applied for the multi-scale characterization of macropores (>50 nm), mesopores (2–50 nm), and micropores (<2 nm) in shale, elucidating the intrinsic relationships among mineral composition, organic matter evolution, and pore fractal features [30–32]. Fractal theory provides a robust framework to describe the geometric properties and structural functionality of solid surfaces, enabling the hierarchical classification and evaluation of shale reservoir heterogeneity [24–26]. The fractal dimension D, a parameter quantifying surface roughness or structural irregularity, serves as a critical tool for characterizing reservoir complexity and assessing shale heterogeneity. Traditional methods for calculating fractal dimensions include box-counting, mercury intrusion porosimetry (MIP), and the Frenkel–Halsey–Hill (FHH) model [22,25–28]. Specifically, the FHH fractal model, based on nitrogen adsorption isotherms, derives surface fractal dimensions from adsorption data [33,34]. The Menger sponge model, applied to high-pressure mercury intrusion curves, calculates pore volume fractal dimensions, while box-counting methods, utilizing SEM or micro-CT images, directly quantify two- or three-dimensional fractal characteristics of pore structures [35]. Studies demonstrate significant correlations between fractal dimensions and shale reservoir properties such as specific surface area, porosity, and permeability, offering key parameters for reservoir quality classification and "sweet spot" prediction [36–38]. Building on this foundation, this study focuses on core samples from the Upper Paleozoic coal-bearing shale in the Yangquan Block of the Qinshui Basin. By integrating FE-SEM, high-pressure mercury intrusion, and low-pressure nitrogen adsorption experiments, we quantitatively analyze the controlling factors of fractal dimensions and clarify the coupled effects of mineral composition (quartz, clay) and organic matter characteristics (TOC, R_o) on pore structure. These insights aim to provide theoretical support for sweet spot prediction and the efficient development of overmature coal-bearing shale gas reservoirs [39–42].

2. Materials and Methods

2.1. Materials

The Qinshui Basin, located in southeastern Shanxi Province, exhibits a NNE-SSW-trending composite syncline structure [43]. This large-scale, nearly N-S-oriented composite synclinal basin, developed on a Paleozoic basement and was formed during the Late Mesozoic. It underwent multi-phase tectonic events, including the Indosinian, Yanshanian, and Himalayan orogenies. The basin features relatively simple internal structures dominated by secondary folds, with sparsely developed faults, indicating tectonic stability. Large-scale faults are primarily concentrated along the basin margins, where tectonic activity intensifies [43]. The Yangquan Block is situated in the northern part of the Qinshui Basin (Figure 1).

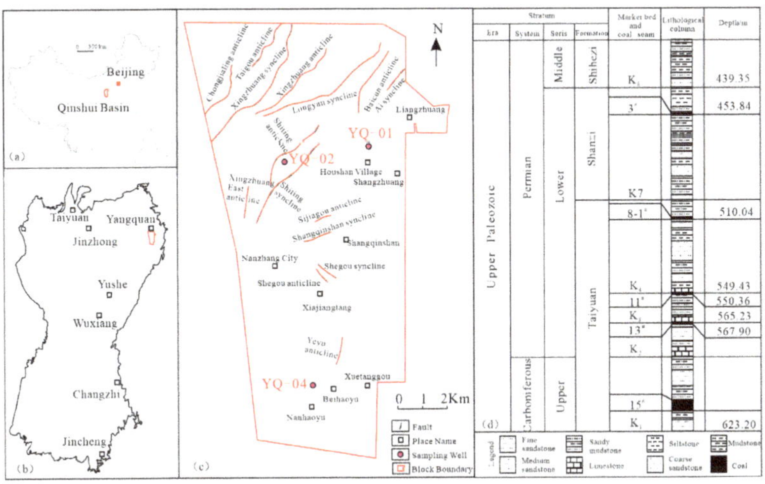

Figure 1. Location map of the sampling wells and composite stratigraphic column of the Upper Paleozoic coal-bearing shale in the Yangquan Block, Qinshui Basin, China. (**a**) Location map of the Qinshui Basin, (**b**) Location map of the Yangquan Block, (**c**) Location map of sampling wells, (**d**) Comprehensive stratigraphic column of the Upper Paleozoic.

This study analyzed 12 core samples collected from three wells (YQ-01, YQ-02, and YQ-04) in the Yangquan Block. The samples originate from coal-bearing shale intervals of the Shanxi Formation and Taiyuan Formation, comprising grayish black to black carbonaceous mudstone and grayish black to black silty mudstone, with sampling depths ranging from 308.2 to 557.2 m.

2.2. Organic Geochemistry Experiments

2.2.1. Total Organic Carbon (TOC) Analysis

The TOC content of shale samples was determined using a CS230HC carbon sulfur analyzer (Wuxi Yinzhuo Analytical Instrument Co., Wuxi, China) following the Chinese National Standard GB/T 19145-2003 [44]. Testing was conducted at the China University of Petroleum (Beijing) under controlled laboratory conditions (temperature: 24 °C; relative humidity: 48%). Experimental parameters included a carrier gas pressure of 0.27 MPa, oxygen purity of 99.5%, a combustion gas flow rate of 2 L/min, and an analytical gas flow rate of 0.5 L/min.

2.2.2. Rock-Eval Pyrolysis

Rock-Eval pyrolysis was performed in accordance with GB/T 18602-2012 [45] using an OGE-II Hydrocarbon Analyzer (Jiangsu Huaan Scientific Equipment, Nantong, China) at the China University of Petroleum (Beijing). Prior to analysis, approximately 100 mg of each sample was ground to <200 mesh. Key parameters, including S_1 (free hydrocarbons), S_2 (hydrocarbons from kerogen cracking), $S_1 + S_2$ (total hydrocarbon potential), and T_{max} (temperature at maximum pyrolysis yield), were obtained to evaluate source rock quality and organic matter maturity.

2.2.3. Vitrinite Reflectance Measurement

Vitrinite reflectance analysis adhered to SY/T 5124-2012 [46] and was conducted using a Leica DM 4500P (Wetzlar, Germany) microspectrophotometer at the China University of Petroleum (Beijing). Samples were crushed to 20~40 mesh, embedded in epoxy resin, and polished into thin sections. Reflectance measurements were performed under 50× oil immersion objectives to ensure accuracy.

2.3. X-Ray Diffraction Analysis

X-ray diffraction experiments were conducted at the China University of Mining and Technology (Beijing) using a Rigaku D_{max}/2000-PC X-ray diffractometer (Rigaku Corporation, Akishima, Japan) in accordance with the industry standard SY/T 5163-2018 [47] to determine the whole-rock mineral composition of coal-bearing shale samples. Key testing parameters included a tube voltage of 40 kV, tube current of 100 mA, scanning range of 2.5–70°, scanning speed of 4°/min, and step size of 0.02°. Samples were ground to 200 mesh, loaded into grooved glass slides, and analyzed to obtain mineral diffraction patterns. Quantitative analysis of minerals was performed using MDI Jade6.5 software by comparing results with standard mineral diffraction databases.

2.4. Reservoir Pore Characterization Experiments

2.4.1. Scanning Electron Microscopy

Microstructural characterization of pore development in shale samples was performed using the following instruments: a conventional ZEISS Merlin scanning electron microscope (Oberkochen, Germany, Carl Zeiss AG), a Hitachi SU8010 field emission scanning electron microscope (FE-SEM) (Tokyo, Japan, Hitachi High-Tech), and a Thermo Fisher Apreo FE-SEM equipped with Ar-ion cross-section polishing (Waltham, MA, USA, Thermo Fisher Scientific), with all analyses conducted in strict compliance with the Chinese petroleum industry standard SY/T 5162-2014 [48]. For SEM analysis, samples (~6 mm in diameter, ~0.5 mm thick) with freshly fractured surfaces obtained by natural spalling were selected to observe the overall pore distribution. For FE-SEM analysis, to precisely characterize pore morphology and size, samples underwent argon ion polishing followed by gold coating of polished surfaces. Sample preparation criteria matched SEM requirements.

2.4.2. Low-Pressure Gas Adsorption

Low-pressure CO_2 adsorption and low-temperature N_2 adsorption–desorption experiments were conducted at the Beijing Physical and Chemical Testing Center using Micromeritics ASAP 2020 Physisorption Analyzer (Micromeritics Instrument Corp., Norcross, GA, USA) following the Chinese National Standard GB/T 19587-2017 [49]. Sample preparation was carried out as follows: samples were ground to 60 mesh and pretreated at 120 °C under a vacuum for 5 h. For CO_2 adsorption, micropore (<2 nm) parameters were derived from adsorption isotherms (relative pressure: 0.0001~0.032) using the density functional theory (DFT) model. For N_2 adsorption, mesopore (2–50 nm) parameters were calculated via the Brunauer–Emmett–Teller (BET) and Barrett–Joyner–Halenda (BJH) models (relative pressure: 0.011~0.995).

2.4.3. High-Pressure Mercury Intrusion

HPMI experiments were performed at the Beijing Physical and Chemical Testing Center using Micromeritics AutoPore IV 9500 Mercury Intrusion Porosimeter (Micromeritics Instrument Corp., Norcross, GA, USA) under GB/T 21650.1-2008/ISO 15901-1:2005 [50,51]. Sample preparation was carried out as follows: cylindrical specimens (diameter $\leq$ 1 cm, height $\leq$ 3 cm) were dried at 110 °C for 4 h. Regarding experimental conditions, pore structure parameters were measured up to 30,000 psia. Notably, excessive intrusion pressure may induce artificial fractures or damage original pore networks, leading to anomalous results.

2.5. Fractal Theory

The pore structure characteristics of shale are critical parameters for evaluating reservoir quality, encompassing storage capacity and fracturability. The storage capacity of shale reservoirs depends on pore heterogeneity, which is closely linked to permeability and fracture initiation potential. Pore heterogeneity refers to the complexity of pore size, abundance, distribution, morphology (including pore type and connectivity), and spatial arrangement. Given the multifractal nature of shale, this study employs experimental data on low-temperature N_2 adsorption and high-pressure mercury intrusion (HPMI) to calculate fractal dimensions [52,53].

The nitrogen adsorption method is widely used to characterize the heterogeneity of irregular pore structures in solids [54,55]. The Frenkel–Halsey–Hill (FHH) equation, extensively applied to porous materials including shale [54,55], offers two computational models: one based on van der Waals forces and the other on capillary condensation mechanisms. The latter is more suitable for studying the heterogeneity of porous media [56,57]. The fractal dimension D derived from the FHH model is expressed as the slope of the linear relationship between $\ln(V)$ and $\ln(\ln(P/P_0))$:

$$\ln V = C + (D - 3)\ln[\ln(P/P_0)] \tag{1}$$

where V is the gas adsorption volume (cm^3/g), P is the equilibrium pressure (MPa), P_0 is the gas saturation vapor pressure (MPa), and D is the dimensionless fractal dimension.

For high-pressure mercury intrusion, the fractal dimension is calculated using the relationship between capillary pressure (P_c) and the non-wetting phase volume fraction (S_{Hg}) [22,23,27]:

$$\lg(1 - S_{Hg}) = (D - 3)\lg P_c - (D - 3)P_{min} \quad (2)$$

where S_{Hg} is the mercury saturation (%), P_c is the capillary pressure (MPa), and P_{min} corresponds to the capillary pressure at the maximum pore throat radius. The fractal dimension D is derived from the slope k of the trend line as $D = k + 3$.

3. Results

3.1. Organic Matter Characteristics

The total organic carbon (TOC) content is a critical indicator for evaluating the gas generation potential of mud shale [15]. The Upper Paleozoic coal-bearing shale in the Yangquan Block, Qinshui Basin, exhibits high organic matter abundance, with an average TOC of 3.82 wt% (Table 1). The Shanxi Formation shale shows higher TOC values (average: 6.16 wt%), influenced by adjacent coal seams, while the Taiyuan Formation shale has a lower TOC (average: 2.65 wt%). The hydrocarbon generation potential (S_1 + S_2) ranges from 0.12 to 3.49 mg/g (mean: 1.06 mg/g) in the Shanxi Formation and from 0.07 to 0.56 mg/g (mean: 0.27 mg/g) in the Taiyuan Formation, indicating that there is superior gas generation capacity in the Shanxi Formation.

Table 1. Basic physical properties of Upper Paleozoic coal-bearing shale samples from Yangquan Block, Qinshui Basin (relative error $\leq$ 2%).

Sample	Depth (m)	Formation	Lithology	TOC (%)	R_o (%)	T_{max} (°C)	$S_1 + S_2$ (mg/g)
YQ-01-12	308.2	Shanxi	Grayish black mudstone	3.29	2.16	550	0.42
YQ-01-22	350.7	Shanxi	Black mudstone	4.63	2.13	545	0.12
YQ-01-26	379.6	Taiyuan	Grayish black mudstone	5.11	2.24	549	0.48
YQ-01-43	461.4	Taiyuan	Black mudstone	2.47	-	558	0.07
YQ-01-48	477.3	Taiyuan	Black mudstone	1.75	-	546	0.18
YQ-02-02	383.7	Shanxi	Black mudstone	1.97	2.22	543	0.22
YQ-02-07	409.6	Taiyuan	Black mudstone	2.67	2.29	564	0.21
YQ-02-11	466.5	Taiyuan	Black mudstone	3.63	2.56	541	0.16
YQ-04-04	452.6	Shanxi	Carbonaceous mudstone	14.73	2.28	551	3.49
YQ-04-08	485.5	Taiyuan	Black mudstone	2.11	-	554	0.29
YQ-04-15	517.3	Taiyuan	Carbonaceous mudstone	2.48	2.4	564	0.56
YQ-04-19	557.2	Taiyuan	Black silty mudstone	0.98	-	574	0.19

Vitrinite reflectance (R_o) and pyrolysis peak temperature (T_{max}) were used to assess thermal maturity. The R_o values of the studied shale range from 2.13 to 2.56% (mean: 2.29%), with T_{max} values ranging between 541 and 574 °C (mean: 553 °C) (Table 1). Maceral composition analysis revealed that the vitrinite content is the highest, followed by inertinite and liptinite, classifying the organic matter as Type III kerogen. These results confirm that the shale is in the overmature stage, favorable for dry gas generation.

3.2. Mineral Composition

Coal-bearing shale exhibits complex mineral compositions, where variations in the mineral content critically influence pore structure, gas adsorption, and storage capacity. Clay minerals, with strong gas adsorption affinity, dominate the mineral assemblage and significantly affect the shale gas adsorption potential [58,59]. X-ray diffraction (XRD) analysis revealed that Upper Paleozoic coal-bearing shale in the Yangquan Block are dominated by clay minerals (23.9–97.8 wt%, average: 58.9 wt%), followed by quartz (2.0–60.0 wt%, average: 33.2 wt%). Other minerals include potassium feldspar (average:

2.4 wt%), plagioclase (1.8 wt%), calcite (1.4 wt%), siderite (3.8 wt%), and pyrite (2.4 wt%) (Table 2).

Table 2. The mineral composition of Upper Paleozoic coal-bearing shale samples from the Yangquan Block, Qinshui Basin.

Sample	Mineral Composition (%)								
	Quartz	Potassium Feldspar	Plagioclase	Calcite	Dolomite	Siderite	Pyrite	Ankerite	Clay Minerals
YQ-01-12	26.2	0	0	0	0	0	2.4	0	71.4
YQ-01-22	20.9	0	1.3	0	0	2.9	0	0	74.9
YQ-01-26	31	0	1.3	0	0	1.1	0	0	66.6
YQ-01-43	2.2	0	0	0	0	0	0	0	97.8
YQ-01-48	41.8	1	2.1	0	0	8	1.3	0	45.8
YQ-02-02	38.6	0	0	0	0	5.3	3.2	0	52.9
YQ-02-07	26.6	0	1.4	0	0	0	1.1	0	70.9
YQ-02-11	31.5	0	1.3	0	0	0	1.5	0	65.7
YQ-04-04	33.5	3.9	0	1.4	0	0	0	0	61.2
YQ-04-08	60.0	0	3.6	1.3	0	0	6.3	0	28.8
YQ-04-15	50.1	0	0	1.7	0	0	1.9	0	46.3
YQ-04-19	36.4	2.4	0	1.0	19	1.6	1.8	13.9	23.9

The brittle mineral content (e.g., quartz and feldspar) is a key factor for successful hydraulic fracturing in shale reservoirs [60–63]. Compared to the Barnett Shale (USA) [63], Yangquan Block shales exhibit a higher clay mineral content and a lower brittle mineral content, suggesting reduced fracability. However, the high clay mineral content enhances gas adsorption capacity, potentially compensating for challenges in reservoir stimulation during shale gas development [37,64].

3.3. Pore Type Characteristics

As a tight reservoir, mud shale exhibits complex pore systems. Porosity, a key parameter for evaluating storage-seepage capacity and free gas content, positively correlates with the total gas content in shale reservoirs [27]. The pore volume (PV) and specific surface area (SSA) are critical structural parameters influencing gas occurrence and resource estimation [27,64]. The Upper Paleozoic coal-bearing shale in the Yangquan Block, Qinshui Basin, displays low porosity (1.1~4.9%, avg. 3.1%) and ultra-low permeability ($0.0008 \sim 0.1910 \times 10^{-3}$ μm^2, avg. 0.0274×10^{-3} μm^2), yet these values are slightly higher than those of the Shanxi Formation shale in the Huainan Coalfield [35,37,38].

Scanning electron microscopy (SEM) provides a direct visualization of pore occurrence states in shale. Following the classification by R. G. Loucks et al. [58], pores are categorized into interparticle pores, intraparticle pores, OM-hosted pores (organic matter, OM), and microfractures. The diversity in pore types and morphologies is influenced by mineral composition, compaction, and dissolution processes [27]. SEM observations of Upper Paleozoic coal-bearing shales in the Yangquan Block confirm the presence of interparticle pores, intraparticle pores, and microfractures (Figure 2a–i). Interparticle and intraparticle pores predominantly develop within the mineral matrix and are controlled by diagenetic processes such as compaction, dissolution, and mineral phase transformation [41,58].

Influenced by mineral composition and compaction–dissolution effects, interparticle pores exhibit flaky, triangular, or irregular morphologies (Figure 2a–c). These pores display favorable connectivity, facilitating shale gas migration. Intraparticle pores are moderately developed between clay mineral layers (Figure 2c–e). These pores are relatively small in size and susceptible to tectonic and diagenetic stresses. However, the layered and flaky structure of clay minerals provide mechanical support, preserving some pores during evolution. OM-hosted pores, characterized by elliptical, circular, or honeycomb shapes, are critical adsorption spaces unique to shale reservoirs. However, they are sparsely observed in the study area (Figure 2f). Their limited development compared to the Barnett Shale (North America) [63] and Shanxi Formation shales in the Huainan Coalfield [35,37,38] likely results from pore collapse and closure due to overmaturity (R_o > 2.0%) [27,64]. Microfractures (Figure 2g–i), primarily tectonic fractures formed along clay mineral laminae, serve as essential pathways for gas storage, seepage, and production [27,64]. In summary, the micron- to nano-scale pores in the Upper Paleozoic coal-bearing shale of the Yangquan Block provide critical storage space for shale gas despite the dominance of low-connectivity pores and limited OM-hosted porosity.

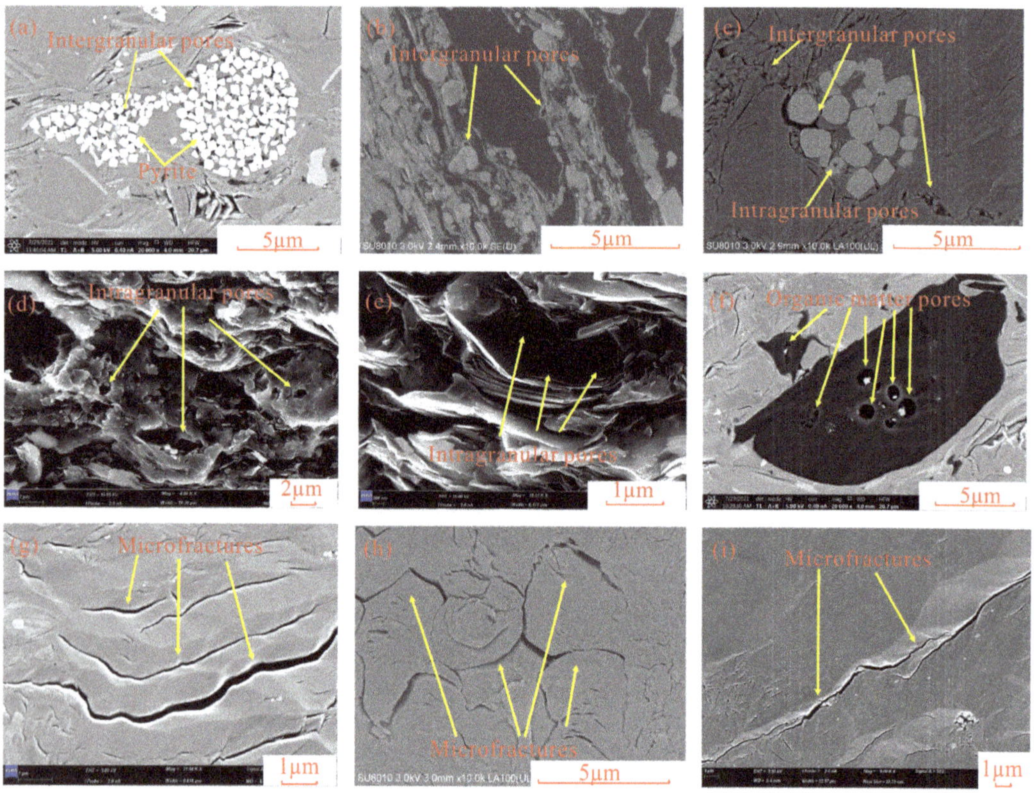

Figure 2. Field emission scanning electron microscopy (FE-SEM) images of Upper Paleozoic coal-bearing shale samples from the Yangquan Block, Qinshui Basin. (**a**) Intergranular pores, (**b**) Intergranular pores, (**c**) Intergranular pores, (**d**) Intragranular pores, (**e**) Intragranular pores, (**f**) Organic matter pores, (**g**) Microfractures, (**h**) Microfractures, (**i**) Microfractures.

3.4. Full-Scale Pore Size Distribution in Shale

The structural characteristics of shale pores critically influence reservoir storage performance [65–67]. Based on the International Union of Pure and Applied Chemistry (IUPAC) classification, nanopores are categorized into three types by diameter (d): macropores (d > 50 nm), mesopores (2 < d < 50 nm), and micropores (d < 2 nm) [68,69]. To comprehensively characterize the pore structure and size distribution of shale, this study employs high-pressure mercury intrusion (HPMI), low-temperature N_2 adsorption, and low-pressure CO_2 adsorption to investigate the structural features of macropores, mesopores, and micropores, respectively.

3.4.1. High-Pressure Mercury Intrusion (HPMI)

The mercury intrusion curves of the Shanxi Formation coal-bearing shale show a gradual increase in the mercury injection volume with rising pressure (0–200 MPa), indicating progressive pore filling. The low-pressure stage (<50 MPa) corresponds to macropore (>1000 nm) filling, the medium-pressure stage (50–150 MPa) to mesopores (2–50 nm), and the high-pressure stage (>150 MPa) to micropores (<2 nm). The incomplete overlap between mercury intrusion and extrusion curves suggests the presence of "ink-bottle" pores (narrow throats with wider interiors), highlighting poor connectivity, particularly in mesopores (Figure 3a). In contrast, the Taiyuan Formation exhibits higher total mercury intrusion volumes, with a notable increase in micropore filling at high pressures (>150 MPa), reflecting greater microporosity development. The pronounced hysteresis in the extrusion curve indicates complex pore structures and poorer connectivity in micropores and mesopores, likely due to the closure of clay mineral (e.g., illite) interlayer fractures (Figure 3b).

Pore volume distributions in the Shanxi Formation display a bimodal pattern, peaking at 10–100 nm (mesopores) and 1–10 μm (macropores), with mesopores dominating (Figure 3c). The high macropore volume (>1000 nm) is attributed to tectonic fractures or dissolution vugs. In the Taiyuan Formation, the pore volume is concentrated at <10 nm (micropores) and 50–100 nm (mesopores), reflecting residual clay mineral interlayer pores after organic pore collapse at high maturity. The macropore volume is lower than that in the Shanxi Formation, likely due to compaction inhibiting fracture development (Figure 3d).

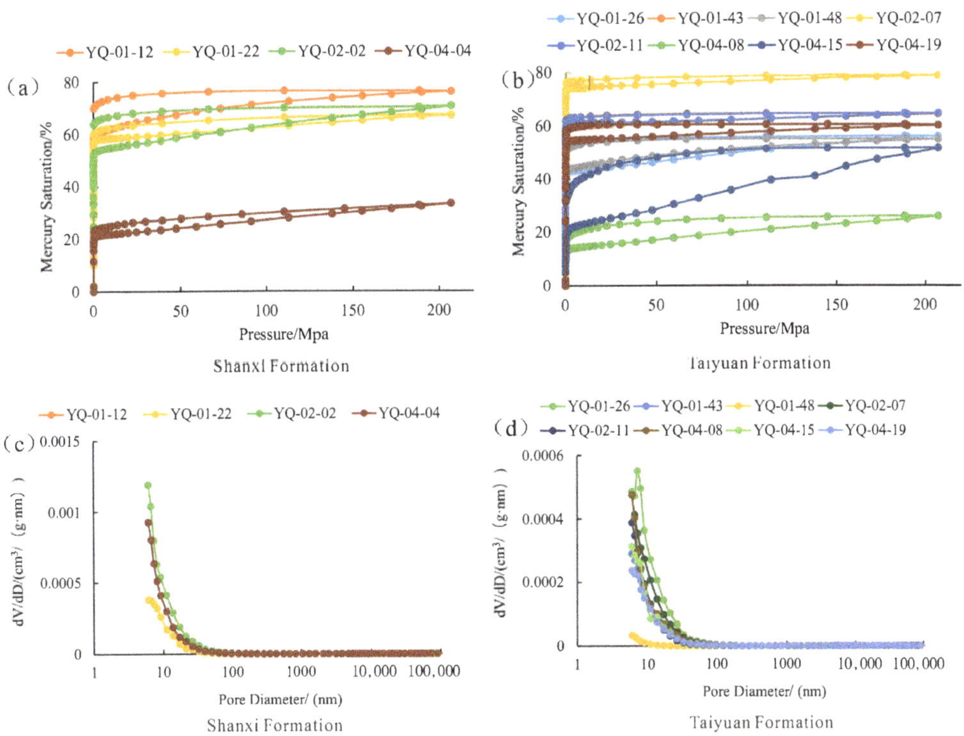

Figure 3. *Cont.*

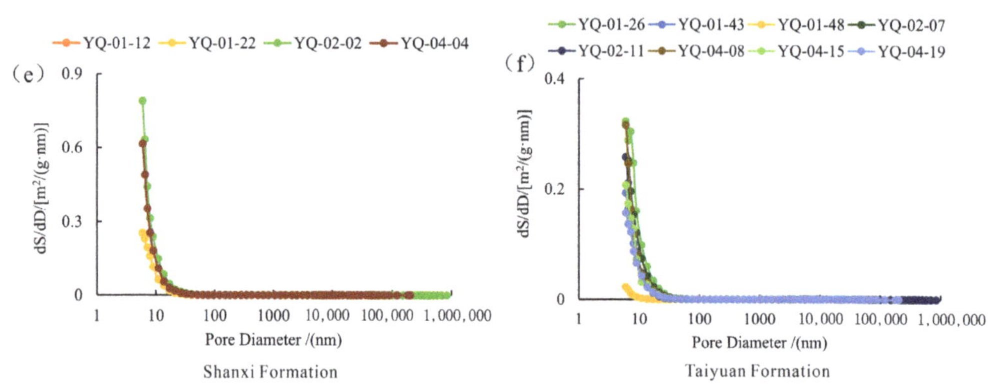

Figure 3. Mercury intrusion–extrusion curves, pore volume distribution, and specific surface area distribution of Upper Paleozoic coal-bearing shale based on mercury intrusion porosimetry (MIP) in Yangquan Block, Qinshui Basin. (**a**) Mercury intrusion and extrusion curves of the Shanxi Formation, (**b**) Mercury intrusion and extrusion curves of the Taiyuan Formation, (**c**) Pore volume distribution of shale in the Shanxi Formation interpreted by mercury intrusion porosimetry (MIP), (**d**) Pore volume distribution of shale in the Taiyuan Formation interpreted by MIP, (**e**) Specific surface area distribution of shale in the Shanxi Formation interpreted by MIP, (**f**) Specific surface area distribution of shale in the Taiyuan Formation interpreted by MIP.

The specific surface area in the Shanxi Formation is predominantly composed of micropores (<10 nm, >70%) and mesopores (2–50 nm, ~20%), with macropores (>1000 nm) having a negligible impact due to their smooth surfaces (Figure 3e). The Taiyuan Formation shows extreme microporosity dominance (<2 nm, >85% contribution), driven by the high specific surface area of clay mineral (e.g., illite) interlayer fractures, while mesopores contribute minimally (<10%) due to organic pore closure (Figure 3f).

3.4.2. Low-Temperature N_2 Adsorption

The N_2 adsorption–desorption isotherms of the Shanxi Formation coal-bearing shale exhibit type IV curves with H3 hysteresis loops, indicative of slit-shaped mesopores (2–50 nm) associated with clay mineral interlayer fractures (Figure 4a). A sharp adsorption increase at high relative pressures ($P/P_0 > 0.9$) suggests the presence of limited macropores (>50 nm) or fractures. Samples YQ-01-12 and YQ-04-04 show elevated adsorption capacities in the high-pressure range, likely due to tectonic fractures or dissolution vugs, while YQ-01-22 and YQ-02-02 exhibit reduced adsorption, attributed to compaction-induced pore closure. In contrast, the Taiyuan Formation coal-bearing shale also displays Type IV isotherms but with narrower hysteresis loops, reflecting poorer pore connectivity dominated by clay mineral interlayer micropores (Figure 4b). Enhanced adsorption at low relative pressures ($P/P_0 < 0.1$) highlights well-developed micropores (<2 nm), consistent with residual clay mineral microporosity after organic pore collapse at high maturity. Samples YQ-01-48 and YQ-04-19 demonstrate significantly higher adsorption capacities, likely due to an elevated illite content and denser microporosity.

Shanxi Formation: A bimodal distribution peaks at 10–50 nm (mesopores, ~60% contribution) and 1–10 μm (macropores). The macropore volume (>1000 nm) in samples like YQ-04-04 is linked to tectonic fractures or dissolution vugs. Micropores (<2 nm) account for <10% of the total pore volume, emphasizing a meso-macropore-dominated system optimized for seepage (Figure 4c). Taiyuan Formation: The core volume concentrates in <2 nm (micropores, >40%) and 2–10 nm (small mesopores), reflecting clay mineral-dominated microporosity post-organic pore closure. The mesopore volume (10–50 nm) is significantly lower than that in the Shanxi Formation due to compaction, while macropores (>1000 nm) are nearly absent, indicating poor connectivity reliant on microporous adsorption (Figure 4d).

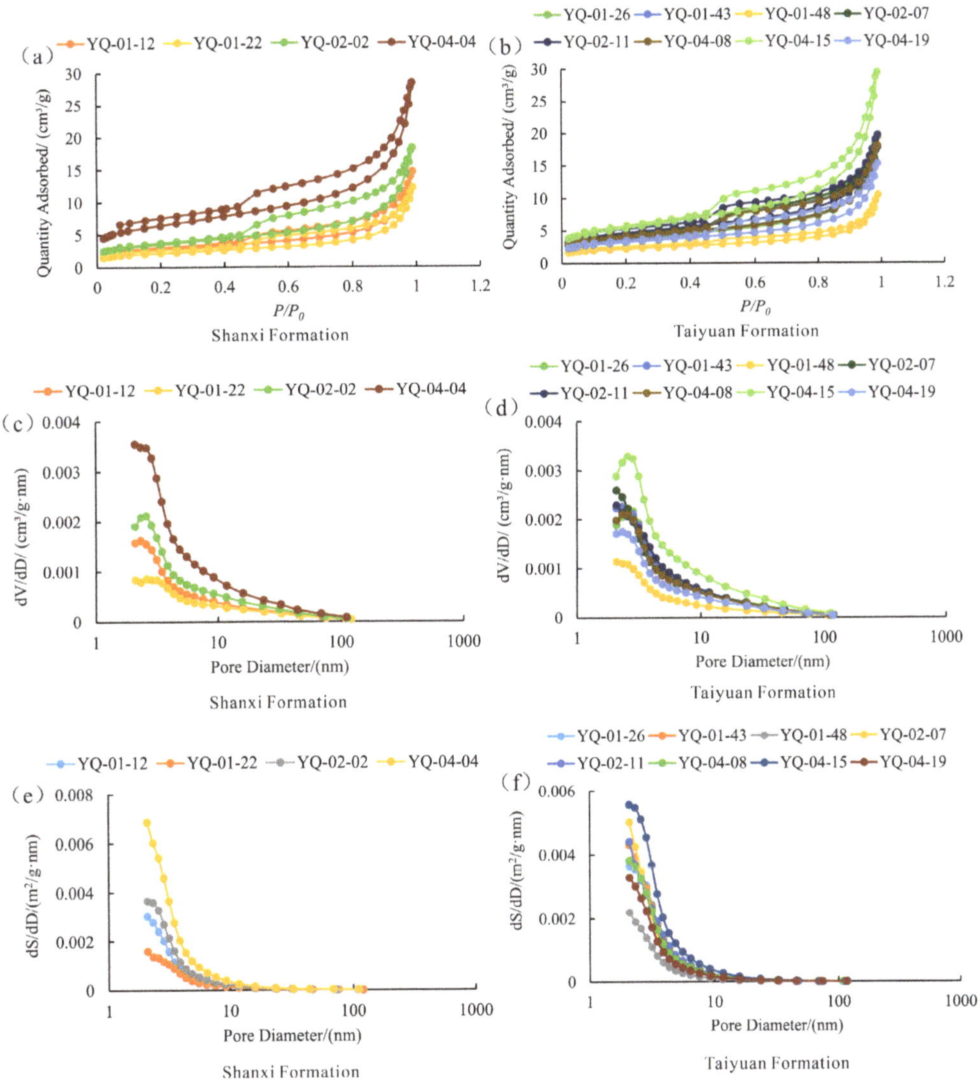

Figure 4. N_2 adsorption–desorption isotherms, pore volume distribution, and specific surface area distribution of Upper Paleozoic coal-bearing shale based on N_2 adsorption analysis in Yangquan Block, Qinshui Basin. (**a**) N_2 adsorption-desorption isotherms of the Shanxi Formation, (**b**) N_2; adsorption-desorption isotherms of the Taiyuan Formation, (**c**) Pore volume distribution of shale in the Shanxi Formation interpreted by N_2 adsorption, (**d**) Pore volume distribution of shale in the Taiyuan Formation interpreted by N_2 adsorption, (**e**) Specific surface area distribution of shale in the Shanxi Formation interpreted by N_2 adsorption, (**f**) Specific surface area distribution of shale in the Taiyuan Formation interpreted by N_2 adsorption.

Shanxi Formation: Micropores (<2 nm) contribute >70% of the total specific surface area, despite their low volumetric share, due to their high surface roughness. Mesopores (2–50 nm) account for ~20%, primarily from clay interlayer fractures and residual organic pores (Figure 4e). Taiyuan Formation: Micropores (<2 nm) dominate with >85% contribution, driven by the nanoscale roughness of clay mineral (e.g., illite) interlayer fractures. Mesopores (2–50 nm) contribute <10%, underscoring limited development post-organic pore closure (Figure 4f).

3.4.3. Low-Pressure CO_2 Adsorption

The CO_2 adsorption isotherms of the Shanxi Formation coal-bearing shale exhibit Type I behavior, indicating microporosity dominance (<2 nm), with rapid adsorption saturation at low relative pressures ($P/P_0 < 0.1$), reflecting the high adsorption capacity of micropores. The absence of hysteresis loops suggests poor connectivity and closed pore structures, likely due to organic pore collapse at high maturity or closure of clay mineral interlayer fractures (Figure 5a). Sample YQ-04-04 shows significantly higher adsorption capacity, attributed to its elevated clay mineral (e.g., illite) content and enhanced microporosity development. In the Taiyuan Formation coal-bearing shale, CO_2 adsorption at low pressures ($P/P_0 < 0.1$) exceeds that of the Shanxi Formation, indicating denser micropores associated with nanoscale interlayer pores in mixed-layer clay minerals (e.g., smectite-illite). Samples YQ-01-48 and YQ-04-19 exhibit exceptional adsorption capacities, reflecting uniform micropore distribution and higher surface roughness (Figure 5b).

Shanxi Formation: A bimodal distribution peaks at 2–10 nm (small mesopores) and 10–50 nm (mesopores), with mesopores accounting for ~60% of the total volume. Macropores (>50 nm) contribute minimally, though sample YQ-01-12 shows an elevated macropore volume, likely linked to localized dissolution vugs or tectonic fractures (Figure 5c). Taiyuan Formation: Micropores (<2 nm) dominate, exceeding 50% of the total pore volume, reflecting clay mineral-controlled microporosity post-organic pore closure. The mesopore volume (10–50 nm) is significantly lower than that in the Shanxi Formation due to compaction, highlighting limited connectivity and reliance on microporous adsorption (Figure 5d).

Shanxi Formation: Micropores (<2 nm) contribute >80% of the total specific surface area, driven by their high surface roughness, while mesopores (2–50 nm) account for <20%, primarily from clay interlayer fractures and residual organic pores (Figure 5e). Taiyuan Formation: Micropores (<2 nm) dominate with >90% contribution, consistent with the nanoscale roughness of clay mineral (e.g., illite) interlayer pores. Mesopores (2–50 nm) contribute < 5%, underscoring restricted development following organic pore closure (Figure 5f).

By integrating high-pressure mercury intrusion (HPMI), low-pressure N_2 adsorption, and low-temperature CO_2 adsorption analyses, this study comprehensively characterized the full-scale pore size distribution of 12 core samples from the Upper Paleozoic coal-bearing shale in the Qinshui Basin. The total pore volume (PV) of the shale ranges from 0.0255 to 0.8171 cm^3/g. Specifically, micropores (PV: 0.0013–0.0072 cm^3/g) contribute 0.31–14.75% (mean 5.18%) to the total PV, mesopores (PV: 0.0102–0.0326 cm^3/g) account for 3.99–76.19% (mean 36.92%), and macropores (PV: 0.0040–0.7820 cm^3/g) dominate with 15.66–95.71% (mean 57.90%) of the total PV. These results demonstrate that the pore volume system in the Upper Paleozoic coal-bearing shale of the Qinshui Basin is predominantly attributed to macropores, highlighting their critical role in controlling the overall pore structure characteristics (Figure 6a, Table 3).

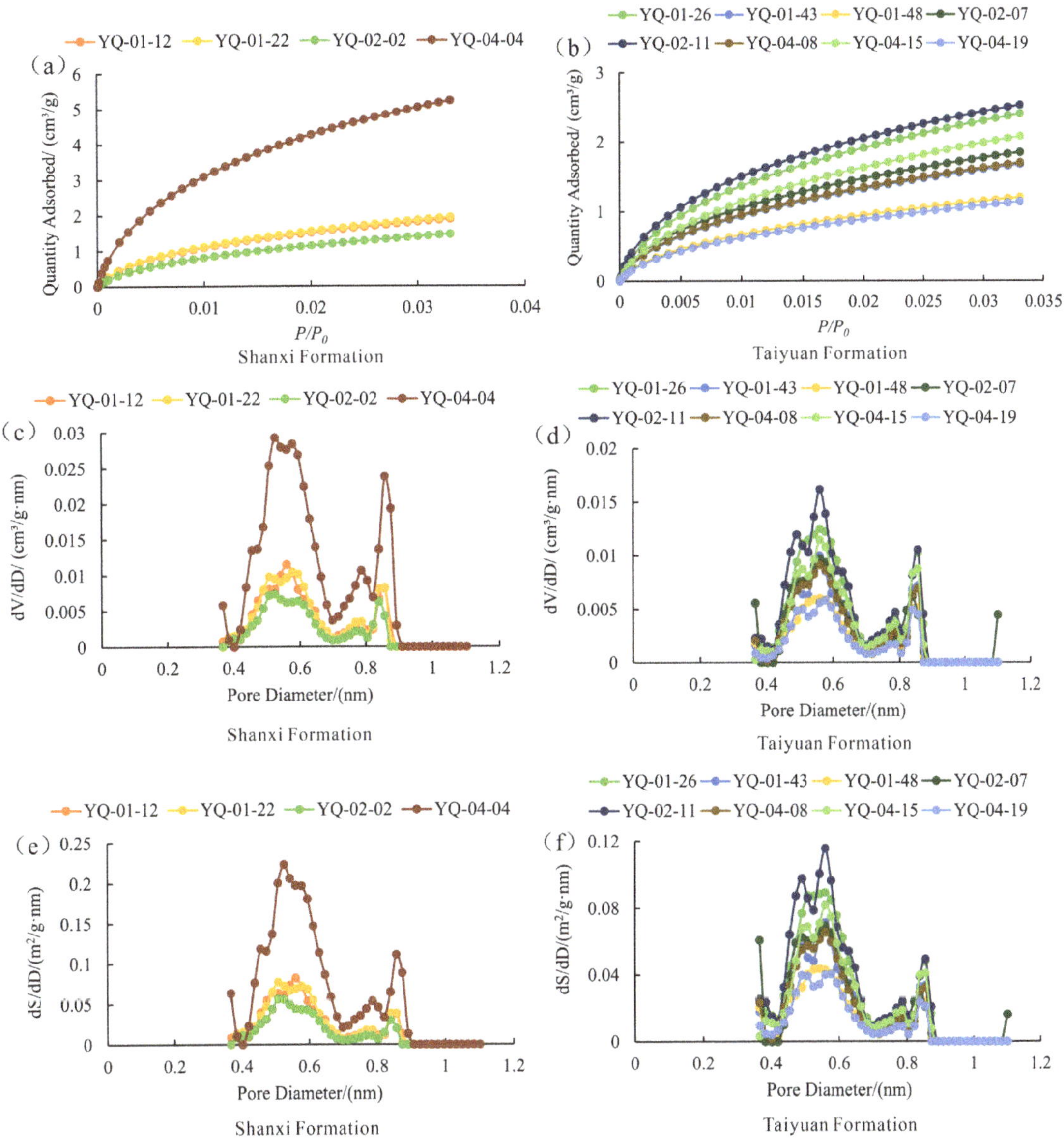

Figure 5. CO_2 adsorption–desorption isotherms, pore volume distribution, and specific surface area distribution of Upper Paleozoic coal-bearing shale based on CO_2 adsorption analysis in Yangquan Block, Qinshui Basin. (**a**) CO_2 adsorption isotherms of the Shanxi Formation, (**b**) CO_2 adsorption isotherms of the Taiyuan Formation, (**c**) Pore volume distribution of shale in the Shanxi Formation interpreted by CO_2 adsorption, (**d**) Pore volume distribution of shale in the Taiyuan Formation interpreted by CO_2 adsorption, (**e**) Specific surface area distribution of shale in the Shanxi Formation interpreted by CO_2 adsorption, (**f**) Specific surface area distribution of shale in the Taiyuan Formation interpreted by CO_2 adsorption.

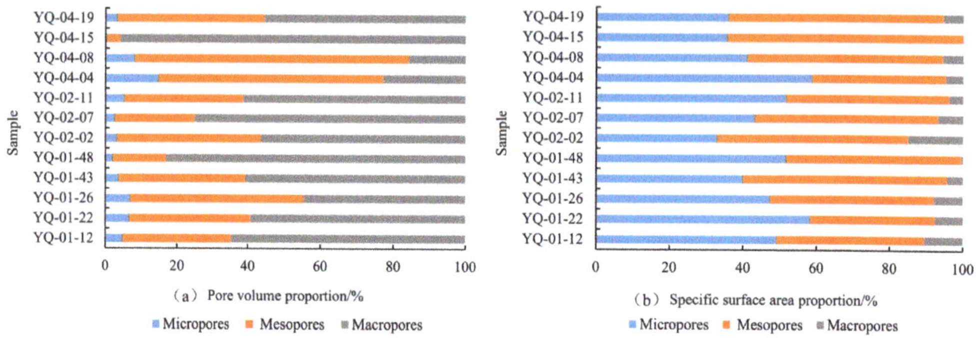

Figure 6. Pore volume and specific surface area characteristics of Upper Paleozoic coal-bearing shale samples from Yangquan Block, Qinshui Basin. (**a**) Shale pore volume characteristics, (**b**) Shale specific surface area characteristics.

Table 3. Full-scale pore distribution characteristics of Upper Paleozoic coal-bearing shale samples from the Yangquan Block, Qinshui Basin.

Sample	Pore Volume/(cm^3/g)				Specific Surface Area/(m^2/g)			
	Micropores	Mesopores	Macropores	Total Pore Volume	Micropores	Mesopores	Macropores	Total Specific Surface Area
YQ-01-12	0.0024	0.0152	0.0322	0.0498	8.261	6.800	1.773	16.834
YQ-01-22	0.0025	0.0127	0.0220	0.0371	8.417	4.943	1.102	14.462
YQ-01-26	0.0030	0.0213	0.0196	0.0439	10.259	9.728	1.719	21.706
YQ-01-43	0.0021	0.0195	0.0333	0.0548	6.955	9.729	0.764	17.448
YQ-01-48	0.0015	0.0102	0.0573	0.0690	4.950	4.598	0.053	9.601
YQ-02-02	0.0017	0.0208	0.0291	0.0516	5.792	9.241	2.672	17.705
YQ-02-07	0.0023	0.0195	0.0660	0.0879	7.948	9.227	1.260	18.435
YQ-02-11	0.0034	0.0212	0.0390	0.0636	11.578	10.026	0.854	22.458
YQ-04-04	0.0072	0.0305	0.0111	0.0488	24.060	14.978	1.943	40.981
YQ-04-08	0.0021	0.0195	0.0040	0.0255	7.082	9.175	0.967	17.224
YQ-04-15	0.0025	0.0326	0.7820	0.8171	8.402	15.225	0.002	23.629
YQ-04-19	0.0013	0.0162	0.0219	0.0395	4.535	7.413	0.691	12.639

By integrating the analytical results of the specific surface area (SSA), the Upper Paleozoic coal-bearing shale in the Qinshui Basin exhibits an SSA range of 9.601–40.981 m^2/g. Specifically, micropores (SSA: 4.535–24.060 m^2/g) contribute 32.71–58.71% (mean 45.38%) to the total SSA, mesopores (SSA: 4.598–15.225 m^2/g) account for 34.18–58.65% (mean 48.57%), and macropores (SSA: 0.002–2.672 m^2/g) show minimal contributions of 0.9–15.09% (mean 6.05%). These findings indicate that mesopores predominantly govern the SSA characteristics of the Upper Paleozoic coal-bearing shale in the Qinshui Basin, followed by micropores (Figure 6b, Table 3).

In the Upper Paleozoic coal-bearing shale samples from the Yangquan Block, a distinct relationship exists between the pore volume and specific surface area (Figure 7). Micropores and mesopores exhibit a strong positive correlation between the pore volume and specific surface area, reflecting their significant contribution to adsorption capacity through high surface roughness and nano-scale complexity. In contrast, macropores show a weak correlation as their larger pore sizes contribute minimally to the surface area despite occupying substantial pore volume. Overall, the total pore volume and specific surface area display a moderate positive correlation, driven predominantly by the dominance of micropores and mesopores in shaping surface properties. This behavior underscores the critical role of nano- to micro-scale pores in enhancing adsorption potential, while macropores primarily influence seepage efficiency without markedly affecting surface interactions. These findings align with the dual "adsorption-seepage" mechanisms in heterogeneous shale reservoirs, where pore-scale heterogeneity governs both gas storage and flow dynamics (Figure 7).

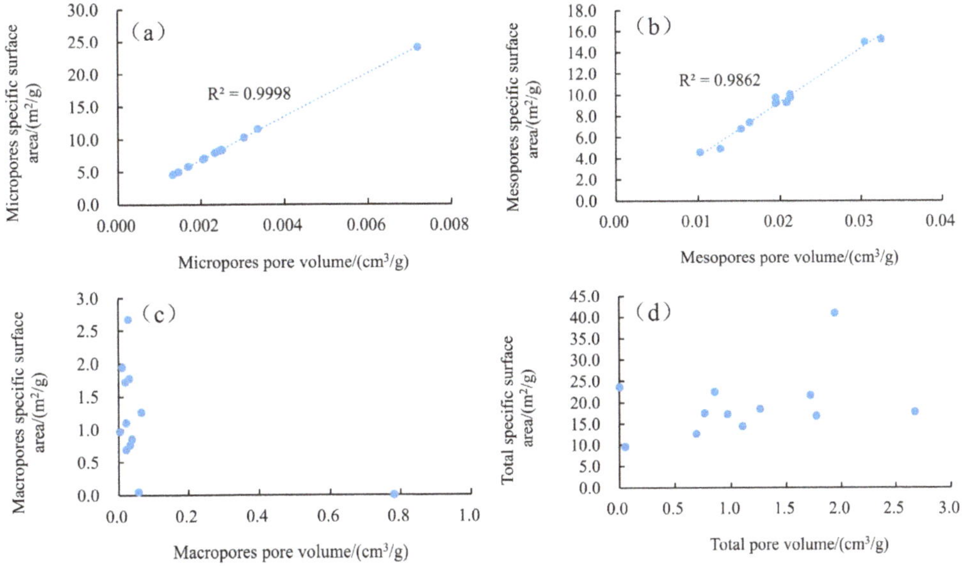

Figure 7. Relationship between pore volume and specific surface area of Upper Paleozoic coal-bearing shale samples from Yangquan Block, Qinshui Basin. (**a**) Relationship between micropore volume and micropore specific surface area, (**b**) Relationship between mesopore volume and mesopore specific surface area, (**c**) Relationship between macropore volume and macropore specific surface area, (**d**) Relationship between total pore volume and total specific surface area.

3.5. Fractal Characteristics of Shale

Through low-temperature N_2 adsorption experiments, the dual logarithmic relationship between adsorption volume and relative pressure (P/P_0) is established using Equation (1). The fractal dimension D is derived from the slope of this curve's linear regression. For coal-bearing shales exhibiting multi-scale pore systems, segmented fractal dimensions quantify heterogeneity across distinct pore size ranges: D_1 (surface fractal dimension) is defined at low relative pressure ($P/P_0 < 0.45$), where multilayer adsorption dominates as gas molecules progressively cover pore surfaces. This regime quantifies surface roughness, with D_1 values approaching 3 indicating highly irregular surfaces (e.g., clay mineral interlayer pores and organic pores) and $D_1 \approx 2$ reflecting smoother surfaces (e.g., quartz intergranular pores). D_2 (pore volume fractal dimension) characterizes the $P/P_0 > 0.45$ range, where capillary condensation occurs in mesopores (2–50 nm). D_2 captures pore network complexity and spatial filling capacity, where values near 3 signify intricate structures (e.g., intertwined nanopores/microfractures), while lower values (closer to 2) indicate enhanced connectivity (e.g., homogeneous mesopore networks). Fractal dimension fitting curves from the FHH model are shown in Figure 8, with the results summarized in Table 4. This integrated approach provides a quantitative framework to assess pore structure complexity and connectivity, offering critical insights into multi-scale heterogeneity and its implications for gas storage/transport mechanisms in coal-bearing shale reservoirs.

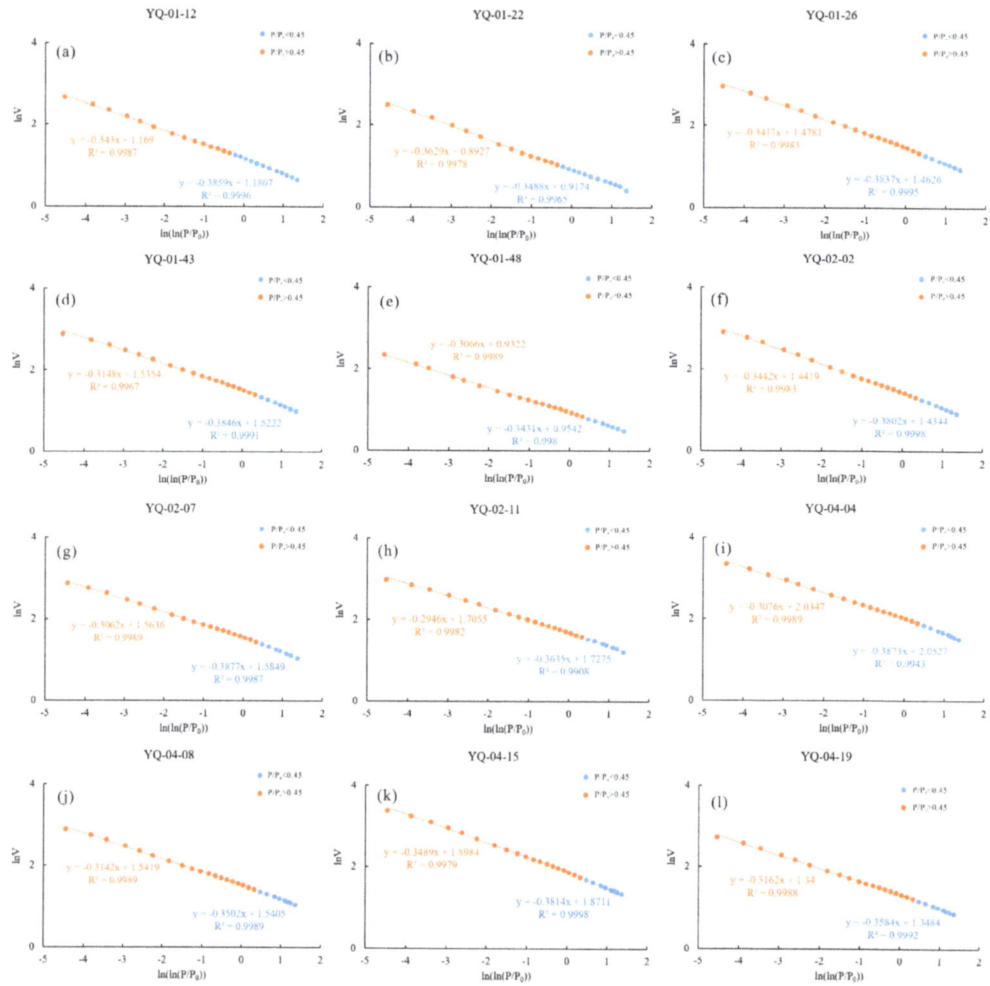

Figure 8. Fitting results of low-temperature N_2 adsorption experimental data for Upper Paleozoic coal-bearing shale samples from Yangquan Block, Qinshui Basin. (**a**) Fractal dimensions of YQ-01-12 Shale based on the FHH model, (**b**) Fractal dimensions of YQ-01-22 Shale based on the FHH model, (**c**) Fractal dimensions of YQ-01-26 Shale based on the FHH model, (**d**) Fractal dimensions of YQ-01-43 Shale based on the FHH model, (**e**) Fractal dimensions of YQ-01-48 Shale based on the FHH model, (**f**) Fractal dimensions of YQ-02-02 Shale based on the FHH model, (**g**) Fractal dimensions of YQ-02-07 Shale based on the FHH model, (**h**) Fractal dimensions of YQ-02-11 Shale based on the FHH model, (**i**) Fractal dimensions of YQ-04-04 Shale based on the FHH model, (**j**) Fractal dimensions of YQ-04-08 Shale based on the FHH model, (**k**) Fractal dimensions of YQ-04-15 Shale based on the FHH model, (**l**) Fractal dimensions of YQ-04-19 Shale based on the FHH model.

Based on the International Union of Pure and Applied Chemistry (IUPAC) pore classification standard and the unique characteristics of coal-bearing shale reservoirs [67,69], pore size ranges correspond to distinct types, formation mechanisms, and associated fractal dimensions derived from high-pressure mercury intrusion (HPMI) experiments: D_3 (macro-mesopore transition), defined for pores of 50–1000 nm, reflects pores primarily controlled by diagenesis (cementation and dissolution) and mechanical compaction, such as intergranular mineral pores and minor microfractures, with its fractal dimension quantifying pore surface roughness and connectivity ($D_3 \rightarrow 3$ indicates complex structures; $D_3 \rightarrow 2$ suggests uniform permeability); D_4 (macropore-fracture transition), covering 1000–10,000 nm, is dominated by tectonic fracturing and dissolution processes, including natural microfractures and dissolution vugs, with its fractal dimension capturing the geometric complexity of intertwined pores and fractures ($D_4 \rightarrow 3$ signifies intricate networks; $D_4 \rightarrow 2$ indicates simplified systems); and D_5 (macro-fracture segment), applying to pores > 10,000 nm, corresponds to macroscopic fractures, oversized vugs, or tectonic fractures formed by intense tectonic activity (e.g., faulting), with its fractal dimension reflecting fracture surface roughness and network heterogeneity ($D_5 \rightarrow 3$ denotes rough, branched fractures; $D_5 \rightarrow 2$ represents straight fractures). This geologically justified segmentation, based on coupled diagenetic–tectonic mechanisms specific to coal-bearing shales, employed a multi-scale HPMI analysis strategy where experimental data were substituted into Equation (2); the linear regression of $\lg(1 - S_{Hg})$ versus $\lg(P_c)$ yielded slopes to calculate D_3, D_4, and D_5 (the results are shown in Table 4, and fitting curves are shown in Figure 9), establishing a framework that bridges pore structure complexity with reservoir performance to predict gas storage heterogeneity, flow dynamics, and hydraulic fracturing efficacy.

Table 4. Fractal dimensions of pores in Upper Paleozoic coal-bearing shale samples from Yangquan Block, Qinshui Basin.

Sample	D_1	R^2	D_2	R^2	D_3	R^2	D_4	R^2	D_5	R^2
YQ-01-12	2.6141	0.9996	2.6570	0.9987	2.9682	0.9814	2.9891	0.9556	2.9629	0.8977
YQ-01-22	2.6512	0.9965	2.6371	0.9978	2.9612	0.9477	2.9897	0.9263	2.9901	0.9655
YQ-01-26	2.6163	0.9995	2.6583	0.9983	2.9852	0.9888	2.9956	0.9748	2.9885	0.9408
YQ-01-43	2.6154	0.9991	2.6852	0.9967	2.9707	0.9638	2.994	0.9657	2.9932	0.9754
YQ-01-48	2.6569	0.9980	2.6934	0.9989	2.9834	0.9802	2.9944	0.9556	2.9816	0.9024
YQ-02-02	2.6198	0.9998	2.6558	0.9983	2.9760	0.9880	2.9891	0.9543	2.9774	0.9659
YQ-02-07	2.6123	0.9987	2.6938	0.9989	2.8955	0.6979	2.9928	0.9916	2.9897	0.9600
YQ-02-11	2.6365	0.9908	2.7054	0.9982	2.9465	0.8104	2.9923	0.9706	2.9942	0.9962
YQ-04-04	2.6127	0.9943	2.6924	0.9989	2.9812	0.7915	2.9958	0.9628	2.9935	0.9260
YQ-04-08	2.6498	0.9989	2.6858	0.9989	2.9777	0.9701	2.9961	0.7884	2.9926	0.9481
YQ-04-15	2.6186	0.9998	2.6511	0.9979	2.9508	0.9927	2.9893	0.7504	2.9805	0.9650
YQ-04-19	2.6416	0.9992	2.6838	0.9988	2.8555	0.9171	2.9806	0.6503	2.9928	0.9688

D_1 (FHH) and D_3 (HPMI) exhibit a weak positive correlation with the clay content, while D_2 (FHH) and D_4 (HPMI) are suppressed in quartz-rich areas due to more homogeneous pore structures. Key differences arise from their operational scales: FHH focuses on the nano-scale (D_1–D_2), whereas HPMI covers the macro-to-fracture scale (D_3–D_5). Consequently, D_2 (mesopore volume complexity) and D_3 (macropore surface roughness) show no direct link, as D_2 is suppressed by organic pore collapse at high maturity, while D_3 is enhanced by dissolution. Technical limitations also contribute to differences: HPMI at high pressures can induce artificial fractures, potentially overestimating D_4–D_5 values, while FHH ignores contributions from macropores.

Significant differences exist between the Shanxi and Taiyuan Formations. The Shanxi Formation is dominated by mesopores, exhibits lower D_2 values indicating more homogeneous mesopores, and has a higher proportion of macropores linked to dissolution and fractures. This results in higher D_4 and D_5 values, suggesting good flow potential. In contrast, the Taiyuan Formation is micropore-dominated, shows higher D_2 values reflecting complex micro-mesopore interweaving, and has lower macropore abundance primarily from clay interlayer pores. Consequently, it exhibits higher D_1/D_2 values favoring gas adsorption but lower D_5 values, indicating a greater need for effective stimulation.

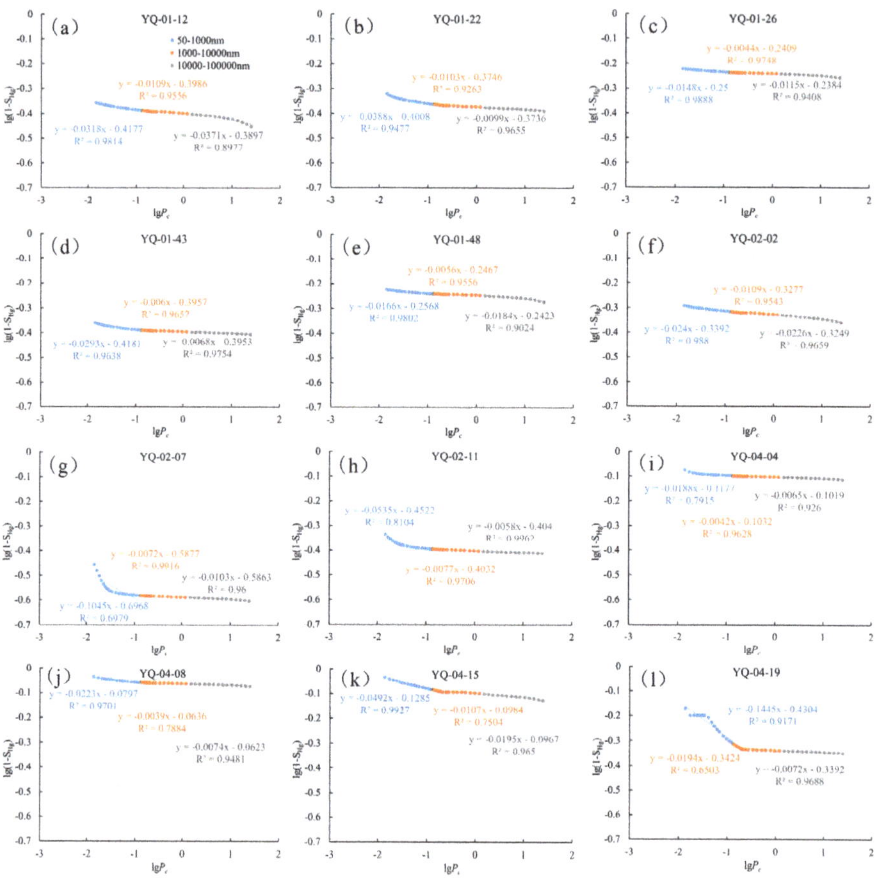

Figure 9. Fitting results of high-pressure mercury intrusion experimental data for Upper Paleozoic coal-bearing shale samples from Yangquan Block, Qinshui Basin. (**a**) Fractal dimensions of YQ-01-12 Shale based on high-pressure mercury intrusion, (**b**) Fractal dimensions of YQ-01-22 Shale based on high-pressure mercury intrusion, (**c**) Fractal dimensions of YQ-01-26 Shale based on high-pressure mercury intrusion, (**d**) Fractal dimensions of YQ-01-43 Shale based on high-pressure mercury intrusion, (**e**) Fractal dimensions of YQ-01-48 Shale based on high-pressure mercury intrusion, (**f**) Fractal dimensions of YQ-02-02 Shale based on high-pressure mercury intrusion, (**g**) Fractal dimensions of YQ-02-07 Shale based on high-pressure mercury intrusion, (**h**) Fractal dimensions of YQ-02-11 Shale based on high-pressure mercury intrusion, (**i**) Fractal dimensions of YQ-04-04 Shale based on high-pressure mercury intrusion, (**j**) Fractal dimensions of YQ-04-08 Shale based on high-pressure mercury intrusion, (**k**) Fractal dimensions of YQ-04-15 Shale based on high-pressure mercury intrusion, (**l**) Fractal dimensions of YQ-04-19 Shale based on high-pressure mercury intrusion.

4. Discussion

4.1. Relationship Between Organic Matter Characteristics and Fractal Dimensions

Organic matter characteristics (e.g., abundance, type, and maturity) act as primary controls on the evolution of shale pore fractal dimensions by regulating the development and spatial arrangement of organic pores and dissolution pores [27,64]. This study analyzed the relationship between total organic carbon (TOC) and fractal dimensions, revealing a slight increase in D_1 with higher TOC, albeit with a weak correlation, while TOC showed negligible associations with D_2, D_3, D_4, and D_5. In the Upper Paleozoic coal-bearing shale of the Yangquan Block (Qinshui Basin), which is in a high-maturity stage (R_o > 2%), organic pores collapse under compaction, diminishing TOC's contribution to pore complexity. Instead, interlayer fractures in clay minerals (e.g., illite) and quartz intergranular pores dominate the pore structure, diluting TOC's influence.

The average vitrinite reflectance (R_o) of 2.29% and T_{max} of 553 °C confirm the shale's overmature dry gas stage, indicating advanced organic matter evolution, likely involving graphitization transformation. The high R_o and T_{max} values suggest complete hydrocarbon generation and thermal degradation, with organic pores largely being collapsed. Notably, R_o exhibits a weak negative correlation with D_4 and D_5 but a positive correlation with D_3. This divergence arises because high maturity (1) promotes the development of interlayer fractures in clay minerals (enhancing microporosity roughness, D_3), (2) suppresses macropore complexity (D_4, D_5) via compaction, and (3) enhances microporosity heterogeneity through organic matter shrinkage. At R_o > 2.0%, organic pore closure is prevalent, though tectonic fractures (D_5) may become more complex due to stress release.

As shown in Figure 10c, T_{max} displays a significant negative correlation with D_2 but no clear relationship with D_1 or D_3. An elevated T_{max} value (>550 °C) corresponds to overmaturity, leading to organic pore closure (D_2 decline) but potentially enhancing mineral dissolution pores (weak positive correlation with D_3). At T_{max} > 560 °C, thermal cracking may generate secondary microfractures (D_4, D_5).

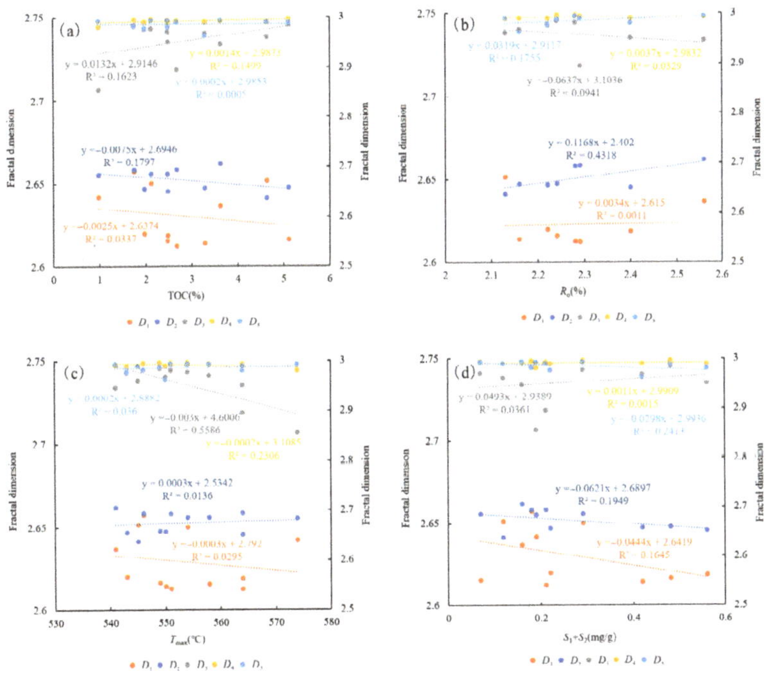

Figure 10. Relationships between fractal dimensions of pores and organic matter characteristics (TOC, R_o) in Upper Paleozoic coal-bearing shale samples from Yangquan Block, Qinshui Basin. (**a**) Relationship between TOC and fractal dimension, (**b**) Relationship between R_o (vitrinite reflectance) and fractal dimension, (**c**) Relationship between T_{max} and fractal dimension, (**d**) Relationship between S_1 + S_2 (hydrocarbon generation potential) and fractal dimension.

Hydrocarbon potential (S_1 + S_2) inversely correlates with fractal dimensions: higher S_1 + S_2 values correspond to reduced D_3 and D_5, suggesting that early hydrocarbon generation may degrade pore connectivity (D_3), while late-stage tectonic fractures (D_5) improve permeability but reduce adsorption capacity. The weak associations between

TOC, $S_1 + S_2$, and fractal dimensions stem from organic pore collapse at high maturity and mineral interference.

In summary, R_o and T_{max} suppress macropore complexity through compaction but may enhance microporosity heterogeneity. Overmaturity ($R_o > 2.0\%$, $T_{max} > 550$ °C) results in organic pore closure, weakening correlations between D_1–D_2 and TOC/R_o. Tectonic fractures (D_4–D_5), governed by stress rather than thermal maturity, may exhibit increased complexity due to post-diagenetic uplift, highlighting the interplay of organic evolution, diagenesis, and tectonic activity in shaping fractal pore networks.

4.2. Relationship Between Mineral Composition and Fractal Dimensions

The type, particle size, and assemblage of minerals significantly influence shale pore morphology and size, thereby shaping fractal characteristics [67]. This study investigated correlations between fractal dimensions and quartz/clay mineral contents. The results show that the quartz content exhibits negligible associations with fractal dimensions D_1–D_4, with only a weak positive correlation to D_5. This limited influence may arise from the simple structure and smooth surfaces of quartz intergranular pores, which suppress fractal complexity, coupled with masking effects from clay minerals or organic pores. Additionally, the narrow quartz content range (0–60%) in the samples likely hindered the development of significant trends.

In the Upper Paleozoic coal-bearing shale of the Yangquan Block, high-quartz zones feature simplified pore structures (low fractal dimensions), favoring superior permeability but weak adsorption. High-clay zones exhibit abundant micropores (elevated D_1, D_2) but poor macropore connectivity (low D_4, D_5), necessitating fractures to enhance seepage. Clay mineral transformations (e.g., smectite-to-illite conversion) may remodel pore structures, amplifying microporosity complexity (D_1, D_2). Tectonic fractures (D_5) emerge as critical factors for improving permeability, particularly in clay-rich intervals. These findings underscore the dual role of mineralogy in balancing adsorption and flow dynamics, with fracture networks serving as key enablers for gas mobility in heterogeneous coal-bearing shale systems (Figure 11a).

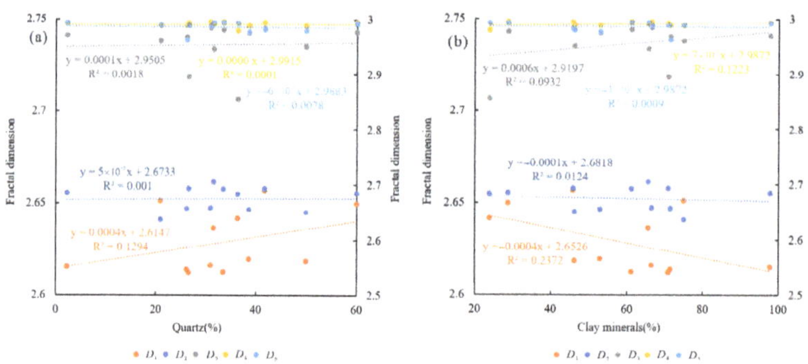

Figure 11. Relationships between fractal dimensions of pores and mineral composition (quartz and clay) in Upper Paleozoic coal-bearing shale samples from Yangquan Block, Qinshui Basin. (**a**) Relationship between quartz content and fractal dimension, (**b**) Relationship between clay mineral content and fractal dimension.

In contrast, the clay mineral content demonstrates a weak positive correlation with D_1 and D_2, no significant link to D_3–D_4, and a moderate negative correlation with D_5. Clay minerals (e.g., illite) likely dominate microporosity complexity through interlayer fractures, contributing to rough surfaces (higher D_1 and D_2) and enhanced adsorption capacity. However, high-clay-content regions exhibit poor macropore connectivity (lower

D_4 and D_5) due to strong plastic deformation, which inhibits tectonic fracture development (D_5 reduction) (Figure 11b).

4.3. Relationship Between Pore Structure and Fractal Dimensions

Pore structure parameters in shale are critical factors influencing fractal dimensions [27]. The relationship between pore volume and fractal dimensions is complex: macropore development generally simplifies pore structures (reducing fractal dimensions), whereas mesopore and micropore proliferation enhances pore network complexity (increasing fractal dimensions).

When the total pore volume increases, the fractal dimension D_1 slightly decreases, albeit with an extremely weak correlation, while D_4 exhibits a weak positive correlation, suggesting that macropore volume growth may marginally amplify pore complexity. No significant associations are observed between the total pore volume and D_2, D_3, or D_5. This weak correlation arises because the total pore volume comprises contributions from micropores, mesopores, and macropores, whose opposing impacts on fractal dimensions may offset one another. For D_4 (macropore range: 1000–10,000 nm), volume increases could stem from isolated dissolution pores (simple structures) or fracture networks (complex structures), leading to contradictory fractal responses. In high-maturity shale, organic pore compaction may reduce the micropore volume, but tectonic fractures (macropores) may partially compensate for total porosity.

As shown in Figure 12b, the total specific surface area (SSA) displays a moderate negative correlation with D_5, a weak positive correlation with D_2, and no significant links to D_1, D_3, or D_4. An elevated SSA is typically attributed to micropores (<50 nm); however, the negative correlation with D_5 (macrofractures > 10,000 nm) suggests that reduced fracture roughness (D_5) may coexist with high SSA from microporosity sources such as organic pores (high D_1) or clay mineral interlayer fractures (high D_2). This spatial decoupling implies that macrofractures (D_5) and microporosity-dominated SSA are inversely related. In overmature shales, organic pore closure reduces the SSA, but tectonic fractures (high D_5) may dominate as primary flow pathways, necessitating a balance between adsorption capacity and seepage efficiency. These findings highlight the intricate trade-offs between pore-scale heterogeneity and reservoir-scale performance in coal-bearing shale systems.

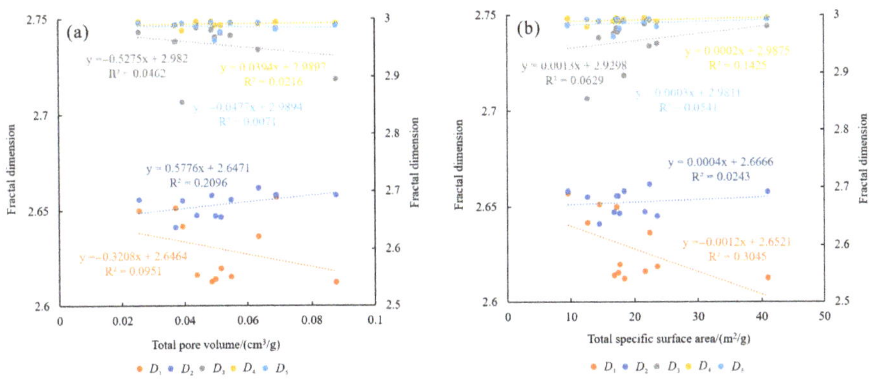

Figure 12. Relationships between fractal dimensions of pores, pore volume, and specific surface area in Upper Paleozoic coal-bearing shale samples from Yangquan Block, Qinshui Basin. (**a**) Relationship between total pore volume and fractal dimension, (**b**) Relationship between total specific surface area and fractal dimension.

4.4. Implications for Shale Gas Exploration in the Upper Paleozoic Coal-Bearing Shale of the Yangquan Block, Qinshui Basin

Previous studies have revealed that the fractal dimension of pores in coal-bearing shale significantly governs the enrichment of shale gas. Specifically, the gas adsorption capacity of coal-bearing shale exhibits a positive correlation with the fractal dimension; therefore, a higher fractal dimension corresponds to greater adsorbed gas content in coal-bearing shale [68,69]. Simultaneously, the fractal dimension also regulates the gas migration efficiency. A higher fractal dimension reflects more complex pore structures and increased tortuosity within the coal-bearing shale reservoir, leading to shale gas having reduced seepage capacity. Diminished flow dynamics enhances gas retention, thereby favoring the preservation of shale gas accumulations in coal-bearing shale [27].

The Qinshui Basin, a significant shale gas enrichment zone in China, exhibits substantial resource potential in the Upper Paleozoic coal-bearing shale of the Yangquan Block, characterized by the following traits: high organic matter abundance, where the total organic carbon (TOC) content ranges from 1 to 5%, with localized peaks of up to 6%, indicating robust hydrocarbon source rock potential; moderate thermal maturity, where vitrinite reflectance (R_o) values predominantly fall between 1.0 and 2.5%, spanning the mature to overmature stages, suitable for shale gas generation and preservation; and multi-scale pore development, involving the coexistence of micropores (<50 nm) and tectonic fractures (>10,000 nm), with fractal dimensions (D_1–D_5) revealing heterogeneity that underscores dual adsorption and seepage capabilities.

A quantitative analysis of fractal dimensions (D_1–D_5), integrating low-temperature N_2 adsorption (D_1, D_2) and high-pressure mercury intrusion (D_3–D_5) experiments, elucidates the heterogeneity of multi-scale pore systems in coal-bearing shale, offering novel metrics for reservoir classification. Overmaturity ($R_o > 2.0\%$) induces organic pore collapse but enhances seepage capacity through tectonic fracture development (D_5), resolving the dynamic balance between "adsorption-seepage" trade-offs. A "sweet spot" model—high D_1 (micropore adsorption) + low D_4 (macropore seepage) + high D_5 (fracture complexity)—is proposed based on correlations between fractal dimensions, mineral composition (quartz and clay), and total specific surface area, refining exploration targeting.

5. Conclusions

(1) The Upper Paleozoic coal-bearing shale in the Yangquan Block of the Qinshui Basin exhibits high organic matter abundance (TOC: avg. 3.82%) and has reached an overmature thermal maturity stage (R_o: avg. 2.29%). Mineralogically, it is characterized by a high clay mineral content (avg. 58.9% by mass) and a low brittle mineral content.

(2) The Upper Paleozoic coal-bearing shale in the Yangquan Block of the Qinshui Basin exhibits low porosity and low permeability characteristics. Porosity ranges from 1.1 to 4.9% (average 3.1%), while permeability varies between 0.0008×10^{-3} and 0.1910×10^{-3} μm^2 (average 0.0274×10^{-3} μm^2). Pore types are dominated by interparticle pores, intraparticle pores, and microfractures, which collectively provide essential storage space for shale gas accumulation.

(3) The Upper Paleozoic coal-bearing shale in the Yangquan Block of the Qinshui Basin exhibits a total pore volume ranging from 0.0255 to 0.8171 cm^3/g (mean: 0.1157 cm^3/g) and a total specific surface area (SSA) of 9.60–40.98 m^2/g (mean: 19.43 m^2/g). Micropores and mesopores demonstrate a strong positive correlation between their pore volumes and SSA, indicating that these pore types serve as the primary storage carriers for shale gas in the Yangquan Block. This micro-mesopore dominance suggests favorable shale gas storage potential. However, the limited fracability of the shale reservoir, likely due to the high clay mineral content and low brittle mineral

proportions, poses significant challenges for gas extraction, potentially hindering commercial development.

(4) The surface fractal dimension (D_1) and volumetric fractal dimension ($D2$) quantify pore surface roughness and the filling capacity of nano-scale pore networks, respectively. A D_1 value approaching 3 (e.g., 2.8–2.9) indicates well-developed organic pores or clay mineral interlayer fractures, signifying strong adsorption potential. $D_2 \rightarrow 3$ reflects complex structures formed by intertwined micropores and mesopores. D_3 (50–1000 nm) characterizes the connectivity of mineral intergranular pores and microfractures, while D_4 (1000–10,000 nm) delineates the spatial distribution of dissolution vugs and fracture networks. D_5 (>10,000 nm) captures the roughness and seepage efficiency of macroscopic fractures. The segmented variations in fractal dimensions (e.g., D_3 = 2.6 vs. D_5 = 2.9) underscore the inherent heterogeneity of the "adsorption-seepage" contradiction in reservoirs, highlighting the dual challenges of balancing gas storage and flow capacity in coal-bearing shale systems.

(5) The weak correlation ($R^2 < 0.2$) between total organic carbon (TOC) and micropore fractal dimensions (D_1 and D_2) suggests that high thermal maturity ($R_o > 2.0\%$) induces organic pore compaction, diminishing TOC's contribution to pore complexity. Increasing R_o suppresses macropore fractal dimensions (D_4 and D_5 reduction) but may indirectly enhance micropore roughness (D_3 weak positive correlation) through clay mineral transformations (e.g., illitization). High-quartz zones (>50%) exhibit uniform pore structures ($D_3 \approx 2$), favoring superior permeability but weak adsorption capacity. In contrast, high-clay zones (>60%) elevate micropore fractal dimensions (D_1 and D_2) while suppressing macropore connectivity (D_4 and D_5 decline). Optimal reservoirs require the synergistic combination of high D_1 (>2.8) + low D_4 (<2.6) + high D_5 (>2.7), ensuring strong micropore adsorption, efficient macropore seepage, and complex fracture networks. This model balances adsorption–seepage trade-offs, offering a targeted framework for identifying high-potential intervals in heterogeneous, overmature coal-bearing shale systems.

Author Contributions: Conceptualization, X.L. and J.Z.; methodology, X.L.; software, J.Z.; validation, X.L., J.Z. and X.Z. (Xueqing Zhang); formal analysis, J.Z.; investigation, X.Z. (Xueqing Zhang); resources, X.L. and S.K.; data curation, X.Z. (Xiaoyan Zou); writing—original draft preparation, J.Z.; writing—review and editing, X.Z. (Xiaoyan Zou), J.Z., and Y.Y.; visualization, J.Z.; supervision, X.L.; project administration, X.L.; funding acquisition, X.L. and S.K. All authors have read and agreed to the published version of the manuscript.

Funding: This research was supported by the Natural Science Foundation Project of China (U1810201) and the Influence and Contribution of Inorganic Fluids on Natural Gas Generation under Deep-Seated Conditions (CSJ-2023-08).

Data Availability Statement: The data that support the findings of this study are available from the corresponding author upon reasonable request. All the data can be found in the paper.

Conflicts of Interest: The authors declare that they have no conflicts of interest.

References

1. Dai, J.; Dong, D.; Ni, Y.; Gong, D.; Huang, S.; Hong, F.; Zhang, Y.; Liu, Q.; Wu, X.; Feng, Z. Distribution Patterns of Tight Sandstone Gas and Shale Gas. *Pet. Explor. Dev.* **2024**, *51*, 767–779. [CrossRef]
2. Guo, X.; Wang, R.; Shen, B.; Wang, G.; Wan, C.; Wang, Q. Geological Characteristics, Resource Potential, and Development Direction of Shale Gas in China. *Pet. Explor. Dev.* **2025**, *52*, 17–32. [CrossRef]
3. Li, S.; Zhou, Z.; Nie, H.; Zhang, L.; Song, T.; Liu, W.; Li, H.; Xu, Q.; Wei, S.; Tao, S. Distribution Characteristics, Exploration & Development, Geological Theories Research Progress and Exploration Directions of Shale Gas in China. *China Geol.* **2022**, *5*, 110–135. [CrossRef]

4. Mcmahon, T.P.; Larson, T.E.; Zhang, T.; Shuster, M. Geologic Characteristics, Exploration and Production Progress of Shale Oil and Gas in the United States: An Overview. *Pet. Explor. Dev.* **2024**, *51*, 925–948. [CrossRef]
5. Nie, H.; Dang, W.; Zhang, Q.; Zhang, J.; Li, P.; Zhang, S.; Ding, J.; Chen, Q.; Feng, Y.; Zhang, X. Evaluation of Gas Content in Organic-Rich Shale: A Review of the History, Current Status, and Future Directions. *Geosci. Front.* **2024**, *15*, 101921. [CrossRef]
6. Zhai, G.; Wang, Y.; Zhou, Z.; Yu, S.; Chen, X.; Zhang, Y. Exploration and Research Progress of Shale Gas in China. *China Geol.* **2018**, *1*, 257–272. [CrossRef]
7. Zhang, J.; Li, Z.; Wang, D.; Xu, L.; Li, Z.; Niu, J.; Chen, L.; Sun, Y.; Li, Q.; Yang, Z.; et al. Shale Gas Accumulation Patterns in China. *Nat. Gas Ind. B* **2023**, *10*, 14–31. [CrossRef]
8. Zou, X.; Li, X.; Zhang, J.; Li, H.; Guo, M.; Zhao, P. Characteristics of Pore Structure and Gas Content of the Lower Paleozoic Shale from the Upper Yangtze Plate, South China. *Energies* **2021**, *14*, 7603. [CrossRef]
9. Zou, C.; Dong, D.; Wang, Y.; Li, X.; Huang, J.; Wang, S.; Guan, Q.; Zhang, C.; Wang, H.; Liu, H.; et al. Shale Gas in China: Characteristics, Challenges and Prospects (II). *Pet. Explor. Dev.* **2016**, *43*, 182–196. [CrossRef]
10. Miao, H.; Jiang, Z.X.; Gong, X.; Li, C.M.; Wang, D.D.; Wu, Q.Z. The shale gas migration capacity of the Qiongzhusi Formation: Implications for its enrichment model. *Phys. Fluids* **2025**, *37*, 056603. [CrossRef]
11. He, X.; Chen, G.; Wu, J.; Liu, Y.; Wu, S.; Zhang, J.; Zhang, X. Deep Shale Gas Exploration and Development in the Southern Sichuan Basin: New Progress and Challenges. *Nat. Gas Ind. B* **2023**, *10*, 32–43. [CrossRef]
12. Zhang, J.; Xiao, X.; Wang, J.; Lin, W.; Han, D.; Wang, C.; Li, Y.; Xiong, Y.; Zhang, X. Pore Structure and Fractal Characteristics of Coal-Bearing Cretaceous Nenjiang Shales from Songliao Basin, Northeast China. *J. Nat. Gas Geosci.* **2024**, *9*, 197–208. [CrossRef]
13. Yun, L. Accumulation and Exploration Enlightenment of Shallow Normal-Pressure Shale Gas in Southeastern Sichuan Basin, SW China. *Pet. Explor. Dev.* **2023**, *50*, 1308–1319. [CrossRef]
14. Yong, R.; Yang, H.; Wu, W.; Yang, X.; Yang, Y.; Huang, H. Controlling Factors and Exploration Potential of Shale Gas Enrichment and High Yield in Permian Dalong Formation, Northern Sichuan Basin, SW China. *Pet. Explor. Dev.* **2025**, *52*, 285–300. [CrossRef]
15. Zhang, L.; He, X.; Li, X.; Li, K.; He, J.; Zhang, Z.; Guo, J.; Chen, Y.; Liu, W. Shale Gas Exploration and Development in the Sichuan Basin: Progress, Challenge and Countermeasures. *Nat. Gas Ind. B* **2022**, *9*, 176–186. [CrossRef]
16. Zhao, Q.; Zhou, T.; Wang, H.; Jin, Y.; Zhou, S.; Duan, X. Determination of Adsorption Parameters in Shale Gas Resource/Reserve Calculation: Case Study of Wufeng Formation–Longmaxi Formation in the Sichuan Basin. *Nat. Gas Ind. B* **2023**, *10*, 304–311. [CrossRef]
17. Lu, C.; Xiao, X.; Gai, H.; Feng, Y.; Li, G.; Meng, G.; Gao, P. Nanopore Structure Characteristics and Evolution of Type III Kerogen in Marine-Continental Transitional Shales from the Qinshui Basin, Northern China. *Geoenergy Sci. Eng.* **2023**, *221*, 211413. [CrossRef]
18. Sun, J.; Xiao, X.; Cheng, P. Methane Absorption of Coal-Measure Shales with and without Pore Water from the Qinshui Basin, North China: Based on High-Pressure Methane Absorption Experiments. *Int. J. Coal Geol.* **2022**, *263*, 104116. [CrossRef]
19. Yang, X.; Tang, S.; Xi, Z.; Zhang, S.; Zhang, Q.; Zhang, K.; Lin, D.; Wang, J. Dynamic Evolution and Differential Enrichment of Deep Coalbed Methane: A Case Study in Qinshui Basin. *Int. J. Coal Geol.* **2025**, *299*, 104696. [CrossRef]
20. Zhang, X.; Li, X.; Yang, J.; Zhang, B.; Sun, J.; Yu, Z. Characterization of the Full-Sized Pore Structure and Controlling Factors of the Coal-Bearing Shale in the Wuxiang Block, South-Central Qinshui Basin, China. *Front. Earth Sci.* **2022**, *9*, 813925. [CrossRef]
21. He, H.; Liu, P.; Xu, L.; Hao, S.; Qiu, X.; Shan, C.; Zhou, Y. Pore Structure Representations Based on Nitrogen Adsorption Experiments and an FHH Fractal Model: Case Study of the Block Z Shales in the Ordos Basin, China. *J. Pet. Sci. Eng.* **2021**, *203*, 108661. [CrossRef]
22. He, T.; Zhou, Y.; Chen, Z.; Zhang, Z.; Xie, H.; Shang, Y.; Cui, G. Fractal Characterization of the Pore-Throat Structure in Tight Sandstone Based on Low-Temperature Nitrogen Gas Adsorption and High-Pressure Mercury Injection. *Fractal Fract.* **2024**, *8*, 356. [CrossRef]
23. Ma, X.; Guo, S.; Shi, D.; Zhou, Z.; Liu, G. Investigation of Pore Structure and Fractal Characteristics of Marine-Continental Transitional Shales from Longtan Formation Using MICP, Gas Adsorption, and NMR (Guizhou, China). *Mar. Pet. Geol.* **2019**, *107*, 555–571. [CrossRef]
24. Ning, S.; Xia, P.; Hao, F.; Tian, J.; Fu, Y.; Wang, K. Pore Fractal Characteristics between Marine and Marine–Continental Transitional Black Shales: A Case Study of Niutitang Formation and Longtan Formation. *Fractal Fract.* **2024**, *8*, 288. [CrossRef]
25. Wang, Y.; Han, D.; Lin, W.; Jia, Y.; Zhang, J.; Wang, C.; Ma, B. Factors Controlling Differences in Morphology and Fractal Characteristics of Organic Pores of Longmaxi Shale in Southern Sichuan Basin, China. *Fractal Fract.* **2024**, *8*, 555. [CrossRef]
26. Mandelbrot, B.B. *Les Objets Fractals: Forme, Hazard et Dimension*; Flammarion: Paris, France, 1975.
27. Zhang, J.; Li, X.; Wei, Q.; Sun, K.; Zhang, G.; Wang, F. Characterization of Full-Sized Pore Structure and Fractal Characteristics of Marine–Continental Transitional Longtan Formation Shale of Sichuan Basin, South China. *Energy Fuels* **2017**, *31*, 10490–10504. [CrossRef]
28. Feng, Y.; Xiao, X.-M.; Wang, E.-Z.; Gao, P.; Lu, C.-G.; Li, G. Gas Storage in Shale Pore System: A Review of the Mechanism, Control and Assessment. *Pet. Sci.* **2023**, *20*, 2605–2636. [CrossRef]

29. Li, X.; Li, Y.; Li, J.; Zou, X.; Guo, M.; Wang, Z.; Zhang, X.; Wang, F. Characteristics of Pore Structures from the Lower Paleozoic Shale Gas Reservoirs in Northern Guizhou, South China. *J. Nat. Gas Geosci.* **2020**, *5*, 241–253. [CrossRef]
30. Liu, Z.; Zhao, S.; Hu, Z.; Shen, B.; Feng, D.; Zhang, J.; Ye, J.; Ma, C.; Dong, Z. Characteristics and Controlling Factors of Coal-Measure Unconventional Reservoirs—A Case Study of the Carboniferous-Permian in the Ordos Basin. *Unconv. Resour.* **2025**, *6*, 100153. [CrossRef]
31. Nie, H.; Jin, Z.; Li, P.; Jay Katz, B.; Dang, W.; Liu, Q.; Ding, J.; Jiang, S.; Li, D. Deep Shale Gas in the Ordovician-Silurian Wufeng–Longmaxi Formations of the Sichuan Basin, SW China: Insights from Reservoir Characteristics, Preservation Conditions and Development Strategies. *J. Asian Earth Sci.* **2023**, *244*, 105521. [CrossRef]
32. Zhang, M.; Fu, X. Influence of Reservoir Properties on the Adsorption Capacity and Fractal Features of Shales from Qinshui Coalfield. *J. Pet. Sci. Eng.* **2019**, *177*, 650–662. [CrossRef]
33. Wang, Q.; Lu, H.; Wang, T.; Liu, D.; Peng, P.; Zhan, X.; Li, X. Pore Characterization of Lower Silurian Shale Gas Reservoirs in the Middle Yangtze Region, Central China. *Mar. Pet. Geol.* **2018**, *89*, 14–26. [CrossRef]
34. Liang, J.; Huang, W.; Wang, H.; Blum, M.J.; Chen, J.; Wei, X.; Yang, G. Organic Geochemical and Petrophysical Characteristics of Transitional Coal-Measure Shale Gas Reservoirs and Their Relationships with Sedimentary Environments: A Case Study from the Carboniferous-Permian Qinshui Basin, China. *J. Pet. Sci. Eng.* **2020**, *184*, 106510. [CrossRef]
35. Xiang, J.; Li, J.; Li, X.; Ma, X.; Sun, J.; Zou, Z.; Qin, Y. Isotopic and Hydrochemical Evidence for the Origins of Unconventional Gases in the Middle and Lower Jurassic Coal Measures in the Fukang Area of Junggar Basin, NW China. *Int. J. Coal Geol.* **2025**, *302*, 104726. [CrossRef]
36. Meng, Z.; Zhang, J.; Wang, R. In-Situ Stress, Pore Pressure and Stress-Dependent Permeability in the Southern Qinshui Basin. *Int. J. Rock Mech. Min. Sci.* **2011**, *48*, 122–131. [CrossRef]
37. Li, J.; Tang, S.; Zhang, S.; Li, L.; Wei, J.; Xi, Z.; Sun, K. Characterization of Unconventional Reservoirs and Continuous Accumulations of Natural Gas in the Carboniferous-Permian Strata, Mid-Eastern Qinshui Basin, China. *J. Nat. Gas Sci. Eng.* **2018**, *49*, 298–316. [CrossRef]
38. Zhang, M.; Fu, X.; Zhang, Q.; Cheng, W. Research on the Organic Geochemical and Mineral Composition Properties and Its Influence on Pore Structure of Coal-Measure Shales in Yushe-Wuxiang Block, South Central Qinshui Basin, China. *J. Pet. Sci. Eng.* **2019**, *173*, 1065–1079. [CrossRef]
39. Peng, C.; Zou, C.; Yang, Y.; Zhang, G.; Wang, W. Fractal Analysis of High Rank Coal from Southeast Qinshui Basin by Using Gas Adsorption and Mercury Porosimetry. *J. Pet. Sci. Eng.* **2017**, *156*, 235–249. [CrossRef]
40. Zhang, J.; Li, X.; Xiaoyan, Z.; Zhao, G.; Zhou, B.; Li, J.; Xie, Z.; Wang, F. Characterization of the Full-Sized Pore Structure of Coal-Bearing Shales and Its Effect on Shale Gas Content. *Energy Fuels* **2019**, *33*, 1969–1982. [CrossRef]
41. Ge, T.; Pan, J.; Wang, K.; Liu, W.; Mou, P.; Wang, X. Heterogeneity of Pore Structure of Late Paleozoic Transitional Facies Coal-Bearing Shale in the Southern North China and Its Main Controlling Factors. *Mar. Pet. Geol.* **2020**, *122*, 104710. [CrossRef]
42. Zhang, Q.; Jin, Z.; Deng, E.; Xu, Z.; Zhu, R.; Liu, Q.; Yan, Z.; Hu, K.; Zheng, T.; Hu, Z. Accumulation and Maturation of Organic Matter in Shales: The Coal-Bearing Permian Longtan Formation at Tucheng Syncline, Southwestern Guizhou, China. *Appl. Geochem.* **2025**, *180*, 106257. [CrossRef]
43. Shao, L.-Y.; Yang, Z.-Y.; Shang, X.-X.; Xiao, Z.-H.; Wang, S.; Zhang, W.-L.; Zheng, M.-Q.; Lu, J. Lithofacies Palaeogeography of the Carboniferous and Permian in the Qinshui Basin, Shanxi Province, China. *J. Palaeogeogr.* **2015**, *4*, 384–412. [CrossRef]
44. *GB/T 19145-2003*; Determination of Total Organic Carbon in Sedimentary Rock. Standardization Administration of China (SAC): Beijing, China, 2003.
45. *GB/T 18602-2012*; Rock Pyrolysis Analysis. Standardization Administration of China (SAC): Beijing, China, 2012.
46. *SY/T 5124-2012*; Microscopy Identification Method of Macerals in Sedimentary Rocks. National Energy Administration (NEA) of China: Beijing, China, 2012.
47. *SY/T 5163-2018*; Analytical Method of Rock Samples by Scanning Electron Microscopy. National Energy Administration (NEA) of China: Beijing, China, 2018.
48. *SY/T 5162-2014*; Analysis Method for Clay Minerals and Common Non-Clay Minerals in Sedimentary Rocks by X-Ray Diffraction. National Energy Administration (NEA) of China: Beijing, China, 2014.
49. *GB/T 19587-2017*; Determination of the Specific Surface Area of Solids by Gas Adsorption Using the BET Method. Standardization Administration of China (SAC): Beijing, China, 2017.
50. *GB/T 21650.1-2008*; Pore Size Distribution and Porosity of Solid Materials by Mercury Porosimetry and Gas Adsorption—Part 1: Mercury Porosimetry. Standardization Administration of China (SAC): Beijing, China, 2008.
51. *ISO 15901-1:2005*; Pore Size Distribution and Porosity of Solid Materials by Mercury Porosimetry and Gas Adsorption—Part 1: Mercury Porosimetry. International Organization for Standardization (ISO): Geneva, Switzerland, 2005.
52. Mahamud, M.M.; Novo, M.F. The Use of Fractal Analysis in the Textural Characterization of Coals. *Fuel* **2008**, *87*, 222–231. [CrossRef]

53. Peter, P.; Avnir, D. Chemistry in noninteger dimensions between two and three. *J. Chem. Phys.* **1983**, *79*, 3558–3565. [CrossRef] [PubMed]
54. Avnir, D.; Jaroniec, M. An Isotherm Equation for Adsorption on Fractal Surfaces of Heterogeneous Porous Materials. *Langmuir* **1989**, *5*, 1431–1433. [CrossRef]
55. Pfeifer, P. Fractal dimension as working tool for surface-roughness problems. *Appl. Surf. Sci.* **1984**, *18*, 146–164. [CrossRef]
56. Tatlıer, M.; Erdem-Şenatalar, A. Method to evaluate the fractal dimensions of solid adsorbents. *J. Phys. Chem.* **1999**, *103*, 4360–4365. [CrossRef]
57. Yao, Y.; Liu, D.; Tang, D.; Tang, S.; Huang, W. Fractal characterization of adsorption-Pores of coals from North China: An investigation on CH_4 adsorption capacity of coals. *Int. J. Coal Geol.* **2008**, *73*, 27–42. [CrossRef]
58. Loucks, R.G.; Reed, R.M.; Ruppel, S.C.; Hammes, U. Spectrum of pore types and networks in mudrocks and a descriptive classification for matrix-related mudrock pores. *AAPG Bull.* **2012**, *96*, 1071–1098. [CrossRef]
59. Reed, R.M.; Loucks, R.L. Imaging nanoscale pores in the Mississippian Barnett Shale of the northern Fort Worth Basin. *AAPG Annu. Conv. Abstr.* **2007**, *16*, 115.
60. Bustin, R.M.; Bustin, A.; Cui, A.; Ross, D.; Pathi, V.M. Impact of Shale Properties on Pore Structure and Storage Characteristics. Presented at the SPE Shale Gas Production Conference, Fort Worth, TX, USA, 16–18 November 2008. [CrossRef]
61. Montgomery, S.L.; Jarvie, D.M.; Bowker, K.A.; Pollastro, R.M. Mississippian Barnett Shale, Fort Worth basin, north-central Texas: Gas-shale play with multi–trillion cubic foot potential. *AAPG Bull.* **2005**, *89*, 155–175. [CrossRef]
62. Jarvie, D.M.; Hill, R.J.; Ruble, T.E.; Pollastro, R.M. Unconventional shale-gas systems: The Mississippian Barnett Shale of north-central Texas as one model for thermogenic shale-gas assessment. *AAPG Bull.* **2007**, *91*, 475–499. [CrossRef]
63. Bowker, K.A. Barnett Shale gas production, Fort Worth Basin: Issues and discussion. *AAPG Bull.* **2007**, *91*, 523–533. [CrossRef]
64. Zhang, J.; Li, X.; Wei, Q.; Gao, W.; Liang, W.; Wang, Z.; Wang, F. Quantitative Characterization of Pore-Fracture System of Organic-Rich Marine-Continental Shale Reservoirs: A Case Study of the Upper Permian Longtan Formation, Southern Sichuan Basin, China. *Fuel* **2017**, *200*, 272–281. [CrossRef]
65. Rowe, H.D.; Loucks, R.G.; Ruppel, S.C.; Rimmer, S.M. Mississippian Barnett Formation, Fort Worth Basin, Texas: Bulk Geochemical Inferences and Mo–TOC Constraints on the Severity of Hydrographic Restriction. *Chem. Geol.* **2008**, *257*, 16–25. [CrossRef]
66. Hu, J.; Tang, S.; Zhang, S. Investigation of Pore Structure and Fractal Characteristics of the Lower Silurian Longmaxi Shales in Western Hunan and Hubei Provinces in China. *J. Nat. Gas Sci. Eng.* **2016**, *28*, 522–535. [CrossRef]
67. IUPAC. Reporting physisorption data for gas/solid systems with special reference to the determination of surface area and porosity. *Pure Appl. Chem.* **1982**, *54*, 2201–2218. [CrossRef]
68. Broekhoff, J.C.P.; Boer, J.H. Studies on pore systems in catalysts XIII: Pore distributions from the desorption branch of a nitrogen sorption isotherm in the case of cylindrical pores: B applications. *J. Catal.* **1968**, *10*, 377–390. [CrossRef]
69. Rouquerol, J.; Avnir, D.; Fairbridge, C.W.; Everett, D.H.; Haynes, J.M.; Pernicone, N.; Ramsay, J.D.F.; Sing, K.S.W.; Unger, K.K. Recommendations for the characterization of porous solids (Technical Report). *Pure Appl. Chem.* **1994**, *66*, 1739–1758. [CrossRef]

MDPI AG
Grosspeteranlage 5
4052 Basel
Switzerland
Tel.: +41 61 683 77 34

Fractal and Fractional Editorial Office
E-mail: fractalfract@mdpi.com
www.mdpi.com/journal/fractalfract

www.ingramcontent.com/pod-product-compliance
Lightning Source LLC
Chambersburg PA
CBHW061454220926
51850CB00004B/87